Isozymes

II

Physiological Function

Isozymes

II

Physiological Function

EDITED BY

Clement L. Markert

Department of Biology
Yale University

ACADEMIC PRESS   New York   San Francisco   London   1975

A Subsidiary of Harcourt Brace Jovanovich, Publishers

ACADEMIC PRESS RAPID MANUSCRIPT REPRODUCTION

ACADEMIC PRESS, INC.
111 Fifth Avenue, New York, New York 10003

United Kingdom Edition published by
ACADEMIC PRESS, INC. (LONDON) LTD.
24/28 Oval Road, London NW1

Library of Congress Cataloging in Publication Data

International Conference on Isozymes, 3d, Yale
    University, 1974.
    Isozymes.

    Bibliography: v. 1, p.   ; v. 2, p.
    Includes indexes.
    CONTENTS: v. 1.   Molecular structure.–v. 2.   Physi-
ological function.
    1.   Isoenzymes–Congresses.   I.   Markert, Clement
Lawrence, (date)   ed.   II.   Title.   [DNLM: 1.   Iso-
enzymes–Congresses.   W3 IN182A 1974i / QU135 1587 1974i]
QP601.148 1974   574.1'925   74-31288
ISBN 0–12–472702–6 (v. 2)

Contents

CONTENTS

CONTENTS

CONTENTS

CONTENTS

Contributors

S. P. Adler   Laboratory of Biochemistry, National Heart and Lung Institute, National Institutes of Health, Bethesda, Maryland 20014

Richard Ainsley   Department of Zoology, The University of Texas, Austin, Texas 78712

Merry B. Alexander   Department of Horticulture, Colorado State University, Fort Collins, Colorado 80521

Bohdan Bakay   Department of Pediatrics, University of California, San Diego, La Jolla, California 92037

B. J. Benecke   Institute fur Physiologische Chemie, 355 Marburg/L, Lahnberge, Bundesrepublik Deutschland

George J. Brewer   Department of Human Genetics, University of Michigan Medical School, Ann Arbor, Michigan 48104

Carolyn S. Brown   Department of Biological Sciences, Purdue University, West Lafayette, Indiana 47907

Patrick J. Carmody   Division of Human Genetics, Department of Pediatrics, Children's Hospital of Buffalo, State University of New York, Buffalo, New York 14222

Mary E. Case   Department of Zoology, University of Georgia, Athens, Georgia 30602

Michael J. Champion   Department of Physiology and Biophysics, University of Illinois, Urbana, Illinois 61801

Y. S. Choi   Guttman Laboratory for Human Pharmacology and Pharmacogenetics, Department of Pharmacology, New York University, New York, New York 10016

J. E. Ciardi   Laboratory of Biochemistry, National Heart and Lung Institute, NIH, Bethesda, Maryland 20014

S. Chung   Department of Chemistry, Biochemistry and Biophysics Program, University of Notre Dame, Notre Dame, Indiana 46556

Godfrey G. S. Collins   Department of Pharmacology, School of Pharmacy, University of London, London WC1N 1AX, United Kingdom

David M. Crisp   Department of Biochemistry, The University of Bristol Medical School, University Walk, Bristol, United Kingdom

Russell Curry   Department of Biology, University of California, Riverside, California 92502

Norman P. Curthoys   Department of Biochemistry, University of Pittsburgh School of Medicine, Pittsburgh, Pennsylvania 15261

Ronald G. Davidson   Division of Human Genetics, Department of Pediatrics, Children's Hospital of Buffalo, State University of New York, Buffalo, New York 14222

D.W. Denna   Department of Horticulture, Colorado State University, Fort Collins, Colorado 80521

J. Donlon   Department of Biochemistry, University College, Galway, Ireland

E. J. Duke   Department of Zoology, University College, Belfield, Stillorgan Road, Dublin 4, Ireland

Hans M. Eppenberger   Institute for Cell Biology, Swiss Federal Institute of Technology, CH-8006, Zurich, Switzerland

Johannes Everse   Department of Chemistry, University of California, San Diego, La Jolla, California 92037

Todor I. Evrev   Department of General Biology, Medical Academy, Sofia, Bulgaria

A. Ferencz   Institut fur Physiologische Chemie, 355 Marburg/L, Lahnberge, Bundesrepublik Deutschland

P.F. Fottrell   Department of Biochemistry, University College, Galway, Ireland

David J. Fox   Department of Zoology, The University of Tennessee, Knoxville, Tennessee 37916

Irene Fuhr   Department of Biology, University of California, Riverside, California 92502

Patrick J. Gaffney   National Institute for Biological Standards and Controls, Holly Hill, London NW3 6RB, England

Peter Georgiev   Department of Pathological Physiology, Veterinary Institute for Infectious and Parasitic Diseases, Sofia, Bulgaria

Norman H. Giles   Department of Zoology, University of Georgia, Athens, Georgia 30602

A. Ginsburg   Laboratory of Biochemistry, National Heart and Lung Institute, National Institutes of Health, Bethesda, Maryland 20014

Fresia Gonzalez   Department of Biochemistry, Medical College of Wisconsin, Milwaukee, Wisconsin 53233

Robert S. Greenfield   Department of Biochemistry, Cornell University Medical College, New York, New York 10021

H. E. Guderley   Department of Zoology, University of British Columbia, Vancouver, B. C. V6T 1W5, Canada

Minoru Hamada   Department of Pathological Biochemistry, Atomic Disease Institute, Nagasaki University School of Medicine, 12-4, Sakamoto-machi, Nagasaki-shi 852, Japan

Christopher O. Hawtrey   Department of Biological Sciences, California State University, Hayward, California 94542

T. J. Hayden   Department of Zoology, University College, Belfield, Stillorgan Road, Dublin 4, Ireland

Rochelle Hirschhorn   Department of Medicine, New York University School of Medicine, New York, New York 10016

P. W. Hochachka   Department of Zoology, University of British Columbia, Vancouver, B. C. V6T 1W5, Canada

H. Hogendoorn   Netherlands Institute for Fishery Investigations, Ymuiden, The Netherlands

Edward W. Holmes, Jr.   Departments of Medicine and Biochemistry, Duke University Medical Center, Durham, North Carolina 27710

Herbert O. Hultin   Department of Food Science and Nutrition, University of Massachusetts, Amherst, Massachusetts 01002

Nathan O. Kaplan   Department of Chemistry, University of California, San Diego, La Jolla, California 92037

Howard M. Katzen   Merck Institute for Therapeutic Research, Rahway, New Jersey 07065

William N. Kelley   Departments of Medicine and Biochemistry, Duke University Medical Center, Durham, North Carolina 27710

Robert G. Kemp   Department of Biochemistry, Medical College of Wisconsin, Milwaukee, Wisconsin 53233

Motoshi Kitamura   Department of Clinical Chemistry, Toranomon Hospital, Tokyo, Japan

G. Barrie Kitto   Clayton Foundation Biochemical Institute, Department of Chemistry, The University of Texas, Austin, Texas 78712

Ellis L. Kline   Department of Biological Sciences, Purdue University, West Lafayette, Indiana 47907

Hans J. Kuhn   Institute for Cell Biology, Swiss Federal Institute of Technology, CH-8006, Zurich, Switzerland

B. N. La Du   Guttman Laboratory for Human Pharmacology and Pharmacogenetics, Department of Pharmacology, New York University, New York, New York 10016

Rosanne M. Leipzig   Department of Human Genetics, University of Michigan Medical School, Ann Arbor, Michigan 48104

Wilhelmina de Ligny   Netherlands Institute for Fishery Investigations, Ymuiden, The Netherlands

Edwin H. Liu   Department of Biology, University of South Carolina, Columbia, South Carolina 29208

Ian D. Longshaw   Department of Biochemistry, University of Bristol, The Medical School, University Walk, Bristol, United Kingdom

Robert Lyons   McArdle Laboratory for Cancer Research, The Medical School, The University of Wisconsin, Madison, Wisconsin 53706

E. M. Lyslova   Sechenov Institute of Evolutionary Physiology and Biochemistry, Academy of Sciences, Leningrad, U.S.S.R.

Clement L. Markert   Department of Biology, Yale University, New Haven, Connecticut 06520

Ronald R. Marquardt   Department of Animal Science, University of Manitoba, Winnipeg, Manitoba R3T 2N2, Canada

M. Martinez-Carrion   Department of Chemistry, Biochemistry and Biophysics Program, University of Notre Dame, Notre Dame, Indiana 46556

Shiro Miwa   Third Department of Internal Medicine, Yamaguchi University School of Medicine, 1144 Kogushi Ube-shi, Yamaguchi-ken 755, Japan

Thomas W. Moon   Department of Biology, University of Ottawa, Ottawa, Ontario, Canada K1N 6N5

Bernardo Nadal-Ginard   Department of Biology, Yale University, New Haven, Connecticut 06520

Asha Naidu   Department of Biological Sciences, California State University, Hayward, California 94542

Koji Nakashima   Third Department of Internal Medicine, Yamaguchi University School of Medicine, 1144 Kogushi Ube-shi, Yamaguchi-ken 755, Japan

Christopher Nelson   Department of Biological Sciences, California State University, Hayward, California 94542

Toshihiro Nishina   Department of Clinical Chemistry, Toranomon Hospital, Tokyo, Japan

Kaoru Nishiyama   Department of Biochemistry, Kobe University School of Medicine, Ikuta-ku, Kobe, Japan

Yasutomi Nishizuka   Department of Biochemistry, Kobe University School of Medicine, Ikuta-ku, Kobe, Japan

William L. Nyhan   Department of Pediatrics, University of California, San Diego, La Jolla, California 92037

Fred J. Oelshlegel Jr.   Department of Human Genetics, University of Michigan Medical School, Ann Arbor, Michigan 48104

Zen-ichi Ogita   Department of Genetics, Medical School, Osaka University, Joancho-33, Kita-ku, Osaka 530, Japan

Masao Ogihara   Department of Internal Medicine, Daisan Hospital, Tokyo Jikeikai University School of Medicine, 106 Izumi, Komae-shi, Tokyo 182, Japan

R. D. O'Brien   Section of Neurobiology and Behavior, Cornell University, Ithaca, New York 14850

G. O'Cuinn   Department of Biochemistry, University College, Galway, Ireland

C. O. Piggott   Department of Biochemistry, University College, Galway, Ireland

Henry C. Pitot   McArdle Laboratory for Cancer Research, The Medical School, The University of Wisconsin, Madison, Wisconsin 53706

Chris I. Pogson   Biological Laboratory, University of Kent, Canterbury, CT2 7NJ, United Kingdom

C. Ladd Prosser   Department of Zoology, University of Illinois, Urbana, Illinois 61801

Mario C. Rattazzi   Division of Human Genetics, Department of Pediatrics, Children's Hospital of Buffalo, State University of New York, Buffalo, New York 14222

A. Relimpio   Department of Chemistry, Biochemistry and Biophysics Program, University of Notre Dame, Notre Dame, Indiana 46556

Jesus Rodriguez   McArdle Laboratory for Cancer Research, The Medical School, University of Wisconsin, Madison, Wisconsin 53706

J. Ruiz-Herrera   Departamento de Microbiologia, Escuela Nacional de Ciencias Biologicas, Instituto Polytecnico Nacional, Mexico 17, D. F. Mexico

A. Segal   Laboratory of Biochemistry, National Heart and Lung Institute, National Institutes of Health, Bethesda, Maryland 20014

K. H. Seifart   Institut fur Physiologische Chemie, 355 Marburg/L, Lahnberge, Bundesrepublik Deutschland

T. P. Serebrenikova, Sechenov Institute of Evolutionary Physiology and Biochemistry, Academy of Sciences, Leningrad, U.S.S.R.

Kenji Shinohara   Third Department of Internal Medicine, Yamaguchi University School of Medicine, 1144 Kogushi Ube-shi, Yamaguchi-ken 755, Japan

P. Z. Smyrniotis   Laboratory of Biochemistry, National Heart and Lung Institute, National Institutes of Health, Bethesda, Maryland 20014

Denis D. Soderman   Merck Institute for Therapeutic Research, Rahway, New Jersey 07065

George N. Somero   Scripps Institution of Oceanography, Box 1529, La Jolla, California 92037

E. R. Stadtman   Laboratory of Biochemistry, National Heart and Lung Institute, National Institutes of Health, Bethesda, Maryland 20014

M. J. Stankewicz   Department of Chemistry, Biochemistry and Biophysics Program, University of Notre Dame, Notre Dame, Indiana 46556

Jisnuson Svasti   Department of Biochemistry, Faculty of Science, Mahidol University, Bangkok, Thailand

Yoshimi Takai   Department of Biochemistry, Kobe University School of Medicine, Ikuta-ku, Kobe, Japan

Suresh S. Tate   Department of Biochemistry, Cornell University Medical College, New York, New York 10021

Irwin P. Ting   Department of Biology, University of California, Riverside, California 92502

R. K. Tripathi   Section of Neurobiology and Behavior, Cornell University, Ithaca, New York 14850

Michael Y. Tsai   Department of Biochemistry, Medical College of Wisconsin, Milwaukee, Wisconsin 53233

David C. Turner   Institute for Cell Biology, Swiss Federal Institute of Technology, CH-8006, Zurich, Switzerland

H. E. Umbarger Department of Biological Sciences, Purdue University, West Lafayette, Indiana 47907

J. M. Varela Central Institute for Brain Research, Ijdijk, 28, Amsterdam, The Netherlands

B. L. Verboom Netherlands Institute for Fishery Investigations, Ymuiden, The Netherlands

N. A. Verzhbinskaya Sechenov Institute of Evolutionary Physiology and Biochemistry, Academy of Sciences, Leningrad, U.S.S.R.

Elliot S. Vesell Department of Pharmacology, Milton S. Hershey Medical Center, Hershey, Pennsylvania 17033

Carl S. Vestling Department of Biochemistry, University of Iowa, Iowa City, Iowa 52242

E. I. Villarreal-Moguel Departamento de Microbiologia, Escuela Nacional de Ciencias Biologicas, Instituto Politecnico Nacional, Mexico 17, D. F. Mexico

Sumalee Viriyachai Department of Biochemistry, Faculty of Science, Mahidol University, Bangkok, Thailand

Theo Wallimann Institute for Cell Biology, Swiss Federal Institute of Technology, CH-8006, Zurich, Switzerland

Daniel Wellner Department of Biochemistry, Cornell University Medical College, New York, New York 10021

Gregory S. Whitt Department of Zoology, University of Illinois, Urbana, Illinois 61801

J. Willemsen Netherlands Institute for Fishery Investigations, Ymuiden, The Netherlands

F. Ray Wilson, Department of Biology, Baylor University, Waco, Texas 76703

Danny Wolfe Department of Biological Sciences, California State University, Hayward, California 94542

James B. Wyngaarden Department of Medicine and Biochemistry, Duke University Medical Center, Durham, North Carolina 27710

Hirohei Yamamura Department of Biochemistry, Kobe University School of Medicine, Ikuta-ku, Kobe, Japan

Moussa B. H. Youdim MRC Clinical Pharmacology Unit, University Department of Clinical Pharmacology, Radcliffe Infirmary, Oxford OX2 6HE, United Kingdom

CONTRIBUTORS

William C. Zschoche   Department of Biology, University of California, Riverside, California 92502

Preface

Isozymes are now recognized, investigated, and used throughout many areas of biological investigation. They have taken their place as an essential feature of the biochemical organization of living things. Like many developments in the biomedical sciences, the field of isozymes began with occasional, perplexing observations that generated questions which led to more investigation and, finally, with the application of new techniques, to a clear recognition and appreciation of a new dimension of enzymology.

The area of isozyme research is only about 15 years old but has been characterized by an exponential growth. Since the recognition in 1959 that isozymic systems are a fundamental and significant aspect of biological organization, many thousands of papers have been published on isozymes. Several hundred enzymes have already been resolved into multiple molecular forms, and many more will doubtless be added to the list. In any event, it is now the responsibility of enzymologists to examine every enzyme system for possible isozyme multiplicity.

Two previous international conferences have been held on the subject of isozymes, both under the sponsorship of the New York Academy of Sciences—the first in February 1961, and the second in December 1966. And now, after a somewhat longer interval, the Third International Conference was convened in April 1974, at Yale University. For many years, a small group of investigators has met annually to discuss recent advances in research on isozymes. They have published a bulletin each year and have generally helped to shape the field; in effect, they have been a standing committee for this area of research. From this group emerged the decision to hold a third international conference, and an organizing committee of five was appointed to carry out the mandate for convening a third conference. This Third International Conference was by far the largest of the three so far held with 224 speakers representing 21 countries, and organized into nine simultaneous sessions for three days on April 18, 19, and 20, 1974. Virtually every speaker submitted a manuscript for publication, and these total almost 4,000 pages. The manuscripts have been collected into four volumes entitled, *I. Molecular Structure; II. Physiological Function; III. Developmental Biology; and IV. Genetics and Evolution.* The oral reports at the Conference and the submitted manuscripts cover a vast area of biological research. Not every manuscript fits precisely into one or another of the four volumes, but the most appropriate assignment has been made wherever possible. The quality of the volumes and the success of the Conference must be credited to the participants and to the organizing committee. The scientific community owes much to them.

Acknowledgments

I would like to acknowledge the help of my students and my laboratory staff in organizing the Conference and in preparing the volumes for publication. I am grateful to my wife, Margaret Markert, for volunteering her time and talent in helping to organize the Conference and in copy editing the manuscripts.

Financial help for the Conference was provided by the National Science Foundation, the National Institutes of Health, International Union of Biochemistry, Yale University, and a number of private contributors:

Private Contributors

American Instrument Company
Silver Spring, Maryland 20910

Canalco
Rockville, Maryland 20852

CIBA-GEIGY Corporation
Ardsley, New York

Gelman Instrument Company
Ann Arbor, Michigan 48106

Gilford Instrument Laboratories, Inc.
Oberlin, Ohio

Hamilton Company
Reno, Nevada 89502

Kontes Glass Company
Vineland, New Hampshire 08360

The Lilly Research Laboratories
Indianapolis, Indiana

Merck Sharp & Dohme
West Point, Pennsylvania

Miles Laboratories, Inc.
Elkhart, Indiana

New England Nuclear
Worcester, Massachusetts 01608

Schering Corporation
Bloomfield, New Jersey

Smith Kline & French Laboratories
Philadelphia, Pennsylvania

Warner-Lambert Company
Morris Plains, New Jersey

MEDICAL USES OF ISOZYMES

ELLIOT S. VESELL
Department of Pharmacology
Milton S. Hershey Medical Center
Hershey, Pennsylvania 17033

ABSTRACT. Since isozymes of lactate dehydrogenase (LDH) have been more extensively employed in clinical contexts than other isozymes, this review focuses on medical applications of LDH isozymes. Isozymes other than LDH have also been used widely in medicine both for diagnostic and experimental purposes. Analysis of isozymes confers a greater specificity in diagnosis and in the search for pathogenesis of various disease states than does analysis of total, unseparated serum activity of an enzyme. Diagnostic specificity can be further increased by analyzing simultaneously several different isozyme systems in biological fluids or tissues.

Various medical uses of isozymes could be extended and expanded by fresh approaches to the functional significance of enzyme heterogeneity. Therefore, the current status of this problem with respect to LDH isozymes is considered. This review stresses that for each metabolic function proposed to explain enzyme heterogeneity, kinetic data on which the theory rests should be gathered under conditions as close as possible to those existing in vivo. In the past, some theories of isozyme function emerged on the foundation of kinetic studies performed under a single set of highly unphysiological conditions. Future approaches to explanations of the metabolic functions of isozymes should be based on experiments performed under a wide variety of different conditions maintained as closely as possible to those prevailing in vivo.

A. *Historical Background*

It is of historical interest that medical applications of multiple forms of an enzyme preceded any other biological use. Prior to their medical use, multiple forms of an enzyme were almost always regarded with annoyance by biochemists who viewed them as an embarrassing artifact, a testimony to poor laboratory technique, without any biological significance.

In 1950 Meister reported heterogeneity of LDH from beef heart, confirmed by Neilands in 1952, but no biological significance of this heterogeneity was suggested. In 1957, two years before the term isozyme was coined, starch block electrophoresis revealed the presence in human serum of multiple

1

forms of lactate dehydrogenase (LDH) (Vesell and Bearn). A
new dimension in medical diagnosis was introduced when the
pattern of serum LDH isozymes, highly reproducible in healthy
subjects, was demonstrated to change dramatically in various
disease states, including myocardial infarction and acute
myelogenous leukemia (Vesell and Bearn, 1957) (Fig. 1). In
healthy volunteers, the LDH activity, now designated LDH-2,

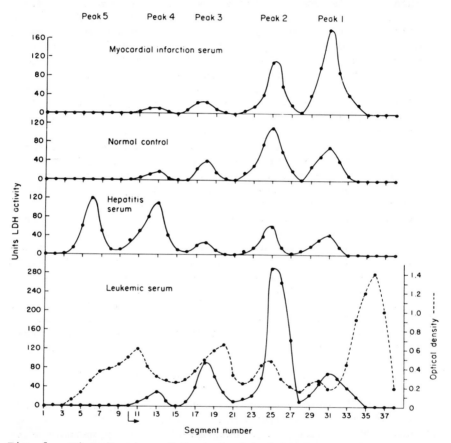

Fig. 1. Distribution of lactate dehydrogenase activity in normal human serum and sera from patients with myocardial infarction, chronic granulocytic leukemia, and infectious hepatitis.
These 4 sera were separated simultaneously on a starch block.
Note that each patient shows alterations in individual LDH
activity peaks from the normal pattern. The curve with the
broken line for the leukemic serum represents the protein concentration. The origin is indicated by the arrow.

that migrated with an electrophoretic mobility like that of
alpha-2 globulin, now predominated. Next came LDH-1; LDH-3
was the third most active fraction (Fig. 1). However, in sera
of patients with myocardial infarction the normal pattern
became distorted, LDH-1, rather than LDH-2, predominated
(Fig. 1). In acute myelogenous leukemia the serum LDH-2 peak
became larger than normal (Fig. 1). Early in hepatitis, LDH-5,
which in normal serum is usually absent or present in trace
amounts, became the most prominent isozyme (Wieme, 1959) (Fig.
1). As Table 1 shows, various disease states in man are associated with certain serum LDH isozyme patterns. It became
clear that much more diagnostic information could be obtained
by electrophoretic separation of serum LDH activity, rather
than by assaying only total unseparated serum LDH activity
which was nonspecifically elevated in many disorders of different organs. Today, electrophoretic separation of serum LDH
activity is one of the most sensitive laboratory tests in the
diagnosis of myocardial infarction and in the assessment of
the response of infarction patients to therapy.

TABLE 1
ENZYMES IN MULTIPLE MOLECULAR FORMS
EMPLOYED IN CLINICAL MEDICINE

ESTERASES

    Acetylcholinesterase
    Cholinesterase

PHOSPHATASES

    Acid Phosphatase
    Alkaline Phosphatase
    5'-Nucleotidase

SACCHARIDASES

    Amylase
    β-Glucuronidase

PROTEASES

    γ-Glutamyl Transpeptidase
    Leucine Aminopeptidase

NUCLEASES

    Ribonuclease

DEHYDROGENASES

    Alcohol Dehydrogenase
    Glucose-6-P Dehydrogenase
    Glutamic Dehydrogenase
    α-Glycerophosphate Dehy-
      drogenase
    Isocitric Dehydrogenase
    Lactate Dehydrogenase
    Malate Dehydrogenase

OTHERS

    Aldolase
    Creatine Phosphokinase
    Glutamic Oxalacetic Trans-
      aminase
    Phosphohexose Isomerase
    Pyruvate Kinase

After this work on LDH isozymes in sera, performed in the late fifties by several laboratories (Hes, 1958; Hill, 1958; Vesell and Bearn, 1961; Wieme and Maercke, 1961; Wroblewski and LaDu, 1961; Cohen et al, 1964), many enzymes other than LDH were shown to exist in multiple forms in human serum (Table 1). For certain disease states, resolution of these enzymes into their component isozymes was also demonstrated to confer greater diagnostic specificity than assay of the total unseparated serum activity of the enzyme. In several instances, simultaneous determination of the isozymic patterns of several different enzymes has been shown to be useful in diagnosis (Nerenberg and Pogojeff, 1969). However, at the present time LDH isozymes remain by far the most commonly used isozymic system in clinical medicine. For this reason, my subsequent discussion concentrates on medical applications of LDH isozymes.

B. *Mechanism of Altered Serum LDH*

Activity in Various Disease States

In several diseases, the particular pattern of serum LDH isozymes associated with the disorder can be explained by reference to the normal LDH isozymic pattern of the pathologically involved tissue. Fig. 2 shows such LDH patterns in 6 human tissues obtained at autopsy. The supernatants of tissue homogenates were separated by starch gel electrophoresis and the bands of LDH activity visualized by immersing the gels in a solution containing lactate, NAD, and nitro blue tetrazolium. This staining technique adapted for starch gels by Markert and Møller (1959) is a particularly important contribution in the history of medical applications of isozymes.

Each of the 6 human tissues contains 5 bands of LDH activity, 4 migrating at pH 7.4 toward the anode, one toward the cathode. Although these 6 tissues resemble one another in the number of LDH isozymes they have and in the electrophoretic mobility of corresponding bands, each tissue differs from the others in the relative distribution of total LDH activity among the 5 isozymes. Heart contains predominantly LDH-1, whereas in the kidney LDH-1, 2, and 3 are closer in the percentage of the total LDH activity accounted for by each anodal band. In liver and skeletal muscle LDH-5 predominates, whereas lung exhibits LDH activity evenly distributed among the 5 bands. Furthermore, within each tissue different cell types have different LDH isozyme patterns. The clinical significance of differences among tissues in distribution of LDH isozymes arises from the particular kind of distortion that occurs in

4

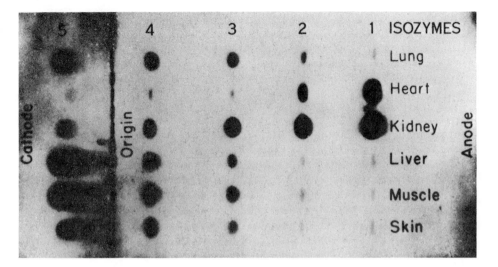

Fig. 2. Photograph of a starch gel treated to reveal the LDH isozymes in homogenates of 6 human tissues obtained at autopsy from a healthy subject killed in an accident. Note the presence of 5 LDH isozymes in each homogenate. Reproduced by permission from Vesell (1965a).

the normal serum LDH isozyme pattern when various disease states ensue. Generally, damage to a tissue is associated with release of its LDH isozymes into serum so that changes in the serum pattern reflect the isozyme pattern of the injured tissue. Because LDH activity occurs at many hundred fold higher concentrations in most mammalian tissues than in serum, damage of only a small portion of a tissue is generally reflected in large elevations of serum LDH activity. When heart muscle is damaged as a result of occlusion of a coronary artery, release of the most prominent isozyme from cardiac muscle, LDH-1, is reflected by a differential elevation of this isozyme in serum. As the damaged heart muscle heals, serum LDH-1 returns to its normal level. Thus far, serum LDH isozymes are the most sensitive indication of myocardial infarction and even when total serum LDH activity has returned to normal, persistence of an abnormal serum LDH isozyme pattern, characterized by elevation of LDH-1, can remain as a sign of residual cardiac damage. However, in cases of pure angina pectoris, the serum LDH isozyme pattern is normal. Where very severe chest pain from repeated attacks of angina pectoris may finally lead to a consideration of acute myocardial infarction as a diagnostic possibility, serum LDH isozymes can provide a useful test to distinguish between the two conditions. The diagnostic utility and sensitivity of serum LDH isozymes as an index of cardiac muscle pathology are illustrated by studies in 17 patients receiving heart transplants (Nora et al., 1969). During the first month when the rejection response was most difficult to detect and the electrocardiogram proved least reliable, alterations of the normal serum LDH isozymes with elevation of serum LDH-1 were the first biochemical indications of the rejection response. However, elevation of serum LDH-1 is by no means specific for myocardial pathology. In such hematological diseases as pernicious anemia and various hemolytic anemias in which erythrocytes lyse, serum LDH isozyme patterns also exhibit predominance of LDH-1 and LDH-2, the main isozymes in red blood cells (Starkweather et al., 1965).

By contrast, early in certain diseases of the liver, such as hepatitis and carbon tetrachloride toxicity, in which widespread damage to hepatic cells occurs, release of the predominant liver isozyme, LDH-5, is reflected by elevated serum LDH-5 activity (Wieme and Van Maercke, 1961). Here again, elevation of serum LDH-5 is far from pathognomonic of liver disease since in acute injuries of skeletal muscle or in diseases of skeletal muscle such as dermatomyositis, LDH-5 and LDH-4, the predominant isozymes of skeletal muscle, are released from damaged tissue to produce elevations of these isozymes in serum.

However, in Duchenne's muscular dystrophy, a severe disorder of skeletal muscle transmitted as an X-linked recessive and characterized by atrophy and fragmentation of certain skeletal muscles, the serum LDH isozyme pattern is not predictable from the previous examples (Dreyfus et al., 1962; Wieme and Lauryssens, 1958). Total serum LDH activity is elevated, but the distribution of this activity among the 5 LDH isozymes is essentially normal. Among the explanations offered are that LDH-5 and LDH-4 may be low in dystrophic muscle due to selective removal of these isozymes from muscle or due to replacement of affected muscle fibers by fibrous tissue with an isozyme pattern different from muscle. Alternatively, dystrophic muscle may resemble embryonic skeletal muscle in being low in these cathodal isozymes or finally the disease may produce a selective alteration in the turnover of certain isozymes in skeletal muscle and of different isozymes in serum.

Thus, not all serum LDH isozyme patterns observed in various disease states can be explained simply by the concept of liberated LDH isozymes from damaged tissue into serum. As an additional example of this problem, in many tumors of rodents and man, affected tissues frequently reveal a shift from the normal

isozyme pattern toward an increase in LDH-5 and LDH-4
(Richterich and Burger, 1963; Goldman et al, 1964). However,
in certain tumors, the serum LDH isozyme pattern may not be
abnormal or may show additional isozymes (Starkweather and
Schoch, 1962). Predominance of cathodal LDH isozymes in tissues with cancer have been interpreted as indications of a
reversion to a more primitive, undifferentiated fetal metabolic pattern in which these cathodal isozymes are known to
be most prominent (Schweitzer et al, 1973). In man cancer
throughout its course is rarely confined to a single organ;
thus various complex changes occurring in other tissues both
from metastases and independently from such tumor-induced processes as infection, infarction, anoxia, or starvation could
account for the occasionally observed aberrant alterations in
serum LDH isozyme pattern during cancer.

With respect to interrelationships among tissues affected
by a pathological process, the following studies are presented
to illustrate how a single environmental insult, in this case
experimental hemorrhagic shock, can simultaneously affect
several different tissues, each of which contributes to an
altered serum LDH isozyme pattern (Vesell et al, 1959). Fig. 3
shows the total serum LDH activity in 8 dogs bled from the
femoral artery into an elevated reservoir to maintain an arterial pressure of 30 mm Hg for 1 to 5 hours before transfusion
of the blood in the elevated reservoir. The normal blood
pressure of the dog is between 110 to 120 mm Hg. On the right
hand side of the figure, the 3 dogs transfused after only 1
hour at 30 mm Hg showed least change in total serum LDH,
whereas the 2 dogs maintained in hypotension for 2 hr before
transfusion had much greater serum LDH elevations, reaching
25 times the initial value. On the left side of the figure
are 3 dogs transfused after 3 to 5 hr of hemorrhagic hypotension. Each dog died, and each exhibited increasing serum LDH
activity until death occurred at 6 to 12 hr after the start
of the study. This experiment suggests that the magnitude of
the elevation in total serum LDH activity is roughly proportional to the extent of the tissue damage produced, that is,
the longer the period of shock, the greater the rise in total
serum LDH activity. Which tissues contribute to elevated
total serum LDH activity? To answer this question, the arterial supply and venous outflow of 3 tissues were tapped and
sequential blood specimens obtained. Fig. 4 shows the results
in 3 dogs, in which the LDH serum activity in the arterial
supply and the venous return of the dog kidney, liver, and tissues of an extremity were measured at various intervals. For
each tissue, the serum LDH activity was similar initially in
venous and arterial blood. However, after several hours of

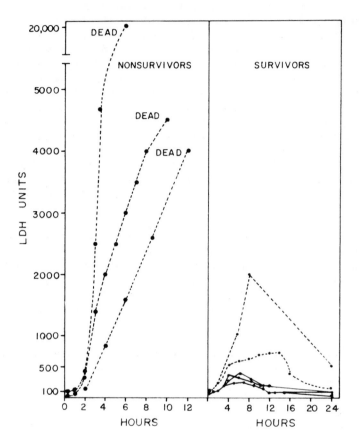

Fig. 3. Plasma LDH activity in 8 dogs bled to arterial pressure 30 mm Hg for 1, 2, 3, and 5 hr and then given replacement
transfusion. Solid lines show activities during hemorrhagic
hypotension, broken lines after transfusion. Reproduced by
permission from Vesell et al, 1959.

shock each tissue exhibited a rise in venous serum LDH activity above that in the arterial supply, suggesting a release
of LDH activity from each of these tissues. The final question
concerns which LDH isozymes in serum become elevated in these
experiments. Fig. 5 shows the 5 serum LDH isozymes of a dog
before and after hemorrhagic hypotension. In this electrophoretic separation, each of the 5 serum isozymes exhibited an
appreciable elevation, as might have been predicted from the
knowledge that shock damages several tissues and releases LDH

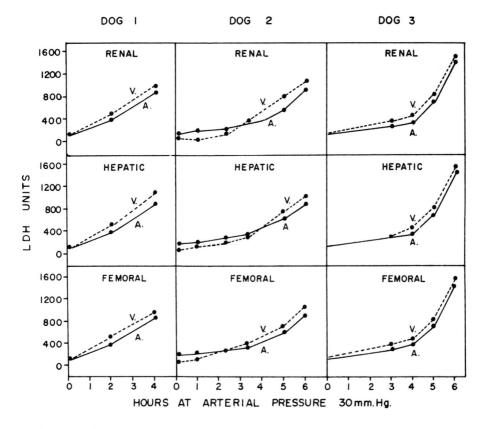

Fig. 4. In all 3 experiments in which the A–V difference in
LDH activity was measured, plasma LDH activity in hepatic,
renal and femoral venous blood attained consistently higher
levels than that of arterial blood. Maximum differences were
200 units in renal and hepatic venous blood and 100 units in
femoral venous blood.

activity from at least 3 different tissues, each with a distinct isozyme pattern (Vesell et al, 1959).

What information of clinical applicability can be gleaned
from these studies? Several direct applications of these experimental studies on shock are relevant to clinical situations.

Firstly, shock may occur in several discrete disease states
in man, including such diverse entities as myocardial infarction, congestive heart failure, pulmonary embolism, infections,

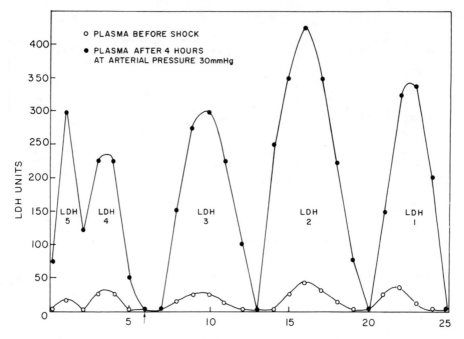

Fig. 5. Electrophoretic pattern of LDH activity in dog plasma before and during hemorrhagic hypotension. Five peaks of LDH activity are present. From the left these are located in gamma globulin, beta globulin, alpha-2 globulin and between alpha-1 globulin and albumin. All are elevated in shock. The origin is at the arrow.

pancreatitis, cerebral vascular accidents, and carcinomatosis. Therefore, serum LDH isozyme patterns associated with an initially uncomplicated case of one of the above disorders may change dramatically if the patient deteriorates through extension of the disease process and the consequent development of shock. Sequential determinations of serum LDH isozyme patterns are thus both useful in diagnosis and prognosis, since they can help in distinguishing between progress toward recovery, as indicated by return of aberrant serum LDH isozyme patterns to normal, and between exacerbation of the disorder, as suggested by an increased abnormality of the serum LDH isozyme or isozymes. As a specific example, patients suffering from acute myocardial infarction can exhibit 3 different types of change in their serum LDH isozyme patterns, each requiring a different therapeutic approach. Return of the elevated serum LDH-1 toward normal indicates healing of the infarcted myocardial tissue and permits the physician to encourage a very gradual resumption of activity customary in treatment of such patients during uncomplicated survival of the first 5 to 6 days after a myocardial infarction. However, if, after several days of such progress, the serum LDH-1 activity suddenly rises, the physician must suspect a fresh myocardial infarction or extension of the old infarction, events unfortunately not uncommon early in the recuperative period.

Finally, elevation of several serum LDH isozymes, particularly LDH-5, after a seemingly uneventful initial post-infarction course can indicate the supervention of shock secondary to heart failure. Also, shock due to heart failure can be present at the time of the myocardial infarction and here, too, elevation of several serum LDH isozymes, in addition to LDH-1, can be a useful diagnostic aide. It must be emphasized that the patient's general condition, the physical examination by the physician, and several other laboratory tests, including the electrocardiogram, must all be assessed to produce a complete picture; reliance on only one or two laboratory tests, however sensitive they may be, can prove to be misleading.

The second point raised by the experimental studies in hemorrhagic shock is that the medical use of isozymes can extend beyond their application to specific diagnostic and prognostic problems in patients to more general and basic investigations of the pathogenesis of various disease processes. Since multiple forms of enzymes perform complex metabolic roles in the cell, disorders that produce a derangement of normal cellular physiology and metabolism should be reflected in aberrant isozyme patterns. Application of this principle has led to the use of many different isozymic systems as tools in medicine to explore the pathogenesis of various disease states and to gain a better understanding of how certain disease processes affect normal cellular function and metabolism. For example, myocardial infarctions of varying extents have been produced in dogs by ligating different portions of the coronary artery to define the relationship between the size of the resultant area of necrotic cardiac muscle and the extent of serum LDH-1 elevation (Nachlas et al, 1964). These studies helped to establish ranges of LDH-1 elevations compatible with good and with poor clinical prognoses. Similarly, dose response relationships have been sought between different doses of hepatotoxic drugs and toxins, such as carbon tetrachloride, and the extent of the rise in serum LDH-5 or renal damage after administration of mercuric chloride and the extent of rise in serum LDH-1 and LDH-2 (Cornish et al, 1970). Effects of the

administration of various hormones on cellular and serum isozyme patterns have been studied (Goodfriend and Kaplan, 1964).
This approach has also been employed to investigate effects of
various experimental tumors and infections. One particularly
interesting experimental study identified reduced clearance of
LDH-5 from mouse serum as the cause of markedly elevated serum
LDH activity in otherwise seemingly normal mice infected with
the Riley virus (Majy et al, 1965; Rowson et al, 1965).
Although no similar example has thus far been unequivocally
established in any human disease to account for elevated serum
LDH activity, it is possible to visualize how reduced clearance
of LDH isozymes either through renal, reticuloendothial, or
cellular catabolic derangements could produce elevations in
serum LDH isozymes in certain human disorders. These are only
isolated experimental uses of isozymes of which there are several hundred additional illustrations in the literature.

Serum and tissue LDH isozymes have also served as tools to
investigate such normal processes as aging (Vesell and Bearn,
1962a; Gerlach and Fegeler, 1973), sex differences (Cohen et
al, 1967), diurnal variations (Starkweather et al, 1966;
Jacey and Schaefer, 1968), genetic heterogeneity (Vesell,
1965a), ontogeny (Markert and Ursprung, 1962; Vesell et al,
1962), and protein synthesis and degradation (Fritz et al,
1969, 1973). Each of these areas has been fruitfully illumi-
nated through the application of the isozyme technique, but
further discussion of them is unwarranted here, other than to
state that such information obtained in healthy subjects may
eventually be applied in medicine when these normal processes
become deranged through the advent of disease.

C. Methodological Considerations

As these examples illustrate, during the past 15 years isozymes have gained widespread acceptance as useful aids in the
diagnosis and follow up care of patients with various disease
states, as well as in experimental studies on the pathogenesis
of certain disorders. Generally serum activity of a particular
enzyme existing in multiple forms in serum is measured after
electrophoretic separation of the proteins. However, for some
serum enzymes existing in multiple forms, electrophoresis can
be eliminated if differential heat stabilities or sensitivity
to inhibitors distinguish among the various isozymes. Serum
has been the most frequently tapped biological source for diagnostic applications of isozymes; other sources, including
urine, cerebrospinal fluid, saliva, sweat, seminal fluid, and
milk have also been examined, as well as pericardial, pleural,
and hepatic effusions and transudates. In addition to these

biological fluids, tissues themselves serve in certain instances as a source of material for the diagnostic use of isozymes.

Certain methodological considerations must be carefully
checked to insure that the assay system is appropriate for the
particular use being contemplated. Such methodological considerations have been described elsewhere (Vesell and Brody,
1964; Fritz et al, 1970) so that they need not be reviewed
here other than to stress the unfortunate consequences of a
failure to check them out carefully. All types of artifact
can ensue if such methodological considerations are ignored.
For example, the intensity of tetrazolium staining on zymograms
is proportional to the LDH activity in biological specimens
only over a limited range of LDH activity. A more unusual
example of artifact is illustrated in Fig. 6 which shows the

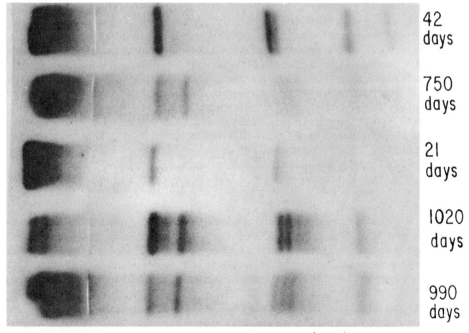

42 days

750 days

21 days

1020 days

990 days

Cathode Origin   Anode

Fig. 6. Photograph of a starch gel showing LDH isozymes in
human skeletal muscles from 5 individuals. These muscles were
treated identically except for length of storage at -25°C.
Note that splitting of bands increases with length of storage
of the whole muscle. Reproduced by permission from Vesell and
Brody (1964).

LDH isozyme pattern of human skeletal muscles stored as the unhomogenized whole tissue for varying periods of time at -70°C (Vesell and Brody, 1964). The figure reveals that the number and intensity of the subbands of each LDH isozyme increase with time of storage. The phenomenon of certain types of subbanding remains incompletely understood, but has been frequently observed in various biological systems.

D. *Functional Significance of Isozymes*

Because isozymes have been employed so extensively in medicine as diagnostic and prognostic aids and as tools in elucidating the pathogenesis of various disease processes, a consideration of their physiological and biochemical function seems appropriate. Better understanding of the functional significance of isozymes could lead to improved and extended medical applications.

Several distinct functions of isozymes in the biochemical modulation of intracellular reactions have been identified. The first example is that of the same biochemical reaction required within morphologically distinct subcellular organelles, each possessing a different electrical charge; under such circumstances, it would be advantageous for several forms of the same enzyme to exist, each having a different charge to permit easier binding to and hence function on differently charged subcellular particles. Such situations have been described for LDH (Allen, 1961; Vesell, 1965) and malate dehydrogenase (Munkres, 1968; Henderson, 1968) isozymes. Even within an organelle, the same catalytic activity might have to function in association with macromolecules of different charge, in which case the existence of differently charged forms of an enzyme might be useful in their attachment.

Hultin and associates observed that in skeletal muscle from various species LDH-5 exists in a free and in a bound form, that certain physicochemical properties of these forms differ markedly, and that pyruvate inhibits the free form of LDH-5 much more than the bound form (Hultin et al, 1966; Melnick and Hultin, 1968; Hultin et al, 1972; Ehmann and Hultin, 1973; Melnick and Hultin, 1973). These kinetic differences between bound and free LDH-5 provide a metabolic regulatory function for the conversion of bound to free LDH-5 in skeletal muscle (Hultin et al, 1972) and suggest other possible controls that could operate through the distribution of different isozymes within different subcellular organelles. According to this hypothesis, in skeletal muscle during exercise as NADH concentrations increase, previously inactive, membrane bound LDH-5 would become solubilized, hence activated by virtue of its capacity to reoxidize NADH in the presence of an appropriate electron acceptor. This mechanism would help to maintain rates of glycolysis under anaerobic conditions while simultaneously minimizing competition for electrons between LDH-5 and mitochondria under aerobic conditions when more LDH-5 would be bound.

Fig. 7 shows the LDH isozyme pattern in different organelles isolated by sucrose density gradient centrifugation of homogenates from rat kidney. Even in undisrupted cells differential staining of isozymes by inhibitors shows localization of LDH isozymes in distinguishable regions of the cell.

A second possible regulatory function of multiple forms of an enzyme involves reversible reactions in which one isozyme is preferentially adapted to one substrate, whereas the other isozyme serves as a better catalyst for the second than for the first substrate. Aldolase isozymes provide an example of this situation. Class A aldolases favor cleavage of fructose 1,6-diphosphate to dihydroxyacetone phosphate and glyceraldehyde 3-phosphate and, therefore, glycolysis. On the other hand, class B aldolases are better suited to fructose 1,6-diphosphate synthesis and, thus, glyconeogenesis (Rutter et al, 1968). As yet, no specialized physiological function has been established for class C aldolases.

A third regulatory function of isozymes involves differential feedback inhibition. The role of isozymes in regulating rates of metabolite flow at branched metabolic pathways in bacteria is one of the most elegant and biologically significant functions yet ascribed to them (Stadtman, 1968). The situation in which several end products are derived from a common precursor controlled by a single enzyme is potentially dangerous for the cell. If the concentration of precursor is regulated by a feedback mechanism, an accumulation of one of the end products could lead to a deficiency of the others. In *E. coli*, aspartyl phosphate is an intermediate in the biosynthesis of methionine, threonine, and lysine. Aspartokinase converts aspartate to aspartyl phosphate in this organism, and the aspartokinase activity in crude extracts is inhibited by threonine, lysine, and methionine. Thus, by feedback inhibition, excessive accumulation of threonine, for example, would be expected to lead to a deficiency of lysine and methionine. However, such a deficiency fails to occur because the organism possesses three aspartokinase isozymes, one sensitive only to threonine, another only to lysine, and a third only to methionine. These observations with aspartokinase were extended to other bacterial enzymes, and it was concluded that many isozymic systems in bacteria function in a similar fashion at branched pathways to regulate the relative flow of metabolites

15

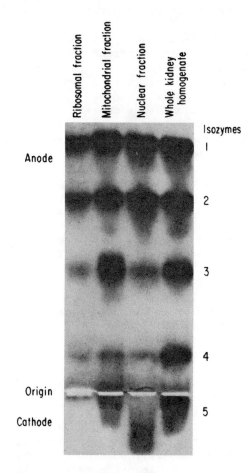

Fig. 7. Photograph of a starch gel showing LDH isozyme patterns in subcellular fractions of rat kidney. Homogenates of
whole kidney, nuclei, and mitochondria prior to application on
the gel each contained 16,000 units/ml; the ribosomal preparation contained 10,000 units/ml. Reproduced by permission from
Vesell (1965).

over each pathway (Stadtman, 1968). Thus far in mammals such
elegant control of branched metabolic pathways has not been
demonstrated.

16

Finally, tissue differences in isozyme distribution could arise from the large differences that exist among tissues in rates of isozyme degradation (Fritz et al, 1969, 1973). If a particular enzymic activity is critical within a tissue where one molecular form is rapidly degraded, an alternative form of the enzyme not subject to such rapid degradation would be required. Isozymes might fulfill this requirement if, as shown for the LDH isozymes (Fritz et al, 1973), various tissues of an organism exhibited wide differences in the rate constant for degradation of a single protein. Accordingly, some isozymic systems might have arisen from the biochemical necessity of maintaining the activity of critical enzymes in the face of rapid rates of degradation in particular tissues. The mechanisms responsible for the intracellular catabolism of enzymes are only now beginning to be understood. In the case of LDH isozymes, it may be pertinent to the degradative process that two intracellular LDH-inhibiting peptides have been identified; one inhibits LDH-1, and the other inhibits LDH-5 (Schoenenberger and Wacker, 1966). Both inhibitors appear in urine. Also in uremic dialysates, 3 low molecular weight LDH inhibitors, two of which appear to be urea and oxalate, were isolated by countercurrent extraction (Wilkinson et al, 1970). Studies performed in vitro revealed differences between LDH-1 and LDH-5 in rates of proteolytic degradation and thermal denaturation (Vesell and Yielding, 1968); furthermore, certain common intracellular metabolites differentially protected LDH isozymes from degradation (Vesell and Yielding, 1968). Large differences in rates of degradation of LDH isozymes suggest that perhaps differences among tissues in certain physicochemical conditions and in concentrations of various metabolites that affect isozyme decay in vitro might play a role in regulating their catabolism in vivo. From an evolutionary standpoint, development of isozymes to protect critical catalytic activities from destruction would represent an interesting, new type of selective pressure in which the goal would be maintenance of a sufficient variety of isozymic forms to permit survival of the enzymic activity required in each tissue.

In the case of the LDH isozymes, it has been suggested that gene duplication with subsequent mutations at both the parent and daughter loci is the biological mechanism for producing multiple molecular forms of an enzyme from a single primordial catalyst (Vesell and Bearn, 1962b). Such a mechanism had previously been proposed for the evolution of the α and β chains of the hemoglobin molecule (Ingram, 1961). The genes coding for the two different A and B subunits of the LDH isozymes apparently reside on different chromosomes (Santachiara et al, 1970; Ruddle et al, 1970).

17

On the basis of the inhibition of LDH-1 by pyruvate concentrations at which LDH-5 is uninhibited, many of us accepted at one time or another a theory designed to explain tissue-specific isozyme patterns (Cahn et al, 1962; Dawson et al, 1964). This hypothesis, termed the "aerobic-anaerobic" theory, maintains that LDH-1 is the main isozyme in so-called "aerobic" tissues, whereas LDH-5 predominates in so-called "anaerobic" tissues. However, quantitative oxygen tensions or pyruvate concentrations required to place a tissue in either the "aerobic" or "anaerobic" category have not been offered. Thus, tissues frequently are designated either "aerobic" or "anaerobic," not because certain pyruvate concentrations have been measured in them under well defined conditions, but rather as a convenience to investigators to suit various LDH isozyme patterns, which are then interpreted as conforming to the predictions of the theory. This circular reasoning is illustrated in a textbook of genetics where liver is designated an anaerobic tissue because it contains almost exclusively LDH-5 (Porter, 1968). Hepatocytes do contain almost exclusively LDH-5, but hepatocytes are one of the most aerobic of cells, containing an abundance of mitochondria. The fact that the isozyme pattern of hepatocytes is contrary to the predictions of the theory must be taken as evidence against the theory. Other tissues whose isozyme patterns do not conform to the predictions of the theory are mature erythrocytes, platelets, and lens fibers (Vesell, 1965b). The diverse biological reasons for a relative abundance or deficiency of mitochondria in different tissues tend to be overlooked by the "aerobic-anaerobic" theory which is limited to a consideration of the relative abundance of "A" and "B" subunits, as though this ratio alone could determine whether a tissue functioned aerobically or anaerobically.

The "aerobic-anaerobic" theory is based on kinetic differences between LDH-1 and LDH-5. At 25°C LDH-1 is inhibited by pyruvate concentrations to which LDH-5 is resistant, but this great difference between LDH-1 and LDH-5 in capacity to withstand substrate inhibition at 25°C is substantially reduced at temperatures more physiological for mammals. This fact has been known since 1961, when the data shown in Fig. 8 were published by Plagemann et al. Although rabbit LDH-1 is very sensitive to substrate inhibition at 6°C, at the more physiological temperature of 40°C, its sensitivity to the concentrations of pyruvate used in the experiment is negligible from the point of view of forming a basis for a theory of biological function (Fig. 8). We confirmed this observation for human LDH-1 (Vesell, 1968).

Another question relating to the "aerobic-anaerobic" theory

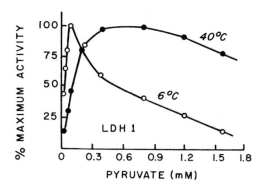

Fig. 8. Effect of temperature on pyruvate inhibition of rabbit LDH-1. Reproduced by permission from Plagemann et al (1961). Note that at physiological temperature for the rabbit substrate inhibition is negligible compared to that observed at 6°C.

is whether, under even the most anaerobic circumstances, concentrations of pyruvate sufficiently high to inhibit LDH-1 are ever attained in vivo. Concentrations of pyruvate and lactate in various tissues have been known for many years (Vesell and Pool, 1966). Concentrations of pyruvate and lactate in canine skeletal muscle even under the most anaerobic conditions rarely exceed 1.0 and 25 mM, respectively; even under unphysiological conditions of temperature (25°C), LDH-1 is not significantly inhibited by these substrate concentrations (Vesell and Pool, 1966).

Furthermore, LDH-1 inhibition is dependent on the degree of dilution of the enzyme. Intracellular concentrations of LDH isozymes have been estimated (Wuntch et al, 1969). At these intracellular isozyme concentrations appreciable pyruvate inhibition of LDH-1 cannot be detected, even at pyruvate concentrations as high as 20 mM (Wuntch et al, 1969, 1970a). It should be emphasized that intracellular concentrations of LDH isozymes, pyruvate and lactate are not uniform. The enzyme and its substrates, as mentioned in the discussion of subcellular isozyme localization, exhibit specific subcellular distributions; hence, estimations of concentrations based on tissue homogenates are at best only approximations. Because the concentrations of LDH isozymes at certain regions within the cell are much higher than at other regions, the concentrations of LDH isozymes in certain subcellular sites are higher than those calculated from tissue homogenates (Wuntch et al, 1969). Similarly, substrate concentrations are also probably higher at certain sites within the cell than calculations made from tissue homogenates would indicate. However, such differences as may exist between estimations based on tissue homogenates and the local concentrations actually existing within cells do not alter the conclusion that no substrate inhibition of LDH-1 occurs under physiological conditions. No LDH-1 inhibition could be detected when the isozyme was present at physiological concentrations (Wuntch et al, 1969, 1970a, 1970b) and the experiments were performed under physiological conditions without prolonged prior incubation (Fig. 9).

Inhibition is not observed with excess pyruvate and lactate at physiological enzyme concentrations possibly because, under these conditions, the abortive ternary complex composed of LDH, NAD, and pyruvate is not formed. It is believed that inhibition of LDH by pyruvate in vitro results from formation of this ternary complex, which is a time-dependent reaction. The time for formation of this complex is considerably longer at the high concentrations of the enzyme that exist in vivo (Everse et al, 1970). Therefore, whether this ternary complex ever forms in vivo or under physiological conditions in vitro is at the core of the dispute concerning the relationship between intracellular pyruvate concentrations and tissue-specific isozyme patterns. When the reaction is run immediately after mixing physiological concentrations of LDH, pyruvate, and NADH without prior incubation, the results suggest that ternary-complex formation and inhibition do not occur (Wuntch et al, 1969, 1970a, 1970b). However, Everse et al (1970) demonstrated that, when LDH and pyruvate are preincubated for 30 min prior to addition of NADH, substrate inhibition does occur. For several reasons this experiment may be unphysiological. It ignores the fact that intracellularly several dehydrogenases and other proteins compete with LDH for NADH (Wuntch et al, 1970b). Accordingly, when incubated in a mixture containing LDH, NADH, and pyruvate, these dehydrogenases, as well as other proteins and the coenzymes themselves, decrease the extent of substrate inhibition (Wuntch et al, 1970b). Furthermore, concentrations of pyruvate and NADH in the cell are in a state of dynamic flux (Chance et al, 1965; Frenkel, 1968a,b). They do not remain at the constant concentrations assumed in the model proposed by Everse et al (1970) where LDH, pyruvate, and NADH are incubated in a cuvette for 30 min unexposed to any of the numerous metabolites normally present in cells. Since NAD and NADH concentrations change continuously within cells (Chance et al, 1965; Frenkel, 1968a,b), their concentrations at actual sites of the LDH isozymes and other NAD-linked dehydrogenases are difficult to determine, as is the extent of abortive ternary-complex formation.

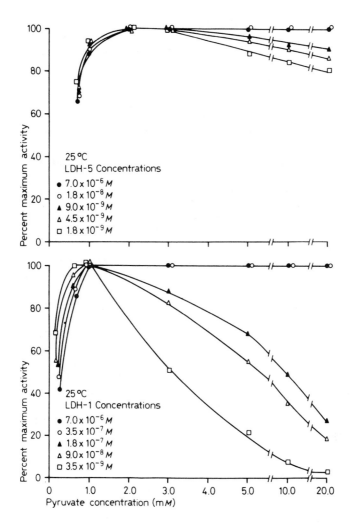

Fig. 9. Effect of increasing pyruvate concentration on the activity of several concentrations of LDH-1 and LDH-5 purified from rat kidney. All reagents were prepared in 0.1 M sodium phosphate buffer, pH 7.0, and the final NADH concentration was 0.56 mM. Molar concentrations of partially purified LDH-1 and LDH-5 were calculated from turnover numbers. The LDH-1 was purified 130-fold, and LDH-5 was purified 98-fold. The maximum specific activity remained constant over the range of enzyme concentrations examined. With 1.0 mM pyruvate, approximately 3.2 µmoles of NADH were oxidized per second per milli-

(*Fig. 9 legend continued*)   gram of LDH-1, and approximately
4.5 μmoles of NADH were oxidized per second per milligram of
LDH-5 with 2.0 mM pyruvate.   Reproduced by permission from
Wuntch et al (1970a).

For different reasons, Griffin and Criddle (1970) believe
that the LDH tetramer is probably not subject to inhibition
in vivo; they reported that the monomeric subunit of LDH, and
not the tetramer, is required for ternary-complex formation
in vitro. Finally, the experiments of Coulson and Rabin (1969)
are pertinent to the subject of substrate inhibition. They
suggested that LDH-1 inhibition by pyruvate is attributable
to the enol form of pyruvate present in commercial preparations
as an impurity (Coulson and Rabin, 1969); the enol and keto
forms exist in vivo in equilibrium but the actual extent of
intracellular LDH inhibition would be restricted by the enol-
keto tautomerization rate of pyruvate. According to this view,
only the intracellular pyruvate present in the enol form could
inhibit LDH, and this amount must be less than the total intracellular pyruvate measured in the experiments described earlier
(Vesell and Pool, 1966).

The problem of LDH isozyme fucntion may be approached from
a different point of view by posing this question: if an isozymic system were to be constructed in order to provide appropriate concentrations of one isozyme uninhibited by substrate
and of another isozyme sensitive to substrate in tissues where
substrate does not accumulate, how many isozymic forms would
be required from a purely functional, teleological point of
view? Clearly, only 2. Since LDH-5 is uninhibited by substrate, it should appear in tissues where lactate accumulates.
Since LDH-5 resists inhibition by high substrate concentrations
better than all other LDH isozymes, it should, therefore, exist
alone in skeletal muscle. By contrast, LDH-1 should be the
only isozyme in cardiac muscle where lactate is rapidly utilized and hence fails to accumulate. Fig. 2 shows that, unlike
predictions based only on this "aerobic-anaerobic" theory, the
biological facts are that LDH isozymes do not appear in isolation in human tissues. Each LDH isozyme in each tissue appears
with several other LDH isozymes. Thus the functional significance of LDH-2, -3, and -4 is ignored by the "aerobic-anaerobic" theory.

A theory designed to explain the functional significance
of the LDH isozymes should address itself to the metabolic
functions of each of the separate isozymes present in most
mammalian tissues, not simply to only two of these forms.
Hopefully, it might even shed some light on the LDH-X located
in sperm. While it is true that LDH-5 is the predominant iso-

zyme in many tissues and LDH-1 is the predominant isozyme in many other tissues, it is also true that in a few tissues, such as lung and pancreas, neither LDH-1 nor LDH-5 predominates but most LDH activity resides in LDH-2, -3, and -4.

In this discussion of the metabolic function of isozymes, the principle has been stressed that any proposed function of an isozymic system should conform as closely as possible to the available biological facts, including the number of the biochemically separable forms of the enzyme and the physiological range of conditions under which these forms exist in different tissues, including physiological concentrations of substrate, enzyme, inhibitors, as well as pH, temperature, and even incubation time in the cuvette. By varying the conditions in the cuvette to keep them as close as possible to those prevailing in vivo, investigators will be more likely than otherwise to develop theories of isozymic function that have biological meaning.

E. Future Medical Applications of Isozymes

In conclusion a glimpse into possible future medical uses of isozymes may be appropriate. Probably the diagnostic uses of isozymes will be greatly expanded in the future both by the introduction for clinical application of new isozyme systems and by simultaneous use of 2 or more isozyme systems in a patient to help establish diagnoses more specifically. Also simpler, more rapid chemical tests to identify isozymes will be achieved. The increasing ease of automation of laboratory tests encourages the expectation that several isozyme systems could be automated with great consequent saving in labor and expense, thereby facilitating more widespread use. At the level of experimental medicine, it may be anticipated that familiarity with the techniques of identifying isozymes in tissues and biological fluids will lead to successful application of the isozyme technique in elucidating some of the secrets that still shroud the pathogenesis of many human diseases. Finally, diagnostic and experimental medical applications of isozymes could be improved and enlarged by development of fresh approaches to the functional significance of the isozyme systems used in medicine.

REFERENCES

Allen, J. M. 1961. Multiple forms of lactic dehydrogenase in tissues of the mouse: their specificity, cellular localization, and response to altered physiological conditions. *Ann. N. Y. Acad. Sci.* 94: 937-951.

Cahn, R. D., N. O. Kaplan, L. Levine, and E. Zwilling 1962. Nature and development of lactic dehydrogenases. *Science* 136: 962-969.

Chance, B., B. Schoener, and S. Elsaesser 1965. Metabolic control phenomena involved in damped sinusoidal oscillations of reduced diphosphopyridine nucleotide in a cell-free extract of *Saccharomyces carlsbergensis*. *J. Biol. Chem.* 240: 3170-3181.

Cohen, L., J. Djordjevich, and V. Ormiste 1964. Serum lactate dehydrogenase isoenzyme patterns in cardiovascular and other diseases, with particular reference to acute myocardial infarction. *J. Lab. Clin. Med.* 64: 355-374.

Cohen, L., J. Block, and J. Djordjevich 1967. Sex-related differences in isozymes of serum lactic dehydrogenase. *Proc. Soc. Exp. Biol. Med.* 126: 55-60.

Cornish, H. H., M. L. Bartb, and V. N. Dodson 1970. Isozyme profiles and protein patterns in specific organ damage. *Toxicol. Appl. Pharmacol.* 16: 411-423.

Coulson, C. J. and B. B. Rabin 1969. Inhibition of lactate dehydrogenase by high concentrations of pyruvate: the nature and removal of the inhibitor. *FEBS Letters.* 3: 333-337.

Dawson, D. M., T. L. Goodfriend, and N. O. Kaplan 1964. Lactic dehydrogenases: functions of the two types. *Science* 143: 929-933.

Dreyfus, J. C., J. Demos, F. Schapira, and G. Schapira 1962. Lactate dehydrogenase in myopathic muscle: apparent persistence of a fetal type. *C. R. Acad. Sci.* (Paris) 254: 4384.

Ehmann, J. D. and H. O. Hultin 1973. Substrate inhibition of soluble and bound lactate dehydrogenase. *Arch. Biochem. Biophys.* 154: 471-475.

Everse, J., R. L. Berger, and N. O. Kaplan 1970. Physiological concentrations of lactate dehydrogenases and substrate inhibition. *Science* 168: 1236-1238.

Frenkel, R. 1968a. Control of reduced pyridine nucleotide oscillations in beef heart extracts. I. Effects of modifiers of phosphofructokinase activity. *Arch. Biochem. Biophys.* 125: 151-156.

Frenkel, R. 1968b. Control of reduced diphosphopyridine nucleotide oscillations in beef heart extracts. II. Oscillations of glycolytic intermediates and adenine nucleotides. *Arch. Biochem. Biophys.* 125: 157-165.

Fritz, P. J., E. S. Vesell, E. L. White, and K. M. Pruitt 1969. The roles of synthesis and degradation in determining tissue concentrations of lactate dehydrogenase-5. *Proc. Natl. Acad. Sci.* 62: 558-565.

Fritz, P. J., E. L. White, K. M. Pruitt, and E. S. Vesell 1973. Lactate dehydrogenase isozymes. Turnover in rat heart, skeletal muscle, and liver. *Biochem.* 12: 4034-4039.

Gerlach, U. and W. Fegeler 1973. Variations with age in the lactate dehydrogenase isoenzyme pattern in the aorta of rats. *Enzyme* 14: 1-12.

Goldman, R. D., N. O. Kaplan, and T. C. Hall 1964. Lactic dehydrogenase in human neoplastic tissues. *Cancer Res.* 24: 389-399.

Goodfriend, T. L. and N. O. Kaplan 1964. Effects of hormone administration on lactic dehydrogenase. *J. Biol. Chem.* 239: 130-135.

Griffin, J. H. and R. S. Criddle 1970. Substrate-inhibited lactate dehydrogenase. Reaction mechanism and essential role of dissociated subunits. *Biochem.* 9: 1195-1205.

Henderson, N. S. 1968 Intracellular location and genetic control of isozymes of NADP-dependent isocitrate dehydrogenase and malate dehydrogenase. *Ann. N. Y. Acad. Sci.* 151: 429-440.

Hess, B. 1958. DPN-dependent enzymes in serum. *Ann. N. Y. Acad. Sci.* 75: 292-303.

Hill, B. R. 1958. Further studies of the fractionation of lactic dehydrogenase of blood. *Ann. N. Y. Acad. Sci.* 75: 304-310.

Hultin, H. O., J. D. Ehmann, and R. L. Melnick 1972. Modification of kinetic properties of muscle lactate dehydrogenase by subcellular associations and possible role in the control of glycolysis. *J. Food Sci.* 37: 269-273.

Hultin, H. O., C. Westort, and J. H. Southard 1966. Adsorption of lactate dehydrogenase to the particulate fraction of homogenized skeletal muscle. *Nature* 211: 853-854.

Ingram, V. M. 1961. Gene evolution and the haemoglobins. *Nature* (London) 189: 704-708.

Jacey, M. and K. E. Schaefer 1968. Circadian cycles of lactic dehydrogenase in urine and blood plasma: response to high pressure. *Aerospace Med.* 39: 410-412.

Mahy, B. W. J., K. E. K. Rowson, C. W. Parr, and M. H. Salaman 1965. Studies on the mechanism of action of the Riley virus. I. Action of substances affecting the reticulo-endothelial system on plasma enzyme levels in mice. *J. Exp. Med.* 122: 967-981.

Markert, C. L. and F. Møller 1959. Multiple forms of enzymes: tissue, ontogenetic and species specific patterns. *Proc. Natl. Acad. Sci.* 45: 753-763.

Markert, C. L. and H. Ursprung 1962. The ontogeny of isozyme patterns of lactate dehydrogenase in the mouse. *Develop. Biol.* 5: 363-381.

Meister, A. 1950. Reduction of α,γ-diketo and α-keto acids catalyzed by muscle preparations and by crystalline lactic dehydrogenase. *J. Biol. Chem.* 184: 117-129.

Melnick, R. L. and H. O. Hultin 1968. Solubilization of bound lactate dehydrogenase by NADH in homogenates of trout skeletal muscle as a function of tissue concentration. *Biochem. Biophys. Res. Commun.* 33: 863-868.

Melnick, R. L. and H. O. Hultin 1973. Studies on the nature of the subcellular localization of lactate dehydrogenase and glyceraldehyde-3-phosphate dehydrogenase in chicken skeletal muscle. *J. Cellular Physiol.* 81: 139-148.

Munkres, K. D. 1968. Genetic and epigenetic forms of malate dehydrogenase in *Neurospora*. *Ann. N. Y. Acad. Sci.* 151: 294-306.

Nachlas, M. M., M. M. Friedman, and S. P. Cohen 1964. A method for the quantitation of myocardial infarcts and the relation of serum enzyme levels to infarct size. *Surgery* 55: 700-708.

Neilands, J. B. 1952. The purity of crystalline lactic dehydrogenase. *Science* 115: 143-144.

Nerenberg, S. T. and G. Pogojeff 1969. Laboratory diagnosis of specific organ diseases by means of combined serum isoenzyme patterns. *Am. J. Clin. Path.* 51: 429-439.

Nora, J. J., D. A. Cooley, B. L. Johnson, S. C. Watson, and J. D. Milam 1969. Lactate dehydrogenase isozymes in human cardiac transplantation. *Science* 164: 1079-1080.

Plagemann, P. G. W., K. F. Gregory, and F. Wróblewski 1961. Die elektrophoretisch trennbaren Lactat-dehydrogenasen des Säugetieres. III. Einfluss der Temperatur auf die Lactat-dehydrogenasen des Kaninchens. *Biochem. Z.* 334: 37-48.

Porter, I. H. 1968. *Heredity and Disease*. McGraw-Hill, New York, 222

Richterich, R. and A. Burger 1963. Lactate dehydrogenase isoenzymes in human cancer cells and malignant effusions. *Enzym. Biol. Clin.* 3: 65-72.

Rowson, K. E. K., B. W. J. Mahy, and M. H. Salaman 1965. Studies on the mechanism of action of the Riley virus II. Action of substances affecting the reticuloendothelial system on the level of viraemia. *J. Exp. Med.* 122: 983-992.

Ruddle, F. H., V. M. Chapman, T. R. Chen, and R. J. Klebe 1970. Linkage between human lactate dehydrogenase A and B and Peptidase B. *Nature* (London) 227: 251-257.

Rutter, W. J., T. Rajkumar, E. Penhoet, and M. Kochman 1968. Aldolase variants: structure and physiological significance. *Ann. N. Y. Acad. Sci.* 151: 102-117.

Santachiara, A. S., M. M. Nabholz, V. Maggiano, A. J. Darlington, and W. Bodmer 1970. Genetic analysis with man-mouse somatic cell hybrids. *Nature* (London) 227: 248-251.

Schoenenberger, G. A. and W. A. C. Wacker 1966. Peptide inhibitors of lactic dehydrogenase (LDH). II. Isolation and characterization of peptides I and II. *Biochem.* 5: 1375-1379.

Schweitzer, E. S., F. Farron, and W. E. Knox 1973. Distribution of lactate dehydrogenase and its subunits in rat tissues and tumors. *Enzyme* 14: 173-184.

Stadtman, E. R. 1968. The role of multiple enzymes in the regulation of branched metabolic pathways. *Ann. N. Y. Acad. Sci.* 151: 516-530.

Starkweather, W. H. and H. K. Schoch 1962. Some observations on the lactate dehydrogenase of human neoplastic tissue. *Biochim. Biophys. Acta* 62: 440-442.

Starkweather, W. H., L. Cousineau, H. K. Schoch, and C. J. Zarafonetis 1965. Alterations of erythrocyte lactate dehydrogenase in man. *Blood* 26: 63-73.

Starkweather, W. H., H. H. Spencer, E. L. Schwarz, and H. K. Schoch 1966. The electrophoretic separation of lactate dehydrogenase isoenzymes and their evaluation in clinical medicine. *J. Lab. Clin. Med.* 67: 329-343.

Vesell, E. S. 1965a. Genetic control of isozyme patterns in human tissues. *Progr. Med. Genet.* 4: 128-175.

Vesell, E. S. 1965b. Lactate dehydrogenase isozyme patterns of human platelets and bovine lens fibers. *Science* 150: 1735-1737.

Vesell, E. S. 1968. Introduction. *Ann. N. Y. Acad. Sci.* 151: 1-13.

Vesell, E. S., and A. G. Bearn 1957. Localization of lactic dehydrogenase activity in serum fractions. *Proc. Soc. Exp. Biol. Med.* 94: 96-99.

Vesell, E. S., and A. G. Bearn 1961. Isozymes of lactic dehydrogenase in human tissues. *J. Clin. Invest.* 40: 586-591.

Vesell, E. S., and A. G. Bearn 1962a. Localization of a lactic dehydrogenase isozyme in nuclei of young cells in the erythrocyte series. *Proc. Soc. Exp. Biol. Med.* 111: 100-104.

Vesell, E. S., and A. G. Bearn 1962b. Variations in the lactic dehydrogenase of vertebrate erythrocytes. *J. Gen. Physiol.* 45: 553-565.

Vesell, E. S., and I. A. Brody 1964. Biological applications of lactic dehydrogenase isozymes. *Ann. N. Y. Acad. Sci.* 121: 544-559.

Vesell, E.S., M.P. Feldman, and E.D. Frank 1959. Plasma lactic dehydrogenase activity in experimental hemorrhagic shock. *Proc. Soc. Exp. Biol. Med.* 101: 644-648.

Vesell, E.S., J. Philip, and A.G. Bearn 1962. Comparative studies of the isozymes of lactic dehydrogenase in rabbit and man. Observations during development and in tissue culture. *J. Exp. Med.* 116: 797-806.

Vesell, E.S. and P.E. Pool 1966. Lactate and pyruvate concentrations in exercised ischemic canine muscle: relationship of tissue substrate level to lactate dehydrogenase isozyme pattern. *Proc. Natl. Acad. Sci. USA.* 55: 756-762.

Vesell, E.S. and K.L. Yielding 1968. Protection of lactate dehydrogenase isozymes from heat inactivation and enzymatic degradation. *Ann. N.Y. Acad. Sci.* 151: 678-689.

Wieme, R.J. 1959. *Studies on Agar-Gel Electrophoresis.* Arscia Uitgaven N.V., Brussels.

Wieme, R.J. and Y. Van Maercke 1961. The fifth (electrophoretically slowest) serum lactic dehydrogenase as an index of liver injury. *Ann. N.Y. Acad. Sci.* 94: 898-911.

Wieme, R.J. and M.J. Lauryssens 1962. Lactate dehydrogenase multiplicity in normal and diseased human muscle. *Lancet* 1: 433.

Wilkinson, J.H., Y. Fujimoto, D. Senesky, and G.D. Ludwig 1970. Nature of the inhibitors of lactate dehydrogenase in uremic dialysates. *J. Lab. Clin. Med.* 75: 109-119.

Wróblewski, F. and K.F. Gregory 1961. Lactic dehydrogenase isozymes and their distribution in normal tissues and plasma and in disease states. *Ann. N.Y. Acad. Sci.* 94: 912-932.

Wuntch, T., E.S. Vesell, and R.F. Chen 1969. Studies on rates of abortive ternary complex formation of lactate dehydrogenase isozymes. *J. Biol. Chem.* 244: 6100-6104.

Wuntch, T., R.F. Chen, and E.S. Vesell 1970a. Lactate dehydrogenase isozymes: kinetic properties at high enzyme concentrations. *Science* 167: 63-65.

Wuntch, T., R.F. Chen, and E.S. Vesell 1970b. Lactate dehydrogenase isozymes: further kinetic studies at high enzyme concentration. *Science* 169: 480-481.

MECHANISMS OF ACTION AND BIOLOGICAL FUNCTIONS
OF VARIOUS DEHYDROGENASE ISOZYMES

JOHANNES EVERSE and NATHAN O. KAPLAN
Department of Chemistry
University of California, San Diego
La Jolla, California 92037

ABSTRACT: A short review is presented pertaining to the considerations that led us to propose the theory that the A-type lactate dehydrogenase is especially geared to serve as a pyruvate reductase in anaerobically metabolizing tissues, whereas B-type enzyme is better suited to serve as a lactate dehydrogenase in aerobically metabolizing tissues. The paper summarizes the evidence suggesting that the activity of the B-type LDH as well as that of the mitochondrial malate dehydrogenase may be regulated *in vivo* by means of abortive ternary complex formation. The advantage of such a regulatory mechanism for the tissues involved is discussed.

Experimental evidence is also presented suggesting that the enol-pyruvate, essential for the formation of the abortive LDH-NAD-pyruvaté complex, may be generated *in vivo* by a non-enzymatic enolization of keto-pyruvate.

About a decade ago we first advanced the theory that the A-type isozyme of lactate dehydrogenase (LDH) is especially geared to serve as a pyruvate reductase in tissues that are dependent on anaerobic glycolysis, whereas the properties of the B-type isozyme make this enzyme more efficient for the oxidation of lactate in tissues with an aerobic metabolism (Kaplan, 1961; Cahn et al., 1962). Note that in alternative terminologies A = M and B = H . This hypothesis has been subjected to some controversy over the ensuing years, especially for the following reasons: (1) It has been stated that differences in the kinetic properties between the two isozymes, and especially the inhibition shown by the A-type LDH in the presence of high pyruvate concentrations, exist only in laboratory experiments and has no physiological significance (Vesell and Pool, 1966; Wuntch et al., 1969, 1970). (2) It has been argued that the theory cannot be generally applied, since there are a number of exceptions to this thesis. These include the liver, which is aerobic but contains the A-type LDH in many animals, and the erythrocytes, which have an anaerobic metabolism, but their LDH isozyme distribution varies greatly from one species to another (see below).

As a result of these arguments the theory has been accepted or partly accepted by certain members of the scientific community, whereas others do not believe that the theory
has any general significance. The position of the non-belie-
vers has been eloquently explained and defended by Dr.
Vesell at this Conference, and it may therefore be useful if
we also utilize this opportunity to restate our theory concerning the function of the two LDH isozymes in somewhat more
detail as well as present the evidence that has accumulated
in support of our views.

Let us first focus our attention on the metabolic pathways involved in the energy production for the contractile
process in heart muscle as compared to that in white skeletal
muscle. White skeletal muscles, which are largely anaerobic,
are almost completely dependent on glycolysis for their ATP
production. Hence, in order to ensure the proper operation
of these muscles it is imperative that the cellular breakdown of glucose to lactate occurs rapidly and effectively.
This is ensured by the large negative free energies of the
pyruvate kinase and the lactate dehydrogenase reactions, together equaling about -13.5 kcal. The complete reduction of
pyruvate to lactate ($\Delta G^{O}{}' = -6.0$ kcal.) thus serves as a
final pulling force in the glycolytic pathway, thereby preventing the accumulation of intermediary products and simultaneously assuring a complete reoxidation of the NADH that
is formed in the triosephosphate dehydrogenase reaction. It
would obviously be disadvantageous for a proper functioning
of the white skeletal muscles if they contained an LDH that
would be partially inhibited from reducing pyruvate to
lactate.

Whereas striated muscles derive their energy almost 100%
from glucose, the heart muscle derives its energy from several sources. Under conditions of complete rest about 65%
of the energy output of the heart is derived from the oxidation of fatty acids and about 35% from the oxidation of
glucose and lactate (Schlant, 1970). During prolonged and
moderately severe exercise, however, the heart may obtain as
much as 60% of its energy from the oxidation of lactate (Keul
et al., 1965; Hirche and Lochner, 1961; Lochner and Nassari,
1959). In fact, it has been shown in several laboratories
that the heart takes up lactate in increasing amounts when
the arterial level of lactate becomes elevated (Bing et al.,
1954; Bing, 1965; Krasnow et al., 1962). This is schematically illustrated in Fig. 1. The lactate produced in the
skeletal muscles is transported via the bloodstream, and part
of it is utilized by the heart for energy production. Under
these conditions the LDH in the skeletal muscles actually

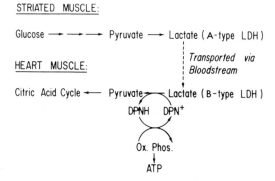

STRIATED MUSCLE:

Glucose —→ —→ —→ Pyruvate —→ Lactate (A-type LDH)

HEART MUSCLE:

Citric Acid Cycle ←— Pyruvate ⇌ Lactate (B-type LDH)

Fig. 1. Biological functions of the lactate dehydrogenase isozymes.

---

serves as a pyruvate reductase, whereas the enzyme in the heart muscle operates as a lactate dehydrogenase.

The concentration of lactate in blood normally varies between 1 and 12 mM. Since the equilibrium constant of the LDH catalyzed reaction is about 5×10^4 at neutral pH, and the $NAD^+/NADH$ ratio in heart cells is about 1000 (Opie and Mansford, 1971; Safer et al., 1971)[1], lactate oxidation would be expected to occur readily when the intracellular levels of lactate are higher than about 5 mM. However, when the blood lactate level is below this value, lactate oxidation could not be expected to occur at a sufficient rate to sustain the heart's activity. Increasing amounts of glucose and fatty acids are then utilized in order to maintain a proper functioning of the heart muscle.

Thus, under conditions of complete rest with concomitant low levels of blood lactate, the heart utilizes significant amounts of glucose as an alternative source of energy. This glucose is metabolized in the heart cells via the glycolytic pathway to pyruvate and NADH. At this point, unlike the reaction in white muscles, where reduction of pyruvate to lactate is imperative in order to regenerate NAD^+, the pyruvate could enter the mitochondria to be oxidized to acetyl-

---

[1]This ratio was calculated from the lactate/pyruvate ratio and the α-glycerophosphate/dihydroxyacetonephosphate ratio in perfused hearts. A lower value (30-50) is usually obtained by direct measurements of the NAD^+ and NADH levels in heart extracts.

CoA, and the NADH (formed from glycolysis) could enter the mitochondria via the malate or the α-glycerophosphate shuttle or be reoxidized via other pathways (Kaplan, 1972; Isaacs et al., 1968). These pathways thus compete with the LDH reaction for the available pyruvate and NADH. This is shown schematically in Fig. 2. On the basis of this competition, one would expect that under normal conditions at least some of the pyruvate would be reduced to lactate in the heart muscle. It is well known, however, that no lactate is produced by the heart muscle even when the level of lactate in the blood is low (Schlant, 1970). Furthermore, in terms of energy production it is obviously far more advantageous for the heart to metabolize the pyruvate and NADH via the mitochondria than to utilize the LDH reaction (reduction of pyruvate to lactate).

These were the underlying considerations which led us to hypothesize that in aerobic tissues such as the heart the LDH reaction may be inhibited in order to prevent the pyruvate from being reduced to lactate.

EVIDENCE IN SUPPORT OF THE ANAEROBIC-AEROBIC THEORY

Much evidence has been accumulated over the years indicating that the inhibition of the B-type LDH at high pyruvate concentrations is caused by the formation of a ternary complex among the enzyme, NAD^+, and the enol-form of pyruvate (Fromm, 1963; Kaplan et al., 1968; Gutfreund et al., 1968; Everse et al., 1971). In this complex the NAD^+ and the enol-pyruvate are covalently linked to each other at the 4 position of the nicotinamide ring as shown in Fig. 3. The structure of this complex has recently been confirmed in our laboratory using high frequency nmr (Arnold and Kaplan, 1974).

A possible mechanism for the formation of the abortive complex was first described by Coulson and Rabin (1969), who also indicated its similarity to the mechanism of lactate oxidation. This is shown in Fig. 4. The fact that the same amino acid residues that are involved in the enzymatic activity are also involved in the formation of the abortive complex accounts for the fact that ternary complexes are formed by both LDH isozymes; differences exist solely in the affinities of the isozymes for NAD^+ and enol-pyruvate (Everse and Kaplan, 1973).

No information has been presented as yet pertaining to the origin of the enol-pyruvate. The equilibrium constant of the pyruvate tautomerism is 4.1×10^{-6} at pH 7 (Albery et al., 1965), indicating that at equilibrium only a very small percentage of pyruvate exists as enol-pyruvate. Furthermore, the rate of enolization is a very slow process ($k = 5 \times 10^{-8} \, sec^{-1}$)

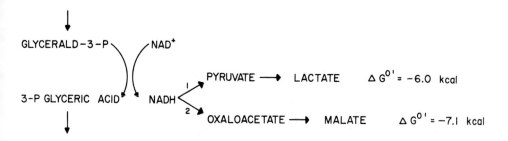

Fig. 2. Scheme illustrating two pathways by which cytoplasmic NADH may be oxidized in aerobically metabolizing tissues.

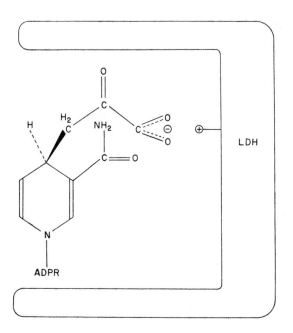

Fig. 3. Structure of the LDH-NAD-pyruvate ternary complex. From Everse et al., 1971.

(Albery et al., 1965; Kosicki, 1968). Enol-pyruvate has been shown to be an intermediate in the pyruvate kinase and the oxaloacetate decarboxylase reactions (Rose, 1960; Kosicki and Westheimer, 1968; Willard and Rose, 1973); however, no evidence has been presented, indicating that enol-pyruvate dis-

33

sociates from these enzymes into the cytoplasm.

We have recently performed some preliminary experiments in an attempt to solve this problem. Previously, we have shown that most of the B-type LDH is converted to the abortive complex when physiological concentrations of enzyme, NAD^+, and commercial pyruvate are incubated for about 60 minutes (Everse et al., 1972). When the commercial pyruvate is replaced by pure ketopyruvate, one observes the same rate of complex formation. This is illustrated by the data in Table I. No increase in rate was observed in the presence of pyruvate kinase or oxaloacetate decarboxylase or when a dialyzed heart extract was added to the reaction mixture. The same data were observed when we followed the increase in optical density at 325 nm instead of the increase in LDH inhibition. These data suggest that the enol-pyruvate required for the formation of the abortive complex is formed by a non-enzymatic enolization of keto-pyruvate and that under our experimental conditions this enolization is probably the rate limiting step in the formation of the abortive complex when physiological concentrations of enzyme are used. The data also indicate that even though the intracellular steady-state concentration of enol-pyruvate may be very low, the rate of ternary complex formation is sufficiently high to be of physiological significance. It should also be borne in mind that enol-pyruvate may be released upon the dissociation of the abortive complex; its rate of conversion to ketopyruvate, however, is sufficiently slow so that it may remain available for some time for subsequent ternary complex formation.

TABLE I

Experiment	Time at which 50% inhibition of LDH was observed (min.)
LDH, NAD^+, commercial pyruvate	24
LDH, NAD^+, commercial pyruvate and dialyzed heart extract	25
LDH, NAD^+, phosphoenolpyruvate, pyruvate kinase, ADP, Mg^{++}	25
LDH, NAD^+, oxoaloacetate, oxaloacetate decarboxylase, Mg^{++}	23
LDH, NAD^+, purified keto-pyruvate	23

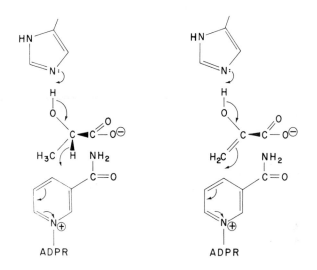

Fig. 4. A comparison of the proposed reaction mechanisms for the LDH-catalyzed oxidation of' lactate and the LDH-catalyzed formation of the NAD-pyruvate binary complex. The imidazole residue is postulated to be the proton-donating group on the lactate dehydrogenase. From Everse and Kaplan, 1973.

The LDH-NAD-pyruvate complex is slowly dissociated in the presence of NADH (Kaplan et al., 1968), whereas the addition of lactate causes a rapid dissociation of the complex (Everse and Kaplan, 1973). These results might be predicted: When the intracellular lactate level rises, it would be of advantage to the heart to have an active LDH and metabolize the lactate. An increase in the NADH level, however, should not *a priori* activate LDH, but rather should stimulate the oxidation of NADH via the mitochondrial shuttle systems. The B-type LDH thus appears to operate as a one-way valve: Under normal metabolic conditions it resists the reduction of pyruvate, but readily catalyzes the oxidation of lactate.

The rate of dissocation of the abortive complex upon dilution is also rather slow. This fact prompted us to attempt to demonstrate its presence *in vivo*. The data shown in Fig. 5 indicate that a considerable increase in LDH activity is found following the disruption of heart tissue, whereas no such activation is observed when chicken breast muscle tissue is used. The activation of the LDH from heart tissue is considerably less in the presence of 0.1 mM NAD[+] or pyruvate, although the same activation is observed following dialysis

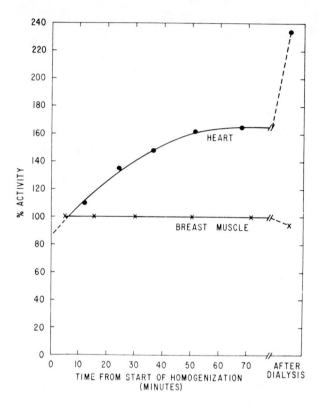

Fig. 5. Typical activity rates of LDH from chicken hearts and
breast muscle following disruption of the tissues. 10 g of
fresh chicken hearts and breast muscles were minced in a blender for 30 sec with 30 ml of 0.1 M phosphate buffer, pH 7.5.
The mixture was filtered over a Buchner funnel and the filtrate
was transferred to syringe 1 of a Stopped Flow Apparatus.
Syringe 2 contained 2.8×10^{-4} M NADH and 1×10^{-3} M pyruvate in the same buffer. The time the blender was started is
indicated as zero minutes. Dialysis was overnight against 1
liter of 0.1 M phosphate buffer, pH 7.5. From Kaplan and
Everse, 1972.

of the extracts. These results strongly indicate that the
LDH may be present largely in an inhibited form in heart
tissue.

As pointed out earlier, our theory states that an increase
in cytoplasmic NADH levels should not promote the reduction
of pyruvate, but rather the reoxidation of the reduced coenzyme by the cytoplasmic malate dehydrogenase and other path-

ways. This implies that the cytoplasmic malate dehydrogenase
(MDH) must be fully active *in vivo*. It is known that the cytoplasmic MDH is strongly inhibited in the presence of high concentrations of malate, whereas the mitochondrial enzyme is inhibited by high levels of oxaloacetate (Kitto and Kaplan, 1966).
This inhibition also appears to be resulting from abortive complex formation by the MDHs (Bernstein et al., 1974). We therefore tested the activity of the two MDH isozymes immediately
following tissue disruption. The results are shown in Fig. 6.
A large increase in activity of the mitochondrial MDH was observed following the disruption of beef heart mitochondria and
subsequent dialysis, whereas comparatively little activation
was observed with the cytoplasmic isozyme.

These results would lend credence to the expectation that
the cytoplasmic MDH is largely present in an active form in
the heart cell and would therefore compete favorably with the
inhibited LDH for the available NADH. The mitochondrial MDH,
on the other hand, appeared to be mostly present in a form that
was inhibited with respect to the reduction of oxaloacetate
to malate. The bovine heart was obtained fresh approximately
three hours before the experiment, and it seems reasonable to
assume that the transport of malate into the mitochondria was
at a minimum at the time of the extraction. The presence of
the strongly inhibited mitochondrial MDH could indicate that
this enzyme also operates as a one-way valve, i.e., that the
operation of the malate dehydrogenase shuttle resists the
transport of NADH out of the mitochondria in a similar manner
as the LDH in heart tissue resists the oxidation of NADH by
pyruvate reduction.

The data that we have presented suggest that the activity
of the B-type LDH as well as that of the mitochondrial MDH
may be under metabolic control. The regulatory mechanisms of
these enzymes are unique in that their activities appear to be
regulated by their own substrates, products, and cofactors.

EFFECT OF OXYGEN DEPLETION

Fig. 7 shows a simplified scheme of the energy-producing
pathways in heart muscle. The two reactions that we have
thus far investigated with respect to the biological function
of the isozymes are marked by a little circle. Under normal
conditions these reactions tend to prevent a reverse flow of
the reducing equivalents. However, when the oxygen supply
becomes limiting or is interrupted, the NADH levels in the mitochondria will rise, which will in turn result in an increase
in the NADH levels in the cytoplasm. In addition the low ATP/
(ADP + P_i) ratio will cause an acceleration in the activity

37

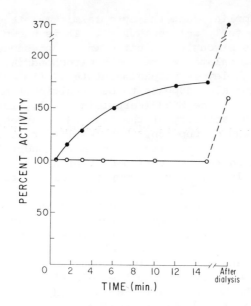

Fig. 6. Changes in the activity of beef heart cytoplasmic (o) and mitochondrial (●) MDH following the disruption of the tissues. The conditions for measuring changes in the cytoplasmic MDH were analogous to those described for LDH in Fig. 5, except that the experiment was performed in 0.25 M sucrose to prevent disruption of the mitochondria. In order to measure changes in the activity of the mitochondrial enzyme, mitochondria were prepared by standard procedures. At time 0 the mitochondria were disrupted by a one-hundred-fold dilution of the preparation into 0.01 M phosphate buffer, pH 7.5, and the activity was measured as indicated in Fig. 5. The activity during the first measurement was arbitrarily set at 100%.

of the glycolytic pathway, thereby augmenting the increase in cytoplasmic NADH. The high concentration of NADH will then cause the LDH ternary complex to dissociate. Hence, during severe hypoxia or anoxia the heart begins to convert glucose to lactate.

THE ROLE OF LDH IN THE PASTEUR EFFECT

The mechanism of the Pasteur Effect has puzzled scientists for many years, especially the complete cessation of lactate production in the presence of oxygen (Krebs, 1972). One would expect that under aerobic conditions the reoxidation of NADH

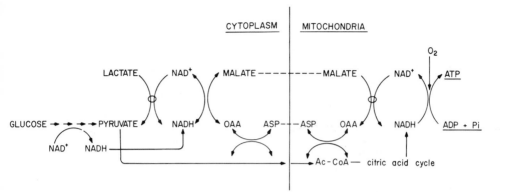

Fig. 7. Schematic illustration of the pathway involving glucose and lactate utilization for the production of ATP in heart muscle. The reactions indicated by a small circle are presumably under metabolic control.

may take place via the LDH as well as via other pathways as shown in Fig. 2. Such a competition for the available NADH would lead to a diminished lactate production, but not a complete cessation. If, however, under aerobic conditions the LDH is inactivated through the formation of abortive complexes, the complete halt in lactate production under the above aerobic conditions may be easily comprehended.

LDH ISOZYMES IN ERYTHROCYTES AND LIVER

Erythrocytes - The type of LDH found in erythrocytes of different species is quite variable. This is illustrated in Table II. The reason for these differences is not quite clear. The metabolic rate in the erythrocyte may be so slow that the type of LDH may not be of significance, since pyruvate is generally not oxidized by erythrocytes. However, NADH can be oxidized in the red blood cells by enzymes other than LDH, such as methemoglobin reductase. Whether there is a relationship of the type of LDH in erythrocytes to the amount of 2,3 diphosphoglyceric acid is unknown.

Liver - The LDH isozyme pattern in liver varies greatly among various animals, although the liver has a strongly aerobic metabolism. Dr. Ogihara (1974) has shown that there is a certain correlation between the type of LDH present in the liver and the dietary habits of the animal. The significance of this correlation appears to be somewhat nebulous at this time. We would like to propose another hypothesis concerning the type

TABLE II

% B-TYPE IN ERYTHROCYTES OF DIFFERENT SPECIES

Rabbit	100%
Human	80%
Chicken	79%
Goat	74%
Dog	42%
Hamster	35%
Rat	12%
Mouse	<5%

TABLE III

% B-SUBUNITS IN "ADULT" LIVER OF DIFFERENT MAMMALS[a]

Man	5%	Guinea Pig	22%	Sheep	90%
Rat	5%	Rabbit	60%	Beef	91%
Mouse	5%	Dog	10%	Deer	95%
Hamster	5%	Cat	8%	Goat	94%
Bat	5%	Pig	78%	Horse	5%

[a]From Fine et al., 1963.

LDH and the animal's behavior, which may perhaps be somewhat more logical. Table III shows the percent B-type LDH that has been found in the livers of various animals. It appears that the animals which either attack or hide when meeting a predator (e.g., cat, rat, mouse) have predominantly the A-type LDH, whereas the animals that are dependent on outrunning their predator, such as sheep, deer, and rabbit have predominantly the B-type enzyme in their livers. Such a correlation could be explained by saying that the running animals need a mechanism to rapidly clear their system of excess lactate, since frequent activity may be necessary for their survival. The B-type LDH in their liver gives them such an advantage since it has a higher affinity for lactate than the A-type enzyme. The attacking or hiding animals can afford the time to rest and therefore do not need the B-type LDH in their livers. Hence, the amount of B-type LDH present in the liver could be a function of the amount of muscular exercise the animal has to resort to in its struggle for survival.

The presence of the A-type LDH in liver and that of the B-type LDH in human erythrocytes, we believe does not violate our original theory, since the function of the enzyme is quite different in these cells than it is in muscle cells. It is imperative that one considers the function of a particular isozyme within the metabolic framework of a given cell in order to interpret the occurrence of one isozyme over another in that cell. It should also be borne in mind that most tissues have a heterogeneous cell population, which could affect the isozyme pattern of a tissue extract. It is our hypothesis that the isozymes of LDH and MDH have evolved to provide the cell with an additional powerful tool to regulate the flow of its metabolites to its best advantage.

ACKNOWLEDGEMENTS

This work was supported in part by grants from the National Institutes of Health (USPHS CA 11683-05) and the American Cancer Society (BC-60-P).

REFERENCES

Albery, W. J., R. P. Bell and A. L. Powell, 1965. Kinetics of Iodination of Pyruvic Acid, *Trans. Faraday Soc.* 61: 1194-1198.

Arnold, L. J. and N. O. Kaplan 1974. The Structure of the Abortive Diphosphopyridine Nucleotide-Pyruvate-Lactate Dehydrogenase Ternary Complex as Determined by Protein Magnetic Resonance Analysis. *J. Biol. Chem.* 249: 652-655.

Bernstein, L. H., J. Everse and N. O. Kaplan 1974. Structural and Functional Significance of Ternary Complex Formation by Lactate and Malate Dehydrogenases. (in preparation).

Bing, R. J. 1965. Cardiac Metabolism. *Physiol. Revs.* 45: 171-213.

Bing, R. J., A. Siegel, A. Vitale, F. Balboni and E. Sparks 1953. Metabolism of Human Heart *in Vivo. J. Clin. Invest.* 32: 556.

Cahn, R. D., N. O. Kaplan, L. Levine and E. Zwilling 1962. Nature and Development of Lactic Dehydrogenases. *Science* 136: 962-969.

Coulson, C. J. and B. R. Rabin 1969. Inhibition of Lactate Dehydrogenase by High Concentrations of Pyruvate: The Nature and Removal of the Inhibitor. *FEBS Letters* 3: 333-337.

Everse, J., R. E. Barnett, C. J. R. Thorne and N. O. Kaplan 1971. The Formation of Ternary Complexes by Diphosphopyridine Nucleotide-Dependent Dehydrogenases. *Arch. Biochem. Biophys.* 143: 444-460.

Everse, J., R. L. Berger and N. O. Kaplan 1972. Complexes of Pyridine Nucleotides and Their Function. In: *Structure and*

Function of Oxidation-Reduction Enzymes. A. Åkeson and A. Ehrenberg, Eds. Pergamon Press. Oxford. p. 691-708.

Everse, J. and N. O. Kaplan 1973. Lactate Dehydrogenases: Structure and Function. *Advances in Enzymology*. Vol. 37. A. Meister, Ed. Academic Press. New York, p. 61-133.

Fine, I. H., N. O. Kaplan and D. Kuftinec.1963. Developmental Changes of Mammaliam Dehydrogenases. *Biochemistry* 2: 116-121.

Fromm, H. J. 1963. Determination of Dissociation Constants of Coenzymes and Abortive Ternary Complexes with Rabbit Muscle Lactate Dehydrogenase from Fluorescence Measurements. *J. Biol. Chem.* 238: 2938-2944.

Gutfreund, H., R. Cantwell, C. H. McMurray, R. S. Criddle and G. Hathaway 1968. The Kinetics of the Reversible Inhibition of Heart Lactate Dehydrogenase through the Formation of the Enzyme-Oxidized Nicotinamide Adenine Dinucleotide-Pyruvate Compound. *Biochem. J.* 106:683-687.

Hirche, H. J. and W. Lochner 1961. Ueber den Stoffwechsel des Herzens bei Vermehrtem Milchsaureangebot, *Verhandl. Deutsch. Gesellsch. Kreislaufforsch.* 27: 207.

Isaacs, G. H., B. Sacktor and T. A. Murphy. 1969. The Role of the α-Glycerophosphate Cycle in the Control of Carbohydrate Oxidation in Heart and in the Mechanism of Action of Thyroid Hormone. *Biochim. Biophys. Acta* 177: 196-203.

Kaplan, N. O. 1961. Regulatory Effects of Enzyme Action. In: *Mechanism of Action of Steroid Hormones*. Pergamon Press. Oxford. p. 247-255.

Kaplan, N. O. 1972. Pyridine Nucleotide Transhydrogenase. *The Harvey Lectures*. 66: 105-133.

Kaplan, N. O., J. Everse and J. Admiraal 1968. Significance of Substrate Inhibition of Dehydrogenases. *Ann. N. Y. Acad. Sci.* 151: 400-412.

Keul, J., E. Doll. H. Stein, U. Fleer and H. Reindell 1965. Ueber den Stoff wechsel des Menschlichen Herzens III. *Pflüger's Arch. Ges. Physiol.* 282: 43-53.

Kitto, G. B. and N. O. Kaplan 1966. Purification and Properties of Chicken Heart Mitochondrial and Supernatant Malic Dehydrogenases. *Biochemistry* 5: 3966 -3980.

Kosicki, G. W. 1968. Oxaloacetate Decarboxylase from Cod. Catalysis of Hydrogen-Deuterium Exchange in Pyruvate. *Biochemistry* 7: 4310-4314.

Kosicki, G. W. and F. H. Westheimer 1968. Oxaloacetate Decarboxylase from Cod. Mechanism of Action and Stereoselective Reduction of Pyruvate by Borohydride. *Biochemistry* 7: 4304-4309.

Krasnow, E. M., W. A. Neill, J. V. Messer and R. Gorlin 1962. Myocardial Lactate and Pyruvate Metabolism. *J. Clin.*

Invest. 41: 2075-2085.

Krebs, H. A. 1972. The Pasteur Effect and the Relationship Between Respiration and Fermentation. In: *Essays in Biochemistry*. Vol. 8. P. N. Campbell and F. Dickens, Eds. Academic Press. London. p. 1-34.

Lochner, W. and M. Nassari 1959. Ueber den Venosen Sauerstoff-druck, die Einstellung der Coronardurchblutung und den Kohlenhydratstoffwechsel des Herzens bei Muskelarbeit. *Pflüger's Arch. Ges. Physiol.* 269: 407-416.

Ogihara, M. 1974. Correlation between Lactate Dehydrogenase Isozymes of the Livers and Dietary Environments in Mammals. *Isozymes II. Physiological Functions*. Academic Press, New York.

Opie, L. H. and K. R. L. Mansford 1971. The Value of Lactate and Pyruvate Measurements in the Assessment of the Redox State of Free Nicotinamide-Adenine-Dinucleotide in the Cytoplasm of Perfused Rat Heart. *Europ. J. Clin. Invest.* 1: 295-306.

Rose, I. A., 1960. Studies on the enolization of Pyruvate by Pyruvate Kinase. *J. Biol. Chem.* 235: 1170-1177.

Safer, B., C. M. Smith and J. R. Williamson 1974. Control of the Transport of Reducing Equivalents across the Mito-chondrial Membrane in Perfused Rat Heart. *J. Mol. Cell. Card.* 2: 111-124.

Schlant, R. C. 1970. Metabolism of the Heart, In: *The Heart*, J. W. Hurst and R. B. Logue, Eds. 2nd Ed. McGraw-Hill Book Co. New York, p. 111-133.

Vesell, E. S. and P. E. Pool 1966. Lactate and Pyruvate Con-centrations in Exercised Ischemic Canine Muscle. Relationship of Tissue Substrate Level to Lactate Dehydrogenase Isozyme Patterns. *Proc. Nat. Acad. Sci. U. S. A.* 55: 756-762.

Willard, J. M. and I. A. Rose 1973. Formation of Enolpyruvate in the Phosphoenolpyruvate Carboxytransphosphorylase Reaction. *Biochemistry* 12: 5241-5246.

Wuntch, T., E. S. Vesell and R. F. Chen 1969. Studies on Rates of Abortive Ternary Complex Formation of Lactate Dehydrogenase Isozymes. *J. Biol. Chem.* 244: 6100-6104.

Wuntch, T., R. F. Chen and E. S. Vesell 1970. Lactate Dehy-drogenase Isozymes. Kinetic Properties at High Enzyme Concentrations. *Science* 167: 63-65.

Wuntch, T., R. F. Chen and E. S. Vesell 1970. Lactate Dehydro-genase Isozymes: Further Kinetic Studies at High Enzyme Concentrations. *Science* 169: 480-481.

USE OF AFFINITY CHROMATOGRAPHY FOR PURIFICATION OF LACTATE DEHYDROGENASE AND FOR ASSESSING THE HOMOLOGY AND FUNCTION OF THE A AND B SUBUNITS

BERNARDO NADAL-GINARD AND CLEMENT L. MARKERT
Department of Biology, Yale University
New Haven, Connecticut   06520

ABSTRACT. A recently reported system of affinity chromatography has been exploited to purify lactate dehydrogenase isozymes from animals of four classes of vertebrates. Purification is achieved within three to four hours and recovery of enzyme approximates 80%. The procedure is based upon binding LDH to the chromophore of blue dextran and subsequently eluting the different isozymes with the oxidized or reduced coenzyme. The pattern of binding and elution of the LDH isozymes from the blue dextran clearly reveals evolutionary homologies among the different subunits. These data on homologies are more direct and reliable than those obtained from immunological or kinetic studies. The B_4 isozyme from all species studied either failed to bind to the blue dextran column or was readily eluted with NAD or AMP. By contrast, the A_4 isozyme from all species was eluted only with NADH. The elution of the heteropolymers composed of both A and B subunits was quantitative. A_1B_3 and A_2B_2 were eluted with increasing amounts of NAD or L-lactate. A_3B_1 was eluted only with NADH. The C_4 isozymes from mammals and fishes behave like the heteropolymers of A and B subunits, but are somewhat more similar to the B_4 than to the A_4 isozyme. Our data demonstrate that the binding site of the enzyme to the blue dextran is not at the catalytically active site of the enzyme and, moreover, that the binding of the oxidized coenzyme induces a conformational change in A subunits that is different from that induced in B subunits. The elution patterns of the heteropolymers from the blue dextran suggests that a cooperative effect exists between the different subunits. From a consideration of the properties of the bound and the free enzyme and from an appreciation of the binding and kinetic properties of the A and B subunits, in particular the binding of LDH-A_4 to cell membranes, we present a comprehensive hypothesis to account for the physiological role of the LDH isozymes. LDH-B_4 predominates in tissues with a constant supply of oxygen and these tissues normally do not produce lactate, but utilize it. Isozymes rich in B subunits exist free in the cytoplasm and, because of high affinity to NAD,

function as true lactate dehydrogenases. The activity of
these isozymes is modulated in the cell through the formation of abortive ternary complexes of enzyme, pyruvate,
and NAD, which serve to prevent the production of excessive amounts of NADH from the oxidation of lactate. Thus
this negative feedback system maintains the NAD-NADH ratio
at an appropriate level in the cell. By contrast, the A_4
isozyme and other isozymes rich in A subunits predominate
in tissues that are subject to periods of anaerobiosis
and/or that do not utilize lactate, but produce it. These
isozymes have a high affinity for NADH; when free in the
cytoplasm, they compete with the Krebs cycle and the cytochrome system for the reduced coenzyme. Deleterious competition is prevented by binding excess LDH A-containing
subunits to membranes thus reducing total catalytic activity. Under anaerobic conditions, as levels of NADH increase,
the bound LDH A_4 is released from the membranes just as
it is from blue dextran when NADH is added. Thus through
this positive feedback system an appropriate NAD-NADH ratio
is again maintained even under conditions of rapid fluctuation in oxygen availability.

INTRODUCTION

L-lactate dehydrogenase (LDH) (E.C.1.1.1.27) catalyzes the
interconversion of lactate and pyruvate in the presence of the
coenzyme, nicotinamide adenine dinucleotide (NAD). In most
mammals and birds LDH exists in five tetrameric isozymic forms
produced by the random association of two different polypeptides, A and B, to compose the five possible tetramers: A_4,
A_3B_1, A_2B_2, A_3B_1, and B_4 (Markert, 1962; 1963). By means of
physical, kinetic, and immunological techniques it has been
demonstrated that the A (or B) subunits of different species
are more homologous to one another than are the A and B subunits of the same species (Markert and Appella, 1963; Cahn et
al., 1962; Markert and Holmes, 1969). Each subunit is encoded
in a separate gene and although they probably evolved from a
single ancestral gene, they are now quite distinct.
The A and B genes are active to different degrees or at
least they have a different degree of expression in different
cell types at different stages of cell differentiation (Markert
and Ursprung, 1962; Cahn et al., 1962) thus giving support to
the belief that the evolutionary divergence of these genes
enables each to fulfill a different physiological function.
But the precise nature of this difference has not been demonstrated. The only generalization that seems relevant to the
different functions is that tissues normally subjected to
periods of anaerobiosis, like skeletal muscle, are richer in

A subunits, and tissues with a constant supply of oxygen are richer in B subunits (Cahn et al., 1962; Kaplan and Goodfriend, 1964).

A third gene, controlling the synthesis of a third type of LDH subunit, designated C, is also present in mammals and birds, giving rise to the so-called X-bands that appear to be formed exclusively in the primary spermatocytes, with biochemical and kinetic properties different from the A and B subunits (Blanco and Zinkham, 1963; Goldberg, 1963; Zinkham et al., 1965; Zinkham, 1968).

Many fish exhibit a more complicated array of LDH patterns, consisting of from as few as two isozymes in some species to as many as 22 in others. These complex LDH patterns suggest the presence of other LDH genes in addition to the A and B genes already discussed. Markert and Faulhaber (1965) explained most of these isozyme patterns on the basis of a third gene specially active in eye and brain tissues of the bony fishes. Recently Sensabaugh and Kaplan (1972) proposed that several gadoid fishes have another gene for a "liver-specific" LDH in addition to the A and B genes. Because the physical, kinetic and immunological properties of the characteristic eye and liver LDHs are quite similar to the B_4 isozyme, it has been suggested that the gene responsible for these isozymes arose during evolution by duplication of the B gene. Recently, Shaklee et al. (1973) presented evidence in support of the monophyletic origin of the LDH genes that function specifically in either eye or liver cells, designating them both as LDH-C genes.

The physiological role of these special isozymes of LDH is not known despite the demonstrated differences in the kinetic properties of the A_4, B_4, and C_4 isozymes (Battelino and Blanco, 1970; Sensabaugh and Kaplan, 1972). The highly specific tissue distribution in the expression of this C gene makes the isozymic product of this gene one of the best targets for an effort to elucidate the physiological roles of the different isozymes of LDH, and of isozymes in general.

LDH has been considered a classical example of a soluble enzyme because it can easily be extracted almost completely with several different homogenization procedures and in many buffers. Recently, however, different investigators have presented data suggesting specific subcellular localizations of the different isozymes with most of the B_4 in the cytosol and most of the A_4 and C_4 attached to subcellular organelles (Amberson et al., 1965; Agostini et al., 1966; Hultin and Westort, 1966; Hultin and Southard, 1967; Hyldgaard and Valenta, 1970; Baba and Sharma, 1971; Melnick and Hultin, 1973). This distribution correlates with the different extractability of the two isozymes (Hyldgaard and Valenta, 1970). The differ-

ences in kinetic properties of the free and bound A_4 (Hultin
et al., 1972) plus its specific localization in a specific
subcellular compartment introduces a new aspect to the problem of assigning a physiological role for each of the different forms of LDH. Despite the significance of such an assignment the data are still far from complete and in some cases
contradictory (Hyldgaard and Valenta, 1970). The specific
physiological roles are still very uncertain. Studies of the
kinetics of LDH covalently bound to cellulose (Wilson et al.,
1968) are difficult to interpret and probably the changes in
enzyme properties do not have any relation to the changes produced by the binding of the enzyme to cellular membranes if
that does in fact occur in vivo.

Ryan and Vestling (1974) discovered that rat LDH-5(A_4)
binds to blue dextran and that it can be specifically eluted
with one mM $NADH_2$. These authors exploited this fact for the
purification of rat LDH A_4. They covalently coupled blue
dextran to sepharose-4B by the cyanogen bromide procedure
(Cuatrecasas, 1970).

During the course of purifying mouse LDH by the procedure
of Ryan and Vestling we found that the A_4 isozyme and all
heteropolymers containing A subunits bound to the blue dextran
resin, but the B_4 isozyme did not. Most of the heteropolymers
could be eluted with NAD, but A_4 could only be eluted with
$NADH_2$. The strikingly different binding and elution behavior
of the A and B containing isozymes resembles their behavior
in binding to biological membranes as reported by Hultin et
al. (1972). These observations prompted us to use this absorption system to study the molecular homology of the LDH
isozymes in vertebrates occupying different positions in the
evolutionary scale in order to obtain more information on the
origin of the third gene, its characteristic behavior with
the oxidized and reduced coenzyme, and with different substrates. At the same time we hoped to gain insight into inter-
subunit interactions that could shed some light on the physiological significance of tissue-specific isozyme patterns.

MATERIALS AND METHODS

Specimens. Tissues from the following organisms were studied
-- mammals: human, mouse, rat, and dog; the chicken (*Gallus
gallus*); frogs: *Rana pipiens* and *Xenopus laevis*; fish: silver
hake (*Merluccius bilinearis*),Atlantic cod (*Gadus morhua*), rainbow smelt (*Osmerus mordax*), Atlantic mackerel (*Scomber scombrus*),
green moray (*Symnothorax funebris*), ballyhoo (*Hemiramphus bras-
iliensis*), rainbow trout (*Salmo gairdneri*), and dogfish shark
Squalus acanthias). All the studies were done using fresh

tissues or tissues stored at -20 C.

Chemicals. Blue dextran and Sepharose-4B were obtained from Pharmacia Fine Chemicals. NAD, NADH$_2$, AMP, Li$^+$-lactate, sodium pyruvate, sodium oxamate, and potassium oxalate were purchased from Sigma; β-mercaptoethanol and cyanogen bromide were obtained from Eastman Organic Chemicals.

LDH activity and protein assays. LDH activity was assayed spectrophotometrically by measuring the formation of NAD from NADH$_2$ per minute at 340 nm in a Beckman Model DU spectrophotometer with a Gilford Model 2000 absorbance recorder. Activity of LDH in international units was calculated according to the following equation: ΔOD per min at 340 nm x dilution factor divided by 6.22 = international units of LDH/ml. Protein concentration was determined by the microbiuret technique (Goa, 1953; Bailey, 1967) or spectrophotometrically by measuring the absorbance at 280 and 260 nm and applying the formula: protein concentration in mg/ml = 1.55 x absorbance at 280 nm minus 0.76 x absorbance at 260 nm.

Electrophoresis. Vertical starch gel electrophoresis was accomplished at 4 C with 12% gels using a Tris-borate-EDTA discontinuous buffer system at pH 8.6 (Boyer et al., 1963). Following electrophoresis the gels were sliced horizontally and stained from LDH at 37 C by a tetrazolium method previously described by Markert and Faulhaber (1965).

Preparation of the blue dextran resin. The sepharose-4B was activated and substituted by the general procedure outlined by Cuatrecasas (1970) and by Ryan and Vestling (1974).

Sepharose-4B was extensively washed with distilled water and resuspended in an equal volume of distilled water. Then 300 mg of cyanogen bromide per ml of packed sepharose was added quickly to the stirred suspension. The pH of the suspension was immediately raised to, and maintained at, pH 11 with dropwise addition of 10N NaOH. The temperature was maintained at about 20 C by adding crushed ice. The reaction was completed in about 10 min and then a large amount of ice was added to the suspension, which was quickly transferred to a Buchner funnel and washed under suction with 10 to 15 volumes of cold 0.1 M NaHCO$_3$. The activated sepharose was immediately suspended in the same volume of 0.4 M sodium carbonate buffer pH 10 containing 20 mg/ml of dissolved blue dextran. The suspension was gently stirred for 18 hr at 4 C and extensively washed with 1 M KCl to remove the unreacted blue dextran. In this way, about 40 mg of blue dextran was covalently attached to a ml of packed sepharose.

Extract preparation and use of the affinity column. The tissues were homogenized in 5 volumes of 20 mM Tris, 1 mM mercaptoethanol, pH 8.6 or 20 mM sodium phosphate buffer pH 7.0 in a

Sorval Omnimixer. The homogenate was centrifuged at 30,000 g
for 30 min in a Sorval RC2B centrifuge. The supernatant was
passed through a cheese cloth and directly applied to the blue
dextran resin which had been previously equilibrated with the
same buffer. Up to 250-300 units of LDH/ml of packed resin
can be applied to the column. The column was washed with the
equilibration buffer at the maximum flow rate of the column
until the eluate had an absorbance at 280 nm of less than 0.02.
The columns were then eluted with twice the bed volume of different concentrations of NAD ranging from 1 mM to 15 mM, with
or without different concentrations of substrates (L-lactate
or pyruvate) or substrate analogs (oxamate and oxalate). In
some cases (see Results) the extract was dialyzed with the
equilibration buffer in order to remove the NAD, $NADH_2$, lactate,
and pyruvate present in the tissue extract; different concentrations of NAD from 1 to 10 mM with or without substrate or
inhibitors were then added to the extract. In such cases the
NAD elution step was omitted. Then the column was washed with
twice the volume of the column with 10 mM Tris or phosphate
buffer and 0.5 mM mercaptoethanol. In other cases the ionic
strength of the buffer was raised by adding 0.2-0.5 M NaCl.
The LDH was detached from the column by eluting with two column
volumes of 0.5 mM $NADH_2$ in 10 mM buffer and 0.5 mM mercaptoethanol, followed by washing with the equilibration buffer
until less than 0.1 units/ml of LDH could be detected in the
eluant. Usually all the LDH activity was eluted in less than
two void volumes.

RESULTS

BEHAVIOR OF THE A_4 ISOZYMES OF LDH

From all species, the A_4 isozyme was completely absorbed
to the column and completely eluted with 0.5 mM $NADH_2$ (Fig.
1-4). the recovery of the enzyme was greater than 95% with a
400 to 600 fold increase in specific activity, depending upon
the species, and with a final purity of about 70%. The partially purified LDH of some species was passed through a DEAE-
Sephadex column and the contaminant proteins removed to yield
a 30% increase in specific activity and to produce a single
band after starch and acrylamide gel electrophoresis. The
gels were stained for enzymatic activity and for protein.
The recovery of the enzyme after the DEAE-Sephadex step was
approximately 85%.

Using a salt gradient, the enzyme was eluted between 0.3
and 0.5 M NaCl depdending upon the species from which the
enzyme was extracted. With this procedure all the isozymes

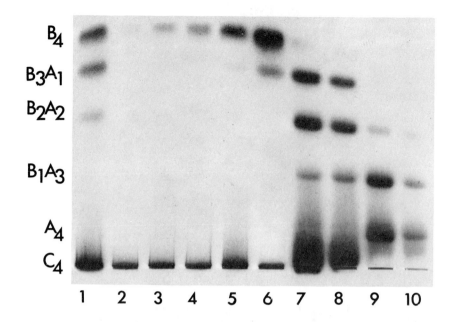

Fig. 1. Elution of LDH isozymes obtained from mouse testis. Channel 1: Control. Channels 2-6: 20 mM phosphate buffer wash, pH 7.0. Channels 7-8: Elution buffer, 1 mM NAD^+, 1 mM Li-lactate. Channels 9-10: 20 mM phosphate buffer, 0.5 mM $NADH_2$.

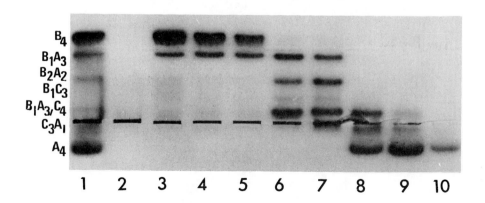

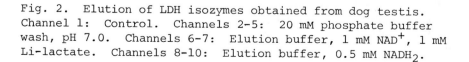

Fig. 2. Elution of LDH isozymes obtained from dog testis. Channel 1: Control. Channels 2-5: 20 mM phosphate buffer wash, pH 7.0. Channels 6-7: Elution buffer, 1 mM NAD^+, 1 mM Li-lactate. Channels 8-10: Elution buffer, 0.5 mM $NADH_2$.

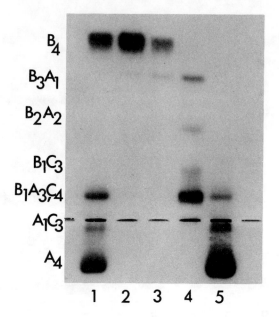

Fig. 3. Elution of LDH isozymes obtained from rat testis. Channel 1: Control. Channels 2-3: 20 mM phosphate buffer wash, pH 7.0. Channel 4: Elution buffer, 1 mM NAD^+, 1 mM Li-lactate. Channel 5: Elution buffer with 0.5 mM $NADH_2$.

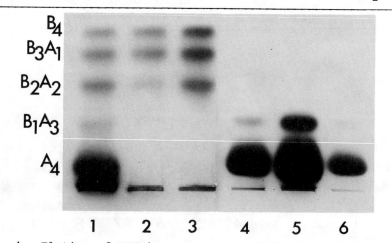

Fig. 4. Elution of LDH isozymes obtained from mouse tissues. Channel 1: Control. Mouse tissue extract dialyzed against 20 mM phosphate buffer, pH 7.0. After dialysis 1 mM NAD^+ was added. Channels 2-3: 20 mM phosphate buffer wash, pH 7.0. Channels 4-6: Elution buffer, 0.5 mM $NADH_2$.

elute in the same fractions and the specific activity is only about 30% of the final specific activity obtained by the complete procedure (washing with 1 mM NAD, 1 mM Lithium-lactate, and finally eluting with 0.5 mM $NADH_2$).

Concentrations of NAD from 1 to 15 mM in the elution buffer do not release the enzyme independently of the presence or absence of 1 mM Lithium lactate. Concentrations of 10 mM of either lactate, pyruvate, oxamate, or oxalate are inefficient in releasing the enzyme in the absence of the coenzyme. When the lithium lactate concentration is increased to 3 to 5 mM in the presence of 1 mM NAD, the enzyme can be completely eluted in about 10 void volumes. Substitution of the 1 mM NAD, 3-5 mM lactate with a normal elution buffer without substrate or coenzyme, or both, immediately stops the release of the enzyme from the column.

Elution with 1 mM NAD plus 1 mM pyruvate produces a slow but eventually complete release of the enzyme from the column. By increasing the concentration of pyruvate to 5 mM the elution can be accelerated and completed in two void volumes.

If NAD and pyruvate are added to the dialyzed extract 20 min before loading the column, the enzyme is not retarded by the column. The addition of 1 mM NAD and 1 mM oxamate or oxalate produces a complete release of the enzyme, if already bound to the column, or prevents binding if added before application to the column.

BEHAVIOR OF THE B_4 ISOZYMES OF LDH

In the three species of mammals studied, mouse, dog and rat, the B_4 isozyme is only slightly retarded by the column and 90% to 95% of the enzyme is eluted with the main bulk of protein or in the first two column volumes of equilibration buffer (Fig. 1-4). That effect is not the result of overloading the column because it occurs whether the column is loaded with 10 or with 300 units of LDH per ml of packed sepharose.

In enzyme preparations from amphibia and fish, the B_4 isozyme is retarded as, for example, from the silver hake, or completely absorbed by the column as is true for the isozyme from the green moray, but in all cases the isozyme is completely eluted by a 1 mM solution of NAD.

When mouse tissue extract was dialyzed against the homogenization buffer for 12 hr, almost all the B_4 isozyme was retained on the column and eluted with 1 mM NAD. If the B_4 isozyme that washed through in the non-dialyzed extract is now dialyzed against the same buffer in an attempt to remove all the coenzyme present and then re-run through the column, 90% is bound to, but released by, the addition of 1 mM NAD.

BEHAVIOR OF THE HETEROPOLYMERS OF A AND B SUBUNITS

All the heteropolymers of A and B subunits bind to the column and only a fraction of the A_1B_3 isozyme of the mouse or rat is not retained (Fig. 1-4). The elution by NAD is quantitative and correlated with the number of B subunits in the heteropolymer. In the absence of lactate, LDH-2 (A_1B_3) is completely eluted, LDH-3 (A_2B_2) is eluted 30%, and LDH-4 (A_3B_1) is eluted 10%. Those results are also reproduced if, instead of NAD, 1 to 10 mM of AMP is used. When 1 to 5 mM lactate are added, the elution of the heteropolymers is increased and eventually all the LDH is eluted from the column at the higher concentrations of substrate. When 1 mM NAD was used without the addition of lactate to the dialyzed extract, no retardation of LDH-1, LDH-2, or LDH-3 could be detected, but LDH-4 was absorbed 90% and LDH-5, 100% (Fig. 4). None of the heteropolymers is retarded by the column if 10 mM NAD is added to the dialyzed extract.

BEHAVIOR OF ISOZYMES CONTAINING C SUBUNITS

The behavior of isozymes with C subunits covers all the spectra of elution patterns and can be classified in three categories: a) In the case of the highly specialized and highly anodal "eye band" of the Atlantic mackerel, the C_4 isozyme is not absorbed to the column and the heteropolymers containing C subunits are completely eluted with NAD. b) In the more "primitive" fishes like the green moray and the smelt in which the C gene has diverged less from the B and the regulation of both the B and C genes is still very parallel, the C isozymes are completely eluted with NAD as is the B_4 isozyme of these fishes. Moreover, the LDH-X (C_4) of mammals and most of the heteropolymers between A and B subunits of mammals elute in the same manner (Fig. 1-3). c) The so-called "liver band" of the gadoid fishes (silver hake and Atlantic cod) is highly cathodal and is eluted only with $NADH_2$, just like the A_4 isozyme.

RELATIONSHIP OF ELUTION PATTERN AND ELECTROPHORETIC MOBILITY

Because of the seeming correlation between anodal migration and elution with NAD, the isozymes of the fish, ballyhoo, were tested. These isozymes exhibit a mobility of the A_4 and B_4 isozymes that is the reverse of that usually observed. Nevertheless, the elution patterns proved to be identical to those of the other species studied. Thus, the more cathodal B_4 was eluted with NAD and the more anodal A_4 was eluted only with $NADH_2$.

The trout is a tetraploid fish with two A and two B genes, each encoding an electrophoretically distinct LDH subunit. Five A-A' tetramers are formed and five B-B' tetramers. One of the A-A' tetramers occupies the same position as one of the B-B' tetramers. Nevertheless all the B-B' isozymes, regardless of electrophoretic mobility, were eluted with NAD, and the A-A' isozymes were eluted only with $NADH_2$.

DISCUSSION

BINDING AND ELUTION OF LDH

Ryan and Vestling (1974) first reported that blue dextran differentially bound the A_4 isozyme of LDH from rat tissues. Although no explanation was provided by these authors for the nature of the binding, it is apparent that the chromophore is involved and not the dextran because LDH does not bind to sephadex. Although the chromophore is positively charged, this appears not to be a critical aspect of the binding between the chromophore and the LDH. Our data show that the different isozymic subunits of LDH bind to blue dextran without regard to the specific net charge characterizing the isozyme. That is, the binding of A subunits of different net charge obtained from different species all bind similarly to the blue dextran even though their net charges vary greatly, ranging all the way from a net negative to a net positive charge at pH 7. Moreover, the binding site on the LDH must be other than the enzymatically active site, because neither the substrate, lactate or pyruvate, nor the cofactor NAD affect the binding of LDH to the chromophore. In fact, enzyme bound to NAD can still bind to the column. Nevertheless, the binding to blue dextran of LDH from species of every class of vertebrates indicates that the binding site has been highly conserved throughout the evolution of the vertebrates, just as conserved, in fact, as the enzymatically active site.

Although Ryan and Vestling (1974) only reported that the A_4 isozyme from rat tissues bound to blue dextran, our investigations have extended their work to demonstrate that all the isozymes of LDH, no matter what the subunit composition, can be bound under the proper conditions to blue dextran. Nevertheless, the binding of these different kinds of subunits is not identical and, in fact, blue dextran can be used to separate the different isozymes of LDH, one from another. Even though A subunits from different vertebrate species have many differences in amino acid composition (Holbrook et al., 1974), they nevertheless are basically very similar and this similarity is clearly apparent in their binding and release from blue dextran. All A subunits attach strongly to blue dextran and are eluted by $NADH_2$, but are not eluted by even much higher concentrations of NAD -- even fifteen times higher. As would be expected, the presence of $NADH_2$ in suitable concentrations prevents the A subunits from binding to the column. The failure of NAD to elute A subunits from the column is not due to a failure of NAD to bind to the enzyme; a combination of NAD and pyruvate (producing the abortive ternary complex) will elute A subunits from the column or, if present initially, will prevent the binding of A subunits. Likewise, the use of inhibitors such as oxamate or oxalate in combination with NAD prevents the binding of A subunits to the column or elutes them if they are already bound. It should be emphasized that even very high concentrations of substrate or inhibitor, in the absence of co-factor, are totally ineffective in eluting bound A subunits from the column. The enzyme must clearly first bind to the co-factor and then to the substrate or inhibitor before the binding properties to blue dextran are affected, in agreement with the compulsory nature of the kinetics of the enzyme (O'Carra and Barry, 1972; Holbrook and Gutfreund, 1973).

In contrast to the A subunits, B subunits bound to blue dextran are eluted by NAD and, of course, also by NADH as are the A subunits. The failure of B subunits in tissue extracts to bind to blue dextran is attributable to the presence of bound coenzyme. If B subunits are exhaustively dialyzed to remove all the coenzyme, then these subunits bind readily to blue dextran. The greater combination of coenzyme with B, as compared to A subunits in tissue extracts, is due to the substantially greater affinity of B subunits for coenzyme (Wieland et al., 1962; Everse and Kaplan, 1973). The fact that B subunits can also be eluted from blue dextran by AMP suggests that only the adenine moiety is significant in the binding or elution of B subunits. With A subunits, AMP is totally ineffective. As with the A subunits, the B subunits, from whatever vertebrate, behave alike, indicating great conservatism in the evolutionary structure of these subunits.

Since LDH exists in five isozymic forms, three of which are heteropolymers of A and B subunits, the behavior of these heteropolymers on binding or eluting from blue dextran should be particularly instructive. As might be expected the heteropolymers are intermediate between A_4 and B_4 in their binding to, and elution from, blue dextran. The characteristics of the heteropolymers are a function of the relative proportion of A and B subunits in the heteropolymer. That is, A_3B_1 behaves very much like A_4, and A_1B_3 behaves more like B_4, with A_2B_2 in between. Crystallographic and fluorometric data demonstrate

that the conformation of A and B subunits changes on binding
to co-factor (Everse et al., 1971; Rossman, 1970; Holbrook et
al., 1974). Furthermore, the binding of NAD as compared to
$NADH_2$ produced a different conformational change as shown by
fluorometric measurements (Vestling and Kunsch, 1968; Everse
and Kaplan, 1973; Holbrook et al., 1974). However, there are
no data to demonstrate conclusively that A and B subunits
undergo different conformational changes when bound to co-
factor. Nevertheless, our data do suggest such a difference.

We find that the A and B subunits bind differently to blue
dextran and are eluted differently by NAD and $NADH_2$. Other
interpretations are possible, but one hypothesis presumes that
the conformation of the subunits is changed in different ways
by binding to co-factor so that NAD will elute B subunits from
the column or prevent their binding whereas only $NADH_2$ will do
the same for A subunits. In the presence of NAD, A subunits
will bind to blue dextran and A subunits will not. The heteropolymers, for example A_2B_2, will bind to the column in the
presence of NAD. One might interpret this result to mean that
the two A subunits in the heteropolymer were bound to the
column and the B subunits were not. However, the A_2B_2 heteropolymer is eluted from the column by a concentration of NAD
that fails to elute the A_4 isozyme. Therefore, the behavior
of A subunits in a heteropolymer is different from their behavior in a homopolymer. We interpret this result to mean
that the binding of NAD to the B subunits in a heteropolymer
alters the conformation of the tetramer so that the A subunits
no longer bind to the chromophore of blue dextran. Thus we
suggest that conformational changes are essential in the binding and elution of LDH from the column, and our results indicate that a kind of cooperativity must exist between the different subunits of LDH so that binding of NAD to a subunit
affects the conformation of associated subunits. This cooperativity appears to have a quantitative parameter since A_3B_1
is not eluted from the column with NAD, A_2B_2 is eluted to some
degree, and A_1B_3 is eluted very readily. Thus, the effect on
A subunits produced by binding NAD to associated B subunits
increases with larger numbers of B subunits in the heteropolymer. By using different methods other authors have presented
evidence of some cooperativity between different LDH subunits
(Rouslin and Braswell, 1968; Kaloustian and Kaplan, 1969;
Ainslie and Cleland, 1969; Ainslie, 1970).

It should be noted that LDH bound to blue dextran still
retains some enzymatic activity although activity is reduced,
again suggesting that the binding of LDH to the blue dextran
is not at the enzymatically active site. Bound LDH does have
a reduced affinity for substrate, probably because of confor-

mational changes associated with the binding. For example,
A subunits bound to NAD and attached to blue dextran are not
eluted by additional amounts of NAD, but if the substrate
lactic acid is added so that enzymatic activity will generate
$NADH_2$, then the A subunits are released from the column. In
this connection, it should be noted that Hultin et al. (1972)
have reported that LDH bound to membranes has different kinetic
properties from unbound LDH and, in particular, that the enzymatic activity is reduced. This behavior, as reported by Hultin,
corresponds with what we find for LDH bound to blue dextran.

HOMOLOGY AMONG LDH SUBUNITS: A, B, AND C

The differential binding of A and B subunits to blue dextran and the great similarity of behavior among all A subunits
as one group and among all B subunits as a different group
made possible a study of the homologies among the different
subunits of LDH as expressed in differential binding to, or
elution from, blue dextran. First, as we mentioned earlier,
all A subunits from whatever vertebrate -- from lampreys to
man -- bind to blue dextran and are not eluted with NAD, but
are eluted with $NADH_2$. All B subunits, on the other hand, are
eluted with NAD as well as with $NADH_2$. This is strong evidence
for the homology of these subunits, perhaps more decisive than
immunochemical similarities.

Among some organisms, particularly fish, the variety of
electrophoretic mobilities characterizing A and B or C subunits
can be very confusing. Ordinarily, LDH-B subunits of vertebrates
have the largest net negative charge, but among some fish either
A or C subunits are more negatively charged (Markert et al.,
1975). Thus, the presumed homologies among the A and among
the B subunits, if based only upon tissue distribution or upon
kinetic properties, could be misleading. One baffling example
is the great multiplicity of LDH isozymes found in salmonid
fish; these are tetraploid and consequently have two A genes
and two B genes. The distinct subunits produced by these
four genes lead to the synthesis of a large number of isozymes
-- in theory, as many as thirty-five; in practice, a somewhat
smaller number can be detected after electrophoretic resolution
(Massaro and Markert, 1968). Now, however, the true nature
of the subunit composition of these isozymes is easily ascertained by virtue of the differential binding of these isozymes
to a blue dextran column. What were previously thought (on
immunochemical evidence) to be isozymes composed of A subunits,
in fact behave in accord with that expectation when tested
against the blue dextran column. The less than 35 isozymes
found in salmonid fish, such as trout, is partly due to

occupation of the same electrophoretic site by isozymes of
different subunit composition. For example, one of the A
tetramers can occupy the same position as a B tetramer, but
these isozymes can nevertheless be distinguished easily from
one another by their differential binding to, and elution from,
a blue dextran column.

Although most lactate dehydrogenase activity is encoded
in two genes, A and B a third gene does exist and codes for a
different subunit - the C subunit. This third gene for LDH
is found in mammals and birds and also in fish. In mammals
and birds, the C gene codes for a special LDH, LDH-X, found
only in primary spermatocytes and descendant cells (Blanco
and Zinkham, 1963; Zinkham, 1968). Tissue extracts of testes
contain large amounts of the B_4 isozyme and smaller amounts of
the C_4 isozyme. When these extracts are passed through a blue
dextran column, the C_4 isozyme binds to the column and the B_4
does not. The bound C_4 isozyme can then be eluted with NAD,
just like B_4 isozymes that have been exhaustively dialyzed so
as to remove NAD. This result suggests either that the C
subunit is intermediate in its behavior between the A_4 and
B_4 isozymes or that it has less affinity for NAD and, in
tissue extracts, is not saturated with NAD as are the B subunits. In any event, the C subunits are by this test clearly
more homologous to B than to A subunits.

Among fish, the third gene for LDH codes for an isozyme
that is found primarily either in the eye tissues of advanced
teleosts or in liver tissues. In primitive teleosts, the C
gene is active in a great variety of tissues, usually the same
tissues in which the B gene is primarily active. On immunochemical grounds, it can be demonstrated that the C subunit is
more closely related to the B than it is to the A (Shaklee et
al., 1973). In terms of the behavior of the C-containing
isozymes on blue dextran columns, however, a different generalization emerges. Depending upon the cellular location and
physiological activity of the C_4 isozyme, the properties of
the isozyme, as tested by a blue dextran column, can range
all the way from those characteristic of the B_4 to those
characteristic of the A_4 isozyme. In tissues in which the C
gene is expressed primarily in the eye, the resulting isozyme
does not bind to blue dextran and is eluted just like the B_4
isozyme. In those fish in which the C gene is active primarily
in the liver, the resulting C_4 isozyme binds and is eluted
exactly like the A_4 isozyme. In the primitive teleost in
which the C gene is active in a wide variety of tissues, the
C-continaing isozymes all attach and elute from the column in
parallel to the B subunits.

These data on the homologies of the A, B, and C subunits
fit previous data except that the behavior of the liver-specific
C isozyme in certain teleost fish is not in accord with the
homologous relationships of the C and B genes as shown by all
other tests. Both the immunochemical tests and kinetic data
suggest that this isozyme is more closely related to the B_4
than to the A_4 (Sensabaugh and Kaplan, 1972), but the behavior
on a blue dextran column is clearly like the A_4 and not the
B_4. This exceptional behavior, in conflict with the rest of
the data on homology, does make sense in terms of the physiological role which these different isozymes play in their
respective tissue locations.

It should be noted that the chromatographic behavior of
the LDH isozymes is not simply a reflection of their electrophoretic mobilities nor of their kinetic properties. The
differential behavior in this type of column chromatography of
isozymes of identical electrophoretic mobility demonstrates
that the net charge on the molecule does not specify chromatographic behavior. Likewise, kinetic properties are not indirectly measured by chromatographic behavior. The liver-specific
C isozyme of fish behaves chromatographically just like A_4
isozymes, but kinetically the C_4 isozymes are much closer to
the B_4 isozymes (Sensabaugh and Kaplan, 1972).

PHYSIOLOGICAL ROLE

The tissue specificity and constancy of LDH isozyme patterns leaves no doubt that they must have physiological significance. But just what this significance is has been unclear.
Perhaps the principal unsolved problem of the LDH isozymes
is the identification of their individual roles in cell physiology. So far, the following generalizations have been offered: 1) tissues with a relatively abundant and constant
supply of oxygen are rich in B subunits and 2) tissues which
experience transient anaerobiosis are rich in A subunits (Cahn
et al., 1962; Markert and Ursprung, 1962). These generalizations tend to fit the isozyme distributions found during ontogeny and in the adult with, however, several notable exceptions, particularly isozyme patterns of the liver. Apparently,
it is not so much the absolute level of available oxygen, but
rather the fluctuating availability of oxygen that determines
the isozyme distribution (Vesell and Poole, 1966). Again, a
constant oxygen supply is associated with B subunits, and a
fluctuating supply is associated with A subunits. From our
data from column chromatography and from some evidence in the
literature on the differential binding of A and B subunits to
biological membranes (Hultin et al., 1972), we have developed

the following comprehensive hypothesis concerning the physiological role of the different isozymes of LDH.

There is an excellent correlation between the presence of A subunits in a tissue and the production of lactic acid; similarly, the presence of B subunits is correlated with the use of lactic acid in producing pyruvate. These are not "one-way" enzymes, of course, but their kinetic properties uniquely fit them for the specific catalytic roles which they play. The significant properties, because of the compulsory ordered kinetics, are the binding constants for NAD and $NADH_2$. These generally correspond with the relative values of the Km's, but have been measured specifically only in a few instances (Holbrook and Gutfreund, 1973; Stinson and Holbrook, 1973; Heck, 1969; Whittaker et al., 1974). The binding constants for the substrates, lactate and pyruvate, are not so significant because these are bound only after NAD or $NADH_2$ has been bound. An additional item of great significance, we believe, is the binding of LDH to membranes, a behavior which we consider analogous to the binding of LDH to blue dextran.

In the simplest case, let us take the abundance of B subunits in heart muscle tissue. Heart muscle is known to consume substantial amounts of lactic acid (Young, 1970), converting it to pyruvate and oxidizing pyruvate to carbon dioxide and water through the tricarboxylic acid cycle. Very little, if any, lactic acid is produced in heart muscle which is normally abundantly supplied with oxygen. The binding constant of B subunits from pig heart muscle has been measured (Stinson and Holbrook, 1973). Given the binding constant for NAD and the high ratio of $NAD/NADH_2$ in the cell, under normal circumstances the B subunits in heart muscle would be bound to NAD. In fact, the LDH in extracts of many tissues, including heart muscle, is normally bound to NAD and, hence, fails to bind to blue dextran unless the LDH is exhaustively dialyzed to remove the NAD. Therefore, heart muscle LDH B_4 functions principally to oxidize lactic acid to pyruvate with the generation of NADH which, in its turn is oxidized through the cytochrome system to generate energy in support of the normal physiology of the heart. If too much ATP is generated by the cell, then glycolysis is inhibited by ATP at the phosphofructokinase step, but still another control mechanism is needed to depress the conversion of lactic acid to pyruvate by LDH. The oxidation of lactic acid could subvert the controls on $NADH_2$ production by glycolysis since the oxydation of lactate also yields $NADH_2$. Thus, in cell environments with high levels of NAD and pyruvate, an abortive ternary complex of NAD, pyruvate, and enzyme would be formed to reduce the titer of active enzyme and to depress the utilization of lactic acid for the production of pyruvate.

The existence and formation in vivo and in vitro of these abortive ternary complexes between LDH-NAD-pyruvate and LDH-NADH-lactate is well documented (Fromm, 1961; 1963; Zewe and Fromm, 1962; Anderson et al., 1964; Everse et al., 1971; Gutfreund et al., 1968; Wuntch et al., 1969). Thus, by this negative feedback system involving LDH-B the ratio of NAD to $NADH_2$ could be kept relatively constant.

The situation with A subunits is quite different. In a well oxygenated muscle very little if any lactic acid is utilized or produced. LDH is not necessary for the production of ATP when the oxygen supply is adequate. In fact, the high affinity of $LDH-A_4$ for $NADH_2$ could be harmful to the cell because the $NADH_2$ would be used up during the conversion of pyruvate to lactic acid, thus hampering oxidative phosphorylation via the Krebs cycle and cytochrome system; this energy-yielding oxidation requires $NADH_2$. When $NADH_2$ is in short supply it would be advantageous for the cell to inhibit LDH activity. That "inhibition" could very well be produced by the binding of the $LDH-A_4$ enzyme to cell membranes in an environment with a high $NAD/NADH_2$ ratio. Increased production of $NADH_2$ would release (or elute) bound LDH to participate in the catalytic conversion of $NADH_2$ to NAD with the concomitant generation of lactic acid from pyruvate. This means that $LDH-A_4$ will function well in a metabolic environment containing abundant pyruvate and depressed supplies of oxygen. Thus, a kind of positive feedback exists in these cells (skeletal muscle and liver) so that increasing amounts of $NADH_2$ liberates increasing amounts of LDH which then leads to the oxidation of this $NADH_2$ in the conversion of pyruvic acid to lactic acid. This is precisely the reaction that occurs in liver tissues and in skeletal muscle during periods of low availability of oxygen.

It is interesting to note that the liver-specific C subunits of fish show exactly the same chromatographic properties with blue dextran as do A subunits; presumably their kinetic properties, with reference to NAD and $NADH_2$ are similar to those of A rather than to those of B subunits. Thus, the kinetic properties of A and B subunits, and probably also C subunits, for binding NAD to $NADH_2$ and for binding to, and eluting from, biological membranes, fits each of these isozymes to perform a specific physiological role appropriate to the cell in which it function - principally B subunits for well-oxygenated tissues such as heart muscle and brain, and largely A subunits for those tissues producing lactic acid as a consequence of continuous or intermittent deprivation of oxygen. Moreover, the heteropolymers of A and B subunits, with properties intermediate to the homopolymers, also have specific roles to play in facilitating precise responses of cellular metabolism to the availability of substrates and oxygen.

Thus, a positive feedback mechanism functions in tissues such as liver and skeletal muscle to make available optimum amounts of active LDH and a negative feedback mechanism achieves the same end result in heart muscle. Both feedback mechanisms serve to maintain optimum ratios of NAD to $NADH_2$. Clearly, the LDH isozyme system has been developed and maintained through the course of evolution because of its specific utility in fulfilling the metabolic requirements of the different cells in which the isozymes function.

ACKNOWLEDGEMENTS

This research was supported by NSF grant # GB 36749. Dr. Bernardo Nadal-Ginard held a fellowship from the Population Council during this research. The assistance of Margarita Vicens de Nadal is gratefully acknowledged.

REFERENCES

Agostini, A., C. Vergani, and L. Villa 1966. Intracellular distribution of the different forms of lactic dehydrogenase. *Nature* 209:1024-1025.

Ainslie, G.R. 1970. Kinetic studies on alcohol and lactate dehydrogenases. *Ph.D. Dissertion* University of Wisconsin.

Ainslie, G.R. and W.W. Cleland 1969. The effect of oligomeric environment on the kinetics of lactate dehydrogenase subunits. *Fed. Proc. Abst., 53rd Meeting* 28:468, Abst. #1176.

Amberson, W.R., F.J. Roisen, and A.C. Bauer 1965. The attachment of glycolytic enzymes to muscle ultrastructure. *J. Cell Physiol.* 66:71-90.

Anderson, S.R., J.R. Florini, and C.S. Vestling 1964. Rat liver lactate dehydrogenase. III. Kinetics and specificity. *J. Biol. Chem.* 239:2991-2997.

Baba, N. and H.M. Sharma 1971. Histochemistry of lactic dehydrogenase in heart and pectoralis muscles of rat. *J. Cell Biol.* 51:621-625.

Bailey, J.L. 1967. *Techniques in Protein Chemistry*, 2nd ed., Elsevier Publishing Co., New York. pp. 406 .

Battelino, L.J. and A. Blanco 1970. Testicular lactate dehydrogenase isozyme: Nature of multiple forms in guinea pig. *Biochim. Biophys. Acta* 212:205-212.

Blanco, A. and W.H. Zinkham 1963. Lactate dehydrogenase in human testis. *Science* 139:601-602.

Boyer, S.H., B.C. Fainer, and E.J. Watson-Williams 1963. Lactate dehydrogenase variants from human blood: evidence for molecular subunits. *Science* 141:642-643.

Cahn, R.D., N.O. Kaplan, L. Levine, and E. Zwilling 1962. Nature and development of lactate dehydrogenase. *Science* 136:962-969.

Cuatrecasas, P. 1970. Protein purification by affinity chromotography. Derivatization of agarose and polyacrylamide beads. *J. Biol. Chem.* 245:3059-3065.

Everse, J., R.E. Barnett, C.J.R. Thorne, and N.O. Kaplan 1971. The formation of ternary complexes by diphosphopyridine nucleotide-dependent dehydrogenases. *Arch. Biochem. Biophys.* 143:444-460.

Everse, J. and N.O. Kaplan 1973. Lactate dehydrogenase structure and function. *Advances in Enzymology* 37:61-133.

Fromm, H.J. 1961. Evidence for ternary-complex formation with rabbit-muscle lactic acid dehydrogenase, diphosphopyridine nucleotide and pyruvic acid. *Biochem. Biophys. Acta* 52:199-200.

Fromm, H.J. 1963. Determination of dissociation constants and abortive ternary complexes with rabbit muscle lactate dehydrogenase from fluorescence measurements. *J. Biol. Chem.* 238:2938-2944.

Goa, J. 1953. A microbiuret method for protein determination. Determination of total protein in cerebrospinal fluid. *Scand. J. Clin. Lab. Invest.* 5:218-222.

Goldberg, E. 1963. Lactic and malic dehydrogenase in human spermatozoa. *Science* 139:602-603.

Gutfreund, H., R. Cantwell, C.H. McMurray, R.S. Criddle, and G. Hathaway 1968. The kinetics of the reversible inhibition of heart lactate dehydrogenase through the formation of the enzyme-oxidized adenine dinucleotide-pyruvate compound. *Biochem. J.* 106:683-687.

Heck, H. d'A. 1969. Porcine heart lactate dehydrogenase. Optical rotatory dispersion, thermodynamics and kinetics of binding reactions. *J. Biol. Chem.* 244:4375-4381.

Holbrook, J.J. and H. Gutfreund 1973. Approaches to the study of enzyme mechanisms. Lactate dehydrogenase. *FEBS Letters* 31:157-169.

Holbrook, J.J., A. Liljas, S.J. Steindel, and G. Rossman 1974. Lactate dehydrogenase. In: *The Enzymes*, in press.

Hültin, H.O. and C. Westort 1966. Factors affecting the distribution of lactate dehydrogenase between particulate and non-particulate fractions of homogenized skeletal muscle. *Arch. Biochem. Biophys.* 117:523-533.

Hultin, H.O. and J.H. Southard 1967. Cellular distribution of lactate dehydrogenase in chicken breast muscle. *J. Food Sci.* 32:503-510.

Hultin, H.O., J.D. Ehmann, and R.L. Melnick 1972. Modification of kinetic properties of muscle lactate dehydrogenase by subcellular associations and possible role in the control of glycolysis. *J. Food Sci.* 37:269-273.

Hyldgaard, J. and M. Valenta 1970. Extraction and localization of porcine lactate dehydrogenase activity and isozymes. *Enzymol. Biol. Clin.* 11:336-359.

Kaloustian, H.D. and N.O. Kaplan 1969. Lactate dehydrogenase of lobster (*Homarus americanus*) tail muscle. II. Kinetics and regulatory properties. *J. Biol. Chem.* 244:2902-2910.

Kaplan, N.O. and T.L. Goodfriend 1964. Role of the two types of lactic dehydrogenase. In: *Advances in Enzyme Regulation* . G. Weber, ed., Pergamon Press, Inc., Oxford. 2:203-254.

Markert, C.L. 1962. Isozymes in kidney development. In: *Hereditary Development and Immunological Aspects of Kidney Disease*, Jack Metcoff, ed. Northwestern University Press, Evanston, Ill. pp. 54-63.

Markert, C.L. 1963. Lactate dehydrogenase isozymes: dissociation and recombination of subunits. *Science* 140: 1329-1330.

Markert, C.L. and E. Appella 1963. Immunochemical properties of lactate dehydrogenase isozymes. *Ann. N.Y. Acad. Sci.* 103:915-929.

Markert, C.L. and H. Ursprung 1962. The ontogeny of isozyme patterns of lactate dehydrogenase in the mouse. *Develop. Biol.* 5:363-381.

Markert, C.L. and I. Faulhaber 1965. Lactate dehydrogenase isozyme patterns of fish. *J. Exptl. Zool.* 159:319-332.

Markert, C.L. and R.S. Holmes 1969. Lactate dehydrogenase isozymes of the flatfish, pleuronectiformes: kinetic, molecular and immunochemical analysis. *J. Expt. Zool.* 171:85-104.

Markert, C.L., J.B. Shaklee, and G.S. Whitt 1975. Evolution of a gene. In press.

Massaro, E.J. and C.L. Markert 1968. Isozyme patterns in salmonid fishes: Evidence for multiple cistrons for lactate dehydrogenase polypeptides. *J. Expt. Zool.* 168:223-238.

Melnick, R.L. and H.O. Hultin 1973. Studies on the nature of the subcellular localization of lactate dehydrogenase and glyceraldehyde-3-phosphate dehydrogenase in chicken skeletal muscle. *J. Cell. Physiol.* 81:139-148.

O'Carra, P. and S. Barry 1972. Affinity chromatography of lactate dehydrogenase. Model studies demonstrating the potential of the technique in the mechanistic investigation as well as in the purification of multi-substrate enzymes. *FEBS Letters* 21:281-285.

Rossman, M.J. 1970. Structure and mechanism of lactate de-hydrogenase. In: *Pyridine Nucleotide-Dependent Dehydro-genases*. H. Sund, ed., Springer, Berlin, Heidelberg, New York, pp. 172-174.

Rosulin, W. and E. Braswell 1968. Analysis of factors affect-ing LDH subunit composition determinations. *J. Theoret. Biol.* 19:169-182.

Ryan, L.D. and C.S. Vestling 1974. Rapid purification of lactate dehydrogenase from rat liver and hepatoma. A new approach. *Arch. Biochem. Biophys.* 160:279-284.

Sensabaugh, G.F. and N.O. Kaplan 1972. A lactate dehydrogenase specific to the liver of gadoid fish. *J. Biol. Chem.* 247:585-593.

Shaklee, J.B., K.L. Kepes, and G.S. Whitt 1973. Specialized lactate dehydrogenase isozymes: the molecular and genetic basis for the unique eye and liver LDHs of teleost fishes. *J. Exptl. Zool.* 185:217-240.

Stinson, R.A. and J.J. Holbrook 1973. Equilibrium binding of nicotinamide nucleotides to lactate dehydrogenase. *Biochem. J.* 131:719-728.

Vesell, E.S. 1966. pH depdendence of lactate dehydrogenase isozyme inhibition by substrate. *Nature* 210:421-422.

Vesell, E.S. and P.E. Poole 1966. Lactate and pyruvate con-centrations in exercised ischemic canine muscle: Relation-ship of tissue substrate level to lactate dehydrogenase isozyme patterns. *Proc. Nat. Acad. Sci. U.S.A.* 55:756-761.

Vestling, C.S. and J. Künsch 1968. Spectrophotometric analysis of rat liver lactate dehydrogenase-coenzyme and coenzyme analog complexes. *Arch. Biochem. Biophys.* 127:568-575.

Whittaker, J.R., D. W. Yates, N.G. Bennett, J.J. Holbrook, H. Gotfreund 1974. The identification of intermediates in the reaction of pig heart lactate dehydrogenase with its substrates. *Biochem. J.* 139:677-697.

Wieland, Th., P. Duesberg, G. Pfleiderer, A. Stock, and E. Sann 1962. An dehydrogenasen gebundene umwandlungsprodukte des diphospho-pyridin-nucleotids. *ABB Supp.* 1:260-265.

Wilson, J.R.H., G. Kay and M.D. Lilly 1968. The preparation and kinetics of lactate dehydrogenase attached to water-insoluble paritcles and sheets. *Biochem. J.* 108:845-853.

Wuntch, T., E.S. Vesell, and R.F. Chen 1969. Studies on rates of abortive ternary complexes formation of lactate dehy-drogenases isozymes. *J. Biol. Chem.* 244:6100-6104.

Young, V.R. 1970. The role of skeletal and cardiac muscle in the regulation of protein metabolism. In: *Mammalian Protein Metabolism*, N.H. Munro, ed., Academic Press, N.Y. 10:585-614.

Zewe, V. and H.J. Fromm 1962.   Kinetic studies of rabbit
    muscle lactate dehydrogenase. *J. Biol. Chem.*   237:1668-
    1675.
Zinkham, W.H., L. Kupchyk, A. Blanco, and H. Isense 1965.
    Polymorphism of lactate dehydrogenase isozymes in pigeons.
    Nature 208:284-286.
Zinkham, W.H. 1968.   Lactate dehydrogenase isozymes of testes
    and sperm: Biological and biochemical properties and
    genetic control. *Ann. N.Y. Acad. Sci.* 151:598-610.

EFFECT OF ENVIRONMENT ON KINETIC CHARACTERISTICS OF CHICKEN LACTATE DEHYDROGENASE ISOZYMES

HERBERT O. HULTIN
Department of Food Science and Nutrition
University of Massachusetts
Amherst, Massachusetts 01002

ABSTRACT. LDH-5 binds to the particulate fraction of homogenized muscle tissue. The distribution is a function of the pH and ionic strength of the medium and the concentration of NADH. The range of pH and NADH concentration over which the solubilization occurs is physiological. Studies under conditions which imitate those *in situ* demonstrate that LDH-5 can bind to the particulate structures of muscle at physiological ionic strengths. There is some specificity of binding of LDH-5 to various subcellular fractions, but it is not very great.

LDH-1 does not bind to the particulate structures of either skeletal or cardiac muscle. LDH-3 binds but less strongly than LDH-5, being particularly susceptible to solubilization by NADH.

Some kinetic characteristics of LDH-5 and LDH-3 at 4, 16, 23, and 40° are modified by binding to the particulate fraction of muscle. V_{max} of LDH-5 is significantly reduced while that of LDH-3 is little affected. The apparent K_M values of both isozymes are generally increased on binding. Binding affords LDH-5 almost complete protection from inhibition by NAD^+ and pyruvate at 40° but less at lower temperatures. LDH-3 is not protected by binding at any temperature.

Based on these results, an hypothesis is presented to explain the role of reversible solubilization-binding of lactate dehydrogenase in the control of cellular metabolism.

INTRODUCTION

Lactate dehydrogenase (LDH) (EC 1.1.1.27) has been considered to exist as a soluble entity in the cytoplasm of the cell. In more recent years, however, evidence has been accumulating that lactate dehydrogenase (LDH) may be associated with subcellular structures in some tissues (Amberson et al., 1965; Agostoni et al, 1966; Domenech et al., 1970; Baba and Sharma, 1971). Since a detailed knowledge of the subcellular location of the LDH isozymes and their relationship

69

to other components of the cell is essential for a clear
understanding of the role this enzyme plays in metabolism, we
undertook to study the question as to whether lactate dehydrogenase is, in fact, a soluble enzyme in vivo and, if not, what
modifications of its activity might be expected by association
with subcellular particulate fractions of muscle cells.

Several questions may be asked related to the question
of whether an enzyme which can be prepared in soluble form
from a tissue does, in fact, exist wholly or in part bound
to the subcellular structures. These are: (1) Can the
enzyme bind to subcellular particulate structures under
physiological conditions? (2) Does the enzyme have a specific distribution among the various subcellular fractions, and
does this distribution affect the specific function of the
enzyme? (3) Do enzyme characteristics differ when bound
and when soluble, and what is the basis for the difference?
(4) Is the enzyme bound *in situ*? (5) Does the enzyme show
reversible binding-solubilization *in situ* under conditions
requiring cellular control of enzyme activity? This paper
describes our attempts to answer the first four of these
questions.

METHODS AND RESULTS

CONDITIONS GOVERNING DISTRIBUTION OF LDH BETWEEN SOLUBLE AND PARTICULATE FRACTIONS OF HOMOGENIZED SKELETAL MUSCLE

The distribution of LDH-5 (A_4) between the soluble and
particulate fractions of homogenized muscle tissue under
various conditions was studied. In Fig. 1 are shown the results as a function of pH in a solution buffered with 15 mM
imidazole. There is an inflection point in the solubility
curve around 7.1 to 7.2 (Hultin and Westort, 1966). This
curve could be shifted up or down depending on the ionic
strength of the environment. The distribution of LDH-5 increased with increasing ionic strength for the chloride salts
of Ca^{+2}, Mg^{+2}, K^+, and Na^+. A straight line relationship was
obtained when the % solubilization was plotted as a function
of the log of the ionic strength. There was no cation specificity among those used. Other factors, such as time of homogenization after death of the chicken and tissue concentration,
could be explained in terms of the two basic parameters of pH
and ionic strength (Hultin and Westort, 1966).

It was also found that the cofactor of the LDH-catalyzed
reaction, NADH, was a very effective solubilizing agent for
the enzyme (Melnick and Hultin, 1968). The hypothesis that
the enzyme must be only functional in the soluble state since

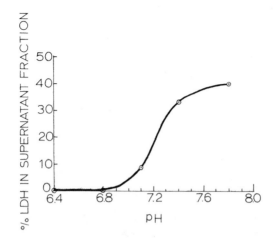

Fig. 1: Solubilization of LDH from the particulate fraction of homogenized chicken breast muscle as a function of pH (Hultin and Westort, 1966).

the physiological amount of NADH solubilized the enzyme under the assay conditions was negated when the solubilizing effect of NADH was examined as a function of tissue concentration (Fig. 2). Solubilization of LDH is expressed as a function of NADH concentration at various levels of tissue particulates at pH 7.0 in 10mM imidazole buffer (Ratner et al., 1974). The tissue concentration here refers to that amount of insoluble protein that would be present in an homogenate containing this percentage of muscle (w/v).

LDH-3 (A_2B_2) was more susceptible than LDH-5 to solubilization by pH and ionic strength, being solubilized at lower values of these two parameters than was LDH-5 (Nitisewojo, 1972). LDH-3 was also much more sensitive to solubilization by NADH than was LDH-5. At pH 6.5 in 10mM imidazole buffer, 0.1 mM NADH solubilized approximately 75% of LDH-3 but less than 1% of LDH-5 (Nitisewojo, 1972). LDH-1 (B_4) does not bind to the particulate fraction of skeletal muscle or heart muscle under any of the conditions that we have examined so far. It is not unreasonable, therefore, to expect that the binding of the hybrid LDH-3 would be less than that of LDH-5. It appears that LDH binds through the A subunit, not the B.

The specificity of solubilization by NADH is high. NADPH is as effective, but NAD^+ and $NADP^+$, as well as several metabolites structurally related to NAD, have limited or negligible effects (Melnick and Hultin, 1968; 1970).

As closely as can be judged, we feel that the conditions

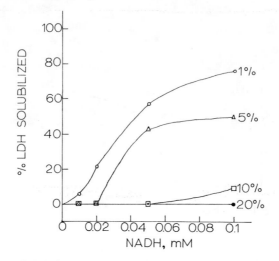

Fig. 2: Solubilization of LDH from the particulate fraction of homogenized chicken breast muscle as a function of NADH concentration at various levels of tissue particulates in 10mM imidazole buffer at pH 7.0. The tissue concentrations were equivalent to those corresponding to (w/v) homogenates. A 1% preparation contains approximately 2 mg of insoluble protein per ml.

involving solubilization represent those which are reasonably physiological. The inflection point of the solubilization curve comes at a pH slightly greater than 7, and the concentration of NADH which solubilizes is within the range of physiological concentrations (Long, 1961). Solubilization does occur at an ionic strength considerably below that considered physiological. However, we shall consider this question later in more depth.

DISTRIBUTION STUDIES OF LACTATE DEHYDROGENASE

Although the results above indicated that LDH might be able to bind under physiological conditions, we had no information at this point as to whether we were dealing with an artifact of the preparative procedure.

We examined this question first by determining the distribution of lactate dehydrogenase. We wished to know if the distribution could be related to a specific function of a particular subcellular fraction. Results shown in Table I indicate that lactate dehydrogenase is relatively uniformly spread among all the major particulate fractions of the muscle (Hultin and Southard, 1967). We also compared the distribution

TABLE I

SPECIFIC LDH ACTIVITIES OF SOME FRACTIONS FROM CHICKEN BREAST
MUSCLE

Fraction	Specific Activity μmole substrate/min/mg protein
Nuclear-myofibrillar	4.0-6.1
Mitochondria	7.2-8.5
[a]Post-mitochondrial fractions	
A	7.8-10.6
B	0.4-1.6
C	1.8-1.9
Sarcolemmae	3.2-39.0
Whole homogenate	4-7

[a]Post-mitochondrial fractions:
  A. Fraction between 8,000g - 20 min and 28,000g - 1 hr
     in 0.5M sucrose, pH 7.0.
  B. Fraction between 15,000g - 10 min and 70,000g - 40 min
     in 0.88M sucrose, pH 7.0.
  C. Fraction between 28,000g - 1 hr and 100,000g - 1 hr in
     0.5M sucrose, pH 7.0.

in freshly homogenized tissue to the distribution in particulate fractions that had been previously washed free of lactate
dehydrogenase and then bound again to soluble enzyme. A different distribution pattern in these two situations might
indicate that one was an artifact, the other a true distribution. Results showed that the distributions were similar
(Hultin, Westort and Southard, 1966).

It was demonstrated that the muscle particulate fraction
could bind over six times the amount of LDH that was normally
associated with it (Hultin, Westort and Southard, 1966). Likewise we have shown that glyceraldehyde-3-phosphate dehydrogenase can compete with lactate dehydrogenase for the available
binding sites (S. Hirway, unpublished observations). We
therefore concluded that interactions of the LDH with the
subcellular particulate fractions were probably non-specific
and any functional significance related to binding per se
rather than to specific location. We do not assume that non-
specific interaction implies no metabolic significance.

The evidence indicates that LDH is probably reversibly
bound to the subcellular particulate fractions *in situ* rather
than associated permanently with them. Thus, any role in
metabolic control would probably be related to the ability to
bind and dissociate reversibly from cellular structures.

HERBERT O. HULTIN

KINETIC PROPERTIES OF BOUND AND SOLUBLE LACTATE DEHYDROGENASE ISOZYMES

The next question which we considered was whether the binding of lactate dehydrogenase affected its kinetic properties. The comparison of the soluble and bound LDH isozymes was accomplished with a stopped-flow technique (Ehmann and Hultin, 1973a). The enzyme concentrations used in comparing the bound and soluble enzymes were the same and were obtained by first assaying the bound and soluble preparations under conditions of pH, ionic strength, and tissue and enzyme concentrations such that all of the enzyme of the bound preparation would be solubilized. Since the LDH does show substrate inhibition, V_{max} was calculated by running the assays at various substrate levels and using the reciprocal plots to extrapolate to a V_{max} value.

Some kinetic properties of LDH-1, LDH-3, and LDH-5 were examined as a function of temperature to determine the effect of modification of the subunit environment on the enzyme. The properties of the various isozymes have been reported to vary markedly with temperature (Vesell, 1965; Maisel and Kerrigan, 1966). We examined LDH-3 and LDH-5, both soluble and bound to the particulate fraction of the muscle. The assays were carried out at pH 6 in 5 mM imidazole buffer using $10^{-8}M$ enzyme and 0.1mM NADH. The buffer was sufficient to maintain constant pH during the reaction. The conditions were chosen to keep most LDH-3 bound to simplify interpretation of the data. Even under these conditions, about 10% of the LDH-3 in the bound preparations was soluble under the conditions of assay, and corrections had to be made for this in all cases. Some investigators have concluded that pH 6 is very close to the pH in the muscle cell (Carter et al., 1967) and that a pH approching this can be attained when the muscle is very active (McLoughlin, 1970).

A summary of the V_{max} values of the three isozymes in the bound and soluble state and at 4, 16, 23, and 40°C are given in Table 2 (Nitisewojo, 1972). It is clear that the V_{max} values do not follow a simple Arrhenius relationship; a very steep change generally occurs between 16 and 23°. There is a greater decrease in the maximal velocity caused by binding for LDH-5 than for LDH-3. The decrease in V_{max} for LDH-5 on binding to the muscle particulate fraction is substantial.

The summary of the apparent K_M values as a function of LDH isozyme and temperature is shown in Table 3 (Nitisewojo, 1972). An increase in the apparent K_M occurred with increasing temperature; the increase was 4-10 fold depending on the

TABLE II

SUMMARY OF V_{MAX} VALUES OF LDH-1 AND BOUND AND SOLUBLE LDH-3 AND LDH-5 AT VARIOUS TEMPERATURES.

Temperature	LDH-1	LDH-3 Soluble	Bound	LDH-5 Soluble	Bound
4°	1.2	3.1	2.4	4.4	1.7
16°	2.3	6.7	4.5	12.9	4.8
23°	8.9	19.1	13.4	20.0	5.9
40°	8.1	24.5	21.4	57.5	18.2

$V_{max} \equiv$ nmoles NADH oxidized per sec. The reactions were carried out at pH 6.0 in a stopped-flow apparatus as previously described (Ehmann and Hultin, 1973a). The enzyme concentration in all cases was 10^{-8}M.

TABLE III

SUMMARY OF APPARENT K_M VALUES OF LDH-1 AND BOUND AND SOLUBLE LDH-3 AND LDH-5 AT VARIOUS TEMPERATURES.

Temperature	LDH-1	LDH-3 Soluble	Bound	LDH-5 Soluble	Bound
4°	0.32	1.1	2.4	1.6	2.7
16°	0.83	1.9	3.3	3.1	7.1
23°	1.9	3.6	4.6	5.2	10.2
40°	2.4	5.0	7.5	18.8	18.5

$K_M \equiv 10^5$ x M pyruvate. The reactions were carried out at pH 6.0 and 10^{-8}M enzyme in a stopped-flow apparatus as previously described (Ehmann and Hultin, 1973a).

specific isozyme. The apparent K_M for LDH-1 was less than that for LDH-3 which was less than that for LDH-5. LDH-3 and LDH-5 showed an increase in apparent K_M on binding with the exception of LDH-5 at 40°. Similar results have been obtained for LDH-5 under different conditions (Ehmann and Hultin, 1973a,b,c).

Pyruvate inhibition of lactate dehydrogenase can occur through the formation of an abortive ternary complex of enzyme, NAD^+, and pyruvate (Zewe and Fromm, 1962). Spectral and fluorescent properties of the complex (Fromm, 1963; Vestling and Künsch, 1968; DiSabato, 1968), dissociation constants (Fromm, 1963), and the kinetics of formation (Gutfreund et al., 1968;

Wuntch, Vesell and Chen, 1969) have been investigated. It
has been suggested that the formation of this abortive ternary
complex is a control mechanism for the regulation of LDH activity *in situ* (Kaplan et al., 1968). We studied the effect
of abortive ternary complex formation on the activity of the
soluble and bound LDH-5 as above with a stopped flow procedure. We incubated the enzyme in the presence of NAD$^+$ rather
than generating the NAD$^+$ during the reaction. We did this
since in the cell the ratio of NAD$^+$ to NADH is high (Long,
1961); thus, pre-incubation should be more physiological than
NAD$^+$ generation during the reaction. We adjusted the incubation time of the enzyme with NAD$^+$ to achieve maximal inhibition. The time required varied as a function of temperature,
being slower at the lower temperatures (Ehmann and Hultin,
1973b). The net results of these studies are summarized in
Fig. 3 which shows a plot of the percentage activity of the
enzyme in the presence of NAD$^+$ to that in its absence vs.
pyruvate concentration (Ehmann and Hultin, 1973b). Inhibition
with the soluble enzyme was much greater than with the bound
enzyme, and there was very little difference as a function of
temperature. In the case of the bound LDH-5, however, there
was a difference depending on the temperature. At 40°C, which
is close to the physiological temperature of the chicken,
there was essentially no inhibition. A significant amount of
inhibition occurred at the lower temperatures, but less than
with the soluble enzyme. It is apparent, therefore, that
bound LDH is somewhat protected from inhibition by pyruvate
and NAD$^+$.

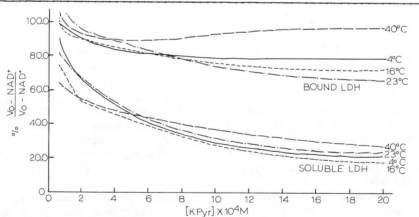

Fig. 3: Percentage of initial velocity of LDH in the presence
of NAD$^+$ compared to that in its absence as a function of pyruvate concentration at 4, 16, 23, and 40°. (Ehmann and Hultin,
1973b).

Similar studies on the protection of bound LDH-3 from inhibition by pyruvate and NAD$^+$ produced no evidence of any protection. Obviously, the presence of the two B type subunits in LDH-3 has a profound effect on the ability of the enzyme to interact with NAD$^+$ and pyruvate.

HYPOTHESIS FOR CONTROL OF LDH BASED ON KINETIC DATA AND REVERSIBLE BINDING-SOLUBILIZATION

Fig. 4 gives the summary of our hypothesis for the control mechanism of lactate dehydrogenase based on the kinetic and binding data which we have obtained to date (Hultin, Ehmann,and Melnick, 1972). Glycogen or glucose is the energy source for muscle via glycolysis. Oxidation of glyceraldehyde-3-phosphate to 1,3-diphosphoglycerate by glyceraldehyde-3-phosphate dehydrogenase requires NAD$^+$ and produces NADH. The latter must be removed since it is a very potent inhibitor of the glyceraldehyde-3-phosphate dehydrogenase. There is competition between the electron transport chain of the mitochondrion and LDH for the electrons of NADH. The former is more efficient and should be the pathway of choice under most conditions, and this is generally so because at the low NADH levels in resting muscles, LDH is bound and relatively inactive. When oxygen is depleted, as in stress, mitochondrial oxidations cease. The NADH level builds up and solubilizes the LDH, which then can reoxidize the NADH more efficiently than does the bound LDH. This mechanism allows the maintenance of glycolysis under anaerobic conditions while at the same time minimizing competition for electrons between LDH and mitochondria under aerobic conditions. There is the added factor that under any particular set of conditions of substrate concentration, pH, etc., a whole range of activities of LDH is possible depending on the amount of enzyme that is bound as illustrated in Fig. 5 by the arrow. Thus the relative activity of the lactate dehydrogenase can be controlled by factors other than the metabolites taking part in the reaction. An example of this would be a change in the ionic environment of the cell in the area where the enzyme is functioning. The data on inhibition by NAD$^+$ and pyruvate would indicate that although the soluble enzyme is potentially more active, it is also subject to more precise control. On the other hand, the bound enzyme, although inherently less active, can function even in the presence of the high levels of NAD$^+$ expected in resting muscle.

An early theory was that LDH-3, which has been shown to be more similar to LDH-1 than to LDH-5 (Anderson, 1971), might

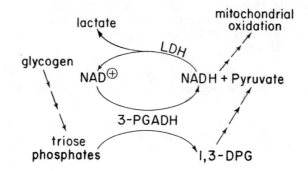

Fig. 4: Proposed scheme for the role of reversible binding-solubilization of LDH in carbohydrate metabolism. Details are given in the text.

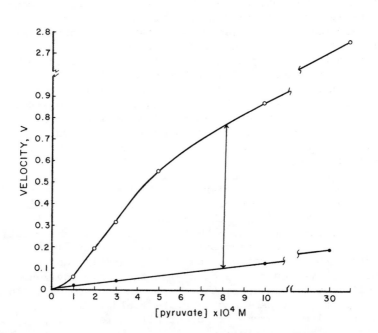

Fig. 5: Initial velocity of LDH as a function of pyruvate concentration in the soluble (o) and bound phase (●). The arrow represents the potential range of activity of the enzyme at that particular pyruvate concentration dependent on the percentage of enzyme that is in the bound phase.

bind and be affected by binding in a manner similar to LDH-5.
Thus the hybrid would have the binding properties of LDH-5 and
the kinetic properties of LDH-1 with all the implications this
would have in terms of finer control. However, based on the
available data, we do not believe this to be true. The V_{max}
of LDH-3 is little affected by binding and no protection is
afforded from substrate inhibition by binding of this particular isozyme. This does not mean that binding of LDH-3 is
not important in control. It may be that the advantage of
binding LDH-3 is related simply to the proximity to other
bound enzymes, like glyceraldehyde-3-phosphate dehydrogenase
(Dagher, 1971).

BINDING OF LACTATE DEHYDROGENASE IN SITU

The next question that was examined was whether the
enzyme was capable of binding *in situ*. Several approaches
were used in this study. In one, the imbibition-centrifuga-
tion technique of Amberson, Roisen, and Bauer (1965) was used
to wash the ultrastructure of muscle with buffer at a pH and
ionic strength which should dissolve all of the enzyme. The
muscle was packed into centrifuge tubes and centrifuged at
high speed (36,000 rpm) to remove the press juice. The
muscle was then suspended in buffer which is absorbed into
the muscle and then centrifuged. This was repeated several
times until the amount of lactate dehydrogenase being removed
from the muscle was very low. At this point the muscle was
homogenized and assayed for lactate dehydrogenase. When
this was done, a considerable amount of enzyme was still
found associated with the muscle tissue. Amberson, Roisen,
and Bauer (1965) assumed that this enzyme was attached to the
ultrastructure.
We did not assume that the presence of the enzyme in the
tissue at the end of the washing cycles was sufficient by
itself to conclude that the enzyme was bound to the ultrastructure. We used the concept of "comparative extraction"
where we compared the extraction of lactate dehydrogenase
with an enzyme comparable in size and structure (Melnick and
Hultin, 1973). This was glyceraldehyde-3-phosphate dehydrogenase. Results of a typical experiment are shown in Fig. 6.
Assuming that the glyceraldehyde-3-phosphate dehydrogenase
that is not removed is still completely soluble in the aqueous
phase of the muscle segments but its removal is diffusion-
limited, the difference between the amount of the glyceralde-
hyde-3-phosphate dehydrogenase remaining and that of LDH represents the minimal amount of lactate dehydrogenase that is
restricted in its removal from the tissue, either due to

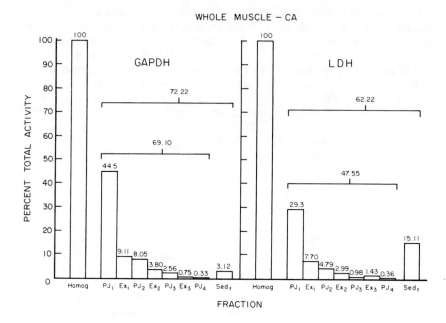

Fig. 6:   The activities of glyceraldehyde-3-phosphate dehydrogenase and lactate dehydrogenase in press juices, extracts,
and final sediments of chicken breast muscle strips.   The
activities of the other fractions are expressed as % of the
homogenate activity.   (Melnick and Hultin, 1973).

binding to the ultrastructure or to compartmentation within
subcellular structures.

   Another approach which we used was to compare the activity of the LDH obtained in the whole muscle to that in the
press juice.   When expressed as activity per ml of water, a
ratio of one would indicate that the concentrations are the
same and the enzyme may be present exclusively in the soluble
phase.   If the ratio of the activity in the whole muscle to
that in the press juice is greater than one, it would indicate
that some of the enzyme was not freely soluble in the muscle.
Results which we obtained for lactate dehydrogenase and
glyceraldehyde-3-phosphate dehydrogenase showned that although
the ratio with glyceraldehyde-3-phosphate dehydrogenase was
one, that with lactate dehydrogenase was considerably greater
than one, indicating that the latter enzyme was retained in
part by the tissue (up to 50%) (Melnick and Hultin, 1973).

   A further approach has been the study of the uptake of
lactate dehydrogenase into the muscle against a concentration
gradient.   The muscle is first centrifuged to remove the press

juice. It is then imbibed with a buffer of pH 7.5, ionic
strength 0.2, which should be enough to solubilize the enzyme
according to our in vitro studies. On a per gram of water
basis the activity of the LDH in the reimbibed muscle is lower than it was in the original muscle and lower than in the
first press juice. This reimbibed muscle now is soaked for
various time periods in the original press juice. Assuming
that the muscle segments are freely permeable to the lactate
dehydrogenase enzyme in the press juice, an equilibrium should
be reached between the muscle and the press juice. This, in
fact, does not happen, but the muscle takes up much more enzyme than would be predicted from an equilibrium distribution,
as shown in Table 4 (Ross, 1973). We interpret the excess
enzyme taken up to be due to binding to the ultrastructure
even though the ionic strength was such that the enzyme
would have been solubilized in vitro. Thus, LDH can bind to
the subcellular particulate structures under conditions of
ionic strength which are similar to those encountered *in situ*.

TABLE IV

REINTRODUCTION OF LDH FROM MUSCLE PRESS JUICE INTO PRESSED
CHICKEN BREAST MUSCLE

Sample	Activity (μmoles/min per ml water in	
	Muscle	Press juice
Whole muscle	1745	
Pressed muscle	2517	1184
Imbibed muscle	793	
Restored muscle		
5 hr	1570	900
24 hr	1945	587

Pressed muscle is after removal of press juice. Imbibed
muscle has been soaked in phosphate buffer, pH 7.5, and
ionic strength 0.2, and the restored muscle is imbibed
muscle that has been soaked in the original press juice.

SUMMARY

We have found that lactate dehydrogenase of chicken breast muscle can bind to subcellular particulate structures under physiological conditions. Although the enzyme does not have a specific distribution, there is evidence that some important kinetic properties of the enzyme are modified when it does interact with the subcellular structures. An hypothesis is presented which relates how these changes might affect the metabolic function of the enzyme. We have also presented evidence that the enzyme can bind under conditions which are similar to those found *in situ*. The next step should be to show that reversible binding-solubilization does occur *in situ* under conditions where such a control mechanism is advantageous to the cell. We are currently working on this problem.

Lactate dehydrogenase is only one of many enzymes that is solubilized by substrate, product, or effector. Others include phosphofructokinase (Mansour, 1965), glyceraldehyde-3-phosphate dehydrogenase (Arnold and Pette, 1970; Dagher, 1971), hexokinase (Rose and Warms, 1967); and aldolase (Masters, Sheedy, and Winzor, 1969; Arnold and Pette, 1970). It seems unlikely that all of these are merely coincidences. The enzymes which demonstrate this behavior for the most part, are important regulatory enzymes with subunit structures. Binding of these components to various parts of the cell may modify their activity, allowing additional control for the cell. In the case where the enzymes exist in isozymic forms, differential interaction of these isozymes with the various subcellular components and the different modifications caused by these interactions may provide further metabolic control to the functioning cell.

Other workers have suggested that LDH-5, but not LDH-1, is a regulatory enzyme based on kinetic data and response to inhibitors (Fritz, 1967; Orleans-Harding and Mahler, 1968). My suggestion that binding-solubilization of LDH-5 may be a control mechanism fits in with these suggestions since LDH-1 does not bind and thus cannot be involved with this type of control.

ACKNOWLEDGEMENTS

This work was supported in part by U. S. Public Health Service Grant GM-12064 and by Grant GB 21262 from the National Science Foundation.

REFERENCES

Agostoni, A., C. Vergani, and L. Villa 1966. Intracellular distribution of the different forms of lactate dehydrogenase. *Nature* 209: 1024-1025.

Amberson, W. R., F. J. Roisen, and A. C. Bauer 1965. The attachment of glycolytic enzymes to muscle ultrastructure. *J. Cell. Physiol.* 66: 71-90.

Anderson, S. R. 1971. Selective adsorption of bis (1-anilino-8-naphthalenesulfonate) to the multiple forms of lactate dehydrogenase. *Biochemistry* 10: 4162-4168.

Arnold, H. and D. Pette 1970. Binding of aldolase and triosephosphate dehydrogenase to F-actin and modification of catalytic properties of aldolase. *Eur. J. Biochem.* 15: 360-366.

Baba, N. and H. M. Sharma 1971. Histochemistry of lactic dehydrogenase in heart and pectoralis muscles of rat. *J. Cell Biol.* 51: 621-635.

Carter, N. W., F. C. Rector, D. S. Campion, and D. W. Seldin 1967. Measurement of intracellular pH of skeletal muscle with pH-sensitive glass microelectrodes. *J. Clin. Invest.* 46: 920-933.

Dagher, S. M. 1971. Association of glyceraldehyde-3-phosphate dehydrogenase with subcellular fractions of chicken breast muscle. Ph.D. Thesis, University of Massachusetts, Amherst.

DiSabato, G. 1968. Complexes of chicken heart lactic dehydrogenase with coenzymes and substrates. *Biochem. Biophys. Res. Comm.* 33: 688-695.

Domenech, E. M. D., C. E. Domenech, and A. Blanco 1970. Distribution of lactate dehydrogenase isozymes in subcellular fractions of rat tissues. *Arch. Biochem. Biophys.* 141: 147-154.

Ehmann, J. D. and H. O. Hultin 1973a. Substrate inhibition of soluble and bound lactate dehydrogenase (isozenzyme 5). *Arch. Biochem. Biophys.* 154: 471-475.

Ehmann, J. D. and H. O. Hultin 1973b. Substrate inhibition of chicken muscle lactate dehydrogenase as a function of temperature. *J. Food Sci.* 38: 1115-1118.

Ehmann, J. D. and H. O. Hultin 1973c. Temperature dependence of the Michaelis constant of chicken breast muscle lactate dehydrogenase. *J. Food Sci.* 38: 1119-1121.

Fritz, P. J. 1967. Rabbit lactate dehydrogenase isoenzymes: effect of pH on activity. *Science* 156: 82-83.

Fromm, H. J. 1963. Determination of dissociation constants and abortive ternary complexes with rabbit muscle lactate dehydrogenase from fluorescence measurements. *J. Biol.*

Chem. 238: 2938-2944.

Gutfreund, H., R. Cantwell, C. H. McMurray, R. S. Criddle, and G. Hathaway 1968. The kinetics of the reversible inhibition of heart lactate dehydrogenase through the formation of the enzyme-oxidized nicotinamide adenine dinucleotide-pyruvate compound. *Biochem. J.* 106: 683-687.

Hultin, H. O., J. D. Ehmann, and R. L. Melnick 1972. Modification of kinetic properties of muscle lactate dehydrogenase by subcellular associations and possible role in the control of glycolysis. *J. Food Sci.* 37: 269-273.

Hultin, H. O. and J. H. Southard 1967. Cellular distribution of lactate dehydrogenase in chicken breast muscle. *J. Food Sci.* 32: 503-510.

Hultin, H. O. and C. Westort 1966. Factors affecting the distribution of lactate dehydrogenase between particulate and non-particulate fractions of homogenized skeletal muscle. *Arch. Biochem. Biophys.* 117: 523-533.

Hultin, H. O., C. Westort, and J. H. Southard 1966. Adsorption of lactate dehydrogenase to the particulate fraction of homogenized skeletal muscle. *Nature* 211: 853-854.

Kaplan, N. O., J. Everse, and J. Admiraal 1968. Significance of substrate inhibition of dehydrogenases. *Ann. N. Y. Acad. Sci.* 151: 400-412.

Long, C., ed. 1961. *"Biochemists' Handbook"*, p. 782. D. Van Nostrand Co., Princeton.

Maisel, H. and M. Kerrigan 1966. Effect of temperature (and substrate concentration) on chicken lactate dehydrogenase activity. *Proc. Soc. Exp. Biol. Med.* 123: 847-848.

Mansour, T. E. 1965. Heart phosphofructokinase. Active and inactive forms of the enzyme. *J. Biol. Chem.* 240: 2165-2172.

Masters, C. J., R. J. Sheedy, and D. J. Winzor 1969. Reversible adsorption of enzymes as a possible allosteric control mechanism. *Biochem. J.* 112: 806-808.

McLoughlin, J. V. 1970. Muscle contraction and postmortem pH changes in pig skeletal muscle. *J. Food Sci.* 35: 717-719.

Melnick, R. L. and H. O. Hultin 1968. Solubilization of bound lactate dehydrogenase by NADH in homogenates of trout skeletal muscle as a function of tissue concentration. *Biochem. Biophys. Res. Comm.* 33: 863-868.

Melnick, R. L. and H. O. Hultin 1970. Factors affecting the distribution of lactate dehydrogenase between particulate and soluble phases of homogenized trout skeletal muscle. *J. Food Sci.* 35: 67-72.

Melnick, R. L. and H. O. Hultin 1973. Studies on the nature

of the subcellular localization of lactate dehydrogenase and glyceraldehyde-3-phosphate dehydrogenase in chicken skeletal muscle. *J. Cell. Physiol.* 81: 139-148.

Nitisewojo, P. 1972. A study of some catalytic properties of bound and soluble lactate dehydrogenase isoenzymes from chicken. Ph. D. Thesis, University of Massachusetts, Amherst.

Orleans-Harding, J. G. and R. Mahler 1968. Preferential activation of M_4-lactate dehydrogenase isoenzyme by citrate. *Biochem. J.* 107: 31P.

Ratner, J. H., P. Nitisewojo, S. Hirway, and H. O. Hultin 1974. A study of some factors involved in the interaction between lactate dehydrogenase isoenzymes and particulate fractions of muscle. *Int. J. Biochem.* (in press).

Rose, I. A. and J. V. B. Warms 1967. Mitochondrial hexokinase; release, rebinding, and location. *J. Biol. Chem.* 242: 1635-1645.

Ross, R. E. 1973. Intracellular distribution of lactate dehydrogenase and glyceraldehyde-3-phosphate dehydrogenase in chicken skeletal muscle. Ph. D. Thesis, University of Massachusetts, Amherst.

Vesell, E. S. 1965. Lactate dehydrogenase isozymes. Substrate inhibition in various human tissues. *Science* 150: 1590-1593.

Vestling, C. S. and U. Künsch 1968. Spectrophotometric analysis of rat liver lactate dehydrogenase-coenzyme and coenzyme analog complexes. *Arch. Biochem. Biophys.* 127: 568-575.

Wuntch, T., E. S. Vesell, and R. F. Chen 1969. Studies on rates of abortive ternary complex formation of lactate dehydrogenase isozymes. *J. Biol. Chem.* 244: 6100-6104.

Zewe, V. and H. J. Fromm 1962. Kinetic studies of rabbit muscle lactate dehydrogenase. *J. Biol. Chem.* 237: 1668-1675.

LACTATE DEHYDROGENASE FROM LIVER,
MORRIS HEPATOMAS, AND HTC CELLS

CARL S. VESTLING
Department of Biochemistry
University of Iowa
Iowa City, Iowa 52242

ABSTRACT: Lactate Dehydrogenase (LDH) from adult rat liver
was isolated in the author's laboratory as a pure enzyme
and characterized in 1953. Since that time many studies
dealing with catalytic function and with the structure of
the enzyme have been reported. The present report concerns
three advances: (1) An improved isolation procedure involving the use of a Sepharose-Blue Dextran affinity column
makes it possible to obtain 100-mg quantities of pure enzyme
within 24 hours; (2) Comparison of pure LDH from liver and
from Morris hepatomas grown *in vivo* and in cell culture
(HTC cells) reveals no structural or functional differences
when catalytic function, immunochemical properties, heat
inactivation, oxamate inhibition, ethanol sensitivity, and
electrophoretic behavior are studied; (3) Further experiments
involving ultraviolet difference spectrophotometry have been
directed toward an understanding of the slow changes in
NAD^+ and NADH chromophoric behavior which result when binary
complexes with LDH are formed and full catalytic activity of
the enzyme is retained.

INTRODUCTION

IMPROVED ISOLATION PROCEDURES

In 1953 before the isozyme era lactate dehydrogenase (LDH)
was isolated from rat liver (Gibson et al., 1953) with the aid
of classical methods of protein fractionation. Some properties
of the pure, crystalline enzyme were reported by Davisson et
al. (1953). It is of interest to note that rat liver LDH was
the fourth enzyme from mammalian liver to be acceptably-charac-
terized as a pure protein. The first three were catalase,
alcohol dehydrogenase, and glutamate dehydrogenase. In the
case of LDH, salt and solvent fractionation steps under rigid
control of temperature, ionic strength, and pH led to very low
yields of pure enzyme and required a long time. We referred to
LDH isolated in this way as "12-week LDH". It is of interest
to note that the properties reported for liver LDH in 1953 have
stood the test of time.
A major advance in liver LDH isolation was made by Hsieh

and Vestling (1966) who used fractionation on carboxymethyl cellulose and diethylaminoethyl cellulose under appropriate conditions to obtain LDH of comparable purity and identical properties but in better yield (30%) and in a much shorter time period. This LDH we came to refer to as "12-day LDH". It became possible to isolate 100- to 200-mg quantities of LDH which showed maximum specific catalytic activity and could be stored at -20° in a suitable medium. It should be mentioned that the presence of 16 SH groups per LDH tetramer (140,000 molecular weight) contributes to a rather healthy instability of pure enzyme. The use of external thiols made it possible to obtain the pure enzyme and store it.

Another point of interest is that the method of extraction of the enzyme from fresh or frozen liver is exceedingly important. When we began our LDH studies, we extracted with 0.5 M NaCl and then quickly lowered the temperature and brought the concentration of ethanol to 20% (vol/vol). We noted that this procedure led to clear, pink solutions for later fractionation steps and that we gained two other advantages: (1) Proteolysis was effectively inhibited, and (2) nucleic acids were precipitated.

METHODS, MATERIALS AND RESULTS

Many of the reports on LDH from our laboratory have dealt with studies of "12-day LDH". We have recently developed an affinity chromatography column (Ryan and Vestling, 1974), which makes available to us what we call "12-hour" or "instant" LDH. These terms represent only slight exaggeration for the purpose of literary emphasis. In a study of possible affinity columns for LDH isolation Ryan used the substance, Blue Dextran (Pharmacia Fine Chemicals, Piscataway, N. J.) as a marker only to find that LDH and Blue Dextran formed a tight ion concentration-sensitive complex, which could be dissociated by increasing the ion concentration or by treatment with NADH at low ionic strength. Accordingly Blue Dextran was coupled covalently to Sepharose-4B by the cyanogen bromide procedure to create a Blue Dextran-resin column. Application of a fresh crude liver or hepatoma extract in 20 mM Tris buffer, pH 8.6, to the column led to quantitative binding of LDH. The extracting buffer was also 1 mM with respect to 2-mercaptoethanol and 1 mM with respect to phenylmethylsulfonyl fluoride, the latter substance being added to inhibit proteolysis. The column was washed with extracting buffer (minus phenylmethylsulfonyl fluoride) and then with 1 mM NAD - 1mM lithium lactate in 10 mM Tris buffer, pH 8.6, to remove loosely bound proteins. Elution was accomplished with 1 mM NADH, followed by concentration by ultrafiltration and passage over a column of DEAE-Sephadex (A-50) either as described by Ryan and Vestling (1974) or equilibrated with 0.02 M potassium phosphate 1 mM 2-mercaptoethanol, pH 7.0. LDH of maximal purity was obtained in about 60% yield and in a very short time. This procedure now makes it possible to isolate pure LDH from liver, hepatomas, and perhaps from other sources in varying quantities and makes possible a variety of continuing studies of the pure protein. A summary of steps for the Ryan and Vestling procedure is shown in Table 1. In Figure 1 are shown polyacrylamide gel electrophoresis patterns of LDH with activity staining (See Carlotti et al., 1974) showing single component behavior for pure enzyme from normal liver and from HTC cells (Morris Hepatoma 7288 C grown in cell culture).

COMPARISON OF PURIFIED LLH'S FROM NORMAL RAT LIVER AND MORRIS HEPATOMAS IN RATS AND IN CULTURE

A major objective of studies of liver LDH has been to determine if it could serve as an indicator of gene function which might have been permanently altered in the process of carcinogenesis. Sufficient detailed knowledge of the structure and properties of the purified enzyme and of its catalytic function has been accumulated to make such an investigation feasible. For example it is of interest to demonstrate whether the isozymic character (A_4) of the liver LDH is preserved in the case of LDH isolated from a hepatoma.

Accordingly a series of studies was recently reported (Carlotti et al.,1974) which indicate that the purified LDH from two Morris hepatomas (Morris,1965) shows several properties which are identical to those exhibited by the enzyme from normal adult rat liver.

For these studies pure "normal" LDH was prepared, as was LDH from Morris Hepatoma 7777 (fast growing (one month), poorly differentiated) grown in Buffalo strain rats; from Morris Hepatoma 7793 (slow growing (four months), well differentiated) also grown in Buffalo strain rats; and from HTC cells (Morris Hepatoma 7288 C, less well differentiated and of intermediate growth rate) grown in culture. The LDH was isolated by the "12-day" and the "12-hour" procedures.

No gross differences in catalytic function could be detected. Such studies as heat inactivation, oxamate inhibition, sensitivity to aqueous ethanol, starch gel and polyacrylamide gel electrophoresis showed no differences as measured under the conditions employed. As indicated above the results in Figure 1 are typical of these experiments.

In order to make an immunochemical comparison a rabbit antiserum was prepared directed against "normal" LDH. Immunodiffusion experiments revealed a very small amount of impurity in our "maximally pure" LDH. The precipitation zone showed a

TABLE 1

SUMMARY OF PRINCIPAL STEPS IN THE RAPID ISOLATION OF LDH (25 g OF STARTING MATERIAL).

Step	Liver			Hepatoma (Morris No. 7777)		
	Protein mg/ml	Specific Activity IEC Units/mg Protein	% Recovery	Protein mg/ml	Specific Activity IEC Units/mg Protein	% Recovery
Extraction 20 mM Tris	62	0.19	(100)	28	0.29	(100)
Blue Dextran Resin NADH Elution	ND*	ND	83	ND	ND	79
DEAE-Sephadex concentrate	1.78	112	61	0.89	107	46

*Not determined.

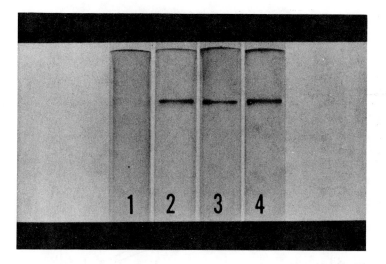

Fig. 1. Polyacrylamide gel electrophoresis of normal rat liver
LDH (2), HTC LDH (3) and a mixture of the two (4). Zones detected by activity staining in 108 mM lithium lactate, 1.56 mM
NAD, 10.8 mM NaCl, 540 µM $MgCl_2$, 135 mM Tris, 320 µM nitroblue
tetrazolium chloride, 86 µM phenazine methosulfate, pH 8.6.
Tube No. 1 is a control with no applied enzyme. (From Carlotti
et al., 1974).

faint fuzziness. Removal of the antibody against the impurity
was accomplished by absorption with a small amount of crude rat
liver homogenate followed by centrifugation (Raffel, 1961).
The resulting antiserum gave a single sharp zone of precipitation in both immunodiffusion and immunoelectrophoresis experiments. As reported by Carlotti et al. (1974) no evidence
against immunochemical identity could be gained. Immunochemical
titration of "normal" and tumor LDH against "normal" antiserum as shown in Figure 2 revealed no differences.

We conclude, therefore, on the basis of these experiments
that the original carcinogenic event(s) did not involve alterations in gene expression concerned with the biosynthesis of
LDH. Those changes which are obvious in the case of the hepatomas must have involved regions of the genes distant from
those concerned with LDH biosynthesis.

DIFFERENCE SPECTRAL STUDIES

Vestling and Künsch (1968) reported a series of studies of
ultraviolet difference spectra involving binary-complexes of
LDH-oxidized coenzyme, LDH-reduced coenzyme, and an LDH-oxidized
and reduced coenzyme analog. Several conclusions emerged from

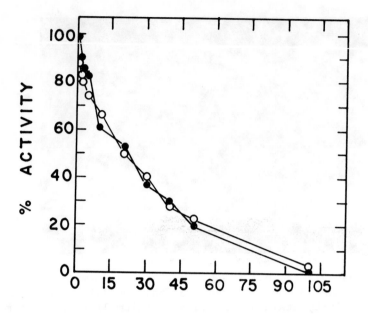

Fig. 2. Immunotitration of normal rat liver LDH (-0-0-0-) and HTC LDH (-●-●-●-) against rabbit antiserum directed against normal LDH. Systems incubated for 40 min at 37° before assay. Solvent was 500 mM NaCl - 18 mM NaHCO₃. (From Carlotti et al., 1974.)

this report.

LDH and either NAD or $_{AcPyAD}$ (the acetyl analog of NAD in which the carboxamide function of the pyridine moiety is replaced by acetyl) formed a binary complex with spectral properties distinctly different from the corresponding LDH-reduced coenzyme complexes. The LDH-NAD and LDH-AcPyAD binary complexes showed relatively low absorbance peaks which were easily measured by difference spectrophotometry in a Cary 15 recording spectrophotometer with cuvettes in tandem. The cuvettes were so arranged that LDH and coenzyme were mixed in the sample beam and in separate cuvettes in the reference beam. Vestling and Künsch (1968) failed to note the time dependence of spectral development as shown in Figure 3 for LDH-NAD binary complexes. A specific effort was made by these authors to detect time changes, and no such changes were found in an early experiment. Recently Vestling and Wiechert (unpublished experiments) noted the striking, slow, temperature dependent spectral changes which are interpreted as reflecting conformational adjustments in the LDH-NAD binary complex which affect the chromophoric

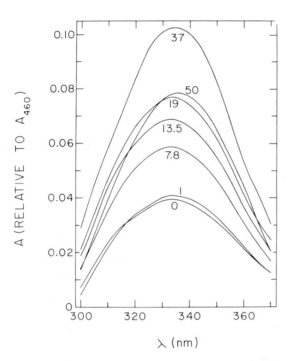

Fig. 3. Difference Spectra. Binary Complex: LDH(1.2×10^{-5}M)
- NAD(4.8×10^{-4}M). Solvent: 0.02 M$(NH_4)_2SO_4$ - 0.02 M
potassium phosphate - 0.001 M 2-mercaptoethanol. pH 7.8. 25°.
Numbers on curves denote hours after mixing. At 50 hours partial loss of enzyme activity was recorded.

behavior of the pyridine moiety. The absorbance maxima of
Figure 3 are plotted against time in Figure 4.

Vestling and Künsch (1968) reported the slow absorbance
changes at 25° when the binary complexes, LDH-NAD and LDH-

93

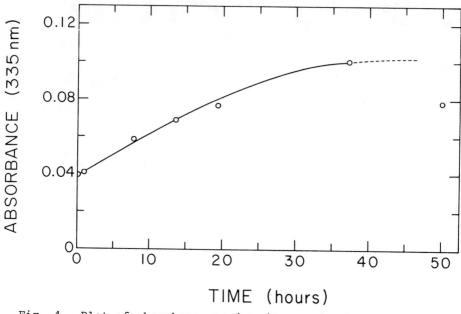

Fig. 4. Plot of absorbance maxima (A_{335} nm) of the curves of Figure 3 vs. time. (At 50 hours partial loss of enzyme activity was recorded.)

AcPyAD, were converted to the corresponding abortive ternary complexes, LDH-NAD-Pyruvate and LDH-AcPyAD-Pyruvate. Time periods of one to three hours were required to reach equilibrium absorbance values. These spectral changes were fully reversible without loss of enzyme activity after prolonged dialysis in the case of AcPyAD-buffer but only partially reversible in the case of dialysis against NAD-buffer.

Vestling and Künsch (1968) also noted slow time changes of absorbance in the case of the binary complexes, LDH-NADH and LDH-AcPyADH. A recent experiment by Vestling and Wiechert (unpublished results) is shown in Figure 5. The spectral curves reflect structural modifications of the chromophoric system of the bound reduced dinucleotides. Further study of these systems is in progress.

For the purposes of this discussion let it be stated that the availability of X-ray crystallographic data (Adams et al., 1972, 1973) for dogfish muscle LDH (A_4) will contribute to an understanding of the conformational adjustments which occur in LDH when various coenzyme complexes are formed. Little is known about the details of the mechanisms which control the spatial coiling of the peptide chains of proteins as they grow on the ribosome and are released. It is possible that very

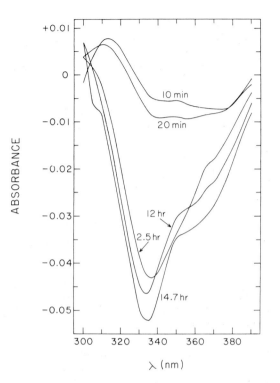

Fig. 5. Difference Spectra. Binary Complex: LDH(6×10^{-6}M) - NADH(2.5×10^{-4}M). Solvent: 0.02 M(NH_4)$_2SO_4$ - 0.02 M potassium phosphate - 0.001 M 2-mercaptoethanol. pH 7.8. The times denote intervals after mixing. System kept at 0° between spectral measurements.

delicate differences in conformation may exist when one compares "normal" LDH with hepatoma LDH. We are looking into this possibility with the aid of various spectral approaches.

To conclude this account the author is pleased to recognize the major step forward which resulted when the early experiments of Markert and Møller (1959), Markert and Appella (1961), Kaplan and Ciotti (1961), and others appeared and touched off the isozyme era.

REFERENCES

Adams, M.J., M. Buehner, K. Chandrasekhar, G.C. Ford, M.L. Hackert, A. Liljas, P. Lentz, Jr., S.T. Rao, M.G. Rossmann, I.E. Smiley, and J.L. White 1972. Subunit interactions in lactate dehydrogenase. *Protein-Protein Interactions*. R. Jaenicke and E. Helmreich, editors. Springer-Verlag, Berlin. pp. 139-158.

Adams, M.J., M. Buehner, K. Chandrasekhar, G.C. Ford, M.L. Hackert, A. Liljas, M.G. Rossman, I.E. Smiley, W.S. Allison, J. Everse, N.O. Kaplan, and S.S. Taylor 1973. Structure-function relationships in lactate dehydrogenase. *Proc. Natl. Acad. Sci. USA.* 70: 1968-1972.

Carlotti, R. J., G. F. Garnett, W. T. Hsieh, A. A. Smucker, C. S. Vestling and H. P. Morris 1974. Comparison of purified lactate dehydrogenases from normal rat liver and Morris hepatomas in rats and in culture. *Biochim. Biophys. Acta* 341: 357-365.

Davisson, E. O., D. M. Gibson, B. R. Ray, and C. S. Vestling 1953. Rat liver lactic dehydrogenase. II. Physico-Chemical characterization of the crystalline enzyme. *J. Phys. Chem.* 57: 609-613.

Gibson, D. M., E. O. Davisson, B. K. Bachhawat, B. R. Ray, and C. S. Vestling,1953. Rat liver lactic dehydrogenase. I. Isolation and chemical properties of the crystalline enzyme. *J. Biol. Chem.* 203: 397-409.

Hsieh, W. T. and C. S. Vestling 1966. Rat liver lactate dehydrogenase. *Biochem.Preparations* 11: 69-75.

Kaplan, N. O. and M. M. Ciotti 1961. Evolution and differentiation of dehydrogenases. *Ann. N. Y. Acad. Sci.* 94: 701-722.

Markert, C. L. and E. Apella 1961. Physiochemical nature of isozymes. *Ann. N. Y. Acad. Sci.* 94: 678-690.

Markert, C. L. and F. Møller 1959. Multiple forms of enzymes: Tissue, ontogenetic, and species specific patterns. *Proc. Nat. Acad. Sci.U.S.A.* 45: 753-763.

Morris, H. P. 1965. Studies on the development, biochemistry and biology of experimental hepatomas. *Adv. Cancer Res.* 9: 227-302.

Raffel, S. 1961. *Immunity,* 2nd Ed., Appleton-Century-Crofts, New York, pp. 516-520.

Ryan, L. D. and C. S. Vestling 1974. Rapid purification of lactate dehydrogenase from rat liver and hepatoma: A new approach. *Arch. Biochem. Biophys.* 160: 279-284.

Vestling, C. S. and U. Künsch 1968. Spectrophotometric analysis of rat liver lactate dehydrogenase - coenzyme and coenzyme analog complexes. *Arch. Biochem. Biophys.* 127: 568-575.

HEREDITARY DEFICIENCY OF SUBUNIT B OF LACTATE DEHYDROGENASE

MOTOSHI KITAMURA AND TOSHIHIRO NISHINA
Dept. Clinical Chemistry
Toranomon Hospital, Tokyo

ABSTRACT: A case of complete deficiency in the synthesis of the B subunit of lactate dehydrogenase is presented. The hereditary basis, incidence, erythrocyte glucose metabolism, and the properties of LDH A_4 have been investigated in this case. A 64-year-old male with mild diabetes drew our attention because of the abnormally low LDH activity present in his serum which had a value of 77 mIU/ml. The analysis of serum, hemolysate, saliva, and specimens of bone marrow aspiration demonstrated the presence of only the A_4 isozyme with complete absence of the faster moving B_4 LDH fraction. A similar study made on family members revealed low LDH activity in the serum and erythrocytes in one of his brothers and all of his five children, corresponding to a decrease in the activity of B_4 fractions. Based on the comparison of the calculated ratio of B/A subunits in normal and affected family members, it is hypothesized that the proband is homozygous and the affected family members are heterozygous. These results strongly indicate that the deficiency of LDH subunit B has a hereditary basis.

The activity of erythrocyte glycolytic enzyme in the patient and his family members was within the normal range except for a marked decrease of LDH activity. On the other hand, among analyzed erythrocyte glycolytic intermediates and adenine nucleotides, the values of dihydroxyacetone phosphate, fructose 1,6-diphosphate, and glyceraldehyde 3-phosphate were remarkably high in the proband, whereas the concentrations of these compounds was normal in the family members. The mechanism of the abnormal metabolism seen in the proband may be explained by the disturbance at the glyceraldehyde 3-phosphate dehydrogenase step due to a relative decrease in NAD caused by the block at the LDH step. The fact that pyruvic acid in erythrocytes did not increase might indicate that the acid diffused out of the erythrocytes without producing a deleterious effect. No evidence for the existence of hemolytic anemia was recognized in the patient. Our survey indicates that the incidence of these LDH heterozygotes in the Japanese population is as low as approximately 1/7,000; thus, this LDH abnormality is extremely rare.

INTRODUCTION

It has been well established that lactate dehydrogenase (LDH) can be separated into five isozymes by means of electrophoresis or column chromatography (Markert and Møller, 1959). These five isozymes are tetramers composed of two randomly associated subunits A and B (Markert, 1963). These subunits are sometimes designated M and H for A and B, respectively. Variant subunits derived from mutant gene locus (Shaw and Barto, 1963) and linkage with immuno-antibody (Biewenga, 1972) have been reported to form electrophoretically abnormal patterns of LDH isozymes. However, complete absence of a subunit resulting from a genetic abnormality has never been reported except for our own presentations (Kitamura et al., 1971; Miwa et al., 1971). In the case we have studied B subunits of LDH are completely lacking; only the A_4 isozyme is present. The purpose of this paper is to report on the incidence and hereditary basis of this LDH B subunit deficiency, the properties of LDH in this case, and on the resulting abnormality of erythrocyte glucose metabolism.

CASE REPORT

The proband, patient J. A., a 64-year-old male from non-consanguineous parents, visited our out-patient clinic for mild diabetes. The patient appeared normal on physical examination. Routine laboratory examination was not remarkable except for moderately impaired glucose tolerance, slight elevation of cholesterol as shown in Table 1, and abnormally low serum LDH activity. The LDH titer was only one-third of the lower limits for normal subjects. Although the diabetic state was brought under good control with a restricted diet and treatment with tolazamide, the abnormally low serum LDH activity was unaffected by this treatment. As shown in Table 2, the patient presents no obvious evidence of hemolysis. That is, no anemia, reticulocytosis, hyperbilirubinemia, or bone marrow hyperplasia were detected, while the half life of ^{51}Cr was 30 days and normal. His family members collaborated with the study. His wife, two brothers, two sons, and three daughters, as indicated in Fig. 1, were similarly examined.

MATERIALS AND METHODS

Freshly obtained sera and saliva of the patient and family members were analyzed. Erythrocytes, leucocytes, and

TABLE I

LABORATORY FINDINGS

J. A. Japanese Male 64-year-old
Diabetes Mellitus on Tolazamide

LDH	77 mIU (203-325 mIU)	
GOT	14 Karmen U.	
GPT	9 Karmen U.	
Protein,	total	8.1 g/dl
	albumin	62.8 per cent
	alpha 1-globulin	4.3 per cent
	alpha 2-globulin	11.6 per cent
	beta globulin	9.4 per cent
	gamma globulin	12.1 per cent
Cholesterol		255 mg/dl
Sodium		140 mEq/l
Potassium		3.8 mEq/l
Chloride		107 meq/l

TABLE II

HEMATOLOGY

Serum LDH	77 mIU (203-325)
Erythrocyte LDH	2.5 U (37-51)
Hemoglobin	13.7 g/dl
Reticulocyte	1.1 per cent
Serum bilirubin	
Total	0.5 mg/dl
Indirect reacting	0.4 mg/dl
Bone marrow	
Erythroid hyperplasia	(-)
Erythrocyte survival	
^{51}Cr, half-life	30 days

platelets prepared from heparinized venous blood were washed with saline, disrupted by freezing and thawing or by adding distilled water, and immediately analyzed. In order to investigate the incidence of subunit B deficiency of LDH, the sera of patients subjected to our laboratory from Jan. 1971 to May 1972 were examined. The number of serum samples was

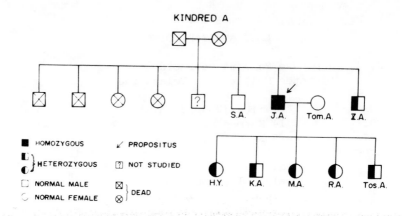

Fig. 1. Pedigree of the family

33,425, which corresponded to approximately 20,000 patients. Total LDH activity was determined by the method described by Hill (1956) with a slight modification for laboratory use. For detailed analysis, the LDH activity of erythrocytes was determined spectrophotometrically with a slight modification of the method of Kubowitz et al. (1943). In this method, 3 ml of incubation mixture contained the following reagents: 1.40 mM of pyruvate, 0.15 mM of NADH, 0.05 M of TEA-HCl at pH 7.4, and 0.02 ml of hemolysate corresponding to approximately 2.5×10^6 erythrocytes. The conversion from NADH to NAD was followed at 37°C and at 340 nm. The LDH activities of serum and saliva were measured at 37°C and expressed in terms of IU/ml, while the LDH of corpuscular elements was expressed in terms of $IU/10^{10}$ cells. For the analysis of LDH isozymes, agar gel electrophoresis with our modification (Yoshida et al., 1966) of Wieme's method (1959) was employed, using lactate for substrate and nitro-blue tetrazolium for enzyme staining. The activity of erythrocyte glycolytic enzymes was determined by the method of Bergmeyer (1963), while glycolytic intermediates and adenine nucleotides were determined by the method described by Minakami et al. (1965).

RESULTS AND DISCUSSION

1. LDH ACTIVITY AND ISOZYMES

Abnormally low activity of serum LDH in this case is shown in Fig. 2. The bell type curve indicates the distribution of LDH activity in normal subjects, while the shaded parts indicate the distribution of activity in the hospital

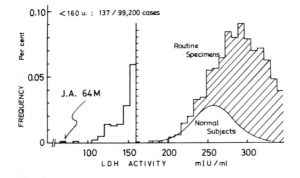

DISTRIBUTION OF SERUM LDH ACTIVITY
IN ROUTINE SPECIMENS

Fig. 2. Distribution of serum LDH activity in routine specimens.

population subjected to routine examination. The value of 150 mIU was found at the rate of 0.05%, while only two cases with 100 mIU or less were detected among approximately 100,000 cases. One of these two cases was the one presented here. He showed the lowest activity of all the cases subjected to routine examination in our hospital.

The pattern of serum LDH isozymes from a normal subject is shown in the upper part of Fig. 3. Normal serum stains strongly for LDH-1, 2, and -3, and slightly for LDH 4 and 5 (LDH-1=B_4, LDH-2=A_1B_3, LDH-3=A_2B_2, LDH-4=A_3B_1, and LDH-5=A_4). On the other hand, only LDH-5 (A_4) was detected in the serum or erythrocytes of the patient J. A. The band located at the LDH-3 position on the zymogram is hemoglobin. In the zymograms of erythrocytes from normal subjects, hemoglobin was not evident, but this result is due simply to the amount present. That is, the LDH activity of the erythrocytes from the patient was so low (3.7 IU/10^{10} RBC; the control, 53.5 IU/10^{10} RBC) that a large amount of hemolysate had to be applied on the agar plate in order to assure a stainable quantity of the isozymes. Thus a large amount of hemoglobin was necessarily also added to the agar plate.

2. HEREDITARY BASIS OF THE DEFICIENCY

The wife of our patient, three brothers, and five children, nine persons in all, are alive. The LDH activity of hemolysates from eight of these individuals has been investi-

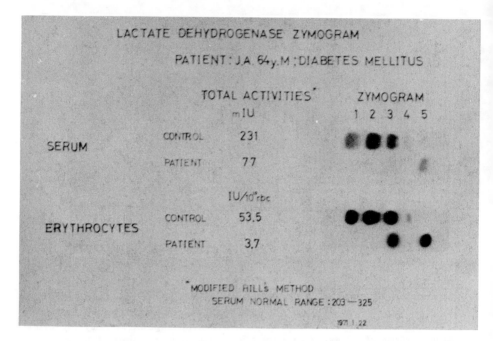

Fig. 3. Serum and erythrocyte LDH zymograms of the patient.
gated. First, as shown in Fig. 4, the total LDH activity was
low in one of the brothers and in all five children, each showed approximately 1/2 to 2/3 the normal value of LDH. Although
five isozymes could be detected in all hemolysates analyzed,
the activity of both the B_4 fraction and the A_1B_3 fraction
was obviously low. The isozyme patterns of the family members
are shown in Fig. 5. This abnormality, namely, the decrease
in total LDH activity and particularly in isozymes B_4 and
A_1B_3, was also detected in the sera of six affected family
members. The ratio of B to A subunits was calculated for the
sera and erythrocytes of the six family members with low LDH
activity. The calculation was artificial, but it helped us to
compare family members with normal subjects. As reported in
Table 3, the ratio of B to A was one half of that of normal
subjects in both erythrocytes and serum. It can be concluded
from these results that the proband is homozygous lacking B
subunits of LDH, while six family members with low LDH activity are heterozygous.

3. PROPERTIES OF LDH

The deficiency of subunit B of LDH in the propositus was

102

ERYTHROCYTE LACTATE DEHYDROGENASE ISOZYME PATTERN
OF THE FAMILY MEMBERS

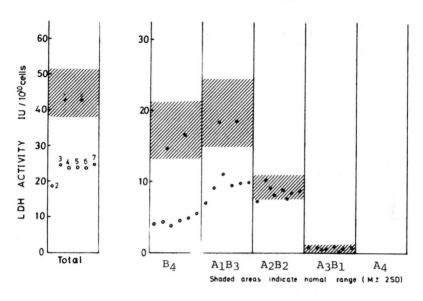

Shaded areas indicate nomal range (M ± 2SD)

Fig. 4. Erythrocyte LDH isozyme pattern of the family
members.

ERYTHROCYTES L D H ZYMOGRAMS OF THE FAMILY

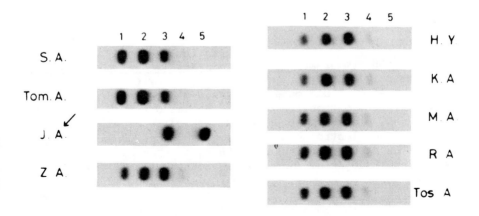

Fig. 5. LDH zymograms of the family members.

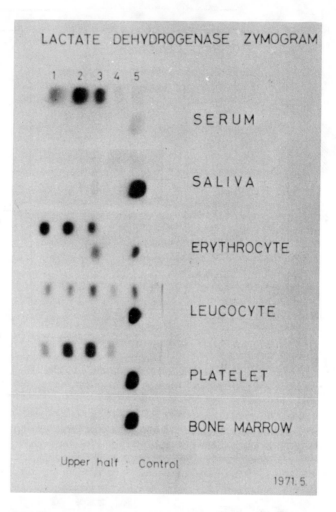

Fig. 6. <u>LDH zymograms from various specimens of the patient.</u>
seen not only in the serum and the erythrocytes as stated
above but also in all other materials studied. As shown in
Fig. 6, in the propositus only LDH A_4 was detected in the
leucocytes, platelets, bone marrow aspiration specimens, and
saliva. The electrophoretic movement of the LDH A_4 from the
propositus agreed with that of LDH A_4 in specimens from normal control subjects. When equal amounts of sera from the
patient and from normal individuals were mixed and analyzed
electrophoretically, the distribution of enzyme activity among
the separated isozymes was compatible with the calculated
expected values without any discrepancy.

TABLE III
B-SUBUNIT: A-SUBUNIT RATIO OF LACTATE
DEHYDROGENASE IN SERUM AND ERYTHROCYTES

Serum

	no. of cases	B/A mean	range
Family members*	6	1.4	1.1-1.6 (min.-max)
Normal subjects	108	2.5	2.0-3.0 ($\pm$ 2SD)

Erythrocytes

	no. of cases	mean	range
Family members*	6	2.2	2.0-2.4 (min.-max)
Normal subjects	11	3.8	3.4-4.3 (min.-max)

*5 children and 1 brother possesing
low LDH

Properties of LDH A_4 were investigated physicochemically and enzymologically. It is well known that the subunit A of LDH is less stable than subunit B. Therefore, the heat stability and urea inhibition of the LDH preparations were investigated. LDH is inactivated on freezing. In the propositus, the loss in LDH A_4 is obvious. As shown in Fig. 7, the LDH activity of the patient's hemolysate decreased to 50% of the original activity after being stored frozen for three days, while the activity of normal hemolysates decreased only approximately 15%. On heating at $60^{\circ}C$ for one hour, the LDH activity obtained from the patient was completely inactivated, whereas only 34% inactivation occurred in normal samples. Also inhibition by urea produced a similar instability in the LDH from the patient (Fig. 8).

In addition to physicochemical instability, the subunit A of LDH is characterized by a low affinity for the substrate. Fig. 9 shows that the Km value of the patient's LDH was 54.1 mM, which is almost five times more than that of normal control hemolysates (11.2 mM). Thus, a remarkably low affinity was demonstrated. A similar tendency is shown with pyruvate. Fig. 10 shows this relationship. The optimal concentration of pyruvate in hemolysates of the patient was higher than that in control hemolysates, and substrate inhibition appeared slowly at high concentration of pyruvate. By analyzing the properties of LDH in various materials from the patient and on the basis of electrophoretic analysis, physicochemical stability, and enzymo-dynamics, the patient's LDH appeared to contain only a single isozyme - LDH A_4.

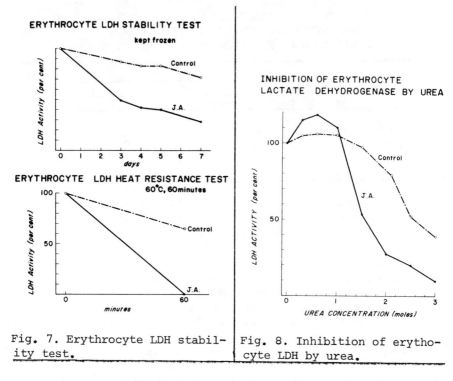

Fig. 7. Erythrocyte LDH stability test.

Fig. 8. Inhibition of erythocyte LDH by urea.

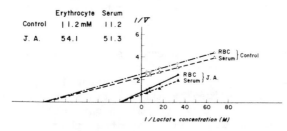

Fig. 9. LDH Km for lactate.

4. *ERYTHROCYTE GLUCOSE METABOLISM*

LDH is an enzyme which functions in the final process of anaerobic glucose metabolism, converting pyruvate to lactate. The absence of all LDH isozymes containing B subunits from the mature erythrocytes of the proband might be expected to affect erythrocyte carbohydrate metabolism to a considerable degree. To investigate this possibility, the activity of erythrocyte glycolytic enzymes was determined. As shown in Table 4, all

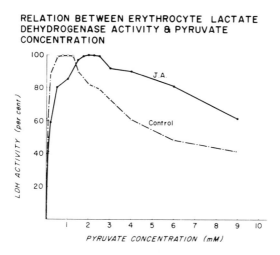

RELATION BETWEEN ERYTHROCYTE LACTATE
DEHYDROGENASE ACTIVITY & PYRUVATE
CONCENTRATION

Fig. 10.   LDH Km for pyruvate.

TABLE IV
ACTIVITIES OF GLYCOLYTIC
ENZYMES IN ERYTHROCYTES

	Control	Patient
	$\mu M/min/10^{10}$ RBC	
Hexokinase	0.19	0.11
Glucosephosphate isomerase	10.5	11.8
Phosphofructokinase	2.16	4.78
Aldolase	0.64	0.98
Triosephosphate isomerase	548	554
Glyceraldehyde 3-phosphate dehydrogenase	39.4	23.4
Phosphoglycerate kinase	42.9	41.1
Phosphoglycerate mutase	9.36	9.59
Enolase	2.78	5.19
Pyruvate kinase	4.32	4.83
Lactate dehydrogenase	40.1	3.1
Glucose 6-phosphate dehydrogenase	2.49	2.20
6-Phosphogluconate dehydrogenase	1.54	1.70
Glutathione reductase	2.07	1.67
Adenulate kinase	64	68

the glycolytic enzyme activities were within the normal
range except for a striking decrease in LDH activity. Above
all, related with the results to be described below, it
should be especially noted that glyceraldehyde-3-phosphate
dehydrogenase was within the normal range. Moreover, glycolytic intermediates and adenine nucleotides were also measured.
The results are shown in Table 5 and also schematically illustrated in Fig. 11. The measured value of the intermediate
was shown relatively by taking one as the median of the
normal values in the figure. The results indicated that no
obvious changes in pyruvate and lactate were observed, whereas fructose 1,6-diphosphate, dihydroxyacetone phosphate, and
glyceraldehyde 3-phosphate increased markedly. The accumulation of pyruvate even when LDH activity was actually decreased was not demonstrated. Jaffee (1970) predicted the existence
of LDH deficiency in his review of hemolytic anemia and assumed that the accumulated pyruvic acid would diffuse through
the erythrocyte membranes, leaving NADH within the erythrocytes.
Our results (Fig. 12) clearly show the participation of pyridine nucleotides in the glycolytic metabolism, confirming
Jaffe's prediction. Moreover, NADH can not diffuse out of
erythrocytes and thus accumulates in them. As a result, a
marked decrease in NAD occurs and the glyceraldehyde-3-phosph-
ate dehydrogenase reaction does not proceed satisfactorily

TABLE V
ERYTHROCYTE GLYCOLYTIC INTERMEDIATES ETC.

	Normal range (mean)	Patient
Glucose 6-phosphate (mμM/ml cells)	21- 44 (33)	32
Fructose 6-phosphate (mμM/ml cells)	4- 16 (11)	10
Fructose 1,6-diphosphate (mμM/ml cells)	6- 26 (16)	114
Dihydroxyacetone phosphate (mμM/ml cells)	8- 22 (15)	143
Glyceraldehyde 3-phosphate (mμM/ml cells)	4- 14 (9)	16
2,3-Diphosphoglycerate (mμM/ml cells)	4,730-6,450 (5,590)	5,020
3-Phosphoglycerate (mμM/ml cells)	43- 103 (73)	63
2-Phosphoglycerate (mμM/ml cells)	3- 30 (17)	10
Phosphoenolpyruvate (mμM/ml cells)	11- 30 (22)	15
Pyruvate (mμM/ml whole blood)	31- 62 (47)	49
Lactate (mμM/ml whole blood)	833-1,453 (1,143)	748
ATP (mμM/ml cells)	1,186-1,524 (1,355)	1,245
Lactate formation (mμM/hr/ml cells)	808-1,992 (1,400)	962
Pyruvate formation (mμM/hr/ml cells)	(355)	528

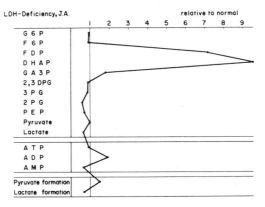

Fig. 11. Erythrocyte glycolytic intermediates and adenine nucleotides, etc.

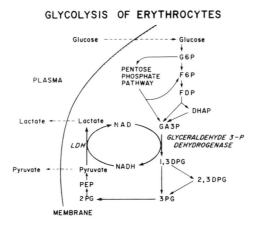

Fig. 12. Glycolysis of erythrocytes.

because of the shortage of NAD. Consequently, metabolic changes similar to those brought about in deficiency of glyceraldehyde-3-phosphate dehydrogenase are induced. The accumulation of glyceraldehyde 3-phophate, dihydroxy acetone phosphate and fructose 1,6-diphosphate may be explained as described above. This metabolic pattern was not observed in the heterozygous family members.

5. INCIDENCE OF LDH DEFICIENCY

In the patient with LDH deficiency, the serum LDH activity was very low and a similar tendency was observed in the heterozygous members of his family. This characteristic might be used to identify patients with an LDH deficiency, we therefore measured the total LDH activity in the sera of patients with non-specific illness in order to screen a large population. Individuals with low LDH activity could easily be recognized by our tests. In this manner, approximately 33,000 sera from 20,000 patients were analyzed in a period of about a year. 78 serum specimens (0.4%) showing LDH activity less than the lower normal limits (M-3SD) were further analyzed for their isozyme patterns. These patterns from sera and erythrocytes demonstrated that three of the subjects might be considered to be heterozygous; the incidence was 1/7,000, indicating that this metabolic abnormality is extremely rare. Recently, Dr. Kanno (1974) of Keio University in Tokyo found the incidence of possible heterozygous LDH B deficiency to be 1/5,000. We found the first deficient homozygote quite accidentally, but in the near future the second case may be found by systematic research with identification of heterozygotes as a first step.

REFERENCES

Bergmeyer, H. V. 1963. Methods of Enzymatic Analysis, *Academic Press*, New York.

Biewenga, J. 1972. Serum lactate dehydrogenase isozymes linked to immunoglobulin A. *Clin. Chim. Acta* 40: 407-414.

Hill, B. R. 1956. Some properties of serum lactic dehydrogenase. *Cancer Res*. 16: 460-468.

Jaffe, E. R. 1970. Hereditary hemolytic disorders and enzymatic deficiencies of human erythrocytes. *Blood* 35: 116-134.

Kanno, T. 1974. Personal communication.

Kitamura, M., N. Iijima, F. Hashimoto, and A. Hiratsuka 1971. Hereditary deficiency of subunit H of lactate dehydrogenase. *Clin. Chim. Acta* 34: 419-423.

Kubowitz, F., and P. Ott 1943. Isolierung und kristallisattion eines garungsferments aus tumoren. *Biochem. Zeitschr.* 314: 94-117.

Markert, C. L. 1963. Lactate dehydrogenase isozymes: dissociation and recombination of subunits. *Science* 140: 1329-1330.

Markert, C. L. and F. Møller 1959. Multiple forms of enzymes: tissue, ontogenetic, and species specific patterns. *Proc. Nat. Acad. Sci.* 45: 753-763.

Minakami, S., C. Suzuki, T. Saito, and H. Yoshikawa 1965. Studies on erythrocyte glycolysis: I. Determination of the glycolytic intermediates in human erythrocytes. *J. Biochem.* 58: 543-550.

Miwa, S., T. Nishina, Y. Kakehashi, M. Kitamura, A. Hiratsuka, and K. Shizume 1971. Studies on erythrocyte metabolism in a case with hereditary deficiency of H-subunit of lactate dehydrogenase. *Acta Haemat. Jap.* 34: 2-6.

Shaw, C. R. and E. Barto 1963. Genetic evidence for the subunit structure of lactate dehydrogenase isozymes. *Proc. Nat. Acad. Sci.* 50: 211-214.

Wieme, R. J. 1959. An improved technique of agargel electrophoresis on microscope slides. *Clin. Chim. Acta* 4: 317-321.

Yoshida, M., K. Ishikawa, and M. Kitamura 1966. Studies on lactate dehydrogenase isozymes of body fluids (I): An improved method of agar-gel electrophoresis on microscopic slides. *Physico-Chem. Biol.* (Japan). 11: 345-350.

THE PROPERTIES OF PURIFIED LDH-C$_4$ FROM HUMAN TESTIS

JISNUSON SVASTI AND SUMALEE VIRIYACHAI
Department of Biochemistry, Faculty of Science
Mahidol University, Rama VI Rd., Bangkok, Thailand

ABSTRACT. Lactate dehydrogenase isozyme-X (LDH-C$_4$) was purified by ammonium sulphate fractionation and DEAE-cellulose chromatography from 224 g of human testis. The purified isozyme had a specific activity of 80 I.U./mg, which represents a 571-fold purification with a yield of 21%. Upon electrophoresis in a polyacrylamide gels this preparation exhibited a high degree of homogeneity, with contamination from other proteins and isozymes totalling 5% or less. The properties of human LDH-C$_4$ were compared with purified LDH-B$_4$ (62 I.U./mg) and partially purified LDH-A$_4$ (102 I.U./mg), isolated from human heart and human liver, respectively.

The kinetic parameters determined include pH optima, K_m values for substrates lactate, pyruvate, and α-ketobutyrate, and K_m values for NAD$^+$ and NAD$^+$-analogues. For the reaction with α-ketoacids, our results confirm previous studies showing that human LDH-C$_4$ can utilize α-ketobutyrate much better than the other isozymes. Oxalate severely inhibited the pyruvate reaction. For the lactate reaction, oxalate inhibition was competitive for all isozymes but much less severe, with LDH-B$_4$ being particularly resistant to inhibition.

Human LDH-C$_4$, unlike mouse LDH-C$_4$, is not significantly more heat stable than the other isozymes. The amino acid composition of human LDH-C$_4$ and human LDH-B$_4$ were determined and compared to published results on mouse LDH-C$_4$ and on mammalian LDH-B$_4$ and A$_4$. This comparison did not reveal any unusual amino acid composition that could be considered characteristic of all LDH-C$_4$ isozymes.

INTRODUCTION

Lactate dehydrogenase isozyme-X (LDH-C$_4$) has been shown by several investigators (e.g. Blanco and Zinkham, 1963; Goldberg, 1963) to occur only in mature testes and spermatozoa. This isozyme appears to be synthesized during the primary spermatocyte stage of the spermatogenic cycle (Zinkham et al., 1964) and in spermatozoa appears to be localized in the midpiece (Clausen, 1969). The LDH-C$_4$ isozyme has been obtained in highly purified form from mice (Wong et al., 1971; Goldberg, 1972), rats (Schatz and Segal, 1969, and bulls (Kolb et al., 1970).

The crystalline mouse enzyme has been shown to differ from the LDH-1 (B_4) and LDH-5 (A_4) isozymes in its kinetic and immunological properties and in its amino acid composition.

The demonstration in female mice (Goldberg and Lerum, 1972) and in female rabbits (Goldberg, 1973a) that immunological inhibition of LDH-C_4 significantly reduces the number of pregnancies makes LDH-C_4 of particular interest to the reproductive biologist. Although antiserum to mouse LDH-C_4 will cross-react with human LDH-C_4 (Goldberg, 1971), extrapolation of studies in other species to the human system may not by entirely justified, in view of the possible metabolic differences between spermatozoa of different species. Since previous studies of the catalytic properties of human LDH-C_4 (Clausen and Olisen, 1965; Wilkinson and Withycombe, 1965) were carried out on relatively impure preparations, we began our studies with the purification of human LDH-C_4. Since human LDH-B_4 and LDH-A_4 have previously been purified (Nisselbaum et al., 1964; Burd and Usategui-Gomez, 1973), the purification of these isozymes will not be described in detail.

MATERIALS AND METHODS

PURIFICATION OF HUMAN LDH ISOZYMES

Human testes obtained at autopsy were used as the starting material for the purification of LDH-C_4. About 200 g of testis were homogenized in 250 ml of 1 mM EDTA and 1 mM β-mercaptoethanol in a Waring blender. The crude extract, obtained after centrifugation and filtration through cheesecloth, was precipitated by adding solid ammonium sulphate to 70% saturation. The precipitate was resuspended in 10 mM potassium phosphate buffer, pH 7.5, 1 mM EDTA, 1 mM β-mercaptoethanol, and again fractionated with ammonium sulphate. The fraction precipitating between 40% - 65% saturation was dialyzed against 10 mM potassium phosphate buffer, pH 7.5, 1 mM EDTA, 1 mM β-mercaptoethanol, and applied to a column (4.5 cm X 40 cm) of DEAE-cellulose equilibrated with the same buffer. Very little LDH activity was retained by this column. Thus the breakthrough peak was precipitated with ammonium sulphate at 70% saturation, dialyzed, and applied to another column (4.5 cm X 60 cm) of DEAE-cellulose equilibrated with the same 10 mM potassium phosphate buffer, pH 7.5. The column was washed with 1.5 liters of starting buffer followed by two liters of 0.2 M NaCl in the same buffer.

Four pools (A,B,C, and D) were precipitated with ammonium sulphate to 70% saturation (Fig. 1). Small samples of each pool were subjected to electrophoresis in 5% polyacrylamide

114

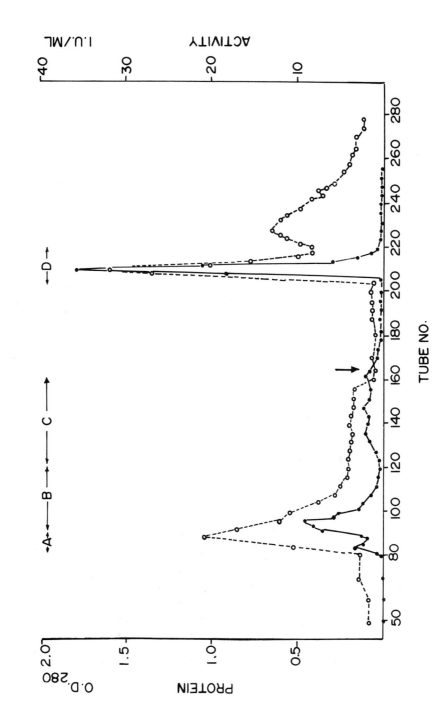

FIGURE I: DEAE CELLULOSE ELUTION PROFILE

Fig. 1 legend. <u>DEAE-cellulose column elution profile (second column)</u>. The sample derived from a 40-65% ammonium sulphate cut of testis crude extract, was loaded to a column (4.5 cm X 60 cm) of DEAE-cellulose, equilibrated with 10 mM potassium phosphate buffer, 1 mM β-mercaptoethanol, 1 mM EDTA, pH 7.5. The column was washed with one liter of the same buffer, followed by two liters of 0.2 M NaCl in the same buffer (indicated by arrow). For tubes 0-165, fractions of 6 ml were collected: thereafter (tubes 166-280), fractions of 12 ml were taken. Protein was estimated by measuring $O.D._{280}$ (indicated by o ----- o). LDH activity I.U./ml (indicated by ●————●), was determined in 50 mM potassium phosphate buffer, pH 7.0, 0.3 mM sodium pyruvate and 0.160 mM NADH at 25° C. Four pools were taken, pool A, pool B, pool C and pool D, as shown.

gels in 10 mM glycine-NaOH buffer, pH 9.6, and stained for LDH activity. To simplify alignment, gels containing mixtures of consecutive pools and a marker mixture of partially purified LDH-B_4 and LDH-A_4 were also run (Fig. 2). Although LDH-C_4 was also present in pool D, it also seems to be the only one present in pool C. An interesting feature of these gels is that other band(s), with a yield too low to be observed in the crude testis homogenate, are located between LDH-4 (A_3B_1) and LDH-5 (A_4). One possible explanation is that in humans there are alleles of the C locus as has been found occasionally in pigeons (Zinkman et al., 1964). A more likely explanation is that these bands represent hybrid forms between the A subunit and the C subunit as found by Goldberg (1973 b) in rat and guinea pig testes. However, in our preparation this may not be an in vivo phenomenon because testes were stored frozen on collection and thawed before use, so that artifacts may have been produced.

The results of the above purification are summarized in Table I. The specific activity of the purified LDH-C_4 is 80 I.U./mg, which represents a purification of 571-fold with a yield of 21.5%. The last step, precipitation with ammonium sulphate yielded considerable purification presumably because the low protein concentration led to incomplete precipitation. This preparation of LDH-C_4 was again run on polyacrylamide gels and stained for both enzyme activity and for protein (Fig. 3). These gels showed that the LDH-C_4 preparation had a high degree of homogeneity. Some minor contaminants were present but only amounted to a total of about 5%. LDH-4 (A_3B_1) was the major LDH contaminant but represented only about 2% of the total LDH activity.

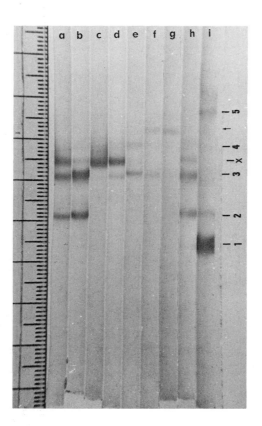

Fig. 2. Polyacrylamide gel electrophoresis of DEAE-cellulose pools. Samples of the DEAE-cellulose pools A,B,C, and D (Fig. 1) were run on 5% polyacrylamide gels in 10 mM glycine-NaOH buffer pH 9.6 and stained for LDH activity. Gel a: pool C + pool D; gel b: pool D; gel c: pool C; gel d: pool B + pool C; gel e: pool B; gel f: pool A + pool B; gel g: pool A; gel h: testis crude extract; gel i: mixture of partially purified LDH-1 and LDH-5. Markings at the side 1,2,3,4,5, and X indicate the positions of the bands corresponding to the respective isozymes (B_4, A_1B_3, A_2B_2, A_3B_1, A_4 and C_4). Arrow ↓ indicates the presence of a band (in gels f and g) not present in the crude extract.

TABLE I

PURIFICATION OF LDH-C_4 FROM HUMAN TESTIS (224 gm)

Fraction	Total Protein Mg	Total Activity I.U. LDH-C_4	Specific Activity I.U. LDH-C_4/Mg	Purification	% Yield
Crude Extract	10576	1520	0.14	-	100
70% $(NH_4)_2SO_4$ PPTE	3740	1520	0.40	2.85	100
40-65% $(NH_4)_2SO_4$ Cut	2125	1100	0.51	3.64	72.3
First DEAE Column	840	1215	1.44	10.3	79.9
Second DEAE Column (Pool C)	18	490	27.2	194	32.2
70% $(NH_4)_2SO_4$ PPTE (Pool C)	4.1	328	80.0	571	21.5

Protein was estimated by using the OD_{280}:OD_{260} method. LDH-C_4 activity was calculated by assaying total LDH activity in 50 mM potassium phosphate buffer, pH 7.5, containing 0.3 mM sodium pyruvate, and 0.15 mM NADH at 25°C, and multiplying this by the % LDH-C_4 content of the sample, as determined by polyacrylamide gel electrophoresis and scanning of the gels. Thus the total LDH activity in the crude extract was 7600 I.U., of which 20% or 1520 I.U. was LDH-C_4.

Fig. 3. Polyacrylamide gel electrophoresis of purified LDH-X.
Samples of purified LDH-X were run on 5% polyacrylamide gels
in 10 mM glycine-NaOH buffer, pH 9.6. Gel 1: stained for
LDH-activity; gel 2: stained for protein.

LDH-B_4 was isolated from 1.1 kg of human heart muscle,
again principally by ammonium sulphate precipitation followed
by DEAE-cellulose chromatography, eluted with a 0-0.3 M NaCl
gradient in 10 mM potassium phosphate buffer, pH 7.0. A final
yield of 210 mg of purified LDH-B_4 was obtained with a specific
activity of 62 I.U./mg. Polyacrylamide gel electrophoresis
(Fig. 4) showed that this preparation was very homogeneous
and that the only visible contaminant, LDH-2 (A_1B_3), represented 1% or less of the total activity or protein. Some problems
were encountered during the purification of LDH-A_4 from human
liver. The preparation used in subsequent experiments was
obtained by ammonium sulphate precipitation, DEAE-cellulose
chromatography, followed by affinity chromatography on NAD$^+$-
linked Sepharose (Mosbach et al., 1972). This preparation

119

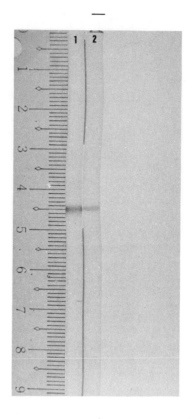

Fig. 4. <u>Polyacrylamide gel electrophoresis of purified LDH-1.</u>
Samples of purified LDH-1 were run on 5% polyacrylamide gels
in 10 mM glycine-NaOH buffer, pH 9.6. <u>Gel 1</u>: stained for
<u>LDH-activity; gel 2 : stained for protein.</u>
had a specific activity of 102 I.U./mg but showed substantial
amounts of contamination when subjected to polyacrylamide gel
electrophoresis, with LDH-A_3B_1 activity representing as much
as 15% of the LDH-A_4 activity.

KINETIC PROPERTIES

The kinetic properties of the purified LDH-X (C_4) and
LDH-1 (B_4) preparations and the partially purified LDH-5 (A_4)
preparation were studied at 37°C with the following results.
a) <u>Lactate reaction</u>: For the reaction of lactate with

120

TABLE II

REACTION WITH LACTATE

	LDH-C$_4$	LDH-B$_4$	LDH-A$_4$
A) With NAD$^+$			
Optimum pH	9.0-9.5	11.0	9.0-9.5
Saturating Concn L-Lactate mM	50	75	75-100
K$_m$ L-Lactate mM	10.2	9.95	13.8
K$_m$ NAD$^+$ mM	0.041	0.091	0.083
B) With NAD$^+$-Analogues			
K$_m$ Thio-NAD$^+$ mM	0.066	0.069	0.125
K$_m$ 3-Ac-Py-NAD$^+$ mM	0.029	0.018	0.051
Ratio $\dfrac{V_{max} \text{ (Thio-NAD}^+)}{V_{max} \text{ (NAD}^+)}$	0.87	0.39	0.39
Ratio $\dfrac{V_{max} \text{ (3-Ac-Py-NAD}^+)}{V_{max} \text{ (NAD}^+)}$	0.056	0.019	0.055

Reaction with lactate. (a) with NAD$^+$; (b) With NAD$^+$-analogues
The optimum pH was determined at 37 C in 50 mM glycine-NaOH
pH 8 to 12 or 50 mM potassium phosphate buffer, pH 7 to 8,
containing 108 mM sodium DL-lactate and 0.15 mM NAD$^+$. Subsequent studies (lactate saturation, K$_m$ and V$_{max}$ determinations)
were carried out in 50 mM glycine-NaOH buffer, pH 9.0 for LDH-
C$_4$ and LDH-A$_4$, and pH 11.0 for LDH-B$_4$. For determination of
the saturating concentration of lactate and K$_m$ for lactate,
the NAD$^+$ concentration used was 0.15 mM NAD$^+$. For K$_m$ and V$_{max}$
data for NAD$^+$ and NAD$^+$-analogues, DL-lactate concentration
was maintained at 200 mM. Kinetic data for L-lactate are
calculated assuming equimolar proportions of the D and L-forms.

NAD$^+$ (Table 2 a), LDH-X has a similar pH optimum (9.0-9.5) to
that of LDH-5, whereas LDH-1 has a pH optimum almost 2 pH
units higher. LDH-X has a similar K$_m$ for lactate to the two
isozymes, but a rather lower K$_m$ for NAD$^+$. Kinetic data for
the raction of lactate with NAD$^+$-analogues are shown in Table
2b. For all three isozymes, the K$_m$'s for 3-acetylpyridine-
NAD$^+$ are lower than the K$_m$'s for NAD$^+$, but the V$_{max}$ for these
isozymes is much lower (1.9-5.6%) than that of NAD$^+$. For thio-
NAD$^+$, LDH-X has a relatively high V$_{max}$ (87% of V$_{max}$ with NAD$^+$)
compared to LDH-1 and LDH-5 (both 39% of V$_{max}$ with NAD$^+$).

   b) Reaction with α-keto acids: The three isozymes had
closely similar pH optima with pyruvate as substrate (pH 7.0

for LDH-B_4 and LDH-C_4, and pH 6.5 for LDH-A_4). For the reaction with α-ketobutyrate, the optimum pH's were approximately 0.5 units higher for each of the three isozymes. Kinetic data for the reaction of pyruvate or α-ketobutyrate with 0.15 mM NADH are shown in Table 3. These data confirm previous results (Clausen and Ovlisen, 1965; Wilkinson and Withycombe, 1965) that human LDH-C_4, like LDH-C_4 from other species, can utilize α-ketobutyrate as a substrate better than do the other isozymes, LDH-B_4 and LDH-A_4. This is demonstrated by the low K_m value for α-ketobutyrate, and the high ratio of V_{max} (α-ketobutyrate) to V_{max} (pyruvate) of LDH-C_4 compared to LDH-B_4 and LDH-A_4.

TABLE III

REACTION WITH α-KETOACIDS

	LDH-C_4	LDH-B_4	LDH-A_4
Optimum Pyruvate mM	0.2	0.3	0.5
Optimum α-ketobutyrate mM	2.0	10	25
K_m (pyruvate) mM	0.053	0.061	0.11
K_m (α-ketobutyrate) mM	1.08	1.96	2.97
Ratio $\dfrac{V_{max} \ (\alpha\text{-ketobutyrate})}{V_{max} \ (\text{pyruvate})}$	1.88	0.96	0.55

Assays were carried out in 50 mM potassium phosphate buffer containing 0.15 mM NADH at the optimum pH for each isozyme and substrate. For sodium pyruvate, the pH's were 7.0 for LDH-B_4 and LDH-C_4, and 6.5 for LDH-A_4. For sodium α-ketobutyrate, the pH's were 7.5 for LDH-B_4 and LDH-C_4, and 7.0 for LDH-A_4.

c) _Inhibition by oxalate_: For the pyruvate reaction, all three isozymes are severely inhibited by oxalate (Fig. 5a) with 60% inhibition being reached at concentrations of oxalate as low as 0.5 mM. Inhibition of the lactate reaction is less severe (Fig. 5b), and the LDH-B_4 isozyme seems particularly resistant to inhibition. For the lactate reaction, inhibition is competitive.
Heat stability
A study of the heat stability of the three isozymes is shown in Fig. 6. Human LDH-A_4 is easily inactivated by heat,

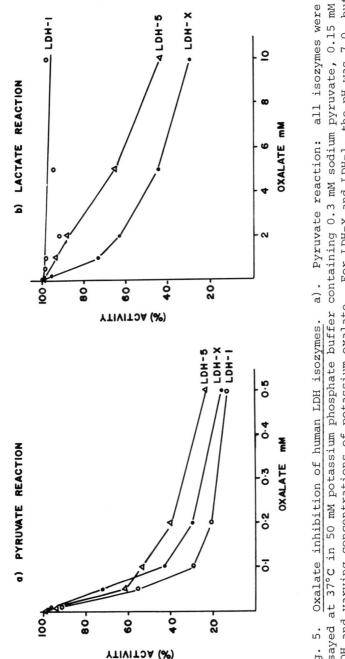

FIGURE 5: OXALATE INHIBITION

a) PYRUVATE REACTION

b) LACTATE REACTION

Fig. 5. Oxalate inhibition of human LDH isozymes. a). Pyruvate reaction: all isozymes were assayed at 37°C in 50 mM potassium phosphate buffer containing 0.3 mM sodium pyruvate, 0.15 mM NADH and varying concentrations of potassium oxalate. For LDH-X and LDH-1, the pH was 7.0, but for LDH-5 the pH was 6.5. b). Lactate reaction: all isozymes were assayed at 37°C in 50 mM glycine-NaOH buffer containing 200 mM sodium DL-lactate, 0.167 mM NAD$^+$ and varying concentrations of potassium oxalate. For LDH-X and LDH-5, the pH was 9.0, but for LDH-1 the pH was 11.0.
●————● LDH-X; ○————○ LDH-1; △————△ LDH-5.

123

FIGURE 6: HEAT STABILITY

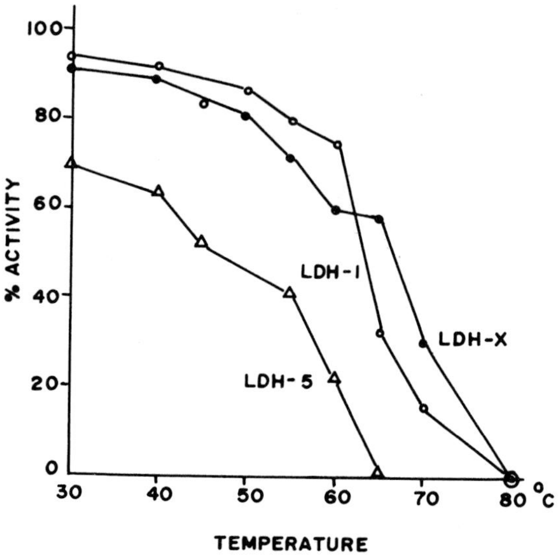

Fig. 6. <u>Heat stability of human LDH isozymes</u>. Isozymes were preincubated at the indicated temperatures for 10 minutes in 5 mM potassium phosphate buffer, pH 7.0 containing 1 mg/ml bovine serum albumin and cooled in an ice-bath. Residual activity was assayed in 50 mM potassium phosphate buffer, con-taining 0.3 mM sodium pyruvate and 0.16 mM NADH at 37°C. For LDH-1 and LDH-X, pH 7.0 was used and for LDH-5, pH 6.5. Re-<u>sults are expressed relative to a control kept in an ice-bath</u>. whereas LDH-C_4 and LDH-B_4 have a similar stability to heat. This result confirms a previous report (Wilkinson and Withy-combe, 1965), that human LDH-C_4, in contrast to mouse LDH-C_4, is not more heat stable than the other human LDH isozymes.
Amino acid composition

Although human LDH-C_4, as purified in this study is not quite homogeneous, the low level of contamination present will probably not significantly affect the amino acid composition data shown in Table 4. Our preliminary results do not indicate any dramatic difference between human LDH-C_4 and human LDH-B_4. In LDH-C_4, the content of arginine, proline, glycine, pheny-lalanine, and possibly threonine seem slightly higher, while the contents of aspartic acid and valine seem to be slightly lower. Comparisons with other data are not straight-forward because different conditions were used for hydrolysis and analysis, and because although all the data were computed for a molecular weight of 140,000, the total number of residues calculated in our study was 1296.5 for human LDH-B_4 and 1301 for LDH-C_4, while the data of Goldberg (1972) on mouse LDH-C_4

124

TABLE IV

AMINO ACID COMPOSITION OF LDH ISOZYMES

	Human LDH-*C4	Human LDH-*B4	Mouse[A] LDH-C4	Human[B] LDH-B4	Mammalian[C] LDH-B4	Mammalian[C] LDH-A4
Lysine	101.3 ± 8.5	109.0 ± 3.1	104	96	98	105
Histidine	28.3 ± 3.6	27.9 ± 0.6	32	25	26	38
Arginine	50.7 ± 2.6	37.5 ± 1.3	41	30	32	39
Aspartic Acid	126.3 ± 4.0	147.1 ± 2.2	155	134	135	121
Threonine	67.0	58.3	80	49	51	48
Serine	90.3	86.0	108	88	89	88
Glutamic Acid	151.7 ± 1.0	152.2 ± 1.5	108	128	126	119
Proline	70.7 ± 2.4	55.4 ± 3.3	48	44	46	44
Glycine	117.3 ± 2.1	95.8 ± 1.1	144	91	92	103
Alanine	93.3 ± 5.3	91.5 ± 2.1	83	85	82	82
Valine	102.0 ± 1.2	133.7 ± 0.1	139	142	135	129
Methionine	24.7 ± 1.8	29.6 ± 2.8	13	33	33	33
Isoleucine	82.8 ± 4.8	81.7 ± 1.8	74	88	88	86
Leucine	117.0 ± 3.0	127.7 ± 2.2	179	135	136	139
Tyrosine	33.6 ± 2.5	30.9 ± 2.9	25	27	27	28
Phenylalanine	44.2 ± 3.0	32.2 ± 0.6	23	22	22	28
TOTAL	1301.2	1296.5	1356	1217	1218	1230

Duplicate samples of purified LDH-C4 and LDH-B4 were hydrolysed in 6 N HCl containing 1 mg/ml phenol. Hydrolysates were analyzed on a Hitachi model KLA-3B amino acid analyzer. The results for proteins purified in this study are denoted by an asterisk (*) and represent arithmetic means of the duplicate samples. However, for serine and threonine the higher of the two values has been taken, and corrected for destruction by dividing by 0.90 and 0.95 repectively. Data

(legend to table IV, continued)
for mouse LDH-C$_4$ (indicated by superscript [A]) are taken from
Goldberg (1972). Data for human LDH-B$_4$ (indicated by superscript [B]) are taken from Pesce et al. (1967). Data for mammalian LDH-B$_4$ and LDH-A$_4$ (indicated by superscript [C]) were average values calculated by Goldberg (1972) from the data of Pesce et al. (1967).

and of Pesce et al. (1967) on human LDH-B$_4$ total 1356 and 1217 residues, respectively. There seem to be no amino acids present in characteristically low or high amounts in both human and mouse LDH-C$_4$. The markedly high values of glycine, leucine, and possibly threonine, and the markedly lower content of methionine may thus be idiosyncrasies of mouse LDH-C$_4$ rather than features common to all LDH-C$_4$ isozymes.

ACKNOWLEDGEMENTS

We thank the Rockefeller Foundation and the Faculty of Science, Mahidol University for providing the facilities for this work. S.V. gratefully acknowledges the support of a research studentship from the Prince of Songkhla University.

REFERENCES

Blanco, A. and W.H. Zinkham 1963. Lactate dehydrogenase in human testes. *Science* 139:601-602.

Burd, J.F. and M. Usategui-Gomez 1973. Immunochemical studies on lactate dehydrogenase. *Biochim. Biophys. Acta* 310: 238-247.

Clausen, J. 1969. Lactate dehydrogenase isozymes of sperm cells and testes. *Biochem. J.* 111:207-218.

Clausen, J. and B. Ovlisen 1965. Lactate dehydrogenase isoenzymes of human semen. *Biochem. J.* 97:513-517.

Goldberg, E. 1963. Lactic and malic dehydrogenases in human spermatozoa. *Science* 139:602-603.

Goldberg, E. 1971. Immunochemical specificity of lactate dehydrogenase-X. *Proc. Nat. Acad. Sci. U.S.A.* 68:349-352.

Goldberg, E. 1972. Amino acid composition and properties of crystalline lactate dehydrogenase X from mouse testes. *J. Biol. Chem.* 247:2044-2048.

Goldberg, E. 1973 a. Infertility in female rabbits immunized with lactate dehydrogenase X. *Science* 181:458-459.

Goldberg, E. 1973 b. Molecular basis for multiple forms of LDH-X. *J. Exp. Zool.* 186:273-278.

Goldberg, E. and J. Lerum 1972. Pregnancy suppression by an antiserum to the sperm specific lactate dehydrogenase. *Science* 176:686-687.

Kolb, E., G.A. Fleisher and J. Larner 1970. Isolation and characterization of bovine lactate dehydrogenase X. *Biochem.* 9:4372-4380.

Mosbach, K., H. Guildford, R. Ohlsson, and M. Scott 1972. General ligands in affinity chromatography. *Biochem. J.* 127:625-631.

Nisselbaum, J.S., D.E. Packer and O. Bodansky 1964. Comparison of actions of human brain liver, and heart lactic dehydrogenase variants on nucleotide analogues and on substrate analogues in the absence and presence of oxalate and oxamate. *J. Biol. Chem.* 239:2830-2935.

Pesce, A., T.P. Fondy, F. Stolzenbach, F. Castillo, and N.O. Kaplan 1967. The comparative enzymology of lactic dehydrogenases. III. Properties of the H_4 and M_4 enzymes from a number of vertebrates. *J. Biol. Chem.* 242:2151-2167.

Schatz, L. and H.L. Segal 1969. Reduction of α-ketoglutarate by homogeneous lactic dehydrogenase X of testicular tissue. *J. Biol. Chem.* 244:4393-4397.

Wilkinson, J.H. and W.A. Withycombe 1965. Organ specificity and lactate-dehydrogenase activity. *Biochem. J.* 97:663-668.

Wong, C., R. Yanex, D.M. Brown, A. Dickey, M.E. Parks, and R.W. McKee 1971. Isolation and properties of lactate dehydrogenase isozyme X from Swiss mice. *Arch. Biochem. Biophys.* 146:454-460.

Zinkham, W.H., A. Blanco, and L.J. Clowry 1964 a. An unusual isozyme of lactate dehydrogenase in mature testes: localization, ontogeny and kinetic properties. *Ann. N.Y. Acad. Sci.* 121:571-588.

Zinkham, W.H., A. Blanco, and R.L. Kupchyk 1964 b. Lactate dehydrogenase in pigeon testes: genetic control by three loci. *Science* 144:1353-1354.

AUTOANTIGENICITY OF LDH-X ISOZYMES

TODOR I. EVREV
Department of General Biology
Medical Academy, Sofia, Bulgaria

ABSTRACT. It has been established that antisera against human spermatozoa decreases the fructolytic activity of human spermatozoa without changing the oxygen uptake. moreover, one ml of these sera produces an average of 50% in the total LDH activity of 10^8 washed human spermatozoa. The decrease in enzyme activity is 7% for LDH-2, 14% for LDH-3, 26% for LDH-X, 32% for LDH-4, and 50% for LDH-5. The activity of LDH-1 is not changed by these antisera.

By comparison, auto- and iso-immunization of guinea pigs with testis homogenates from sexually mature animals yield antisera that completely inhibit the activity of $LDH-X_1$, $LDH-X_2$, and $LDH-X_3$ from the testes of guinea pigs without changing the activity of the other LDH isozymes.

On the basis of these results, it may be concluded that: 1) LDH-X isozymes are auto-antigens like other spermatozoan auto-antigens, and 2) anti-LDH-X antibodies probably play a role as immunological factors in some forms of male sterility.

INTRODUCTION

In 1963 Markert and Appella showed that the five isozymes of beef lactate dehydrogenase (LDH) stimulate the formation of at least four kinds of antibodies when injected into rabbits. In the same year, Blanco and Zinkham (1963) and Goldberg (1963) found an LDH-X isozyme that was specific for spermatozoa and for mature testes in men as well as in many other mammals. Moreover, in our investigations (Popivanov and Evrev, 1964) as well as in the studies of Ackerman (1968a, 1968b), it was proven that induced and normally-occurring antibodies can reduce the fructolytic activity of human spermatozoa. Nevertheless, these sera did not reduce oxygen uptake by human spermatozoa as shown by Warburg measurements according to the techniques of Salisbury and Lodge (1962).

These results encouraged us to investigate the inhibitory effect of sperm antibodies upon the fructolytic enzymes of spermatozoa and especially upon the LDH isozymes. In our initial studies we investigated the LDH isozymic content of human testes (prepubertal and mature), monolayer testicular cultures, long-term cultivated seminal tubules, spermatozoa, seminal plasma, blood sera, and erythrocyte hemolysates. Our results confirmed the cell specificity of LDH-X (Evrev et al., 1970) as first postulated by Blanco and Zinkham (1963).

RESULTS AND DISCUSSION

FRUCTOLYTIC INHIBITION BY ANTIBODIES

Some of our results on fructose utilization by human sperm suspended in normal rabbit serum and in rabbit antiserum to human sperm are shown in Table I. Fructolytic activity was reduced significantly by antisera in each of the seven sperm samples tested. No correlation was evident between the ABO blood group of the sperm donor and the amount of fructolysis in the sperm suspension. Although these results demonstrate that antibodies to sperm can reduce the metabolic activity of sperm, the results do not identify the specific enzymes that are affected.

TABLE I

FRUCTOLYTIC ACTIVITY OF HUMAN SPERMATOZOA
SUSPENSIONS (10^8 SPERM) AFTER TREATMENT
WITH RABBIT ANTISERA TO HUMAN SPERMATOZOA

Blood group of sperm donor	Residual quantity of fructose after incubation with 10^8 human spermatozoa, expressed in µg determined according to Mann's technique (1964)		
	1.0 ml immune serum. Residual fructose in µg	1.0 ml normal serum. Residual fructose in µg	difference in µg (ave. 40.3 µg)
A	96	62	34
A	88	34	54
B	94	60	34
B	89	62	27
O	98	52	46
O	86	32	54
AB	81	48	33

CELL SPECIFICITY OF LDH-X

In order to study specific anti-enzyme effects of antisera
elicited by sperm, we selected the isozymes of lactate dehydrogenase, particularly the sperm-specific LDH-X isozymes.
Our results show that LDH-X appears only in spermatozoa,
testes, and in organ cultures of testes from mature individuals (Figure 1). In agreement with the results of previous
authors, we failed to detect the LDH-X isozyme in the testes
of prepubertal individuals. We also failed to find LDH-X in
monolayer testicular cultures and in the other tissues and
fluids we examined (Figure 1).

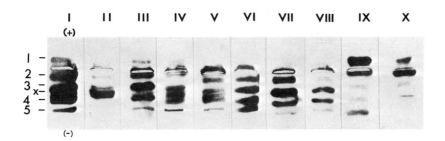

Figure 1. Zymograms of LDH isozymes. I Parenchyma of testes
of mature man. II Spermatozoa. III Parenchyma of testes
of immature boy. IV Seminal tubules cultured for one week.
V Seminal tubules cultured for 2 weeks. VI Monolayer testicular culture after 18 days. VII Monolayer testicular culture
after 76 days in culture. VIII Seminal plasma. IX Serum.
X Erythrocyte hemolysate. LDH-1 = B_4, LDH-2 = A_1B_3, LDH-3 =
A_2B_2, LDH-4 = A_3B_1, LDH-5 = A_4.

Judging by the morphological features exhibited by most
of the cells of the monolayer cultures (see Fig. 2) we conclude that chiefly Sertoli cells grow in monolayer testicular
cultures even from mature individuals. After 3-4 days, the
LDH-X band of these cultures disappears along with all obvious spermatogenic cells. Evidently the LDH-X isozyme is not
produced by Sertoli cells. The absence of LDH-X in monolayer
testicular cultures and in the testes of prepubertal individuals indicates that LDH-X is specific for spermatocytes and
post-meiotic cells. Prepubertal testes contain only spermatogonia and the nonspermatogenic interstitial cells as determined by histological examination.

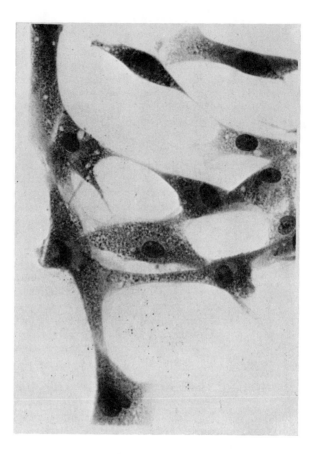

Figure 2.   Monolayer testicular culture.   Sertoli cells predominate.

AUTO-IMMUNITY TO LDH-X

The cell specificity of LDH-X is further supported by the
work of Zinkham (1968) and also by the results of our own work
(Evrev et al., 1971) on enzymatic changes in guinea pig testes
that accompany the development of experimental auto-immune
aspermatogenic orchitis (EAAO).   Figure 3 shows a zymogram of
the LDH isozymes in testis homogenates of non-immunized control guinea pigs and in four guinea pigs with EAAO, two with
partial exfoliation of the spermatogenic cells and two with
almost complete exfoliation.   The figure shows that the homogenate of the control animal contains 7 bands (after agar gel
electrophoresis).   Two bands are LDH-X, one band located between LDH-3 and LDH-4, and the other between LDH-4 and LDH-5.

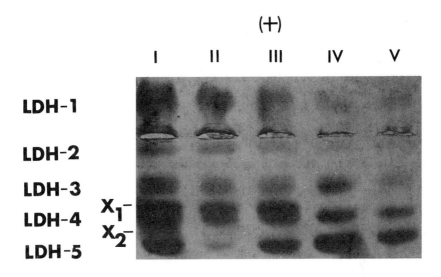

(+)

I   II   III   IV   V

LDH-1

LDH-2

LDH-3

LDH-4 X_1^-

LDH-5 X_2^-

(−)

Figure 3. Zymogram of LDH isozymes. I Control - mature guinea pig non immunized. 7 LDH isozymes are seen and two of them are LDH-X's. II and III Experimental animals with incomplete exfoliation of the cells from the spermatogenic cell line. LDH-X_2 is missing. IV and V Experimental animals with almost complete exfoliation of the spermatogenic cells. Both LDH-X_2 and LDH-X_1 isozymes are missing.

In the homogenates of the two experimental animals with partial exfoliation of the spermatogenic cells, LDH-X_1 persists while LDH-X_2 is missing (Figure 3, channels II and III). In two other animals, with almost complete exfoliation of the spermatogenic cells, both LDH-X_1 and LDH-X_2 are missing (see Figure 3, channels IV and V) and the intensity of LDH-1 and LDH-2 is diminished.

The total LDH activity in testis homogenates of the animals with partial alteration of the spermatogenic cells decreases by 57%. For the animals with almost complete exfoliation of these cells the decrease is 76%. These results confirm that LDH-X isozymes are specific for spermatozoa and meiotic and post-meiotic spermatogenic cells, and that the LDH-X isozymes can be used as biological markers.

$$LDH-X = LDH-C$$

Zinkham, Blanco, and Kupchyk (1963) assumed that LDH-X is a tetramer composed of C subunits, and that its synthesis is

133

controlled by a third LDH gene -- the C gene. These authors
and other investigators (Markert, 1968; 1970; Goldberg, 1971;
Blackshaw et al., 1968; Blackshaw and Elkington, 1970) are in
general agreement that the LDH-C gene is active only in the
primary spermatocytes and inactive in all other cells. The
LDH-A and LDH-B genes are both active in nearly every cell,
including, presumably, the spermatogonia -- the cell precursors of the primary spermatocytes.

Wherever more than one LDH-X band is present, as in guinea
pigs, only one of these isozymes is the C_4 tetramer. The
others are heteropolymers of C and A or C and B subunits.
Goldberg (1971) investigated the immunochemical specificity
of LDH-X with hetero-immune anti-LDH-X sera and presented
immunochemical evidence confirming the hypothesis that the
C subunit is encoded in a separate gene. We have also investigated the antigenic specificity of the LDH-X isozyme (Evrev
et al., 1971).

ANTI-LDH-X AND MALE STERILITY

All these facts led us to conclude that anti-enzymes could
be formed and directed against specific spermatozoan isozymes
and that LDH-X could possess auto-antigenic properties similar
to those of other spermatozoan auto-antigens which play a role
in male sterility. To test this conclusion, we studied the
inhibitory effect of anti-testes sera and immune lymphocytes
(obtained by auto- and iso-immunization of sexually mature
guinea pigs) on the LDH isozymes of the testes of sexually
mature guinea pigs. Simultaneously, we studied the inhibitory
effect of hetero-immune anti-spermatozoa sera on the LDH isozymes of human spermatozoa.

Homogenates from testes of sexually mature and immature
guinea pigs were obtained by mechanical crushing of seminal
tubules. Human spermatozoa were obtained from ejaculates of
healthy donors by centrifugation and triple washing with buffered saline. Anti-testes sera were prepared by immunization
of guinea pigs with autologous and homologous testes homogenates, following Freund's technique (Freund et al., 1953).
Hetero-immune anti-sperm sera were obtained by immunizing
rabbits with washed human spermatozoa. Cells of testes and
lymph nodes were obtained from guinea pigs by fragmenting the
corresponding organs with scissors, adding buffered saline,
and filtering through gauze.

Preparation 1. Testes homogenates of sexually mature guinea
pigs were centrifuged for 30 minutes at 5,000 rpm at 4°C. The
supernatants were divided into three equal parts. An equal

volume of serum from an auto-immunized animal was added to the first part; a cell suspension of lymph nodes of an immunized animal was added to the second part; and an equal volume of normal serum of guinea pigs was added to the third. The material was then incubated for 16 hours at 4°C.

Preparation 2. Anti-testis sera, obtained by auto- and iso-immunization of sexually mature guinea pigs, were separately mixed with equal volumes of testis cell suspensions of sexually mature and immature guinea pigs, and with washed human spermatozoa. The supernatants were added to an equal volume of testis homogenates from sexually mature guinea pigs, and then stored for 16 hours at 4°C.

Preparation 3. Human spermatozoa, washed three times, were separated into 2 equal suspensions. An equal volume of rabbit antiserum to human spermatozoa was added to the first suspension, and normal rabbit serum to the second. The samples and controls, thus prepared, were incubated for 16 hours at 4°C.

The material from preparations 1, 2, and 3, following incubation at 4°C, were centrifuged at 5,000 rpm for 30 min in a refrigerated centrifuge at 10°C, and the supernatants subjected to electrophoresis of two types: a) electrophoresis in agar gel performed by the method previously described (Evrev et al., 1970); b) disc polyacrylamide gel electrophoresis carried out according to the procedure of Toubs (1967).

The agar plates and columns of polyacrylamide were analyzed after Van der Helm (1961) for LDH isozymes, and the stain produced by each separate isozyme was recorded as a percentage of the total stain with the aid of an integrating densitometer.

The LDH isozymes of human spermatozoa (from 10 healthy donors), treated with rabbit anti-spermatozoa serum, showed a decrease in the activity of LDH-2 of 7%; LDH-3, 14%; LDH-X, 26%; LDH-4, 32%; and of LDH-5, 40%, as compared with the controls (spermatozoa of the same donors and normal rabbit serum; cf. Figure 4).

The inhibitory effect of the hetero-immune anti-spermatozoa serum increased progressively from LDH-2 to LDH-5. The LDH-1 is not inhibited at all, while LDH-5 is inhibited the most strongly (cf. Figure 4). In view of the concept of Appella and Markert (1961) and Markert (1963) setting forth the tetrameric structure of the LDH isozymes, we conclude that the hetero-immune anti-spermatozoa serum, obtained in our laboratory, contains anti-enzymes, which inhibit the A and C monomers and do not inhibit the B monomers (LDH-1 consists only of B subunits). This means that under the conditions of hetero-immunization it is not possible to obtain pure anti-LDH-X antibodies.

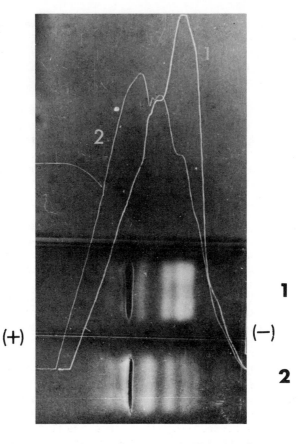

Figure 4. Zymogram of LDH isozymes and superimposed density tracing of resolved bands. 1) homogenate of washed human spermatozoa and normal rabbit serum (control), and 2) homogenate of washed human spermatozoa and hetero-immune rabbit antiserum to human spermatozoa. Some bands may be rabbit serum isozymes but the shift in pattern reflects the removal of human sperm isozymes by the antiserum.

The results of the studies on the inhibiting effect of anti-testes sera from auto-immunized guinea pigs on the LDH isozymes of sexually mature guinea pigs are presented in Figure 5. Clearly, LDH-X_1 , LDH-X_2, and LDH-X_3, which are specific for the testis of mature guinea pigs, are totally inhibited after incubation with anti-testis sera from auto-immune guinea pigs.

The same effect is achieved with antisera from iso-immune guinea pigs (see Figure 6) as well as with lymphocytes from auto- and iso-immune animals. Under these conditions

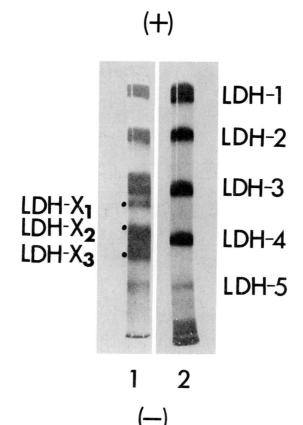

(+)

LDH-1

LDH-2

LDH-3

LDH-X$_1$
LDH-X$_2$
LDH-X$_3$

LDH-4

LDH-5

1   2

(—)

Figure 5. Zymogram of LDH isozymes after polyacrylamide gel electrophoresis of: 1) Testis of sexually mature guinea pig and normal serum of guinea pig. 2) Anti-testis serum from auto-immune guinea pig.

the other LDH isozymes do not change at all. This means that the anti-testis sera, as well as lymphocytes from auto- and iso-immune guinea pigs, contain antibodies with specific inhibitory effects against LDH-X isozymes.

The results of the studies of the inhibitory effect of anti-testis sera, previously absorbed with cells of seminal tubules of sexually mature and immature guinea pigs, on the LDH isozymes are shown in Figure 7. It can be seen that the anti-testis sera, following absorption with testis of sexually mature guinea pigs, lose their ability to inhibit LDH-X isozymes. This indicates that the inhibiting factors in the anti-testis sera act as immune antibodies and are not enzyme

137

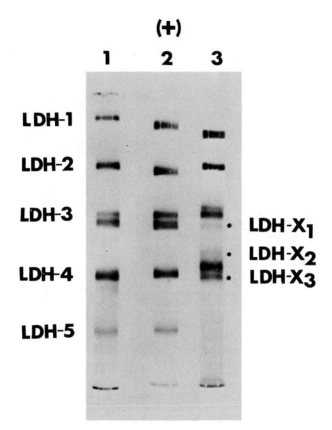

Figure 6. Zymogram of LDH isozymes after polyacrylamide gel electrophoresis of: 1) Testis of sexually immature guinea pig. LDH-3 consists of two bands. No X bands are present. 2) Testis of sexually mature guinea pig and lymphocytes of immune guinea pig. LDH-X isozymes are missing. LDH-3 is a doublet. 3) Testis of sexually mature guinea pig.

inhibitors of another nature. Moreover, these antisera lose their inhibitory effect after thermal inactivation.

Anti-testis sera, absorbed with testes of sexually immature guinea pigs, does not lose its ability to inhibit LDH-X isozymes. This is obviously due to the absence of the LDH-X$_1$, LDH-X$_2$, and LDH-X$_3$ antigens in the testes of sexually immature animals.

(+)

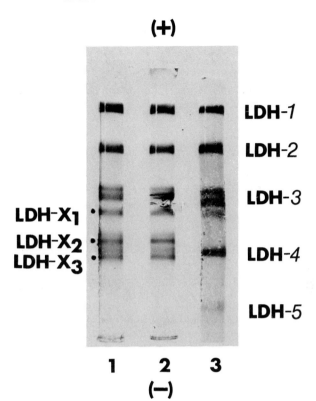

Figure 7.   Zymogram of LDH isozymes after polyacrylamide gel
electrophoresis.   1) Testis of sexually mature guinea pig.
2) Testis of sexually mature guinea pig and anti-testis serum,
absorbed with testis of a sexually mature guinea pig.   3) Testis of sexually mature guinea pig and anti-testis serum, absorbed with testis of a sexually immature guinea pig.

CONCLUSIONS

    The results of our investigations permit us to make the
following conclusions:
    1) LDH-X isozymes are specific markers of spermatozoa and
of the meiotic and post-meiotic spermatogenic cells.   This is
in agreement with the results of Allen (1961), Blanco and
Zinkham (1963), and of Goldberg (1963).
    2) During the development of experimental auto-immune
aspermatogenic orchitis, the LDH-X isozymes disappear when
exfoliation of spermatozoa and other spermatogenic cells
takes place in the testes.
    3) LDH-X is immunologically distinct from the other LDH
isozymes and this confirms the concept that the C subunits

are encoded in a separate gene, different from the LDH-A and LDH-B genes.

4) The anti-sera against auto- and isologous testes in guinea pigs exert selective inhibitory effects upon LDH-X isozymes and do not influence the other LDH isozymes.

5) LDH-X isozymes are auto-antigens, similar to the other spermatozoan auto-antigens.

REFERENCES

Ackerman, D.R. 1968a. Antibodies of the ABO system and the metabolism of human spermatozoa. *Nature* 219: 1159-1160.

Ackerman, D.R. 1968b. Freezing alters the glycolytic function of human sperm in the presence of ABO isoantibodies. *Nature*. 219:

Allen, J.M. 1961. Multiple forms of LDH in tissues of mouse. Their specificity, cellular localization, and response to altered physiological conditions. *Ann. N.Y. Acad. Sci.* 94: 937-951.

Appella, E. and C.L. Markert 1961. Dissociation of lactate dehydrogenase into subunits with guanidine hydrochloride. *Biochem. Biophys. Res. Commun.* 6: 171-176.

Blackshaw, A.W., O.C. Miller, and C.N. Craves 1968. Lactate dehydrogenase isozymes of calf and adult bull testes. *J. Dairy Sci.* 51: 950.

Blackshaw, A.W. and J.S.H. Elkington 1970. Developmental changes in LDH isoenzymes in the testis of the immature rat. *J. Reprod. Fertil.* 22: 69- 75.

Blanco, A. and W.H. Zinkham 1963. Lactate dehydrogenases in human testes. *Science* 139: 601-602.

Evrev, T., R. Popivanov, I. Podoplelov, and S. Zhivkov 1971. Antigenic specificity of the LDH-X isozyme. *Compt. rend. Acad. Bulgare Scie.* 24: 1571.

Evrev. T., S. Zhivkov, and L. Russev 1970. LDH isozymes in testicular cultures and human testes. *Human Heredity* 20: 70

Evrev, T., R. Popivanov, I. Kehayov, I. Podoplelov, and S. Zhivkov 1971. Enzyme changes in guinea pig testes accompanying the development of auto-immune aspermatogenic experimental orchitis. *Compt. rend. Acad. Bulgare Scie.* 24: 1283.

Freund, J., M.M. Lipton, and G.E. Thompson 1953. Aspermatogenesis in the guinea pig induced by testicular tissue and adjuvant. *J. Exp. Med.* 97: 711.

Goldberg, E. 1963. Lactic and malic dehydrogenases in human spermatozoa. *Science* 139: 602-603.

Goldberg, E. 1971. Immunochemical specificity of lactate dehydrogenase-X. *Proc. Natl. Acad. Sci. U.S.A.* 68: 349-352.

Mann, T. 1964. *Biochemistry of Semen and of the Male Reproductive Tract.* Methuen, London.

Markert, C.L. 1970. Isozymes and cellular differentiation. *Advances in the Biosciences* 6: 511-528.

Markert, C.L. 1968. The molecular basis for isozymes. *Ann. N.Y. Acad. Sci.* 151: 14-40.

Markert, C.L. 1963. Lactate dehydrogenase isozymes: dissociation and recombination of subunits. *Science* 140: 1329-1330.

Markert, C.L. and E. Appella 1963. Immunochemical properties of LDH isozymes. *Ann. N.Y. Acad. Sci.* 103: 915-929.

Popivanov, R. and T. Evrev 1964. Detecting of the group of human spermatozoa according to their fructolytic activity. *Exp. Med. i. Morf.* 1: 8.

Salisbury, G.W. and R. Lodge 1962. Metabolism of spermatozoa. In: *Advances in Enzymology,* F.F. Nord, editor, John Wiley & Sons, New York 24: 56.

Toubs, M.P. 1967. Shandou Instrument Application.

Van der Helm, H.I. 1961. Simple method of demonstrating lactic acid dehydrogenase isozymes. *Lancet* 11: 108.

Zinkham, W.H. 1968. Lactate dehydrogenase isozymes of testis and sperm: biological and biochemical properties and genetic control. *Ann. N.Y. Acad. Sci.* 151: 598-610.

Zinkham, W.H., A. Blanco, and L. Kupchyk 1963. Lactate dehydrogenase in testis: dissociation and recombination of subunits. *Science* 142: 1303-1304.

RELATIONSHIP BETWEEN THE CHANGE IN THE TERTIARY STRUCTURE
OF HUMAN TUMOR LACTATE DEHYDROGENASE AND CARCINOGENESIS

MINORU HAMADA
Department of Pathological Biochemistry,
Atomic Disease Institute,
Nagasaki University School of Medicine, 12-4, Sakamoto-machi,
Nagasaki-shi, 852, Japan

ABSTRACT. Comparative studies on highly purified lactate
dehydrogenase (LDH) isozymes (A_4, A_2B_2, and B_4) of human
uterus, uterine myoma, uterine myosarcoma, and cervical
cancer indicated a marked increase of the A_4 isozyme in
cancerous tissues. The conformational structures of
purified A_4 isozymes from these neoplastic tissues were
compared to investigate their relationship to carcinogenesis. The A_4 isozyme in neoplasms differs from the
normal noncancerous tissue isozyme by its greater resistance to heat inactivation and its less precise conformational arrangement or unfolding as shown by its marked
decrease in right-handed α helices. The A_4 isozyme in
neoplasms is like other A_4 isozymes in its molecular
weight (140,000), subunit size (35,000), amino acid
composition, and immunochemical cross reactivity with
other A_4 isozymes.

INTRODUCTION

In human uterine tissues the isozymes of lactate dehydrogenase (L-lactate: NAD^+ oxidoreductase E.C. 1.1.1.27) may be
separated electrophoretically into at least five bands (Okabe
et al., 1968). In neoplastic diseases LDH isozymes have been
characterized by a shift (Starkweather et al., 1962; Richterich
et al 1963; Goldman et al., 1963; Nisselbaum et al., 1963;
Dawson et al., 1964; Okabe et al., 1968; and Schapira et al.,
1970) in the pattern, with a marked increase in the A_4 isozyme
in neoplasms as compared to both normal tissues and benign
tumors.

In this paper the pure forms of the isozymes, A_4, A_2B_2,
and B_4 from human uterus, uterine myoma, and uterine leiomyosarcoma were studied by immunological and biochemical methods.
I describe the structure of the A_4 isozyme, which characteristically unfolds in cancerous tissues.

MATERIALS AND METHODS

The study was undertaken with the following human samples:

a) surgically removed tissues from uterine myoma and uterine myosarcoma, and b) uterine tissues from autopsies (within 6 hours *post-mortem*).

Purification and assay of the enzyme. The materials were stored at -20°C, and prepared by the method of our previous report (Okabe et al., 1968). Purification of LDH isozymes from the uterine tissues yielded homogeneous preparations, as indicated by chromatographic, electrophoretic, immunochemical, and ultracentrifugal criteria. LDH enzyme assay, the reaction with lactate and NAD, was measured according to the method of Neiland (1955), and specific activity of LDH is expressed as micromoles of NADH formed per minute per milligram of enzyme protein. The protein concentration was determined by the phenol method (Layne 1957).

Electrophoresis. Thin layer electrophoresis on 5% acrylamide (Raymond et al., 1959) gel was performed with a pH 8.8, Tris-HCl buffer system, at 4°C, in a horizontal electrophoresis cell (Ogita et al., 1966). Enzyme activity was localized by staining as described by the method of Fine et al. (1963). The quantitative determination of each isozyme of LDH was performed by the method previously reported (Okabe et al., 1968). The molecular weight of the native enzymes was estimated by disc gel electrophoresis using the method of Hedrick et al. (1968). Polyacrylamide gel electrophoresis in sodium dodecyl sulfate (SDS) was carried out in 5% gels as described by Shapiro et al. (1967) and by Weber et al. (1969). The denaturation and carboxymethylation of the proteins were performed by the method of Inoue et al., (1972).

Immunochemical techniques. The double diffusion test in agar was carried out at 5°C by a modification of the method of Ouchterlony (1962). Antisera were produced in rabbits against the purified A_4, A_2B_2, and B_4 isozymes from uterine myoma, and the gamma globulin fraction was isolated from serum by precipitation with ammonium sulfate by the method of Campbell et al. (1964).

Amino acid analysis. S-Carboxymethylated (Craven et al., 1965) and performic acid oxidized (Moore 1963) samples of the LDHs were prepared prior to acid hydrolysis (Moore et al., 1963). The subsequent amino acid analysis was performed by the method of Spackman et al. (1958), using a Beckman model 116 amino acid analyzer. Tryptophan and tyrosine were estimated spectrophotometrically as described by Goodwin et al.

144

(1946).

Circular dichroism spectroscopy. Circular dichroism (CD) spectra were obtained at 25°C with a JASCO spectropolarimeter model ORD/UV/CD-5; light path length varied from 0.5 to 1.0 cm. Recordings were made in the range 190 to 260 nm and 260 to 300 nm, and all spectra were corrected for base-line shifts of the dialysate buffer. Protein concentrations were determined at the completion of CD studies. The molecular ellipticity, $(\underline{\Theta})$, was calculated from $(\underline{\Theta})_\lambda = 3,300\ (\underline{\varepsilon}_L - \underline{\varepsilon}_R)$, where $\underline{\varepsilon}_L$ and $\underline{\varepsilon}_R$ are the absorptivities at wavelength $\overline{\lambda}$ of $\overline{\text{left}}$ and right circular polarized light. The helical contents were calculated from $(\underline{\Theta})_{222}$ according to Holzwarth et al. (1965). The instrument was calibrated with D-10-camphor sulfonate in the CD, at 290 nm to give an $\underline{\varepsilon}_L - \underline{\varepsilon}_R$ of 2.20 (DeTar et al., 1969; Cassim et al., 1969; and Urry et al., 1967). All experiments were conducted at 25°C in 0.1 M sodium phosphate buffer (pH 7.0)

RESULTS AND DISCUSSION

In our previous paper (Okabe et al., 1968), we suggested that the A_4 isozymes from neoplastic tissues are different from the A_4 isozyme in normal tissues. These differences were most marked in the catalytic action and elevated thermal stability as observed also by Südi (1970).

Electrophoresis. The complexity of the electrophoretic pattern of human uterine LDH is shown in Fig. 1. The isozymic deviation in cancerous tissues is shown by the markedly increased A_4 isozyme. This pattern also shows the loss of organ specificities of the enzyme activity in cancerous tissues.

Purification of the LDH isozymes. Table 1 shows a summary of the degree of purification of A_4, A_2B_2, and B_4 isozymes from each tissue. There was no significant difference in each specific activity and no qualitative changes in the A_4 isozymes in neoplasms, nor was there any change in its specific activity.

Under certain conditions various types of isozymes have distinctive susceptibilities to inactivation by heating. This distinctive susceptibility to heat was found in the myosarcoma LDH isozyme (Okabe et al., 1968). From these data it was inferred that since the A_4 isozyme in neoplasms had increased resistance to heat inactivation, there must be some

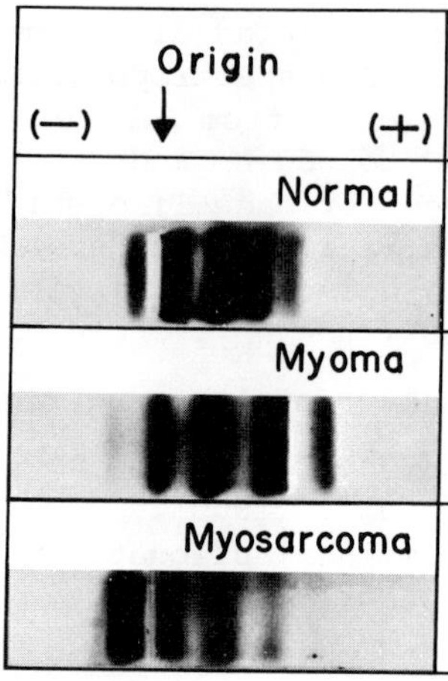

Origin (—) ↓ (+)	Percentage of Isozymes				
	A_4	A_3B_1	A_2B_2	A_1B_3	B_4
Normal	2	26	45	21	6
Myoma	13	32	35	16	4
Myosarcoma	38	34	21	5	3

Fig. 1. Electrophoretic patterns and percentages of the isozymes of human lactate dehydrogenase in water extracts of normal uterus and its neoplasms. Electrophoretic procedures and the quantitative determination of each enzyme are described in the text.

---

modification in its sterical structure (Südi,1970).

Immunochemical studies. The Ouchterlony double diffusion test in agar is illustrated in Fig. 2. The immunological reaction between the A_4 isozyme in myosarcoma and anti-A_4 antibody formed against the A_4 isozyme in myoma suggests a sequence homology between them. However, when reacted with anti-B_4, no lines of identity or cross reaction were noted. The A_4 and B_4 isozymes have distinct antigenic specificities as described by other reports (Markert and Holmes,1969).

Determination of molecular weight. Like other A_4 isozymes, the A_4 isozymes in neoplasms have a molecular weight of 140,000 as determined by polyacrylamide gel disc electrophoresis (Fig. 3). A subunit size of 35,000 was determined by SDS-polyacrylamide gel electrophoresis with an alkylated sample after reduction in 8 M urea (Fig. 4). These observations are similar to those for other species as reported by Fosmire et al. (1972), Allison et al. (1969), and Huston et al. (1972).

Amino acid composition. The amino acid composition of

TABLE 1

Purification of the isozymes of human lactate
dehydrogenase from uterus, its myoma and myosarcoma

Purification step		Total protein (mg)	Specific* activity	Total** activity	Yield (%)
(Myoma)	600g				
Water extract		35,540	0.17	6,180	100
Hydroxyl-apatite	A_4	10.0	13.90	140	2.2
	A_2B_2	28.3	27.75	780	12.5
	B_4	12.0	16.20	200	3.3
Myosarcoma	35g				
Water extract		1,479	0.18	234	100
Hydroxyl-apatite	A_4	1.2	12.15	12.3	5.0
	A_2B_2	2.1	26.00	40.0	16.5
	B_4	1.5	14.26	21.3	8.7
(Normal)	200g				
Water extract		6,865	0.19	1,300	100
Hydroxyl-apatite	A_4	1.7	14.3	25.0	1.9
	A_2B_2	3.3	29.20	1,000	7.7
	B_4	2.0	15.01	30.0	2.3

(*μmoles of NADH/min·mg;   **μmoles of NADH/min)

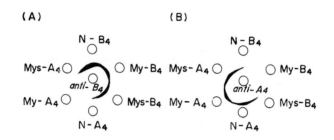

Fig. 2. Comparisons of human uterine lactate dehydrogenase by Ouchterlony double diffusion. A, the antiserum, in the center well, was prepared against the purified B_4 isozyme in myoma. The antigens are in the surrounding wells: $N-B_4$, normal B_4; $My-B_4$, myoma B_4; $Mys-B_4$, myosarcoma B_4; $N-A_4$, normal A_4; $My-A_4$, myoma A_4; $Mys-A_4$, myosarcoma A_4; and B, the antiserum in the center well was prepared against the A_4 isozyme in myoma. The antigen assignments in the outer wells are as in A.

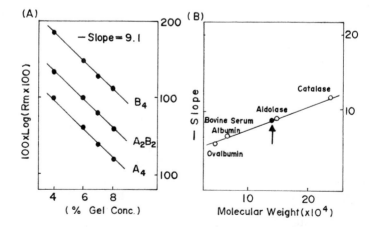

Fig. 3. Separation and estimation of molecular weights of proteins by disc gel electrophoresis. A, plots of the log of protein mobility relative to the dye front <u>versus</u> acrylamide gel concentration; B, plots of slope by % gel concentration <u>versus</u> molecular weight (X 10^4).

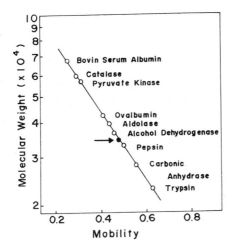

Fig. 4. Determination of the apparent molecular weight of the dissociated subunits of lactate dehydrogenase isozymes by polyacrylamide gel electrophoresis in SDS. Standard protein samples and lactate dehydrogenase itself were subjected to electrophoresis as described in "Methods." Mobility relative to the tracking dye was plotted against the known subunit molecular weight of each protein as described by Weber et al. (1969). The mobility of the subunits of lactate dehydrogenase is indicated by an <u>arrow</u>.

human uterine LDH isozymes was calculated on the assumption that the enzyme is composed of four identical polypeptide chains each with a molecular weight of 35,000 (Table 2). Significant differences were not expected between the A_4 isozymes from cancerous and normal tissue. These analyses, however, provide clear indications of the differences in the composition of A_4 and B_4 isozymes as described by Wieland et al. (1961), and Pesce et al. (1964). There were four residues of half cystine in each subunit from the A_4, A_2B_2, and B_4 isozymes. I have been studying the function of active thiols of the human LDH isozymes in relation to the electrophoretic pattern of LDH, its quaternary structure and subunit organization, since it has been claimed that some human LDH isozymes can exist as conformers (Markert, 1968; Pfleiderer et al., 1970; Levi et al., 1971; and Dudman et al., 1973).

   In summary, one may assume that the A_4 isozyme in neoplasms probably has the same primary structure as that of the A_4

TABLE 2
Amino acid composition of LDH

	Myoma		Myosarcoma		Normal	
	A_4	B_4	A_4	B_4	A_4	B_4
	(residues/35,000 g subunit)					
Try[c]	8.5	8.8	9.2	8.6	9.5	8.0
Lys	20.6	18.5	17.7	19.9	18.4	21.0
His	4.3	3.9	4.7	4.4	4.2	4.1
Arg	7.2	9.7	8.1	9.4	9.1	10.1
Asp	31.7	34.0	32.3	33.5	31.7	33.3
Thr[a]	13.2	15.2	13.4	14.7	13.2	15.6
Ser[a]	17.8	16.9	17.7	16.3	17.8	17.4
Glu	32.4	40.5	33.6	42.0	32.2	42.2
Pro	11.6	10.7	12.0	12.4	11.4	11.5
Gly	24.6	18.0	24.4	19.3	23.0	17.5
Ala	20.2	20.2	17.0	19.8	20.0	20.5
1/2 Cys[d]	3.9	4.0	3.9	3.9	3.9	3.9
Val[b]	35.6	33.0	34.5	34.0	35.5	34.5
Met[d]	7.4	7.6	7.5	8.5	7.5	8.5
Ileu[b]	21.5	19.7	21.9	22.6	21.5	21.9
Leu	27.8	34.8	28.0	33.7	28.0	34.0
Tyr[c]	2.0	2.0	2.0	2.0	1.8	2.2
Phe	4.2	4.5	4.0	4.6	4.2	4.7

a) Corrected for destruction during acid hydrolysis.
b) Corrected for slow release during acid hydrolysis.
c) Determined spectrophotometrically (Goodwin et al., 1946).
d) Determined as cysteic acid (Moore 1963).

isozyme in normal tissues, but is structurally less organized, and can, therefore, give rise to slightly different molecular forms.

Circular dichroism studies. Figure 5 shows a preliminary survey of the data. The CD spectrum in the wavelength range

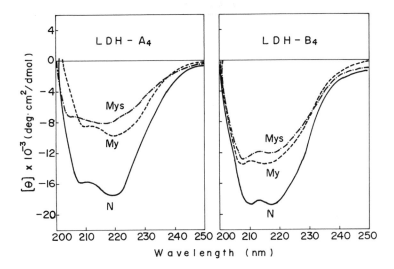

Fig. 5. Circular dichroism spectra of native uterine A_4 and B_4 isozymes. Samples were in 0.1 M sodium phosphate buffer, pH 7.0. N, normal; My, myoma; and Mys, myosarcoma. Enzyme concentration was 1.5×10^{-7} M.

200 to 250 nm shows the α helical structure by a corresponding negative maximum at 209 and 220 nm. The α helical contents as calculated from a molecular ellipticity for these isozymes are shown in Table 3. Very interestingly, the α helical contents of the A_4 isozyme in myosarcoma were lower than those of the A_4 isozyme in normal tissues and myoma. The A_4 isozyme in cancerous tissues is, therefore, markedly unfolded and may be considered as being "immature," in its tertiary structure, as compared to the A_4 isozyme from normal tissues for other species, since Fondy et al., (1965) have observed 30% alpha helix formation in pig heart and 35% in pig muscle. It is interesting that a value of about 40% helicity has been found in other species such as chicken, dogfish, and rabbit (Pesce et al., 1967). Perhaps this amount of helix is necessary for the homologous enzyme to maintain its tetrameric structure. In the CD spectrum (Fig. 6) from the wavelength

TABLE 3
Circular dichroism characteristics of LDH isozymes

		$-(\underline{\theta})_{222}^{220} \times 10^{-3}$	α Helix (%)
Myoma	A_4	9.7	32
	A_2B_2	10.0	25
	B_4	12.6	35
Myosarcoma	A_4	7.6	19
	A_2B_2	11.6	24
	B_4	11.8	35
Normal	A_4	17.4	40
	A_2B_2	8.9	25
	B_4	18.5	46

range 260 to 300 nm, the A_4 isozyme in cancerous tissues shows different peaks, troughs, and crossover points than does the A_4 isozyme from normal tissues. Therefore, there are detectable differences in the steric conformation at aromatic side chains of the A_4 isozyme from cancerous and normal tissues.

To explain the findings it is not necessary to assume that the A_4 isozyme in neoplasms would differ from the A_4 isozyme in normal tissues in its primary structure, but only in its tertiary structure. The degree of activity of an enzyme such as LDH appears to be highly dependent upon its tertiary structure. However, a precise conformation might not be too critical for the occurrence of specific catalytic activity. Optical rotatory dispersion spectroscopy and titration studies of the functional residues of the isozyme should provide more detailed information about the structural characteristics of the A_4 isozyme in neoplasms. These studies are still in progress.

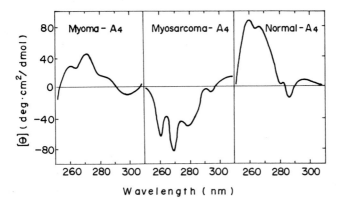

Fig. 6. Circular dichroism spectra in the wavelength range 260-300 nm of native A$_4$ isozymes. Samples were in 0.1 M sodium phosphate buffer, pH 7.0. The sample assignments in the figure are as in Fig. 5.

ACKNOWLEDGEMENTS

The author is indebted to Prof. Masahiko Koike for his encouragement, helpful suggestions, and discussions throughout this work. I also thank Dr. Kyoko Ogasahara for her assistance in optical property measurements. I wish to thank Dr. Keiichiro Okabe in supplying human leiomyosarcoma.

REFERENCES

Allison, W.S., J. Admiraal, and N.O. Kaplan 1969. The subunits of dogfish M$_4$ lactate dehydrogenase. *J. Biol. Chem.* 244: 4743-4749.

Campbell,D.H., J.S. Garvey, N.E. Cremer, and D.H. Sussdoy 1964. *Methods in Immunology.* W.A. Benjamin Inc., N.Y. pp. 263.

Cassim, J.Y., and J.T. Yang 1969. A computerized calibration of the circular dichrometer. *Biochem.* 8: 1947-1951.

Craven, G.R., E. Steers, Jr., and C.B. Anfinsen 1965. Purification, composition, and molecular weight of the β-galactosidase of *Escherichia coli* K12. *J. Biol. Chem.* 240: 2468-2477.

Dawson, D.M., T.M. Goodfriend, and N.O. Kaplan 1964. Lactic dehydrogenases: Functions of the two types. *Science*

143: 929-933.

DeTar, D.F. 1969. Suggested primary standards for calibration of optical rotatory and circular dichroism instruments. *Anal. Chem.* 41: 1406-1408.

Dudman, N.P.B. and B. Zerner 1973. Lactate dehydrogenase: Electrophoretic behaviour, electron microscopy and structure. *Biochim. Biophys. Acta* 310: 248-263.

Fine, I.H. and L.A. Costello 1963. The use of starch electrophoresis in dehydrogenase studies. *Methods Enzymol.* 6: 958-972.

Fondy, T.P., A. Pesce, I. Freedberg, F.E. Stolzenbach, and N.O. Kaplan 1965. The comparative enzymology of lactate dehydrogenase, II. Properties of the crystalline HM_3 hybrid from chicken muscle and of H_2M_2 hybrid and H_4 enzyme from chicken liver. *Biochem.* 3: 522-530.

Fosmire, G.J. and S.N. Timasheff 1972. Molecular weight of beef heart lactate dehydrogenase. *Biochem.* 11: 2455-2460.

Goldman. R.D. and N.O. Kaplan 1963. Alterations of tissue lactate dehydrogenase in human neoplasms. *Biochim. Biophys. Acta* 77: 515-518.

Goodwin, T.W. and R.A. Morton 1946. Spectrophotometric determinations of tyrosine and tryptophan in proteins. *Biochem. J.* 40: 628-632.

Hedrick, J.L. and A.J. Smith 1968. Size and charge isomer separation and estimation of molecular weights of proteins by disc gel electrophoresis. *Arch. Biochem. Biophys.* 126: 155-164.

Holzwarth, G. and P. Doty 1965. The ultraviolet circular dichroism of polpyeptides. *J. Amer. Chem. Soc.* 87: 218-228.

Huston, J.S. , W.W. Fish, K.G. Mann, and C. Tanford 1972. Studies on the subunit molecular weight of beef heart lactate dehydrogenase. *Biochem.* 11: 1609-1612.

Inoue, H. and J.M. Lowenstein 1972. Acetyl coenzyme A carboxylase from rat liver. *J. Biol. Chem.* 247: 4825-4832.

Layne, E. 1957. Spectrophotometric and turbidimetric methods for measuring proteins. *Methods Enzymol.* 3: 447-454.

Levi, A.S. and N.O. Kaplan 1971. Physical and chemical properties of reversibly inactivated lactate dehydrogenase. *J. Biol. Chem.* 246: 6409-6417.

Markert, C.L. 1968. The molecular basis for isozymes. *Ann. N.Y. Acad. Sci.* 151: 14-40.

Markert, C.L. and R.S. Holmes 1969. Lactate dehydrogenase isozymes of the flatfish, *pleuronectiformes:* kinetic, molecular, and immunochemical analysis. *J. Exp. Zool.*

171: 85-103.

Moore, S. 1963. On the determination of cystine as cysteic acid. *J. Biol. Chem.* 238: 235-237.

Moore, S. and W.H. Stein 1963. Chromatographic determination amino acids by the use of automatic recording equipment. *Methods Enzymol.* 6: 819-831.

Neiland, J.B. 1955. Lactic dehydrogenase of heart muscle. *Methods Enzymol.* 1: 449-454.

Nisselbaum, J.S. and O. Bodansky 1963. Purification, kinetic, and immunochemical studies of the major variants of lactic dehydrogenase from human liver, hepatoma, and erythrocytes; comparison with the major variant of human heart lactic dehydrogenase. *J. Biol. Chem.* 238: 969-974.

Ogita, Z., M. Hashinotsume, and Y. Kosugi 1966. The poly-acrylamide gel electrophoresis. *SABCO J.* 2; 58-65.

Okabe, K., T. Hayakawa, M. Hamada, and M. Koike 1968. Purification and comparative properties of human lactate dehydrogenase isozymes from uterus, uterine myoma, and cervical cancer. *Biochem.* 7: 79-90.

Ouchterlony, Ö. 1962. Diffusion-in-gel methods for immuno-logical analysis. *Prog. Allergy* 6: 30-77.

Pesce, A., R.H. McKay, F.E. Stolzenbach, R.D. Cahn, and N.O. Kaplan 1964. The comparative enzymology of lactate dehydrogenases, I. Properties of the crystalline beef and chicken enzymes. *J. Biol. Chem.* 239: 1753-1761.

Pesce, A., T.P. Fondy, F. Stolzenbach, F. Castillo, and N.O. Kaplan 1967. The comparative enzymology of lactate dehydrogenases, III. Properties of the H_4 and M_4 enzymes from a number of vertebrates. *J. Biol. Chem.* 242: 2151-2167.

Pfleiderer, G. and D. Jeckel 1970. Structure and action of lactate dehydrogenase. *FEBS Symposium* 18: 157-161.

Raymond, S. and L. Weintraub 1959. Acrylamide gel as a supporting medium for zone electrophoresis. *Science* 130: 711.

Richiterich, R., P. Schfroth, and H. Aebei 1963. A study of lactate dehydrogenase isozyme pattern of human tissues by adsorption-elution on Sephadex-DEAE. *Clin. Chim. Acta* 8: 178-192.

Schapira, F., J. C. Dreyfus, and G. Schapira 1970. Ontogenic evolution and pathological modification of molecular forms of some isozymes. *FEBS Symposium* 18: 305-320.

Schapiro, A.L., V. Viñuela, and J.V. Maizel 1967. Molecular weight estimation of polypeptide chains by electrophoresis in SDS-polyacrylamide gels. *Biochem. Biophys. Res. Commun.* 28: 815-820.

Spackman, D.H., W.H. Stein, and S. Moore 1958. Automatic recording apparatus for use in the chromatography of amino acids. *Anal. Chem.* 30: 1190-1206.

Starkweather, W.H. and H.K. Schock 1962. Some observations on the lactate dehydrogenase of human neoplastic tissue. *Biochim. Biophys. Acta* 62:440-442.

Südi, J. 1970. Temperature dependence of the mechanism of heat inactivation of pig lactate dehydrogenase isozyme H_2M_2. *FEBS Symposium* 18: 169-176.

Urry, D.W. and J.W. Pettegrew 1967. Model systems for interacting heme moieties, II. The ferriheme octapeptide of cytochrome c. *J. Amer. Chem. Soc.* 89: 5276-5283.

Weber, K. and M. Osborn 1969. The reliability of molecular weight determinations by dodecyl sulfate-polyacrylamide gel electrophoresis. *J. Biol. Chem.* 244: 4406-4412.

Wieland, T. and G.P. Pfleiderer 1961. Chemical differences between multiple forms of lactic acid dehydrogenases. *Ann. N.Y. Acad. Sci.* 94: 691-700.

CORRELATION BETWEEN LACTATE DEHYDROGENASE ISOZYME PATTERNS OF MAMMALIAN LIVERS AND DIETARY ENVIRONMENTS

MASAO OGIHARA

Department of Internal Medicine
Daisan Hospital, Tokyo Jikeikai
University School of Medicine
106 Izumi, Komae-shi
Tokyo 182, Japan

ABSTRACT. The study of the lactate dehydrogenase (LDH) isozymes in the livers of mammals shows a very close relationship between liver metabolism and dietary habits. The liver LDH patterns of twenty-two species of mammals were investigated. Several species (e.g. humans, dogs, cats, etc.) possess high activity of LDH-5 (A subunits). In contrast, a high activity of LDH-1 (B subunits) is found in the order Artiodactyla (such as pigs and cows), while the hetropolymers LDH-2, LDH-3, and LDH-4 are predominant in rabbits. A comparison of liver LDH isozyme patterns with dietary habits of mammals reveals that high levels of A subunits are found in the livers of carnivores. On the other hand, high levels of B subunits are found in the livers of herbivores. Chipmunks and guinea pigs have higher levels of B subunits than rats and mice, all in the order Rodentia, and this is closely related to their vegetable dietary habits. A few differences of LDH isozyme activities found in human livers were also shown to be correlated with differences in dietary habits. In an experiment with rats maintained on a high starch diet, liver LDH-1, LDH-2, and LDH-3 activities (predominately B subunits) were found to have increased in comparison with those of the control rats. It appears that an increase in the B subunits of lactate dehydrogenase was induced in the livers of the experimental rats as a result of their adaptation to a high starch diet.

INTRODUCTION

Many investigations have demonstrated that mammals contain five major isozymes of lactate dehydrogenase formed by the random combination of two different subunits into tetramers (Appella and Markert, 1961; Cahn et al., 1962; and Markert, 1963). The organ and species specificities of the lactate dehydrogenase isozymes were established by the comparison of the lactate dehydrogenase (LDH) isozymes from different organs of the same species or from the same organ in different species (Markert and Møller, 1959; Appella and Markert, 1961; Kaplan and Cahn, 1962; Vesell and Philip, 1963; and Wroblewski and Gregory, 1961). In regard to the physiological role of the isozymes, it has been proposed that the LDH B_4 isozyme is best suited for function in an aerobic environment, whereas the LDH A_4 isozme has kinetic properties suitable for functioning in an anaerobic environment (Cahn et al., 1962). The primary focus of this investigation concerns the lactate dehydrogenase isozymes that are found in the livers of different species of mammals. In the present paper a correlation is demonstrated between the lactate dehydrogenase isozymes in the liver and the dietary habits of twenty-two species of mammals belonging to six orders. In addition, a change in the liver lactate dehydrogenase isozyme pattern was observed when the rats were fed on a high starch diet.

MATERIALS AND METHODS

Twenty-two species of mammals belonging to six orders were employed for this research:

a. Primates: fourteen Japanese adult human males and two Japanese Macaques (*Macaca fuscata*).

b. Carnivora: four domestic dogs (*Canis familiaris*), one raccoon dog (*Nycterutes procyonoides*), one masked palm civet (*Paguma larvata*), one Japanese red fox (*Vulpes vulpes*), six domestic cats (*Felis catus L.*), one Siberian weasel (*Mustela sibirica*), and one Japanese marten (*Martes melampus*).

c. Rodentia: five mice (*Mus musculus*), eighty rats (*Rattus rattus*), one Asian chipmunk (*Seiurus vulgaris lis T.*), one large flying squirrel (*Petaurista*), and two guinea pigs (*Cavia cobaya*).

d. Lagomorpha: four rabbits (*Oryctolagus cuniculatus L.*).

e. Perissadoctyla: three domestic horses (*Equus caballus L.*).

f. Artiodactyla: four oxen (*Bos taurus*), three sheep (*Ovis ammon*), three goats (*Capra hircus L.*), five Japanese deer (*Cervus nippon T.*), six pigs (*Sus scrofa var. domesticus* Brisson), and one wild boar (*Sus scrofa L.*).

The livers of normal adult human males were obtained by autopsy within ten hours after death. Death was caused by either cerebral hemorrhage or by traffic accident. Each adult of the other twenty-one species of mammals was killed by decapitation and its liver stored immediately at -20°C. Samples of 0.5g of the liver tissues were washed in 2.5 ml of 0.9% NaCl and then ground in a Potter-Elvehjem homogenizer. Then

the homogenate was centrifuged at 2°C for forty-five minutes
at 5000 x g in a Spinco Model L centrifuge, and the supernatant was used directly for electrophoresis.

A total of eighty Sprague-Dawley rats were used to investigate the effects of dietary differences on liver LDH patterns.
Forty experimental rats were kept in separate cages and maintained on a diet that contained 80 percent starch (to which
a few vitamins were added) for twelve weeks beginning twelve
days after birth. The forty control rats were maintained on
an ordinary diet for the same period of time. After decapitation of the rats, the livers, other organs, and gastrointestinal tracts were rapidly removed and carefully weighed. In
addition, the lengths of the gastrointestinal tracts were
measured. Liver tissues were embedded in paraffin, sectioned,
stained with hematoxylin-eosin and Periodic acid-Schiff's
reagent (PAS) and examined by light microscopy.

Method of electrophoresis. Twenty μl of supernatant was
subjected to horizontal polyacrylamide gel electrophoresis.
A pH 8.6 tris-citrate buffer was employed in the gel. A
voltage gradient of 6 V per cm was employed for four hours
at 4°C. Six different supernatants were electrophoresed at
the same time. These gels were stained for LDH activity by
incubating them for one hour at 37°C in the dark in a solution
containing DL-lactate 70% (0.3 ml), phosphate buffer pH 7.4,
0.1 M (10 ml), sodium cyanide 0.1 M (0.1 ml), phenazine methosulphate 1 mg/ml (1 ml), nitroblue tetrazolium 1 mg/ml (5 ml),
and nicotinamide adenine dinucleotide (NAD) 10 mg/ml (1 ml).
A purple formazan dye was deposited at the sites of LDH activity on the polyacrylamide gel.

The absorbance of each band at 500 mμ was determined by
spectrophotometric analysis using a linear gel transport
device. The areas under the peaks obtained were then calculated and enzyme activity estimated according to the amount
of the enzyme initially placed on the polyacrylamide gel.

RESULTS

The LDH isozyme compositions of the livers of the twenty-
two species of mammals are summarized in Table 1. The livers
from both species of primates investigated (humans and Japanese macaques) showed predominantly LDH-5 activity. However,
the LDH-2 and LDH-3 isozymes were more prominent in the Japanese macaque liver than in the human liver. LDH-5 is the most
intensely staining isozyme in the livers of dogs, raccoon
dogs, and the Japanese red fox, belonging to the family Cani-
dae. Although LDH-4 is prominent in the liver of the raccoon
dog and Japanese red fox, it is present in very low amounts

TABLE 1

LDH ISOZYME COMPOSITIONS OF THE LIVERS OF TWENTY-TWO MAMMALS

Mammals	LDH Isozymes (%)				
	LDH-5	LDH-4	LDH-3	LDH-2	LDH-1
1) Primates					
man	48.5	29.4	10.5	7.8	5.0
Japanese Macaque ..	50.3	9.5	22.0	16.0	2.1
2) Carnivora					
dog	46.8	5.0	24.5	14.5	9.0
raccoon dog	37.0	37.0	21.0	3.4	1.6
red fox	49.5	27.6	13.6	5.8	3.9
cat	61.7	29.0	4.0	3.1	2.3
masked palm civet .	45.0	35.5	12.5	3.5	3.5
weasel	50.0	37.6	11.0	1.5	-
Japanese marten ...	76.0	6.2	7.0	7.8	3.0
3) Rodentia					
rat	89.0	11.0	-	-	-
mouse	88.4	12.6	-	-	-
Asian chipmunk	49.0	-	21.5	21.5	8.0
flying squirrel ...	73.5	0.5	0.5	0.5	25.0
guinea pig	38.0	38.0	19.7	3.3	1.0
4) Lagomorpha					
rabbit	12.8	26.8	30.5	14.1	15.8
5) Artiodactyla					
ox	4.0	11.6	25.8	36.9	22.2
sheep	3.1	1.1	15.4	28.5	52.1
goat	6.2	1.5	10.0	16.8	59.2
deer	-	-	29.5	48.7	21.8
pig	1.0	4.6	23.8	39.3	32.3
wild boar	8.8	3.4	28.4	42.6	16.8
6) Perissodactyla					
horse	32.2	26.7	17.0	3.0	1.0

in the liver of the domestic dog. The livers of other carnivorous mammals, such as cats, masked palm civet, and weasel, possess large amounts of LDH-4 and LDH-5, whereas the Japanese marten exhibits predominantly LDH-5 activity. The livers of mice and rats show similar patterns with very high LDH-5 activity and no LDH-1, LDH-2, or LDH-3 activity. The flying squirrel possesses a high level of LDH-1 and LDH-5 whereas LDH-2, and LDH-3, and LDH-4 are present in lower concentrations.

The chipmunk liver possesses equally high activity of

LDH-2 and LDH-3. In the guinea pig livers, the LDH-4 and LDH-5 activities are equally high whereas the LDH-3 activity is slightly lower. Rabbit livers showed an interesting pattern in that the LDH-3 isozyme was present in highest amounts, LDH-4 the next highest, but LDH-1 and LDH-2 showed only moderate activity. In horse livers LDH-5 and LDH-4 are most prevalent whereas LDH-3 is somewhat less abundant.

The livers from deer, goats, sheep, ox, pigs, and wild boar exhibit predominantly LDH-1 and LDH-2 whereas LDH-3 showed only moderate activity. These LDH isozyme patterns are almost completely reversed in comparison with those present in the fifteen other mammals, except rabbits (Table 1).

LDH Isozymes of Human Livers. It was observed that LDH isozymes in Japanese human livers differed as the dietary habits differed. It was further shown that the LDH isozymes in the liver from males who ate meat regularly possessed the highest levels of LDH-5 (Table 2). On the other hand, males who maintained a diet of rice, vegetables, and fruit had livers high in LDH-4 and LDH-5 with lower levels of LDH-1, LDH-2, and LDH-3.

Effect of High Starch Diet on Rat Liver LDH. The effects of the control diet and the high starch diet on the LDH isozyme of the body and organs are shown in Table 3. The weights of the rats in the control group were slightly less than those of the animals in the experimental group. The stomach length was nearly the same for both groups, whereas the lengths of the small intestine, caecum, and colon of the rats

TABLE 2
LDH ISOZYMES OF THE LIVERS OF HUMAN MALES

	LDH-5	LDH-4	LDH-3	LDH-2	LDH-1
Mean Value (14)*	48.5	29.4	10.5	7.8	5.2
	±4.6	±3.3	±0.5	±0.6	±0.2
Meat Diets (3)**	62.5	27.0	5.5	3.5	1.5
	±3.4	±1.3	±0.3	±0.3	±0.1
Vegetable	43.3	33.6	14.0	11.1	13.0
Diets (2)***	±3.1	±3.2	±0.6	±0.15	±0.3

* Number of males.
** Those who ate meat as a regular part of their diet.
***Those who did not eat meat due to religious reasons.

TABLE 3
ANATOMICAL MEASUREMENTS OF THE RATS FED ON THE
STANDARD DIET AND THE HIGH STARCH DIET

Measurements	Standard Diet		High Starch Diet	
	Male (20)*	Female (20)	Male (20)	Female (20)
Body Weight (g)	338 ±2.7	245 ±1.9	260 ±2.8	211 ±1.5
Body Length (cm)	22.2 ±0.1	21.0 ±0.05	20.7 ±0.04	19.4 ±0.05
Tail Length (cm)	19.3 ±0.02	18.3 ±0.09	15.7 ±0.25	17.2 ±0.02
Length of Gastrointestinal Tracts (cm)				
Stomach	6.0 ±0.05	4.7 ±0.07	5.7 ±0.18	5.2 ±0.15
Small Intestine	128 ±3.3	121 ±0.4	164 ±0.3	148 ±1.2
Caecum	9.0 ±0.14	7.8 ±0.14	15.0 ±0.4	15.0 ±0.93
Large Intestine	18.2 ±0.45	16.2 ±0.13	21.3 ±0.8	23.0 ±0.4
Total Length	161.2 ±3.4	149.7 ±0.6	209.7 ±1.0	196.2 ±1.3
Weight of the Organs				
Liver	13.0 ±0.08	8.7 ±0.04	8.3 ±0.13	7.9 ±0.34
Heart	1.08 ±0.02	0.78 ±0.1	0.86 ±0.03	0.64 ±0.08
Pancreas	0.70 ±0.02	0.50 ±0.02	0.80 ±0.04	0.60 ±0.05
Kidney	1.10 ±0.1	0.80 ±0.03	0.70 ±0.1	0.60 ±0.03

*Number of rats

in the experimental group were, on the average, longer (approximately 26 cm, 6.6 cm, and 5 cm, respectively) than those of the control group.

All rats in the experimental group had longer intestines (some by as much as 47 cm) than did the rats in the control group. The weight of the livers, hearts, and kidneys of the rats in the experimental group were slightly less than those of animals in the control group.

The histological findings were as expected. PAS positive granules in the intrahepatic cells of the experimental rats were observed to have increased more than those of the control group.

The livers of animals in the control group possessed primarily LDH-5 whereas LDH-4 was present at the level of 11%, and LDH-3, LDH-2, and LDH-1 were present in only trace amounts. The experimental group was arbitrarily divided into four types according to the relative activities of LDH-1, LDH-2, LDH-3, LDH-4, and LDH-5 compared to those of the control group (Table 4).

Type   I (seven rats)   The LDH-1, LDH-2, and LDH-3 activities were significantly increased but were still present at relatively low levels (Figure 1).

Type II (thirteen rats)   A marked increase was observed in LDH-2 and LDH-3.

Type III (sixteen rats)   The LDH-3 isozyme increased distinctly more than in the control group.

Type IV (four rats)   The LDH isozyme patterns are the same as those of the control group.

DISCUSSION

All mammals have five principle tetrameric LDH isozymes composed of A and B subunits. The two types of subunits are under the control of separate genetic loci (Shaw and Barto, 1963; Appella and Markert, 1961; Markert, 1963; and Markert and Møller, 1959). Many investigators (Bailey and Wilson, 1969; Cahn et al., 1962; Fine et al., 1963; Kaplan and Cahn, 1962; Kaplan and Goodfriend, 1964; Markert and Møller, 1959; Vesell and Philip, 1963; and Wroblewski and Gregory, 1961) who have compared the LDH isozymes of organs and tissues of various species have demonstrated a marked species variation in the number, distribution, and electrophoretic mobilities of these isozymes. Humans and mice, for example, possess primarily LDH-5 in their livers, which is in complete contrast to

TABLE 4

LDH ISOZYMES OF THE LIVERS OF RATS FED ON A
STANDARD DIET AND A HIGH STARCH DIET

Diet	LDH-5	LDH-4	LDH-3	LDH-2	LDH-1
Standard Diet					
Male (20)*	86.85** $\pm$2.01	12.95 $\pm$1.11	0.2	–	–
Female (20)*	87.00 $\pm$2.20	12.90 $\pm$1.50	0.1	–	–
High Starch Diet					
Type 1 (7)	86.60 $\pm$2.20	7.4 $\pm$1.01	1.2 $\pm$0.02	1.8 $\pm$0.01	3.0 $\pm$0.03
Type 2 (13)	85.20 $\pm$2.30	10.3 $\pm$1.20	3.9 $\pm$0.07	3.6 $\pm$0.10	–
Type 3 (16)	87.90 $\pm$2.50	10.4 $\pm$1.04	1.7 $\pm$0.14	–	–
Type 4 (4)	87.50 $\pm$2.10	12.5 $\pm$1.20	–	–	–

* Number of rats
** Percentage of LDH isozymes $\pm$ S.D.

that of the livers in ruminants, which contain predominantely
LDH-1 activity, as has already been reported (Kaplan and Goodfriend, 1964; and Masters and Hinks, 1966).

Fatty acids have been observed in the liver of ruminant
animals which contain predominantly LDH-1 activity (Kaplan
and Goodfriend, 1964). These fatty acids are converted to
glucose in the liver. The possibility exists that oxidative
metabolism in the ruminant livers is greater than that of
other mammals (i.e., horse, rats, mice) and this may be related to the conversion of fatty acids to glucose (Cahn et al.,
1962; Fine et al., 1963; Kaplan and Cahn, 1962; and Kaplan and
Goodfriend, 1964). Although the pigs, rabbits and wild boars
do not belong to the ruminants their livers contain predominantly LDH-1 activity.

The comparison of LDH isozyme patterns in the livers from
twenty-two mammals revealed large differences between species.
Fifteen mammals, including some carnivores, possessed high

(+)

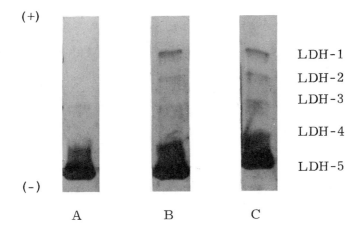

LDH-1
LDH-2
LDH-3
LDH-4
LDH-5

(-)

A   B   C

A. Control Rat Liver
B. Experimental Rat Liver No. 1
C. Experimental Rat Liver No. 2

Fig. 1.   LDH Isozyme Compositions of Control and Experimental
Rat Liver.   Photograph of zymograms showing LDH-1, LDH-2, and
LDH-3 activities which appeared in livers of rats No. 1 and
No. 2 fed on a high starch diet.   Note that the LDH-1, LDH-2,
and LDH-3 isozymes of the control rat liver are greatly dim-
inished.

LDH-5 activity, whereas, seven other mammalian species, i.e.,
oxen, goats, sheep, deer, pigs, rabbits, and wild boars, had
predominantly LDH-1 activity.
    The percentages of meat and vegetables in the diets of the
mammals studied is shown in Table 5.   A high correlation was
found between the LDH isozyme patterns of the liver and the
dietary habits of the mammalian species examined.   In the
order Rodentia, LDH isozymes of the livers of the chipmunk and
guinea pig showed high B subunits levels in comparison with
the isozymes of mouse and rat livers.   It should be noted that
the staple food of the chipmunk and guinea pig contained
almost all vegetable material and that an adequate amount of
fish meat was given to the mice and rats.   It is questionable
whether or not the differences in liver LDH isozymic patterns
mentioned above were influenced by their dietary habits over
a long period of time.   However, the different patterns of
LDH isozymes in different mammalian livers may well account
for the metabolic role of each liver.

TABLE 5

PERCENTAGES OF STAPLE FOOD IN THE DIET OF TWENTY-TWO MAMMALS

	Carnivorous (flesh and eggs)		Harbivorous (plants or vegetables)	
	percentage	kind	percentage	kind
man	40-60	fish, meat, flesh	40-60	rice, potato, fruit, berry
Japanese macaque	10-20	small, animals, insects	80-90	fruit, corn, nut, seed
dog	50-70	bird, meat	30-40	rice, corn, potato
raccoon dog	60-80	fish, bird, reptile, small, animals, crab	20-30	fruit, berry, nut
red fox	60-80	bird, egg, insect, rodent	20-30	fruit, berry, potato
cat	50-60	fish, bird, egg, meat, small animals	30-60	rice, potato, bread
masked palm civet	50-60	fish, bird, reptile, rodent	30-40	rice, corn, potato, radish
weasel	60-80	fish, bird, egg, reptile	20-40	fruit, berry
Japanese marten	60-90	fish, bird, egg, reptile	10-40	berry, nut
rat	35	fish	65	corn
mouse	35	fish	65	corn
flying squirrel	20-30	bird, insects	70-80	berry, nut, leaf, bud sprout
Asian chipmunk	10-15	egg	80-90	fruit, berry, nut, seed
guinea pig	5-8	fish	92-95	corn, potato, radish, leaf, grass

TABLE 5 (cont.)

rabbit	5-8	fish	92-95	corn, potato, radish, grass, weed
horse	0		100	corn, potato, grass, weed
ox	0		100	corn, potato, radish, leaf, grass, weed
sheep	0		100	"
goat	0		100	"
deer	0		100	"
pig	20-30	fish, meat	70-80	corn, potato, fruit
wild boar	20-30	reptile	70-80	potato, corn, grass, root, nut

The small differences found among the LDH isozyme patterns of human livers appeared to be correlated with differences in dietary habit (Table 2). This observation was consistent with the result from the study of the liver LDH isozymes in rats maintained on a high starch diet. It was found that the activities of LDH-1, LDH-2, and LDH-3 increased in the livers of the rats on this high starch diet. The mechanism by which the increased LDH-1, LDH-2, and LDH-3 activities occur in the experimental rat livers can not be easily explained. However, it has now been well established that in animal tissues, levels of enzyme activity can be increased by the direct administration of hormones or by the alteration of the nutritional state of the animal.

Several examples of enzymatic adaptation in rat livers are already known to us (Cahn et al., 1962; Schimke 1964; Schimke et al., 1964; Sharma et al., 1963; Weber et al., 1966; and Weber et al., 1965). For example Schimke (1964) and Schimke et al. (1964) reported the role and importance of synthesis and degradation in the control of tryptophan pyrrolase and arginase levels in rat livers. Sharma et al. (1963) reported the effects of diet and insulin on glucose-adenosine triphosphate phosphotransferase of rat livers. These investigations have demonstrated the physiological function of induced and repressed enzymes in mammalian tissue.

The increased LDH-1, LDH-2, and LDH-3 isozyme activities in the experimental rat livers may be induced by adaptation to a high starch diet. The morphological changes in the lengths of the intestinal tracts or weights of the livers in the experimental rats, as shown in Table 3, are distinctly related to the effects of the high starch diet.

Increased PAS positive granules in the intrahepatic cells may be promoted by the glycogen metabolism in the livers of the experimental rats. These morphological and histological changes may be related to the increase of LDH-1, LDH-2, and LDH-3 activities in the livers of experimental rats that were maintained on high starch diets.

ACKNOWLEDGEMENT

I wish to thank Professor C. L. Markert for his guidance in this investigation. The original research reported here is in collaboration with Dr. Hiroshi Kobayashi of my research group.

I also wish to thank Dr. Tetsuro Kikushima and Dr. Toji Kikushima for supplying me with many mammalian livers, and Professor Ichiro Ohira for his most helpful advice.

REFERENCES

Appella, E. and C. L. Markert 1961. Dissociation of lactate dehydrogenase into subunits with guanidine hydrochloride. *Biochem. Biophys. Res. Commun.* 6: 171-176.

Bailey, G. S. and A. C. Wilson 1968. Homologies between isozymes of fishes and those of higher vertebrates. *J. Biol. Chem.* 243: 5843-5853.

Cahn, R. D., N. O. Kaplan, L. Levine, and E. Zwilling 1962. Nature and development of lactic dehydrogenases. *Science* 136: 962-969.

Fine, I. H., N. O. Kaplan, and D. Kuftinec 1963. Developmental changes of mammalian lactic dehydrogenases. *Biochemistry* 2: 116-121.

Kaplan, N. O. and R. D. Cahn 1962. Lactic dehydrogenase and muscular dystrophy in the chicken. *Biochemistry* 48: 2123-2130.

Kaplan, N. O., and T. L. Goodfriend 1964. Role of the two types of lactic dehydrogenases. *Advance Enzyme Reg.* 2: 203-213.

Markert, C. L. 1968. Molecular basis for isozymes. *Ann. N.Y. Acad. Sci.* 151: 14-40.

Markert, C. L. 1963. Lactate dehydrogenase isozymes: Dissociation and recombination of subunits. *Science* 140: 1329-1330.

Markert, C. L. and F. Møller 1959. Multiple forms of enzyme; tissue, ontogenic, and species specific patterns. *Proc. Nat. Acad. Sci.* 45: 753-763.

Masters, C. J. and M. Hinks 1966. The regulation of lactate dehydrogenase biosyntheis during morphogenesis. *Biochim. Biophys. Acta* 113: 611-613.

Schimke, R. T. 1964. The importance of both synthesis and degradation in the control of arginase level in rat liver. *J. Biol. Chem.* 239: 3808-3817.

Schimke, R. T., E. W. Sweeney, and C. M. Berlin 1964. The roles of synthesis and degradation in the control of rat liver tryptophan pyrolase. *J. Biol. Chem.* 240: 322-331.

Sharma, C., R. Manjeshwar, and S. Weinhouse 1963. Effects of diet and insulin on glucose-adenosine triphosphate phosphotransferase of rat liver. *J. Biol. Chem.* 238: 3840-3845.

Shaw, C. R. and E. Barto 1963. Genetic evidence for the subunit structure of lactate dehydrogenase isozymes. *Proc. Nat. Acad. Sci.* 50: 211-214.

Vesell, E. S. and J. Philip 1963. Isozymes of lactic dehydrogenase: sequential alterations during development. *Ann. N.Y. Acad. Sci.* 151: 14-40.

Weber, G., R. L. Singhal, N. B. Stamm, M. A. Lea, and E. A. Fisher 1966. Synchronous behavior pattern of key glycolytic enzyme: glucokinase, phosphofructokinase, and pyruvate kinase. *Advance in Enzyme Regulation* 4: 59-81.

Weber, G., R. L. Singhal, N. B. Stamm,and S. K. Srivastava 1965. Hormonal induction and supression of liver enzyme biosythesis. *Fed. Proc.* 24: 745-754.

Wroblewski, F. and K. F. Gregory 1961. Lactic dehydrogenase isozymes and their distribution in normal tissue and plasma in disease state. *Ann. N.Y. Acad. Sci.* 94: 912-931.

THE PHYSIOLOGICAL SIGNIFICANCE OF LDH-X

CHRISTOPHER O. HAWTREY, ASHA NAIDU,
CHRISTOPHER NELSON, and DANNY WOLFE
Department of Biological Sciences
California State University, Hayward

ABSTRACT. Purified LDH-1, LDH-5, and LDH-X from mice were analysed for their response to substrates and coenzymes. The affinities of the isozymes (Km) for lactate, pyruvate, and NAD were highest for LDH-X, intermediate for LDH-1, and lowest for LDH-5. NADH differed in that LDH-1 had greater affinity than did LDH-X. A model is proposed to explain this observation. The environment of the fallopian tube contains significant quantities of lactate. Because of its high affinity for lactate the LDH-X could produce energy by converting lactate to pyruvate with concomitant conversion of NAD to NADH. The relatively low affinity of LDH-X for NADH would make the reduced coenzyme available to the electron transport system for ATP production. The continuous operation of the system requires that NADH be available to the mitochondria and that LDH-X have the greatest access to NAD and lactate.

Of all the enzymes that exist as multiple molecular catalysts of chemical reactions, the lactate dehydrogenase (LDH) isozymes are by far the most thoroughly studied in nearly every aspect. They were the first isozymes to be discovered, and their evolutionary, developmental, genetic, biochemical, and physiological significance has been the basis for many investigations. While this statement is true for the LDH's formed in most tissues of the body, the group of special LDH isozymes composed of C subunits have not been as thoroughly studied (McKee et al, 1972, Goldberg, 1972, Battellino et al, 1968). This report will deal with the biochemical response and physiological significance of one of the special isozymes, the sperm specific LDH (or LDH-X) that is produced in mouse testes.

Of course, the ultimate physiological question and probably one of the most important questions that can be asked about isozymes is: "Why does the cell produce so many different kinds of molecules that do essentially the same thing for the cell?" This question is not new and for LDH-1 and LDH-5 the answer or answers have been discussed and codified until most new physiology and biochemistry texts present some explanation. The essence of these explanations is that LDH, which "sits" at the junction between the glycolytic pathway and the TCA

cycle, regulates cell metabolism in accord with anaerobic or
aerobic conditions. The enzyme achieves this result by having
different affinities for substrate and consequently producing
different concentrations of lactate and pyruvate under the
conditions existing within the cell. While this explanation
is acceptable, there is still much speculation about whether
the formation of abortive ternary complexes at high substrate
concentrations has physiological influence on cellular activity, (see Vesell and also Everse this conference, also, Vesell
and Pool, 1956, Wuntch et al, 1969).

One need not be a physiologist, a biochemist, or a histol-
ogist to appreciate that sperm cells with their LDH-X isozyme
differ from the heart muscle and skeletal muscle cells that
are typical sources of LDH-1 and LDH-5 respectively. Three
of these differences are of particular importance in our investigation. First, sperm cells have a minimum of cytoplasm
and therefore have little internal environment to regulate.
Second, by the nature of their biological function, sperm
live in one of the most constantly changing of all cellular
environments. Therefore, it appears that sperm LDH does not
regulate the internal cellular milieu by sequestering lactate
nor does it regulate the external environment as it migrates
through the reproductive pathway. Finally, in previous reports
(Hawtrey and Goldberg, 1970) we demonstrated that in most of
its measured biochemical parameters LDH-X resembles LDH-1 more
than LDH-5. However, availability of O_2 and lactate in the
environment through which the sperm must travel resembles
the intracellular environment of skeletal muscle more closely
than that found in heart muscle (see discussion). From these
observations it is reasonable to conclude that the physiological role of sperm LDH is different from that of either the
heart (LDH-1) or muscle (LDH-5) isozymes. With the sperm
specific LDH we must again ask: "Why does this particular
isozyme exist? What does it do for the sperm that is special?
What function does it perform that is not done by the other
isozymes?" To answer these questions we decided to analyze
the association of the enzyme with all of the molecules affected during its catalytic activity.

MATERIALS AND METHODS

Purified LDH isozymes were prepared as previously described
(Hawtrey and Goldberg, 1970) with the following modifications
to improve the purity of the fractions. Chromatography of
LDH-1 was done on a glass bead column (Sigma G-370-100) which
retained all of the other LDH isozymes and the major contaminating proteins such as hemoglobin and myoglobin. The glass

bead chromatography enabled us to obtain LDH-1 isozymes from all three tissue sources, i.e. testes, heart muscle, and skeletal muscle in sufficient quantities for analytical use. Kinetic analyses of these fractions indicated no difference in their response, regardless of source. All of the chromatographed samples were further purified by slab gel electrophoresis using the 6 mm slab Gelman chamber with a single slot. The band containing each isozyme was cut from the gel after comparison with stained strips cut from each side of the gel, homogenized in a Waring blender at 4°C, and placed in 0.05M pH 7.4 phosphate buffer overnight in the refrigerator. The samples were filtered to remove the acrylamide chunks and either used then for analysis or concentrated using Aquacide II (Cal Bio Chem). Substrates and coenzymes were obtained from Sigma Chemical Co. Spectrophotometry and electrophoresis were performed as previously described (Hawtrey and Goldberg, 1970; Goldberg, 1965).

RESULTS

Studies of initial reaction velocities for the purified LDH isozymes using pyruvate as substrate are presented as Lineweaver Burk plots in Fig. 1. Although each isozyme was tested at several concentrations of coenzyme, the points on the graph were calculated using saturating levels of coenzyme. At lower levels of coenzyme there were no substantial differences in the Km values for any of the isozymes. These data indicate that LDH-5 has the least affinity for pyruvate, LDH-1 is intermediate, and LDH-X has the greatest affinity for pyruvate. Similar experiments involving lactate and nicotinamide adenine dinucleotide (NAD) reactions with the enzyme are presented in Figs. 2 and 3 respectively. The enzyme interaction with these molecules is essentially the same as with pyruvate. The response of the isozymes to reduced coenzyme (NADH), however, is significantly different from the other substrates (Fig. 4). In this case LDH-1 shows the greatest affinity for the NADH substrate while LDH-X exhibits intermediate values. In this case, we have also presented data on the response to NADH at pyruvate levels that did not saturate the enzyme. Low substrate levels, which might be more physiologically realistic, did not affect the Km values for either purified isozyme. Fig. 5 is a tabulation of the data from all the Lineweaver Burk plots. In each case the isozymes display internal consistency of reaction, that is, their affinity for lactate is least, follwed by NAD, pyruvate, and NADH. This is consistent with the compulsory sequence of enzyme to coenzyme to substrate binding mechanism proposed by many authors (reviewed by

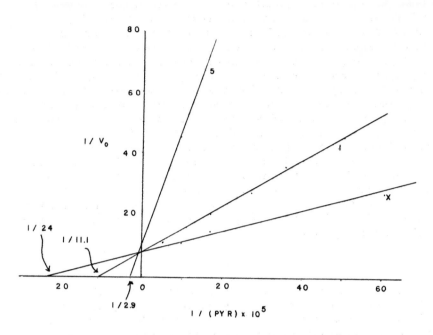

Fig. 1. Lineweaver Burk plots for the response of LDH-1, LDH-5, and LDH-X to pyruvate. NADH concentration for each isozyme is greater than saturating, i.e. 0.01 mM or more.

Vestling and Kunsch, 1968). There is also the consistency previously noted between isozymes in that for all substrates LDH-5 has the least affinity for the molecules involved in its catalytic activity, followed by LDH-1 and finally LDH-X; the only exception being the case of the reaction of the isozymes with reduced coenzyme.

DISCUSSION

How then are these biochemical parameters related to the physiological role of LDH-X in the sperm cell? To understand this question we must first examine some basic parameters of sperm physiology.

Although sperm have greatly reduced cytoplasmic components they still have the basic metabolic functions of cells, that is, glycolysis, the Krebs tricarboxylic acid cycle, and electron transport. Bishop (1962) in his review article on sperm motility has made a number of interesting observations. Most significant is his distinction between the metabolism carried

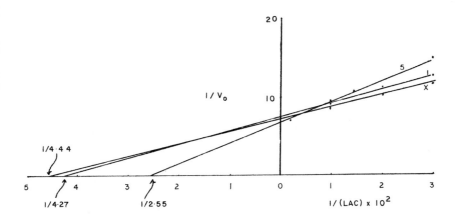

Fig. 2. Lineweaver Burk plots for the response of LDH-1, LDH-5, and LDH-X to lactate. NAD concentrations for each isozyme is greater than saturating, i.e. 3.0 mM or more.

on by avian and mammalian sperm, as opposed to all other sperm types. This correlates well with the occurence of LDH-X only in the testes of birds and mammals. This is probably a reflection of the internal fertilization which occurs in these animals. Bishop emphasizes the capability of sperm to metabolize aerobically and he also points out that they can function well in anaerobic environments. Sperm (of bulls, at least) accumulate lactate in vitro even in aerobic conditions. The oxygen concentration of the uterus and fallopian tubes is 40 mm Hg (Bishop, 1956) or roughly that of the blood in the venous system. Moreover, sperm oxidize glucose completely to CO_2 and H_2O, and the flagella-midpiece can accomplish this metabolic activity without the nucleus or acrosome. Chemical analysis of the flagella-midpiece indicates that it contains no polysaccharide or other stored substrate (Mann, 1954). When the sperm is mixed with other components of the seminal fluid they receive a rich supply of fructose (Mann, 1954) which is then metabolized. As the sperm travels farther in the reproductive tract, fructose, the energy source is steadily diminished until in the environment of the fallopian tube the only substantial substrate remaining is lactate (Hammer and Williams, 1965). Rothschild, 1959, had pointed out that glycolysis alone may not produce sufficient energy in the form of ATP to support sperm function. The cells of the uterus are rich in LDH-5 and this would be consistent with an abun-

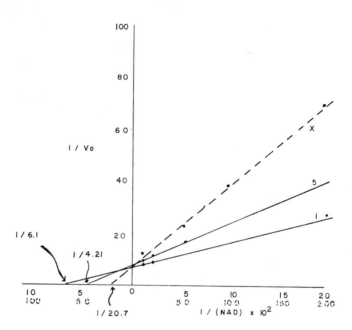

Fig. 3. Lineweaver Burk plots of the response of LDH-1, LDH-5, and LDH-X to NAD. Lactate concentration for each isozyme is greater than saturating, i.e. 8.3 mM or more.

dance of lactate in the environment (Markert and Ursprung, 1962).

All of these observations suggest that sperm cells are capable of considerable metabolic function. When glucose is available, normal glycolysis provides energy for the cell. As the reproductive journey of the sperm nears completion, the cell enters an environment that is poor in oxygen and six carbon substrates. Even if the substrates were available there is some question as to the metabolic pathway which could be utilized. Glycolysis alone, because of low oxygen availability, might not provide enough energy.

We would like to propose the following model as an explanation of the unusual biochemical properties of LDH-X. The proposal includes an energy producing pathway that is 1) short and simple and 2) provides 50% more energy to the sperm than does glycolysis (Fig. 6).

The cells in the upper part of the reproductive tract are

176

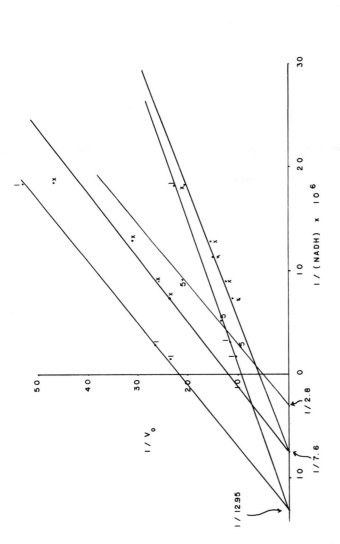

Fig. 4. Lineweaver Burk plots for the response of LDH-1, LDH-5, and LDH-X to NADH. Pyruvate concentrations for the lower line for each isozyme was greater than saturating, i.e. 0.3 mM or more. The pyruvate concentration for the upper line for LDH-1 was 0.05 mM and the upper line for LDH-X was 0.01 mM.

Km

	LDH-I	LDH-5	LDH-X
NAD	1.64×10^{-4}	2.37×10^{-4}	4.83×10^{-5}
NADH	7.76×10^{-6}	3.57×10^{-5}	1.34×10^{-5}
LAC	2.34×10^{-3}	3.92×10^{-3}	2.25×10^{-3}
PYR	9×10^{-5}	3.45×10^{-4}	4.17×10^{-5}

Fig. 5. Table of calculated Michaelis constants for the several isozymes and their substrates as determined from the Lineweaver Burk plots.

rich in LDH-5 and have high concentrations of lactate which is in part transported to the fluids of the fallopian tubes (Hammer and Williams, 1965). Because the LDH-X of sperm has a higher affinity for lactate than do the other isozymes, it readily converts part of the lactate into pyruvate with a concomitant conversion of NAD into NADH. It is important to this model that the sperm enzyme have a greater affinity for its substrates than is true for the competing isozymes so that the excess lactate will force the reaction in the direction of pyruvate formation. The only problem is NADH. The reduced coenzyme must not be so tightly bound to the enzyme that it becomes unavailable for oxidation by mitochondria. If the LDH-X did not have this low affinity for coenzyme then the reaction would quickly reach a new equilibrium value in the vicinity of the sperm and further lactate oxidation would stop. The model is consistent with all of the data. The available substrate is utilized. The affinity constants of the isozymes indicate that LDH-X would be preferentially saturated with coenzyme and substrate; the mitochondria could function as usual. The primary virtue of the system is its simplicity. In the model only one enzyme other than those in the electron transport system is required to increase the energy supply to the cell by 50%. Only a small amount of O_2

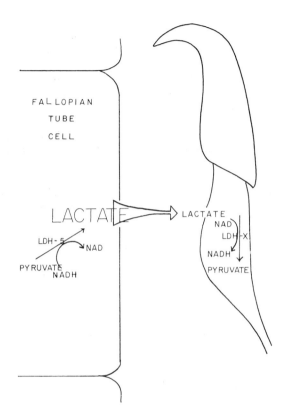

Fig. 6. Model system for the physiological role of LDH-X in
the sperm. Lactate produced in the cells of the fallopian
tubes is released into the lumen of the tube. This lactate
is converted to pyruvate in the sperm and the NADH produced
is used in the mitochondria to produce ATP.

---

is available in the environment of the cell so it seems reasonable that the electron transport system is limited by the
available oxygen.

Thus, LDH-X seems to serve the sperm in a unique physiological manner. Rather than being an aerobic enzyme in an anaerobic environment it has a high affinity for the one metabolic lactate that is available for energy production. The
evolutionary advantage of this system would be very high
since sperm with this isozyme whould have an increased ability
to fertilize an egg.

REFERENCES

Battellino, L. J., F. R. Jaime, and A. Blanco 1969. Kinetic properties of rabbit testicular lactate dehydrogenase isozymes. *J. Biol. Chem.* 243: 5185-5192.

Bishop, D. W. 1956. Oxygen concentrations in the rabbit genital tract. In *Proceedings of the Third International Congress on Animal Reproduction*. London: Brown, Knight and Truscott, Ltd. pp. 53-55.

Bishop, D. W. 1962. Sperm motility. *Phys. Rev.* 42: 1-59.

Goldberg, E. 1965. Lactate dehydrogenase in spermatozoa: Subunit interactions *in vitro*. *Arch. Biochem. Biophys.* 109: 134-141.

Goldberg, E. 1972. Amino acid composition and properties of crystalline lactate dehydrogenase X from mouse testes. *J. Biol. Chem.* 247: 2044-2048.

Hammer, C. E. and W. L. Williams 1965. Composition of rabbit oviduct secretions. *Fert. Ster.* 16: 170-176.

Hawtrey, C. O. and E. Goldberg 1970. Some kinetic aspects of sperm specific lactate dehydrogenase in mice. *J. Exp. Zool.* 174: 451-462.

Mann, T. 1954. *The biochemistry of semen*. London: Methuen and Co. Ltd. 240 pp.

Markert, C. L. and H. Ursprung 1962. The ontogeny of isozyme patterns of lactate dehydrogenase in the mouse. *Dev. Biol.* 5: 363-381.

McKee, R. W., E. Lonstaff, and A. L. Latner 1972. Lactate dehydrogenase in human testes. *Clin. Chim. Acta* 39: 221-227.

Rothschild, Lord 1959. Anaerobic heat production of bull spermatozoa. II. The effects of changes in the colligative and other properties of the suspending medium. *Proc. Roy. Soc.* (London), B 151: 1-22.

Vesell, E. S. and P. E. Pool 1966. Lactate and pyruvate concentrations in exercised ishemic canine muscle: Relationship of tissue substrate level to lactate dehydrogenase isozyme pattern. *Proc. Natl. Acad. Sci. USA* 55: 756-762.

Vestling, C. S. and U. Kunsch 1968. Spectrophotometric analysis of rat liver lactate dehydrogenase-coenzyme and coenzyme analog complexes. *Arch. Biochem. Biophys.* 127: 568-575.

Wuntch, T., E. S. Vesell and R. F. Chen 1969. Studies on rates of abortive ternary complex formation of lactate dehydrogenase isozymes. *J. Biol. Chem.* 244: 6100-6104.

STUDIES ON THE ACTIVITY OF SELECTED ENZYMES
AND ISOZYMES IN EXPERIMENTAL HEART INFARCT IN DOGS

PETER GEORGIEV
Department of Pathological Physiology
Veterinary Institute for Infectious and Parasitic Diseases
Sofia, Bulgaria

ABSTRACT. Myocardial infarction was experimentally induced in dogs and subsequent changes of enzyme activity in
the blood serum measured. Isozymic patterns of lactate
dehydrogenase and malate dehydrogenase were determined;
significant changes in pattern occurred soon after the
infarction, and during recovery the patterns gradually
returned to normal. Three additional enzymes (GOT, GPT,
and CPK) also increased in the blood serum following infarction. No changes in the zymogram showing non-specific
esterases nor in the total amount of esterase activity
was detected as a consequence of infarction. Experimental
infarcts were produced by application of algal tents of
Laminaria digitata to branches of the coronary artery.
Infarcts were confirmed by cytological examination and
the enzymatic changes in the blood serum correlated closely with the size of the infarct. All of these enzyme and
isozyme measurements might prove to have practical value
in medical diagnosis and in the treatment of coronary
artery disease.

INTRODUCTION

In 1954, La Due et al. first reported characteristic changes
in the activity of serum glutamic oxalacetic transaminase
(GOT) after myocardial infarction. Since that time, many
investigators have measured the titers of enzymes and of
various isozymes in blood serum to determine the utility of
such measurements as diagnostic indicators of myocardial
infarction and of other diseases as well (Korovkin, 1965;
Latner, 1968; Rodina, 1971; Wilkinson, 1962, 1965; Wieme, 1963;
Wroblewski, 1961, 1963; Wroblewski et al., 1956, 1960).
With the development of the concept of isozymes (Markert
and Møller, 1959) and after the resolution of lactate dehydrogenase (LDH) into its constituent isozymes and the identification of species specific patterns of LDH during development
and in each adult tissue (Markert and Ursprung, 1962; Markert,
1963), the LDH isozyme system became a focus of attention as
a diagnostic tool in myocardial infarction (Vesell, 1975).
LDH-1 (B_4) and LDH-2 (A_1B_3) are the predominant isozymes in

the cardiac muscle of most mammals, and LDH-5 (A_4) is the predominant isozyme in most skeletal muscles and in the liver of many mammals. The isozyme patterns in the serum, whether of LDH or of other enzymes, are generally believed to be the result of contributions from many cells of different tissues. Thus if any particular cell type suffered injury that brought about the release of its isozymes into the blood serum, then the isozyme pattern of the serum should change so as to reflect the tissue source of the isozymes. Thus, increases in LDH-1 and LDH-2 might well indicate damage to cardiac muscle, and increases in LDH-5 might well indicate damage to skeletal muscle or to the liver.

In recent years, the zymogram technique (Hunter and Markert, 1957) has become widely used since this technique provides a quick and definitive assessment of changes in enzyme and isozyme patterns. The aim of this present investigation was to use the zymogram technique to elucidate the diagnostic significance of changes in the isozyme patterns of LDH in blood serum and also to examine isozymic forms of several other enzymes, including malate dehydrogenase (MDH), non-specific esterases, glutamic oxalacetic transaminase (GOT), glutamic pyruvic transaminase (GPT), and creatine phospho-kinase (CPK). Serum isozyme patterns were examined in normal control dogs and in dogs after induced myocardial infarction.

MATERIALS AND METHODS

The experiments were carried out on 20 clinically healthy dogs with body weights between 9 and 12 kg. Experimental myocardial infarction was induced in 18 of the dogs; two dogs were subjected only to thoracotomy and pericardiotomy in order to record the effects of operative trauma on the activity of enzymes and patterns of isozymes in blood serum. Access to the heart muscle was obtained by means of a left thoracotomy through the fourth to fifth intercostal space after a preliminary intra-tracheal ether-oxygen intubation. Branches of the coronary artery were occluded by the method of Maros et al. (1969) by using the alga *Laminaria digitata*. This technique resulted in a high survival rate among the dogs after the infarction was induced.

Blood samples were taken from the dogs prior to the operation, as a control, 6 hours after the operation, and every day thereafter for ten days. Measurements of enzymes in these samples made it possible to monitor changes in enzyme patterns before infarction, after infarction, and during the recovery period. During the same time periods, electro-cardiograms (ECG) and vectorcardiograms (VCG) were also

182

recorded. Out of the 18 operated dogs, three animals died
by the third day, but the remaining dogs survived from 15
to 20 days after the infarction. Then they were killed by
decapitation for a detailed pathoanatomic examination. The
isozyme patterns of LDH, MDH, and non-specific esterase were
determined by the zymogram technique using the vertical starch
gel electrophoresis method of Smithies (1959). The hydrolyzed
starch was produced by the Electrostarch Company, Madison,
Wisconsin, USA. Electrophoretic resolution was obtained in
16-18 hr at a current strength of 10-15 mA. Isozyme patterns
were then visualized on the gels by standard staining techniques (Shaw and Prasad, 1970). The activity of LDH, GOT,
GPT, and CPK was measured by spectrophotometry using ready
enzyme tests obtained from Boehringer. A VSU-2 spectrophotometer (Zeiss) was used for the spectrophotometric determinations
at a wave length of 366 nm. All enzyme measurements are
expressed in International Units of activity.

RESULTS AND DISCUSSION

Enzyme activity. The activity in serum of LDH, GOT, GPT, and
CPK up to the tenth day after the experimentally induced heart
infarct in the dogs is given in Table 1. The table shows
that a considerable increase in the activity of all of these
enzymes occurs by the sixth hour after infarction. Between
the fourth and fifth day, the activity of GOT, GPT, and CPK
returned to the initial control level. The activity of LDH
showed a clear pattern of change with somewhat greater individual variability among the dogs. In particular, dog # 10
represented an exception in that a decrease of LDH activity
was observed in the serum. Two dogs, # 4 and # 6, showed
unusually high activities in the normal serum, 95 units of
LDH per ml of serum compared to an average value of 79. These
dogs also exhibited much greater initial activities after
infarction - at six hours showing 1,041 units compared to an
average value for the other dogs of 515 units. It should also
be noted that the dogs with unusually high LDH activity after
infarction also displayed a correspondingly larger infarct as
revealed by pathoanatomical examination. This observation
supports the concept that the high enzyme activity found in
the blood serum is due to the release of enzymes from necrotic
cells of the myocardium (Wilkinson, 1962).

The induction of experimental infarction by operative in-
sertion of *Laminaria digitata* on a branch of the corony artery
not only affects the cells of the myocardium but also other
tissues may be injured and caused to release their enzymes.
Consequently, two dogs were subjected to thoracotomy only.

TABLE I

ACTIVITY OF SERUM LDH, GOT, GPT, AND CPK IN EXPERIMENTALLY-INDUCED CARDIAC INFARCTION

Enzymes	Prior to the in-farction	After Operatively-Induced cardiac infarction				
		6th hour	1st day	2nd day	3rd day	4th day
LDH	78.10 ±13.00	313.30 ±68.00	208.90 ±23.00	178.50 ±26.00	88.50 ±12.00	86.40 ±19.50
GOT	14.20 ± 2.20	64.00 ±10.00	118.90 ±26.00	112.20 ±20.00	71.40 ± 8.00	18.20 ± 4.50
GPT	20.60 ± 3.50	48.20 ± 5.00	17.50 ± 1.50	19.20 ± 2.00	24.50 ± 4.00	13.50 ± 2.50
CPK	1.16 ± 0.50	3.30 ± 0.90	7.60 ± 1.40	5.10 ± 1.20	5.50 ± 1.45	0.65 ± 0.20

Enzymes	5th day	6th day	7th day	8th day	9th day	10th day
LDH	67.40 ±9.00	96.00 ±11.00	121.40 ±23.00	86.00 ±12.00	80.50 ±13.00	61.10 ±9.00
GOT	22.80 ±3.50	18.00 ± 4.50	19.20 ± 3.00	5.00 ± 0.50	4.50 ± 1.00	8.50 ±2.00
GPT	7.70 ±1.00	8.30 ± 2.00	7.20 ± 1.50	13.00 ± 3.00	9.20 ± 0.50	13.50 ±1.50
CPK	0.45 ±0.10	1.85 ± 0.40	1.30 ± 0.40	0.55 ± 0.15	0.28 ± 0.10	0.00 ±0.00

The results from these two dogs revealed some increase in the activity of the enzymes tested, but this increase was insignificant in comparison with the dogs in which a myocardial infarction was induced. The activity of LDH, GOT, GPT, and CPK after infarction increased by as much as 300% by the sixth hour after the infarction, whereas the dogs subjected only to thoracotomy showed increases ranging between 40 and 60%. From these findings, we suggest that the changes in the activity of LDH, GOT, GPT, and CPK are primarily attributable to the experimentally induced cardiac infarction and are only slightly affected by the operative trauma, itself. In addition, the changes in ECG and VCG corresponded quite well to the changes in enzyme and isozyme activity measured in the blood serum (Figures 1 and 2).

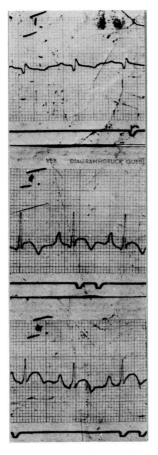 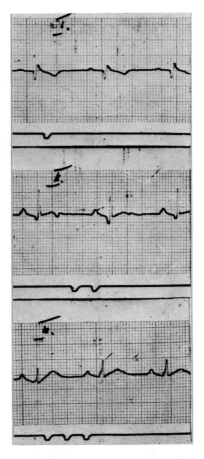

Fig. 1. Electrocardiographic recordings from a dog 6 hours after experimentally induced infarction.

Fig. 2. Electrocardiographic recording from a dog 10 days after experimentally induced infarction.

ISOZYME ACTIVITY

LDH isozymes. Figure 3 is a zymogram showing the relative activity of the different isozymes of LDH. This zymogram reveals that, by the sixth hour after an experimentally-induced infarct, a sharp increase in the activity of LDH-1 and LDH-2 occurred. This increased activity remained until the third day and then gradually declined toward normal levels. Measurements of total LDH activity in the blood serum paralled the changing amounts of LDH-1 and LDH-2 shown on zymo-

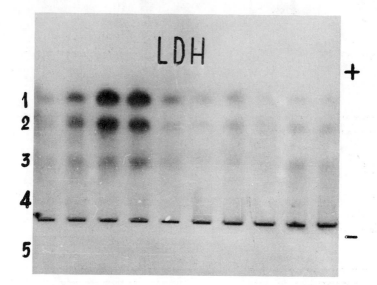

Fig. 3. Zymogram of LDH activity from dogs with experimental-ly-induced myocardial infarction. From left to right - first channel prior to the infarction; remaining channels from the 6th hour to the 8th day after the infarction.

grams since these are the principal isozymes in the serum after an infarct. A close correlation was found at post mortem examination between the size of the infarct and the level of LDH in the serum. The electrocardiographic and the vectorcardiographic studies also revealed the development of a myocardial infarction after the application of the *Laminaria* tent on a branch of the coronary artery (see Figures 1 and 2).

Histopathological investigations revealed that all dogs treated with *Laminaria* tents did, in fact, develop foci of infarction but of different sizes. It seems clear that the levels of LDH in the blood serum parallel closely the amount of injury to the heart muscle and thus measurements of LDH, particularly of the individual isozymes, provides an excel-lent diagnostic tool (Korovkin, 1965; Rodina, 1971; Wilkinson, 1962; and Wust and Thus, 1966) (See figures 4 and 5).
MDH Isozymes. Figure 6 presents a zymogram of the isozymes of malate dehydrogenase. These, too, change considerably in activity in the blood serum after experimentally induced myo-cardial infarction. Prior to the infarction only one isozyme of MDH was evident, and this isozyme showed a very low activity. As the infarction developed, a new isozyme appeared and there

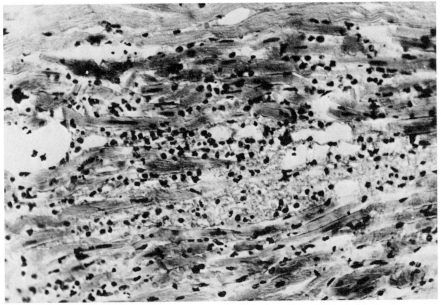

Fig. 4. Photomicrograph of heart from dog dying 18 hours after experimental infarction: parenchymatosis dystrophy of the muscular fibers; dissociation of muscular fibers: interstitial edema.

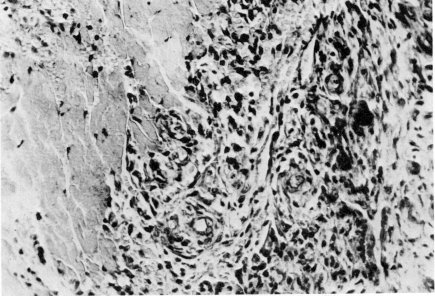

Fig. 5. Photomicrograph of heart from dog killed 12 days after experimentally induced heart infarction: organization of an infarction.

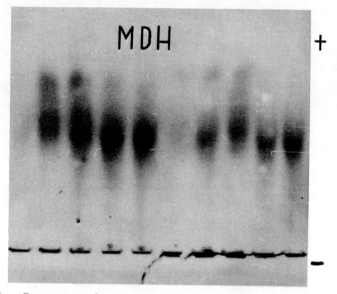

Fig. 6. Zymogram of MDH activity in the two dogs with experimentally-induced myocardial infarction. From left to right - first and sixth channels prior to the infarction; the remaining channels after the infarction.

was also an increase in the initial isozyme. It is known that MDH is mainly a mitochondrial enzyme and thus the appearance of increased amounts of MDH in the blood serum after infarction suggests that the mitochondria of the cardiac muscle cells are releasing enzymes into the blood serum. *Isozymes of non-specific esterase.* Figure 7 presents a zymogram of non-specific esterases. Five different isozymic forms are evident, but none of these undergo any significant change during the development of a myocardial infarct. Our studies reveal that only the isozyme spectrum of LDH and MDH changes with myocardial infarction. But these isozyme patterns can be of considerable importance in the diagnosis of acute infarction in man as well as in dogs and should prove, therefore, to be a valuable source of information for the physician responsible for diagnosing and treating coronary artery disease.

SUMMARY

1. Experimentally induced cardiac infarction in dogs results in an increase in the activity of serum LDH, GOT, GPT,

Fig. 7. Zymogram of non-specific esterase activity in the
dog with experimentally-induced myocardial infarction. From
left to right, first channel prior to the infarction; remaining channels 1st, 2nd, 3rd, and 4th day after the infarction.

and CPK by 200 to 300% by the sixth hour after the infarct.

2. In all dogs with experimentally induced myocardial
infarction a sharp increase in the activity of the isozymes
LDH-1 and LDH-2 occurred and this increase returned to normal after the third day.

3. During the development of a cardiac infarction a second
isozyme of malate dehydrogenase appeared in the serum and the
first isozyme increased in amount.

4. The isozymes of non-specific esterase did not change
either in amount or in kind as a consequence of myocardial
infarction.

All these enzymatic changes in blood serum might be of
diagnostic significance in the treatment of coronary artery
disease in human beings.

ACKNOWLEDGEMENTS

I wish to express my gratitude to Dr. Luben Lozanov for the pathoanatomical examinations, to Dr. Ilija Popiliev for the electrocardiogramic measurements, and to Dr. Todor Buchvarov, Dr. Kiril Tenev, and Dr. Dimiter Golemandv for the help given me in producing the experimental induced infarctions. I want also to express my gratitude to Prof. Metodi Petrichev, head of the Department of Pathological Physiology, who supported this work.

REFERENCES

Atanasov, N. 1968. Po vaprosa za metabolitnoto znachenie na takanite laktatdehidrogenazni izozimi u podopitni jivotni. Disertatzia. Plovdiv.

Hunter, R.L. and C.L. Markert 1957. Histochemical demonstration of enzymes separated by zone electrophoresis in starch gels. *Science* 125: 1294-1295.

Kaplan, N.O. 1964. Lactate dehydrogenase: structure and function. *Brookhaven Symp. Biol.* 17: 131-139.

Korovkin, F. 1965. Fermenti v diagnostike infarkta miokarda. Moskva.

LaDue, S., F. Wroblewski, and A. Karmen 1954. Serum glutamate oxalacetate transaminase activity in human transmural acute myocardial infarction. *Science* 120: 497-499.

Latner, A. 1968. Isoenzymes in Biology and Medicine. London.

Markert, C.L. 1963. Epigenetic control of specific protein synthesis in differentiating cells. In: *Cytodifferentiation and Macromolecular Synthesis*. Academic Press, N.Y. pp. 65-84.

Markert, C.L. and F. Møller 1959. Multiple forms of enzymes: tissue, ontogenetic, and species specific patterns. *Proc. Natl. Acad. Sci. USA* 45: 753-763.

Markert, C.L. and H. Ursprung 1962. The ontogeny of isozyme patterns of lactate dehydrogenase in the mouse. *Develop. Biol.* 5: 363-381.

Maros, T. et al. 1969. Experimentalnii model na serdechni infarkt. *Cor. Vasa.* 11: 93-94.

Rodina, F. 1971. Kompensatornie vozmojnosti serdechni sosudisti sistemi pri ostroi lokalnoi ishemii miokarde v experimente. *Minks.*

Shaw, C.R. and R. Prasad 1970. Starch gel electrophoresis of enzymes - compilation of recipes. *Biochem. Genet.* 4: 297-320.

Smithies, O. 1959. An improved procedure for starch gel electrophoresis: Further variations of the serum protein of normal individuals. *Biochem. J.* 71: 585-587.

Vesell, E.S. 1975. Medical uses of isozymes. *II. Isozymes: Physiology and Function.* C. L. Markert, editor, Academic Press, New York.

Wieme, R. 1963. Multiple molecular forms of enzymes and their use in clinical diagnosis. *Nature* 199: 437-439.

Wilkinson, H. 1965. *Isoenzymes.* London.

Wilkinson, H. 1962. *An Introduction to Diagnostic Enzymology* London.

Wroblewski, F. 1961. Enzymes in medical diagnosis. *Sci. Amer.* 205: 97-107.

Wroblewski, F. 1963. Serum enzyme isoenzyme alterations in myocardial infarction. *Progress in Cardiovascular Disease* 6: 63-83.

Wroblewski, F., F. Ross, and K. Gregory 1960. Isoenzymes in myocardial infarction. *New Engl. J. Med.* 263: 521-536.

Wroblewski, F., P. Ruegseger, and S. La Due 1956. Serum lactic dehydrogenase activity in acute transmural myocardial infarction. *Science* 123: 1122-1123.

Wust, H. and G. Thuss 1966. Vergeichende untersuhungen von LDH-isoenzym bestimungemethoden. *Artz. Lab.* 12: 97-108.

ISOZYME PATTERNS IN TISSUES OF TEMPERATURE-ACCLIMATED FISH

F. RAY WILSON, MICHAEL J. CHAMPION
GREGORY S. WHITT and C. LADD PROSSER
Department of Physiology and Biophysics
and
Department of Zoology
University of Illinois
Urbana, Illinois 61801

ABSTRACT. Lactate dehydrogenase and malate dehydrogenase isozyme patterns from goldfish (*Carassius auratus* L.) tissues were examined by starch gel electrophoresis. Electrophoretic analyses reveal no detectable qualitative or quantitative alterations in the isozyme patterns due to the effects of temperature acclimation at 5°, 15°, and 25°C. The only variations of LDH isozyme patterns observed were those attributable to genetic polymorphism and to the different isozyme patterns of white and red skeletal muscle fibers. Specific activity of LDH was similar for all acclimation temperatures. Various glycolytic and oxidative isozymes from green sunfish (*Lepomis cyanellus*) tissues were examined by starch gel electrophoresis and spectrophotometric activity analysis. No qualitative alterations were noted in the isozyme patterns due to the effects of temperature acclimation at 5° and 25°C although quantitative alterations were observed. Warm-acclimated sunfish display an increase in glycolytic enzymes (LDH, aldolase, phosphoglucomutase, G-3-PDH) and a decrease in an oxidative enzyme (MDH). Cold-acclimated sunfish display increased oxidative enzyme activity and decreased glycolytic enzyme activity. We conclude that qualitative isozyme alterations of these enzyme systems occur as a direct result of temperature acclimation only in trout, while other groups of fish show either quantitative alterations or no changes in isozyme patterns.

INTRODUCTION

Eurythermal poikilotherms are capable of metabolic alterations which permit them to acclimate or acclimatize over a wide range of temperature, the limits for such acclimations are genetically determined. Mechanisms for individual acclimation have been extensively studied at the molecular level. Concepts and mechanisms are reviewed by Prosser, (1963, 1967a, 1967b, 1969, 1973); Hazel and Prosser, (1974); and Hochachka and Somero, (1973).

Two general approaches have been used in studying the effects of temperature on poikilotherms: (1) resistance adaptations involving tolerance at temperature limits and (2) capacity adaptations involving the measurement of many rate functions at the temperature of acclimation or at intermediate temperatures. In addition to these two approaches, the adaptive properties associated with environmentally induced stress may be considered for three time periods: direct responses, compensatory acclimation for days or weeks, and long-term genetic changes evidenced after generations.

Studies of capacity adaptations reveal that a relatively constant energy liberation is maintained at different acclimation temperatures. Cold acclimated animals generally have increased oxygen consumption (Kanungo and Prosser, 1959), altered glycolytic flux (Hochachka, 1968) and changes in utilization of the pentose shunt (Hochachka and Hayes, 1962, Freed, 1969). These metabolic effects may be explained by changes in a wide variety of enzymes, both oxidative and glycolytic, which show altered activities during acclimation. These changes led Hazel and Prosser (1970) to conclude that, in general, enzymes which show positive compensatory acclimation are involved in metabolic pathways leading to energy liberation and those that show no, or inverse, acclimation patterns are related to the breakdown of metabolic intermediates.

Because many enzymatic changes have been observed during compensatory acclimation, their underlying mechanisms have become the focal point of numerous investigations. Molecular mechanisms have been postulated to explain the changes in enzyme activity observed during temperature acclimation. These include: (1) changes in inorganic ion concentration, (2) changes in lipids and other cofactors (Hazel, 1972), and (3) qualitative and quantitative changes in enzymes (both isozymal and non-isozymal) (Wilson, 1973).

It is the purpose of this report to investigate the extent to which qualitative and quantitative alterations of isozymes (both glycolytic and oxidative) occur in response to altered environmental temperatures in two different fish: the common goldfish (*Carassius auratus* L.) and the green sunfish (*Lepomis cyanellus*).

MATERIALS AND METHODS

Goldfish (*Carassius auratus* L.) were obtained from the Auburndale Goldfish Company of Chicago and green sunfish (*Lepomis cyanellus*) were provided by Dr. William F. Childers of the Illinois Natural History Survey (Urbana, Illinois). Fish were held for four weeks after arrival in a constant temperature room at 15°C in 10 gallon aquaria containing aerated natural well water. They were then transferred to constant-temperature rooms of 5° and 25°C (12 hour photoperiod).

Fish maintained at 5°C were fed on alternate days, those at 15°C once daily, and those at 25° twice daily, a schedule previously found to provide a state of uniform nutrition at each temperature. Goldfish received Conditioner Goldfish food (Wardley), and green sunfish were fed meal worms.

Fish placed at altered environmental temperatures were considered to be acclimated after a period of at least four weeks (Sidell et al, 1973). The preparation of the tissue extracts and the starch gel electrophoretic procedures are described in Wilson et al, (1973).

Staining procedures for electrophoresis gels The staining was accomplished according to procedures listed in Shaw and Prasad, (1970).

Polyacrylamide Gel Electrophoresis Electrophoresis of the enzyme preparations was performed on 7% polyacrylamide gels (pH 8.3) according to Davis, (1964).

Spectrophotometric assessment of LDH activity Activity calculations for LDH were performed according to the method described by the Worthington Biochemical Corporation. All assays were done at a constant temperature of 15°C in a Beckman DB spectrophotometer with a Sargent SRL recorder.

Myoglobin was assayed spectrophotometrically by absorption at 418 nm.

RESULTS

Lactate dehydrogenase and malate dehydrogenase isozyme patterns of goldfish tissues The lactate dehydrogenase (LDH) isozyme patterns from various goldfish tissues revealed considerable tissue specificity (Fig. 1A). Almost all tissues have a group of five anodal bands with the skeletal muscle possessing only these bands. The most anodal band is the LDH B_4; the most cathodal of the five, the LDH A_4. In the goldfish brain, heart, and kidney there also appears to be a group of more cathodally migrating bands and the liver contains a unique, highly cathodal isozyme, the LDH C_4.

All of the goldfish tissues examined (Fig. 1B) contained three anodal supernatant malate dehydrogenase isozymes (S-MDH) and some cathodal bands with slight activity (mitochondrial malate dehydrogenase, M-MDH). Considerable variation was observed in the amount of each isozyme present in the different tissues.

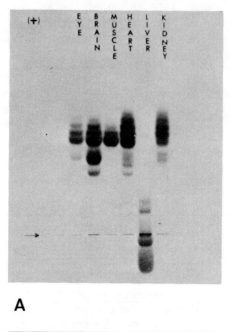

A

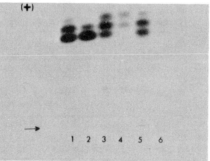

B

Fig. 1. (A) Lactate dehydrogenase isozymes of goldfish tissues. EBT buffer system. (B) NAD-malate dehydrogenase isozymes of goldfish tissue: (1) eye, (2) brain, (3) skeletal muscle (epaxial), (4) heart, (5) liver, (6) kidney. EBT buffer system.

Effects of temperature acclimation on LDH and MDH isozymes in different goldfish tissues Shown in Fig. 2A, B, C are the LDH isozymes from tissues of the brain, skeletal muscle, and cardiac muscle, respectively, of different fish acclimated for a minimum of four weeks at 5°, 15°, and 25°C. There appear to be no gross differences in isozyme activity in the gels on the basis of relative staining intensities for different acclimation temperatures. In each of the figures, considerable variability is observed within acclimation groups but there were no trends observed for isozyme phenotypes to correlate with a specific temperature of acclimation. LDH activity was assayed at three temperatures for muscle samples from 10 fish acclimated to each of three temperatures. When expressed on a protein basis, no significant differences were observed except that the mean for 5°-acclimated fish was less than that for 25°-acclimated fish.

The cytosol malate dehydrogenase was also examined during temperature acclimation. Fig. 2D, E, F show brain, epaxial skeletal muscle, and cardiac muscle, respectively. The MDH isozyme patterns, both in intensity and number of bands, appear to show no changes attributable to the temperature regime. The MDH isozyme phenotypes were not as polymorphic as the LDH isozyme phenotypes.

Goldfish LDH polymorphism In view of the extensive heterogeneity of LDH isozyme patterns between individuals at the same temperature of acclimation, the extent of polymorphism for LDH isozyme patterns from different tissues of goldfish kept at an intermediate temperature of 15°C was examined. Examination of skeletal muscle, heart muscle, and brain of eight different goldfish shown in Fig. 3A, B, C reveals considerable variation in the intensity of staining and in the relative distribution of these particular LDH isozyme bands. The electrophoretic mobilities of the various isozymes did not differ.

Cellular heterogeneity of isozyme synthesis within tissues In addition to phenotypic variability between individuals within a population, another source of variability was observed, namely, the distribution of heterogeneous cell populations within a tissue. Because the red and white muscles are so different in their enzymatic complement we determined the relative contributions of the red and white fibers to the total isozyme pattern (Fig. 3D). The percentage of red muscle contaminating the white muscle was small and similar in each sample. The LDH isozymes from red muscle contain a much higher ratio of LDH B to LDH A subunits than the white skeletal muscle, which contains predominantly A subunits. These results were also obtained for fish acclimated at 5°, 15°, and 25°C.

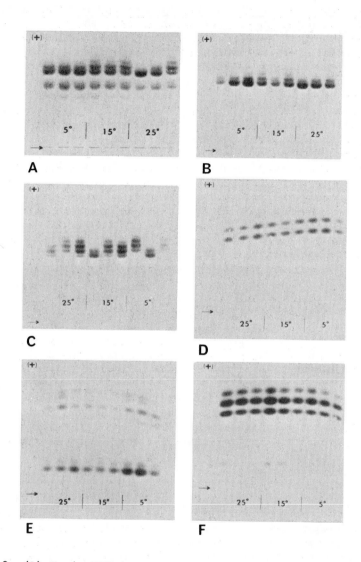

Fig. 2. (A) Brain LDH isozyme patterns in temperature-acclimated goldfish. EBT buffer system. (B) Skeletal muscle (epaxial) LDH isozyme patterns in temperature-acclimated goldfish. EBT buffer system. (C) Cardiac muscle LDH isozyme patterns in temperature-acclimated goldfish. EBT buffer system. (D) Brain NAD-MDH isozyme patterns in temperature-acclimated goldfish. Tris-citrate buffer system. (E) Epaxial skeletal muscle NAD-MDH isozyme patterns in temperature-acclimated

Fig. 2 legend continued
goldfish. Tris-citrate buffer system. (F) Cardiac muscle
NAD-MDH isozyme patterns in temperature-acclimated goldfish.
Tris-citrate buffer system.

Effects of temperature acclimation upon quantitative alterations of glycolytic enzymes in the green sunfish Because we
observed no effects of temperature acclimation upon the gross
quantitative or qualitative changes in the enzyme patterns of
the goldfish we then investigated the isozyme repertory of
tissues of the green sunfish (Lepomis cyanellus) to determine
whether the enzyme patterns were affected by temperature
acclimation. Because the sunfish appear to have a lower level
of genetic polymorphism than the goldfish we felt that the
green sunfish would be a more sensitive indicator of effects
of temperature acclimation. In addition the epaxial muscle
of this species appears to be uniformly composed of white
fibers.

Fig. 4A, B, C are representative zymograms showing the effects of temperature upon the levels of various glycolytic
enzymes (epaxial skeletal muscle phosphoglucomutase, epaxial
skeletal muscle aldolase, and liver glyceraldehyde-3-phosphate
dehydrogenase, respectively) within tissues of acclimated
green sunfish. In these three figures, the slots containing
the most intensely staining bands are from tissues of fish
subjected to the 25°C acclimation temperature. The lowest band
density was observed for fish acclimated at the cold temperature (5°C).

In an attempt to correlate staining intensity with enzyme
activity, the data obtained from polyacrylamide gels of 10%
muscle homogenates from green sunfish acclimated to 5°, 15°,
and 25°C were compared with data obtained from spectrophotometric analysis of the activity of the homogenates. Fig. 5A
shows that a linear correlation exists between area of the gel
stained for LDH and resulting activity. In general, a contin-
uum exists from 5°C to 25°C in that the enzyme from colder
acclimated animals was present in a lower amount and showed
less total activity than the corresponding enzyme from the
warmer acclimated animals.

Effects of acclimation upon an oxidative enzyme In an
attempt to compare the changes in glycolytic enzymes with
those of an oxidative enzyme during temperature acclimation,
malate dehydrogenase was examined from the brain homogenates
of animals subjected to 5°, 15°, and 25°C (Fig. 5B). The slots
on the zymogram with the most intense dye deposition were those
of the cold acclimated (5°C) animals.

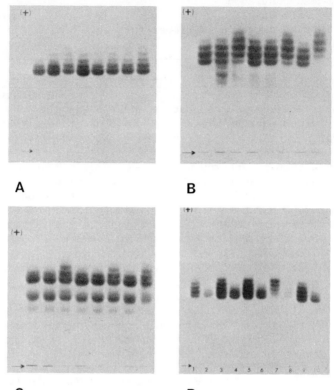

Fig. 3. (A) LDH isozyme phenotypes from skeletal muscle of goldfish maintained at 15°C. EBT buffer system. (B) LDH isozyme phenotypes from cardiac muscle of goldfish kept at 15°C. EBT buffer system. (C) LDH isozyme phenotypes from brain of goldfish kept at 15°C. EBT buffer system. (D) LDH isozyme patterns of red (odd-numbered slots) and white (even-numbered slots) skeletal muscle of goldfish acclimated at 15°C. EBT buffer system. Percentage of myoglobin of red muscle contaminating white muscle preparation calculated by (Mb) red muscle/(Mb) white muscle x 100. Slot No. 2, 3.2%; No. 4, 5.1%; No. 6, 5.5%; No. 8, 3.2%; No. 10, 5.1%.

DISCUSSION

Within the past decade many reports of differences in the distributions of isozymes in temperature-compensated fish have appeared in the literature. In the first study of the effect

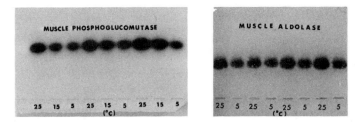

A   B

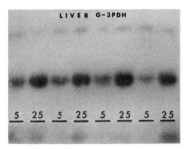

C

Fig. 4. (A) Green sunfish epaxial muscle phosphoglucomutase
from animals acclimated to 5° and 25°C. Tris-citrate buffer
system. (B) Epaxial muscle aldolase from *Lepomis cyanellus*
acclimated at 5° and 25°C. Tris-citrate buffer system. (C)
Green sunfish liver glyceraldehyde - 3-phosphate dehydrogenase
from 5° and 25°C acclimated animals. Tris-citrate buffer
system.

---

of temperature compensation upon LDH isozyme synthesis in
goldfish tissues, Hochachka (1965) reported dramatic differences in levels of the skeletal muscle isozymes from warm-
and cold-adapted goldfish with the cold-acclimated individuals
exhibiting an increase in activity over the warm-acclimated.
Temperature-induced alterations in the isozyme pattern of
tissues of acclimated fish have also been reported for trout
brain acetylcholinesterase (Baldwin and Hochachka, 1970),
trout liver citrate synthases (Hochachka and Lewis, 1970),
and trout liver isocitrate dehydrogenase (Moon and Hochachka,
1972a, 1972b).

In the present examination of the LDH and MDH isozyme patterns of several tissues from goldfish acclimated to 5°, 15°,
and 25°C for a minimum of four weeks, the substantial altera-

201

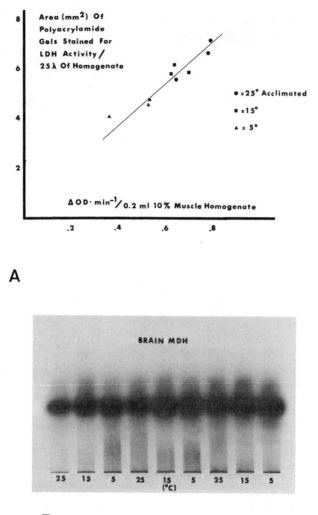

A

B

Fig. 5. (A) Comparison of area obtained from scanning polyacrylamide gels stained for green sunfish LDH (ordinate) with
data obtained from spectrophotometric analysis of kinetic
activity (abscissa). (● = 25°C acclimated, ■ = 15°C acclimated,
▲ = 5°C acclimated). (B) Brain malate dehydrogenase from
green sunfish acclimated to 5° and 25°C. Tris-citrate buffer
system.

tion of isozyme patterns reported earlier for goldfish LDH
was not observed. The LDH and MDH isozyme patterns in the
goldfish were observed to show no significant differences,
either qualitative or quantitative, due to temperature acclimation. However the extensive isozyme polymorphisms in the
population of goldfish may have masked slight changes in the
isozyme patterns due to temperature acclimation. The variation of the LDH isozymes in skeletal muscle of 15°C acclimated
goldfish was not due to differences in the electrophoretic
mobility of the subunits but rather due to differences generated by altered ratios of A to B subunits. Also no changes
in LDH activity were found associated with temperature
acclimation.

In addition to the phenotypic variability of isozymes within a population of goldfish, an additional source of variability was observed - that of distribution of heterogeneous cell
populations within a tissue (i.e., red and white skeletal
muscle). The red muscle contained a much higher ratio of LDH
B to LDH A subunits whereas the white skeletal muscle contained
predominantly A subunits. No effect of temperature acclimation was observed upon the LDH isozyme patterns of either red
or white goldfish epaxial muscle.

Therefore, in *Carassius auratus* the qualitative alterations
reported in LDH isozyme patterns are not due to an effect of
temperature acclimation but are attributable rather to genetic
polymorphism as well as to the contributions of heterogeneous
cell populations in tissues. Cytochrome oxidase, as measured
by antibody inactivation, shows an increase in amount in 5°-
acclimated and a decrease in 25°-acclimated goldfish as compared with 15°-acclimated fish (Wilson, 1973). Failure to
find corresponding changes in MDH may be due to (1) the lower
sensitivity of electrophoretic methods for quantitating the
enzyme or (2) the possibility that oxidation is rate-limited
by the cytochromes rather than by NADH from the TCA cycle.

Since our work indicated no gross quantitative or qualitative alterations in isozyme patterns from tissues of temperature acclimated *Carassius auratus* other than those due to genetic polymorphism, a survey of representative isozymes (both
glycolytic and oxidative) was made on the green sunfish (*Lepomis cyanellus*) which exhibits a lower level of genetic polymorphism.

Data obtained from green sunfish acclimated to 5° and 25°C
indicate that the warm-acclimated green sunfish have enhanced
glycolytic activity over the cold-acclimated fish. This increase in activity was observed for several enzymes (PGM, aldolase, G-3-PDH, and LDH). The increase of one of these enzymes, LDH, was attributable to a quantitative increase in

the enzyme as shown by comparing the intensity of dye deposition on acrylamide gels with spectrophotometric analysis of
the specific activity.

Comparable effects of temperature acclimation upon the isozymes of green sunfish tissues were observed for the oxidative
enzyme, malate dehydrogenase. However, for this enzyme there
was enhanced MDH activity in the cold (5°C) acclimated individuals.

Goldfish apparently show little or no change in glycolysis
according to temperature as measured by LDH but a marked increase in oxidative activity in the cold, whereas green sunfish show a reduction in glycolysis in the cold and an increase at warm temperatures. Green sunfish are adapted to
annual cycles of greater temperature change than are goldfish.
The increase of glycolytic activity observed in warm-acclimated animals and increased oxidative activity in the cold-acclimated animals is perhaps best explained on the basis of oxygen availability (Somero, 1973; Wilson et al., 1974) rather
than upon the direct effects of temperature. There is decreased
oxygen solubility as the temperature of the water increases.
Thus it is adaptive for fishes to increase their glycolytic
activity under warmer conditions and at the same time to decrease the activity of the enzymes associated with the more
aerobic metabolism and to increase oxidative activity in the
cold.

The question still remains whether altered environmental
temperatures induce dramatic alterations in isozyme patterns
of some fish. From the data presented here and by various
other contributors to this conference (G. N. Somero and T.
Moon) it appears that rainbow trout are the only species of
fish which have been found to undergo a qualitative change
in their isozyme patterns. It is clear that future work must
explore such issues as the effect of polyploidy upon isozyme
changes during temperature acclimation.

ACKNOWLEDGMENTS

This research was supported by NSF GB 16425 and GB 43995
to G. S. Whitt and NSF GB 35240X to C. L. Prosser.
The assays of LDH activity were performed by Bruce Sidell.

REFERENCES

Baldwin, J. and P. W. Hochachka 1970. Functional significance
of isozymes in thermal acclimation - acetylcholinesterase
from trout brain. *Biochem. J.* 116: 883-887.
Davis, E. F. C. 1964. Simplified gel electrophoresis. *Comp.
Biochem. Physiol.* 14: 336-348.

Freed, G. H., M. P. Schreibman, and D. Kaltman 1969. Enzymatic activities in tissues of teleosts. *Comp. Biochem. Physiol.* 28: 771-776.

Hazel, J. R. 1972. The effect of temperature acclimation upon succinic dehydrogenase activity from the epaxial muscle of the common goldfish (*Carassius auratus* L.). I. Properties of the enzyme and the effect of lipid extraction. II. Lipid reactivation of the soluble enzyme. *Comp. Biochem. Physiol.* 43B: 837-882.

Hazel, J. and C. L. Prosser 1970. Interpretation of inverse acclimation to temperature. *A. vergl. Physiol.* 67: 217-228.

Hazel, J. R. and C. L. Prosser 1974. Molecular mechanisms of temperature compensation in poikilotherms. *Physiol. Rev.* 54: 620-677.

Hochachka, P. W. 1965. Isoenzymes in metabolic adaptation of a poikilotherm; subunit relations in lactic dehydrogenases of goldfish. *Arch. Biochem. Biophys.* 111: 96-103.

Hochachka, P. W. 1968. Action of temperatures on branch points in glucose and acetate metabolism. *Comp. Biochem. Physiol.* 25: 107-119.

Hochachka, P. W. and F. R. Hayes 1962. The effect of temperature acclimation on pathways of glucose metabolism in the trout. *Can. J. Zool.* 40: 261-270.

Hochachka, P. W. and J. K. Lewis 1970. Enzyme variants in thermal acclimation. Trout liver citrate synthesis. *J. Biol. Chem.* 245: 6567-6573.

Hochachka, P. W. and G. N. Somero 1973. *Strategies of Biochemical Adaptation*. Philadelphia: W. B. Saunders.

Kanungo, M. S. and C. L. Prosser 1959. Physiological and biochemical adaptation of goldfish to cold and warm temperatures - II. Oxygen consumption of liver homogenate; oxygen consumption and oxidative phosphorylation of liver mitochondria. *J. Cell.Comp. Physiol.* 54: 265-274.

Moon, T. W. and P. W. Hochachka 1972a. Temperature and kinetic analysis of trout isocitrate dehydrogenases. *Comp. Biochem. Physiol.* 42B: 725-730.

Moon, T. W. and P. W. Hochachka 1972b. Temperature and enzyme activity in poikilotherms: isocitrate dehydrogenase in rainbow trout liver. *Biochem. J.* 123: 695-705.

Prosser, C. L. 1963. Perspectives of adaptations: Theoretical aspects. In *Handbook of physiology, adaptation to the environment*, B. Dill, ed., pp. 11-25, Washington, D.C.: Amer. Physiol. Soc.

Prosser, C. L. 1967a. *The Cell and Environmental Temperature*. Oxford: Pergamon Press.

Prosser, C. L. 1967b. *Molecular Mechanisms of Temperature*

Adaptation. Washington, D.C.: Amer. Assoc. Adv. Sci.

Prosser, C. L. 1969. Principles and general concepts of adaptation. *Env. Res.* 2: 404-416.

Prosser, C. L. 1973. *Comparative Animal Physiology.* Ed. 3 Philadelphia: W. B. Saunders.

Shaw, C. R. and R. Prasad 1970. Starch gel electrophoresis of enzymes - a compilation of recipes. *Biochem. Genet.* 4: 297-320.

Sidell, B. D., F. R. Wilson, J. Hazel, and C. L. Prosser 1972. Time course of thermal acclimation in goldfish. *J. Comp. Physiol.* 84: 119-127.

Somero, G. N. 1973. Thermal modulation of pyruvate metabolism in the fish *Gillichthys mirabilis:* the role of lactate dehydrogenases. *Comp. Biochem. Physiol.* 44B: 205-209.

Wilson, F. R. 1973. Quantitating changes of enzymes of the goldfish (*Carassius auratus* L.) in response to temperature acclimation: An immunological approach. Ph.D. Thesis, University of Illinois.

Wilson, F. R., G. Somero, and C. L. Prosser 1974. Temperature-metabolism relations of two species of *Sebastes* from different thermal environments. *Comp. Biochem. Physiol.* 47B: 485-491.

Wilson, F. R., G. S. Whitt, and C. L. Prosser 1973. Lactate dehydrogenase and malate dehydrogenase isozyme patterns in tissues of temperature-acclimated goldfish (*Carassius auratus* L.). *Comp. Biochem. Physiol.* 46B: 105-116.

TEMPERATURE ADAPTATION:
ISOZYMIC FUNCTION
AND THE
MAINTENANCE OF HETEROGENEITY

THOMAS W. MOON

Department of Biology
University of Ottawa
Ottawa, Ontario
Canada K1N 6N5

ABSTRACT. The current status of knowledge concerning isozymic function in the thermal adaptation of ectotherms,
and principally in salmonid fish, is reviewed. Two basic
isozymic strategies are found: 1. the "on-off" synthesis
of unique isozymic forms; and, 2. relative changes in specific isozyme members of a complex isozymic system. These
multiple enzyme strategies are compared to that of maintaining a single enzyme species, the specific conformational
state of which is mediated through alterations in the internal milieu. Although each strategy is as likely to occur
as another, the high degree of enzyme heterogeneity in salmonid and certain other fishes results in favoring the isozymic strategy. Examples are given for each strategy and
discussed in terms of their similar end result, i.e., reaction rate stabilization and optimal control capabilities.
It is concluded that the maintenance of enzyme heterogeneity
in organisms such as salmonids increases flexibility and
adaptability in the face of fluctuating thermal regimes.

INTRODUCTION

Molecular kinetic energy, or temperature, has general consequences for biological systems. But not all organisms are
equally susceptible to thermal change. Endothermic organisms,
by highly evolved mechanisms, maintain internal body temperatures independent of environmental temperatures. Ectothermic
organisms, however, do not have the necessary complex homeo-
static repertoire of the endotherms, but must rely on more basic
behavioral, physiological, and biochemical mechanisms to counteract or avoid thermal effects. Nevertheless, related ectotherms
living at very different temperatures do exhibit similar metabolic rates. This apparent paradox is the result of evolutionary modifications and acclimation, a compensatory change in rate
of a physiological process in response to a change in temperature (Prosser, 1973).
The basic mechanisms associated with thermal compensation

in ectotherms have come to light during the past few years and recent publications have dealt with the broad implications of these mechanisms (Hochachka and Somero, 1971; 1973; Somero and Hochachka, 1971). It seems appropriate at this time to review the specific role of isozymes in these compensatory responses. Indeed, I should also like to propose that the physiological benefits of isozymes are such that they have provided in the past an evolutionary selective force which has maintained and perhaps extended enzyme heterogeneity.

THERMAL HOMEOSTASIS IN ECTOTHERMS

We can group the effects of temperature on biological processes into two broad categories: kinetic and "weak"-bond-structural (Hochachka and Somero, 1973).

The kinetic effects are the most obvious and early work dealt with establishing a relationship between habitat preference and the Arrhenius number or activation energy, E_a. ($E_a = \dfrac{R(T_1T_2) \, (\ln k_2 - \ln k_1)}{(T_2 - T_1)}$, where k values represent velocity constants at each temperature (T_1 and T_2), R is the gas constant, and T is absolute temperature). The recent investigations of Low, Bada, and Somero (1973) have suggested a more complex interpretation of this literature; thus, any established correlation may be fortuitous and of no physiological importance.

Further, since E_a was initially derived from empirical observations, the conditions of the experiment will dictate its magnitude and affect the apparent temperature dependency of the reaction. For example (Fig. 1), adjusting the concentration of glucose-6-phosphate (G-6P) and TPN (NADP) for glucose-6-phosphate dehydrogenase (G-6PD) from the mullet fish, *Mugil cephalus*, changes the temperature dependency of the reaction (Hochachka and Hochachka, 1973). At physiological substrate levels (0.05 mM G-6P) the reaction velocity above 25°C is temperature-independent and below shows reduced sensitivity compared to saturating substrate levels (0.5 mM). Similar observations have been made for many other systems (Hochachka and Somero, 1968; Somero, 1969).

These differential kinetic effects at saturating and subsaturating substrate concentrations lead to the important conclusion that E-S affinity parameters, as defined by the reciprocal of the Michaelis constant, or K_m, vary as a function of temperature for some enzymes isolated from ecto-and endotherms. This aspect of temperature is termed the "weak"-bond-structural effect and has been reviewed (Hochachka and Somero, 1973).

From a physiological point of view the importance of control by E-S affinity is mediation of Q_{10}. ($Q_{10} = (k_2/k_1)^{10/(T_2 - T_1)}$,

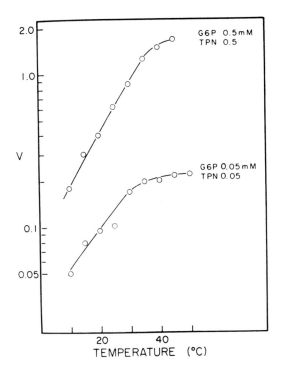

Figure 1. Semilog plots of velocity (V) versus temperature for adult mullet *(Mugil cephalus)* liver G-6PDH under conditions of high (upper curve) and low (lower curve) levels of G-6P and TPN (NADP). Modified from Hochachka and Hochachka (1973).

where k is as above, and T is $^\circ$C.) Thus, at physiological levels of substrate, which are 2- to 10- fold lower than those found to substrate the enzyme, values of Q_{10} are often reduced to 1.0 or below, indicating temperature-independent enzyme function. Were substrate levels not this low, thermal modulation of reaction rates by E-S affinity parameters would be meaningless.

The role of E-S affinity parameters in cellular regulation is well known and the important evolutionary implications have been discussed (Atkinson, 1969). Since temperature and other environmental stresses can be important selective forces, the mediation of these effects through E-S affinity is not surprising. Recent data, to be reviewed when AChE is discussed, are suggestive of the basic mechanisms involved with these thermal perturbations of enzyme structure.

Certain enzymes of ectotherms show complex U-shaped K_m - temperature plots (Fig. 2; see Hochachka and Somero, 1971;

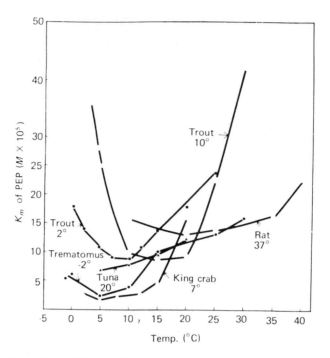

Figure 2. The effect of temperature on the $K_{m(PEP)}$ of pyruvate kinase enzymes from differently adapted species. After Hochachka and Somero (1971).

1973; Somero and Hochachka, 1971). Therefore, the range over which a direct K_m-temperature correlation exists and therefore, Q_{10} modulation, is rather narrow; in many cases, a large increase in K_m is seen at the lower extreme, which, coupled to the loss of available thermal energy can increase Q_{10} values to 20 and above. This large thermal dependency would restrict the enzyme from functioning over this range.

Many ectotherms are eurythermal and have the capacity after a short adaptation period to live within large thermal ranges. What mechanisms eliminate this negative-thermal modulation of particular enzymes at these low temperature extremes?

Numerous enzyme systems have been investigated in the salmonid fishes, in the particular the eurythermal rainbow trout, *Salmo gairdneri*. Certain salmonids are unique in that they arose from a tetraploid ancestor, and much of their reported enzyme heterogeneity can be linked to this evolutionary event (Ohno, 1970). A tetraploid organism initially maintains four homologous chromosomes of each linkage group and through evolutionary time new gene loci develop by diploidization (Ohno, 1970). This process increases the potential for genetic diversification, which could result in enzyme variants "tailored" by natural selection to function under a specific set of environmental conditions. Thus, one strategy open to ectotherms to nullify the detrimental negative-thermal modulation of enzymes would be the expression of enzyme forms better suited for function under the prevailing thermal conditions.

ISOZYMES IN THERMAL ADAPTATION

Although the tetraploid hypothesis (Ohno, 1970) does not apply to all ectotherms, within the salmonids and certain other fishes it could account for the existence of so many isozymic systems. Also, since the processes of acclimation and seasonal acclimatization (Prosser, 1973) rely on existing genetic material, this enzyme heterogeneity may increase the selective advantage of these organisms under fluctuating environmental conditions.

Two mechanisms can make available to the ectothermic cell enzymes with the greatest functional capacity under a given set of conditions (modified from Hochachka and Somero, 1973). First, the isozymic mechanism which consists of two strategies: 1. the "on-off" synthesis of unique isozymic forms; and, 2. relative changes in specific isozymic members of a complex system. Second, a single enzymic form may be present, the conformation of which can be directly modified by conditions of the internal milieu. This system is analogous to the conformational isozymes first identified by Kitto, Wassarman, and Kaplan (1966). Examples of each strategy are numerous, but with the extensive enzyme heterogeneity found in most organisms, and in particular salmonids, the first strategy predominates. Thus, emphasis will be placed on this strategy with mention of the important details of the conformational strategy.

"On-Off" Synthesis of Unique Isozymes

The importance of acetylcholinesterase (AChE) in neural function has prompted intensive study of both the physiological and physiochemical properties of this enzyme from a variety of sources (see Froede and Wilson, 1971). Baldwin and Hochachka (1970) and Baldwin (1971) have reported that brain AChE from temperate, tropical, and polar zone fishes display complex K_m-temperature relationships (Fig. 3). A direct K_m-temperature response is observed at high temperatures, a levelling-off where K_m is temperature-independent, and then a reversal of the K_m response at the low temperature extreme.

Trout acclimated to the cold ($2^{\circ}C$) exhibit an altered kinetic pattern with the minimal K_m value shifted towards the new acclimation temperature. By disc gel electrophoresis and specific staining techniques these kinetic differences are found to

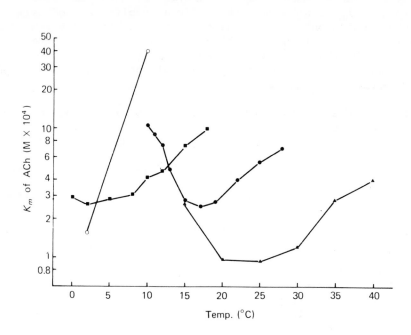

Figure 3. The effect of temperature on the $K_{m(ACh)}$ for AChE enzymes of the electric eel, *Electrophorus electricus* (▲); cold-(2°C) acclimated trout (■); warm-(18°) acclimated trout (●); and the antarctic fish *Trematomus borchgrevinki* (○). After Hochachka and Somero (1971).

represent distinct enzyme variants or isozymes. Acclimating trout to an intermediate temperature gives both isozymes, a result consistent with that observed in a natural trout population when water temperatures were approximately 12°C (Baldwin and Hochachka, 1970). The production of this enzyme variant at low temperatures maintains optimal catalytic function over the entire thermal range of the organism thereby assuring central nervous system integrity and eliminating the detrimental effects of negative-thermal modulation inherent in the kinetics of the warm-enzyme variant.

The well established active site structure and reaction mechanism of AChE permits a detailed study of how the two trout isozymes may differ structurally. Work by Hochachka and Kelly (1974) and Hochachka, Storey, and Baldwin (1974) with defined AChE inhibitors has established the effects of temperature on the contribution of various non-covalent or "weak" chemical bonds to substrate binding. Hydrophobic interactions, which are important at the anionic or substrate binding site of AChE, are less stable at low termperatures and high pressures compared to electrostatic interactions between opposite charges (see Brandts, 1967). Both types of "weak"-interactions are possible at the AChE anionic site; therefore one might predict that stabilization of AChE structure at low temperature could occur with an isozymic variant which relies more heavily on electrostatic interactions than hydrophobic ones. Although this question has not specifically been approached by Hochachka and coworkers, some preliminary comparisons suggest this may be true.

Using low temperature and high pressure (a few hundred atmospheres) perturbations, a correlation is established between the K_m(acetylthiocholine) and the habitat of the organism. Thus, AChE from the abyssal fish, *Antimora rostrata*, has K_m values which are reduced by low temperatures and high pressures compared to the enzyme of the surface dwelling dolphin fish where K_m values increase. This result is suggestive of a greater contribution of non-hydrophobic interactions in the substrate binding to AChE in the abyssal fish. Couple this to a low ΔS (entropy of activation) value in the presence of a good "hydrophobic" inhibitor, phenyltrimethyl ammonium ion, the implication is that a modified anionic pocket exists in the enzyme from the abyssal, cold-adapted fish.

These data of Hochachka and coworkers seem to predict the necessity of enzyme variants at low temperature due to substrate binding constraints. The trout brain AChE system could be used as a model to substantiate this hypothesis.

The synthesis of unique isozymes during thermal acclimation and seasonal acclimatization represents an important, but so far, little identified strategy. Other systems similar to trout brain AChE undoubtedly exist; in fact, recent data on adductor muscle LDH of a freshwater bivalve *(Elliptio complanatus R.)* are consistent with this strategy (Moon, Huner, and Hulbert, unpublished data). Other examples await investigation.

Complex Isozymic Systems

Isozymic changes during thermal adaptation resulting in the relative change in activity of particular members of a complex isozymic set are numerous. Examples include mullet liver G-6PD (Hochachka and Hochachka, 1973), salmonid liver cytoplasmic NADP-isocitrate dehydrogenase (IDH) (Moon and Hochachka, 1971a, b; 1972), salmonid and goldfish LDH (Hochachka, 1967), and brook trout alkaline phosphatases (Whitmore and Goldberg, 1972). Other systems undoubtedly exist, but the lack of electrophoretic and/or isoelectric focusing data does not allow their inclusion here.

Not all complex isozymic systems, however, change with

thermal adaptation. Wilson, Whitt, and Prosser (1973) have evidence that individual variation, not thermal acclimation, is responsible for LDH isozymic changes in goldfish. Also, isozymes of goldfish MDH (Wilson, Whitt, and Prosser, 1973) and succinate dehydrogenase (Hazel, 1972), and loach *(Misgurnus fossilis L.)* G-6PD, LDH, NAD- and NADP-IDH (Mester, Scripcariu, and Niculescu, 1972) show no changes.

Salmonid liver cytoplasmic NADP-linked isocitrate dehydrogenase (IDH) has been extensively studied and found to be coded by at least two and possibly more gene loci (Moon and Hochachka, 1971a). In an electrophorectic study of hatchery reared brook, lake, and splake trout liver IDH (Moon and Hochachka, 1971a) brook trout were found to significantly increase the quantitative expression of the faster moving IDH isozymes in the warm- (17°C) compared to the cold- (4°C) acclimated state. It was suggested that temperature resulted in a shift in the expression of a given subunit type; however, in the absence of kinetic information it is difficult to appreciate the significance of this alteration.

Rainbow trout liver IDH is more complex. Six phenotypes were found in a hatchery population but the relative distribution of these depended upon season (Moon and Hochachka, 1972). A detailed electrophoretic and kinetic study of the $A_2B_2C_2$ phenotype indicated two IDH variants (Moon and Hochachka 1971b): with cold- (2°C) acclimation, an IDH variant displaying minimal K_m (isocitrate) values approaching the acclimation temperature; the warm- (18°C) variant had K_m values coincident with this acclimation temperature, a result consistent with trout brain AChE (see Fig. 3).

Unlike brain AChE, these variants showed quantitative rather than qualitative changes in particular isozymic bands, i.e., there is a five-fold increase in the relative amount of the low mobility isozyme in the cold-acclimated trout, whereas approximately 90% of all the activity in the warm is in the fastmoving band (Moon and Hochachka, 1971b). The contribution of each isozyme to the kinetic pattern could not be determined, but by using other IDH phenotypes this is possible.

The single-band A_2-IDH phenotype has kinetic properties unlike the other phenotypes (Fig. 4; Moon and Hochachka, 1972). A direct K_m-temperature response is not seen for A_2-IDH, but is when the number of slow-mobility subunits increase. Thus, Q_{10} stabilization with decreasing temperature can not occur in the A_2-IDH phenotype, the importance of which has been previously discussed.

Therefore, with trout liver IDH the importance of maintaining enzyme heterogeneity is related to stabilization of reaction rates. In the absence of multiple forms an individual may be

214

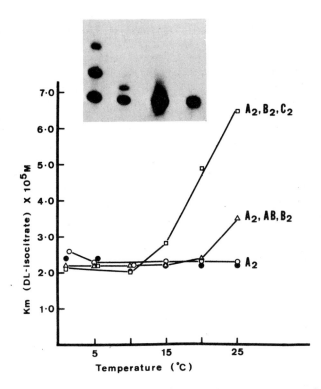

Figure 4. The effect of tissue isozymal content on the K_m (isocitrate)-temperature response for three liver NADP-IDH phenotypes from rainbow trout. The curve for A_2-IDH is given for both crude (O) and partially purified (●) enzyme. Electrophoretic insert (origin at top and anode at bottom): 1. A_2,B_2,C_2. 2. A_2,AB, B_2. 3. Partially purified A_2. 4. Crude A_2. After Moon and Hochachka (1972).

at a disadvantage in a fluctuating thermal environment. A precedence for this hypothesis exists; lake trout IDH is a single enzyme form and this trout has a restricted thermal tolerance (Moon and Hochachka, 1971a). Thus, in a eurythermal animal such as the rainbow trout the isozymic strategy may be important to maintain metabolic integrity. In its absence, other strategies must be considered (Hochachka and Somero; 1973).

Conformational Isozymes

The remaining strategy does not involve changes in the primary structure of enzyme proteins, or classical isozymes. Instead, by specific temperature and/or milieu-directed changes, enzyme properties better suited for a particular environmental condition may prevail. These enzymic alterations have often been termed "conformational isozymes". That these changes are reversible and must occur is a result of the already noted effects of temperature on "weak"-bond interactions.

The best examined ectothermic example is king crab muscle pyruvate kinase (PK) (Somero, 1969). The kinetic characteristics of PK are identical to systems we have already discussed. As seen in Fig. 5A, activity curves as a function of phosphoenolpyruvate concentrations are unpredictably "bumpy" at intermediate temperatures. For these temperatures, and below, two values of K_m (PEP) can be estimated from Hill plots. These are plotted in Fig. 5B. Two kinetic forms of PK, exhibiting altered K_m-temperature responses are indicated. As before, the "cold"-variant extends the thermal range of K_m independency and eliminates the large upswing in the "warm"-variant at low temperatures.

Unlike the other strategies, however, there is no evidence for more than one form of crab muscle PK. Instead, Somero (1969) postulated the existence of a temperature-dependent conformational shift between two kinetic states:

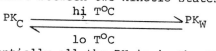

$$PK_C \xrightleftharpoons[\text{lo T}^oC]{\text{hi T}^oC} PK_W$$

At 15°C, essentially all the PK is in the PK_W conformation, and below 5°C all the physiologically active enzyme is in the PK_C conformation.

The inclusion of this example is important, since it indicates that classical isozymes relying on amino acid sequence changes need not occur to insure the presence of an enzyme in the cell which functions optimally under a given thermal regime. Instead, single enzyme forms, maintaining structural flexibility, can constitute an important biochemical strategy.

DISCUSSION

THE ROLE OF ISOZYMES IN THERMAL HOMEOSTASIS

We have seen that isozymes can be involved in the biochemical response of ectotherms to thermal alterations and that these isozymic variants are important to:

1. eliminate the high Q_{10} values resulting from simultaneous increases in K_m or decreases in E-S affinity and the reduction in available thermal energy at low temperatures; and,

2. the maintenance of controlled catalytic function (K_m-modulation) throughout the animal's thermal range.

Thus, metabolic homeostasis is paramount during thermal homeostasis. Johnson (1974) in a more general argument has arrived at the same conclusions, i.e., that the maintenance of heterogeneity may provide a means for metabolic compensation

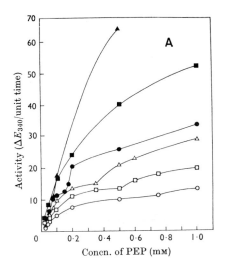

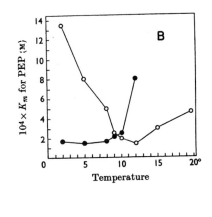

Figure 5. A. Substrate (PEP) saturation curves for king crab leg-muscle PK at 2°C (○), 5°C (□), 8°C (△), 10°C (●), 15°C (■), and 20°C (▲). B. Estimates of the $K_{m(PEP)}$ of the two forms of king crab PK; "cold" PK (●) and "warm" PK (○). After Somero (1969).

under varying environmental conditions. However, isozymic variants, per se, have not been found to result in rate compensation, which is the most consistent outcome of thermal acclimation (Prosser, 1973). Instead, either increases in enzyme concentrations and/or changes in the cellular milieu must accompany these isozymic changes if rate compensation is to occur (Hochachka and Somero, 1973).

Also, from studies by Hebb, Stephens, and Smith (1972) the most stable enzyme structure is that with the highest E-S affinity; this is consistent with the constraints implicit in the effects of temperature on "weak"-bond-structural parameters. Therefore, the existence of temperature-dependent isozymic changes leading to maximal E-S affinities near the new environmental temperature is a necessary outcome of optimizing enzyme structure and metabolic control, but only secondarily, catalytic rates.

It is not necessary, however, for isozymes to have unique amino acid sequences and thus distinctive electrophoretic properties. Instead, cellular demands may be met equally well by a single enzyme protein with stable conformational states of differing kinetic characteristics. These two strategies must be considered mechanistically dissimilar but functionally similar. The salmonids, with a high degree of enzyme heterogeneity, can "afford" to maintain complex isozymic systems, the final isozymic pattern being dictated by the particular environmental conditions. Other ectotherms, and possibly other gene loci, which are not as heterogeneous must rely instead on the conformational isozymic strategy or an equivalent strategy (see Somero, 1974). Thus, one must consider not only the animal, but also the catalytic process when the selective nature of isozymes is questioned.

Caution must be taken in the interpretation of isozyme data. The considerable individual variations observed in the salmonids and other fishes can lead to illusory thermal effects (Wilson, Whitt, and Prosser, 1973), so large numbers of fish must be sampled. Incubation temperatures for gel staining can markedly alter the isozymic pattern, as has been noted by Mester, Scripcariu, and Niculescu (1972). Even so, the examples discussed in this paper illustrate changes modifying specific isozymic systems in an apparently adaptive direction.

ACKNOWLEDGEMENTS

The author is indebted to Drs. P. W. Hochachka and G. N. Somero for their critical examination and discussion of the ideas incorporated into this review; their keen interest and dedication opened an area of research which will lead them and those that follow to a better understanding of enzymes and enzyme characteristics of ectothermic organisms. I would also like to acknowledge Dr. J. C. Fenwick for his critical reading of this manuscript.

REFERENCES

Atkinson, D. E. 1969. Limitation of metabolite concentration and the conservation of solvent capacity in the living cell. In *Current Tropics in Cellular Regulation* (Horecker, B. L. and Stadtman, E. R., eds.) Vol. 1, pp. 29-43, Academic Press, New York.

Baldwin, J. 1971. Adaptation of enzymes to temperature: Acetylcholinesterases in the central nervous system of fishes. *Comp. Biochem. Physiol.* 40B: 181-187.

Baldwin, J. and P. W. Hochachka 1970. Functional significance of isoenzymes in thermal acclimation. Acetylcholinesterase from trout brain. *Biochem. J.* 116: 883-887.

Brandts, J. F. 1967. Heat effects on proteins and enzymes. In *Thermobiology* (Rose, A. H. ed.) pp. 25-72, Academic Press, New York.

Froeda, H. C. and I. B. Wilson 1971. Acetylcholinesterase. In *The Enzymes* (P. Boyer, ed.) 3rd. Ed., Vol. 5, pp. 87-114, Academic Press, New York.

Hazel, J. R. 1972. The effect of temperature acclimation upon succinic dehydrogenase activity from the epaxial muscle of the common goldfish (Carassius auratus L.) --I. Properties of the enzyme and the effect of lipid extraction. Comp. Biochem. Physiol. 43B: 837-861.

Hebb, C., T. C. Stephens, and M. W. Smith 1972. Effect of environmental temperature on the kinetic properties of goldfish brain choline acetyltransferase. Biochem. J. 129: 1013-1021.

Hochachka, P. W. 1967. Organization of metabolism during temperature compensation. In Molecular Mechanisms of Temperature Compensation (Prosser, C. L. ed.) Publ. No. 84, pp. 177-203, Am. Assoc. Advanc. Sci., Washington, D.C.

Hochachka, P. W. and B. C.-Hochachka 1973. Glucose-6-phosphate dehydrogenase and thermal acclimation in the mullet fish. Mar. Biol. 18: 251-259.

Hochachka, P. W. and J. Kelly 1974. Acetylcholinesterase: Temperature adaptation of the binding site. Submitted for publ.

Hochachka, P. W. and G. N. Somero 1968. The adaptation of enzymes to temperature. Comp. Biochem Physiol. 27: 659-668.

Hochachka, P. W. and G. N. Somero 1971. Biochemical adaptation to the environment. In Fish Physiology (Hoar, W. S. and Randall, D. J., eds.) Vol. 6, pp. 99-156, Academic Press, New York.

Hochachka, P. W. and G. N. Somero 1973. Strategies of Biochemical Adaptation. W. B. Saunders Co., Philadelphia.

Hochachka, P. W., K. B. Storey, and J. Baldwin 1974. In Preparation.

Johnson, G. B. 1974. Enzyme polymorphism and metabolism. Science 184: 28-37.

Kitto, G. B., P. M. Wassarman, and N. O. Kaplan 1966. Enzymatically active conformers of mitochondrial malate dehydrogenase. Proc. Nat. Acad. Sci. U.S.A. 56: 578-58.

Low, P., J. Bada, and G. N. Somero 1973. Temperature adaptation of enzymes: Roles of the free energy, the enthalpy, and the entropy of activation. Proc. Nat. Acad. Sci. U.S.A. 70: 430-432.

Mester, R., D. Scripcariu, and S. Niculescu 1972. Effect of temperature on the isoenzymic pattern of loach (Misgurnus fossilis L.). I. Glucose-6-phosphate dehydrogenase, lactate dehydrogenase, NAD- and NADP-isocitrate dehydrogenase. Rev. Roum. Biol.-Zool. 17: 205-217.

Moon, T. W. and P. W. Hochachka 1971a. Effect of thermal acclimation on multiple forms of the liver-soluble NADP-linked isocitrate dehydrogenase in the family Salmonidae. Comp. Biochem. Physiol. 40B: 207-213.

Moon, T. W. and P. W. Hochachka 1971b. Temperature and enzyme activity in poikilotherms. Isocitrate dehydrogenases in rainbow trout liver. *Biochem. J.* 123: 695-705.

Moon, T. W. and P. W. Hochachka 1972. Temperature and the kinetic analysis of trout isocitrate dehydrogenases. *Comp. Biochem. Physiol.* 42B: 725-730.

Ohno, S. 1970. *Evolution by Gene Duplication.* Springer-Verlag, New York.

Prosser, C. L. 1973. *Comparative Animal Physiology.* W. B. Saunders Co., Philadelphia.

Somero, G. N. 1969. Pyruvate kinase variants of the Alaskan king crab; evidence for a temperature-dependent interconversion between two forms having distinct and adaptive kinetic properties. *Biochem. J.* 114: 237-241.

Somero, G. N. 1974. The role of isozymes and allozymes in adaptation to varying temperatures. This volume.

Somero, G. N. and P. W. Hochachka 1971. Biochemical adaptation to the environment. *Amer. Zool.* 11: 157-165.

Whitmore, D. H. and E. Goldberg 1972. Trout intestinal alkaline phosphatases. II. The effect of temperature upon enzymatic activity in vitro and in vivo. *J. Exp. Zool.* 182: 59-68.

Wilson, F. R., G. S. Whitt, and C. L. Prosser 1973. Lactate dehydrogenase and malate dehydrogenase isozyme patterns in tissues of temperature-acclimated goldfish *(Carassius auratus L.). Comp. Biochem. Physiol.* 46B: 105-116.

THE ROLES OF ISOZYMES IN ADAPTATION
TO VARYING TEMPERATURES

GEORGE N. SOMERO
Scripps Institution of Oceanography
Box 1529
La Jolla, California 92037

ABSTRACT. Enzyme variants may serve an adaptive role (1)
by providing the correct vectorial properties for the metabolism of a tissue or an organelle; or (2) by broadening
the environmental tolerance range of an organism. To determine the importance of multiple locus isozymes and allelic isozymes (allozymes) in adaptation to temperature, we
examined the electrophoretic and kinetic properties of enzymes from teleost fishes adapted (1) to widely different
temperatures and (2) to thermal regimes which varied greatly in stability. Fishes from stable and highly variable
thermal regimes did not differ in the extent of enzyme
polymorphism. No isozyme changes were observed during
seasonal acclimitization or laboratory acclimation, except
in the tetraploid rainbow trout. Thus, with one exception,
no correlation was observed between environmental (thermal)
variability and genetic-protein variability. Kinetic
studies revealed that eurythermal fishes possess enzymes
capable of functioning over very broad ranges of temperature, relative to the homologous enzymes of stenothermal
fishes. Recent tetraploids such as the rainbow trout
differ from most fishes in having an additional set of
genes added to an already established family of isozyme
loci. This additional genetic information may provide the
raw material for evolution of environmentally-specific
isozymes.

INTRODUCTION

Two types of functional significance have been ascribed to
multiple enzyme forms, i.e., isozymes. It is now generally
accepted that many tissue- and organelle-specific isozymes
perform regulatory roles in controlling metabolic activity.
When we consider the kinetic properties of a tissue- or organelle-specific isozyme in the context of the metabolic function
of the tissue or organelle, we often find that the isozyme's
functional traits seem especially well-suited for fostering
the rates and directions of metabolic flow characteristic of
the tissue or the organelle.
Whereas this regulatory role of isozymes is well established, a second proposed role of multiple enzyme forms is still

221

a matter of some controversy. This is the suggested role of allelic isozymes (allozymes) and multiple locus isozymes in adapting organisms to environments which vary in space and time. It has been argued by several workers (Johnson, 1973, 1974; Levins, 1968; Selander and Kaufman, 1973; see also the paper by Allard, et al., 1974) that allozymic and multiple locus isozymic forms of enzymes may serve as an important mechanism for broadening the environmental tolerance ranges of organisms, e.g., to temperature, salinity, types of food plant, etc. A single form of each and every type of enzyme may not be adequate to insure success or survival over the entire spectrum of environmental conditions faced by the organism. Environmental adaptations of this type would not involve different isozymes as in the case of tissue- and organelle-specific isozymes. Rather, the "goal"of isozymic adaptations to variable environments is one of insuring that each type of enzymic function necessary for success and survival is maintained under all environmental conditions that the organism is likely to experience.

RESULTS AND DISCUSSION

THE "MULTIPLE VARIANT" AND "EURYTOLERANT PROTEIN" STRATEGIES OF ADAPTATION

If a particular functional or structural trait of an enzyme is subject to environmental perturbation, e.g., under the influence of temperature, hydrostatic pressure, and salt content or composition, and if this trait must be maintained at all times then two basic types of adaptive strategies seem possible (Fig. 1).

If an organism utilizes two or more isozymes to buffer its metabolic functions against the effects of environmental change, the organism uses what can be termed a "multiple variant" strategy. In this situation, no single isozyme is capable of functioning satisfactorily over the entire range of temperature, pressure, etc., experienced by the organism (and the enzyme). Success over the entire range of environmental conditions is possible only through the concerted action of two or more relatively stenotolerant enzymes. Figure 1 illustrates a system in which three enzyme variants are necessary to insure proper function of the enzyme over the entire range of habitat conditions faced by the organism.

An alternative strategy (Figure 1) is termed the "eurytolerant protein" strategy. In this pattern of adaptation to variable environmental conditions, a single form of a protein is capable of maintaining its structural and functional charac-

222

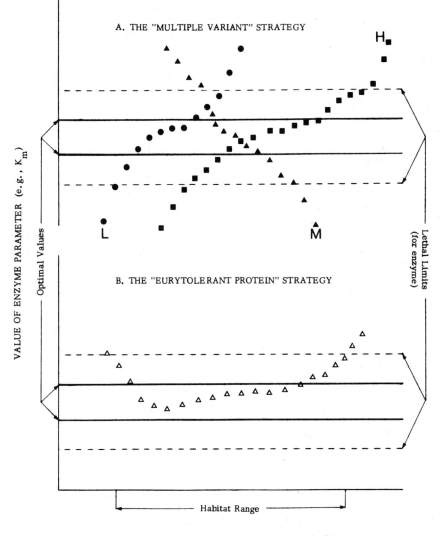

Fig. 1.   Alternative strategies for maintaining enzymic
function and structure over a broad range of an environmental
parameter which perturbs the function or structure.   In the
"multiple variant" strategy three isozymes are required to
maintain the enzymic parameter with a tolerable range: a low-
range (L) variant works well at low temperature (or salinity
or hydrostatic pressure, etc.); a mid-range (M) and a high-
range (H) variant function optimally at other regions of the
environmental spectrum.

Fig. 1. (cont.) In the "eurytolerant protein" strategy a single protein form is able to maintain its functional and structural characteristics within a tolerable range over the entire span of environmental conditions faced by the organism (protein).

---

istics within an acceptable range over the entire spectrum of environmental conditions faced by the organism (enzyme). This strategy can be appreciated as being the simplest and, in many ways, the optimal mechanism for coping with variable environmental conditions. By using only a single, eurytolerant enzyme (or respiratory protein, for example), the cell can (1) reduce the complexity of its genome (fewer isozyme loci, lower genetic load), (2) maintain fewer enzyme molecules within its boundaries (at least in the case of multiple isozyme forms), and (3) avoid the time-lag entailed in "turning-on" new isozyme forms in response to what may be a rapid environmental change.

Each of these two strategies has *a priori* appeal. To determine the importance of these alternate ways of adapting to variable habitat conditions, two types of studies seem necessary. First, it should prove insightful to estimate the levels of isozymic variability in organisms which inhabit either highly stable or extremely variable habitats. Second, to permit sound conclusions about the functional aspects of different isozymes, it is essential that the functional and structural traits of the enzymes under study be scrutinized. The second type of study is necessary if we are to determine the degrees of "stenotolerance" and "eurytolerance" of variants of particular enzymes.

EXPERIMENTAL DESIGN REQUIREMENTS

Before reviewing the relevant data dealing with the relationship between environmental variability and enzymic adaptation strategies, it seems prudent to discuss briefly certain of the experimental design criteria which must be met for a meaningful investigation of phenomena of this nature. The criteria discussed below seem absolutely vital for studies of this class. Unfortunately, there are many studies in which one or more of these experimental design requirements have not been met.

An especially critical aspect of experimental design involves the selection of an environmental parameter and a set of enzymes for investigation. One must focus attention on an environmental parameter which is "felt" by the proteins of interest. For example, it would be illogical to study polymorphisms of Krebs citric acid cycle enzymes in relationship to variability in environmental food sources. The metabolites, such as AcetylCoA, which are "seen" by the enzymes of the citric acid cycle "look" the same regardless of the foodstuff from which they were derived. The foodstuff-enzyme interface is thus very limited, consisting of only the digestive enzymes involved in the initial processing of nutrients.

Since one ideally wishes to examine an environmental-protein interface which includes as many proteins as possible, we believe that temperature represents the optimal parameter for study, at least in the case of ectothermic (poikilothermic) organisms. With the possible exception of hydrostatic pressure, temperature is the only environmental parameter which is "felt" by all proteins of an organism. Changes in temperature lead to (1) alterations in the rates of enzymic activity, (2) changes in the higher orders of protein structure, (3) variations in ligand binding properties, and (4) important changes in the physical state of the lipid molecules to which many enzymes are bound (Hochachka and Somero, 1973). If temperature is used as the environmental parameter of interest, one can therefore select any group of proteins for study, knowing that changes in habitat temperature will affect the functions and structures of the proteins examined. (This statement should not be interpreted to mean that all proteins will be equally affected by temperature changes, however.)

Another vital component of experimental design is the requirement that the same set of enzymes be studied in all groups of organisms to be compared. Different classes of enzymes may exhibit distinctly different levels of enzyme polymorphism (Johnson, 1974; Selander and Kaufman, 1973), and one could very easily end up comparing "apples to oranges" if, for example, he compared esterase polymorphisms in one species to dehydrogenase polymorphisms in a second species.

Interspecific comparisons are another major source of trouble in experimental design, since different species exhibit different levels of polymorphism (Johnson, 1974; Selander and Kaufman, 1973). To determine the effects of different environments on protein variability, one must therefore use species which, aside from their environmental preferences, are otherwise very similar, e.g., in a phylogenetic sense.

TEMPERATURE AND ENZYME POLYMORPHISM: ALLOZYME VARIABILITY IN MARINE TELEOSTS

With these experimental design requirements in mind, we initiated a study to determine if the levels of isozymic variability in ectothermic species which inhabit highly variable thermal habitats are greater than the variabilities

TABLE I

LEVELS OF GENETIC VARIATION IN TELEOST FISHES FROM DIFFERENT THERMAL REGIMES

SPECIES (Family)	COLLECTION SITE (Habitat)	ANNUAL[1] THERMAL RANGE (°C)	n^2	L^3	P^4	$P^5_{0.05}$	$H^6 \pm$ S. E.
Trematomus borchgrevinki (*Nototheniidae*)	McMurdo Sound (sub-ice, pelagic)	1	9	21	4.8	0	0.5 ± 0.5
Trematomus hansoni (*Nototheniidae*)	McMurdo Sound (benthic)	1	26	26	18.5	11.1	2.5 ± 0.4
Trematomus bernacchii (*Nototheniidae*)	McMurdo Sound (benthic)	1	30	26	42.3	15.4	3.3 ± 0.6
Dascyllus reticulatus (*Pomacentridae*)	Manila (tropical reef, shallow)	3	10	29	34.5	---	10.7 ± 1.3
Amphiprion clarkii (*Pomacentridae*)	Manila (tropical reef, shallow)	3	11	27	22.2	--7	9.1 ± 1.1
Halichoeres sp. (*Labridae*)	Manila (tropical reef, shallow)	3	10	28	21.4	--7	5.7 ± 1.5
Coryphaenoides acrolepis (*Macruridae*)	San Diego Trough (deep benthic)	5	18	6	66.7	--7	11.0 ± 2.8
Abudefduf troschelii (*Pomacentridae*)	Revillagigedo Islands (tropical, shallow)	7	16	20	35.0	15.0	5.0 ± 1.1
Leuresthes tenuis (*Atherinidae*)	San Diego (inshore, pelagic)	10	20	33	18.2	15.2	3.6 ± 0.7

TABLE I (cont.)

		[1]	[2]	[3]	[4]	[5]	[6]
Gibbonsia metzi (Clinidae)	San Simeon, California (benthic, intertidal)	10	28	28	28.6	17.9	4.3 ± 0.8
Bathygobius ramosus (Gobiidae)	Revillagigedo Islands (tropical, shallow)	12	16	23	4.3	4.3	0.5 ± 0.4
Mugil cephalus (Mugilidae)	Mission Bay, San Diego (inshore, pelagic, tropical subtropical)	15	20	30	36.7	20.0	7.1 ± 1.1
Gillichthys mirabilis (Gobiidae)	San Diego (benthic, estuarine)	20	30	29	30.0	20.0	4.6 ± 1.3

[1] The annual thermal range is an approximation of the variation in water (body) temperature a member of the sampled population would experience in its native habitat. Our estimates are based on published water temperature data and personal observations.

[2] sample size

[3] number of loci surveyed

[4] percent of loci polymorphic

[5] percent of loci polymorphic when commonest allele has a frequency less than 0.95

[6] mean individual heterozygosity determined by actual count of heterozygous genotypes, plus or minus the standard error of the estimate (values given as "per cent")

[7] samples too small to calculate parameter

(Data from Somero and Soulé, 1974)

characteristic of ectotherms which live under extremely stable temperature conditions. If temperature *per se* is a major determinant of the extent of enzyme polymorphism of a species, then fishes living in such thermally stable environments as McMurdo Sound, Antarctica, the deepsea, and in tropical reef habitats ought to have far less isozymic polymorphism than fishes living in such thermally variable habitats as temperate zone estuaries and tide pools.

The results of our initial electrophoretic studies (Somero and Soulé, 1974) are given in Table I. The enzyme systems examined in this study and the isozyme studies discussed below are listed in Table II. Electrophoretic procedures are detailed in Somero and Soulé (1974).

There is clearly no correlation between the amount of enzyme polymorphism characteristic of a fish and the extent of thermal variation in its habitat (Table I). Statistical analysis of these data showed that, for example, fishes experiencing annual temperature variations of 5°C and greater were no more or no less polymorphic than fishes which experienced temperature variations of less than 5°C (Somero and Soulé, 1974). Variation in other habitat conditions did not appear to correlate with enzymic variability either. For example, *Gillichthys mirabilis* experiences large diurnal and/or seasonal changes in oxygen content and salinity as well as temperature. Yet this fish is only average in terms of enzymic polymorphism.

Alternative rationalizations for the patterns observed in in these studies are discussed by Somero and Soulé (1974) and will not be discussed in this paper, other than to point out that a "time-divergence" model of genetic change (Soulé, 1972, 1973) is consistent with our findings that levels of polymorphism are highest in fishes from habitats which have been most stable over geological time periods (the deepsea and tropical reef assemblages).

TEMPERATURE AND ENZYME POLYMORPHISM: THE ROLE OF ISOZYMES IN TEMPERATURE ACCLIMATION

In the studies just described, we were primarily interested in measuring levels of allelic polymorphism, albeit we did look for--and did not find--differences in the number of gene loci (number of isozymes) present in the 13 species studied. One might raise an objection to our approach, however, since the populations studied were sampled at only one time in their seasonal cycles. Thus, even though the eurythermal and stenothermal fishes did not differ in levels of polymorphism in our samples, the results in Table I can say nothing about the

TABLE II
PROTEINS EXAMINED ELECTROPHORETICALLY

DEHYDROGENASES

α-glycerophosphate dehydrogenase
ethanol dehydrogenase
glucose-6-phosphate dehydrogenase
isocitrate dehydrogenase
lactate dehydrogenase
malate dehydrogenase
octanol dehydrogenase
6-phosphogluconate dehydrogenase
xanthine dehydrogenase

OTHER

acid phosphatase
esterases
fumarase
aspartate aminotransferase
leucine aminopeptidase
peptidases
phosphoglucomutase
phosphohexose isomerase
general protein

---

possibility that the eurythermal fishes might synthesize season-specific isozyme forms. In this case, polymorphism could be detected only by sampling over time.

To investigate this latter possibility, we examined seasonal and acclimational effects on electrophoretic patterns of enzymes in several freshwater and marine teleosts (Table III). The results of these studies can be stated most simply: except in the case of the rainbow trout, where acclimation is known to induce a number of isozyme changes (Baldwin and Hochachka, 1970; Moon and Hochachka, 1971; Hochachka and Somero, 1973; Somero and Hochachka, 1971), no differences in enzyme electrophoretic patterns were noted between populations of the different species acclimated or acclimitized to different temperatures. These studies were particularly thorough in the case of the three *Gibbonsia* species, as shown by Ms. Bonnie Jean Davis of our laboratory who examined natural populations throughout the year as well as laboratory acclimated specimens.

With the exception of the rainbow trout data, the possible significance of which we discuss below, our findings again are inconsistent with the prediction that variable habitat

229

TABLE III
FISH SPECIES EXAMINED FOR CHANGES IN ISOZYME
ELECTROPHORETIC PATTERNS DURING SEASONAL
ACCLIMITIZATION OR LABORATORY ACCLIMATION

Species	Acclimation Temperatures	Isozyme Changes
Gillichthys mirabilis	8° and 28°C	None
Amieurus nebulosus	10° and 20°C	None
Gibbonsia elegans[1]	10° and 20°C	None
Gibbonsia metzi[1]	10° and 20°C	None
Gibbonsia montereyensis[1]	10° and 20°C	None
Salmo gairdneri	2-4°C and 17-18°C	Numerous[2]

[1]Davis and Somero, unpublished

[2]For a review of the trout isozyme changes, see Hochachka and Somero (1973).

temperatures and variable enzyme systems go hand-in-hand.

"EURYTOLERANT" AND "STENOTOLERANT" PROTEINS: KINETIC STUDIES

To obtain a complete account of the roles of enzyme variants in environmental adaptation, one must combine kinetic and structural studies with electrophoretic experiments. The mere observation that zymogram banding patterns differ either between species or among populations of a single species can tell us nothing about the functional significance of the enzyme variants.

We have conducted comparative kinetic studies of the effects of temperature on enzyme-substrate interactions in hopes of determining whether differences such as those shown schematically in Figure 1 exist either between the enzymes of different species or between the enzymes of warm- and cold-acclimated individuals of a single species.

Perhaps the best existing data on which answers to questions about the roles of "eurytolerant" and "stenotolerant" enzymes can be based are the findings of Dr. John Baldwin (Baldwin and Hochachka, 1970) who has studied brain acetylcholinesterase (AChE) enzymes of a wide variety of teleost fishes (Figure 2). For each of the AChE enzymes of Figure 2, only a single protein species was evident on electrophoretic media. Note, however, the vastly different temperature sensitivities of the apparent Michaelis constants (Km) of the enzymes. One can certainly term the AChE of the mullet fish (*Mugil cephalus*)

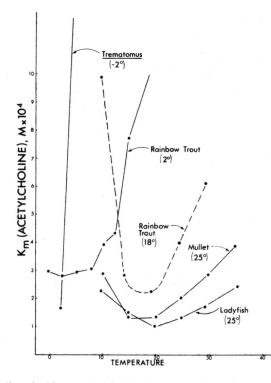

Fig. 2. The influence of assay and habitat temperatures on the apparent Michaelis constant (K_m) of acetylcholine for acetylcholinesterase enzymes from brain tissue of diverse fish species. (From Hochachka and Somero, 1973, with permission of the authors.)

---

a "eurytolerant" or "eurythermal" enzyme, for the substrate binding ability of the enzyme varies extremely little over the wide range of temperatures encountered by this species. In contrast, the AChE of the Antarctic fish, *Trematomus borch-grevinki*, clearly deserves the title, "stenothermal". This AChE binds substrate very well at the low temperatures experienced by the cells of this fish (approximately -1.8 to +2°C), but rapidly loses substrate binding ability at temperatures more than a few degrees above 0°C. The unusually high temperature dependence of substrate binding ability by this critical brain enzyme may be in large measure responsible for setting the low upper incipient lethal temperature of this species, approximately 7°C (Somero and DeVries, 1967).

In contrast to the AChEs of the other species, the enzymes

of the rainbow trout fit the "multiple variant" strategy shown in Figure 1. Both kinetic and electrophoretic evidence (Baldwin and Hochachka, 1970) indicates that warm- and cold-acclimated trout have different AChE isozymes. Trout acclimated to low temperatures (approximately 2-4°C) have a single AChE isozyme with the kinetic properties shown in Figure 2. Warm-acclimated trout have a single AChE isozyme with temperature-Km characteristics which are markedly - and seemingly adaptively (Moon, 1974) - different. Interestingly, trout acclimated to intermediate temperatures (10-12°C) contain both the "warm" and "cold" AChE isozymes (Baldwin and Hochachka, 1970). Thus rainbow trout may employ two variations on the theme of the "multiple variant" strategy: (1) at seasonal extremes of temperature, only the isozyme which is "needed" is synthesized, whereas (2) at intermediate temperatures when both isozymes may be necessary, both isozymes are produced. The observation that trout synthesize only the isozymes actually required under the thermal regime they experience emphasizes a major advantage of using multiple locus isozymes rather than allelic isozymes, i.e., allozymes: in the latter case, both enzyme variants must be synthesized simultaneously (excluding such phenomena as the Lyon effect), while in the case of isozymic variants, only the form of the enzyme which works well is produced. The cell can thus save energy and avoid "cluttering" its limited solvent and membrane systems by using the multiple locus isozyme pattern of adaptation.

SUMMARY AND SPECULATIONS

If we temporarily exclude the rainbow trout from our field of vision, we can state with some confidence that temperature variation in and of itself does not appear to be a sufficient cause for complicated isozyme systems, at least in fishes (for an alternative view regarding plants, see Allard, 1974). The "feat" of developing eurythermal enzymes during evolutionary adaptation to variable habitat temperatures seems eminently achievable as judged by the kinetic properties of enzymes from such eurythermal fishes as the mullet (Figure 2) and *Gillichthys mirabilis* (Hochachka and Somero, 1973). It should be noted that pyruvate kinase enzymes from stenothermal and eurythermal fishes display kinetic properties highly comparable to those shown by the different AChE isozymes (Somero and Hochachka, 1971; Low and Somero, in preparation). Thus we conclude that the "eurytolerant" strategy of adaptation is, in general, optimal for the reasons we have given above: the organism can exist with simpler genetic and protein systems, it will always have the "right" protein in its cells, and energy need not be expended synthesizing enzyme variants which do not function well over at least part of the organism's environment range.

If we bring the rainbow trout into the picture, we are faced with having to rationalize a very different adaptive strategy, one which involves season-specific isozymes-- a "multiple variant" strategy. What types of rationalizations can we raise?

The most convincing explanations for the different behavior of the trout can be based on findings of Ohno (1970) and other workers who have examined the chromosomal and genetic characteristics of this species. Rainbow trout and other salmonid fishes appear to be recent tetraploids. For many enzymes there appears to be enough genetic information to code for twice the number of isozyme forms present in so-called "normal" (non-tetraploid) fishes. An important aspect of this additional DNA is that it is added onto a genome which already carries the necessary information for tissue-, cell- and organelle-specific isozymes. This latest increase in ploidy can be regarded, therefore, as the addition of a "frosting" on an already well-developed genetic "cake". Having the types of isozymes which are required to perform tissue-, cell-, and organelle-specific functions, the trout has been able to utilize the latest addition to its genetic repertory to develop environmentally-adaptive isozymes.

The observation that one tetraploid organism has developed a multiple isozyme strategy of environmental adaptation raises the question as to whether other polyploid species rely on similar mechanisms. In the case of fishes, Wilson (1974) has shown that the tetraploid goldfish, *Carassius auratus*, does not synthesize different isozymes during warm- and cold-acclimation. The closest parallels to the trout system may, in fact, be found in plant species. Unlike animals, plants do not face complex genetic "problems" deriving from sex determination mechanisms when increases in ploidy occur. Thus changes in ploidy are relatively widespread among plants, and in these cases we may well find that increases in genetic information have been utilized to develop environmentally-adaptive isozymes akin to those found in the rainbow trout.

REFERENCES

Allard, R. W., A. L. Kahler, and M. T. Clegg 1974. Isozymes in plant population genetics. *I Isozymes: Molecular Structure.* C. L. Markert, editor, Academic Press, N.Y.

Baldwin, J. 1971. Adaptation of enzymes to temperature: acetylcholinesterases in the central nervous system of fishes. *Comp. Biochem. Physiol.* 40: 181-187.

Baldwin, J. and P. W. Hochachka 1970. Functional significance of isoenzymes in thermal acclimation: acetylcholinesterases from trout brain. *Biochem. J.* 116: 883-887.

Hochachka, P. W. and G. N. Somero 1973. *Strategies of Biochemical Adaptation.* W. B. Saunders Co., Philadelphia.

Johnson, G. 1973. Enzyme polymorphism and biosystematics: the hypothesis of selective neutrality. *Ann. Rev. Ecology and Systematics* 4: 93-116.

Johnson, G. 1974. Enzyme polymorphism and metabolism. *Science* 184: 28-37.

Levins, R. 1968. *Evolution in Changing Environments.* Princeton University Press, Princeton, N.J.

Moon, T. W. 1974. Temperature adaptation: Isozymic function and the maintenance of heterogeneity. *II Isozymes: Physiology and Function.* C. L. Markert, editor, Academic Press, N.Y.

Moon, T. W., and P. W. Hochachka 1971. Temperature and enzyme activity in poikilotherms: isocitrate dehydrogenases in rainbow trout liver. *Biochem. J.* 123: 695-705.

Ohno, S. 1970. *Evolution by Gene Duplication.* Springer-Verlag, Berlin.

Selander, R. K., and D. W. Kaufman 1973. Genic variability and strategies of adaptation in animals. *Proc. Nat. Acad. Sci. USA.* 70: 1875-1877.

Somero, G. N. and A. L. DeVries 1967. Temperature tolerance of some Antarctic fishes. *Science* 156: 257-258.

Somero, G. N. and P. W. Hochachka 1971. Biochemical adaptation to the environment. *Amer. Zool.* 11: 159-167.

Somero, G. N. and M. Soulé 1974. Genetic variation in marine teleosts: a test of the niche-variation hypothesis. *Nature,* in press.

Soulé M. 1972. Phenetics of natural populations. III Variation in insular populations of a lizard. *Amer. Naturalist* 106: 429-446.

Soulé, M. 1973. The epistasis cycle: a theory of marginal populations. *Ann. Rev. Ecology and Systematics* 4: 93-116.

Wilson, F. R., M. J. Champion, G. S. Whitt, and C. L. Prosser 1974. Isozyme patterns in tissues of temperature-acclimated fish. *II. Isozymes: Physiology and Function.* C. L. Markert, editor. Academic Press, New York.

CYCLIC NUCLEOTIDE-DEPENDENT PROTEIN KINASES FROM EUKARYOTIC ORGANISMS

HIROHEI YAMAMURA, YOSHIMI TAKAI, KAORU NISHIYAMA,
and YASUTOMI NISHIZUKA
Department of Biochemistry
Kobe University School of Medicine
Ikuta-ku, Kobe, Japan

ABSTRACT. Cyclic AMP-dependent protein kinases distributed in a wide variety of eukaryotic organisms exhibit closely similar catalytic properties and apparently lack tissue-as well as species-specificities, at least with respect to their functional activities. In contrast, cyclic GMP-dependent protein kinase appears to belong to another category and shows entirely different substrate specificity from that of the class of cyclic AMP-dependent protein kinases.

Cyclic AMP (adenosine 3', 5'-monophosphate) is well known as a key intracellular regulator of a number of biological processes in a variety of living cells (Robison et al., 1968). Cyclic AMP-dependent protein kinase has been found in rabbit skeletal muscle by Walsh, Perkins, and Krebs (1968). Subsequently, Greengard and other investigators have reported wide distribution of this enzyme, and biochemical and physiological effects elicited by the cyclic nucleotide are proposed to be mediated through activation of this class of enzymes (Kuo and Greengard, 1969; and Yamamura et al., 1971a). Cyclic AMP-dependent protein kinase is shown to be composed of two subunits, a catalytic unit and a regulatory unit. The cyclic nucleotide is selectively bound to the regulatory unit; this causes the kinase to dissociate and the free catalytic subunit exhibits full activity. The reaction is reversible. This mode of action of cyclic AMP on protein kinase has been clarified by this and other laboratories (Kumon et al., 1970; Gill and Garren, 1970; Tao et al., 1970; and Reimann et al., 1971). In most mammalian tissues, apparently multiple species of this class of enzymes are distinguished (Tao et al., 1970; Yamamura et al., 1971b; Gill and Garren, 1971; Reimann et al., 1971; Miyamoto et al., 1971; and Kumon et al., 1972); but systemic analysis in this laboratory has shown that these multiple kinases are composed of a common catalytic unit and apparently differ from each other in their associated regulatory units. In addition, this class of enzymes is found in a wide variety of vertebral and invertebral tissues, and also in unicellular eukaryotic cells such as yeast (Takai et al., 1974).

The enzyme has not been detected in prokaryotic organisms such as *Escherichia coli.*

Fig. 1 shows a working hypothesis for a possible mechanism of cyclic AMP actions in eukaryotic organisms. Some trigger such as a hormone stimulates adenylate cyclase and produces cyclic AMP. The protein kinase, which is activated by the cyclic nucleotide in a specific manner, reveals broad substrate specificity and phosphorylates pleiotropically various enzymes and proteins, such as phosphorylase kinase, glycogen synthetase, lipase, histone, ribosomal proteins, and membrane-associated proteins. The evidence seems to assign a role of crucial importance to the protein kinase in the simultaneous control of several biological reactions in each tissue. Table I shows subunit structure of protein kinases obtained from rat liver and yeast. All cyclic AMP-dependent protein kinases thus far obtained are composed of catalytic and regulatory units, as mentioned above. The molecular weight of rat liver holoenzyme is different from that of yeast holoenzyme. However, the liver and yeast catalytic units have rather similar molecular weight, 35,000 and 30,000, respectively. The regulatory unit of the liver enzyme has a molecular weight of 150,000. Preliminary analysis suggests that the mammalian regulatory unit is composed of smaller subunits with molecular weight of approximately 40,000. In contrast, the yeast regulatory unit shows the molecular weight of about 28,000 and appears to be composed of two smaller subunits of molecular weight about 14,000. Nevertheless, these enzymes are activated in an essentially similar manner and the activation process is reversible. The catalytic and regulatory units from different origins cross-react; recombination of these units from heterologous sources produces a hybrid cyclic AMP-dependent protein kinase. For example, the yeast catalytic unit is inhibited progressively by the addition of increasing amounts of the rat liver regulatory unit as shown in Fig. 2. This inhibition is completely overcome by the addition of cyclic AMP, and the original catalytic activity is fully restored. The regulatory unit alone is essentially free of kinase activity. Thus the mode of action of cyclic AMP appears to follow a universal mechanism.

Cyclic AMP-dependent protein kinases obtained from various sources appear to lack tissue-and species-specificities and the enzymes exhibit similar properties and substrate specificities. Fig. 3 shows the autoradiograms of tryptic peptides of histone, which was phosphorylated separately with liver or yeast cyclic AMP-dependent protein kinase. The radioactive histone preparations were digested by trypsin, and the tryptic peptides obtained were subjected to paper chromatography followed by high voltage paper electrophoresis

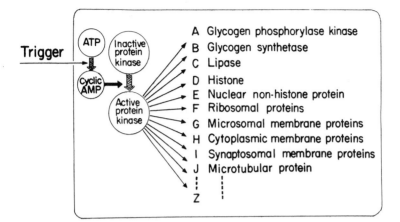

Fig. 1: A proposed mechanism of action of cyclic AMP in eukaryotic organisms.

TABLE I

POSSIBLE SUBUNIT STRUCTURE OF CYCLIC AMP-DEPENDENT
PROTEIN KINASES

	Molecular weight	
	Liver	Yeast
Holoenzyme	180,000	58,000
Catalytic unit	35,000	30,000
Regulatory unit	150,000	28,000

Rat liver and yeast cyclic AMP-dependent protein kinases were obtained under the conditions described previously (Kumon et al., 1972; Takai et al., 1974).

and autoradiography. Closely similar patterns were observed for these enzymes. Under these conditions both histone preparations were equally digested to produce identical sets of more than 38 spots, as visualized by the ninhydrin reaction. The results suggest that the enzymes from different sources phosphorylate the same specific sites of substrate proteins and show identical catalytic properties.

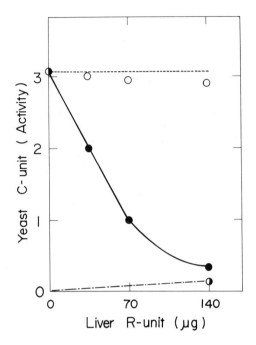

Liver R-unit (µg)

Fig. 2:   Effects of rat liver regulatory unit on yeast cata·
lytic unit. Rat liver regulatory unit was added to the assay
mixture containing 2.5 µmoles of Tris-Cl at pH 7.5, 1.25
µmoles of magnesium acetate, 2.5 µmoles of $[\gamma-^{32}P]$ ATP (5 x
10^4 cpm per nmole), 100 µg of histone and 14 µg of yeast
catalytic unit. Where indicated cyclic AMP (0.4 µM) was added. Other assay conditions were described previously (Takai
et al., 1974). Rat liver regulatory unit was prepared by the
method of Kumon, Yamamura,and Nishizuka (1970) and yeast
catalytic unit was prepared by the method of Takai, Yamamura,
and Nishizuka (1974). C-unit, catalytic unit; R-unit, regulatory unit. o and ● : Protein kinase activity with and without
cyclic AMP (0.4 µM), respectively. ◑ : Protein kinase
activity of regulatory unit alone.

Further evidence for the catalytic identity of the protein
kinases from different sources may be provided by their functional abilities to phosphorylate several enzymes and proteins.
For example, both mammalian and yeast enzymes can phosphorylate
glycogen phosphorylase kinase from rabbit skeletal muscle.
Walsh, Perkins, and Krebs (1968) have shown that muscle glycogen phosphorylase is activated by the protein kinase; cyclic
AMP-dependent protein kinase phosphorylates glycogen phosphory-

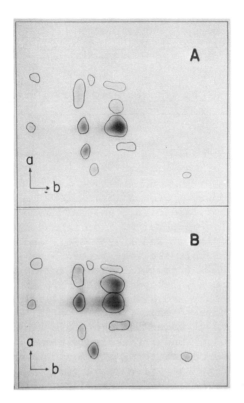

Fig. 3: Autoradiography of tryptic digests of radioactive histone preparations phosphorylated by yeast and rat liver cyclic AMP-dependent protein kinases. Paper chromatography (Direction a) and subsequent paper electrophoresis at pH 3.5 (Direction b) were carried out under the conditions specified earlier (Kumon et al., 1972). A, yeast protein kinase; B, rat liver protein kinase. The data were taken from Takai, Yamamura and Nishizuka (1974).

lase kinase and phosphorylase kinase thus activated is effective in the conversion of inactive phosphorylase to its active form. As shown in Table II, rabbit muscle, rat liver, and yeast cyclic AMP-dependent protein kinases were equally able to phosphorylate the muscle glycogen phosphroylase system and regulate its enzymic activity. The activity observed in the absence of protein kinase appeared to be due to autocatalytic phosphorylation of glycogen phosphorylase kinase as described by Krebs (1966), or to the activated form of glycogen phosphory-lase kinase that slightly contaminated the preparation.

239

TABLE II

EFFECTS OF MAMMALIAN AND YEAST CYCLIC AMP-DEPENDENT PROTEIN
KINASES ON MUSCLE GLYCOGEN PHOSPHORYLASE KINASE

Protein kinase added	^{32}Pi incorporated into glucose 1-phosphate	
	-ATP	+ATP
	cpm	*cpm*
None	154	1,476
Rabbit skeletal muscle	–	3,575
Rat liver	95	3,590
Yeast	86	3,592

Glycogen phosphorylase b (8.5 μg) and glycogen phosphorylase
kinase (0.6 μg) were obtained from rabbit skeletal muscle by
the methods of Fischer and Krebs (1962) and Cohen (1973),
respectively. Rabbit skeletal muscle, rat liver and yeast
cyclic AMP-dependent protein kinases were prepared by the
method of Yamamura, Nishiyama, Shimomura,and Nishizuka (1973)
and Takai, Yamamura,and Nishizuka (1974). The amounts of
kinases used in these experiments were in equivalent quantities and, when glycogen phosphorylase kinase was replaced
by histone (100 μg) under these conditions,each kinase transferred 24 pmoles of the terminal phosphate of ATP to histone
per min at 30°C in the presence of cyclic AMP (7 x 10^{-6} $\underline{M}$).
Where indicated, ATP ($10^{-5}\underline{M}$) was added. Glycogen phosphorylase was assayed by the method of Yamamura, Nishiyama,
Shimomura, and Nishizuka (1973) and other experimental conditions were identical with those given previously (Yamamura
et al., 1971a).

Similarly, the protein kinases are equally able to phosphorylate muscle glycogen synthetase. Bishop and Larner (1969) have
shown that muscle glycogen snythetase is inhibited by the
protein kinase; thus cyclic AMP-dependent protein kinase
phosphorylates glycogen synthetase resulting in the conversion
of I-form to D-form. As shown in Table III, cyclic AMP-dependent protein kinases isolated from the homologous as well
as from the heterologous sources were equally active in the
phosphorylation of muscle glycogen synthetase and inhibited
the enzymic activity. Essentially similar results were
obtained with protein kinases purified from many other vertebrate and invertebrate tissues. Thus the evidence indicates
that cyclic AMP-dependent protein kinases in general show
identical spectra for substrate proteins and lack tissue-
as well as species-specificity regardless of the enzyme
source. Quantitative and qualitative analyses of the substrate proteins within the cell may provide important clues

TABLE III

CONVERSION OF I-FORM TO D-FORM OF MUSCLE GLYCOGEN SYNTHETASE
BY MAMMALIAN AND YEAST CYCLIC AMP DEPENDENT PROTEIN KINASES

Protein kinase	^{14}C-Glucose incorporated into glycogen	Relative Activity
	cpm	%
None	1,482	100
Rabbit skeletal muscle	260	17
Rat liver	196	13
Yeast	270	18

Rabbit skeletal muscle glycogen synthetase was purified by
the method of Villar-Palasi, Rosell-Perez, Hizukuri, Huijing,
and Larner (1966). This preparation was a mixture of I- and
D-forms— initially 22% and 78%, respectively. Rabbit
skeletal muscle, rat liver and yeast cyclic AMP-dependent
protein kinases were prepared as described previously (Yamamura et al., 1973; and Takai et al., 1974). The protein
kinases employed were in same quantities and each transferred
450 pmoles of the terminal phosphate of ATP per min when calf
thymus histone was used as phosphate acceptor. Assay of
glycogen synthetase and other experimental conditions were
identical with those specified earlier (Yamamura et al., 1973).
for better understanding the molecular basis of cyclic AMP
actions in each tissue.

Recently, cyclic GMP (guanosine 3', 5'-monophosphate) was
proposed to be an additional chemical mediator of some hormonal actions (Goldberg et al., 1973), but the metabolic
effects of cyclic GMP are apparently opposite to those of
cyclic AMP in some tissues. George, Polson, O'Toole, and
Goldberg (1970) have found that perfusion of rat heart with
acetylcholine produces an elevation of cyclic GMP concentration, and Illiano, Tell, Siegel, and Cuatrecasas (1973) have
proposed that insulin increases the cyclic GMP content of
isolated fat cells. Cyclic GMP-dependent protein kinase has
been found in lobster tail muscle (Kuo and Greengard, 1970),
mammalian brain, bladder and uterine tissue (Greengard and
Kuo, 1970), anthropods (Kuo et al., 1971), rat cerebellum
(Hofman and Sold, 1972), and in rat pancreas (Van Leemput-
Coutrez et al., 1973). In order to explore the detailed
properties as well as the functional specificity of this
class of enzymes, subsequent studies were undertaken to
compare cyclic GMP-dependent and cyclic AMP-dependent protein
kinases that were partially purified from the silkworm.

Fig. 4 shows that the resolution of cyclic AMP-dependent
and cyclic GMP-dependent protein kinases by calcium phosphate

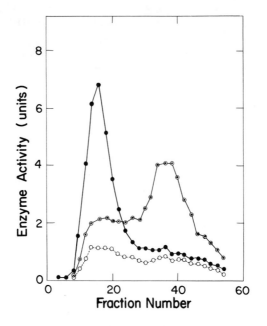

Fig. 4: Separation of cyclic AMP-dependent and cyclic GMP-dependent protein kinases from silkworm by calcium phosphate gel cellulose column chromatography. Pupae of silkworm (30 g) were homogenized at 0°C with 4 volumes of 10 mM potassium phosphate buffer, pH 7.0, containing 4 mM EDTA, 5 mM magnesium acetate and 4 mM 2-mercaptoethanol using Waring blendor for one min at 0°C. The homogenate was centrifuged for 30 min at 12,000 x g. To the supernatant solution ammonium sulfate was added to a final concentration of 60% saturation. After centrifugation the precipitate was dissolved in 20 ml of 10 mM potassium phosphate buffer at pH 7.0 containing 4 mM EDTA and 4 mM 2-mercaptoethanol, and dialyzed overnight against the same buffer. The dialysate was applied to a calcium phosphate gel cellulose column (diameter, 3 cm; length, 13 cm) equilibrated with 0.01 M potassium phosphate buffer at pH 7.0 containing 6 mM 2-mercaptoethanol and 1 mM EDTA. The gel cellulose was prepared by the method of Koike and Hamada (1971). After washing the column with 480 ml of the same buffer, elution was carried out with 800 ml of a linear concentration gradient of potassium phosphate buffer at pH 7.0 (0.01 M to 0.35 M) containing 6 mM 2-mercaptoethanol and 1 mM EDTA. ●, ⊙ and O ; Protein kinase with cyclic GMP, cyclic AMP and without cyclic nucleotide, respectively.

gel cellulose column chromatography. Both kinases were
assayed with calf thymus histone as substrate. The enzymes
were extracted from pupae of silkworm, fractionated by
ammonium sulfate, and then chromatographed under the conditions specified in the legend of Fig. 4. Two protein kinases
appeared; the first peak was dependent on cyclic GMP, whereas the second peak was stimulated by cyclic AMP. The first
peak was able to bind only cyclic GMP and the second could
bind only cyclic AMP. Each enzyme was purified further by
DEAE-cellulose column chromatography. The properties of
these kinases are briefly summarized in Table IV. Although
both cyclic AMP-dependent and cyclic GMP-dependent protein
kinases reacted with histone and protamine, the physical and
kinetic properties of these enzymes differed from each other.
In fact, these enzymes were shown to phosphorylate different
seryl and threonyl residues of histone and protamine as
judged by the fingerprint procedure (Fig. 5). Under these
conditions both histone preparations were digested equally
to produce identical sets of more than 30 spots visualized by
the ninhydrin reaction. The evidence indicates that these
protein kinases phosphorylate specific sites in the substrate
protein and that the natural substrates of these enzymes are
probably different.

TABLE IV

PROPERTIES OF CYCLIC AMP-DEPENDENT AND CYCLIC GMP-DEPENDENT
PROTEIN KINASES PURIFIED FROM SILKWORM.

Properties	Cyclic AMP-dependent Protein kinase	Cyclic GMP-dependent Protein kinase
K_m value for ATP	0.45×10^{-5} $\underline{M}$	3×10^{-5} $\underline{M}$
K_a value for cyclic nucleotide	1.3×10^{-8} $\underline{M}$ (cyclic AMP)	7.5×10^{-9} $\underline{M}$ (cyclic GMP)
Optimum Mg^{++} concentration	3.0 $\underline{mM}$	100 $\underline{mM}$
Optimum pH	pH 7.0	pH 7.5-8.0
Isoelectric point	pH 5.4	pH 5.4
Molecular weight	180,000	140,000

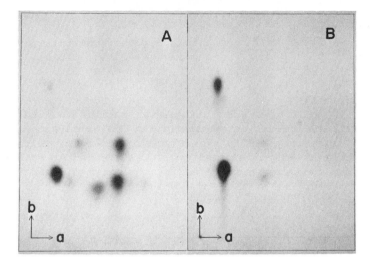

Fig. 5: Autoradiography of tryptic digests of radioactive histone preparations fully phosphorylated by cyclic AMP- and cyclic GMP-dependent protein kinases from silkworm. Calf thymus whole histone was fully phosphorylated separately with the cyclic AMP-dependent or cyclic GMP-dependent enzyme, and then digested with trypsin. The tryptic digestion, paper chromatography and high-voltage paper electrophoresis were carried out under the condition described previously (Takai et al., 1974). A, cyclic AMP-dependent protein kinase. B, cyclic GMP-dependent protein kinase.

Another set of experiments was designed to test whether the insect cyclic GMP-dependent protein kinase was able to phosphorylate muscle phosphorylase kinase. As shown in Table V, the cyclic AMP-dependent enzyme phosphorylated phosphorylase kinase and stimulated glycogenolysis as was found for mammalian and yeast enzymes. In contrast, the cyclic GMP-dependent enzyme was inactive in this capacity under the same conditions. Similarly, the cyclic AMP-dependent enzyme cross-reacted with muscle glycogen synthetase and inhibited its enzymic activity, whereas the cyclic GMP-dependent enzyme did not react with glycogen synthetase (data are not shown). These results clearly indicate that cyclic AMP-dependent and cyclic GMP-dependent protein kinases have different substrate specificities. Presumably, the cyclic GMP-dependent enzyme phosphorylates another class of substrate proteins and regulates their biological activities.

244

TABLE V

EFFECTS OF CYCLIC AMP- AND CYCLIC GMP-DEPENDENT PROTEIN KIN-
ASES FROM SILKWORM ON MUSCLE GLYCOGEN PHOSPHORYLASE KINASE

Protein Kinase	Cyclic nucleotide		^{32}Pi incorporated into glucose 1-phosphate
	Cyclic AMP	Cyclic GMP	
			cpm
None	+	-	360
	-	+	100
Cyclic AMP-dependent protein	+	-	3,550
	-	+	844
Cyclic GMP-protein	+	-	490
kinase	-	+	120

Experimental conditions were identical with those given in
Table II. Where indicated, 5.6 x 10^{-8} M of cyclic nucleotide
was added. Cyclic AMP-dependent and cyclic GMP-dependent
protein kinases added in this experiment each transferred
40 pmoles of the terminal phosphate of ATP into acid-pre-
cipitable material per min at 30°C when calf thymus histone
(100 μg) was used as substrate instead of glycogen synthetase.

Based on the experimental results described above, it may
be emphasized again that cyclic AMP-dependent protein kinases
from mammalian tissues as well as from other sources such as
yeast and silkworm exhibit similar catalytic and functional
properties, and play a central role in regulating various
biological reactions in a variety of eukaryotic cells. In
contrast, in mammals cyclic GMP-dependent protein kinases
have been found thus far only in the brain, pancreas and
uterine tissue. Natural substrate proteins as well as
tissue-and species-specificities of this class of enzymes
are unknown. Nevertheless, the cyclic GMP-dependent enzyme
obtained from silkworm is now shown to exhibit clearly different substrate specificity from that of cyclic AMP-dependent
enzymes, and neither muscle phosphorylase kinase nor glycogen
synthetase serves as substrate for the cyclic GMP-dependent
enzyme. Further studies are currently underway in an effort
to identify the natural substrate proteins as well as the
biological roles of cyclic GMP-dependent protein kinases.

ACKNOWLEDGEMENTS

This investigation has been supported in part by the research grants from the Jane Coffin Childs Memorial Fund for Medical Research, the Toray Science Foundation and the Scientific Research Fund of the Ministry of Education of Japan. The authors are grateful to Miss Reiko Shimomura, Mr. Yorihiko Morishita, Mr. Hideki Katakami, and Mr. Kuniyasu Sakai for their skillful technical assistance.

REFERENCES

Bishop, J. S. and J. Larner 1969. Presence in liver of a 3',5'-cyclic AMP stiumlated kinase for the I form of UDPG-glycogen glucosyltransferase. *Biochim. Biophys. Acta* 171: 374-377.

Cohen, P. 1973. The subunit structure of rabbit skeletal muscle phosphorylase kinase and the molecular basis of its activation reactions. *Eur. J. Biochem.* 34: 1-14.

Fischer, E. H. and E. G. Krebs 1962. Glycogen phosphorylase b from rabbit skeletal muscle. *Methods Enzymol.* 5: 369-373.

George, W. J., J. B. Polson, A. G. O'Toole, and N. D. Goldberg 1970. Elevation of guanosine 3',5'-cyclic phosphate in rat heart after perfusion with acetylcholine. *Proc. Nat. Acad. Sci. U.S.A.* 66: 398-403.

Gill, G. N. and L. D. Garren 1970. A cyclic 3',5'-adenosine monophosphate-dependent protein kinase from the adrenal cortex: Comparison with a cyclic AMP binding protein. *Biochem. Biophys. Res. Commun.* 39: 335-343.

Gill, G. N. and L. D. Garren 1971. Role of the receptor in the mechanism of action of adenosine 3',5'-cyclic mono-phosphate. *Proc. Nat. Acad. Sci. U.S.A.* 68: 786-790.

Goldberg, N. D., R. F. O'Dea, and M. K. Haddox 1973. Cyclic GMP. In *Advances in Cyclic Nucleotide Research*, Vol. 3., P. Greengard and G. A. Robison, eds., Raven Press, New York: 155-223.

Greengard, P. and J. F. Kuo 1970. On the mechanism of action of cyclic AMP. In *Advances in Biochemical Psychopharmacology*, Vol. 3., P. Greengard and E. Costa, eds., Raven Press, New York: 287-306.

Hofman, F. and G. Sold 1972. A protein kinase activity from rat cerebellum stimulated by guanosine 3',5'-monophosphate. *Biochem. Biophys. Res. Commun.* 49: 1100-1107.

Illiano, G., G. P. E. Tell, M. Siegel, and P. Cuatrecasas 1973. Guanosine 3',5'-cyclic monophosphate and the action of insulin and acetylcholine. *Proc. Nat. Acad. Sci. U.S.A.*

70: 2443-2447.

Koike, M. and M. Hamada 1971. Preparation of calcium phosphate gel deposited on cellulose. *Methods Enzymol.* 22: 339-342.

Krebs, E. G. 1966. Phosphorylase b kinase from rabbit muscle. *Methods Enzymol.* 8: 543-546.

Kumon, A., K. Nishiyama, H. Yamamura, and Y. Nishizuka 1972. Multiplicity of adenosine 3',5'-monophosphate-dependent protein kinases from rat liver and mode of action of nucleoside 3',5'-monophosphate. *J. Biol. Chem.* 247: 3726-3735.

Kumon, A., H. Yamamura, and Y. Nishizuka 1970. Mode of action of adenosine 3',5'-cyclic phosphate on protein kinase from rat liver. *Biochem. Biophys. Res. Commun.* 41: 1290-1297.

Kuo, J. F. and P. Greengard 1969. Cyclic nucleotide-dependent protein kinases. IV. Widespread occurrence of adenosine 3',5'-monophosphate-dependent protein kinase in various tissues and phyla of the animal kingdom. *Proc. Nat. Acad. Sci. U.S.A.* 64: 1349-1355.

Kuo, J. F. and P. Greengard 1970. Cyclic nucleotide-dependent protein kinase. VI. Isolation and partial purification of a protein kinase activated by guanosine 3',5'-monophosphate. *J. Biol. Chem.* 245: 2493-2498.

Kuo, J. F., G. R. Wyatt, and P. Greengard 1971. Cyclic nucleotide-dependent protein kinases. IX. Partial purification and some properties of guanosine 3',5'-monophosphate-dependent and adenosine 3',5'-monophosphate-dependent protein kinases from various tissues and species of arthropoda. *J. Biol. Chem.* 246: 7159-7167.

Miyamoto, E., G. L. Petzold, J. S. Harris, and P. Greengard 1971. Dissociation and concomitant activation of adenosine 3',5'-monophosphate-dependent protein kinase by histone. *Biochem. Biophys. Res. Commun.* 44: 305-312.

Reimann, E. M., C. O. Brostrom, J. D. Corbin, C. A. King, and E. G. Krebs 1971. Separation of regulatory and catalytic subunits of the cyclic 3',5'-adenosine monophosphate-dependent protein kinase(s) of rabbit skeletal muscle. *Biochem. Biophys. Res. Commun.* 42: 187-194.

Robison, G. A., R. W. Butcher, and E. W. Sutherland 1968. Cyclic AMP. *Annual Review of Biochemistry* 37: 149-174.

Takai, Y., H. Yamamura, and Y. Nishizuka 1974. Adenosine 3',5'-monophosphate-dependent protein kinase from yeast. *J. Biol. Chem.* 249: 530-535.

Tao, M., M. L. Salas, and F. Lipmann 1970. Mechanism of activation by adenosine 3',5'-cyclic monophosphate of a protein phosphorylase from rabbit reticulocytes. *Proc. Nat. Acad.*

Sci. U.S.A. 67: 408-414.

Van Leemput-Coutrez, M., J. Camus, and J. Christophe 1973. Cyclic nucleotide-dependent protein kinases of the rat pancreas. *Biochem. Biophys. Res. Commun.* 54: 182-190.

Villar-Palasi, C., M. Rosell-Perez, S. Hizukuri, F. Huijing, and J. Larner 1966. Muscle and liver UDP-glucose: α-1, 4-glucan α-4-glucosyltransferase (Glycogen synthetase). *Methods Enzymol.* 8: 374-384.

Walsh, D. A., J. P. Perkins, and E. G. Krebs 1968. An adenosine 3',5'-monophosphate-dependent protein kinase from rabbit skeletal muscle. *J. Biol. Chem.* 243: 3763-3765.

Yamamura, H., A. Kumon, and Y. Nishizuka 1971a. Cross-reactions of adenosine 3',5'-monophosphate-dependent protein kinase systems from rat liver and rabbit skeletal muscle. *J. Biol. Chem.* 246: 1544-1547.

Yamamura, H., A. Kumon, K. Nishiyama, M. Takeda, and Y. Nishizuka 1971b. Characterization of two adenosine 3',5'-monophosphate-dependent protein kinases from rat liver. *Biochem. Biophys. Res. Commun.* 45: 1560-1566.

Yamamura, H., K. Nishiyama, R. Shimomura, and Y. Nishizuka 1973. Comparison of catalytic units of muscle and liver adenosine 3',5'-monophosphate-dependent protein kinases. *Biochemistry* 12: 856-862.

Yamamura, H., M. Takeda, A. Kumon, and Y. Nishizuka 1970. Adenosine 3',5'-cyclic phosphate-dependent and independent histone kinases from rat liver. *Biochem. Biophys. Res. Commun.* 40: 675-682.

SINGLE REACTIONS WITH MULTIPLE FUNCTIONS:
MULTIPLE ENZYMES AS ONE OF THREE PATTERNS IN MICROORGANISMS

CAROLYN S. BROWN, ELLIS L. KLINE,
and H. E. UMBARGER
Department of Biological Sciences
Purdue University
West Lafayette, Indiana   47907

ABSTRACT: Diverging and converging pathways are frequently encountered in the study of cell metabolism and therefore a single chemical transformation is often found to serve multiple functions. Control of metabolite flow through multipurpose anabolic transformations can be achieved by concerted inhibition by the multiple endproducts, sequential inhibition by each branch point intermediate, or by uniquely controlled multiple enzymes (isozymes). Five distinct functional classes of the latter are discussed here. The catabolic and biosynthetic threonine deaminases and acetohydroxy acid synthases are considered in some detail as typical examples.

INTRODUCTION

Viewed very simply, cell metabolism can be divided into three kinds of pathways: the central metabolic routes from which intermediates are drawn via a great many highly diverging routes (usually with an anabolic function) and into which compounds are directed from a variety of highly converging routes (usually with a catabolic function). In both the diverging and the converging pathways, reactions necessarily occur that have multiple functions. Very often the enzymes catalyzing these multifunctional reactions are controlled both in amount and in activity in ways that make them highly suited to the roles they play.

Considering at this point only the control of carbon flow through the multifunctional reaction, three patterns have been encountered. These patterns are represented in Fig. 1, in which a common pathway diverges into two distinct branches. One mode of control is by concerted inhibition in which the first reaction in the pathway is inhibited by the concerted action of both endproducts (Datta and Gest, 1964, and Paulus and Grey 1968). A second mode of control is by sequential inhibition in which the branch point intermediate is the inhibitor of the first common reaction and the first reactions in the specific pathways are inhibited by the corresponding endproducts (Jensen and Nester, 1963). In the third mode of

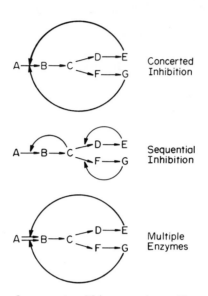

Fig. 1. Patterns for controlling carbon flow in the multi-functional reaction, A → B.

control two enzymes catalyze the first step in the common pathway, one enzyme is inhibited by one endproduct and the second is inhibited by the other (Stadtman et al., 1961).

SOME EXAMPLES OF MULTIPLE ENZYMES IN BACTERIA

In microorganisms many multifunctional reactions are catalyzed by multiple enzymes. Table 1 lists a few chosen to illustrate the way in which one of a pair may differ from the other. One category would be that in which one enzyme has a catabolic role and the other a biosynthetic role. Relatively early in the study of branched-chain amino acid biosynthesis, two clearly defined examples were encountered. One pair was the biosynthetic and the catabolic threonine deaminases studied in *E. coli* (Umbarger and Brown, 1957) and the other was the biosynthetic and catabolic acetohydroxy acid synthases of *Aerobacter aerogenes* (Halpern and Umbarger, 1959). In each case the biosynthetic enzymes are controlled by the repression mechanism affecting branched-chain amino acid biosynthesis and by endproduct inhibition. The dependence upon cyclic AMP for the formation of the catabolic threonine deaminase and upon AMP for its activity clearly indicates its role as a

catabolic enzyme (Shizuta et al., 1970). Again, the degradative acetohydroxy acid synthase which leads to acetoin
synthesis is formed and is active only after the pH of the
medium has fallen as a result of sugar fermentation (Juni,
1952). The enzymes in both of these pairs are clearly distinguished from each other in terms of effector ligands and
regulation of formation. By the Enzyme Commission rules, they
would seem to fall into the formal definition of isozymes.

More subtle in different functions are the multiple enzymes
that have biosynthetic roles but are repressed and inhibited by
different endproducts. The three DAHP synthases of *E. coli*
constitute one example. One enzyme is inhibited and repressed
specifically by tryptophan, another is inhibited and repressed
by tyrosine, whereas the third is inhibited by phenylalanine
but multivalently repressed by phenylalanine and tryptophan
(Brown and Somerville, 1971 and Im et al., 1971).

Somewhat analogous are the three aspartokinase activities
(Cohen, 1969). In *E. coli* K-12, two are exhibited by proteins
that also have homoserine dehydrogenase activities. Aspartokinase I--homoserine dehydrogenase I is inhibited by threonine and, like the other two products of the thr operon, is
multivalently repressed by threonine and isoleucine. Aspartokinase II--homoserine dehydrogenase II is repressed by methionine. The strongest inhibition of the third aspartokinase
(which exhibits no other activity) by a single amino acid is
by lysine, but there is significant but weaker inhibition also
by leucine, methionine, and several other amino acids. Furthermore, there is a strong synergistic inhibition that results
from the concerted action of lysine and one of the weaker inhibitors (Lee et al., 1966). Lysine specifically represses
aspartokinase III. In *E. coli* B, however, aspartokinase II
has no associated homoserine dehydrogenase activity. Which of
the three multiple enzymes in the two strains are to be
considered isozymes?

Another class of multiple enzymes are those that have different catabolic functions and which under normal circumstances
could not substitute for each other. Kornberg and Malcovati
(1973) have described two pyruvate kinases in *E. coli*. One is
activated by fructose-1, 6-diphosphate and is the functionally
predominate one when glycolysis is a major metabolic route.
When gluconeogenesis (from TCA cycle intermediates for example)
is important, the AMP-activated enzyme is predominant.

Exactly reciprocal to the general pattern of diverging
pathways in biosynthesis is the pattern of converging pathways
exhibited by catabolic pathways. A typical example is the
separately induced enzymes transferring the CoA group between
succinate and β-ketoadipate (Hoet and Stanier, 1970). The

251

TABLE 1

Some Selected Examples of Multiple Enzymes in Bacteria

Type	Enzyme Activity	Typical Organism	Distinguishing Features	Reference
	Acetohydroxy Acid Synthase	*A. aerogenes*	Biosynthetic: Required FAD, inhibited by valine, pH 7 optimum, multivalently repressible Catabolic: pH 6 optimum, FAD not required, formed at low pH during carbohydrate formation	Halpern (1959) Huseby et al. (1971)
Multiple Enzymes with Biosynthetic and Catabolic Functions	Threonine Deaminase	*E. coli*	Biosynthetic: Isoleucine inhibits, multivalently repressed Catabolic: AMP required; inducible, catabolite repressed	Burns and Zarlengo (1968) Shizuta and Hayaishi (1970) and Gerlt et al. (1973)
	Ornithine Trans- carbamylase	*P. aeruginosa*	Biosynthetic: Optimal pH 8.5; repressed by arginine Catabolic: Optimal pH 7.3; cooperative kinetics; inhibited by putrescine	Stalon (1972) and Ramos et al. (1965)

252

TABLE 1 (continued)

3-Deoxy-D-Arabino Heptulosonic acid-7-Phosphate Synthase	*E. coli*	Tyrosine inhibited enzyme; repressed by tyrosine Phenylalanine inhibited enzyme; multivalently repressed by phenylalanine plus tryptophan Tryptophan inhibited enzyme; repressed by tryptophan	Brown and Somerville (1971) and Im et al. (1971)
Multiple Enzymes with Different Biosynthetic Functions			
Aspartokinase	*E. coli* K-12	Aspartokinase I--Homoserine dehydrogenase I: inhibited by threonine; multivalently repressed by threonine plus isoleucine Aspartokinase II--Homoserine dehydrogenase II; repressed by methionine	Cohen (1969)
	E. coli B	Aspartokinase II: Same as for K-12, except that aspartokinase II not associated with a homoserine dehydrogenase II activity Aspartokinase III: inhibited by lysine, repressed by lysine	

253

TABLE 1 (continued)

Type	Enzyme Activity	Typical Organism	Distinguishing Features	Reference
	Glutamine-amidotransferases			
	Carbamyl phosphate Synthetase	E. coli	A 42,000 mol. wt. subunit of carbamyl phosphate synthetase: allows glutamine to be utilized in reactions	Pinkus and Meister (1972)
	P-Aminobenzoate Synthetase	E. coli	A 9,000 mol. wt. subunit required for PABA synthesis	Huang and Gibson (1970)
Multiple Enzymes with Different Catabolic Functions	Pyruvate Kinase	E. coli	Pyk-F-activated by fructose-1,6-diphosphate Pyk-A-activated by AMP	Kornberg and Malkovati (1973)
	β-Ketoadipate Succinyl CoA Transferase	P. fluorescens	Transferase I: Induced by growth on aromatic substrates Transferase II: Induced by growth on C_6 and higher dicarboxylic acids	Hoet and Stanier (1970)
	Chorismate Mutase	E. coli	Chorismate Mutase T--part of phrephenate dehydrogenase; only the second activity is inhibited by tyrosine	Koch et al. (1971)
Similar Enzyme Activities Covalently			Chorismate Mutase P--part of prephenate dehydratase; inhibited by phenylalanine	Dopheide et al. (1972)

TABLE 1 (continued)

Linked to Proteins Bearing Different Activities	Glutamine amido-transferases	E. coli	
	Anthranilate Synthase, Component II		Part of same protein that has phosphoribosyl anthranilate transferase activity	Ito and Yanofsky (1966)
	Cytidine tri-phosphate		Allows glutamine to be used in CTP formation	Levitzki and Koshland (1971)
Multiple Enzymes with Apparently Identical Functions	Ornithine trans-carbamylase	E. coli	argI and argF products separable by chromatography	Legrain et al. (1972)

enzyme induced by aromatic compounds has the physiological function of acylating β-ketoadipate derived from aromatic catabolism. The enzyme induced by the higher dicarboxylic acids has the physiological role of acylating adipic acid.

An example of enzymes normally not considered as multiple enzymes is the glutamine-amidotransferases that constitute subunits of enzymes exhibiting glutamine-dependent amination reactions. *E. coli* carbamyl phosphate synthetase and p-aminobenzoate (PABA) synthetase are the two examples cited in Table 1. Some glutamine amidotransferase activities are a still more special case because they are carried on polypeptide chains that make them in fact different enzymes. Nevertheless, they may differ from some of the examples already cited only by the covalent linkage to the portion of the protein carrying the second activity. Presumably, the glutamine amidotransferase subunits for carbamyl phosphate formation and PABA formation are specifically associated with their corresponding larger subunits. Does this specific association eliminate them from consideration as isozymes? If so, perhaps some of the other examples of multiple enzymes are not isozymes. Within the cell, there may be an association of some enzymes that catalyze sequential reactions. Such an association contributes to their fitness for the functions they serve. For example, we are currently entertaining the possibility of an as yet undemonstrated association of the biosynthetic threonine deaminase with the next enzyme in the sequence, the acetohydroxy acid synthase. We would, however, certainly not anticipate a similar association on the part of the degradative threonine deaminase even if they were to appear in the cell at the same time (and they normally do not).

Another common activity that is associated with two distinctly different proteins in *E. coli* is chorismate mutase. One activity, chorismate mutase T, is catalyzed by the protein carrying prephenate dehydrogenase activity and is repressed by tyrosine (Dopheide et al., 1972). Tyrosine does not inhibit chorismate mutase T activity but it does inhibit prephenate dehydrogenase activity. The second activity, chorismate mutase P, is catalyzed by the protein carrying prephenate dehydratase and is repressed and inhibited by phenylalanine (Koch et al., 1971).

Finally in Table 1, is an example in *E. coli* K-12 (but not *E. coli* B) that is perhaps most nearly analogous to the A and B lactate dehydrogenases that first stimulated interest in isozymes. Two genes specify ornithine transcarbamylase activity, argF and argI (Syvanen and Roth, 1972). So far as we know they are functionally equivalent. They may be the result of gene duplication but, if so, the two genes have undergone

some degree of evolutionary modification. The enzyme appears
to be a trimer in each case and when both genes are active,
chromatography of the extracts yields four clearly separated
isozymes as expected of two subunits randomly associating as
a trimer (Legrain et al., 1972).

Of the limited number of examples reviewed above, that of
the two biosynthetic ornithine transcarbamlyases is clearly
different from the rest. There is no question that the four
separable species are isozymes of ornithine transcarbamylase.
Fortunately, we are spared here a further dilemma in *E. coli*
for it does not contain the degradative ornithine transcarbamylase (which catalyzes the phosphorylytic cleavage of
citrulline) that is found in some pseudomonads. The latter
is clearly as functionally distinct from the arginine biosynthetic enzyme as the two threonine deaminases are from each
other. A question of biochemical semantics that does arise is
whether such functionally different multiple enzymes should be
considered isozymes. A question that arises at the other
extreme in differences between enzymes is whether the conven-
tion recommended by the Enzyme Commission that allelic forms
of an enzyme be considered isozymes is a justifiable one.
Would an inactive gene product be an isozymic form of the wild
type enzyme? Would one with 1% of the wild type activity or
with 25% of the wild type activity be considered isozymic?
Again, if the mutant form of the enzyme has full activity in
one enzyme test but not in another would the mutant form and
parental proteins be isozymic? Is there an accomodation in
the formal definition for proteins that appear to be isozymes
by one test but not by another?

*THE MULTIPLE ENZYMES ENCOUNTERED IN THE STUDY OF
BRANCHED-CHAIN AMINO ACID BIOSYNTHESIS.*

The two kinds of enzymes occurring in multiple forms that
were encountered early in the study of branched-chain amino
acid biosynthesis were the **threonine** deaminases and the acetohydroxy acid synthases. Actually, there is no problem in distinguishing between the two forms in either case and, for all
practical purposes it appears that the forms are not inter-
changeable. It might be useful, however, to summarize properties of these **multiple enzymes as they are now understood.**
The two enzymes differ from each other not only in function
but also in their response to effectors and in the way their
synthesis is regulated. The biosynthetic enzyme has been
studied in a variety of microbial and plant forms and all cannot be reviewed here. Several have been purified and studied
extensively, but none more extensively with respect to struc-

ture and function than those from *E. coli* (Calhoun et al., 1973) and *S. typhimurium* (Hatfield and Burns, 1970 and Burns and Zarlengo, 1968). The very important distinctive characteristic of the biosynthetic enzyme is its inhibition by isoleucine which results in marked cooperative binding of threonine. In addition, there have been a number of observations on the biosynthetic threonine deaminase from *E. coli* and *S. typhimurium* that suggest that it may have a role in the control of all the products of the ilv gene cluster (Hatfield and Burns, 1970; Levinthal et al., 1973;and Calhoun and Hatfield, 1973). An analysis of this possibility is currently underway in several laboratories and with several organisms.

The biodegradative threonine deaminase has been studied most extensively in two orgainsms, *E. coli* (Gerlt et al., 1973) and *Clostridium tetanomorphum* (Nakazawa and Hayaishi, 1967). The enzymes are similar in both. The distinctive characteristic is that the enzyme requires AMP (ADP for the Clostridial one). It is not only unaffected by isoleucine but is normally formed and functions in media that would invariably contain isoleucine. Like many catabolic enzymes it is repressed by glucose and the repression is overcome by cyclic AMP (Shizuta et al., 1970). However, there is not much information on the nature of the induction or the repression of the enzyme in any organism.

From the above, it is clear that the two enzymes need not be confused during enzyme assay and indeed the two enzymes would usually not be encountered in the same extract, except by use of rather ingenious physiological or genetic manipulations. Nevertheless, the potential for forming two kinds of threonine deaminase does exist in some organisms.

The two forms of acetohydroxy acid synthetase that have been studied are the degradative and biosynthetic enzymes of *Aerobacter areogenes* (Halpern and Umbarger, 1969). By the "biosynthetic enzyme" is meant an enzyme that will form both acetolactate and acetohydroxy butyrate and is inhibited by valine. At this time we are not prepared to say that there is a single biosynthetic enzyme in *A. aerogenes*. By analogy with the recent studies on *S. typhimurium* that we shall describe later, we would expect that there are at least two.

The valine-sensitive enzyme is characterized in most organisms including *A. aerogenes* by a requirement for FAD (Störmer and Umbarger, 1964). The reason for the requirement is not clear but is is of interest that a biochemically homologous reaction catalyzed by glyoxylate carboligase also requires FAD (Gupta and Vennesland, 1964). The regulation of the enzyme can best be considered when the regulation of the ilv cluster is reviewed. The enzyme was originally observed

to have a pH optimum for activity between pH 7.5 and 8.0
(Umbarger and Brown, 1958). However, this was probably an
artifact due, as in *E. coli* extracts, to the presence in the
extracts of a very small amount of valine which led to an apparent lack of activity in the pH 6.0 to 7.0 range where the
sensitivity to valine is greatest. Actually, the pH range of
optimal activity is quite broad with a maximum near 7 (C. S.
Brown, unpublished observations).

The degradative enzyme has been purified and crystallized
(Störmer, 1968). It is not a flavoprotein and is not stimulated by FAD. It is not inhibited by valine. Its pH optimum
is low, pH 6.0. It appears in the medium only after the pH
drops below 6.0. It is obviously suited to its role as a
means for *A. aerogenes* to route carbohydrate fermentation to
acetoin production via acetolactate (Juni, 1952).

Here, too, the two enzyme forms can be readily distinguished from each other. It would be possible to design enzyme
assays in which both would be detected as well as assays in
which one or the other would be detected.

ARE THERE MULTIPLE FORMS OF THE BIOSYNTHETIC ACETOHYDROXY ACID SYNTHASE?

The idea that more than one enzyme can form acetohydroxy
butyrate and acetolactate (see Fig. 2 for the pathway leading
to isoleucine and valine) has only emerged in the last few
years. We can now explain why the inhibition of acetohydroxy
acid synthase by valine does not block isoleucine biosynthesis
and inhibit growth of *E. coli* B and of *S. typhimurium*. However, we are still embarrassingly unable to account for the
fact that, in *E. coli* K-12, in which valine does inhibit the
(apparently) single acetohydroxy acid synthase with a resulting
inhibition of growth, we have not found mutants lacking that
presumed single enzyme. To consider the problem in context
requires a brief summary of the way we currently think the ilv
gene cluster is controlled. In considering the control mechanism, we can, owing to considerable uncertainty concerning
the nature of the regulatory elements, refer only to repression
or derepression signals, to induction signals and to the response of the ilv genes to those signals.

The isoleucine and valine forming enzymes in *E. coli*
strain K-12 appear to be specified by a cluster of five genes
arranged on the chromosome as shown in Fig. 2. While in general the control of gene expression in the cluster is multivalent, only the genes ilvA, D and E require all three
branched-chain amino acids (isoleucine, valine and leucine) to
be in excess for repression to occur. The ilvB gene, which is

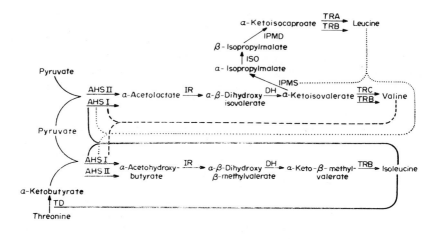

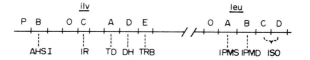

Fig. 2.   Gene-enzyme relationships in branched-chain amino acid biosynthesis.   Top -- the biosynthetic steps.   Bottom -- the arrangement of the structural gene clusters in *E. coli* K-12.   TD, threonine deaminase (isoleucine-sensitive); AHAS I, acetohydroxy acid synthase I (valine-sensitive); IR, acetohydroxy acid isomeroreductase; DH, dihydroxy acid dehydrase; TRA, B, C, transaminases A, B and C; IPMS, α-isopropylmalate synthase; ISO, isopropylmalate isomerase; IPMD, β-isopropylmalate dehydrogenase.   For additional details, see text.

thought to specify the valine-inhibited acetohydroxy acid synthase, is repressed when only valine and leucine are in excess. Thus, when valine is added to the culture medium of *E. coli* K-12, the formation of acetohydroxy butyrate is blocked as well as that of acetolactate.   The resulting isoleucine restriction leads to derepression only of genes ilvA, D and E. There is not only repression of the ilvB gene but an actual decay in the activity of the rather unstable enzyme.   The B gene product is derepressed when valine is limiting and, since the activity of acetohydroxy acid synthase would be largely uncontrolled with limiting valine, there would be an induction of the ilvC gene product, the isomeroreductase, which is a substrate induced enzyme (Ratzkin et al., 1972).   The limitation of valine would also result in derepression of the ilvA, D and E gene products. Limiting leucine also derepresses the four repressible genes ilvB, A, D and E but because valine would be in excess (by definition) when leucine is limiting, the acetohydroxy acid synthase, while elevated, would be inhibited and there would be very little internal induction of the ilvC product. Thus, in the K-12 strains, three regulatory groups can be readily distinguished within the ilv **gene cluster by independently restricting the supply of isoleucine, valine or leucine.**

In *S. typhimurium,* all five enzyme activities are elevated when any one of the three branched-chain amino acids is limiting. The reason for this was clarified when it was realized that there was a gene specifying a second acetohydroxy acid synthetase that was derepressed by the same combination of derepression signals that caused derepression of the ilvA, D and E genes (Blatt et al., 1972). The second acetohydroxy acid synthase is insensitive to valine. This enzyme is specified by the ilvG gene. O'Neill and Freundlich (1972) have presented evidence for its occurrence between ilvE and ilvD. However, three point mapping experiments have not been reported so that this location must be considered tentative. Thus with leucine and isoleucine restriction in this organism, even in excess valine, there was sufficient acetolactate formed whenever the ilvG gene was derepressed to induce the isomeroreductase.

In the past, the ilvA, D and E genes, which responded to a common multivalent repression signal, were considered to be a single operon under the control of the ilvO region which was thought to be the operator for those three genes (Ramakrishnan and Adelberg, 1965). We now find that the ilvC gene lies between ilvA and the ilvO region (Kline et al., 1974). (Its relative position with ilvB is tentative since we have no ilvB mutants in the K-12 strain.) The nature of the ilvO region is not clear at this time but mutations in it do result in elevated expression of the ilvA, D and E genes but reduced expression of the ilvC and B genes. We suspect that ilvE has a separate operator-promoter region and that ilvA and D alone may constitute an operon.

If there were, in fact, only a single gene specifying acetohydroxy acid synthase in *E. coli* K-12, one would expect that ilv- mutants lacking this enzyme would be readily found. For *E. coli* B and *S. typhimurium,* each of which are thought to have an ilvG gene as well as an ilvB gene, mutants that are valine-sensitive (like strain K-12) and double mutants that are devoid of acetohydroxy acid synthase activity can be obtained in two sequential mutagenic steps. We cannot explain the absence of acetohydroxy acid synthase-negative derivatives of *E. coli* K-12.

Let us, however, consider the kind of evidence that led us to think there is only a valine-sensitive acetohydroxy synthetase in wild type *E. coli* K-12 and both a sensitive and an insensitive one in *E. coli* B and *S. typhimurium*.

The realization that two acetohydroxy acid synthases occurred in *E. coli* W and *S. typhimurium* came from some experiments in collaboration with Drs. Blatt and Pledger (Blatt et al., 1972) a few years ago. Until that time it had been assumed that the acetohydroxy acid synthases of these organisms were also specified by an ilvB gene as in *E. coli* K-12 but that in these strains the ilvB gene was derepressed by limiting isoleucine as well as by limiting valine or leucine.

Blatt et al,(1972) found, however, that when isoleucine was limited in these strains the acetohydroxy acid synthase activity was virtually insensitive to valine. Further analysis showed that the valine-sensitive activity behaved exactly as it did in *E. coli* K-12 -- it was repressed by only valine and leucine acting in concert. These observations and those reported at the same time by O'Neill and Freundlich (1972) led to the idea that *S. typhimurium* contained a functional ilvG gene that was controlled in the same way as were the ilvA, D and E genes and the *E. coli* K-12 did not. Whether the non-expression of the ilvG gene in *E. coli* K-12 was due to its absence, to its occurrence in the null state or to restricted expression of the entire ilvADE (and G) region of the ilv cluster is not known. However, all attempts to detect valine-insensitive acetohydroxy acid synthase activity upon derepressing the ilvA, D and E genes by isoleucine limitation or by mutants leading to increased derepression have so far been unsuccessful. Indeed, the ilvO mutations, described by Ramakrishnan and Adelberg (1965) and which lead to increased ilvADE function and to valine resistance, might well be explained if those mutations did result in sufficient expression of putative ilvG gene to permit acetohydroxy butyrate formation in the presence of valine. O'Neill and Freundlich (1972) cited experiments with valine-resistant derivatives of the K-12 strain of *E. coli* which behaved in this very way. However, examination of numerous isogenic pairs of strains with or without authentic ilvO lesions has until now revealed no evidence for an ilvG function (increased resistance of acetohydroxy acid synthesis to valine) in our laboratory.

To examine this question more thoroughly, a series of experiments was undertaken in which hydroxylapatite chromatography (O'Neill and Freundlich, 1972) was used to separate the sensitive and insensitive enzymes. The fractions eluted from the columns obtained were assayed for acetolactate for-

mation in the presence and absence of 10^{-2} M valine. This
concentration of valine has been found to yield maximal inhibition in all extracts tested. The organisms examined were
either prototrophic derivatives of *E. coli* B and K-12, of
S. typhimurium LT-2 or of ilv-leu auxotrophs which allowed us
to examine the hydroxylapatite elution profile of acetohydroxy acid synthase activity formed under conditions of
physiological derepression. In addition, several valine-
resistant derivatives of the K-12 strain and a valine-sensitive
derivative of the B strain were examined.

As Fig. 3. shows, *S. typhimurium* contained essentially only
a valine-insensitive acetohydroxy acid synthase when the cells
were grown with limiting isoleucine. When either valine or
leucine were limiting, however, there was a heterogeneity in
the acetohydroxy acid synthase activity. A large portion of
the activity was eluted from the column at a later time than
that observed with the isoleucine-limited extract and that in
later fractions was sensitive to valine. While the separation
was not a sharp one, owing presumably to the large amounts of
extract placed on the column, the hydroxylapatite elution profiles from this organism were in complete accord with the
earlier findings using density gradient centrifugation: a
valine-insensitive acetohydroxy acid synthase that appeared
when any of the three branched-chain amino acids were limiting
and a valine-sensitive enzyme appearing only when leucine or
valine were limiting.

In contrast were the results (Fig. 4) obtained with an
E. coli K-12 strain when it was grown under the three conditions of limitation. With limiting valine or leucine, the
enzyme was entirely of the sensitive type; there was no indication of an early eluted, valine-insensitive enzyme. The
"activity" in the presence of valine in these extracts shown
in the figure is probably due more to an inadequate blank than
to any residual acetolactate formation in the presence of
valine, for there was essentially no red color in the tubes.
The isoleucine-limited culture exhibited only a markedly
repressed level of valine-sensitive acetohydroxy acid synthase
activity. Only the activity in the absence of valine has been
plotted in the top of Fig. 4. It should be pointed out that
the extract was prepared from a culture that had been limited
for isoleucine for a three hour period following growth in
excess branched-chain amino acids. During this time there was
very little increase in cell protein and the acetohydroxy acid
synthase activity actually dropped below the (repressed) level
it had been at the start of the limitation procedure. While
the data are not shown, there was marked derepression of the
ilvA, D and E gene products during this period.   In the

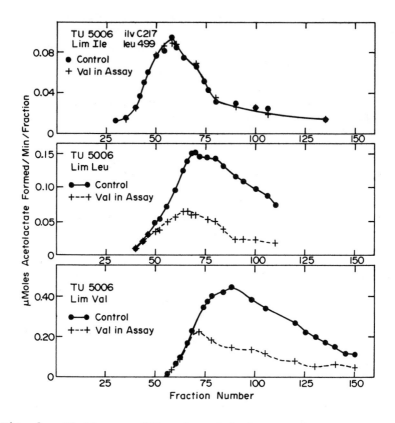

Fig. 3. Elution profile of acetohydroxy acid synthase activities of *S. typhimurium* under various conditions of derepression. Cells from 1 liter cultures grown under limiting conditions and harvested as described by Kline et al. (1974), were suspended in 5 ml buffer (10^{-2} M potassium phosphate, pH 7.1, 20% glycerol, 5×10^{-3} M $MgCl_2$) containing 1 mg thiamine pyrophosphate and 0.1 mg FAD. The extract was loaded on a hydroxylapatite column (1 x 20 cm) equilibrated with the same buffer. Elution was with 220 ml of the same buffer containing a linear concentration gradient of potassium phosphate between 2×10^{-2} M and 4×10^{-2} M. Fractions of about 1.5 ml were collected every two minutes. All fractions were kept cold and were assayed with and without 10^{-2} M **valine as described** by Kline et al. (1974).

S. typhimurium **ilv-leu** auxotroph, limiting isoleucine also resulted in disappearance of the valine sensitive enzyme but the valine-insensitive enzyme was formed at a derepressed rate.

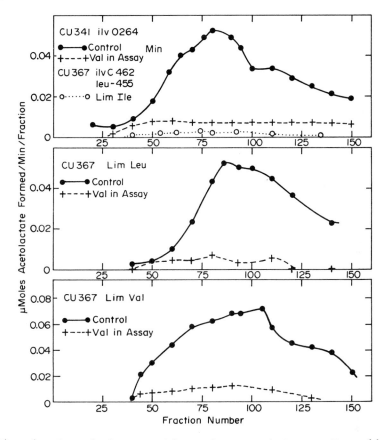

Fig. 4.   Acetohydroxy acid synthase activity of *E. coli* K-12 under various conditions of derepression and an ilvO (valine-resistant) mutant grown in minimal medium.   Other conditions as in Fig. 3.

Thus, in a K-12 derivative, there does not seem to be any valine-insensitive acetohydroxy acid synthase activity that can be derepressed when the ilvA, D and E genes are derepressed.   It would appear therefore that there is no functional ilvG gene.

Quite compatible with this view is the acetohydroxy acid synthase profile obtained with an *E. coli* K-12 strain bearing the ilvO264 lesion described by Ramakrishnan and Adelberg (1965).   The profile, shown also in the top panel of Fig. 4, is of an extract of cells grown in minimal medium.   Such extracts show elevated levels of the ilvA, D and E gene products but no evidence of an ilvG gene product.

With *E. coli* B, the elution profile (Fig 5A) of an extract prepared from an isoleucine auxotroph grown with isoleucine limiting, revealed a valine-insensitive enzyme that was eluted without any valine-sensitive enzyme. When the extract of a valine-sensitive derivative of *E. coli* B grown in minimal medium was chromatographed on hydroxylapatite, no evidence for a valine-insensitive enzyme was observed (Fig. 5B). It should be noted that the activity was very low, and it appears that less enzyme activity was eluted early with the Val^S extract than with the wild type extract but the enzyme that did appear early was entirely sensitive. Thus, it may be that a small amount of the ilvG product is still formed by the Val^S strain but, if so, it has gained a sensitive character. On the other hand, the difference could be due to a lack of reproducibility of the elution pattern. If so, we would conclude that the Val^S derivative of *E. coli* B lacks the ilvG product. The elution profile (Fig. 6A) of a normal *E. coli* B extract prepared from minimal medium-grown cells showed both insensitive and sensitive activities. Both activities were also revealed when a leucine auxotroph of *E. coli* B/r was limited on leucine (Fig. 6B).

Two other Val^r derivatives of *E. coli* K-12 have been examined in our effort to account for valine resistance. One strain (CU389) carries a lesion linked to the *leu* region. In crude extracts, there is no derepression of the enzymes in the isoleucine and valine pathway but the acetohydroxy acid synthase activity is less sensitive to valine than it is in the parent strain. Upon chromatography, however, only sensitive enzyme was observed (Fig. 7). Interestingly, the profile is broader than the profile shown by a normal strain or by the ilvO (Val^r) strain (Fig. 4). It is also of interest that the storage of an extract of this strain at the same temperature and in the same buffer as that employed in the chromatographic procedure increased the sensitivity to valine. Whether the broad elution pattern was due to more than a single species or to heterogeneity within a single species is, of course, not clear. Significantly, the entire complement of acetohydroxy acid synthase activity in such extracts, while relatively low, was sensitive to valine. Strain CU5117, the ilvF derivative, showed, however, a complete conversion of the activity to a valine-resistant state with no region of the chromatographic elution pattern showing a remnant of the valine-sensitive enzyme that appears in the wild type organism.

These results indicate that at least in *E. coli* B and *S. typhimurium* there are at least two acetohydroxy acid synthases, one that is inhibited by valine and one that is not. The assumption is that the inhibited and non-inhibited enzymes

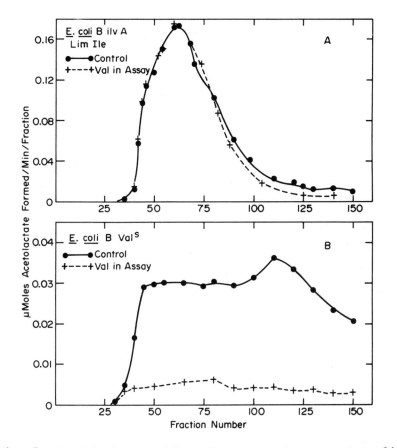

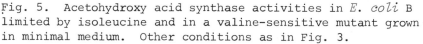

Fig. 5.   Acetohydroxy acid synthase activities in *E. coli* B limited by isoleucine and in a valine-sensitive mutant grown in minimal medium.   Other conditions as in Fig. 3.

are the products of the <u>ilvB</u> and <u>ilvG</u> genes, respectively. The insensitive enzyme appears to respond to the same pattern of repression and derepression as do the <u>ilvADE</u> gene products, a finding compatible with the preliminary evidence that <u>ilvG</u> is near <u>ilvE</u>.   It should be pointed out, however, that a satisfactory genetic and biochemical correlation of the two genes to the two enzymes has not been achieved.   Neither enzyme has yet been purified but based upon the separation achieved it appears that both enzymes appear when either valine or leucine are limiting.

No such heterogeneity with respect to sensitive and insensitive acetohydroxy acid synthases has been found in the

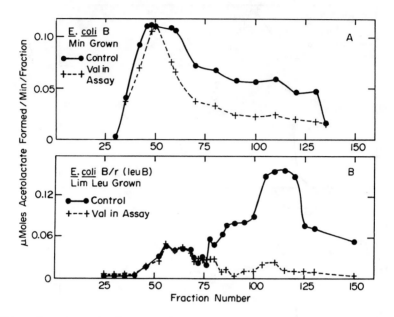

Fig. 6.   Acetohydroxy acid synthase activities in *E. coli* B
grown in minimal medium and in *E. coli* B/r limited on isoleucine.   Other conditions as in Fig. 3.

K-12 strain of *E. coli*.   With an isoleucine limitation, the
acetohydroxy acid synthase is completely repressed and the
activity present at the beginning of the limitation actually
decays.

Of particular interest was the finding that the ilvO
derivative showed no evidence of a valine-insensitive enzyme.
From that finding, we concluded that, while the ilvO mutation somehow increased the expression of the ilvA, D and E
genes, it did not trigger expression of a putative ilvG gene
that might be latent in the K-12 strain.

It was therefore quite startling for us to find only recently that when the ilvO (Val^r) strain, CU341, was grown in
valine-supplemented minimal medium, there was a considerable
fraction (∿50%) of the acetohydroxy acid synthase activity in
a crude extract resistant to 10^{-2} M valine.   This activity
appears very early after transfer to a valine medium and is not
present in large amounts.   We have had no opportunity to
examine the chromatographic behavior of the resistant enzyme
on a hydroxylapatite column.   These experiments and studies

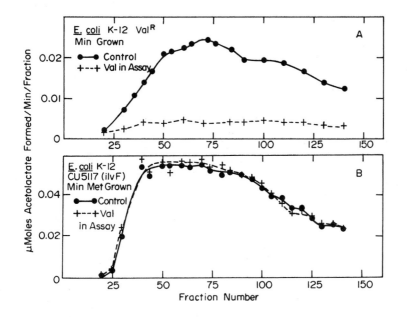

Fig. 7.   Acetohydroxy acid synthase activities in two valine resistant mutants of *E. coli* K-12.   Other conditions as in Fig. 3.   The strain designated Val[r] is strain CU389.

to determine whether the valine insensitive enzyme is formed <u>from</u> the sensitive enzyme or de novo as a separate species are scheduled for the near future.   In any event, it now appears that, at long last, an explanation of the Ramakrishnan and Adelberg (1965) <u>ilvO</u> mutants is very near at hand.

Very pressing now is the need to develop procedures that will allow purification of the acetohydroxy acid synthases so that we can decide by more than intuition whether we are dealing with true isozymes or merely modified products of a single gene.

ACKNOWLEDGEMENTS

Supported by research grant GM12522 from the National Institute of General Medical Sciences.

REFERENCES

Blatt, J. M., W. J. Pledger, and H. E. Umbarger 1972. Isoleucine and valine metabolism in *Escherichia coli*. XX.

Multiple forms of acetohydroxy acid synthetase. *Biochem. Biophys. Res. Commun.* 48: 444-450.

Brown, K. D., and R. L. Somerville 1971. Repression of aromatic amino acid biosynthesis in *Escherichia coli* K-12. *J. Bacteriol.* 108: 386-399.

Burns, R. O., and M. H. Zarlengo 1968. Threonine deaminase from *Salmonella typhimurium*. I. Purification and properties. *J. Biol. Chem.* 243: 178-185.

Calhoun, D. H., and G. W. Hatfield 1973. Autoregulation: A role for a biosynthetic enzyme in the control of gene expression. *Natl. Acad. Sci.* 70: 2757-2761.

Calhoun, D. H., R. A. Rimerman, and G. W. Hatfield 1973. Threonine deaminase from *Escherichia coli*. I. Purification and properties. *J. Biol. Chem.* 248: 3511-3516.

Cohen, G. N. 1969. The aspartokinase and homoserine dehydrogenases of *Escherichia coli*. In: *Current Topics in Cellular Regulation,* Vol. 1 (B. L. Horecker and E. R. Stadtman, Eds.), Academic Press, pp. 183-231.

Datta, P., and H. Gest 1964. Control of enzyme activity by concerted feedback inhibition. *Proc. Natl. Acad. Sci. U.S.A.* 52: 1004-1009.

Dopheide, T. A. A., P. Crewther, and B. E. Davidson 1972. Chorismate mutase-prephenate dehydratase from *Escherichia coli*. II. Kinetic properties. *J. Biol. Chem.* 247: 4447-4452.

Gerlt, J. A., K. W. Rabinowitz, C. P. Dunne, and W. A. Wood 1973. The mechanism of action of 5'adenylic acid-activated threonine dehydrase. V. Relation between ligand-induced allosteric activation and the promoter-oligomer inter-conversion. *J. Biol. Chem.* 248: 8200-8206.

Gupta, N. K., and B. Vennesland 1964. Glyoxylate carboligase of *Escherichia coli:* a flavoprotein. *J. Biol. Chem.* 239: 3787-3789.

Halpern, Y. S., and H. E. Umbarger 1959. Evidence for two distinct enzyme systems forming acetolactate in *Aerobacter aerogenes*. *J. Biol. Chem.* 234: 3067-3071.

Hatfield, G. W., and R. O. Burns 1970. Threonine deaminase from *Salmonella typhimurium*. III. The intermediate substructure. *J. Biol. Chem.* 245: 787-791

Hatfield, G. W., and R. O. Burns 1970. Specific binding of leucyl transfer RNA to an immature form of L-threonine deaminase: Its implications in repression. *Proc. Natl. Acad. Sci. U.S.A.* 66:1027-1035.

Hoet, P. P., and R. Y. Stanier 1970. Existence and functions of two enzymes with β-ketoadipate: succinyl-CoA transferase activity in *Pseudomonas fluorescens*. *Eur. J. Biochem.* 13: 71-76.

Huang, M., and F. Gibson 1970. Biosynthesis of 4-aminobenzoate in *Escherichia coli*. *J. Bacteriol*. 102: 767-773.

Huseby, N.-E., T. B. Christensen, B. J. Olsen, and F. C. Störmer 1971. The pH 6 acetolactate-forming enzyme from *Aerobacter aerogenes*. Subunit structure. *Eur. J. Biochem*. 20: 209-214.

Im, S. W. K., H. Davidson, and J. Pittard 1971. Phenylalanine and tyrosine biosynthesis in *Escherichia coli* K-12: Mutants derepressed for 3-deoxy-D-arabinoheptulosonic acid 7-phosphate synthetase (phe), 3-deoxy-D-arabinoheptulosonic acid 7-phosphate synthetase (tyr), chorismate mutase T-prephenate dehydrogenase, and transaminase A. *J. Bacteriol*. 108: 400-409.

Ito, J., and C. Yanofsky 1966. The nature of the anthranilic acid synthetase complex of *Escherichia coli*. *J. Biol. Chem*. 241: 4112-4114.

Jensen, R. A., and E. W. Nester 1965. The regulatory significance of intermediary metabolites: Control of aromatic acid biosynthesis by feedback inhibition in *Bacillus pubtilis*. *J. Mol. Biol*. 12: 468-481.

Juni, E. 1952. Mechanisms of formation of acetoin by bacteria. *J. Biol. Chem*. 195: 715-726.

Kline, E. L., C. S. Brown, W. G. Coleman, Jr., and H. E. Umbarger 1974. Regulation of isoleucine-valine biosynthesis in and ilvDAC deletion strain of *Escherichia coli* K-12. *Biochem. Biophys. Res. Commun*. 57: 1144-1151.

Koch, G. L. E., D. C. Shaw, and F. Gibson 1971. The purification and characterization of chorismate mutase-prephenate dehydrogenase from *Escherichia coli*. *Biochim. Biophys. Acta* 229: 795-804.

Kornberg, H. L., and M. Malcovati 1973. Control *in situ* of the pyruvate-kinase activity of *Escherichia coli*. *FEBS Letters* 32: 257-259.

Lee, N. M., E. M. Lansford, Jr., and W. Shive 1966. A study of the mechanism of lysine-sensitive asparto-kinase control of lysine. *Biochem. Biophys. Res. Commun*. 24: 315-318.

Legrain, C., P. Halleux, V. Stalon, and N. Glansdorff 1972. The dual genetic control of ornithine carbamoyltransferase in *Escherichia coli*. A case of bacterial hybrid enzymes. *Eur. J. Biochem*. 27: 93-102.

Levinthal, M., L. S. Williams, M. Levinthal, and H. E. Umbarger 1973. Role of threonine deaminase in the regulation of isoleucine and valine biosynthesis. *Nature New Biol*. 246: 65-68.

Levitzki, A., and D. E. Koshland, Jr. 1971. Cytidine triphosphate synthetase. Covalent intermediates and mechanisms of action. *Biochem*. 10: 3365-3371.

Nakazawa, A., and O. Hayaishi 1967. On the mechanism of activation of L-threonine deaminase from *Clostridium tetanomorphum* by adenosine diphosphate. *J. Biol. Chem.* 242: 1146-1154.

O'Neill, J. P. and M. Freundlich 1972. Two forms of biosynthetic acetohydroxy acid synthetase in *Salmonella typhimurium*. *Biochem. Biophys. Res. Commun.* 48: 437-443.

Paulus, H., and E. Gray 1967. Multivalent feedback inhibition of aspartokinase in *Bacillus polymyxa*. *J. Biol. Chem.* 242: 4980-4986.

Pinkus, L. M., and A. Meister 1972. Identification of a reactive cysteine residue at the glutamine binding site of carbamyl phosphate synthetase. *J. Biol. Chem.* 247: 6119-6127.

Pledger, W. J., and H. E. Umbarger 1973. Isoleucine and valine metabolism in *Escherichia coli*. XXI. Mutations affecting derepression and valine resistance. *J. Bacteriol.* 114: 183-194.

Ramakrishnan, T., and Adelberg, E. A. 1965. Regulatory mechanisms in the biosynthesis of isoleucine and valine. II. Identification of two operator genes. *J. Bacteriol.* 89: 654-660.

Ramos, R., V. Stalon, A. Pierard, and J. M. Wiame 1965. Sur l'existence de deux ornithine transcarbamylases chez un Pseudomonas. *Arch. Internat. Physiolog. Biochim.* 73: 155-157.

Ratzkin, B., S. Arfin, and H. E. Umbarger 1972. Isoleucine and valine metabolism in *Escherichia coli*. XVIII. Induction of acetohydroxy acid isomeroreductase. *J. Bacteriol.* 112: 131-141.

Shizuta, Y., and O. Hayaishi 1970. Regulation of biodegradative threonine deaminase synthesis in *Escherichia coli* by cyclic adenosine. *J. Biol. Chem.* 245: 5416-5423.

Stadtman, E. R., G. N. Cohen, G. LeBras, and H. de Robichon-Szulmajster 1961. Feedback inhibition and repression of aspartokinase activity in *Escherichia coli* and *Saccharomyces cerevisiae*. *J. Biol. Chem.* 236: 2033-2038.

Stalon, V. 1972. Regulation of the catabolic ornithine carbamoyltransferase of *Pseudomonas fluroescens*. A study of the allosteric interactions. *Eur. J. Biochem.* 29: 36-46.

Störmer, F. C. 1968. The pH 6 acetolactate-forming enzymes *Aerobacter aerogenes*. II. Evidence that it is not a flavoprotein. *J. Biol. Chem.* 243: 3740-3741.

Störmer, F. C., and H. E. Umbarger 1964. The requirement for flavine adenine dinucleotide in the formation of acetolactate by *Salmonella typhimurium* extracts. *Biochem. Biophys. Res. Commun.* 17: 587-592.

Syvanen, J. M., and J. R. Roth 1972. Structural genes for ornithine transcarbamylase in *Salmonella typhimurium* and *Escherichia coli* K-12. *J. Bacteriol.* 110: 66-70.

Umbarger, H. E., and B. Brown 1958. Isoleucine and valine metabolism in *Escherichia coli*. VII. A negative feedback mechanism controlling isoleucine biosynthesis. *J. Biol. Chem.* 233: 415-420.

COMPARATIVE STUDIES OF MULTIPLE PEPTIDE
HYDROLASES FROM GUINEA PIG SMALL INTESTINAL
MUCOSA AND THE IMPLICATIONS OF THESE STUDIES FOR
THE "MISSING PEPTIDASE" HYPOTHESIS FOR COELIAC DISEASE

P.F. FOTTRELL, J. DONLON, G. O'CUINN AND C.O. PIGGOTT
Department of Biochemistry, University College
Galway, Ireland

ABSTRACT. Three cytoplasmic peptide hydrolases from guinea-
pig small intestinal mucose, termed α,β_1, and β_2, have been
purified and characterized. These were active against a
broad and overlapping range of peptides. The β_1 and β_2
peptide hydrolases had very similar properties, although
they differed in their affinities for substrates. In
contrast, α distinctly differed from these in its substrate
specificity, metal requirement,and other properties.
     A comparision of purified α peptide hydrolase prepared
from cytosol and brush borders showed that both activities
were immunologically and kinetically similar. Various
methods of solubilizing brush border peptide hydrolases
were compared. Sonication was found to be most effective,
while high concentrations of papain appeared to give rise
to artifacts.
     These studies and related investigations have failed
to lend support to the "missing peptidase" hypothesis for
coeliac disease. The overall results, however, do indicate
that these enzymes may be involved in the final stages of
protein digestion within the enterocyte.

INTRODUCTION

The final stages of mammalian protein digestion are now
thought to be mediated by peptide hydrolases (peptidases)
present in the absorptive epithelial cells (enterocytes) of
small intestinal mucosa (Peters, 1970a). In contrast with
intestinal disaccharidases, which are responsible for the
final stages of carbohydrate digestion (Gray, 1971), relatively
little detailed information is available on the properties
of intestinal peptidases. Such information is essential to
a clearer understanding of mammalian protein digestion and
may also help to explain the etiology of malabsorption conditions such as coeliac disease, which is also known as gluten-
induced enteropathy (Booth, 1970).
     Previous studies from this laboratory showed that multiple
forms of intestinal peptidases were detectable in several
mammalian species including guinea-pig and human (Dolly and

Fottrell, 1969; Dolly et al., 1971; Donlon and Fottrell, 1972).
In agreement with the results obtained by other workers (Peters,
1970; Kim et al., 1972) peptide hydrolases were detected in
both brush border and cytosol fractions of the intestinal
mucosal cells.

The objectives of our studies were (a) to purify and characterize the multiple intestinal peptidases; (b) to determine
the relationship between brush border and cytosol peptidases
and (c) to examine possible implications of this study to the
missing peptidase hypothesis for coeliac disease (Frazer, 1956).

MATERIALS AND METHODS

Animals
The guinea-pig was selected for this study for the following
reasons: - (i) to avoid complications due to the binding of
pancreatic proteolytic enzymes to the brush borders (Woodley
and Kenny, 1969; Goldberg et al., 1971) (ii) the lesser amounts
of mucus present in guinea-pig intestine (cf. Forstner, 1968)
allows gentler preparation of brush borders and (iii) to serve
as a corollary to the quantitative studies of Peters (1970b)
on the subcellular localization of intestinal peptide hydrolases.

Starch-gel electrophoresis
Horizontal starch-gel electrophoresis (Smithies, 1955), at
pH 8.4, was employed in the classification of the mucosal
peptide hydrolases. Peptide hydrolases were detected in situ
on the gels by (i) the chromogenic stain of Lewis and Harris
(1967) and (ii) the starch-iodide method of Murphy (1970)
as described previously (Donlon and Fottrell, 1972).

Assay of peptide hydrolase activities
The method used for the assay of peptide hydrolase activities
depends on the measurement (in a coupled assay with L-amino
acid oxidase and peroxidase) of the amino acids liberated as
a result of hydrolysis of a suitable peptide (Donlon and
Fottrell, 1971).

Purification of α-peptide hydrolase
One of the cytoplasmic peptide hydrolases, termed "α" peptidase
on the basis of its electrophoretic mobility on starch gels
(Donlon and Fottrell, 1972), was purified to apparent homogeneity by a combination of gel filtration upon Sephadex G-200,
acetone fractionation and chromatography on DEAE-Sephadex A-25
(Donlon and Fottrell, 1973). A 50-fold purification with a
yield of 12% of original activity was obtained. The identity
of the enzyme being purified was monitored by starch-gel electrophoresis.

Purification of β_1 and β_2 peptide hydrolases

Two other cytoplasmic peptide hydrolases β_1 and β_2 were purified by subjecting crude guinea-pig mucosal extract to salt
fractionation, chromatography on DEAE-cellulose and preparative
polyacrylamide gel electrophoresis (Piggott and Fottrell,
unpublished observations). The elution profile from the
final stage is depicted in Fig. 2.

Purification of brush-border "α" peptide hydrolase

Brush-borders were prepared by the method of Hubscher et al.
(1965). Brush-border enzymes were released by sonication
with an M.S.E. disintegrator at 1.3 A for 20 sec. The brush-
border "α" peptide hydrolase was purified by the method used
for preparation of the cytoplasmic "α" peptide hydrolase.

Immunological studies

An antibody to cytoplasmic "α" peptide hydrolase was raised
in rabbits by the method of Goudie et al. (1966). γ-globulins
were prepared from specific antiserum and from control serum
by batch adsorption with DEAE-cellulose and precipitation with
40% saturated ammonium sulphate.

Solubilization of brush-border peptide hydrolases

Peptide hydrolases were released from brush-borders by a)
sonication as above, or b) treatment with 0.1% Triton X-100
for 60 min. at 37°, or c) treatment with various concentrations
of papain for 60 min. at 37°in a 0.1 M Phosphate buffer, pH
7.0 containing 3.17 mM cysteine hydrochloride. In each case
the solubilized enzymes were separated from particulate matter
by centrifugation at 22,300 g for 15 mins. The supernatant
and resuspended pellet were then assayed for peptide hydrolase
activity using both L-Leu-L-Leu and L-Leu-L-Leu-L-Leu (Donlon
and Fottrell, 1971).

RESULTS

Properties of α, β_1 and β_2 peptidases

Extracts from guinea-pig small intestinal mucosa contained
seven electrophoretically distinct peptide hydrolases (Donlon
and Fottrell, 1972). The most active zones were termed α,β,
and γ, as shown in Fig. 1. With some substrates it was possible to distinguish between two sub-zones contained in the
broad and extremely active β zone (Fig. 2); these were termed
β_1 and β_2. Preliminary studies, using starch-gel electrophoresis, showed that these enzymes had broad and separate
but overlapping substrate specificities, as tested with a
large number of dipeptides and some tripeptides. In summary,
dipeptides were mainly hydrolysed by α,β,and γ whereas a, b
and α were mainly responsible for tripeptide hydrolysis.

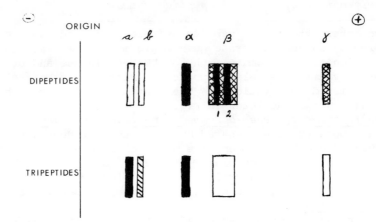

Fig. 1. Zymogram patterns of peptide hydrosase from guinea-pig small intestinal mucosal homogenates.

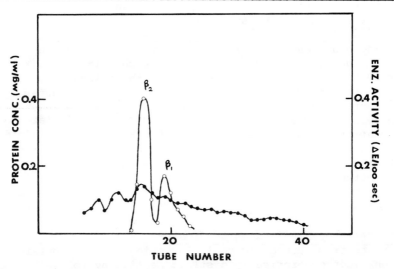

Fig. 2. Elution profile from preparative polyacrylamide-gel electrophoresis of pooled concentrated DEAE fractions, containing β_1 and β_2 peptide hydrolases. The gels were prepared according to a scaled-up modification of the analytical procedure of Davis (1964). Protein (●————●) was monitored by recording the absorbance of 280 nm. Peptide hydrolase activity (O————O) was assayed as described previously (Donlon and Fottrell, 1971).

These patterns have been confirmed using purified preparations of these enzymes. Table 1 shows that α, β_1 and β_2 peptide hydrolases can hydrolyze a broad range of peptides and

278

TABLE I

SUBSTRATE SPECIFICITY OF PURIFIED α, β_1 and β_2
PEPTIDE HYDROLASES

Peptide	Peptide Hydrolase		
	α	β_1	β_2
Gly-L-Leu	−	+	+
Gly-L-trp	−	+	+
Gly-L-tyr	−	+	+
Gly-L-phe	−	+	+
α-L-Glu-L-tyr	+	+	+
γ-L-Glu-L-leu	+	−	−
L-Leu-L-leu	+	+	+
L-Leu-Gly	+	+	+
L-Leu-L-pro	−	−	−
L-Leu-L-tyr	+	+	+
L-Leu-L-tyramide	+	+	+
L-Leu-L-ala	+	+	+
L-Leu-p-nitroanilide	−	−	−
L-Leucinamide	+	−	−
L-Lys-L-phe	+	−	−
L-phe-L-pro	+	−	−
L-pro-L-leu	−	−	−
L-Pro-L-tyr	−	+	+
L-Val-L-Leu	+	+	+

that many of the peptides are hydrolyzed by more than one of
these enzymes. In contrast some of the peptides quoted are
not hydrolyzed, in particular peptides having N-terminal
glycine were not hydrolyzed by α peptide hydrolase.

Table 2 shows that the affinities of these enzymes for
peptides are generally quite distinct. The α and β peptide
hydrolases also differ with respect to their metal requirements,
pH optima and their behavior in various buffer systems, as
shown in Table 3.

Brush border peptide hydrolases

An anitserum was raised in rabbits to the guinea-pig intestinal mucosa cytosol "α" peptide hydrolase. γ-globulins prepared from this antiserum were capable of precipitating the
purified "α" peptide hydrolase prepared from guinea-pig intestinal brush borders (Fig. 3). Comparision of the purified "α"
peptide hydrolases prepared from cytosol and brush borders
indicated that both enzymes had similar metal ion requirements,

TABLE II

MICHAELIS CONSTANTS (mM) OF α, β_1 and β_2
PEPTIDE HYDROLASES

Peptide	Peptide Hydrolase		
	α	β_1	β_2
L-Leu-L-leu	0.47	1.66	0.5
L-Val-L-leu	0.25	0.13	2.5
L-Leu-L-ala	7.1	0.84	1.0
L-Leucinamide	2.0	-	-
Gly-L-tyr	-	3.3	2.5
Gly-L-phe	-	4.0	0.8
Gly-L-trp	-	1.66	1.66
L-Phe-L-pro	0.53	-	-

TABLE III

COMPARISON OF OTHER PROPERTIES OF α, β_1 AND β_2
PEPTIDE HYDROLASES

Property	Peptide hydrolase		
	α	β_1	β_2
Metal Requirements	Activated by divalent manganese	No apparent metal requirements	
pH optimum	9.2	8.2	8.2
Stability in buffer systems	Unstable in Tris. Stable in veronal.	Stable in Tris. Less stable in veronal and borate.	

substrate specificities and K_m values for different substrates
(O'Cuinn, et al., 1974).

It is worth noting that the properties of peptide hydrolases
released from brush-borders and the efficiency of their release depended on the nature of the solubilizing agent. Tables
4a and 4b show that sonication is very effective in releasing
both dipeptide and tripeptide hydrolase activities. Triton
X-100 treatment is somewhat less efficient while low concentrations of papain were ineffective at releasing dipeptide or

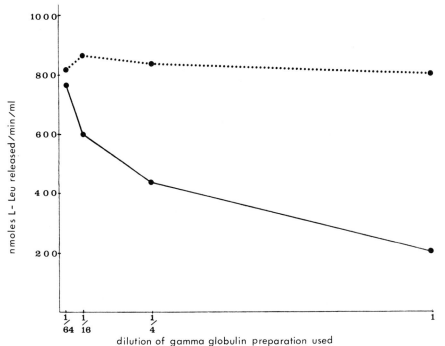

Fig. 3. Brush-border "α" peptide hydrolase activity after treatment with various dilutions of (a) rabbit anti-guinea-pig intestinal mucosa cytosol "α" peptide hydrolase gamma globulin (●——————●) and of (b) normal rabbits gamma globulin (●- - - -●). In both cases the supernatants were tested for activity following centrifugation at 30,000 g for 15 mins.

tripeptide hydrolase. With higher levels of papain a tripeptide hydrolase activity alone was solubilized. When solubilized peptide hydrolases were subjected to starch-gel electrophoresis, the activities released by sonication and treatment with Triton X-100 were found to possess similar electrophoretic mobilities to the cytosol peptide hydrolases (Fig. 4). In contrast papain released a tripeptide hydrolase activity which differed in electrophoretic mobility from the cytoplasmic activities. However, a similar tripeptide hydrolase was detected on starch gels after treatment of the cytosol fraction with papain (Fig. 4).

These results indicate that the tripeptide hydrolase activity released from brush border preparations by papain (Fig. 4) may be an artifact and demonstrates that considerable caution must be taken in the interpretation of experimental results where relatively high concentrations of papain are used to solubilize brush border peptide hydrolases.

TABLE IVa

THE RELEASE OF LEU-LEU HYDROLASE ACTIVITY FOLLOWING
TREATMENT OF BRUSH-BORDER PREPARATIONS WITH
SOLUBILIZING AGENTS

The total activity associated with untreated controls was
taken to be 100%.
Activities in both supernatants and pellets following the
various treatments were expressed as a percentage of the
untreated control.

	Untreated Control	10^{-2} units papain	10^2 units papain	0.1% Triton X-100	Sonication
Supernatant	30	27	0	60	78
Pellet	70	49	70	31	37
Recovery	100	76	70	91	115

TABLE IVb

THE RELEASE OF LEU-LEU-LEU-HYDROLASE ACTIVITY FOLLOWING
TREATMENT OF BRUSH-BORDER PREPARATIONS WITH
SOLUBILIZING AGENTS

	Untreated	10^{-2} units papain	10^2 units papain	0.1% Triton X-100	Sonication
Supernatant	30	14	65	58	100
Pellet	70	72	78	23	15
Recovery	100	86	143	81	115

Activities in supernatnats and pellets were expressed as in
Table IVa.

Implications of these studies in Coeliac Disease

In order to verify Frazer's (1956) hypothesis, that coeliac
disease is due to a genetically deficient intestinal peptidase,
it is necessary to show that the activity of such an enzyme is
markedly reduced in treated subjects (i.e. those on gluten-free
diets for an extended period). Various studies (Dolly and
Fottrell, 1969; Dahlqvist et al., 1970) have shown that peroral
intestinal biopsies from patients with coeliac disease have
reduced peptidase activities. However, it appears that this

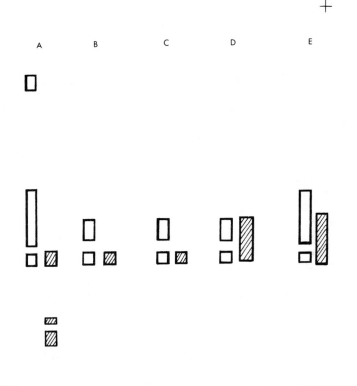

Fig. 4. Zymograms of guinea-pig intestinal mucosal peptide hydrolases from untreated cytosol (A) and cytosol treated with 10^2 units of papain (E); from brush border preparations which were sonicated (B), incubated with 0.1% Triton X-100 (C) and treated with 10^2 units of papain (D). Clear rectangles indicate activities observed when L-leu-L-leu was included in the agar overlay, while hatched rectangles indicate activities observed when L-leu-L-leu-L-leu was present in the overlay. reduction is a secondary response due to villous-atrophy, since the activities return to normal levels after treatment with a gluten-free diet.

Likewise, when the multiple peptidases of peroral biopsies from coeliac patients were examined, it was found that none were absent although the activities with some substrates were reduced (Dolly et al., 1971). These reductions were not found in biopsies from treated patients. In addition, peptide maps obtained after incubating peptic-tryptic-chymotrypic digests of insulin with either coeliac biopsies of biopsies from normal individuals were identical (J.J. Phelan, personal communication).

These different approaches, therefore, have failed to obtain any evidence in support of the "missing peptidase" hypothesis.

DISCUSSION

In the present studies three peptide hydrolases from guinea-pig intestinal mucosa were purified to homogeneity and characterized. In confirmation of previous electrophoretic studies (Donlon and Fottrell, 1972) all three peptidases had relatively broad and overlapping substrate specificities. There were however some interesting differences in specificity between the "α" peptide hydrolase and the β_1 and β_2 peptide hydrolases (see Table 1). For instance, glycine peptides were good substrates for β_1 and β_2 peptidases but were not hydrolysed by α peptidase. The differences in metal requirements between α peptidase and β_1 and β_2 are also noteworthy (Table 3). It is obvious therefore from the aforementioned results and from other findings (see Tables 2 and 3) that an α peptide hydrolase is quite distinct from β_1 and β_2 peptide hydrolases.

In contrast, the two latter peptide hydrolases have very similar electrophoretic mobilities, substrate specificities and metal requirements (Fig. 1 and Tables 1 and 3). The only difference in substrate specificity detected between β_1 and β_2 was in the hydrolysis of L-phe-L-ala, which was hydrolysed by β_2 but not by β_1. Although the latter two enzymes have very similar properties further immunological, chemical and kinetic data are required to determine the extent of this similarity in order to classify them as multiple forms or as isozymes!

As stated earlier the results obtained with a wide range of dipeptides and tripeptides have shown that mammalian intestinal mucosa contain a limited number of peptide hydrolases having activity with a broad spectrum of substrates. Under such circumstances the presence of a relatively specific enzyme , which could be defective in coeliac disease, is extremely unlikely. Nevertheless, Kowlessar et al. (1970) have suggested that there may be a specific enzyme defect in the mucosa of coeliac patients leading to an inability to hydrolyze N-pyrrolidone peptides derived from the enzymic digestion of wheat proteins. However, Woodley (1972) in this laboratory, found no evidence of a deficiency of pyrrolidonecarboxylyl peptidase in coeliac patients.

The results of the studies on the comparison of peptide hydrolases from brush-border and cytosol fractions of intestinal mucosa showed that great caution must be exercised in the interpretation of experiments where papain is used to release peptide hydrolases. It appears from the present studies that papain, at certain concentrations, altered the electrophoretic mobility and substrate specificity of the β peptide hydrolases (Fig. 4). Further studies on the effects of papain on purified preparations of peptide hydrolases are now in progress. It can however be concluded from these studies that there is a striking similarity in the properties of α peptide hydrolase from brush border and cytosol fractions of guinea-pig intestinal mucosa.

In view of these findings and the ease with which some dipeptides are transported intact into the enterocytes (Matthews, 1972), the present enzymes appear to have an important role in the final stages of protein digestion in addition to their presumed role in normal cellular turnover (Peters, 1970a).

REFERENCES

Booth, C.C. 1970. Enterocyte in coeliac disease. *Brit. Med. J.* 3:725-731; 4:14-17.

Dahlqvist, A., T. Lindberg, G. Meeuwisse, and M. Åkerman 1970. Intestinal dipeptidases and disaccharidases in children with malabsorption. *Acta Paediat. Scand.* 59:621-630.

Davis, B.J. 1964. Simplified "disc" electrophoresis. *Ann. N.Y. Acad. Sci.* 121:404-427.

Dolly, J.O., Ann Dillon, M.J. Duffy, and P.F. Fottrell 1971. Further studies on multiple forms of peptidases in mammalian tissues including intestinal mucosa from children with treated and untreated coeliac disease. *Clin. Chem. Acta.* 31:55-62.

Dolly, J.O. and P.F. Fottrell 1969. Multiple forms of dipeptidases in normal human intestinal mucosa and in mucosa from children with coeliac disease. *Clin. Chim. Acta.* 26:555-558.

Donlon, J. and P.F. Fottrell 1971. Quantitative determination of intestinal peptide hydrolase activity using L-amino acid oxidase. *Clin. Chim. Acta.* 33:345-350.

Donlon, J. and P.F. Fottrell 1972. Studies on substrate specificities and subcellular location of multiple forms of peptide hydrolases in guinea-pig intestinal mucosa. *Comp. Biochem. Physiol.* 41B:181-193.

Donlon J. and P.F. Fottrell 1973. Purification and characterization of one of the forms of peptide hydrolases from guinea-pig small intestinal mucosa. *Biochim. Biophys. Acta.* 327:425-436.

Forstner, G.G., S.M. Sabesin, and K.J. Isselbacher 1968. Rat intestinal microvillus membranes. *Biochem. J.* 106:381-390.

Frazer, A.C. 1956. Discussion on some problems of streator-rhoea and reduced stature. *Proc. Roy. Soc. Med.* 49:1009-1013.

Goudie, R.B., C.H.W. Horne, and P.C. Wilkinson 1966. A simple method for producing antibody specific to a single selected diffusible antigen. *The Lancet* 2:1224-1226.

Goldberg, D.M., R. Campbell, and A.D. Roy 1971. The inter-action of trypsin and chymotrypsin with intestinal cells in man and several animal species. *Comp. Biochem. Physiol.* 38B:697-706.

Gray, G.M. 1971. Intestinal digestion and maldigestion of dietary carbohydrates. *Ann. Rev. Med.* 22:391-413.

Hübscher, G., Gwen R. West, and D.N. Brindley 1965. Studies on the fractionation of mucosal homogenates from the small intestine. *Biochem. J.* 97:629-642.

Kim, Y.S., W. Birtwhistle, and Y.W. Kim 1972. Peptide hydro-lases in the brush border and soluble fractions of small intestinal mucosa of rat and man. *J. Clin. Invest.* 51:1419-1430.

Kowlessar, O.D., R.E. Warren, and H.D. Bronstein 1970. Celiac disease: Enzyme defect or immune mechanism? *Progress in Gasteroenterology*, Vol. 2, Groen and Strattan, New York, pp. 409-429.

Lewis, W.H.R. and H. Harris 1967. Human red cell peptidases. *Nature* 215:351-355.

Matthews, D.M. 1972. Rates of peptide uptake by small intest-ine. Peptide transport in Bacteria and Mammalian Gut. *Ciba Foundation Symposium*, Associated Scientific Publishers, Amsterdam, pp. 71-92.

Murphy, M.J. 1970. Starch-iodide as a stain for peroxidases. *J. Chromatogr.* 50:539-540.

O'Cuinn, G., J. Donlon, and P. F. Fottrell 1974. Similarities between one of the multiple forms of peptide hydrolase purified from brush border and cytosol fractions of guinea-pig intestinal mucosa. *FEBS Letts.* 39:225-228.

Peters, T.J. 1970a. Intestinal peptidases. *Gut* 11:720-725.

Peters, T.J. 1970b. The subcellular localization of di- and tri-peptide hydrolase activity in guinea-pig small intestine. *Biochem. J.* 120:195-203.

Smithies, O. 1955. Zone electrophoresis in starch gels: Group variations in the serum proteins of normal human adults. *Biochem. J.* 61:629-641.

Woodley, J.F. 1972. Pyrrolidonecarboxylyl peptidase activity in normal intestinal biopsies and those from coeliac pat-ients. *Clin. Chim. Acta.* 42:211-213.

Woodley, J.F. and A.J. Kenny 1969. The presence of pancreatic proteases in particulate preparations of rat intestinal mucosa. *Biochem. J.* 115:18p.

GENETIC CONTROL AND EPIGENETIC MODIFICATION
OF HUMAN SERUM CHOLINESTERASE

ZEN-ICHI OGITA

Department of Genetics, Medical School, Osaka University
Joancho-33, Kita-ku, Osaka 530, Japan

ABSTRACT. The qualitative and quantitative changes in
the isozymes of human serum cholinesterase are more complicated than those produced by mutations of structural
genes or by differential function of genes during cell
differentiation, whether normal or neoplastic.

The principal results reported in this paper concern
the epigenetic modification of cholinesterase molecules
in the circulation. During disease, such factors as
abnormalities in the release of isozymes into the circulation and epigenetic modification of the circulating
isozyme molecules generate qualitative and quantitative
changes in isozyme patterns.

We previously reported that several extra components
of serum cholinesterase were produced by incubating a
major component of serum cholinesterase (C_4) or complete
normal serum with the serum derived from a 67 year old
patient with a large peritoneal leiomyoma. The patient
possessed very low activities of serum cholinesterase
and experienced prolonged post-surgical apnea. Cholinesterase activity increased to one third that of a normal
person about one year later. The zymograms during this
period showed that the patient's serum contained the C_4
component and also extra components, C_6 and C_7.

The results from genetic data on the family suggested
that a neuraminidase-like enzyme in the patient's serum
might be controlled by genetic factors and that the production of certain extra components depended on the level
of the neuraminidase-like activity. The initial very low
level of serum cholinesterase activity was brought about
by a combination of proteolytic and neuraminidase-like
enzymatic destruction of the C_4 component under the pathological conditions. Certain of the extra components of
the serum cholinesterase can be attributed to epigenetic
modifications of the initial enzyme.

INTRODUCTION

The relationship between the activity of serum enzymes and
diseases is important to clinical biochemists because the
activity of certain serum enzymes changes during illness.
Since the recognition of the biological significance of

289

isozymes (Markert and Møller, 1959), the relationship among isozyme patterns, physiological activity, and various diseases has been widely investigated.

The serum isozymes are synthesized by various organs and cellular organelles and then released into the circulation. Therefore, it should be possible to identify the organ or the tissue involved in a disease by changes in serum isozyme patterns, if the organ specific isozyme pattern is known.

In view of such a possibility, methods of isozyme resolution have been devised and are being used for clinical diagnoses. However, the organs of synthesis and the physiological significance of many serum isozymes are not known. In addition, certain isozymes undergo further modification after their release into the circulation. Thus, it appears that the qualitative and quantitative changes in serum isozymes are even more complicated than the changes occurring within organs or tissues during development or in disease.

Changes in isozyme patterns can be brought about by the following genetic and pathological events:

1) The qualitative and quantitative changes resulting from mutations of genes responsible for the synthesis of the isozymes.

2) Changes in the pattern of gene activation, resulting from normal developmental processes or from aberrant neoplastic differentiation of cells.

3) Activation or deactivation of genes by physiologically active substances such as hormones.

4) Activation or inhibition of enzymes by physiologically active substances such as hormones, metabolic products, or drugs.

5) Abnormalities in the release of isozymes into the circulation due to cell damage, particularly damage to cell membranes, resulting in the leakage of isozymes into the circulation.

6) Changes in the molecular structure of isozymes resulting from epigenetic modification of enzymes in the circulation.

The principal focus of the present report is on the epigenetic modification of enzymes in the circulation. We have investigated the modes of genetic control as well as the epigenetic modifications that have brought about qualitative and quantitative changes in the isozyme patterns of serum cholinesterase (serum Ch E).

MATERIALS AND METHODS

Serum Ch E was resolved by thin layer (1 mm) polyacrylamide gel electrophoresis. Because such thin gel layers are easily

broken in the process of staining for enzyme activity, a
supporting sheet of uncoated cellophane was used. Electrophoresis with polyacrylamide gel as supporting medium was
carried out according to the following procedures.

Stock Solutions. 1) Solution A: 47.5 g of acrylamide and
2.5 g of BIS (N,N'-methylene bisacrylamide) were dissolved in
deionized water to make 250 ml of solution.
  Solution B: 22.8 g of Tris (hydroxymethyl amino methane)
were dissolved in deionized water to which 30.0 ml of 1 N HCl
aqueous solution was added and enough deionized water (or 10%
aqueous glycerol solution) to make 250 ml.
  Solution C: 1 ml of TEMED (N,N,N',N'-tetra methyl ethylene
diamine was dissolved in 100 ml of deionized water.
  Solution D: 120 mg of ammonium persulfate were dissolved
in 100 ml of deionized water.
  2) Working solution: solutions A:B:C:D mixed in the ratio
of 1:1:1:1.
  3) Electrode buffer solution: 4 g of NaOH and 37 g of
boric acid were dissolved in enough deionized water to make
two liters of buffer solution.

Preparation of Thin Gel Layer. Thin, uncoated cellophane of
a quality comparable to dialysis tubing of 124 PD cellophane
(DuPont Co.) was selected because of its high degree of adhesion to the gel. Before use, cellophane sheets are immersed
in deionized water or 5% aqueous glycerol solution until the
cellophane is completely impregnated. The cellophane sheet
is then spread over a glass plate and smoothed out with a
damp sponge to remove any air bubbles. This glass plate with
the cellophane sheet on its upper side may be used as the
bottom support for the gel. On top of this cellophane-glass
plate is placed a frame cut from a sheet of lucite (115 x 165
x 1 mm). A lucite cover plate with the same outside dimensions as the frame is then placed over the mold after the
working solution has been added. The cover plate bears several
lucite ridges 0.8 x 1.0 x 5 mm on its lower surface to make
specimen application slots in the gel layer.
  Gelation commences in approximately 20 to 30 minutes at
25°C. The mold can be removed from the gel plate with the aid
of a stainless steel spatula or a razor blade but should be
left in place for one or two hours.

Electrophoresis. Excess water on the surface of the gel layer
is removed by careful blotting with filter paper. Serum or
any other sample solution, is pippetted into the preformed
specimen slots. Horizontal electrophoresis is carried out

at 5° to 10° C; to prevent overheating the voltage should be adjusted to deliver a constant current of 0.5 to 1.0 mA per cm width of the gel layer. The current is applied for 90 to 120 min.

Enzyme Staining and Plasticization. Serum Ch E is revealed by staining with a mixture consisting of 20 ml of phosphate buffer, pH 6.8, 1 ml of 1% α-naphthyl acetate in acetone solution, and 20 mg of naphthanil diazo blue B.

After enzyme staining, the stained gel layer is washed by running water and the gel layer immersed in 5% aqueous glycerol solution for about two hours. The gel layer is next inserted between sheets of cellophane which have previously been immersed in 5% aqueous glycerin solution; the gel layer is plasticized by incubation at room temperature for about two days.

Determinations of Serum Ch E Activity. Serum Ch E activity was measured by ultraviolet spectrophotometry using benzoylcholine as a substrate. Phenotypes were determined by inhibition of activity by dibucaine and sodium fluoride under standard conditions. A second method that may be used for serum Ch E assay is a manometric method using acetylcholine as a substrate (Kalow and Lindsay, 1955; Kalow and Genest, 1957; Harris and Whittaker, 1961).

RESULTS AND DISCUSSION

ISOZYMES OF SERUM Ch E AND THEIR NOMENCLATURE

Harris et al. (1962) described the appearance and enzymatic behavior of serum Ch E in normal serum as seen after two dimensional electrophoresis, first on paper and then on starch gel. After completion of the runs, four components of serum Ch E activity were found in the gels corresponding to at lease four isozymes which have been designated C_1, C_2, C_3, and C_4. The slowest component (C_4) is the major enzyme, containing most of the serum Ch E activity. In a few sera, a fifth, slower-moving component was found by Harris et al. (1972). This component is designated the C_5 component. Other additional components such as C_6 (Ashton and Simpson, 1966), C_{7a} and C_{7b} (Van Ros and Drvet, 1966) were named on the basis of a four-component serum Ch E system. These extra components were apparently not artifacts because they were observed in fresh specimens. All these components are believed to be isozymes of serum Ch E because of their substrate specificity and their response to inhibitors.

Various nomenclature systems for serum Ch E isozymes have been used such as CHE_1, CHE_2, etc. (Juul, 1968) and CHE-1,

CHE-2, etc. (La Matta et al., 1968; Saeed et al., 1970; Gaffney, 1970). The system of nomenclature used by these workers is based on a sixth and seventh component of serum Ch E. These can be demonstrated in purified enzyme preparations or in Ch E from normal human serum resolved on starch gel or polyacrylamide gels (10 ~ 7.5% gel concentration) with butyryl thiocholine iodide as developer. In these methods the C_4 isozyme described by Harris et al. (1962) separated into three components with decreasing intensity from CHE-5 to CHE-7. The CHE-5 component stained much more intensely than the other six isozymes and is homologous to the C_4 isozyme recognized in other preparations. The nomenclature used by Harris et al., namely, C_1, C_2, etc. will be used to name the serum Ch E isozymes described in this paper.

GENES THAT CONTROL SERUM Ch E

Serum cholinesterase (serum Ch E) is also called plasma cholinesterase or pseudocholinesterase. It rapidly hydrolyzes choline esters of long carbon chain acids such as butylcholine or benzylcholine but may also readily hydrolyze non-choline esters such as α-napthylacetate. Therefore, it is also called non-specific cholinesterase.

Serum Ch E splits suxamethonium (succinyl dicholine) into succinyl monocholine and choline. Since suxamethonium is often used as a muscle relaxant during surgery, this reaction catalyzed by the serum Ch E has an important practical significance. Ordinarily suxamethonium, when injected into the blood stream, is quickly broken down by the action of the serum Ch E and the drug effect disappears. However, the effect of the drug persists for a prolonged period in the presence of inactive or atypical enzyme because of the lowered hydrolytic capacity of these enzymes. As a result, the patient will not recover the ability to breathe.

Mutant genes were discovered during pedigree analyses that included suxamethonium-sensitive patients. At present, family studies have shown that at least six genes control the synthesis of serum cholinesterase.

The enzymatic properties of these esterases are determined by alleles at the E_1 locus. Mutant genes at the E_1 locus that have so far been described are: 1) the normal allele ($E_1{}^u$), which controls the synthesis of the normal enzyme, 2) an atypical gene ($E_1{}^a$), which increases the resistance of the enzyme toward dibucaine (Kalow and Genest, 1957), 3) a fluoride resistance gene ($E_1{}^f$), which increases the resistance of the enzyme toward sodium fluoride (Harris and Whittaker, 1961), and 4) a silent gene ($E_1{}^s$) which yields no enzyme activity

(Liddell et al., 1962). Enzymes of individuals homozygous for these mutant genes have either lower or deficient enzyme activity. Therefore, both homozygotes and heterozygotes carrying these mutant genes are sensitive to suxamethonium.

An additional component, C_5, migrates more slowly in unidimensional starch gel electrophoresis than does the major normal component, C_4. Studies of families led to the hypothesis that the C_5 component was probably inherited as an autosomal dominant; it maps to another locus different from the previously described E_1 locus. This second locus is designated E_2. Persons with this extra component are said to be of C_5^+ phenotype and those lacking it, C_5^-. Harris et al. (1963c) reported that the extra component, C_5, increased the activity of the enzyme by 30% in sera of individuals classified as C_5^+. Neitlich (1966) reported a genetic variant that increased the serum Ch E two to three times the normal mean value with the appearance of a slow-moving band on zymograms obtained by disc-electrophoresis with acrylamide gels as the supporting medium. A study of these families showed that this component was probably inherited as an autosomal dominant. From results of the in vitro and the in vivo studies, Neitlich suggested that the propositus was heterozygous for an enzyme molecule that is more than three times as active as the usual serum Ch E. Yoshida and Motulsky (1969) have shown that the increased enzyme activity of the variant described by Neitlich (1966) could be attributed to an increase in the number of enzyme molecules which have the same specific enzyme activity as the normal enzyme but which are structurally different from the C_5 enzyme component.

Existence of Pre-Enzyme of Serum Ch E. Other factors also have been shown to change quantitatively the serum Ch E activity without altering the above mentioned genetic types. For example, serum Ch E activity is decreased in various disorders of the liver (Vorhaus et al., 1950). The number of cells in the liver with Ch E activity was found to be significantly smaller when the liver was functioning subnormally. These observations suggest that serum Ch E may be synthesized in the liver.

Svensmark (1963), using DEAE cellulose chromatography, showed that the liver contains the pre-enzyme of the serum Ch E - a pre-enzyme with a different mobility and not combined with sialic acid. Thus, pre-enzymes are synthesized in the liver with mobilities different from those found in blood serum. This suggests that the pre-enzyme synthesized in the liver is subsequently modified when it enters the blood stream. In order to understand the quantitative changes of certain isozymes in the serum, it is necessary to examine changes in these isozymes in the organ in which they are first synthesized.

Svensmark (1961a; 1961b) showed that the serum Ch E is a glycoprotein containing several sialic acid residues and that sialidase (neuraminidase, receptor destroying enzyme) treatment causes a decrease in electrophoretic mobility without loss of enzyme activity. The molecular weight of the Ch E was reported to be 348,000 by Haupt et al. (1966).

SUXAMETHONIUM-SENSITIVE PROPOSITI

Propositus No. 1. The patient, a 54 year old male Japanese, was born in Shiga prefecture. He had no significant medical history and came to Kyoto University Hospital for kidney stone treatment. He was subjected to surgery. The physician routinely administered a total of 320 mg of suxamethonium in small aliquots during the operation which lasted one and one half hours. Spontaneous respiration did not begin to return until two hours later and then returned to completion only gradually over a span of five and one half hours from the initial treatment.

Serum was sent to our laboratory for evaluation from Dr. T. Asari and Dr. A. Inamoto of the Medical School, Kyoto University. We found that the patient possessed no serum Ch E activity as determined by the manometric method using acetylcholine and butylcholine as substrates. These results suggested that the patient was a homozygote for the silent gene (Asari et al., 1969).

Family Pedigree and Zymograms. Figure 1 and Figure 2 illustrate the family pedigree and zymograms of serum Ch E through two generations. An examination of the zymogram of the patient's serum Ch E revealed a complete lack of the enzyme. The patient (I-2) has six living children, individuals II - 1, 2, 3, 4, 5, and 7 who are obligate heterozygotes for the same gene, thus requiring subject I-2 to be homozygote for the "silent gene". Enzyme activity levels of obligate heterozygotes range from 54 to 183 μ moles/ml/hr with acetylcholine, and range from 63 to 23 μ moles/ml/hr with butyl-choline as the substrate, respectively. The enzymatic activities of these children were only about one half of the normal level. This can be seen in the zymogram in which the staining intensity is much weaker than that of normal serum Ch E.

Propositus No. 2. A 40 year old male Japanese was referred to Nissei Hospital, Osaka, on October, 1971 for a gall stone

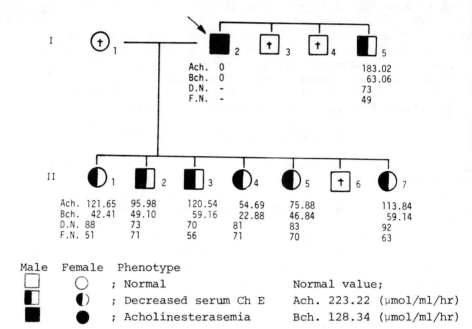

Figure 1. Pedigree of propositus No. 1, who is homozygous for a silent gene for serum Ch E. The figures indicate the cholinesterase activities as estimated by the manometric method using acetylcholine and butylcholine as substrates; the dibucaine and fluoride numbers were determined by standard methods. The arrow indicates the propositus.

operation. He had been suffering from cholelithiasis. Since his initial serum Ch E activity was found to be low during preparative examinations, GOF anesthesia without suxamethonium was performed to avoid prolonged apnea.

Serum was brought to our laboratory for evaluation from Dr. S. Kitamura and Dr. K. Ogli of the Department of Anesthesiology, Medical School, Osaka University. The patient's serum Ch E activity was completely deficient. We estimated the total enzyme activity, dibucaine number, and chloride number. We also resolved the serum Ch E isozymes, using thin layer polyacrylamide gel electrophoresis, from 24 family members distributed in three generations.

Family Pedigree and Zymograms. Figure 3 shows the family pedigree and zymograms of serum Ch E representing three genera-

C_4 Alb-E

I - 2(propositus)

I - 5

II- 1

II- 2

II- 3

II- 4

II- 5

II- 7

Control Serum

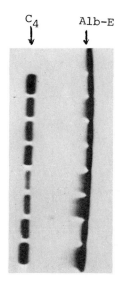

Origin

Figure 2. Zymograms of serum Ch E of members of the family shown in Figure 1. The zymogram of the patient's serum Ch E shows a complete lack of the enzyme for α-naphthyl acetate as substrate. The staining intensity in the zymograms of his children (II-1, II-2, II-3, II-4, II-5, and II-7) is much weaker than that of normal serum Ch E. Alb-E: albumin esterase; C_4: C_4 component.

tions. There were two cases of acholinesterasemia: the propositus (II-5) and his elder sister (II-3). The enzyme activities were respectively, 0.0 and 1.9 μ moles/ml/hr with benzoylcholine as the substrate.

Judging from the absence of serum Ch E activity and from the fact that their parents were consanguineous, we conclude that these individuals may be homozygotes for the "silent gene" of serum Ch E. In addition, the C_5 variant was observed in the spouse (II-6) and daughter (III-5) of the propositus (II-5). As shown in Figure 3, the daughter (III-5) exhibited a higher serum Ch E activity level than did her parents (Kitamura et al., 1973).

It is reported by Harris et al. (1963b) that the activity of serum Ch E is generally high in C_5 variants. Thus, the high enzyme activity of this daughter may be due to the C_5

297

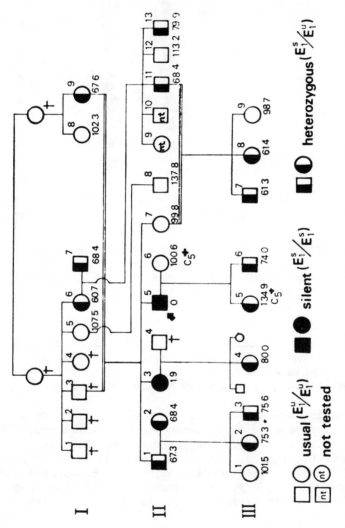

Figure 3. Family distribution of decreased serum Ch E activity and the additional serum cholinesterase electrophoretic component. The figures indicate the serum Ch E activity which was measured by ultraviolet spectrophotometry using benzoylcholine as a substrate. The arrow indicates the propositus. Note that III-5, who carries a silent gene in the heterozygous state, shows increased serum Ch E activity compared to III-6, who is in the heterozygous state with a silent gene. III-5 also has an additional serum Ch E component, C_5, which is evident after thin layer electrophoresis.

298

variant. These findings suggest that the genotypes of the
family members studied were as follows:

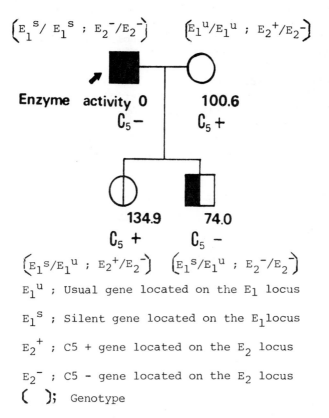

$\left(E_1{}^S/E_1{}^S ; E_2{}^-/E_2{}^-\right)$ $\left(E_1{}^u/E_1{}^u ; E_2{}^+/E_2{}^-\right)$

Enzyme activity 0   100.6

$C_5 -$ $C_5 +$

134.9   74.0

$C_5 +$ $C_5 -$

$\left(E_1{}^S/E_1{}^u ; E_2{}^+/E_2{}^-\right)$ $\left(E_1{}^S/E_1{}^u ; E_2{}^-/E_2{}^-\right)$

$E_1{}^u$; Usual gene located on the E_1 locus

$E_1{}^S$; Silent gene located on the E_1 locus

$E_2{}^+$; C5 + gene located on the E_2 locus

$E_2{}^-$; C5 - gene located on the E_2 locus

( ); Genotype

Figure 4

Propositus No. 3. A 67 year old male Japanese with leiomyoma
was referred in July, 1968, to Tottori University Hospital
for excessive bleeding as observed in feces. An emergency
operation was performed for a supposed gastric ulcer. The
patient was pre-medicated with 0.4 mg atropine and 250 mg
thioamytal. This treatment was followed by 50 mg of suxamethonium. Spontaneous respiration did not return to the
patient for 29 minutes, and this necessitated intermittent
positive pressure respiration. The doctor routinely administered GOF anesthesia with suxamethonium during the operation.
For the second, third, and fourth treatments, 20 mg of suxamethonium were given and the resulting apneas lasted for an
average of 25 minutes. The patient received transfusions of
2400 ml of blood before and after the operation. A large
peritoneal leiomyoma was found but not removed because of

extensive adhesions to neighboring tissues.

Serum was sent to our laboratory for evaluation from Dr. T. Sato and Dr. R. Matsuura of the Medical School, Tottori University. The patient, in October, 1968, possessed very low esterase activities for acetylcholine and butylcholine as determined by the manometric method. The activities were 0.0 and 2.2 µ moles/ml/hr, respectively. These results initially suggested that this patient was a homozygote for the "silent gene", but further investigations with the serum Ch E obtained from the propositus and from 5 living members of his family indicated that the deficiency of serum Ch E activity was not the result of homozygosity for the "silent gene" $(E_1{}^S)$ (Matsuura et al., 1969; Ogita et al., 1969).

Family Pedigree and Zymograms. Figure 5 and Figure 6 show the family pedigree and zymograms of serum Ch E through two generations. The patient had nine children. Two of the nine had

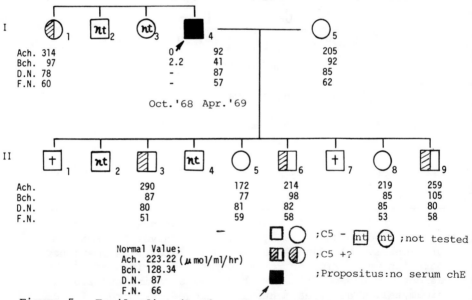

Figure 5. Family distribution of serum Ch E and the C_5-like component. The figures indicate the cholinesterase activity using acetylcholine and butylcholine as substrates; dibucaine and fluoride numbers were determined by using standard methods.

The enzyme activities of the propositus were measured with the serum obtained in October, 1968, and in April, 1969. In April, 1969, the serum Ch E activity of the patient increased to one third to one half of normal activity. Note that each family member who has an increased acetylcholine activity also

300

Fig. 5 (continued). shows a C5-like component after electrophoresis, except for one of the children, II-6.

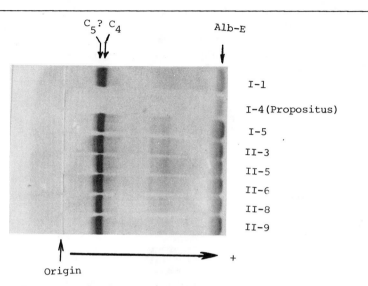

Figure 6. Zymograms of serum Ch E of members of a family
shown in Figure 5. The zymogram of the serum Ch E which was
obtained from the propositus from July through September, 1968,
shows a complete lack of the enzyme for α-naphthyl acetate
as substrate. The propositus's sister (I-1) and three of
his children, II-3, II-6, and II-9 all show a C_5-like component. His elder sister has a C_5-like component which is much
greater than that found in the propositus's children. Alb-E:
albumin esterase; C_4: C_4 component; C_5?: C_5-like component.

died and two were not available for examination. Therefore,
only five children were used in this investigation. Strangely,
the serum Ch E levels of these five children were all normal.
We expected a significant reduction in the enzyme level judging from our previous experience with the "silent gene", if
indeed the propositus were homozygous for this gene. I-4 is
the zymogram of serum Ch E in serum obtained from this patient.
No serum Ch E is evident. I-5 is the zymogram of his "normal"
wife. II-3, 5, 6, 8, and 9 are zymograms of several of his
children. I-1 is the zymogram of his elder sister. They
were all normal. However, it was noted that three of five
children (II-3, II-6, and II-9) who were examined, and his
elder sister (I-1) had the C_5-like component. The propositus's
sister and two of three children (II-3 and II-9) who have the
extra component, has slightly elevated enzyme activities for

acetylcholine as the substrate, but the esterase activity of the children (II-6) was not increased. His elder sister showed higher acetylcholine esterase activity than any of his children.

Figure 7 contains zymograms prepared with the serum taken from this patient at intervals during the span of one year from July, 1968, through July, 1969. Slot 1 contained sera obtained from this patient from July through September, 1968.

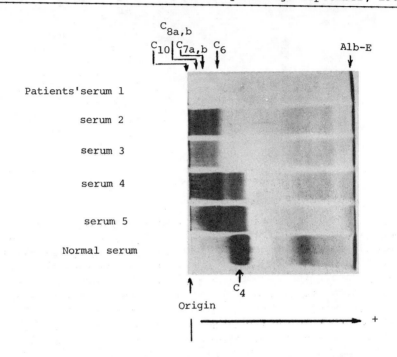

Figure 7. Changes in the zymogram patterns of the patient's serum by using α-naphthyl acetate as the substrate during the span of one year from July of 1968 through July of 1969. Patient's serum 1: Patient's serum for the period from July through September, 1968. Serum 2: for January, 1969. Serum 3: for March, 1969. Serum 4: from the end of March through April, 1969. Serum 5: for July, 1969, just before his death. Alb-E: albumin esterase; C_6, $C_{7a,b}$, $C_{8a,b}$, and C_{10}: extra components.

There was no C_4 component and no, or very low, serum Ch E activity. We have designated such sera as "serum 1". However, the serum of the patient began to show some serum Ch E activity in January of the following year. Zymograms of the serum Ch E for January and March are shown in channels 2 and

3, respectively. Stained areas representing C_6 and $C_{7a,b}$ are
evident. We designated these sera as serum 2 and 3, respectively. In the period from the end of March through April of
the same year the serum Ch E activity of the patient reached
almost one-third that of normal activity. The zymogram for
this period is shown in channel 4. The origin and the C_4, C_6,
and $C_{7a,b}$ isozymes are easily recognized. In addition, close
examination reveals a vague C_5-like band. The zymogram for
July of the same year, just before the death of the propositus,
is shown in channel 5. It is very similar to that of April,
but the intensity of staining is much greater. The zymogram
for a normal serum is shown in channel 6.

Estimation of Modifying Factor in the Patient's Serum. In
order to determine whether there is a constituent in the
patient's serum that modifies the normal C_4 component of serum
Ch E, we carried out the following experiments. The patient's
serum from each stage was mixed with the purified C_4 and incubated. Then the Ch E was resolved by electrophoresis. The
results of these experiments are shown in Figure 8. The first
channel is the zymogram of the patient's serum 1, which was
taken from July to September, 1968, and which had previously
been found to contain no serum Ch E activity. Channel 2 is
the zymogram of the mixture of this patient's serum 1 and the
C_4 component. As can be seen, not only C_4 but C_6, C_{7a}, C_{7b},
C_{8a}, C_{8b}, C_9, and the origin (C_{10}) appeared on the zymogram.
Channel 4 shows the serum 2 from the patient taken during
January, when the serum showed a small amount of Ch E activity.
The zymogram showed a complete lack of the C_4, but small
amounts of the C_6 and $C_{7a,b}$ components. When this serum was
mixed with the purified C_4 component, again C_6 and $C_{7a,b}$ showed
up in addition to C_4, but the C_4 was even more intense. On
the other hand, activity at the origin was reduced. Channel
6 shows the serum taken from the patient during the end of
March through April, 1969. The last channel shows the mixture
of this serum and purified C_4. The result was similar to the
preceeding one, but the trend observed in the preceeding zymogram was even more accentuated. It was also noted that not
only the origin (C_{10}) and C_9, but also the C_8 component was
somewhat reduced. In general, there was a gradual shift of
serum Ch E components from left to right during the course
of these changes.

The question now becomes: What is the factor present in
the patient's serum that modifies the C_4 component?

Modifying Factor as a Neuraminidase-like Enzyme. In developing
a hypothesis, we postulate the existence of substances X and

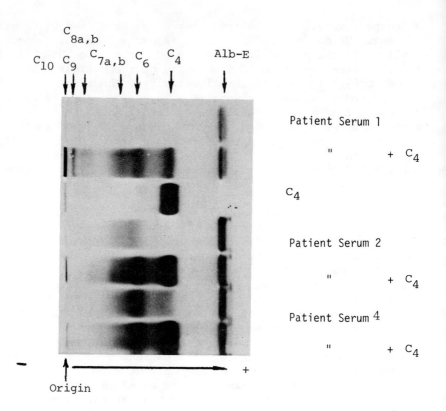

Figure 8. Estimation of a modifying factor in the patient's serum. The electrophoretic analysis was carried out with a mixture of the patient's serum from each stage with purified C_4, which is the main component of serum Ch E; the mixture was incubated. Note that the zymogram of the mixture of this patient's serum with the C_4 component was changed by incubation.

Channel 1 is a zymogram of patient's serum 1; sera obtained from this patient from July through September, 1968. There was no C_4 component and no Ch E activity. Channel 2 is a zymogram of the mixture of serum 1 and the C_4 component, after incubation. Channel 3 is a zymogram of the C_4 component which was purified from normal human sera. Channel 4 is a zymogram of patient's serum 2; patient's serum in January, 1969. Channel 5 is a zymogram of the mixture of serum 2 and the C_4 component, after incubation. Channel 6 is a zymogram of patient's serum 4; patient's serum at end of March through April, 1969. Channel 7 is a zymogram of the mixture of serum 4 and the C_4 component, after incubation. Alb-E: albumin esterase; C_4: C_4 component; C_6, $C_{7a,b}$, $C_{8a,b}$, C_9, and C_{10}: extra components.

Y, X to modify the C_4 component to C_6, $C_{7a,b}$, $C_{8a,b}$, C_9, and C_{10}, and Y to reduce the Ch E activity. From this viewpoint, the patient had abnormal amounts of both X and Y substances. The patient's serum showed no serum Ch E activity during the early stage, from July through October, presumably because of the activity of the Y substance. The Y activity in the patient's serum, for unexplained reasons, began to drop later, from January of the following year, revealing hitherto undetected serum Ch E activity. There was a concomitant drop in the activity of the Y substance during the corresponding period, shifting the zymogram from left to right as shown in Figure 8.

In the following experiment we attempted to identify one of these postulated substances, particularly the X substance. As shown in Figure 9 we first analyzed the patient's serum

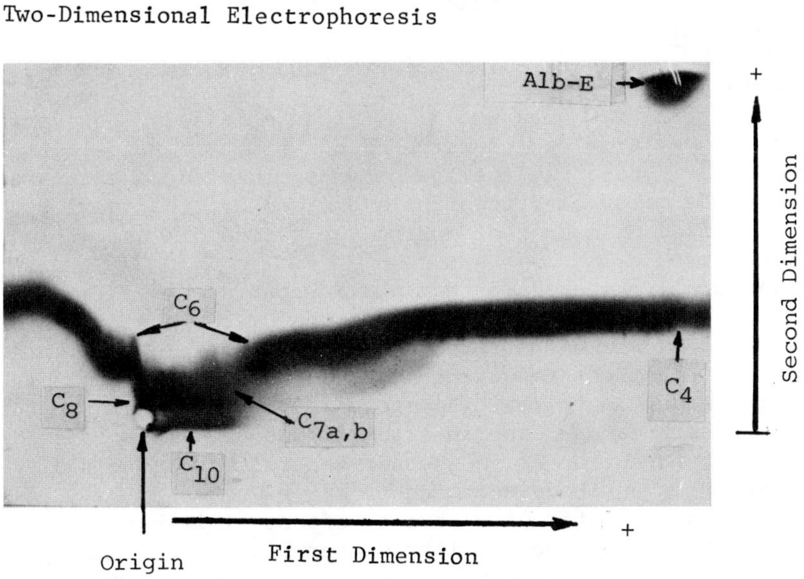

Two-Dimensional Electrophoresis

Figure 9. Two-dimensional electrophoresis for estimation of modifying factors in the serum of the propositus, No. 3.

In this experiment we tried to identify one of these unknown substances, particularly the X substance. A concentrated sample of the patient's serum is placed in the origin of an acrylamide gel layer. Migration directions are marked with arrows; positions of electrodes in the first run and in the second run are designated with + and -. Origin is indicated. After the first run, a thin strip of paper was soaked

305

Fig. 9 (continued). with the purified C_4 and this strip of paper was placed across the field from left to right. After incubation, the C_4-soaked paper was removed and the second electrophoresis was run from the bottom to the top and Ch E activity visualized. The result shows that the X factors are separated near the origin by the first run. Note that changes in mobility of C_4 indicate that extra components are produced by the incubation. Alb-E: albumin esterase; C_6, $C_{7a,b}$, C_8, and C_{10}: extra components.

obtained during the early stage when there was no electrophoretically distinguishable Ch E activity. The electrophoretic separation was achieved in the first dimension. Next, a thin strip of paper was soaked with purified C_4 and this strip of paper was placed across the previously separated channel, and the gel incubated. The C_4-soaked paper was then removed and the underlying gel subjected to a second electrophoretic separation in the perpendicular dimension after which the serum Ch E activity was visualized.

If the patient's serum had contained no substance that interfered with the mobility of the C_4 component, then a straight line zymogram would have been obtained. However, there was a large dip near the origin of the first electrophoresis. Apparently the X-substance migrated to this area during the first electrophoresis and reduced the mobility of the purified C_4 component during the second dimension electrophoresis.

Naturally, we assume that the Y substance was also present, but we assume that the quantity of the C_4 component was large enough to overcome the effect of Y, or that the Y was spread too thin to affect the zymogram. The X factors, responsible for the reaction to the C_4 component, were located near the origin where gamma-globulin fractions occur. This coincidence raised the question as to whether the factors are a modifying enzyme or an antibody reacting to serum Ch E.

Strangely, antibodies have an effect similar to neuraminidase in modifying the mobility of serum Ch E. As shown in Figure 10, varying concentrations of antibody against C_4 (anti-C_4), which were obtained from rabbits, were incubated with the C_4 component and the mixtures subjected to electrophoresis. At suitable concentrations, the zymograms showed additional bands very much like the extra components.

In order to test the effect of antibodies that would counteract the effect of the antibodies used in the preceeding experiment, anti-rabbit gamma globulin was obtained from commercial sources. When mixtures which contained the C_4 component and anti-C_4 antibody were treated with anti-rabbit gamma globulin, the effect of the anti-C_4 antibody was completely

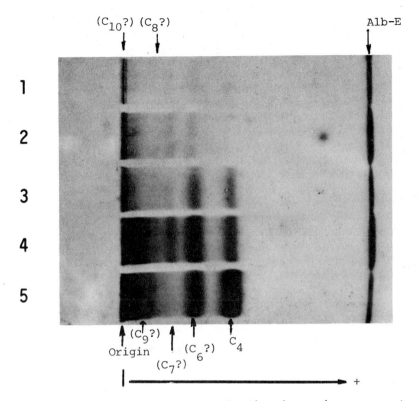

Figure 10. The effects of C_4 antibodies in various concentrations on the C_4 component. Diluted C_4-antibody solutions have effects similar to neuraminidase in modifying the mobilities of serum Ch E components.

1: mixture of 0.1 ml of C_4-antibody and 0.1 ml of normal serum; 2: mixture of 0.1 ml of 2 times diluted antibody and 0.1 ml normal serum; 3: mixture of 0.1 ml 4 times diluted antibody and 0.1 ml normal serum; 4: mixture of 0.1 ml of 16 times diluted antibody and 0.1 ml of normal serum; 5: mixture of 0.1 ml of 32 times diluted antibody and 0.1 ml of normal serum. Alb-E: albumin esterase; C_4: C_4 compound; (C_6?), (C_7?), (C_8?), (C_9?), and (C_{10}?): These products are just like extra components.

eliminated, leaving only the C_4 component.

Therefore, the patient's serum was treated with anti-human gamma globulin, anti-IgM, anti-IgG, and anti-IgA to determine whether these antibodies would eliminate the C_6, C_{7a}, and C_{7b} components as they did in the preceeding experiment. Anti-human gamma globulins had no effect upon the zymograms of the patient's serum. These results demonstrate that the modifica-

307

tion of the serum Ch E of the patient was not due to antibodies, such as autoimmune antibodies in the patient's serum. Thus, we tentatively identify this X substance as a neuraminidase-like enzyme (sialidase). The products in the mixture of the C_4 and the patient's serum were then analyzed by the thiobarbiturate method to test for an increase in sialic acid. Indeed it did increase. And there was a gradual decrease in the neuraminic acid during the course of our investigation, indicating a gradual decrease in the neuraminidase activity. This observation is in agreement with the changes in the serum Ch E patterns during this period (Ogita and Sakoyama, 1971). This finding suggests that the X substance may be an enzyme very similar to neuraminidase (sialidase) because the serum Ch E is known to be an acidic glycoprotein containing several sialic acid residues per molecule (Svensmark, 1961a, 1961b).

Svensmark (1961b) has observed that in certain neurological diseases serum Ch E shows a decreased mobility due to the presence of neuraminidase, Augustinsson and Ekedahl (1962) demonstrated that the treated enzyme had the same kinetic properties as the untreated enzyme and it has been suggested that naturally occurring neuraminidase may give rise to limited cholinesterase heterogeneity.

GENES RESPONSIBLE FOR NEURAMINIDASE

The next question is concerned with whether the increase in neuraminidase-like activity in the serum was determined by the patient's genotype. The results of a family study showed that the relevant structural genes of this patient were normal. All of his children had normal serum Ch E levels. Three of five children who were examined and the sister had an extra band, the C_5-like component, which stained faintly and accompanied a slight increase in serum Ch E activity.

If extra components are generated from modifications of the C_4 component by neuraminidase, then one possibility is that the patient has a repressed neuraminidase gene. The family inheritance data may suggest that the gene is located on an autosomal chromosome and that the production of extra components depends on the level of neuraminidase activity.

OTHER MODIFYING FACTORS AS PROTEOLYTIC ENZYMES

The last question is concerned with the Y. substances. Although we could not obtain definitive data from analyzing the patient's serum, we believe that the Y substances may be proteolytic enzymes similar to trypsin and plasmin. Our preliminary data show that purified C_4 was destroyed by treatment

with these proteolytic enzymes, and also that all serum Ch E
activity disappeared from the patient's serum when reacted
with the proteolytic enzymes and the neuraminidase-like
enzymes simultaneously under the pathological conditions (Ogita
and Sakoyama, 1973). The genetico-biochemical analysis of
this patient will be reported in detail elsewhere.

From these results, we propose the following working
hypothesis.

HYPOTHESIS FOR GENETIC CONTROL OF EXTRA COMPONENTS

We propose that some of the extra components of the serum
Ch E, with electrophoretic mobilities slower than the C_5 component, are modified enzymes formed as a result of the activity of a neuraminidase-like enzyme which is produced in accord
with the genetic background of an individual under appropriate
physiological or pathological conditions. This hypothesis
may be called the *"Epigenetic modification"* hypothesis. Genes
responsible for a neuraminidase-like enzyme are assumed to be
at the N locus, which is located on an autosomal chromosome
and are unrelated to the structural genes for serum Ch E.
This gene is tentatively designated as N^R and its normal
allele is N^u. The N^R gene 'may be activated by changed physiological conditions and thus raise the neuraminidase activity
beyond the normal range. As a result, the C_4 component would
lose neuraminic acids and be transformed into C_5-like components of slower mobility. With further increases in neuraminidase activity, the C_6 or the $C_{7a,b}$ and additional slow
components would be formed.

The hypothesis suggests that the N^R gene should be inherited as an autosomal dominant, but individuals carrying the
N^R gene are not always detectable since the expression of the
gene is dependent upon physiological or pathological conditions. Perhaps the N^R gene is a regulatory gene responsible
for the physiological conditions.

Genotype of Propositus No. 3. The patient (propositus No. 3)
may be interpreted as follows by this hypothesis. The normal
C_4 and C_5-like components may be present in the serum of the
patient under conditions, because his three children and
elder sister have the C_5-like component. However, the neuraminidase activity of this patient's serum was elevated above
the C_5 + (?) phenotype by pathological changes which altered
gene expression. As a result, the C_4 component was converted
into extra components with slower mobility. Furthermore,
degradation of serum Ch E in the patient's serum could be
brought about by reaction with proteolytic enzymes.

The hypothesis suggests that the genotypes of the family

members of the patient (propositus No. 3) are as follows:

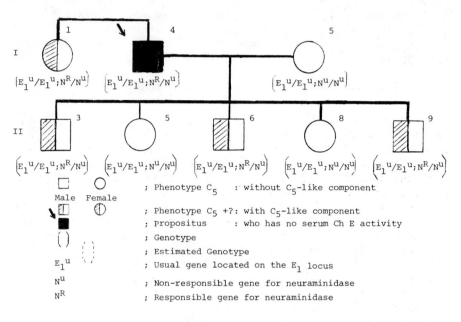

Figure 11

Possible Genetic Backgrounds for the Extra Components. The genetic backgrounds for the extra components which have been reported by other authors may be explained by our hypothesis. Ashton and Simpson (1966) described the electrophoretic pattern of one serum sample from a Brazilian population. This sample revealed an unusual zone tentatively designated C_6, located between the origin and the C_4 component and accompanied by a faint additional zone located between the origin and the C_6 component. They reported that sera from all members of the family were C_5 negative and that only the propositus showed the C_6 zone; the Ch E activity of his serum was just below the average for the usual C_5^- phenotype (Simpson, 1966).

Van Ros and Drvet (1966) described two unusual electrophoretic patterns of serum Ch E which showed additional components distinct from C_5. These have so far been found only in African subjects. The first type of unusual pattern shows the normal zones C_1-C_4 and an additional component called C_6, which could be differentiated from C_5 by its slower migration rate. The second type of unusual pattern also shows the four normal zones in addition to two unusual slow components called C_{7a} and C_{7b}. Unfortunately, it was not possible to study the inheritance of either of these phenotypes.

Although the genotypes of these propositi could not be investigated by the authors, the genetic backgrounds may have been the same as for the genotype of propositus No. 3, who was presumed to be heterozygous for a gene producing a neuraminidase-like enzyme. This is a reasonable assumption to explain the Brazilian family with the C_6 and faint additional components reported by Ashton and Simpson (1966). We assume, of course, that the expression of the N^R gene is dependent on specific physiological or pathological conditions.

With regard to the C_5 component, the family studies by Harris et al. (1963b) indicated that the C_5 component is probably inherited as an autosomal dominant and that the occurrence and properties of the C_5 are determined by genes at two loci, although the molecular basis of the enzymatic variations remains unknown. Furthermore, the existence of a second locus E_2, different from the E_1, was established by Motulsky (1964). Whether the E_2 locus is different from the N locus cannot be decided at this time. This alternative hypothesis may be called the *"Two loci"* hypothesis in contrast to the foregoing *"Epigenetic modification"* hypothesis in which two independent loci are assumed to control the serum Ch E. Genes at the second locus E_2 appear to influence the amount of enzyme and the electrophoretic mobility independently from alleles at the first locus E_1.

However, in occasional samples from individuals apparently carrying the E_2^+ gene, the C_5 component was either very weakly stained or not visible at all. Harris et al. (1963c) pointed out that either some of the heterozygotes did not exhibit the C_5 component or else that it was not detectable by their procedure. They suggested that the genetic mechanism may be more complex. A few exceptions were also noted in the study of over one thousand Brazilian families by Ashton and Simpson (1966).

These inconsistencies in the segregation data may be easily dismissed if the assumption is made that the exceptional individuals with the C_5^+ phenotype may have the same genotype as propositus No. 3. According to this assumption, certain genetic heterogeneities in the expression of the C_5 trait should be established by pedigree analysis; Type 1 is controlled by genes at at least two loci, E_1 and E_2; and Type II is controlled by genes at at least two loci, E_1 and N.

Although genes that increase the neuraminidase-like enzyme have not been identified as yet, the *"Epigenetic modification"* hypothesis, if proven, should complement the *"Two loci"* hypothesis for the extra components as proposed by Harris et al. (1963b). These hypotheses are not necessarily mutually exclusive. The *"Epigenetic modification"* hypothesis, however, is more inclusive and explains our observations more fully.

ACKNOWLEDGEMENTS

The author would like to express his indebtedness to his present and past colleagues who have contributed in so many ways to these studies: Dr. Y. Asari, Dr. A. Inamoto, Dr. S. Kitamura, Dr. K. Ogli, Dr. R. Matsuura, Dr. S. Nozaka, Dr. T. Sato, Mr. Y. Sakoyama, and Miss S. Aotami. And the author is grateful for helpful suggestions of Dr. C.H. Yoon in the early stage of preparation with the manuscript. And the author is also indebted to Dr. C. L. Markert for his available suggestions and encouragement, and for critically reviewing the manuscript.

The original investigations were supported by the Scientific Fund from the Ministry of Education and the Research Grant from the Ministry of Health and Welfare, Japan; and also, in part, by the Research Grant from the Scientific Foundations of the Chiyoda Seimei. The manuscript was prepared with the help of National Science Foundation research grant 36749, while the author was a research associate in the Department of Biology, Yale University, New Haven, Connecticut, U.S.A.

REFERENCES

Asari, Y., A. Inamoto, Z. Ogita, Y. Sakoyama, and S. Aotani 1969. A family of acholinesterasemia which found by prolonged apnea with suxamethonium. *Masui* 18: 1404 (in Japanese).

Ashtone, G.C. and N.E. Simpson 1966. C_5 types of serum cholinesterase in a Brazilian population. *Am. J. Hum. Genet.* 18: 438-447.

Augustinsson, K-B. and G. Ekedahl 1962. The properties of neuraminidase-treated serum cholinesterase. *Biochim. Biophys. Acta* 56: 392-393.

Ecobichon, D.J. and W. Kalow 1963. Effects of sialidase on pseudocholinesterase types. *Can. J. Biochem.* 41: 969-974.

Harris, H. and M. Whittaker 1961. Differential inhibition of human serum cholinesterase with fluoride. Recognition of two new phenotypes. *Nature* (Lond.) 191: 496-498.

Harris, H., D.A. Hopkinson, and E.B. Robson 1962. Two dimensional electrophoresis of pseudocholinesterase components in normal human serum. *Nature* (Lond.) 196: 1296-1298.

Harris, H. and E.B. Robson 1963a. Fractionation of human serum cholinesterase components by gel filtration. *Biochim. Biophys. Acta.* 73: 649-652.

Harris, H., E.B. Robson, A.M. Glen-Bott, and J.A. Thornton 1963b. Evidence from non-allelism between genes affecting human serum cholinesterase. *Nature* 200: 1181-1187.

Harris, H., D.A. Hopkinson, E.B. Robson, and M. Whittaker 1963c. Genetical studies on a new variant of serum cholinesterase detected by electrophoresis. *Ann. Hum. Genet.* (Lond.) 26: 359-382.

Haupt, H., K. Heido, O. Zwisler, and H.G. Schwick 1966. Isolierung und physikolisch-chemische characterisierung der cholinesterase aus Human serum. *Blut.* 14: 65-76.

Juul, P. 1968. Human plasma cholinesterase isozymes. *Clin. Chim. Acta* 19: 205-213.

Gaffney, P.J., Jr. 1970. Human serum cholinesterase partial purification and nature of the heterogeneity of this system. *Biochim. Biophys. Acta* 207: 465-476.

Kalow, W. and H.A. Lindsay 1955. A comparison of optical and manometric methods for the assay of human serum cholinesterase. *Canad. J. Biochem.* 33: 568-574.

Kalow, W. and K. Genest 1957. A method for the detection of atypical forms of human serum cholinesterase. Determination of dibucaine numbers. *Canad. J. Biochem.* 35:339-346.

Kitamura, S., K. Ogli, Z. Ogita, and Y. Sakoyama 1973. Two cases of acholinesterasemia and C_5 variant of cholinesterase isozyme. *Masui.* 22: 667-671 (with English abstract).

Liddell, J., H. Lehmann, and E. Silk 1961. A "silent" pseudocholinesterase gene. *Nature* (Lond.) 193: 561-562.

LaMatta, R.V., R.B. McComb, and H.J. Wetstone 1965. Isozymes of serum cholinesterase: A new polymerization sequence. *Canad. J. Physiology and Pharmacology* 43: 313-318.

LaMatta, R.V., R.B. McComb, C.R. Noll, Jr., H.J. Wetstone, and R.F. Reinfrank 1968. Multiple forms of serum cholinesterase. *Arch. Biochem. Biophys.* 126: 299-305.

Markert, C.L. and F. Møller 1959. Multiple forms of enzymes: Tissue, ontogenetic, and species specific patterns. *Proc. Natl. Acad. Sci. U.S.A.* 45: 753-763.

Matsuura, R., S. Nozaka, T. Sato, and Z. Ogita 1969. One case of acholinesterasemia. *Masui.* 18: 1404-1405 (in Japanese).

Motulsky, A.G. 1964. Pharmacogenetics. *Prog. Med. Genet.* 3: 49-74.

Neitlich, H.W. 1966. Increased plasma cholinesterase activity and succinylcholine resistance: A genetic variant. *J. Clin. Invest.* 45: 380-387.

Ogita, Z., Y. Sakoyama, T. Sato, R. Matsuura, H. Asari, and A. Inamoto 1969. Genetical and biochemical studies on the suxamethonium sensitivity. *Jap. J. Human Genet.* 14: 246-247 (in Japanese).

Ogita, Z. and Y. Sakoyama 1971. Genetic control systems of abnormal pseudocholinesterase components. *Jap. J. Human Genet.* 15: 291.

Ogita, Z. and Y. Sakoyama 1973. Genetic control systems of human serum cholinesterase isozymes. *Jap. J. Human Genet.* 18: 132 (in Japanese).

Saeed, S.A., G.R. Chadwick, and P.J. Mill 1971. Action of proteases on human plasma cholinesterase isoenzymes. *Biochim. Biophys. Acta* 229: 186-192.

Simpson, N.E. 1966. Factors influencing cholinesterase activity in a Brazilian population. *Amer. J. Human Genet.* 18: 243-252.

Svensmark, O. 1961a. Effect of sialidase on the electrophoretic properties of human serum cholinesterase. *Dan. Med. Bull.* 8: 28-29.

Svensmark, O. 1961b. Human serum cholinesterase as a sialoprotein. *Acta Physiol. Scand.* 52: 267-275.

Svensmark, O. 1963. Precursors of serum cholinesterase in human liver. *Acta Physiol. Scand.* 59: 148-149.

Van Ros, G. and R. Drvet 1966. Uncommon electrophoretic patterns of serum cholinesterase (Pseudocholinesterase). *Nature* 212: 543-544.

Vorhaus, L.J., II, H.H. Scudamore, and R.M. Kark 1950. Measurement of serum cholinesterase activity in the study of diseases of the liver and biliary system. *Gastroaenterology* 15: 304-315.

Yoshida, A. and A.G. Motulsky 1969. A pseudocholinesterase variant (*E. cynthiana*) associated with elevated plasma enzyme activity. *Amer. J. Human Genet.* 21: 486-498.

POTENTIALS IN EXPLORING THE PHYSIOLOGICAL ROLE
OF ACETYLCHOLINESTERASE ISOZYMES

J.M. VARELA
Central Institute for Brain Research,
Amsterdam, The Netherlands

ABSTRACT. Saline and Triton X-100 extracts of both
brain and muscle of rat revealed 4 isozymes of AChE by
polyacrylamide gel electrophoresis and gel filtration.
The kinetic properties of AChE as well as the relative
percentage of activity of its isozymic species differ
in the two tissues. The interconversion of the two
native states (free and bound) of AChE, whose isozymes
are likely to be distributed in distinct cell compartments,
might have some implication for the nerve impulse. The
actual relevance of the differential Hill coeficient of
AChE isozymes for repacking of the synaptic vesicles and
for pacing the firing of the synapse, and the relationship (spatial and/or structural) of the isozymes and the
so-called inactivating and activating particles (or
gates) of the ionic channels are discussed. A one isoreceptor-one isozyme hypothesis is considered and preliminary evidence is presented for involvement of
isozymes 1 and 3 in the permeabilities to K^+ and Na^+,
respectively.
    The presence of the enzyme, as revealed by electron
microscopy and biochemical methods, in a few cell
compartments (nuclear envelope, endoplasmic reticulum,
cytosol) suggests a role in cell function unconnected with
synaptic activity. In ths context preliminary exper-
mental findings indicate that participation of soluble
AChE in the regulation of the cytosolic part of the
Krebs cycle deserves exploration.

INTRODUCTION

It is generally accepted that hydrolysis of acetylcholine
(ACh) by acetylcholinesterase (AChE) is the main mechanism
of termination of the action of the mediator in cholinergic
synapses (cf. Nachmansohn, 1972). Until now, involvement in
synaptic transmission seems to be the only function ascribed
by good experimental evidence to this enzyme. However, the
presence of AChE, as shown by ultrastructural histochemistry,
at cellular loci other than the synaptic membranes, namely

the nuclear envelope, the endoplasmic reticulum, and the cytosol, may be an indication that the enzyme is implicated in cell activities other than the synaptic ones, an assumption occasionally found in the literature.

The demonstration that AChE exists in multiple molecular forms (Barron and Bernsohn, 1968; Ho and Ellman, 1969; Massoulie and Rieger, 1969; Varela, 1969, 1973a; Dudai et al., 1972; Knutsen et al., 1973; Rieger et al., 1973) has placed the matter in a new perspective, for one may think that, if the enzyme were to have several functions in the cell, this could well be achieved through distinct isozymic species.

It appears then that any decisive approach to the possible physiological role of AChE and of its isozymes must take into account the above pattern of AChE compartmentation and must determine whether different isozymes are involved.

ACETYLCHOLINESTERASE OF THE RAT

1. Kinetic properties. Before turning to its multi-molecular forms, we shall briefly consider the enzyme. Since the method of preparation of the extracts was described in detail elsewhere (Varela, 1973c), here we shall only indicate that 0.9% NaCl and 1% (v/v) Triton X-100 buffered to pH 7.0 in 0.1 M Tris-HCl were used to obtain soluble (or free) and insoluble (or bound) AChE, respectively.

Brain, heart, and skeletal muscle of rat were employed in these investigations. In all these organs, AChE was found to be chiefly a membrane-bound enzyme, though in both the brain and muscle the ratio of bound to free enzyme is higher than in the heart. It should be remarked, parenthetically, that BChE activity is higher in the latter than in the former (about 30% and less than 10% of the total cholinesterase activity, respectively).

The discussion hereafter, unless otherwise stated, refers to the rat brain and muscle.

Free and bound enzymes of each tissue were found to differ in several kinetic properties: Km, reactivity towards 284C51, cholinergic agonists and antagonists such as carbachol(CARB), atropine(ATR), hexamethonium(HEX), phenyltrimethylammonium(PTM), d-tubocurarine(dTB), sensitivity to temperature of inactivation (55°C) and to storage at 4°C.

Both soluble and insoluble AChE from brain were less sensitive to temperature (55°C) and more sensitive to 284C51 and to carbachol than the enzyme from muscle which was, in turn, more sensitive than the former to PTMA and dTB (Valrela, 1973c) and to storage at 4°C.

The apparent Kms for brain AChE were 4.46 x 10^{-5} $\pm$ 0.09

(n=14) and 2.8 x 10^{-5} ± 0.08 (N=12), for the insoluble and the soluble enzymes, respectively, and 6.9 x 10^{-5} ± 0.19 (N=12) and 9.3 x 10^{-5} ± 0.38(N=9) for the free and bound AChE from muscle. The Km differences between the enzyme from brain and muscle (both the soluble and the insoluble forms) were signifi- cant (p<0.005); the possible physiological implication of this has been discussed (Varela, 1973c). The pH optima were 7.6 and 7.8 for brain and muscle enzymes, respectively. The enzyme from both sources showed considerable decrease of activity at pH 7.0, which was, however, more marked for muscle than for brain AChE.

2. *Isozymes.* Both the soluble and the insoluble forms of AChE were separated into 4 isozymic species with 7.5% poly- acrylamide gels and Sephadex G-200. The electrophoresis was carried out for 3h with a current of 2.5mA per tube. The Sephadex G-200 columns were eluted at 4°C, at a flow rate of 7ml/h with 0.1 M HCl-Tris buffer (pH 7.7) for the soluble enzyme, and 0.1 M phosphate buffer (pH 7.7) containing 1% Triton X-100 for the insoluble enzyme. Fractions of 1-1.5ml were collected and assayed for AChE at 25°C with the method of Ellman et al., (1961), a technique used throughout these studies to measure the enzymic activity. Proteins were determined according to Lowry et al., (1951) with bovine serum albumin as standard, and at 280nm.

It was found that an isozyme (isozyme 2) abounds in the brain and a different one (isozyme 3) predominates in muscle (Figs. 1, 2, 3, and 4). It appears likely that such differences in AChE isozymic population underlie somehow the aforementioned kinetic differences displayed by the enzymes of the two tissues (Varela, 1973a, 1973c).

If one compares the chromatographic profiles of free and bound AChE from brain and one considers the isozymes by the (decreasing) order of magnitude of their activities, one may write 2,1,4,3 for the soluble and 2,1,3,4, for the insoluble enzyme. Similarly, in the case of the muscle, one may write 3,1,4,2 and 3,2,4,1 for the soluble and insoluble AChE, respectively. But it should be stressed that these differences in the percentage of the activities of the isozymes 3 and 4 for the the soluble and insoluble enzymes from brain, and of isozymes 1 and 2 for their homologous forms from muscle remain questionable, for they do not seem to be fully re- producible; we are at present carrying on more work to ascertain this.

The four isozymes showed differences in Km, Hill coefic- ient, and behaviour towards cholinergic modifiers (see below).

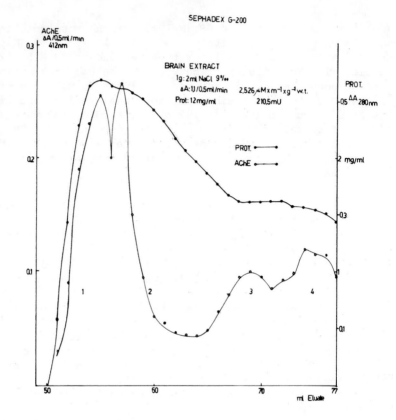

Fig. 1. Brain soluble AChE. Note that isozyme 1 displays the highest activity.

No clear-cut differences in pH could be observed. These results will be published in detail elsewhere.

The molecular weights (MW) of the insoluble AChE isozymes from brain and muscle were estimated, using ferritin (MW: 540,000), gammaglobulin (MW: 160,000), and bovine albumin (MW: 67,000) to standardize the columns, and blue dextran (MW: 2 x 10^6) to determine the void volume. The following values were found: brain, isozyme 1-582,000, isozyme 2-407,500, isozyme 3-162,100, isozyme 4-81,300; muscle, isozyme 1-457,000, isozyme 2-288,000, isozyme 3-202,000, isozyme 4-104,800.

Many reports on the multiple molecular forms of AChE have appeared in the past few years. The number of isozymes described varies from 2 to 5. For instance, three isozymes were found in the nervous system of lobster (Maynard, 1964)

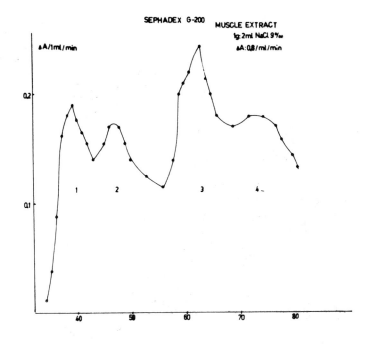

Fig. 2.   Muscle soluble AChE.   Note that isozyme 3 predominates.

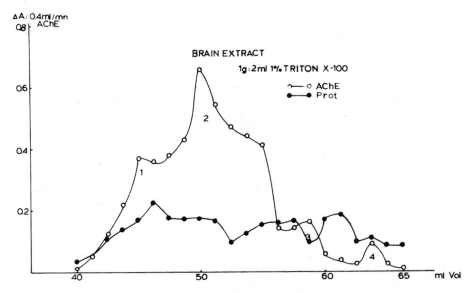

Fig. 3. Brain insoluble AChE Higher activity of the isozyme 2.

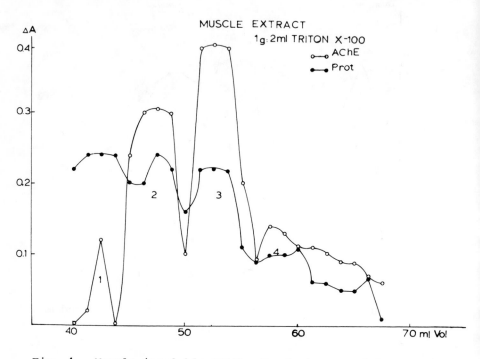

Fig. 4. Muscle insoluble AChE. Predominance of isozyme 3.

by polyacrylamide gel electrophoresis of aqueous extracts.
In an electrophoretic purification of an n-butanol extract
of caudate AChE of calf performed on Sephadex G-25, the
enzyme was "fractioned into three active components (and
perhaps a fourth) with corresponding protein peaks" (Jackson and Aprison, 1966). With barbital buffer extract of rat
brain three isozymes were separated by agar gel electrophoresis
by Bajgar and Zizkovsky (1971), while four forms were found
with polyacrylamide in Triton X-100 extracts of chick brain
by Iqbal and Talwar (1971). Knutsen et al., (1973) have
also reported the separation, both by gel filtration and
electrofocusing, of 4 isozymes from AChE extracted with
EDTA from ox caudate nucleus. Rieger et al., (1973) have
characterized with sucrose gradient centrifugation and with
the electron microscope three asymmetric forms of AChE; they
were found to give rise, under certain conditions, to two
smaller, globular species. Following purification by affinity chromatography, Dudai et al., (1972) detected, again

with sucrose gradient centrifugation, 3 isozymic forms in saline extracts of the same organs.

The differences in number of AChE isozymes observed in these and other studies are probably related more to the techniques (separation methods, extraction media, purification schemes) than to the species employed. It is indeed noteworthy that 4 isozymes were found in the brain of species as different as the cockroach (Kerkut et al., 1972) and the ox (Knutsen et al., 1973) or the chick (Iqbal and Talwar, 1971). A further plausible reason for these differences in the number of isozymic forms is to be found in the tendency shown by the molecules of AChE to form, under low ionic strength, aggregates of high molecular weight (Massoulié et al., 1970; Dudai et al., 1972),which may not enter the gels. Also, if interconversion of the isozymic forms can occur spontaneously (see Massoulié and Rieger, 1969; Rieger et al., 1973), it could be responsible for some of these separation patterns, by facilitating, under certain conditions, the transformation of two or three isozymic species into lower molecular weight forms, the ensuing possible combinations leading to a varying number of AChE isozymes.

In any case, from the above it can be concluded that either saline extracts or extracts obtained with surface active agents show multiple molecular forms of AChE, four free and four bound isozymic forms being the number that has been most often reported. Most workers have used only one medium of extraction, so that the reports deal with the soluble or with the insoluble enzyme. Among the few authors who have studied both forms, one may quote Barron and Bersohn (1968) who characterized three soluble and three insoluble (Triton X-100 extracted) isozymes in human brain extracts, following agar gel electrophoresis, and McIntosh and Plummer (1973) who described the existence of at least 4 soluble and 4 insoluble AChE isozymes in pig brain. The actual relationship between the free and the bound isozymic species remains to be elucidated.

3. *Cellular compartmentation.* As previously indicated, electron microscopy has shown a AChE reaction in the nuclear envelope and in the cisternae of the endoplasmic reticulum of the nerve cells (Brzin et al., 1966; Duffy et al., 1967; Reale et al., 1971; and others) as well as in the synaptic region of neural and myoneural junctions. Activity in the Golgi complex seems to be inconsistent. As for the synaptic distribution, there is no general apreement as to whether the enzyme is present only in the synaptic cleft (Duffy et al., 1967), in both the pre-and postsynaptic membranes

321

(Nyberg-Hansen et al., 1971; Koelle et al., 1974) or in all
the three synaptic loci (synaptic cleft, pre- and postsynaptic membranes) (Lewis and Shute, 1964; Smith and Treherne,
1965). Our own studies (Varela, 1969; and unpublished) tend
to confirm the latter view, i.e., the enzyme is observed
in the gap, the pre- and postsynaptic membranes, apart
from the nuclear envelope and the endoplasmic reticulum.
The localization in the gap is often rejected, but the presence in the synaptic membranes is acknowledged by most
authors, though it is currently stated that quantitative
differences in activity are found between the pre- and
postsynaptic membranes. For example, the enzyme would be
predominantly presynaptic in autonomic ganglia and exclusively,
or almost exclusively, postsynaptic in muscle endplate (cf.
Koelle, 1970; see discussion in Csillik, 1965, and Couteaux,
1973).

While the extrasynaptic active sites of AChE viz. the
nuclear membrane and the endoplasmic reticulum were demonstrated in both the embryonic and mature nerve cell(Duffy
et al., 1967; Pannese et al., 1971), they were found in the
developing but not in the normal adult muscle by Tennyson
et al., (1973). However, it cannot be excluded that this
was due to the conditions of fixation, that is, susceptibility
of the adult enzyme to the fixative could be different from
that of the fetal enzyme. The issue needs clarification.
It should be added, in this connection, that the observations
by Duffy et al., (1967) on neurons of the hypothalamus of
rabbits indicate that the AChE reaction becomes more marked
in the endoplasmic reticulum and in the nuclear envelope as
the maturation of the nerve cells proceeds.

From the foregoing, it may be inferred that ultrastructural
histochemistry points to the existence of 4 AChE sites in the
cell (nuclear envelope, endoplasmic reticulum, pre- and postsynaptic membranes), or 6 AChE compartments, if one accepts
(as not due to a diffusion artifact) the activity in the
cytosol (Tennyson et al., 1973) and in the synaptic gap.

Whether the AChE compartments represent distinct isozymes
is a question, as indicated earlier, to be answered in the
near future with appropriate methods. Meanwhile, it might
be rewarding to try to evaluate or indirectly test one or
more of the most plausible hypotheses that can be advanced
to provide guidelines for such a search. For example, it
does not seem too far-fetched to think that at least two of
the four insoluble AChE isozymes might be present at the
synaptic junction - one localized at the presynaptic membrane
and another at the postsynaptic membrane - and that the remaining two could account for the enzymic staining of the

322

nuclear envelope and the endoplasmic reticulum. The possibility that the latter two organelles, due somehow to their common origin and/or relationships (cf. Watson, 1955), form a single AChE compartment and therefore could contain the same isozyme should be envisaged. Moreover, in addition to the two isozymes distributed at the pre- and post-synaptic membranes as postulated above, a third isozyme may actually be present in the synaptic area, namely in the synaptic cleft (Varela, 1969).

In the above assumptions, only the insoluble AChE isozymes have been considered, in view of the evidence that AChE is present chiefly in a membrane bound form in those cellular loci (Bernsohn et al., 1966; Nachmansohn, 1972; Tennyson et al., 1973). But one is reminded that histochemical (Eränkö et al., 1964; Koelle et al., 1970; Tennyson et al., 1973) and biochemical data (Barron and Bernsohn, 1968; Barron et al., 1968; Varela, 1969, 1973a; McIntosh and Plummer, 1973) have revealed that both free and bound AChE occur in central and peripheral nervous tissues and muscle, though the number of their isozymic forms, as discussed previously, is still open to much debate. If one accepts (see p.3) that there are 4 soluble isozymes, the question may be raised as to their localization, which is seemingly more difficult to attach than that of their bound counterparts. At present, one can only speculate. And probably the simplest hypothesis is to assume that all these soluble forms are distributed in cytosolic pools surrounding the bound AChE isozymic compartments; that is, each bound AChE isozyme would be in a sort of dynamic equilibrium with a free corresponding isozymic species located in the adjacent cytosol. It is worth mentioning, in this context, that a rapid exchange between membrane bound enzymes and a finite pool of their dissociated forms has recently been proposed by Schimke and Dehlinger (1972).

FREE AND BOUND ACETYLCHOLINE HYDROLYSIS AND ITS PROBABLE
CONTRIBUTION TO CELLULAR HOMEOSTASIS

The relative percentage of free and bound AChE present in excitable tissues seems to vary from species to species. Massoulié et al.,(1970) have reported extracting most of the enzyme from the electric eel's electric organ with tris buffer. This result indicates that in these fishes, for example, the bulk of the enzyme could exist in a free state. The view has been expressed that in most mammals, on the contrary, the enzyme is mainly bound (Ho and Ellman, 1969). Our studies on rat brain and muscle agree with this (Varela, 1973c),

since the rat brain and muscle extracts show, respectively
as much as x6-10 and x2-6 more activity in Triton as compared to the 0.9% NaCl extracts. The much higher activity
of the detergent extracts was not due to activation by
Triton but to release of membrane bound enzyme. Three successive contrifugations were made under the same conditions
(see details in Varela, 1973c). Before adding Triton X-100
to obtain insoluble AChE, the pellet resulting from the
extraction of the soluble enzyme (first centrifugation) was
recentrifuged. When samples of the supernatants obtained
by the second centrifugation were subjected to polyacrylamide
electrophoresis, no enzymic reaction could be observed.
There was, as a matter of fact, some activity in these
extracts, as revealed spectrophometrically, but it was
in general below the sensitivity of the enzymophoresis. Such
activity was 6 to 30 times lower than that of the first
(soluble enzyme) and the third (insoluble enzyme) supernatants,
respectively. To rule out the possibility that the action
of Triton could be due to an activation process (Crone, 1971),
the soluble enzyme was incubated with Triton for several periods of time and the activity was measured. No significant
differences were found. McIntosh and Plummer (1973) have
reached a similar conclusion with the pig enzyme.

    The above conclusions are drawn from more than 50 exper-
iments (most of them in duplicate), but we also carried out
a few using four centrifugations; i.e., the pellet after
the soluble extraction was resuspended in NaCl and centrifuged
twice before adding Triton. This procedure was extended
to rat liver, spleen, kidney, and heart as well as to rabbit
brain and muscle. In all these experiments, as was the
case in the previous ones, the enzyme was found to predominate
in a bound form. It should be noted that our findings in
the rabbit are at variance with those of Tennyson et al.,
(1973) who detected more free than bound AChE in rabbit
muscle.

    It has been known for some time that AChE occurs in
tissues or cells (e.g. platelets, red blood cells) which
undoubtedly are not involved in synaptic transmission. The
same is true of ACh (Bhoola et al., 1961; Morley and Schachter, 1962). Further, both the enzyme and its substrate have
been found in cell populations in early stages of development
before any differentiation, and consequently, before the
formation of nervous structures takes place. This has led
to speculations concerning the actual role of ACh and AChE
in cell development and/or morphogenesis (cf. Ryberg, 1973).
There is, on the other hand, much indication that the appearance of AChE in neuroblasts is not dependent on the onset of

the synaptic function (see discussion in Pannese et al., 1971). The list of data suggesting that there is a need for the search for a role of AChE in cell metabolism is rather impressive and cannot be dealt with here. We shall only discuss briefly one of the paths that deserves some attention: actual relationships between ACh/AChE system (which is responsible for synthesis and degradation of ACh and, therefore, utilizes some of the free acetyl groups in excitable cells) and the Krebs cycle (which also uses these groups, but to yield energy in the form of phosphate bonds). As a first approach, the action of several intermediates of the citric acid cycle (citrate, isocitrate, oxalacetate, fumarate, malate, succinate, and 2-oxo-glutarate) on the rate of deacetylation were scanned. These studies are still too preliminary to permit definitive conclusions, but they tend to indicate that the soluble enzyme (of both brain and muscle) is affected (either activated or inhibited depending on the concentration of these ligands) by citrate and isocitrate, whereas the insoluble enzyme is not.

There is now good evidence that acetyl-CoA cannot cross the mitochondria, where it is formed. To explain its presence in the cytosol, it is supposed, on good experimental grounds, that it is condensed in the mitochondria with oxalacetate to form citrate by the same reaction that initiates the citric acid cycle, but the citrate instead of undergoing oxidation is released into the cytoplasmic matrix to be converted back to acetyl-CoA and oxalacetate by ATP citrate lyase (Greville, 1969). (This enzyme occurs in the brain (Williamson and Buckley, 1973) but to our knowledge its presence in other excitable tissues has not yet been determined). One of the fates of these acetyl groups is to form acetylcholine, which is synthesized in the cytosol by choline acetylase (ChAc) and partially stored in the synaptic vesicles (Fonnum, 1973; Schwartz, 1974).

It should be noted that some of the enzymes of the Krebs cycle have a dual distribution between the mitochondrial and the soluble fractions (see details in Rodyn, 1967) and that, accordingly, part of the cycle (from citrate to 2-oxo-glutarate and from fumarate to oxalacetate) can occur in duplicate in the cell, i.e., in both the mitochondria and the cytosol (Greville, 1969).

In view of the fact that some citric acid cycle intermediates appear to influence AChE activity, one might wonder whether the ACh/AChE system could not be inserted in a device controlling the function of the tricarboxilic acid cycle, and, to some extent, the redox potential systems of the cell. Indeed, a sort of balance could exist between these acetyl groups used in the Krebs cycle and those which are converted to ACh and fatty acids. Once in 'suitable' amount in the cytosol, those intermediates could activate soluble AChE, which would tend to hydrolyze free ACh, so that an increase of acetyl groups could prevent (by an allosteric mechanism) more citrate from crossing the mito-chondria to be reconverted to acetyl-CoA; instead, it would enter the citric acid cycle. Conversely, inhibition of AChE would favour availability of ACh and blocking of the cycle.

It is not possible at this stage to envision the level at which such a controlling mechanism could take place, i.e., which soluble enzymes of the Krebs cycle might be involved.

In conclusion, one of the functions of soluble AChE would be, in the above view, to control the level of cyto-solic acetyl groups (through regulation of the cytosolic level of the intermediates of the Krebs cycle, and fatty acid formation). Further, the enzyme may serve as a scavenging mechanism to cope with actual errors of diffusion of ACh, in excitable cells, as postulated by others.

We have already discussed the possibility of different subcellular localizations for the AChE isozymes. The fact that they have distinct kinètic properties could enable them to be related to different metabolic pathways within the cell, including different Krebs cycles.

Evidence has been presented for the existence of at least three different Krebs cycles (Cheng, 1973) in brain, each one linked to a different pool of acetyl-CoA and in relation-ship with a different substrate (pyruvate, citrate, acetate) for ACh synthesis. Two of these acetyl-CoA pools are admittedly located in neurons (pericarya, nerve endings) but their exact distribution and significance remain to be ascertained. The demonstration of the existence of at least three multiple molecular forms of choline acetylase (Fonnum, 1973) is worthy of mention here. It would not be too surprising if future research reveals that each ACh 'species' thus formed is preferentially hydrolyzed by a distinct AChE isozyme.

Finally, it could be interesting to know whether the ACh/AChE system is linked to the acylases systems thought to be responsible for the acetylation of the amino acids such as phenylalanine implicated in the initiation of polyipeptidic chains (Chapeville and Haenni, 1974), and such as lysine and arginine, possibily involved in derepression by histones (cf. De Lange and Smith, 1971). Such plausible roles could explain the presence of the enzyme in the endoplasmic reticu-lum and the nuclear envelope referred to earlier. Although it can be argued that the presence of AChE in the endoplasmic reticulum indicates synthesis of the enzyme, the high amount of bound AChE in these organelles may call for an alternative interpretation. The fact that chloramphenicol and cyclo-heximide, known inhibitors of protein synthesis, show anti-ChE action (Hubard and Quastel, 1973) may not be just a coin-cidence. All the above subjects, properly investigated will no doubt unravel actions of AChE isozymes other than the synaptic ones, or will at least clarify several issues which presently can only raise conjectures. And, as it might have been apparent from the first part of this section, among the overwhelming problems to be understood stands in the fore-ground the significance of the occurrence of AChE (and its isozymes) chiefly in a free state in some species and mainly in a bound form in others.

IONIC CONDUCTANCE AND ACETYLCHOLINESTERASE ISOZYMES

1. Views on cell permeability. As already noted, the pre-sence of AChE in non-innervated or non-nervous structures or cells has led to a number of speculations. For example, its involvement in cationic permeability of red blood cells has been envisaged (Holland and Greig, 1950; cf. Nachmansohn, 1974, p. 134). A role in the permeability of embryonic cells, of the intestinal mucosa, and placenta has also been post-ulated (see discussion in Gerebtzoff, 1959).

In the structures containing ACh, AChE and choline acetyl-ase, in which no nerve cells or axons can be found to be responsible for their manufacture, hydrolysis of ACh by AChE is linked to processes which obviously are not implicated in neurotransmission. Some authors have not hesitated in interpreting generally these and other facts which cannot be summarized here as an indication that AChE is an enzyme main-ly concerned with cell permeability. Synaptic transmission is then viewed as consequence of these permeability phenomena (thereby the enzyme is required) or as one aspect of the more general problem of how ions are transported across cell membranes.

The presence of AChE in axonal membranes in regions other than the synaptic area, including the Ranvier node, may seem to strengthen such a view or to suggest somehow that AChE is involved in nerve conduction (Machmansohn, 1974), besides being implicated in synaptic transmission. Future research shall throw much light on these matters which remain rather controversial or obscure, specially on whether impulse con-duction and synaptic transmission depend on the same mechan-ism (Neumann et al., 1973) or on different mechanisms (Katz, 1969, p. 31).

2. Synaptic transmission: facts and hypotheses. It is currently held that as the impulse reaches the nerve terminal, ACh is released and binds to specific receptors. This leads to conformation changes of the latter and to permeability to small ions, namely Na^+, K^+, and Cl^-. Termination of the action of the mediator and subsequent restoration of the resting condition of the receptor molecules are largely due to hydrolysis of ACh by AChE. Except for the conformation changes which cannot yet be an object of experimental scrutiny, these assumptions have been tested in certain conditions and proved to be generally correct (Changeux, 1972).

Kohoe (1972) has shown that there are distinct cholinergic receptors in Aplysia for the triggering of Na^+, K^+, and Cl^- permeabilities. She acknowledged that the same situation might occur in other phyla. In keeping with this reasoning, one wonders whether a fourth cholinergic receptor might not be envisioned, as, apart from its usual action in ACh release, calcium is known to be the only ion responsible for the spike in certain cholinergic synapses (Karczmar et al., 1972).

Purification studies of a cholinergic receptor (from electrical eel electroplax), which is said to trigger both Na^+ and K^+ movements, tend to show a single protein, though apparently formed by two polypeptidic chains (Klet et al., 1974; Lindström and Patrick, 1974) but the present status of knowledge is still too rudimentary (Klet et al., 1974) to allow any firm conclusion to be drawn regarding the existence of multiple cholinergic receptor proteins in these fishes. The only evidence for heterogeneity of receptor molecules appears to have come from a study by Porter et al. (1973) with alphabungarotoxin in frog and rat, in which they could distinguish two classes of endplate receptors, besides extrajunctional receptors. However, the role of these endplate receptors in action potential and ion permeability has not yet been worked out.

Kohoe's assumption appears attractive and deserves consideration even though, as just seen, evidence for it has still to be provided. If, for example, there were distinct receptors (or ionophores) which would trigger selectively Na^+, K^+ and Cl^- fluxes, one would be tempted to believe that the receptor (or ionophores)-mediators "linkages" responsible for the selective permeabilities to Na^+, K^+, and Cl^- would have to be specifically "broken" by distinct isozymes. Thus, one of the reasons for the occurrence of AChE isozymes could become apparent.

Such ionic fluxes or conductances can be blocked by pharmacological agents. K^+ and Na^+ permeabilities are selectively blocked by PTMA and HEX, respectively. No selective blocking drug is available for Cl^- permeability: dTB, which is effective in blocking Cl^- flux, blocks the K^+ movement as well (Kohoe, 1972).

Since the above agents are known to act on both the cholinergic receptor and AChE (Varela, 1973c), we designed experiments to test their actions on the isozymes. Besides PTMA and HEX, we used lanthanum, for it has been shown to prevent calcium inward movement in nerve terminals (Miledi, 1971). We found that PTMA and HEX have inhibitory actions chiefly on the insoluble isozymes 1 and 3, respectively; lanthanum affected mainly isozyme 1. At high concentrations of the substrate ($5x10^{-4}$ - $5x10^{-3}$) the inhibitory action was occasionally found to change into activation.

It is not known which relationship, if any, the inhibition (or activation) of AChE may have with the aforementioned pharmacological blockage of these ionic fluxes (see below). Yet, the fact that the selectivity found with these drugs in the case of the receptors seems also to exist to a degree in the case of the isozymes is worthy of further exploration. Therefore, after the purification of the isozymes, now under way, we shall carry out binding studies to obtain more direct evidence. Some conclusions may then be possible regarding the above one isoreceptor-one isozyme hypothesis, which poses another problem: since there is evidence that the pores for K^+ and Na^+ are localized on different sides of the membranes, and that both pores are present in presynaptic and post-synaptic membranes of neuromuscular junctions (and probably of central synapses as well) (Eccles, 1973), one wonders whether, in keeping with this hypothesis, instead of the distribution referred to earlier, more than one isozyme could be present in each synaptic membrane, though still spatially differentially distributed, i.e., located at distinct faces of the membrane.

So far, we have been discussing transmission of the im-pulse in terms of operation of 3 molecules: the transmit-ter, the receptor, and the catabolic enzyme. Other entities involved are channels, and the inactivating and activating particles (or gates), but though they have been identified and recognized as distinct on physiological and pharmacologi-cal bases (Amstrong et al., 1973; Keynes and Rojas, 1974), it is still far from clear how they are related to, or de-pendent on, the above molecules.

The different types of channels and associated gates are presently admitted depending on whether they are opened by the transmitter (transmitter channels) or by depolarization (depolarization channels) (Eccles, 1973). Two classes of channels and associated gates for Na^+, K^+ (and for Cl^- and

Ca ?) are accordingly considered, depending on whether they are related to the propagation or to the transfer of the impulse. Their localization is also dissimilar. The transmitter channels are said to be sharply restricted to the postsynaptic membrane, while the depolarization ones are located beyond this zone and in the presynaptic membrane, which is devoid of transmitter channels. This is supposed to hold for both central and neuromuscular synapses (Eccles,1973).

As tetrodotoxin (TTD) can block the depolarization but not the transmitter channels (Eccles, 1973), we tried to see whether sites for TTD could be found on ACh, as is the case with other agents known to influence nerve action (Changeux, 1966; Varela, 1973c). In the concentrations used $(1.5 \times 10^{-4}$ and $10^{-5})$, which are even x100-1000 than those employed by the physiologists, TTD did not affect significantly either the soluble or the insoluble enzyme.

Although, as stated, it remains to be decisively established whether apart from its role in transmission the ACh/AChE system is also involved in nerve conduction, this seems very likely (Nachmansohn, 1974). The fact that most drugs which influence conduction also affect AChE as much as those known to act on transmission would seem to agree with this view. The problem is to investigate whether distinct species or properties of the enzyme are involved in the two processes as, it should be recalled, the enzyme is present in both the conducting and the transmitting parts of the excitable cells.

Our negative results with TTD do not rule out this possibility. We may have not used the right concentrations to elicite reactivity on the part of the enzyme in vitro. Another explanation could, however, be offered for such lack of action. As it has been shown that TTD specifically reacts with the molecules in the channels but has no effect upon their associated gates (Armstrong et al., 1973; Keynes and Rojas, 1974), one might believe that AChE is somehow structurally related to the gates but not to the channels and that drugs such as TTD which bind so specifically to the channels cannot bind to AChE. Conversely, those which bind to the gates could find binding sites on the enzyme molecule. Indeed, we found that potassium iodate, known to interfere with the inactivating particle (or the closing of the gate) (Stämfli, 1974), inhibits AChE when used in the concentration (120 mM) that prolongs the action potential. On the other hand, ethanol and procaine (1%), which alter chiefly the activating particle (the opening of the gate) (Hille, 1970; Keynes and Rojas, 1974), also inhibit AChE. However, we found that, while potassium iodate inhibits mainly the

soluble enzyme from both muscle and brain, alcohol and procaine preferentially inactivate the insoluble enzyme. All
these drugs affect conduction. These studies will be extended to the pruified isozymes.

Obviously none of this is adequate to answer our questions. Most, if not all, of the drugs which react with AChE
(and differentially, whether it is in free or bound state)
are found to influence one or another parameter of the biopotentials. No general explanation can be provided at present for this, and in fact, little can be concluded from
experiments of this sort as long as we ignore how the enzyme,
the receptor, and the ionophore are (spatially and/or structurally) interelated. They can however, contribute to the
elimination of a number of hypotheses in the field. Although
other alternative assumptions can be made, in order that
the hydrolysis of ACh by AChE should trigger the termination
of the ionic fluxes, the enzyme is very likely to be closely
related in space or in structure to the inactivating particle.
This particle, according to the available evidence, seems
to be a protein in the case of the axonal gates (Armstrong
et al., 1973). The enzyme may also be related to the activating particles, as well. In any case, we believe that facts
such as those reported here warn one about the necessity of
looking for a possible involvement of free (and not only
bound) AChE in the modulation of bioelectricity, for it
cannot be excluded that the interplay (or interconversion)
of such forms could be linked somehow to part of the cycle
of the generation of the biopotentials.

Another question which has been difficult to understand
in molecular terms is the fact that acetylcholine, like a
few other transmitters, can trigger excitation in certain
synapses and inhibition in others (ref. in Bloom, 1973). It
is generally argued that this is dependent upon the organization or the properties of the postsynaptic membranes. It is
possible that AChE contributes to these features, and one
may wonder whether the kinetic properties of the enzyme and/
or its isozymic patterns are the same in both cases. Also,
one may ask whether the enzyme shows any differences in
nicotinic compared to muscarinic synapses.

The heart seems to be good material to provide answers
to such questions. The cholinergic synapses it encloses are
purely inhibitory; the cholinergic receptor is muscarinic in
character. The situation contrasts with that of striated
muscle which, in mammals, has only nicotinic excitatory
synapses.

In the following, we shall summarize a few experiments
we have done, in this connection, with heart AChE. Surpri-

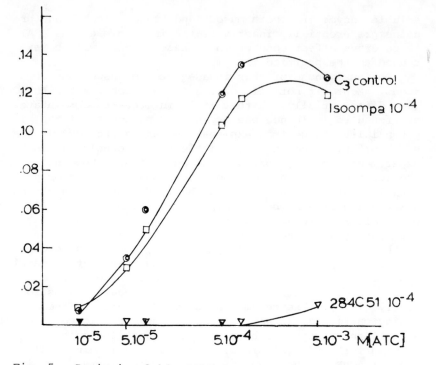

Fig. 5. Brain insoluble AChE is strongly inhibited by
284C51, whereas iso-OMPA has a very weak action. Inhibition
of the enzyme by high concentrations of the substrate
(5.10^{-4} - 5.10^{-3}) observed in the control is not apparent
with the iso-OMPA.

singly, kinetic studies revealed that both the free and
bound enzymes are inhibited by iso-OMPA and are almost insensitive to 284C51, which strongly inhibits the enzyme
from brain and muscle (Figs. 5 and 6). 284C51 shows more
aptitude for inhibiting the insoluble than the soluble form
of AChE of brain and muscle. While almost unreactive towards
heart AChE, again with the bound enzyme, it shows very slight
effect indeed as compared to the inhibition elicited by
iso-OMPA. (While using 284C51, a small inactivation occurs
only, or almost exclusively, with the insoluble BChE; iso-
OMPA strongly inhibits BChE in all the three organs).

The enzyme (both the free and bound forms) is sensitive
to HEX but not to PTMA, and feebly reactive with lanthanum.
(Final concentrations of the reagents in all the studies
reported here: 2.5 mM). It is also more sensitive to
atropine than is either the muscle or the brain enzyme.

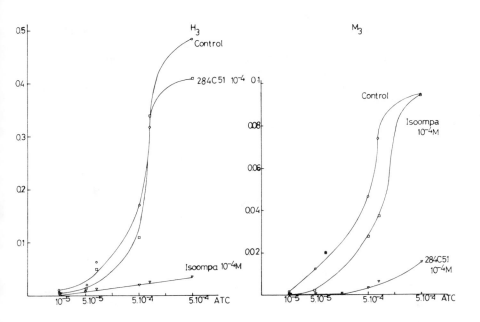

Fig. 6. Muscle insoluble AChE(M_3) behaves towards 284C51 and iso-OMPA like the brain enzyme (previous figure), but substrate inhibition is less commonly found, under the conditions of our experiments, as shown here; an inverse pattern is observed with heart insoluble AChE(H_3). Again no substrate inhibition of the control is seen. Only the insoluble enzymes are presented, but the same features are observed with the soluble ones. Final concentration of the reagents: 2.5 mM.

24 rats were used in these experiments. For each experiment 4 hearts were pooled and the assay for the enzyme was done in duplicate.

The pH optimum for bound enzyme was 7.5-7.6 and the Km 7.2×10^{-5}. The Hill coefficient was higher for heart(2.2) than for brain and muscle (< 2). Polyacrylamide gel electrophoresis revealed 4 isozymes, in the heart extracts, and little activity of isozyme 1. We have not carried out chromatographic analysis, so that we are not able to confirm the impression gained from inspection of the enzymograms, that the isozymic populations differ in a relevant fashion from those of brain and muscle enzymes. AChE from liver, which like heart, receives parasympathetic (vagal) innervation is also sensitive to iso-OMPA, HEX, and atropine.

It would seem essential to study, on the same kinetic basis, AChE from a tissue with ganglionic innervation of nicotinic type to see whether the above differences can be viewed as having any significance in relationship to the nature (nicotinic or muscarinic) of the receptors which accompany AChE in the junctional membranes and of the synaptic communication (excitation or inhibition) the tissue performs.

CONCLUDING REMARKS

In the foregoing pages, we have tried to indicate certain paths where research can be undertaken in order to provide answers to a number of questions regarding the function of AChE and its isozymes. Other directions were left out, because they do not belong to our direct interest or experience or because there is less experimental data or objective knowledge to support them. They raise at present even more speculations and lead to more tentative assumptions than those treated here.

Synpases can be viewed as a particular, highly specialized feature of a general phenomenon in the behavior of cell populations-adhesion or selective contact. As is well known, this phenomenon is the outcome of two others - migration and recognition- and they are all important for the understanding of cell differentiation. The nerve cell has apparently developed these abilities more than any other cell of multicellular systems. It can be assumed that the understanding of synapses will throw much light upon these problems probably even more than the study of embryonic cells, with which developmental biologists have long been struggling. In other words, it seems that any adequate molecular explanation of synaptic events should give some clues for the elucidation of the mechanisms underlying directed cell locomotion, recognition, and adhesive selectivity. We are, at present, far from this goal. Nevertheless, the importance of the molecules lying essentially in the plasma membranes or in the submembrane cytoplasm is fully realized nowadays, and among such molecules, the glycoproteins have raised much interest. The finding that both the cholinergic receptor and AChE are glycoproteins (Powell et al., 1973) is probably of some relevance in this context.

One wonders whether the presence of acetylated groups in molecules having admittedly key functions and occupying strategic positions in the cell, e.g., cell surface (glycoproteins: adhesion and/or recognition), nucleus (histones: derepression), and endoplasmic reticulum (amino acids such as phenylacetylalanine: translation) might be related somehow to the occurrence of AChE in such cellular loci and be indicative of an essential involvement of the ACh/AChE system in the control of the distribution or movement of acetyl groups throughout the cell.

The way a cell recognizes its partner and subsequently establishes a definite contact, for excitation or inhibition of the nerve signals, cannot exclude the (total or partial) participation of the machinery responsible for such synaptic communication, and it seems inconceivable that the ACh/AChE system does not take part in the genesis of the cholinergic synapses.

It has been demonstrated that AChE isozymic populations in neonatal nerve (Iqbal and Talwar, 1971) and muscle tissues (Barron et al., 1968; Wilson et al., 1971) are different from the adult. As previously discussed, the enzyme appears in these tissues before any synaptic function can be detected. Their significance is unclear, but it is scarcely difficult to admit that these neonatal isozymes have no function, and even more difficult to understand why some of them disappear with maturation. Equally, it is not easy to explain the relatively widespread distribution of the enzyme in the nervous system and, consequently, which role it is playing in the structures or synapses which are not cholinergic.

We have suggested in this presentation actions for both soluble and insoluble AChE isozymes, unconnected with either transmission or conduction of the impulse. We shall now take a further risk and embark on one of the directions, which, as indicated above, are plagued with more hypotheses than experimental evidence. We would like to suggest that a differential location (or make up) of isozymes at the synaptic membranes, whether determined by a sort of"isozymic complementarity" (Varela, 1969), could also contribute to the phenomenon of selective adhesion or recognition, and presumably precede synaptic function. It is difficult to guess how this could be achieved. But it is noteworthy that the presently available knowledge on the structure of AChE (from electrical eel electroplax), though far from complete, reveals a molecule extremely complex in organization, formed by 4 to 10 subunits arranged in clusters, and 2 different (light and heavy) polypeptidic chains. Further, some isozymic species are associated with a semi-rigid tail 2-3 nm thick as revealed by electron microscopy (Powell et al., 1973; Rieger et al., 1973).

If all these features are relevant for the function of AChE, one is trying to understand a highly complicated protein, which probably can accomodate several types of regulatory sites for binding of a large range of compounds, and to trigger opposite responses as their concentrations vary. Alternatively (or concomitantly) a molecule with such an heterogeneous composition may be endowed with more than just the hydrolytic capabilities that are conventionally attributed to it.

Finally, a remark should be made about another feature of AChE isozymes - the Hill coefficient - which may prove of some physiological significance, as is the case with hemoglobin. Both AChE (Changeux, 1966; Varela, 1973c) and the cholinergic receptor (Meunier et al., 1973) are allosteric proteins. It should be interesting to investigate whether the differential Hill coefficient displayed by the isozymes could determine a different rate of repacking of the synaptic vesicles and, therefore, contribute to some extent to the pacing of firing of the synapse.

REFERENCES

Amstrong, C.M., F. Bezanilla,and E. Rojas 1973. Destruction of sodium conductance in squid axons perfused with pronase. *J. Gen. Physiol.* 62:375-391.

Bajgar, J. and V. Zizkovsky 1971. Partial characterization of soluble acetylcholinesterase isoenzymes of rat brain. *J. Neurochem.* 18:1609-1614.

Barron, K.D. and J. Bernsohn 1968. Esterases of developing human brain. *J. Neurochem.* 15:273-284.

Barron, K.D., A.T. Ordinario, J. Bernsohn, A.R. Hess, and M.T. Hedrick 1968. Cholinesterases and nonspecific esterases of developing and adult (normal and atrophic) rat gastrocnemius. 1. Chemical assay and electrophoresis. *J. Hist. and Cytochem.* 16:346-361.

Bernsohn, J., K.D. Barron, P.F. Doolin, A.R. Hess, and M.T. Hedrick 1966. Subcellular localization of rat brain esterases. *J. Histochem and Cytochem.* 14:455-472.

Bhoola, K.D., J.D. Calle, and M. Schachter 1961. Identification of acetylcholine, 5-hydroxytryptamine, histamine, and a new kinin in hornet venom (*V. crabro*). *J. Physiol.* 159:167-182.

Bloom, F.E. 1973. Dynamic synaptic communication: finding the vocabulary. *Brain Res.* 62:299-305.

Brzin, M., V.M. Tennyson, and P.E. Duffy 1966. Acetylcholinesterase in frog sympathetic and dorsal root ganglia. *J. Cell Biol.* 31:215-242.

Changeux, J.P. 1966. Responses of acetylcholinesterase from *Torpedo marmorata* to salts and curarizing drugs. *Mol. Pharmacol.* 2:369-392.

Changeux, J.P. 1972. Etudes sur le mechanisme moléculaire de la réponse d'un membrane excitable aux agents cholinergiques. In: *Le système cholinergique en anêsthesiologie et en réanimation* , G.G. Nahas, J.C. Salamagne, P. Viars, and G. Vourc'h, eds., Librairie Arnette, Paris, pp. 99-112.

Chapeville, F. and A.L. Haenni 1974. Biosynthèse des protéines. Traduction génétique. Hermann, Paris.

Cheng, S.C. 1973. Compartmentation of tricarboxylic acid cycle intermediates and related metabolites. In: *Metabolic compartmentation in the rat brain* , R. Balazs and J.E. Cremer, eds., McMillan, London, pp. 107-118.

Couteaux, R. 1973. Structure and cytochemical characteristics of the neuromuscular junction. In: *Neuromuscular blocking and stimulating agents,* J. Cheymol, ed., Pergamon Press, Oxford, pp. 7-56.

Crone, H.D. 1971. The dissociation of rat membranes bearing acetylcholinesterase by the non-ionic detergent triton x-100 and an examination of the product. *J. Neurochem.* 18:489-497.

Csillik, B. 1965. Functional structure of the post-synaptic membrane in the myoneural junction. Akadémiai Kiadó, Budapest.

De Lange, R.J. and E.L. Smith 1971. Histones: structure and function. *Ann. Rev. Biochem.* 40:279-314.

Dudai, Y., I. Silman, N. Kalderon, and S. Blumberg 1972. Purification by affinity chromatography of acetylcholinesterase from electric organ tissue of the electric eel subsequent to tryptic treatment. *Biochim. Biophys. Acta.* 268:138-157.

Duffy, P.E., V. Tennyson, and M. Brzin 1967. Cholinesterase in adult and embryonic hypothalamus. A combined cytochemical electron microscopic study. *Arch. Neurol.* 16:385-403.

Eccles, J.C. 1973. The understanding of the brain. McGraw-Hill, New York.

Ellman, G.L., K. Diane Courtney, V. Andres, Jr., and R.M. Featherstone 1961. A new and rapid colorimetric determination of acetylcholinesterase activity. *Biochem. Pharmacol.* 7:88-95.

Eränkö, O., M. Härkönen, A. Kokko, and L. Räisänen 1964. Histochemical and starch gel characterization of desmo- and lyo-esterases in the sympathetic and spinal ganglia of the rat. *J. Histochem. Cytochem.* 12:570-581.

Fonnum, F. 1973. Molecular aspects of compartmentation of choline acetyl transferase. In: *Metabolic compartmentation in the rat brain,* R. Balazs and J.E. Cremer, eds., McMillan, London, pp. 35-45.

Gerebtzoff, M.A. 1959. *Cholinesterases.* Pergamon Press, Oxford.

Greville, G.D. 1969. Intracellular compartmentation and the citric acid cycle. In: *Citric acid cycle, control and compartmentation,* J.M. Lowenstein, ed., Marcel Dekker, New York, pp. 1-136.

Hille, B. 1970. Ion channels and nerve membranes. In: *Progress in biophysics and molecular biology,* J.A.V. Butler and D. Noble, eds., Pergamon Press, Oxford, pp. 1-32.

Ho, I.K. and G.L. Ellman 1969. Triton solubilized acetylcholinesterase of brain. *J. Neurochem.* 16:1505-1513.

Holland, W.C. and M.E. Greig 1950. Studies on the permeability of erythrocites. III. The effect of physostigmine and acetylcholine on the permeability of dog, cat and rabbit erythrocites to sodium and potassium. *Amer. J. Physiol.* 162:610-618.

Hubbard, J.I. and D.M.J. Quastel 1973. Micropharmacology of vertebrate neuromuscular transmission. *Ann. Rev. Pharmacol.* 13:199-216.

Iqbal, Z. and G.P. Talwar 1971. Acetylcholinesterase in developing chick embryo brain. *J. Neurochem.* 18:1261-1267.

Jackson, R.L. and Aprison, M.H. 1966. Mammalian brain acetylcholinesterase. Purification and properties. *J. Neurochem.* 13:1351-1365.

Karczmar, A.G., S. Nishi, and L.C. Blaber 1972. Synaptic modulations. In: *Brain and human behaviour,* A.G. Karczmar and J.C. Eccles, eds., Spriger, Berlin, pp. 63-92.

Katz, B. 1969. *The release of neural transmitter substances.* Liverpool University press, Liverpool.

Kerkut, G.A., P.C. Emson, R.W. Brimblecombe, P. Beesly, G.W. Oliver, and R.J. Walker 1972. Changes in the properties of acetylcholinesterase in the invertebrate central nervous system. In: *Biochemical and pharmacological mechanisms underlying behaviour,* P.B. Bradley and R. W. Brimblecombe, eds., Elsevier, Amsterdam, pp. 65-77.

Keynes, K.D. and E. Rojas 1974. Courant 'portier' de sodium dans l'axone géant du calmar. In: *Actualités neurophysiologiques*, A.M. Monnier, ed., Masson & Cie., Paris, 10:59-68.

Klett, K.P., B.W. Fulpius, D. Cooper, and E. Reich 1974. The nicotinic receptor; characterization and properties of a macromolecule isolated from *Electrophorus electricus*. In: *Synaptic transmission and neuronal interaction*, M.V.L. Bennett, ed., Raven Press, New York, pp. 179-190.

Knutsen, P., S.L. Chan, and E. Gardner 1973. Separation of multiple forms of mammalian acetylcholinesterase by isoelectric focusing. *Proc. West. Pharmacol. Soc.* 16:170-174.

Koelle, G.B. 1970. Neurohumoral transmission and the autonomic nervous system. In: *The pharmacological basis of therapeutics*, 4th edition, L.S. Goodman and A. Gilman, eds., MacMillan, London, p. 417.

Koelle, G.B., R. Davis, and W.A. Koelle 1974. Refinement of the bis-(thioacetoxy) aurate (I) method for the electron microscopic localization of acetylcholinesterase and nonspecific cholinesterase. *J. Histochem. Cytochem.* 22:252-259.

Koelle, W.A., K.S. Hossaini, P. Akbarzadeh, and G.B. Koelle 1970. Histochemical evidence and consequences of the occurrence of isoenzymes of acetylcholinesterase. *J. Histochem. Cytochem.* 18:812-819.

Kohoe, J. 1972. Three acetylcholine receptors in *Aplysia* neurons. *J. Physiol.* 225:115-146.

Lewis, P.R. and C.C.D. Shute 1964. Demonstration of cholinesterase activity with the electron microscope. *J. Physiol.* 175:5P-7P.

Lindström, J. and J. Patrick 1974. Purification of acetylcholine receptor by affinity chromatography. In: *Synaptic transmission and neuronal interaction*, M.V.L. Bennett, ed., Raven Press, New York, pp. 191-216.

Lowry, O.H., N.J. Rosebrough, A.L. Farr, and R.J. Randall 1951. Protein measurement with the Folin phenol reagent. *J. Biol. Chem.* 193:265-275.

Massoulié, J. and F. Rieger 1969. L'acetylcholinestérase des organes électriques de poissons (torpille et gymnote); complexes membranaires. *European J. Biochem.* 11:441-455.

Massoulié, J., F. Rieger, and S. Tsuji 1970. Solubilisation de l'acetylcholinestérase des organes électriques de gymnote. Action de la trypsine. *Eur. J. Biochem.* 14:430-439.

Maynard, E.A. 1964. Esterases in crustacean nervous system.
I. Electrophoretic studies in lobsters. *J. Exp. Zool.*
157:251-266.

McIntosh, C.H.S. and D.T. Plummer 1973. Multiple forms of
acetylcholinesterase from pig brain. *Biochem. J.*
133:655-665.

Meunier, J.C., H. Sugiyama, J. Cartaud, R. Sealock, and J.P.
Changeux 1973. Functional properties of the purified
cholinergic receptor protein from *Electrophorus electricus*.
62:307-315.

Miledi, R. 1971. Lanthanum ions abolish the 'calcium response'
of nerve terminals. *Nature* 229:410-411.

Morley, J. and M. Schachter 1962. Acetylcholine in non-
nervous tissues of the garden tiger (*Arctia caja*) and
other moths. *J. Physiol.* 162:12P-13P.

Nachmansohn, D. 1972. Biochemistry as part of my life.
Ann. Rev. Biochem. 41:1-28.

Nachmansohn, D. 1974. Importance of structure and organization for chemical reactions in excitable membranes.
In: *Central nervous system. Studies on metabolic regulation and function,* E. Genazzani and H. Herken, ed.,
Springer, Berlin, pp. 121-137.

Neumann, B., D. Nachmansohn, and A. Katchalsky 1972. An
attempt at an integral interpretation of nerve excita-
bility. *Proc. Nat. Acad. Sci.* 70:727-731.

Nyberg-Hansen, R., E. Rinvik, P. Aarseth, and J.A.B. Barstad
1969. Electron microscopic localization of cholinesterase at the neuromuscular junction by a quaternary carbon
analogue of acetylthiocholine as substrate. *Histochem.*
20:40-45.

Pannese, E., L. Luciano, S. Iurato, and E. Reale 1971.
Cholinesterase activity in spinal ganglia neuroblasts:
a histochemical study with the electron microscope.

Porter, C.W., T.H. Chiu, J. Wieckowski, and E.A. Barnard
1973. Types and locations of cholinergic receptor-like
molecules in muscle fibres. *Nature New Biol.* 241:3-7.

Powell, J.T., S. Bon, F. Rieger, and J. Massoulié 1973.
Electrophorus acetylcholinesterase: a glycoprotein;
molecular weight of its subunits. *FEBS Letters* 36:17-
22.

Reale, E., L. Luciano, and M. Spitznas 1971. The fine structural localization of acetylcholinesterase activity in
the retina and optic nerve of rabbits. *J. Histochem.
Cytochem.* 19:85-96.

Rieger, F., S. Bon, J. Massoulié, and J. Cartaud 1973.
Observation par microscopie électronique des formes
allongées et globulaires de l'acétylcholinesterase de

gymnote (*Electrophorus electricus*). *Dur. J. Biochem.* 34:539-547.

Rodyn, D.B. 1967. The mitochondrion. In: *Enzyme cytology*, D.B. Rodyn, ed., Academic Press, London, pp. 103-180.

Ryberg, E. 1973. The localization of cholinesterases and non-specific esterases in echinopluteus. *Zoologica Scripta* 2:163-170.

Schimke, R.T. and P.J. Dehlinger 1972. Turnover of membrane proteins of animal cells. In: *Membrane research*, C.F. Fox, ed., Academic Press, London, pp. 115-133.

Schwartz, J.H. 1974. Synthesis, axonal transport and release of acetylcholine by identified neurons of *Aplysia californica*. In: *Synaptic transmission and neuronal interaction*, M.V.L. Bennett, ed., Raven Press, New York, pp. 239-257.

Smith, D.S. and J.E. Treherne 1965. The electron microscopic localization of cholinesterase activity in the central nervous system of an insect, *Periplaneta americana 1. J. Cell Biol.* 26:445-465.

Stämfli, R. 1974. Recent work on Ranvier nodes. In: *Actualités neurophysiologiques*, A.M. Monnier, ed., Masson & Cie, Paris, 10:59-68.

Tennyson, V.M., M. Brzin, and L.T. Kremzner 1973. Acetylcholinesterase activity in the myotube and muscle satellite cell of the fetal rabbit. An electron microscopic-cytochemical and biochemical study. *J. Histochem. Cytochem.* 21:634-652.

Varela, J.M. 1969. Acetylcholinesterase isoenzymes and nerve transmission. Abstr. VIth FEBS Meeting, p. 295, *Spanish Biochem. Soc., Madrid.*

Varela, J.M. 1973a. Acetylcholinesterase of muscle and brain of rat. Some kinetic differences. *Abstr. IVth Intern. Congr. Neurochem.*, Tokyo, p. 412.

Varela, J.M. 1973b. Properties and physiological significance of acetylcholinesterase isoenzymes of mammalian muscle. *VIIth Symp. Physio. Morph. Neuromusc. Junction*, Polish Acad. Sci., Krakow, pp. 27-28.

Varela, J.M. 1973c. Kinetic characterization of acetylcholinesterase of muscle and rat brain. *Experientia* (Basel) 1347-1349.

Watson, M.L. 1955. The nuclear envelope. Its structure and relation to cytoplasmic membranes. *J. Biophys. Biochem. Cytol.* 1:257-270.

Williamson, D.H. and B.M. Buckley 1973. The role of ketone bodies in brain development. In: *Inborn errors of metabolism*, F.A. Hommes and C.J. van den Berg, eds., Academic Press, London, pp. 81-96.

Wilson, B.W., J.L. Schenkel, and D.M. Fry 1971. Acetyl-cholinesterase isozymes and the maturation of normal and dystrophic muscle. In: *Cholinergic ligand interactions*, D.J. Triggle, J.F. Moran, and E.A. Barnard, eds., Academic Press, London, pp. 137-174.

MOLECULAR PATHOLOGY OF THE MULTIPLE FORMS OF
FIBRINOGEN AND ITS FRAGMENTS

PATRICK J. GAFFNEY
National Institute for Biological Standards and Controls
Holly Hill, London NW3 6RB, England

ABSTRACT. Heterogeneity of human fibrinogen and that of the
various groups of plasmin induced fibrinogen degradation
products (X, Y, D, and E) is demonstrated. The structure-
function relationships of the heterogeneous fibrinogen degradation products in their ability to inhibit the clotting
of fibrinogen is elaborated using sodium dodecyl sulphate
polyacrylamide gel electrophoresis and N-terminal amino
acid analyses.

The occasional pathological bleeding associated with
streptokinase infusion was related to molecular forms of
fibrinogen lacking the carboxy ends of their Aα chains.
Patients undergoing streptokinase treatment for lower limb
venous occlusion were found to contain fibrinogen which
clotted slowly and was crosslinked inadequately by the action
of Factor XIII. In vivo thrombi and emboli were found to be
essentially stabilized by their crosslinked α chain polymers
and the latter were presented as the rate limiting structural
features to the lysis of the fibrin in in vivo thrombi and
pulmonary emboli.

There are various reports concerning the heterogeneity of
human fibrinogen (Hartley and Waugh, 1960; Mosesson and Sherry, 1966; Gaffney, 1971; Mosesson et al, 1972; Mosesson et al,
1972a; inter alia) and its fragments   (Jamieson and Gaffney,
1966; Nilehn, 1967; Arnesen, 1971; Pizzo et al, 1972; Fisher
et al, 1967; inter alia). Hartley and Waugh (1960) first suggested on the basis of solubility curve data that fibrinogen
must be a pauci-disperse system of molecules. Mosesson and
Sherry (1966) supported this view in their isolation from
plasma of high solubility and low solubility fibrinogen and it
is suggested that the high solubility material represented a
pathway of fibrinogen catabolism (Mosesson, 1973). Fibrinogen
is a strategic molecule in blood being the major common sus-
strate of the plasmin (EC 3.4.4.14) and thrombin (EC 3.4.4.13)
enzyme systems and thus is implicated in many aspects of
hemostatic function. The most understood functional property
of fibrinogen is its ability to clot. As a disulfide bonded
(Blomback, 1970) dimer (Blomback and Yamashina, 1958), each
unit of which is composed of three disulfide bonded (Henschen,
1964) polypeptide chains (Aα, Bβ, γ), clotting is initiated

(Bettleheim and Bailey, 1952) by the cleavage of fibrinopeptides A and B by thrombin (EC 3.4.4.13). The subsequent fibrin monomer, $(\alpha, \beta, \gamma)_2$ polymerizes to form a polymer $(\alpha, \beta, \gamma)_{2p}$ which is crosslinked in the presence of Ca^{++} ions and Factor XIII (activated by thrombin) to form urea-insoluble fibrin (Lorand and Konishi, 1964). These crosslinks form rapidly between the γ chains to form γ dimers (γ-γ) and more slowly between the α chains to form α chain polymers (α^p) of over 400,000 molecular weight (McKee et al, 1970). These covalent crosslinks have been shown to be a peptide-like bond between the lysine and glutamine residues of adjacent chains (Matacic and Loewy, 1968; Pisano et al, 1968). Interference in any of the above reactions of the clotting process can cause impaired blood coagulation. The main source of this interference will be shown in this report to be the activity of plasmin on fibrinogen.

This report also presents data on the heterogeneity of fibrinogen and its fragments, the possible subunit basis of some of this heterogeneity and some relationships between structure and function of fibrinogen fragments during hemostatic disfunction episodes. Parts of this report have been published elsewhere (Gaffney, 1971; Gaffney, 1973a; Gaffney et al, 1974).

MATERIALS AND METHODS

Human fibrinogen (Kabi Grade L) and plasmin were obtained from Kabi, Stockholm, Sweden. Human single donor blood was taken into citrate (9 ml blood to 1 ml 3.8% sodium citrate), spun hard, and the plasma was stored at -70° C for further use. The donors were usually members of the staff of the National Institute for Biological Standards and Control (London). Single donor and multi donor fibrinogens were isolated from their respective plasmas by the method of Laki (1951). Bovine thrombin from Parke-Davis was used to isolate non-crosslinked fibrin from plasma (in 10mM EDTA) while crosslinked fibrin was isolated from plasma using bovine thrombin (in 40mM $CaCl_2$). Sulphitolysis of the chains of fibrinogen (SsF fibrinogen) was achieved by the method of Pechere, et al (1958).

Isoelectric focusing in polyacrylamide (IEF-PA) gels (7%) was according to Wrigley (1968) with slight modifications (Gaffney, 1971), and the gels were stained with bromophenol blue (Awdeh, 1969). Polyacrylamide (PA) gel electrophoresis and sodium dodecyl sulphate polyacrylamide (SDS-PA) electrophoresis were according to Davis (1964) and Weber and Osborn (1969), respectively. The marker proteins used to estimate the molecular weights of single chain components were pyruvate kinase (57,000), lactic dehydrogenase (33,500), chymotrypsinogen (25,700) and lysozyme (14,300). The cyanate method for quantitative N-terminal amino acid analysis was used according to Stark and Smyth (1963), and qualitative N-terminal data were obtained by the dansyl-chloride method of Gray and Hartley (1963), using polyamide layers according to Woods and Wang (1967). Fibrinogen digests of various ages were prepared by the addition of plasmin to fibrinogen and stopping the reaction at various times with ε-aminocaproic acid (ε-ACA) as described elsewhere (Jamieson and Gaffney, 1968; Gaffney and Dobos, 1971). Thrombin clotting times were determined by the method of Latallo et al (1962) using 200 μg of fibrinogen digestion products together with 600 μg of fibrinogen in a total volume of 0.2 ml. Under these conditions the clotting time after zero minutes of plasmin digestion was 10 seconds. Fibrinogen core fragments D and E were prepared by DEAE-cellulose chromatography (Nussenzweig et al, 1961) and also by preparative polyacrylamide gel electrophoresis (Gaffney and Brownstone, unpublished work). Fibrinogen fragment X (Marder et al, 1969) was prepared by Sephadex G-200 chromatography of slightly degraded fibrinogen (60 i.u. of the plasminogen activator, streptokinase, to 9 mgs fibrinogen in 1 ml 0.4% citrate/ 0.90% saline, pH 7.4).

Plasma taken into citrate (0.38%) - 0.2M ε-ACA from patients undergoing streptokinase therapy for lower limb venous occlusion was supplied by Dr. C. N. Chesterman (Radcliffe Infirmary, Oxford, England). Rabbit antiserum specific for human fibrinogen was prepared by the regime of Nussenzweig et al (1961), and immunoprecipitation of fibrinogen and its fragments from plasmas has been described elsewhere (Gaffney et al, 1974). Experimental canine lower limb venous thrombi and pulmonary emboli were prepared in vivo using an electrode-stasis model described by Strachan et al (1974), and human pulmonary emboli were obtained by embolectomy. These thrombi and emboli were washed in the following solvents to obtain purified fibrin: ε-ACA (0.2M), 8M urea, 5M guanidine - HCl and 1% SDS solution. The purified fibrin was reduced with 0.2M-β-mercaptoethanol in 4M urea - 1% SDS before examination on SDS-PA gel electrophoresis for polypeptide chain composition. In-gel staining for carbohydrate was according to Zaccharius et al (1969) and occasionally by the procedure of Glossman and Neville (1971) to ensure the absence of non-specific carbohydrate staining of protein. Albumin was used on accompanying SDS-PA gels to insure that only glycopeptide chains stained for carbohydrate.

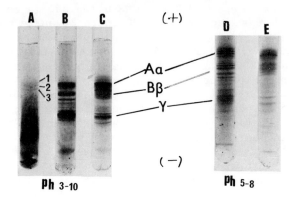

Fig. 1. IEF-PA gel patterns of fibrinogen and its polypeptide chains in 8M urea. (A) Pooled plasma fibrinogen showing three components; S-sulpho-fibrinogen (SsF) from (B) pooled plasma and (C) single donor plasma in a pH 3-10 gradient; S-sulpho-fibrinogen (SsF) from (D) pooled plasma and (E) single donor plasma in a pH 5-8 gradient.

NOMENCLATURE

The three chains of human fibrinogen are named as Aα, Bβ and γ (nomenclature subcommittee of the International Committee for Thrombosis and Haemostasis, Washington, D.C. 1972). The combined nomenclatures of Nussenzweig et al (1961) and Marder et al (1969) are used to describe the various families of fragments released during plasmin (EC 3.4.4.14) degradation of fibrinogen as follows; fibrinogen-> fragments X-> fragments Y-> plasmin resistant core fragments D and E. Individual polypeptide chains are named according to their presence in a particular fragment (i.e. in fragments X, Y, D, or E) and their fibrinogen chain origin as has been described elsewhere (Gaffney, 1972). The fibrinogen/fibrin chain origin and the molecular weight can also be used to characterize and name a polypeptide chain. Where the fibrinogen or fibrin chain origin of a polypeptide chain is not known it is denoted by molecular weight only. Efforts will be made to relate this nomenclature to that used by Mosesson et al (1973) and Pizzo et al (1972). The individual electrophoretic components of core fragment D (Jamieson and Gaffney, 1966) are designated D_1, D_2, D_3, etc. in order of decreasing mobility on polyacrylamide (PA) gel electrophoresis while the two

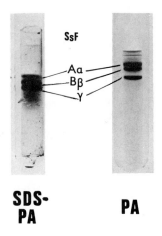

SDS-PA   **PA**

Fig. 2. S-sulpho-fibrinogen (SsF shown in Fig. 1, B) on 8M
urea-PA gel electrophoresis and SDS-PA gel electrophoresis.
The homogeneity of the chains with respect to molecular
weight and charge is evident by these techniques.

---

major molecular weight components of fragment D are named
D_g and D_p (Gaffney, 1972).
The empirical formula of fibrinogen is $(A\alpha, B\beta, \gamma)_2$ and
the fibrin monomer formula $(\alpha, \beta, \gamma)_2$ denotes the loss of
fibrinopeptides A and B by thrombin cleavage. During clotting
the fibrin monomer polymerizes to fibrin polymer, denoted $(\alpha, \beta, \gamma)_{2p}$ and, if Factor XIII and Ca^{++} are present, the γ chains
crosslink to form γ chain dimers (denoted γ-γ) and the α
chains crosslink to form α chain polymers (denoted α^p).

RESULTS AND DISCUSSION

A. Structural Data

Fig. 1 shows the IEF-PA gel patterns of SsF-fibrinogens
from multi-donor and single-donor plasmas in pH gradients
of 3-10 and 5-8. The intact fibrinogen in 8M urea shows two
major components and one minor component (denoted 1, 2, 3 in
Fig. 1, A) while the SsF-chains of fibrinogen evidenced about
six components in each of the Aα, Bβ, and γ chain binding areas.
The heterogeneity of the single donor fibrinogen Aα and γ chains
was slightly less than that of multi-donor fibrinogen. The res-

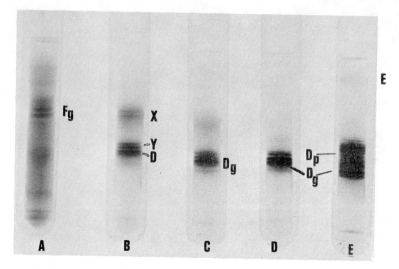

Fig. 3. IEF-PA gel patterns (pH gradient 5-8) of fibrinogen and its plasmin induced degradation products in 8M urea. Identification of the major fragments (X, Y, D, E from the description of Marder et al, 1969) was achieved by running the purified fragments as controls. The various patterns show (A) fibrinogen and fibrinogen treated with plasmin for (B) 2 min (C) 10 min (D) 1 hour and (E) 16 hours.

olution of the components with both types of SsF fibrinogen was greater with the pH 5-8 gradient (Fig. 1, D, E) than with the pH 3-10 gradient (Fig. 1, B, C). The SsF fibrinogen chains were examined on 8M urea-PA gel electrophoresis and SDS-PA gel electrophoresis (Fig. 2). The molecular size and charge homogeneity of the SsF fibrinogen chain preparation was evident. When the chains were eluted from the urea-PA gels and run on IEF-PA gels a heterogeneity similar to that shown in Fig. 1 was seen. Double sulphitolysis or S-carboxymethylation of fibrinogen did not significantly affect the chain pattern on IEF-PA gel electrophoresis. Heterogeneity of fibrinogen has been reported by Blomback et al (1963), who showed that some Aα chains contain a phosphorylated serine at residue 3 and some exhibit an N-terminal alanine deletion. Mosesson et al (1966), while demonstrating two chromatographic fractions in various mammalian fibrinogens, could not detect any differences between the tryptic peptide compositions of their constituent Aα and Bβ chains (Mosesson et al, 1972), but observed differences in their γ chains (Mosesson et al, 1972a). Multiple molecular forms of the γ chains has also been reported elsewhere

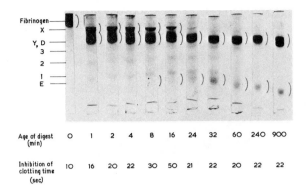

Fig. 4. Polyacrylamide (PA) gel (7%) electrophoresis of fibrinogen and various of its plasmin induced digests. The gels
were stained with 1% amido black in 7% acetic acid. The nomenclature of Marder et al (1969) and data from Fig. 5 were
used to identify the fragments in various staining regions
of the gels. All the major fragments reacted with rabbit
anti-human fibrinogen serum and the lines of identity are
drawn in their appropriate experimental locations. The
ability of these various timed digests of fibrinogen to pro-
long the coagulation time of fibrinogen is shown (in seconds)
in the bottom line, e.g. the figure of clotting time 50 sec
indicates that a digest made up of the fragments shown in the
corresponding PA gel inhibits a standard clotting test
(Latallo et al, 1962) by 40 seconds.

(Gerbeck et al, 1969; Henschen and Edman, 1972). Whether
these reports of chain heterogeneity bear any relationship to
that shown in Fig. 1 remains unknown. The ease of lysis of
the carboxy end of the Aα chain of fibrinogen (Gaffney and
Dobos, 1971) and the reported numerous high molecular weight
early derivatives of fibrinogen in plasma (Mosesson et al,
1967) may explain some of the heterogeneity observed.

Fig. 3 shows the IEF-PA gel patterns of fibrinogen and various timed digests of fibrinogen with plasmin. Again at least
two types of fibrinogen are evident, while with increasing
digestion time the isoelectric point and the heterogeneity of
the various fragments of fibrinogen changes. Using purified
fragments X, D and E it was possible to locate the components
in the gel patterns shown in Fig. 3. It is evident that fragments X, possibly Y and the earliest form of fragment D are
heterogeneous by this technique. Fragment E contains
separate components which do not resolve very well on the 5-8

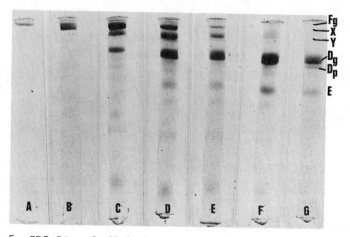

Fig. 5. SDS-PA gel (5%) electrophoretic patterns of (A) fibrin-ogen and its plasmin induced degradation products, whose di-gestion was stopped at various times from (B) 2 minutes to (G) 900 minutes.

pH gradient. Fig. 4 shows a similar timed digest run on PA gel electrophoresis. All the major components react with rab-bit anti-human fibrinogen serum. Under the conditions of this digest the fibrinogen is rapidly reduced in size to frag-ment X and a mixture of fragments Y-D and E (accepting the sequence of degradation and the nomenclature of Marder et al, 1969). The high mobility fragments (named as 1, 2 and 3) are probably fragments from the carboxy region of the Aα chain and the N-terminal end of the Bβ chains (Mosesson et al, 1973) which do not persist beyond 1 hour of digestion. The digests shown on the various gels in Fig. 4 were tested for their anticoagulant activity and the data is shown in the bottom line of Fig. 4. The maximum anticoagulant activity was associated with the 16 minute digest, i.e. soon after the appearance of fragment E (detected by immunodiffusion against rabbit anti-human fibrinogen serum) and when the Y-D complex was the heaviest staining area on the gel. The anti-coagulant activity of a one hour + digest was due to fragments D and E. Since purified fragments D and E can be shown to have only low anticoagulant activity the component of maximum anticoagulant activity is assumed to be fragment Y in agree-ment with findings of Marder and Shulman (1969).

Fig. 5 shows fibrinogen and a series of its digests on SDS-PA gel electrophoresis. The early formation of fragment X (Fig. 5, B) and its digestion with a concomitant increase

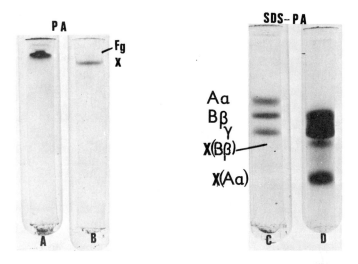

Fig. 6. PA gel (7%) electrophoresis of (A) fibrinogen and
(B) purified fibrinogen fragment X. SDS-PA gel (7%) electrophoresis of the polypeptide chains of (C) fibrinogen and (D)
purified fibrinogen fragment X. The polypeptide chains
X(Bβ) and X(Aα) have molecular weights of 42,000 and 27,000
respectively. See Nomenclature.

of fragments Y and D (Fig. 5, C-D) is followed by a further
increase in the concentration of fragments D and E (Fig. 5,
F). This sequence of release of the major staining fragments
of fibrinogen agrees with that suggested by Marder et al
(1969) namely, fibrinogen -> fragment X -> fragment Y ->
fragments D and E. By comparison of Figs. 4 and 5 it is evident that fragment Y, while having a larger molecular size,
migrates in polyacrylamide gel electrophoresis (Fig. 4) in
the same location as fragment D. The decrease in concentration of fragment Y with an increase in that of the high mobility fragment E and the lower mobility fragment D (Figs. 4 and
5) support the proposal of Marder et al (1969) that 1 mole of
fragment Y degrades to 1 mole of fragment D and 1 mole of fragment E. Table I shows the amino acid analyses of fragments D
and E in this and two other laboratories. The ratio of acidic
to basic residues in fragment E is about twice that of fragment D (our data) which explains why the higher molecular
weight fragment Y (D + E) migrates on polyacrylamide in the
same location as the lower molecular weight but less charged
fragment D. Considerable differences in amino acid analyses
were observed suggesting either different stages of lysis of
the different fragments D and E or differences in the types of

TABLE I

| Amino Acid | Moles of amino acid/10^5 g proteins | | | | | |
| | Fragment D | | | Fragment E | | |
	L1	L2	L3	L1	L2	L3
Lysine	54	56	40	42	58	15
Histidine	12	10	9	16	9	21
Arginine	34	43	25	24	64	25
Aspartic Acid	84	109	72	103	127	98
Threonine	27	49	28	31	61	58
Serine	40	56	36	51	75	108
Glutamic Acid	94	113	64	88	87	85
Proline	20	33	14	29	60	32
Glycine	42	87	48	44	85	120
Alanine	26	32	25	32	30	47
Half-Cysteine	18	14	12	38	38	34
Valine	46	54	24	36	40	35
Methionine	15	17	12	11	2	10
Isoleucine	41	53	25	22	31	13
Leucine	51	64	37	56	59	31
Tyrosine	26	42	20	23	23	6
Phenylalanine	14	28	16	26	31	45

Tryptophan was not determined

L1   From Mills and Triantaphyllopoulos, 1969
L2   From Dudek et al, 1970
L3   From this laboratory

Amino Acid Analyses of core fragments D and E compared to those obtained in two other laboratories.

fibrinogen starting materials from different populations.

Purified fragments X, D and E were examined for their polypeptide chain and N-terminal amino acid compositions with a view tc elucidating aspects of their heterogeneity and function. Fig. 6 shows the PA gel electrophoretic homogeneity of purified fragment X, having a slightly higher mobility than fibrinogen (Fig. 6, A and B). The polypeptide chain composition of fibrinogen and fragment X shows that the latter fragment contains intact Bβ and γ chains with Aα chain fragments denoted X (Aα), of 27,000 molecular weight (Fig. 6, D) in contrast to the intact Aα chains of fibrinogen (Fig. 6, C). Assuming that, like fibrinogen (Blomback and Yamashina, 1958), fragment X is a dimer the molecular weight of its major com-

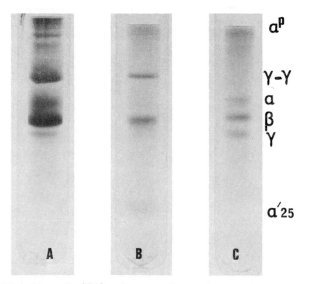

Fig. 7. SDS-PA gel (7%) electrophoretic patterns of the polypeptide chains of the clotted material obtained by thrombin-
Ca^{++} from (A) fibrinogen and (B) fibrinogen fragment X. The
α, β and γ chains of non-crosslinked fibrin (C) were used as
a reference mixture.

ponent is 264,000 (using molecular weights of 58,000 and
47,000 for the Bβ and γ chains of fibrinogen according to
Gaffney and Dobos, 1971). A minor chain fragment (denoted
X (Bβ) and having a molecular weight of 42,000) suggests that
fragment X, homogeneous on PA gel electrophoresis (Fig. 6, B),
contains at least two components of molecular weights,
264,000 and 232,000. Further heterogeneity is suggested by
the fact that some components of the fragment X family clot
following the removal of fibrinopeptide A by thrombin and
some do not (Gaffney, 1973). The purified fragment X is 80%
clottable with thrombin and Fig. 7 shows the polypeptide composition of the reduced fibrin formed compared to that formed
from intact fibrinogen using non-crosslinked fibrin as a reference chain mixture. The polypeptide chain composition of
clotted fibrinogen contains intact β chain, γ-γ dimers and α
polymers (α^P) as described by McKee et al (1970), while the
clotted fragment X was made up of the same chains except that
the α chain fragments of 25,000 molecular weight did not
crosslink. It is noteworthy that X(Aα) in its thrombin mediated cleavage to α'25 decreased 2000 in molecular weight
indicating the loss of the intact fibrinopeptide A from frag-

TABLE II

Material	A.A	Molar Ratios*
Fibrinogen	ALA	2.00
	TYR	1.75
	ASP	1.25
D	ASP	2.70
	GLY	trace
	VAL	trace
E	ALA	0.50
	TYR	2.00
	ASP	0.47
	VAL	1.80
	LYS	1.25

*Calculated on the basis of molecular weights
of 340,000, 80,000, and 45,000 for fibrinogen,
fragment D and fragment E, respectively,
(Marder et al, 1969; Gaffney and Dobos, 1971).

Molar quantities of N-terminal amino acids per mole of
fibrinogen and fragments D and E.

ment X. Factor XIII (the inactive fibrin crosslinking enzyme
in plasma) was present in both preparations as a slight impurity. It would seem that the carboxy half of the Aα chain
of fibrinogen contains the residues involved in the final
stage of stable clot formation, e.g. the glutamyl-lysyl
(Matacic and Loewy, 1968; Pisano et al, 1968) crosslinks catalyzed by Factor XIII following activation by thrombin and
calcium. Indeed the clotting of fragment X in the presence
of whole fibrinogen suggests that the carboxy half of the Aα
chains contain both the donor and acceptor aminoacid residues
and that this fragment can inhibit the crosslinking of the α
chains of normal fibrin. It has been shown that at least one
potential crosslinking region of the α chain is at the N-
terminal end (Finlayson et al, 1972); however this work suggests
that when the crosslinking regions at the carboxy ends of the
Aα chains are removed those at the N-terminal ends may not
function because the formation of the various crosslinks are
probably interdependent.

Considerable difficulty was experienced in the isolation
in quantity of purified fragment Y. Thus its anticoagulant
effect (Marder and Shulman, 1969) was examined on the assumption of Marder et al (1969) that fragment E together with frag-

ment D comprise fragment Y. Furthermore fragment E has been shown immunologically (Marder, 1971) and chemically (Kowalska-Loth et al, 1973) to be similar to the cyanogen bromide derived N-terminal part of fibrinogen. However there is dispute whether and how much fibrinopeptide A is associated with fragment E (Mills, 1972; Pizzo et al, 1973; Mosesson et al, 1973; Peyer and Staub, 1973; Marder et al, 1972). None of the later studies were quantitative and did not relate the presence or absence of fibrinopeptide A in a molar manner to fragment E. One quantitative study (Kowalska-Loth et al, 1973) has suggested on the basis of N-terminal data that both fibrinopeptides A are contained in fragment E. Table II shows that, by the cyanate method, fragment E contains a total of about 6 moles of N-terminal amino acids supporting the notion that this fragment is a dimer containing 6 polypeptide chains. The dansyl method using polyamide layers to separate the N-terminal amino acids supported the quantitative cyanate data but identification of the individual amino acids was sometimes equivocal. By comparison with fibrinogen, which has 2 residues each of alanine and tyrosine and 1 residue of aspartic acid, it seems that fragment E contains some new N-terminal residues. The presence of 1 mole of aspartic acid in human fibrinogen was also reported by von Korff et al (1963). An interpretation of this N-terminal data (Table II) is that one N-terminal end (e.g. one mole of alanine) of the two fibrinogen Aα chains and two N-terminal ends of the two fibrinogen γ chains are conserved in the fragment examined in this study. Treatment of fragment E with thrombin (1 N.I.H. unit bovine thrombin per 100 μg of fragment for 1 hour at 37°C) caused a decrease in its molecular weight as measured on SDS-PA gels which was in keeping with a loss in molecular weight of about 2000. Thus if fragment E represents the N-terminus of fragment Y the anticoagulant effect of the latter can be explained in terms of the exposure of one polymerization site by the cleavage of one mole of fibrinopeptide A from one side of the fibrinogen fragment Y. In this rationale it is assumed that the cleavage by thrombin of two fibrinopeptides A from fibrinogen expose two polymerization sites and subsequent polymerization to fibrin clot visualization can proceed. The introduction of a fragment with one intact and one destroyed polymerization site (namely fragment Y) could be seen to effectively inhibit polymerization. Such an assymetric cleavage of the N-terminal of fibrinogen by plasmin is compatible with the assymetric cleavage of fragment X (Marder et al, 1969) into fragment Y (MW 150,000) and D (MW 82,000) and again emphasizes that the two units of the fibrinogen dimer are most likely different.

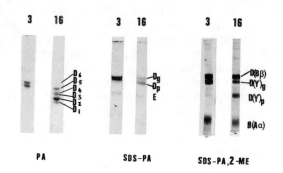

Fig. 8. Electrophoresis of three hour (3) and sixteen hour
(16) fibrinogen + plasmin digests on prolonged PA gel electrophoresis and SDS-PA gel electrophoresis. The polypeptide chain
compositions of fragments D from 3-hour and 16-hour digests
are shown (SDS-PA, 2-ME) following reduction with 0.2M
β-mercaptoethanol (2-ME). See Nomenclature for explanation
of symbols.

Fibrinogen core fragment D has been shown to contain between 6 (Jamieson and Gaffney, 1966) and 9 (Nilehn, 1967)
electrophoretic components. IEF-PA patterns of various digests
of fibrinogen (Fig. 3) show that this heterogeneity increases
with time suggesting that enzymatic degradation towards the
end of the digestive process plays a major role in the heterogeneity. Fragment D from 3-hour and 16-hour plasmin digests
of fibrinogen were examined on PA and SDS-PA gel electrophoresis and the results are shown in Fig. 8. The 3-hour fragment
D contains two electrophoretic components (D_6 and D_5) having
the same molecular weight (one component, D_g) and was composed
of polypeptide chains: D (Aα), MW 12,000; D^g(Bβ), MW 37,000;
and D (γ)$_g$, MW 33,000. Inspection of the molecular weight
data of Pizzo et al (1973) suggests that their fragment D subunit named β" has a molecular weight of 37,000 rather than
44,500 as they reported and is analogous to the D(Bβ) chain
in Fig. 8. The evidence to support the contention of Mosesson
et al (1973) that D(Bβ), called γ' in their nomenclature has
its origin in the γ chain of fibrinogen is unconvincing. The
D_6 and D_5 possibly originated in the two forms of fibrinogen
present in the starting material or from the two different
sides of the fibrinogen dimer molecule. The 16-hour fragment
D contained 6 electrophoretic components on prolonged PA elec-

trophoresis (D_1, D_2, D_3, D_4, D_5, and D_6), having two major molecular weight groupings (D_g, MW 82,000 and D_p, MW 73,000) on SDS-PA gel electrophoresis, and was composed of four poly-peptide chains. From Fig. 8 it is evident that the degrada-tive conversion of fragment D_g to fragment D_p involved the lysis of the polypeptide chain $D(\gamma)_g$ to form a smaller poly-peptide chain $D(\gamma)_p$. This degradation process and the hetero-geneity of the fibrinogen starting material accounts for four of the 6 major electrophoretic components of fragment D. Ex-haustively digested fragment D showed small amounts of another lower molecular weight (60,000) fragment which crossreacted poorly with rabbit anti-human fibrinogen serum but no D frag-ments with the low molecular weights reported by Furlan and Beck (1972) were observed. However IEF-PA gel electrophoresis (Fig. 3) shows about twelve components in fragment D. Since 1 mole of fragment D contains 2.7 moles of aspartic acid (Table II) much of the heterogeneity must be due to degrada-tion of the carboxy terminal parts of the molecule during the final stages of digestion. Whether the predominance of aspar-tic acid as the N-terminal amino acid of fragment D might form the basis of a useful and sensitive test of renal insufficiency is under investigation.

Fragment D inhibits polymerization of fibrin but its overall anticoagulant effect is far less than that of fragment Y on a molar basis (Marder and Shulman, 1969). The mechanism of its polymerization inhibition is not understood but may involve interference of the 2-3 moles of N-terminal aspartic acid of fragment D with the hydrophobic polymerization sites of fibrin since it is believed that the acidic-hydrophilic fibrinopep-tides A maintain the solubility of fibrinogen in aqueous solvents.

B. Pathology Section

1. Fibrinogen, fragment X and a haemorrhagic defect

Whether the heterogeneity of human fibrinogen has any clin-ical significance in haemostasis is as yet unresolved. How-ever the presence of high solubility fibrinogen derivatives (probably plasmin and thrombin induced) in plasma (Mosesson et al, 1967) may be reflected in impaired haemostasis. It has been shown (Gaffney, 1973) that streptokinase (SK) in vitro and in vivo activates plasminogen to convert plasma fibrinogen to large molecular weight fragments by degradation of those regions of the Aα chain involved in their Factor XIII catalyzed crosslinking. To examine the condition of plasma fibrinogen

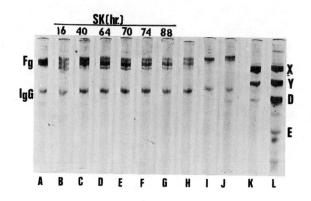

Fig. 9. SDS-PA gel (5%) electrophoretic patterns of the immuno-
precipitates obtained with rabbit anti-human fibrin from a
patient's plasma (A) before, (B-G) during, (H) 1 hr after, and
(I) 18 hr after treatment with SK. Pattern (J) shows the composition of the immunoprecipitate from normal plasma. Patterns
(K) and (L) show the size components in a mixture of fibrinogen degradation products (X, Y, D, E) before (L) and after (K)
treatment with the rabbit anti-human-fibrin serum. The duration of SK treatment is marked in hours above the relevant
gels. Other details are described in the text.

over a prolonged period of SK infusion three patients were
bled at various intervals during conventional treatment with
SK for the relief of lower limb venous occlusion (600,000 IU/
30 min followed by 100,000 IU/hour). The plasma samples were
absorbed with rabbit anti-human fibrin serum and the immuno-
precipitates examined on SDS-PA gel (5%) electrophoresis. The
results for one patient are shown in Fig. 9. Fig. 10 shows
the chain composition of the clotted materials extracted from
the same plasma samples. The pretreatment plasma (Fig. 9, A)
shows a major and minor component again emphasizing the heterogeneity of normal plasma fibrinogen and Fig. 10 (A) shows that
on clotting the plasma the expected (McKee et al, 1970) composition of totally crosslinked fibrin following reduction is
seen on SDS-PA gel electrophoresis, i.e. α chain polymers
(α^p), β chains and γ chain dimers ($\gamma-\gamma$). The fibrinogen-like
material in the plasma 16 hours after SK treatment was begun
was represented by three major components of differing molecular size (Fig. 9, B). These components have been identified
as fibrinogen, 'pre-X' and X on the basis of their electrophoretic comparison with the marker mixture in Fig. 9 (1).

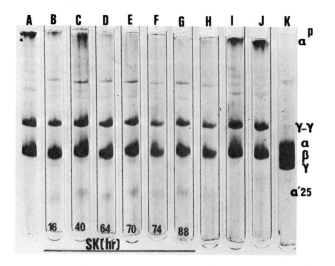

Fig. 10. The SDS-PA gel (5%) patterns of the polypeptide chains of the reduced clots obtained from a patient's plasma (A) before, (B-G) during, (H) 1 hr after, and (I) 18 hr after SK treatment. The duration of SK treatment is marked in hours below the relevant gels. The fibrin chains (α, β and γ) were used as a marker mixture (pattern K). The expected (McKee et al, 1970) chain composition of Factor XIII catalyzed cross-linked fibrin from pre-treatment patient plasma (pattern A) and normal plasma control (pattern J) show α polymers (α^p), β chains and γ dimers (γ-γ) and act as another marker mixture. The $\alpha'25$ denotes a fragment of mol wt about 25,000 derived from α chains (Gaffney, 1973a).

When this plasma was clotted, the polypeptide composition of the washed fibrin was β chains, γ dimers and α chain fragments of molecular weight between 24,000 and 27,000 which are assumed to be similar to the α chain fragments named elsewhere as α' 25 (Gaffney, 1973a). The plasma samples taken at various intervals during the 88 hour SK treatment showed similar SDS-PA gel patterns for their fibrinogen-like material (Fig. 9, B-G) and for the chain composition of their clotted fibrin derivatives (Fig. 10, B-G). Plasma samples examined one and 18 hours after cessation of SK treatment (Fig. 9, H, I) showed that the plasma fibrinogen had returned to the pre-treatment condition (Fig. 9, A) after 18 hours and crosslinked normally (Fig. 10, I). The rapid clearance of these high molecular weight, yet damaged forms of fibrinogen (named 'pre-X' and X) support the recent contention (Mosesson, 1973) that fibrinogenesis repre-

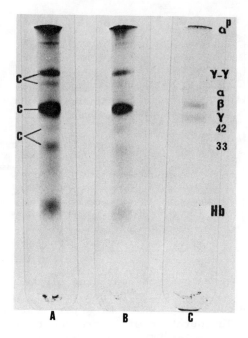

Fig. 11. SDS-PA gel (7%) electrophoretic patterns of the reduced (0.2 β-mercaptoethanol) polypeptide chains of the fibrin
in exhaustively washed experimentally (in vivo) produced dog
(A) thrombus and (B) embolus. The chains of human fibrin
(α, β, γ) act as a marker mixture (C). The polypeptide chains
staining for carbohydrate are denoted by C. Residual amounts
of the subunit of hemoglobin (Hb, MW 16,000) act as a convenient molecular marker on the gels. The polypeptide chains
denoted 42 and 33 have molecular weights of 42,000 and 33,000
respectively, and their chain origin is not known (however
they most likely originate in the β chains of fibrin).

---

sents a major pathway of fibrinogen catabolism. Since incompletely crosslinked fibrin is more prone to lysis than crosslinked fibrin (McDonagh et al, 1971; Gaffney and Brasher, 1973)
the 'damaged' fibrinogen generated during SK treatment should
partially explain hemmorrhagic complications encountered during
SK infusion for the relief of thrombosis. Some patients,
during SK infusion, evidenced some fibrinogen fragments similar to fragment Y, whose anticoagulant properties (Marder and
Shulman, 1969) would further predispose a patient to bleed.

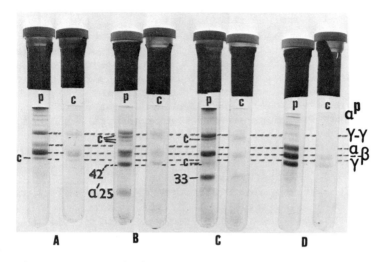

Fig. 12. SDS-PA gel (7%) patterns of the polypeptide chains
in the reduced (0.2M β-mercaptoethanol) fibrin isolated from
(A) plasma-thrombin-Ca^{++}, (B) plasma- streptokinase-thrombin-
Ca^{++}, (C) human pulmonary embolus and (D) plasma-thrombin-
EDTA. Duplicate gels were run, one stained with coomassie
blue for protein (P) and one stained for carbohydrate (C).
Polypeptide bands which stained for carbohydrate but are not
evident in the above photograph are indicated by the symbol
C to the left of the relevant protein stained gel. The horizontal broken lines help to relate the protein bands on the
different gels. The symbols 42 and 33 denote protein bands
of 42,000 and 33,000 molecular weight respectively, whose
fibrinogen/fibrin chain origin is uncertain. Other aspects
of labelling explained in Nomenclature.

2. Fibrin crosslinking, thrombosis and pulmonary embolism

It is accepted that fibrin plays an important role in the
development of thrombosis in vivo, while fibrinolysis is
regarded as a natural protective mechanism against the deposition of fibrin. The fibrinolytic enzyme, plasmin (EC 3.4.4.
14), attacks the chains of intact fibrinogen in the order of
Aα, Bβ, and finally γ (Gaffney and Dobos, 1971) while non-
crosslinked fibrin chains are degraded in the order of α, β
and γ in the intact fibrin (Gaffney, 1973a). However when
the α chains are crosslinked, the fibrin is degraded by plasmin with difficulty and the sequence of attack of the chains
is β and γ chains first and finally the crosslinked α chains
(Gaffney, 1973a; Pizzo et al, 1973). It seems that cross-

linked α chains are the rate limiting step in the lysis of
fibrin in vitro. Histologically viable thrombi (Sevitt, 1973)
from the deep veins of a dog and their subsequent embolized
fragments were examined on SDS-PA for the polypeptide chain
composition of the fibrin. Fig. 11 shows the results obtained.
The major chains of both were α chain polymers (α^p), β chains,
γ chain dimers and two lower molecular weight chain fragments.
The two later fragments stained for carbohydrate and had molecular weights by this SDS-PA gel technique of 42,000 and
33,000. Low levels of the hemoglobin subunit (Hb, molecular
weight 16,000) were also present. Since the Aα chains of fibrinogen do not stain for carbohydrate (Gaffney, 1972; Pizzo
et al, 1972) the two chain fragments (MW 42,000 and 33,000)
are not fragments of the α chain or its crosslinked polymers.
This suggests that any fibrinolytic attack on the dog thrombus
or embolus lysed either the β chains or the γ dimers (most
probably the β chains) leaving the α chain polymers unaffected.
Slight heterogeneity of the γ-γ dimer was also in evidence.
This data, obtained with experimentally induced thrombi and
emboli in dogs, was compared to in vivo emboli in man. Material was obtained by embolectomy and washed exhaustively before
SDS-PA gel examination of its fibrin subunits with appropriate
fibrin controls. The results are shown in Fig. 12. The β
and γ chains of non-crosslinked fibrin stain for carbohydrate
while the α chains do not (Fig. 12, D) and the chains of
crosslinked plasma fibrin (Fig. 12, A) contain α chain polymers,
β chains and γ chain dimers, the latter two staining for carbohydrate. The pulmonary embolus fibrin (Fig. 12, C) contains
α polymers, β chains, heterogeneous γ chain dimers, and chain
fragments of molecular weights 42,000 and 33,000, both staining for carbohydrate and thus not having their chain origin
in the α chain polymers. These 42,000 and 33,000 molecular
weight fragments probably originate in the β chains rather
than in the γ-γ chain dimers. No fragments of the α chains
or their polymers are evident. Fig. 12, B shows that chains
of fibrin obtained from plasma by the simultaneous addition
of streotokinase and thrombin for 10 minutes. The presence
of α chain fragments of molecular weight 25,000 (α' 25) and
the lack of α chain polymers (α^p) was evident as was the
presence of intact β chains and a number of bands in the γ-γ
dimer region. The presence of some carbohydrate staining
fragments of molecular weight 42,000 suggests that if fibrin
is formed during enhanced fibrinolysis some γ-γ dimers and
β chains are degraded in the intact fibrin. No α chain polymers (α^p) are formed since all the α chains have been degraded
to α fragments of molecular weight, 25,000. In contrast the
in vivo formed human embolus contains only intact α chain

polymers with some degraded β chains (molecular weight, 33,000) and heterogeneous γ-γ chain dimers, suggesting that fibrinolytic attack and degradation of the embolus is mainly inhibited by the α chain polymers of the fibrin, while the β and γ chains are prone to limited lysis.

Summary

Using electrophoresis in various media it has been demonstrated that fibrinogen contains either two or three molecular types and that the heterogeneity of its fragments, obtained by digestion with plasmin, increases with time. The subunit basis of some of this heterogeneity has been elaborated and the molecular properties of some of the fragments X, Y, D and E have been described. Relationships between some structural data and functional properties of the fragments were proposed. The release of the Factor XIII crosslinking regions of the Aα chains of fibrinogen during its plasmin induced conversion to fragment X was used as a partial explanation of occasional hemorrhagic complications during the use of streptokinase in thrombolytic therapy. The anticoagulant effect of fragment Y was explained in terms of the asymmetric cleavage of the N-terminal part of the fibrinogen molecule. The heterogeneity of core fragment D from fibrinogen was related to the heterogeneity of fibrinogen. The 2-3 moles of N-terminal aspartic acid in fragment D suggest the monitoring of N-terminal aspartic acid as a possible diagnostic tool for fragment D in plasma and urine, though what clinical significance this might have is unclear beyond measuring renal insufficiency. Knowledge of some of the subunit features of the degradation of in vivo formed thrombi and emboli suggest that inhibitors of Factor XIII activity might play a useful role in the prevention of the deposition of "pathological fibrin." There is need for further research to develop reagents which might preferentially degrade the α chain polymers in fibrin since it has been shown that these crosslinked chains inhibit the lysis of fibrin. Such reagents could play a useful role in thrombolytic therapy.

ACKNOWLEDGMENTS

I wish to express gratitude to Mrs. M. Brasher for expert technical assistance in all aspects of this work. The work on the in vivo effect of streptokinase on fibrinogen was done in conjunction with Drs. C. N. Chesterman and M. J. Allington (Department of Haematology, Radcliffe Infirmary, Oxford, England) while the human and dog embolus studies were conduct-

ed with the cooperation of Drs. C. J. L. Strachan, M. F.
Scully and V. V. Kakkar (King's College Hospital Medical
School, Denmark Hill, London, England).

REFERENCES

Arnesen, H. 1971. Isoelectric focusing of fibrinogen degradation products. *Scand. J. Haemotol.* 13: (suppl.) 43-47.
Awdeh, Z. L. 1969. Staining method for proteins after isoelectric focusing in polyacrylamide gel. *Sci. Tools*
16: 42-43.
Bettleheim, E. R. and K. Bailey 1952. The products of the
action of thrombin on fibrinogen. *Biochim. Biophys. Acta*
9: 578-579.
Blomback, B. 1970. Selectional trends in the structure of fibrinogen of different species. *Symp. Zool. Soc.* (London)
27: 167-187. Academic Press (London).
Blomback, B., M. Blomback, R. F. Doolittle, B. Hessell, and
P. Edman 1963. On the properties of a new human fibrinopeptide. *Biochim. Biophys. Acta* 78: 563-566.
Blomback, B. and I. Yamashina 1958. On the N-terminal amino
acids in fibrinogen and fibrin. *Arkiv. Kemi* 12: 299-319.
Davis, B. J. 1964. Disc electrophoresis II. Method and application to human serum proteins. *Ann. N.Y. Acad. Sci.*
121: 404-427.
Dudek, G. A., M. Kloczewiak, A. Z. Budzynski, Z. S. Latallo,
and M. Kopec 1970. Characterisation and comparison of
macromolecular end products of fibrinogen and fibrin
proteolysis by plasmin. *Biochim. Biophys. Acta* 214: 44-
51.
Finlayson, J. S., M. W. Mosesson, T. J. Bronzert, and J. J.
Pisano 1972. Human fibrinogen heterogeneities II.
Crosslinking capacity of high solubility catabolic intermediates. *J. Biol. Chem.* 247: 5220-5222.
Fisher, S., A. P. Fletcher, N. Alkjaersig, and S. Sherry 1967.
Immuno-electrophoretic characterisation of plasma fibrinogen derivatives in patients with pathological plasmin
proteolysis. *J. Lab Clin. Med.* 70: 903-922.
Furlan, M. and E. A. Beck 1972. Plasmin degradation of human
fibrinogen I. Structural characterisation of degradation
products. *Biochim. Biophys. Acta* 263: 631-644.
Gaffney, P. J. 1971. Heterogeneity of human fibrinogen.
Nature New Biology 230: 54-56.
Gaffney, P. J. 1972. Localisation of carbohydrate in the subunits of human fibrinogen and its plasmin induced fragments. *Biochim. Biophys. Acta* 263: 453-458.
Gaffney, P. J. 1973. The molecular and functional condition

of plasma fibrinogen during thrombolytic therapy with streptokinase (SK). *Thrombosis Res.* 2: 105-114.

Gaffney, P. J. 1973a. Subunit relationships between fibrinogen and fibrin degradation products. *Thrombosis Res.* 2: 201-218.

Gaffney, P. J. and M. Brasher 1973. Subunit structure of the plasmin-induced degradation products of crosslinked fibrin. *Biochim. Biophys. Acta* 295: 308-313.

Gaffney, P. J., C. N. Chesterman, and M. J. Allington 1974. Plasma fibrinogen and its fragments during streptokinase treatment. *Brit. J. Haemat.* 26: 287-295.

Gaffney, P. J. and P. Dobos 1971. A structural aspect of human fibrinogen suggested by its plasmin degradation. *Febs. Lett.* 15: 13-16.

Gerbeck, C. M., T. Yoshikawa, and R. Montgomery 1969. Bovine fibrinogen-heterogeneity of the γ-chains. *Arch. Biochem. Biophys.* 134: 67-75.

Glossman, H. and D. H. Neville 1971. Glycoproteins of cell surfaces. A comparative study of three different cell surfaces of the rat. *J. Biol. Chem.* 246: 6339-6346.

Gray, W. R. and B. S. Hartley 1963. The structure of a chymotryptic peptide from pseudomonas cytochrome C-551. *Biochem J.* 89: 379.

Hartley, R. W. and D. F. Waugh 1960. Solubility, denaturation and heterogeneity of bovine fibrinogen. *J. Amer. Chem. Soc.* 82: 978-986.

Henschen, A. 1963. S-sulpho derivatives of fibrinogen and fibrin: preparations and general properties. *Arkiv. Kemi* 22: 1-28.

Henschen, A. and P. Edman 1972. Large scale preparation of S-carboxymethylated chains of human fibrin and fibrinogen and the occurrence of chain variants. *Biochim. Biophys. Acta* 263: 351-367.

Jamieson, G. A. and P. J. Gaffney 1966. Heterogeneity of fibrin polymerisation inhibitor. *Biochim. Biophys. Acta* 121: 217-220.

Jamieson, G. A. and P. J. Gaffney 1968. Nature of the high molecular weight fraction of fibrinolytic digests of human fibrinogen. *Biochim. Biophys. Acta* 154: 96-109.

Kowalska-Loth, B. B. Garlund, N. Egberg, and B. Blomback 1973. Plasmic degradation products of human fibrinogen, II. Chemical and immunological relation between fragment E and N-DSK. *Thrombosis Res.* 2: 423-450.

Laki, K. 1951. The polymerisation of proteins: the action of thrombin of fibrinogen. *Arch. Biochem. Biophys.* 32: 317-324.

Latallo, Z. S., A. P. Fletcher, M. Alkjaersig, and S. Sherry

1962. Inhibition of fibrin polymerisation by fibrinogen proteolysis products. *Am. J. Physiol.* 202: 681-686.

Lorand, L. and K. Konishi 1964. Activation of the fibrin stabilising factor of plasma by thrombin. *Arch. Biochem. Biophys.* 105: 58-67.

Marder, V. J. 1971. Identification and purification of fibrinogen degradation products produced by plasmin. Considerations on the structure of fibrinogen. *Scand. J. Haemat.* (suppl.) 13: 21-36.

Marder, V. J., A. Z. Budzynski, and H. L. James 1972. High molecular weight derivatives of human fibrinogen produced by plasmin. III. Their NH_2-terminal amino acids and comparison with "NH_2-terminal disylphide knot." *J. Biol. Chem.* 247: 4775-4781.

Marder, V. J. amd N. R. Shulman 1969. High molecular weight derivatives of human fibrinogen produced by plasmin. II. Mechanism of their anticoagulant activity. *J. Biol. Chem.* 244: 2120-2124.

Marder, V. J., N. R. Shulman, and W. R. Carroll 1969. High molecular weight derivatives of human fibrinogen produced by plasmin. I. Physiochemical and immunological characterisation. *J. Biol. Chem.* 244: 2111-2119.

Matacic, S. and A. G. Loewy 1968. The identification of isopeptide crosslinks in insoluble fibrin. *Biochem. Biophys. Res. Commun.* 30: 356-362.

McDonagh, R. P., J. McDonagh, and F. Duckert 1971. The influence of fibrin crosslinking on the kinetics of urokinase-induced clot lysis. *Brit J. Haemat.* 21: 323-332.

McKee, P. A., P. Mattock, and R. L. Hill 1970. Subunit structure of human fibrinogen, soluble fibrin and crosslinked insoluble fibrin. *Proc. Natl. Acad. Sci. USA* 66: 738-744.

Mills, D. A. 1972. A molecular model for the proteolysis of human fibrinogen by plasmin. *Biochim. Biophys. Acta* 263: 619-630.

Mills, D. A. and D. C. Triantaphyllopoulos 1969. Distribution of carbohydrate among the polypeptide chains and plasmin digest products of human fibrinogen. *Arch. Biochem. Biophys.* 135: 28-35.

Mosesson, M. W. 1973. The fibrinogenolytic pathway of fibrinogen catabolism. *Thrombosis Res.* 2: 185-200.

Mosesson, M. W., N. Alkjaersig, B. Sweet, and S. Sherry 1967. Human fibrinogen of relatively high solubility. Comparative biophysical, biochemical and biological studies with fibrinogen of lower solubility. *Biochem.* 6: 3279-3292.

Mosesson, M. W., J. S. Finlayson, and D. K. Galanakis 1973.

The essential covalent structure of human fibrinogen evinced by analysis of derivatives formed during plasmic hydrolysis. *J. Biol. Chem.* 248: 7913-7929.

Mosesson, M. W., J. S. Finlayson, and R. A. Umfleet 1972a. Human fibrinogen heterogeneities. III. Identification of γ-chain variants. *J. Biol. Chem.* 247: 5223-5227.

Mosesson, M. W., J. S. Finlayson, R. A. Umfleet, and D. Galanakis 1972. Human fibrinogen heterogeneities. I. Structural and related studies of plasma fibrinogen on high solubility catabolic intermediates. *J. Biol. Chem.* 247: 5210-5219.

Mosesson, M. W. amd S. Sherry 1966. The preparation and properties of human fibrinogen of relatively high solubility. *Biochem.* 5: 2829-2835.

Murano, G., B. Wiman, M. Blomback, and B. Blomback 1971. Preparation and isolation of S-carboxymethyl derivative chains of human fibrinogen. *Febs. Lett.* 14: 37-41.

Nilehn, J. E. 1967. Split products of fibrinogen after prolonged interaction with plasmin. *Thromb. Diath. Haemorrh.* 18: 89-100.

Nussenzweig, V., M. Seligman, J. Pelmont, and P. Grabar 1961. Les products de degradation du fibrinogen humain par la plasmine. I. Separation et proprietes physico-chemiques. *Ann. Inst. Pasteur* 100 : 377-387.

Pechere, J. F. G. H. Dixon, R. H. Maybary. and H. Neurath 1958. Cleavage of disulphide bonds in trysinogen and α-chymotrypsinogen. *J. Biol. Chem.* 233: 1364-1372.

Peyer, A. and P. W. Staub 1973. Release of fibrinopeptides from fibrinolytic fibrinogen fragment E. *Thromb. Diath. Haemorrh.* 29: 300-312.

Pisano, J. J., J. S. Finlayson, and M. P. Peyton 1968. Cross-link in fibrin polymerised by factor XIII: ε(γ-glutamyl) lysine. *Science* 160: 892-893.

Pizzo, S. V., M. L. Schwartz, R. L. Hill, and P. A. McKee 1972. The effect of plasmin on the subunit structure of human fibrinogen. *J. Biol. Chem.* 247: 636-645.

Pizzo, S. V., M. L. Schwartz, R. L. Hill, and P. A. McKee 1973. The effect of plasmin of the subunit structure of human fibrin. *J. Biol. Chem.* 248: 4574-4583.

Sevitt, S. 1973. Pathology and pathogenesis of deep vein thrombosis in: *Recent advances in thrombosis*. L. Poller (ed.) pp. 17-38. London - Churchill Livingstone.

Stark, G. R. and D. G. Smyth 1963. The use of cyanate for the determination of NH_2-terminal residues in protein. *J. Biol. Chem.* 238: 214-226.

Strachan, C. J. L, P. J. Gaffney, M. F. Scully, and V. V. Kakkar 1974. An experimental model for the study of venous thrombosis in vivo. *Thrombosis Res.* (in press).

Von Korff, R. W., B. Pollara, R. Coyne, J. Runquist, and R. Kapoor 1963. Application of radioisotopic yield to the quantitation of the N-terminal amino acids of fibrinogen. *Biochim. Biophys. Acta* 74: 698-708.

Weber, K. and M. Osborn 1969. The reliability of molecular weight determinations by dodecyl-sulphate-polyacrylamide gel electrophoresis. *J. Biol. Chem.* 244: 4406-4412.

Woods, K. R. and K. T. Wang 1967. Separation of dansyl-amino acids by polyacrylamide layer chromatography. *Biochim. Biophys. Acta* 133 : 369-370.

Wrigley, C. 1968. Gel electrofocusing - a technique for analysing multiple protein samples by isoelectric focusing. *Sci. Tools* 15: 17-23.

Zaccharius, R.M., T.E. Zell, J.H. Morrison, and J.J. Woodlock 1969. Glycoprotein staining following electrophoresis on acrylamide gels. *Anal. Biochem.* 30:148-152.

MALATE DEHYDROGENASE ISOZYMES IN PLANTS:
PREPARATION, PROPERTIES, AND BIOLOGICAL SIGNIFICANCE

IRWIN P. TING, IRENE FÜHR, RUSSELL CURRY,
and WILLIAM C. ZSCHOCHE
Department of Biology
University of California
Riverside, California 92502

ABSTRACT. Vascular plants may have three distinct classes
of isozymes of NAD malate dehydrogenase (EC 1.1.1.37);
cytosol-malate dehydrogenase, mitochondrial-malate dehydrogenase, and microbody-malate dehydrogenase. These organelle specific isozymes function in distinctly different
metabolic sequences. The cytosol form has a role in nonautotrophic CO_2 fixation to form malate which may function
in ionic, osmotic, or redox regulation. The mitochondrial
form functions in the citric acid cycle. Two functional
types of the microbody form exist: a peroxisomal form in
leaf microbodies or peroxisomes functioning in photorespiration, and a glyoxysomal form in glyoxysomes functioning in
the glyoxylate cycle. The kinetic properties of the isozymes are quite similar and we assume that metabolic regulation is largely effected by compartmentation of the isozymes.

INTRODUCTION

Within the wide array of known metabolic pathways, the same
enzymic reaction is frequently repeated. In procaryotic cells
with little or no compartmentation, Umbarger (1961) proposed
that if an enzyme is a component of two or more metabolic
sequences, and if in one of the sequences it is under metabolic control by feedback inhibition, then isozymes must exist.
In eucaryotic cells, however, metabolic pathways with common
enzymic steps and common intermediates may be localized in
different subcellular compartments. We propose that the common
enzymic steps occurring in the separate metabolic pathways
will be catalyzed by isozymes. This does not mean to imply
that in those situations where common reactions occur in the
same compartment, the function of isozymes suggested by
Umbarger and others (Wroblewski, 1961) is not applicable.

In the malate dehydrogenase (MDH) (1-malate: NAD oxidoreductase, EC 1.1.1.37) system discussed here, each isozyme is
known to exist in a different subcellular compartment and each
participates in a different metabolic sequence. Hence, regulation and control of the common metabolic intermediates (i.e.,

oxaloacetate and malate) is largely by spatial compartmentation of *organelle specific isozymes* rather than by complex
feedback or feedforward events as has been described in detail
for microbial systems (Umbarger, 1961; Stadtman, 1968). The
following discussion outlines the evidence for this hypothesis
for higher plant malate dehydrogenases.

Others have proposed that isozymes are localized in different subcellular compartments where they participate in different metabolic sequences (Vesell, 1968).

Localization and Preparation

Localization. The method by which plant malate dehydrogenase isozymes are assigned to specific subcellular organelles
involves separation and purification of organelles and demonstration of electrophoretic correspondence of specific isozymes
with the organelles (Longo and Scandalios, 1969; Rocha and
Ting, 1970a). The organelles are prepared from freshly collected plant tissue by gently chopping with an electric knife
in a pH 7.5, 0.05 M Tris buffer containing 0.5 M sucrose, 1
mM ethylenediamine tetraacetate, 1 mM dithiothreitol, and 0.1%
bovine serum albumin. The organelle preparation is centrifuged at 250 g for 90 sec to remove cellular debris and then
at 1000-3000 g for 5-15 min to pellet intact organelles. The
pellet is collected and layered on a 40-80% (w/v) linear
sucrose gradient and centrifuged at 25,000 rpm in a swing-out
rotor to equilibrium. The gradient is fractionated into 30
or more equal parts, detergent is added to each fraction to
insure rupture of organelles and to release bound enzymes,
and then assayed for organelle marker enzymes plus malate dehydrogenase. Organelle markers most frequently used are: cytochrome c oxidase for mitochondrial membranes and fumarase for
intact mitochondria, catalase and/or glycolate oxidase for
microbodies, NADP cytochrome c reductase for endoplasmic
reticulum, chlorophyll for chloroplast membranes, and NADP
triose phosphate dehydrogenase for intact chloroplasts.

Fig. 1 shows the separation and purification of organelles
plus the distribution of malate dehydrogenase in the gradient.
Two distinct peaks of malate dehydrogenase are evident.

Routinely we electrophorese malate dehydrogenase on starch
gels with a phosphate-citrate buffer, pH 7.0, according to
Fine and Costello (1963).

Fig. 1 shows a zymogram of malate dehydrogenase from spinach
leaf tissue. In this particular tissue, there are three major
isozymes plus a small fast moving band. When the malate dehydrogenase from the microbody and mitochondrial regions of the
isopycnic sucrose density gradient is collected, concentrated,

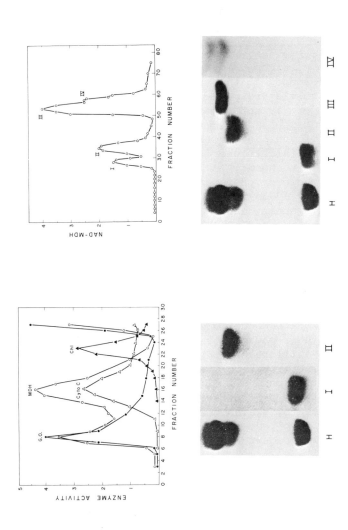

Fig. 1. *Left upper:* Separation of spinach leaf organelles by sucrose density gradient centrifugation. GO = glycolate oxidase, microbody (peroxisome) marker. Cyto C = cytochrome C oxidase, mitochondrial marker. Chl = chlorophyll, chloroplast marker. MDH = NAD malate dehydrogenase. *Left lower:* H = starch gel electrophoresis of malate dehydrogenase from spinach leaf homogenate. I = MDH from microbody region of gradient. II = MDH from mitochondrial region of gradient.

(Fig. 1 legend continued) Right upper: DEAE cellulose column chromatography of malate dehydrogenase from spinach leaf homogenate. Peak I = microbody protein; Peak II = mitochondrial protein; Peaks III and IV are cytosol forms. *Right lower*: H = starch gel electrophoresis of spinach leaf homogenate. I, II, III, and IV are the DEAE-cellulose peaks from above right. Data are from Rocha and Ting (1970a).

and subjected to electrophoresis, it is observed that the slow moving band corresponds to the microbody form and the second anodally moving form corresponds to the mitochondrial form. The fast moving form is not found associated with any membrane or organelle, thus it is assigned as a soluble or cytosol form.

In a variety of green plant tissues which we have investigated, the order of electrophoretic migration has been cytosol-MDH most anodal, mitochondrial-MDH second, and the microbody-MDH the slowest; however, some exceptions have been noted (Zschoche and Ting, 1973a).

In some plant systems, other investigators have reported that the mitochondrial and soluble forms are controlled by distinct genetic loci (Longo and Scandalios, 1969; Grimwood and McDaniel, 1970).

Preparation. A fresh preparation of total malate dehydrogenase can be precipitated between 30 and 80% saturation $(NH_4)_2SO_4$, dialyzed against 5 mM sodium phosphate pH 7.0 buffer, and applied to a DEAE-cellulose column (1.5 x 15 cm). Elution with a linear phosphate (or KCl) gradient, pH 7.0, from 20 to 200 mM will result in fractionation of the various isozymes (Ting, 1968). The elution patterns will vary with the experimental tissue and will require some adjustment (Fig. 1). Each peak can be precipitated once again with ammonium sulfate and checked by starch gel electrophoresis to insure complete separation of each form (Fig. 1). This type of preparation is usually adequate for a variety of comparative studies.

Previously, we had purified to homogeneity the three forms of malate dehydrogenase. The cytosol and mitochondrial forms from maize tissue were copurified by a course of acetone fractionation, and chromatography on hydroxylapatite and sieving gels. Subsequent separation of the forms prior to crystallization was performed on DEAE-cellulose and acrylamide gel electrophoresis columns (Curry and Ting, 1973). The microbody form from spinach leaves was separated from the other forms by 0-15% ammonium sulfate precipitation and then purified to homogeneity by a course of calcium phosphate gel, DEAE cellulose, and hydroxylapatite chromatography (Zschoche and Ting, 1973b).

Distribution. Boser and Pawelke (1961) separated two forms of plant malate dehydrogenase by elution from calcium phosphate gels; cytosol and mitochondrial forms were demonstrated by Ting et al (1966) and Yue (1966). We have now surveyed a variety of vascular plants ranging from primitive rootless forms to advanced monocots and all tissues have had readily detectable isozymes of malate dehydrogenase suggestive of the mitochondrial and soluble forms. In the green tissue and leaves of many higher plants studied, three forms occur; the cytosol form, the mitochondrial form, and a microbody form (Yamazaki and Tolbert, 1969; Rocha and Ting, 1970b) localized in green leaf microbodies or peroxisomes (Tolbert, 1971), the organelles functioning in photorespiration. *Psilotum* (subphylum Psilopsida) the most primitive living vascular plant and lacking true roots, has three forms of MDH with electro-phoretic properties corresponding to microbody-, mitochondrial- and soluble-MDH in both above and below ground tissues (Fig. 2). Green tissue of *Equisetum* and *Selaginella* of the sub-phyla Sphenopsida and Lycopsida, have the three forms (Fig. 2). Ferns (in the subphylum Pteropsida, the same as conifers and flowering plants) also have the presumed microbody form in leaves, but it is absent or much reduced in roots.

Those plant tissues with an active glyoxylate cycle local-ized in microbodies (i.e., glyoxysomes, Beevers, 1969) convert-ing fats to carbohydrates also have a microbody form of MDH (Fig. 2) (Breidenbach, 1969). We have investigated the possi-bility that the glyoxysomal-MDH is different from the peroxi-somal-MDH (Fig. 2), but have not detected any physical or kinetic differences. Thus it appears that the microbody form may function in two different metabolic sequences, but within the same subcellular compartment.

In maize leaf, the microbody form is not readily detectable (Ting, 1968), but Hayden and Cook (1972) have reported a micro-body (glyoxysomal) form in maize endosperm.

Studies with algae and fungi have been minimal, but to date no observation of an MDH with properties of the microbody form have been reported. Benveniste and Munkres (1970) studied the regulatory and enzymic properties of mitochondrial and cytosol forms in the fungus, Neurospora, and Stromeyer et al (1971) found a particulate and nonparticulate form in Chlorella. Saccharomyces has a cytosol and mitochondrial form (Witt et al, 1966). In bacteria, such as *Escherichia coli* and *Bacillus subtilis* only a single form is known (Murphey et al, 1967).

Previously, we reported on the kinetic and regulatory prop-

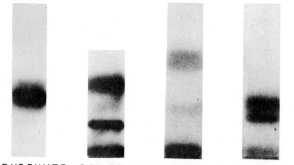

BRYOPHYTE PSILOTUM EQUISETUM SELAGINELLA

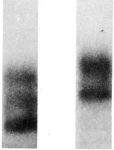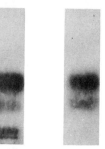

LEAF ROOT LEAF ROOT
ADIANTUM   GOMPHRENA

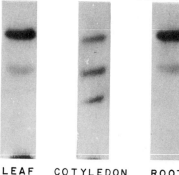

LEAF COTYLEDON ROOT
   CUCUMIS

Fig. 2. Starch gel electrophoresis of malate dehydrogenase from a variety of plants and plant tissues. The slow moving band is assumed to be the microbody form and is absent in most lower plants (the Bryophyte for example) and absent in roots. Adiantum is a fern, and Gomphrena and Cucumis are dicotyledonous plants.

erties of the isozymic forms of mitochondrial and cytosol
malate dehydrogenase in Euglena (Peak et al, 1973). No evidence was found for a microbody form in either auto- or heter-
otrophically grown cells; however, Davis and Merritt (1973)
have recently reported a peroxisomal form in photoautotrophic
Euglena.

Hence our tentative conclusion at this time concerning the
distribution of organelle specific plant MDH isozymes is that
the green photosynthetic tissues of Tracheophytes may have
three forms, a cytosol-MDH, a mitochondrial-MDH, and a microbody-MDH. The latter may occur in nongreen tissues in specialized metabolic microbodies such as glyoxysomes; true roots have
a much reduced or lack completely the microbody form of MDH.
The lower, non-vascular eucaryotic plant phyla including algae
and fungi have the cytosol and mitochondrial forms, but probably lack the peroxisomal microbody form. There is some evidence, however, of a glyoxysomal microbody form in some, e.g.,
Euglena.

Kinetic and Regulatory Properties

All of the MDH isozymes favor formation of malate. The
Michaelis constants (oxaloacetate) are all quite similar
(Table 1) and on the order of 0.03 to 0.06 mM at a pH of about
7.5 (Rocha and Ting, 1971). Comparison with the literature
indicates that these data are largely representative. The
Michaelis constants (oxaloacetate) are highly pH dependent,
and the unpurified cytosol and microbody forms show pH dependent substrate inhibition with more inhibition at low pH. Substrate inhibition of the mitochondrial form is relatively pH
independent. These substrate inhibition properties are largely
lost after purification (unpublished). The Michaelis constants
determined with malate and the pyridine nucleotides (NAD or
NADH) as the variable substrates are not markedly different.
With the exception of the pH dependent substrate inhibition
and activity in the presence of NAD analogs, the kinetic properties of the NAD isozymes are remarkably similar. These
observations are taken to mean that the isozymes catalyze the
same reaction *in situ*, i.e., oxaloacetate $\rightleftharpoons$ malate.

Antiserum prepared against purified maize mitochondrial-
MDH does not cross react with the cytosol form from the same
tissue (Curry and Ting, 1973). Anti-microbody MDH serum cross
reacts about 10% with the mitochondrial form but not with the
cytosol form (Zschoche and Ting, 1973b).

The purified microbody form shows a complex pH-ionic
strength interaction. It is more active at low pH (5.6) when
at high ionic strength (pI = 5.6) and at high pH (7.0) when

TABLE 1

SOME PROPERTIES OF SPINACH LEAF MALATE DEHYDROGENASE*

	c-MDH	m-MDH	mb-MDH
Km OAA (pH 7.5)	0.06 mM	0.04	0.04
Km MAL (pH 8.5)	0.8	6.0	2.8
Km NADH (pH 7.5)	0.02	0.05	0.02
Km NAD (pH 8.5)	0.2	0.6	0.4
TN-NAD/NAD	2.2	0.21	0.15
3-AP-NAD/NAD	1.2	3.0	2.5
Mol. wt. (pig heart MDH standard)	66,000–70,000	66,000–70,000	66,000–70,000
Antiserum identity (mb-MDH)	0%	10%	100%

* Data from Rocha and Ting (1971) and Zschoche and Ting (1973)

at low ionic strength (Zschoche and Ting, 1973b).

Estimates of the molecular size are about 66-70,000 for all forms with the exception of certain freshly prepared microbody forms which chromatograph as if they were complexes. Breidenbach (1969), for example, found that the glyoxysomal microbody form of castor bean endosperm was in a high molecular weight complex but could be broken down to the 66,000-70,000 species by deoxycholate treatment. The glyoxysomal form from cucumber cotyledons elutes from a sieving gel column with an estimated molecular size of about 3 x 70,000, but can be broken down to the 70,000 (unpublished data). The electrophoretic mobility is the same regardless of the size.

Limited data suggest that the microbody-form is less heat stable than the other forms but can readily be stabilized in glycerol solutions.

Relatively little is known regarding the in vivo metabolic regulation of higher plant NAD-MDH isozymes. Several small metabolite molecules when in concentrations of 10 mM or greater tend to inhibit all of the NAD-MDH isozymes. For example, α-ketoglutarate, cis-aconitate, isocitrate, succinate, fumarate, pyruvate, aspartate, glutamate, and asparagine inhibit all the isozymes about equally (Mukerji and Ting, 1969). Glyoxylate, but not glycolate, is known to inhibit the spinach microbody form (Zschoche and Ting, 1973b). Pyruvate and α-ketoglutarate also inhibit. Adenine nucleotides are known to inhibit the mitochondrial-MDH and may play a role in metabolic regulation (Kuramitsu, 1968; Abou-Zamzam and Wallace, 1970). In general, however, allosteric control similar to that reported for microbial malate dehydrogenase proteins is not known (Sanwal, 1969).

These kinetic data are taken to mean that regulation of the common metabolic reactions with the same substrate requirement in different metabolic pathways is largely by spatial compartmentation rather than by complex feedback and feedforward mechanisms. The possibility of substrate level regulation within compartments, however, is not eliminated.

In lower plant forms and bacteria, the level of activity can be regulated by metabolites. For example, isocitrate represses in *Bacillus subtilis* (Fortnagel and Lopez, 1971). The cytosol form in Neurospora (Benveniste and Munkres, 1970), Yeast (Duntze et al, 1968), and Euglena (Peak et al, 1972) can be derepressed or otherwise regulated by catabolites or light. There is no known analogous situation in higher plants yet it appears that the peroxisomal microbody form is repressed in roots and the glyoxysomal microbody form is produced in response to glyoxylate cycle activity.

Biological function and significance

Metabolic roles. i. Soluble-malate dehydrogenase. An important metabolic function in plant tissues is the generation of malate which plays a role in ionic regulation (Ulrich, 1941; MacDonald and Laties, 1964), osmotic adjustment (Bernstein, 1963), and perhaps NADPH production (Ting and Dugger, 1965). Malate for these functions is known to be isolated from mitochondrial metabolism (Lipps and Beevers, 1966) and is most likely generated through the nonautotrophic CO_2 fixation pathway (Ting, 1971). Hence at least one function of the cytosol-MDH is in the nonautotrophic CO_2 fixation pathway (Fig. 3).

ii. Mitochondrial-MDH. Plants,like virtually all other living eucaryotic organisms,have a functional citric acid cycle localized in mitochondria. There can be little doubt, therefore, that a major role of the mitochondrial-MDH is the oxidation of malate to oxaloacetate in the citric acid cycle (Fig. 3).

iii. Microbody-MDH. Our understanding of microbody metabolism in plants is relatively recent. Leaf microbodies or peroxisomes are known to play an important role in photorespiration (Tolbert, 1971), the photosynthesis-related metabolic pathway oxidizing glycolate. Here, Tolbert has proposed a specific role for MDH in the regeneration of reducing power,

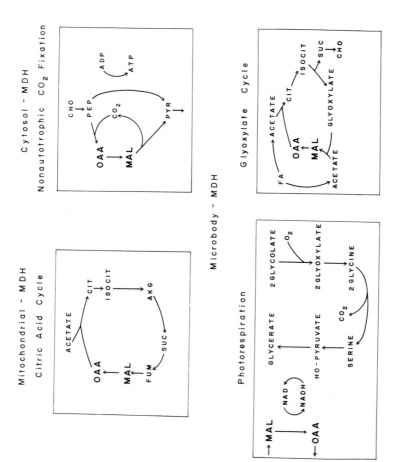

Fig. 3. Presumed metabolic roles of the organelle specific isozymes of malate dehydrogenase. The mitochondrial form functions in the citric acid cycle. The cytosol form functions in nonautotrophic CO_2 fixation in couple with P-enolpyruvate carboxylase. The microbody malate dehydrogenase protein in glyoxysomes functions in the glyoxylate cycle and in peroxisomes it

(*Fig. 3 legend continued*) functions in photorespiration. To date, there is no real evidence that the microbody form in glyoxysomes and peroxisomes of the same tissue is different.

---

i.e., NADH, for the reduction of hydroxypyruvate to glycerate (Fig. 3). In glyoxysomes, with the glyoxylate cycle, a specific role for malate dehydrogenase is known in the oxidation of malate to oxaloacetate similarly to the citric acid cycle (Fig. 3). It may be of significance that algae appear to lack the peroxisomal microbody form of MDH and also do not have glycolate oxidase, but glycolate dehydrogenase (Nelson and Tolbert, 1970). These two proteins catalyze the identical reaction but the latter does not consume O_2. Hence in these plants glycolate dehydrogenase and not malate dehydrogenase could be the redox couple with hydroxypyruvate reductase (Fig. 3).

Chloroplast-MDH. It is interesting and probably significant that a specific chloroplast protein catalyzing the reversible reduction of oxaloacetate to malate occurs in plants (Ting and Rocha, 1971). This protein, although the kinetic properties are similar to the NAD-MDH isozymes, its physical properties are distinctly different, and the reaction is specific for NADP rather than NAD. NADP, of course, is the functional pyridine nucleotide of chloroplasts. A specific role for NADP-MDH in photosynthesis has been proposed (Johnson and Hatch, 1970; Ting and Rocha, 1971).

Although this protein is not an isozyme in series with the NAD-MDH forms, functionally it meets the hypothesis of different proteins catalyzing a common enzymic step in different metabolic pathways localized in different subcellular compartments.

The Shuttle Hypothesis. A possible metabolic role for the various MDH isozymes localized in different subcellular compartments is ionic and redox regulation by malate shuttles (or NAD, NADH, oxaloacetate shuttles) among compartments. This notion was proposed in general for dehydrogenases by Delbruck et al (1959), Heber and Santarius (1965), and Williamson et al (1967), and specifically for plant MDH isozymes by Ting and Rocha (1971) and Johnson (1971).

The metabolic role proposed above for the peroxisomal microbody-MDH is in fact a shuttle. The malate needed to drive the hydroxypyruvate-malate dehydrogenase couple would come from one of the other malate dehydrogenase isozymes.

SUMMARY AND CONCLUSIONS

The available data indicate that vascular green plants have at least three different, site specific proteins catalyzing the NADH dependent reversible reduction of oxaloacetate to malate. Each is localized in a different subcellular compartment, viz., microbodies, mitochondria, and the cytosol. These organelle specific isozymes participate in different metabolic functions or sequences: the glyoxylate cycle and the photorespiration, glycolate pathways for the microbody-MDH, the citric acid cycle for the mitochondrial-MDH, and the nonautotrophic CO_2 fixation pathway for the cytosol-MDH. There is evidence that redox shuttles are important roles for the isozymes as well. The latter statements should not necessarily be taken to mean that other roles or functions are excluded. Since these organelle specific isozymes are coded for by different genetic loci, multiple forms may occur within each organelle.

Because no dramatic substrate level regulation of the various MDH isozymes is known, we propose that metabolic regulation of the same reaction in the different metabolic pathways is largely by spatial compartmentation of organelle specific isozymes. Hence an important consequence of organelle specific isozymes is metabolic regulation by compartmentation.

Although the NADP-malate dehydrogenase does not fit into the set of isozymes with the NAD-malate dehydrogenases, physiologically the four must be considered together.

Perhaps, as good an example of organelle specific isozymes comes from the work of Anderson and Advani (1970). They have shown that three of the enzymes which are common to both the photosynthetic reductive pentose cycle, localized in chloroplasts, and to the glycolytic sequence are, in fact, not the same, but are isozymes. Schnarrenberger et al (1973) demonstrated the same for glucose-6-phosphate and phosphogluconate dehydrogenase. Here, compartmentation and different metabolic roles are clearly indicated. A variety of other organelle specific isozymes are known from higher plants including malate enzyme (Mukerji and Ting, 1968), and NADP-isocitrate dehydrogenase (Tolbert and Yamazaki, 1969).

ACKNOWLEDGMENTS

The development of this work was supported by the US National Science Foundation (NSF GB 2547, GB 6735, and GB 25878) and the University of California Intramural Research Fund. Many others participated directly including Dr. Jean Danner, Dr. S. K. Mukerji, Dr. V. Rocha, Dr. M. Peak,

Jenny Peak, Dr. N. Ahmadi, Paula Frederick, Megan Foster,
Janet Norris, Penny Reid, Polly Hilsabeck, Kay Jolly, and
Kay Adams.

LITERATURE CITED

Abou-Zamzam, A. M. and A. Wallace 1970. Some characteristics
of the mitochondrial and soluble forms of malate dehydrogenase in lemon fruits. *Biochim. Biophys. Acta* 220:
396-409.

Anderson, L. E. and V. R. Advani 1970. Chloroplast and cytoplasmic enzymes. Three distinct isoenzymes associated
with the reductive pentose phosphate cycle. *Plant
Physiol.* 45: 583-585.

Beevers, H. 1969. Glyoxysomes of castor bean endosperm and
their relation to gluconeogenesis. *Ann. N. Y. Acad. Sci.*
168: 313-324.

Beneviste, K. and K. D. Munkres 1970. Cytoplasmic and mitochondrial malate dehydrogenases of Neurospora. Regulatory
and enzymic properties. *Biochim. Biophys. Acta* 220: 161-
177.

Bernstein, L. 1963. Osmotic adjustments of plants to saline
media. II. Dynamic phase. *Am. J. Botany* 50: 360-370.

Boser, H. and G. Pawelke 1961. Zur heterogenität von Enzymen
II: Reindarstellung von zwei Individuen der Malicodehy-
drogenase (MDH) aus Kartoffelknollen. *Naturwissenschaften*
48: 572.

Breidenbach, R. W. 1969. Characterization of some glyoxysomal
proteins. *Ann. N. Y. Acad. Sci.* 168: 342-347.

Curry, R. A. and I. P. Ting 1973. Purification and crystallization of three isoenzymes of malate dehydrogenase from
Zea mays seed. *Arch. Biochem. Biophys.* 158: 213-224.

Davis, B. and M. J. Merrett 1973. Malate dehydrogenase isoenzymes in division synchronized cultures of Euglena.
Plant Physiol. 51: 1127-1132.

Delbrück, A., E. Zebe, and T. H. Bücher 1959. Über Verteil-
ungsmuster von Enzymen des Energie liefernden Stoffwechs-
els im Flugmuskel, Sprungmuskel, und Fettkörper von
Locusta migratoria und ihre cytologische Bedeutung.
Biochem. Z. 331: 273-296.

Duntze, W., D. Neumann, and H. Holzer 1968. Glucose induced
inactivation of malate dehydrogenase in intact yeast
cells. *Europ. J. Biochem.* 3: 326-331.

Fine, I. H. and L. A. Costello 1963. The use of starch gel
electrophoresis in dehydrogenase studies. in *Methods in
Enzymology, Vol. VI* (Colowick and Kaplan, eds). Academic
Press, N. Y., 958-972.

Fortnagel, P. and J. Lopez 1971. The regulation of malate dehydrogenase in sporulation mutants of Bacillus subtilis blocked in the citric acid cycle. *Biochim. Biophys. Acta* 237: 320-328.

Grimwood, B. G. and R. G. McDaniel 1970. Variant malate dehydrogenase isoenzymes in mitochondrial populations. *Biochim. Biophys. Acta* 220: 410-415.

Hayden, D. B. and F. S. Cook 1972. Malate dehydrogenase in maize endosperm: The intracellular location and characterization of the two major particulate forms. *Can. J. Biochem.* 50: 663-671.

Johnson, H. S. 1971. NADP-malate dehydrogenase: photoactivation in leaves of plants with Calvin Cycle photosynthesis. *Biochem. Biophys. Res. Commun.* 43: 703-709.

Johnson, H. S. and M. D. Hatch 1970. Properties and regulation of leaf nicotinamide-adenine dinucleotide phosphate-malate dehydrogenase and malic enzyme in plants with the C_4-dicarboxylic acid pathway of photosynthesis. *Biochem. J.* 119: 273-280.

Kuramitsu, H. K. 1968. Adenine nucleotide inhibition of pea malate dehydrogenase. *Arch. Biochem. Biophys.* 125: 383-384.

Lips, S. H. and H. Beevers 1966. Compartmentation of organic acids in corn roots. II. The cytoplasmic pool of malic acid. *Plant Physiol.* 41: 713-717.

Longo, G. P. and J. G. Scandalios 1969. Nuclear gene control of mitochondrial malic dehydrogenase in maize. *Proc. Natl. Acad. Sci.* 62: 104-111.

MacDonald, I. R. and G. G. Laties 1964. A comparative study of the influence of salt type and concentration on $^{14}CO_2$ fixation in potato slices at 25°C and 0°C. *J. Exp. Bot.* 15: 530-537.

Mukerji, S. K. and I. P. Ting 1968. Malate dehydrogenase (decarboxylating) (NADP) isoenzymes of Opuntia stem tissue. Mitochondrial, chloroplast, and soluble forms. *Biochim. Biophys. Acta* 167: 239-249.

Mukerji, S. K. and I. P. Ting 1969. Malic dehydrogenase isoenzymes in green stem tissue of Opuntia: Isolation and characterization. *Arch. Biochem. Biophys.* 131: 336-351.

Murphey, W. H., C. Barnaby, F. J. Lin, and N. O. Kaplan 1967. Malate dehydrogenases. II. Purification and properties of *Bacillus subtilis, Bacillus stearotheromophilus,* and *Escherichia coli* malate dehydrogenases. *J. Biol. Chem.* 242: 1548-1559.

Nelson, E. B. and N. E. Tolbert 1970. Glycolate dehydrogenase in green algae. *Arch. Biochem. Biophys.* 141: 102-110.

Peak, M. J., J. G. Peak, and I. P. Ting 1972. Isoenzymes of
    malate dehydrogenase and their regulation in *Euglena*
    gracilis Z. *Biochim. Biophys. Acta* 284: 1-15.
Rocha, V. and I. P. Ting 1970a. Preparation of cellular plant
    organelles from spinach leaves. *Arch. Biochem. Biophys.*
    140: 398-407.
Rocha, V. and I. P. Ting 1970b. Tissue distribution of micro-
    body, mitochondrial, and soluble malate dehydrogenase
    isoenzymes. *Plant Physiol.* 46: 754-756.
Rocha, V. and I. P. Ting 1971. Malate dehydrogenases of leaf
    tissue from *Spinacia oleracea:* properties of three
    isoenzymes. *Arch. Biochem. Biophys.* 147: 114-122.
Sanwal, B. D. 1969. Regulatory mechanisms involving nicotina-
    mide adenine nucleotides as allosteric effectors. *J. Biol.*
    Chem. 244: 1831-1837.
Schnarrenberger, C., A. Oeser, and N. E. Tolbert 1973. Two
    isoenzymes each of glucose-6-phosphate dehydrogenase and
    6-phosphogluconate dehydrogenase in spinach leaves. *Arch.*
    Biochem. Biophys. 154: 438-448.
Stadtman, E. R. 1968. The role of multiple enzymes in branched
    metabolic pathways. in *Multiple molecular forms of*
    enzymes (Vesell,E.,ed)*Ann. N. Y. Acad. Sci.* 151: 516-530.
Stromeyer, C. T., F. E. Cole, and P. C. Arquembourg 1971.
    Purification and properties of malate dehydrogenase from
    Chlorella pyrenoidosa. Catalytic mechanism of the par-
    ticulate form. *Biochem.* 10: 729-735.
Ting, I. P. 1968. Malic dehydrogenases in corn root tips.
    Arch. Biochem. Biophys. 126: 1-7.
Ting, I. P. 1971. Nonautotrophic CO_2 fixation and Crassulacean
    Acid Metabolism. in *Photosynthesis and Photorespiration*
    (Hatch, Osmond, and Slatyer, eds). Wiley-Interscience,
    N. Y., 169-185.
Ting, I. P. and W. M. Dugger 1965. Transhydrogenation in root
    tissue: mediation by carbon dioxide. *Science* 150: 1727-
    1728.
Ting, I. P. and V. Rocha 1971. NADP-specific malate dehydrogen-
    ase of green spinach leaf tissue. *Arch. Biochem. Biophys.*
    147: 156-164.
Ting, I. P., I. W. Sherman, and W. M. Dugger 1966. Intracellu-
    lar localization and possible function of malic dehy-
    drogenase isozymes from young maize root tissue. *Plant*
    Physiol. 41: 1083-1084.
Tolbert, N. E. 1971. Microbodies--peroxisomes and glyoxysomes.
    Ann. Rev. Plant Physiol. 22: 45-74.
Tolbert, N. E. and R. K. Yamazaki 1969. Leaf peroxisomes and
    their relation to photorespiration and photosynthesis.
    Ann. N. Y. Acad. Sci. 168: 325-341.

Ulrich, A. 1941. Metabolism of non-volatile organic acids in excised barley roots as related to cation-anion balance during salt accumulation. *Amer. J. Bot.* 28: 526-537.

Umbarger, H. E. 1961. in *Control mechanisms in cellular processes* (Bonner,D.M.,ed). The Ronald Press Co.,N.Y.,67-86.

Vesell, E. S. 1968. Ed. *Multiple molecular forms of enzymes.* Ann. N. Y. Acad. Sci. 151. 689 pp.

Williamson, D. H., P. Lund, and H. A. Krebs 1967. The redox state of free nicotinamide-adenine dinucleotide in the cytoplasm and mitochondria of rat liver. *Biochem. J.* 103: 514-527.

Witt, I., R. Kronan, and H. Holzer 1966. Isoenzyme der malatdehydrogenase und ihre regulation in Saccharomyces cerevisiae. *Biochim. Biophys. Acta* 128: 63-73.

Wroblewski, F. 1961. Ed. *Multiple molecular forms of enzymes.* *Ann. N. Y. Acad. Sci.* 94. 655 pp.

Yamazaki, R. K. and N. E. Tolbert 1969. Malate dehydrogenase in leaf peroxisomes. *Biochim. Biophys. Acta* 178: 11-20.

Yue, S. B. 1966. Isoenzymes of malate dehydrogenase from barley seedlings. *Phytochemistry* 5: 1147-1152.

Zschoche, W. C. and I. P. Ting 1973a. Malate dehydrogenases of *Pisum sativum.* Tissue distribution and properties of the particulate forms. *Plant Physiol.* 51: 1076-1081.

Zschoche, W. C. and I. P. Ting 1973b. Purification and properties of microbody malate dehydrogenase from *Spinacia oleracea* leaf tissue. *Arch. Biochem. Biophys.* 159: 767-776.

MULTIPLE MOLECULAR FORMS OF THE PURINE PHOSPHORIBOSYL TRANSFERASES

WILLIAM L. NYHAN and BOHDAN BAKAY
Department of Pediatrics
University of California, San Diego
La Jolla, California

ABSTRACT. In man hypoxanthine guanine phosphoribosyl transferase catalyzes the conversion of hypoxanthine and guanine to their respective nucleotides, inosinic acid and guanylic acid. Adenine phosphoribosyl transferase catalyzes the conversion of adenine to adenylic acid. A considerable degree of genetically determined heterogeneity has been demonstrated in the case of HGPRT. An electrophoretic method has been developed for HGPRT and APRT which depends for detection on the quantitative precipitation of a labeled nucleotide reaction product by lanthanum. In this system variants have been detected with faster and slower anodal migration than normal. So far, in each instance where a variant enzyme has been recognized, there has been an associated clinical disease. These observations provide an important example of the relationship between altered molecular form of an enzyme and altered physiological function.

INTRODUCTION

The phosphoribosyl transferases are important enzymes of purine interrelation. They are involved in the catalysis of the conversions of the purine bases in the presence of phosphoribosyl pyrophosphate (PRPP) to their respective nucleotides. Hypoxanthine guanine phosphoribosyl transferase (HGPRT) (E.C. 2.4.2.8) (Fig.1) catalyzes the formation of inosinic acid (IMP) or guanylic acid (GMP) from hypoxanthine or guanine. Xanthine also serves as a substrate for this enzyme, and so do the purine analogs 6-mercaptopurine, 6-thioguanine, and 8-azaguanine. The latter property permits the utilization of selective media containing one of these analogs to distinguish cells that contain HGPRT activity from those that do not. Adenine phosphoribosyl transferase (APRT) (E.C.2.4.2.7) catalyzes the conversion of adenine to adenylic acid (AMP). Diaminopurine is also a substrate for this enzyme, providing the basis for a selective medium. These enzymes have been referred to as salvage enzymes. They do provide a mechanism for the reutilization of purines. However, it is clear at least in the case of HGPRT that they occupy a highly important place in the ecology of the cell.

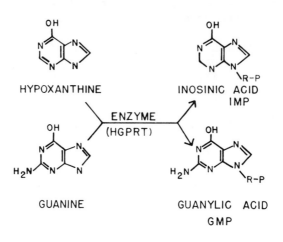

Figure 1. Hypoxanthine guanine phosphoribosyl transferase (HGPRT).

Inherited variation in the case of HGPRT first became evident through clinical observations. The first information that there might be human variation in HGPRT came with the definition of the Lesch-Nyhan syndrome. Patients with this syndrome present an extraordinary clinical picture by which the phenotype can be readily recognized (Lesch and Nyhan, 1964; Nyhan, 1973). They are all severely mentally retarded, with IQs less than 50. They have spastic cerebral palsy and a choreoathetoid movement disorder. They regularly manifest bizarre, compulsive, aggressive behavior. Its most striking feature is self mutilation through biting. Metabolic abnormality in this syndrome is characterized by increased amounts of uric acid in blood and urine, a consequence of oversynthesis of purine *de novo*. As a result of this metabolic abnormality these patients also develop a symptomatology reminiscent of patients with gout. These manifestations include renal stone disease, nephropathy, tophi, and acute arthritis. This clinical picture provides a background against which phenotypic variation can be distinguished.

Patients with the Lesch-Nyhan syndrome have virtually no activity of HGPRT (Seegmiller, et al. 1967; Sweetman and Nyhan, 1972). This enzyme is active in every cell in every mammalian system studied. It is most readily measured quantitatively in the erythrocyte, for in other tissues active nucleotidases variably break down IMP newly formed in the quantitative assay. In those tissues most investigators measure HGPRT in a radiochemical assay in which labeled hypoxanthine is the precursor and radioactivity in both inosine and IMP are measured. This, of course, raises the possibility that some of the inosine formed from hypoxanthine might result from the action of enzymes other than HGPRT such as nucleoside phosphorylase. Therefore, most investigators have preferred to characterize patients on the basis of the assay in the erythrocyte where there are no nucleotidases and one can measure directly the conversion of hypoxanthine to IMP with very little variation about the mean. In patients with the Lesch-Nyhan syndrome the quantitative assay for HGPRT in the erythrocyte cannot be distinguished from zero.

It soon became evident that there were other clinical phenotypes among patients with deficient activity of HGPRT. Some of these are adult patients with clinical gout (Kelley et al. 1969). Others are children who present with renal stone disease (Kogut et al. 1970). Obviously these two populations, both very different from those of the Lesch-Nyhan syndrome, tend to merge together. They all have increased rates of synthesis of purine *de novo*, and they have increased amounts of uric acid in blood and urine. They may come first to attention with renal failure or an acute attack of renal colic, or they may gradually accumulate deposits of urate in the body and ultimately develop acute attacks of arthritis, gouty tophi, and urate nephropathy.

Molecular distinctions from those of patients with the Lesch-Nyhan syndrome were evident in the fact that in all of these patients there was some erythrocyte activity of HGPRT. Furthermore, in any one kindred the quantitative amount of activity among different individuals tended to be the same. Variation in the amount of activity from kindred to kindred has suggested the possibility that among this population there is a considerable number of different genetically determined variants. The amounts of activity have varied some dependent on the condition of the assay method but in general the activity level has been between one and ten percent of the normal level (Kelly et al., 1969; Kogut et al., 1970).

METHODS AND RESULTS

ELECTROPHORETIC ANALYSIS OF THE PHOSPHORIBOSYL TRANSFERASES

In order to pursue the question of genetic heterogeneity in the HGPRT molecule, it was of interest to develop a method for the separation of different molecular forms of this enzyme. We have developed a method which employs electrophoretic separation of the purine phosphoribosyl transferases on polyacrylamide gel (Bakay and Nyhan, 1971). The method can utilize erythrocyte lysates or extracts of tissues, fibroblasts, or amniotic fluid cells. The proteins are separated by electrophoresis on 8 percent 5 x 100 mm acrylamide gel rods. Separation is run in tris-glycine buffer, pH 8.3, using 3 milliamperes per gel. It is complete in about 210 minutes. At this pH red cell proteins have anodal migration. Following electrophoresis the enzyme reaction is run by incubating the gels at 37°C in the presence of PRPP and ^{14}C-hypoxanthine or ^{14}C-guanine for HGPRT and ^{14}C-adenine for APRT.

The critical principle in the detection is the quantitative precipitation of the nucleotide products of these reactions by lanthanum (Bakay, et al., 1969). Following incubation the gels are overlaid with 0.1 $\underline{M}$ lanthanum chloride in 0.1 $\underline{M}$ tris HCl buffer (pH 7.0) and placed for 4 hours in a refrigerator. The labeled nucleotide precipitates are immobilized tightly in the gel and are not removed by 18 hours of washing with water. Unutilized radioactive substrate is well removed by washing for 8-10 hours in running deionized water.

Initially the gels were sliced manually, and each slice digested in H_2O_2, mixed with Bray's solution, and counted in a liquid scintillation spectrometer. The technique has now been completely automated (Bakay, 1971) by the development of a gel syringe fractionator fitted with a constant speed. In this apparatus as the gel is slowly squeezed out it is mixed with 1 mM EDTA-tris (pH 9) buffer, delivered by a pump to a mixing chamber, and mixed with scintillation mixture delivered simultanously by another pump. The mixture is then sent through a flow cell which is monitored by a Beckman β-Mate scintillation spectrometer and the activity is recorded on a strip chart and digitally. The responses are linear, and each enzyme may be quantitated using this method.

In the course of these studies an individual was identified in whom the APRT enzyme migrated more rapidly than the normal APRT (Bakay and Nyhan, 1971). The subject was a 4-year-old boy with overproduction hyperuricemia and normal HGPRT (Nyhan et al., 1969). His quantitative level of APRT activity was elevated. His enzyme migrated about 10-12 percent faster than the normal APRT.

Automated precedure resulted in an increased sensitivity more than five-fold over that of the manual gel slicing method. The application of this technique to the study of patients with the Lesch-Nyhan syndrome revealed evidence of HGPRT activity in every patient tested (Bakay and Nyhan, 1972) (Figure 2). In the patient illustrated, thirty times as much erythrocyte lysate was added to the gel as in the normal individual, and furthermore the lysate was 4 times more concentrated.

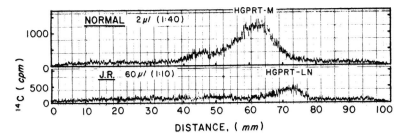

Figure 2. Radioelectropherograms of HGPRT. HGPRT-M has been employed to designate the enzyme from normal individuals and HGPRT-LN the enzyme from patients with the Lesch-Nyhan syndrome.

Enzyme can be demonstrated in patients with less than 120 times the amounts used in controls, but it does take at least 40 times the load. The enzyme variant detected in the patients migrated some 15 percent faster than the normal enzyme.

These observations indicate that the Lesch-Nyhan syndrome is due to a structural gene mutation. This is consistent with data obtained by Rubin and colleagues (1971) and Arnold and Kelly (1971) that in these patients there is a protein which exhibits cross reactivity (CRM+) with antibody prepared against purified normal HGPRT. Altered electrophoretic mobility provides evidence that the variant protein in these individuals has an altered chemical structure. Presumably a genetically determined amino acid substitution leads to an altered primary structure causing a difference in charge or in molecular radius.

Electrophoretic analysis has also been carried out on some individuals with partial activity of erythrocyte HGPRT. Fig. 3 illustrates the patterns obtained in a family in which 4 affected males had hyperuricemia and 5 percent of normal activity of HGPRT in their erythrocytes (Bakay et al., 1972). The variant in this family has been designated HGPRT-L⁻. It migrates about 15 percent faster than the normal enzyme. The shape of the zone of enzyme activity was also different from that of the normal, enzyme. The leading front of the variant enzyme was always steeper than the rest of the profile. The shape was reminiscent of some of the cliffs overlooking the Southern California coastline. Variation from normal in rate of migration in the gel, in distinctive profile and in activity, indicate that this enzyme is a distinct protein. It would be expected to have a greater negative charge or a decreased

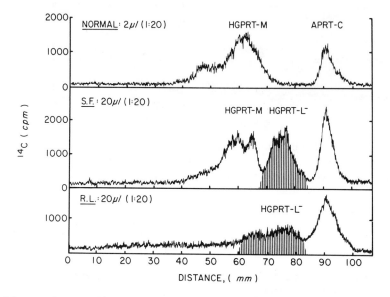

Figure 3. Radioelectropherograms of HGPRT from individuals with normal and partially deficient enzyme activity. R. L. was a hyperuricemic patient whose erythrocytes had 5% of normal activity. S. F. was his heterozygous sister.

molecular radius than the normal HGPRT protein.

The middle curve in Figure 3, designated S. F., represents the half sister of the patient R. L. Her total erythrocyte activity in the test tube assay was 34 percent of normal. In the gel her hemolysate has always given the interesting picture shown. There are two zones of activity in the profile, one in the area of the normal enzyme and one in the area of the variant. However, the zone in the variant area was very different from that seen in her siblings and cousins because it contained a large amount of enzyme activity. We have concluded that she is heterozygous at the HGPRT locus. We have also concluded that the large activity in the variant area represents an activation of the variant enzyme by the normal enzyme, for coelectrophoresis of lysates from hemizygous patients and normal individuals on the same gel led to an increase in activity of the variant enzyme and a pattern like that seen in S. F. (Bakay and Nyhan, 1972-A).

Enzymes with more rapid anodal migration in starch gel than normal HGPRT have been reported by Kelley and colleagues (1969)

HGPRT AND APRT OF HUMAN ERYTHROCYTES

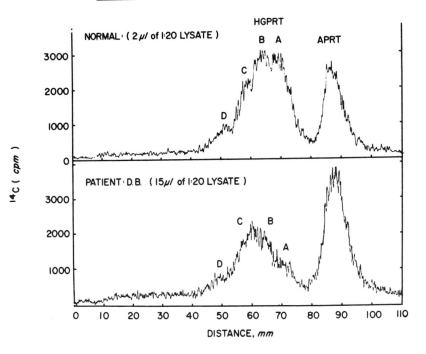

Figure 4.   Radioelectropherograms of HGPRT.   D. B. was an adult male with gout whose blood sample was kindly provided by Dr. Michael Becker.

in two   families in which enzyme activity approximated one and ten percent of normal activity. Our first experience with an enzyme that migrates more slowly than normal is shown in Figure 4. This enzyme from a man with overproduction hyperuricemia and gout also had partial HGPRT activity. His enzyme activity in erythrocyte lysates was 50 percent of normal. The shape of the profile was highly reproducible and characteristic. He appears to be very special and the work-up of this enzyme is in progress.

These approaches to the electrophoretic analysis of isozymes do not require the automated equipment we have developed. Tischfield, Bernard, and Ruddle (1973) have reported that the phosphoribosyl transferases may be separated in this way using flat slab acrylamide gels and lanthanum precipitation followed by autoradiographic detection. The technique has also been applied to other enzymes, such as thymidine kinase and

391

glycerol kinase.

Other approaches to molecular heterogeneity. The elucidation
of different molecular forms of an enzyme in different human
individuals is most easily pursued using electrophoretic
methods. However, a variety of other physical and chemical
methods have been useful in the delineation of genetic heterogeneity. In the family shown in Figure 3 (Bakay et al., 1972)
the K_m for hypoxanthine was 0.022 mM, some two times the value
for this kinetic property in normal individuals (0.011mM).
Kelley and colleagues (1969) have reported a variant with
partial activity of HGPRT in which study of heat stability of
the enzyme showed it to be more sensitive to inactivation by
heat than the normal enzyme, and in another kindred a variant
more resistant to inactivation by heat than normal.

Heterogeneity within the Lesch-Nyhan syndrome has been reported by McDonald and Kelley (1971) who studied a 10-year-
old patient with the classic features of the syndrome who had
an unusual enzyme. In the usual assay his erythrocyte activity
of HGPRT was about 0.2 percent of normal. However, at higher
concentrations of PRPP activity it ranged as high as 8 percent
of normal with guanine as substrate and 34 percent of normal
with hypoxanthine as substrate. With saturating concentrations
of hypoxanthine, activity approximated normal. The kinetic
curves for this enzyme were sigmoid, as contrasted with the
usual hyperbolic curve. The K_m for PRPP was some 13 times
the normal values and that for hypoxanthine was some 10 times
normal.

We have recently studied a patient who displayed many of
the features of the Lesch-Nyhan syndrome. He was spastic and
athetoid and he did have a problem with self destructive behavior, although he had learned to cope with it better than
most. On the other hand he was smart, and in spite of his
motor limitations his IQ appeared normal. This patient's enzyme is currently under evaluation. It is already clear that
it is an unusually unstable enzyme. Activities were as high
as 20 percent of normal in erythrocytes within the first 24
hours after the blood was drawn and decreased steadily to
zero within 4 to 6 days.

Another interesting variant was reported by Benke and
Herrick (1972). This patient had borderline intelligence,
mild spasticity, and an episode of self-mutilation in which he
tore out a large section of hair from his head for no apparent
reason. Assay of HGPRT in erythrocytes and fibroblasts was
normal. However, the patient's synthesis of uric acid from
single precursors in vivo was resistant to azathioprine, and
his cells were resistant to 8-azaguanine just like cells from
patients with the Lesch-Nyhan syndrome. In contrast to such

392

cells, cells of this patient grow normally in hypoxanthine aminopterin thymidine (HAT) medium. This patient has been found to have an abnormal HGPRT with altered kinetic properties.

CONCLUSION

The phosphoribosyl transferases, and in particular HGPRT, are clearly important enzymes in purine metabolism in man. A considerable degree of polymorphism has now been recognized. In each instance to date in which a variant HGPRT molecule has been found, a clearcut clinical illness has been present. Interesting correlations between molecular structure and physiological function are emerging.

ACKNOWLEDGEMENT

This work was supported by Grants No. GM 11702 from the National Institute of General Medical Sciences, National Institutes of Health, United States Public Health Service; and No. CRBS 242, National Foundation Research.

REFERENCES

Arnold, W. J. and W. N. Kelley 1971. Human hypoxanthine-guanine phosphoribosyltransferase. Purification and subunit structure. *J. Biol. Chem.* 246: 7398-7404.

Bakay, B. 1971. Detection of radioactive components in poly-acrylamide gel disc electropherograms by automated mechanical fractionation. *Analyt. Biochem.* 40: 429-439.

Bakay, B. and W. L. Nyhan 1971. The separation of adenine and hypoxanthine-guanine phosphoribosyl transferases isoenzymes by disc gel electrophoresis. *Biochem. Genet.* 5: 81-90.

Bakay, B. and W. L. Nyhan 1972. Electrophoretic properties of hypoxanthine-guanine phosphoribosyl transferase in erythrocytes of subjects with Lesch-Nyhan syndrome. *Biochem. Genet.* 6: 139-146.

Bakay, B. and W. L. Nyhan 1972-A. Activation of variants of hypoxanthine-guanine phosphoribosyl transferase by the normal enzyme. *Proc. Nat. Acad. Sci.* 69: 2523-2527.

Bakay, B., W. L Nyhan, N. Fawcett, and M. D. Kogut 1972. Iso-enzymes of hypoxanthine-guanine-phosphoribosyl transferase in a family with partial deficiency of the enzyme. *Biochem. Genet.* 7: 73-85.

Bakay, B., M. A. Telfer, and W. L. Nyhan 1969. Assay of hypo-xanthine-guanine and adenine phosphoribosyl transferase. *Biochem. Med.* 3: 230-243.

Benke, P. J. and N. Herrick 1972. Azaguanine-resistance as a manifestation of a new form of metabolic overproduction of uric acid. *Am. J. Med.* 52: 547-555.

Kelley, W. N., M. L. Greene, F. M. Rosenbloom, J. F. Henderson, and J. E. Seegmiller 1969. Hypoxanthine-guanine phosphoribosyltransferase deficiency in gout: A review. *Ann. Int. Med.* 70: 155-206.

Kogut, M. D., G. N. Donnell, W. L. Nyhan, and L. Sweetman 1970. Disorder of purine metabolism due to partial deficiency of hypoxanthine guanine phosphoribosyl transferase. A study of a family. *Am. J. Med.* 48: 148-161.

Lesch, M., and W. L. Nyhan 1964. A familial disorder of uric acid metabolism and central nervous system function. *Am. J. Med.* 36: 561-570.

McDonald, J. A. and W. N. Kelley 1971. Lesch-Nyhan syndrome: Altered kinetic properties of mutant enzyme. *Science* 171: 689-690.

Nyhan, W. L. 1973. The Lesch-Nyhan syndrome, in *Annual Review of Medicine*, Vol. 24, Creger, W. P., Editor, Annual Reviews, Inc., Palo Alto, Calif. pp. 41-60.

Nyhan, W. L., J. A. James, A. J. Teberg, L. Sweetman, and L. G. Nelson 1969. A new disorder of purine metabolism with behavioral manifestations. *J. Pediat.* 74: 20-27.

Rubin, C. S., J. Dancis, L. C. Yip, R. C. Nowinski, and M. E. Balis 1971. Purification of IMP: Pyrophosphate phosphoribosyltransferase, catalytically incompetent enzymes in Lesch-Nyhan disease. *Proc. Nat. Acad. Sci.* 68: 1461-1464.

Seegmiller, J. E., F. M. Rosenbloom, and W. N. Kelley 1967. Enzyme defect associated with a sex-linked human neurological disorder and excessive purine synthesis. *Science* 155: 1682-1684.

Sweetman, L., and W. L. Nyhan 1972. Further studies of the enzyme composition of mutant cells in X-linked uric aciduria. *Arch. Intern. Med.* 130: 214-220.

Tischfield, J. A., H. P. Bernhard, and F. H. Ruddle 1973. A new electrophoretic-autoradiographic method for the visual detection of phosphotransferases. *Analyt. Biochem.* 53: 545-554.

THE SIGNIFICANCE OF MULTIPLE MOLECULAR FORMS
OF ACETYLCHOLINESTERASE IN THE SENSITIVITY
OF HOUSEFLIES TO ORGANOPHOSPHORUS POISONING

R. K. TRIPATHI AND R. D. O'BRIEN

Section of Neurobiology and Behavior
Cornell University
Ithaca, New York
14850

ABSTRACT. Multiple forms of acetylcholinesterase from various strains of houseflies, *Musca domestica* (L.) were separated by polyacrylamide gel electrophoresis combined with
specific substrate staining and gel scanning procedures.
The head enzyme was studied most extensively and four isozymes were noted in fresh tissue. The isozymes differed
in their kinetic properties: Michaelis constants (K_m) for
acetylthiocholine varied over a 5-fold range and bimolecular rate constants (k_i) for various inhibitors varied
over a 2-fold range. The isozymes also differed in the
degree of inhibition by various organophosphates in vivo.
Thoracic isozymes were found to be more sensitive to inhibition than head isozymes at the LD_{50} dose of several organophosphates. The soluble isozyme mixture from a resistant strain of housefly exhibited a decrease in sensitivity
towards all five organophosphates examined and this decrease
in sensitivity was attributable to differences in their
affinity for the enzyme, as measured by the dissociation
constant K_d. Evidence was provided that the isozymal variations were epigenetic rather than genetic.

We shall use the term "isozymes" to mean multiple molecular
forms of enzymes as separated by electrophoresis. The existence
of isozymes of acetylcholinesterase (AChE) from vertebrate sources has been amply demonstrated (Hall, 1973; Massoulie and Rieger, 1969; Wilson et al., 1971; Bajgar and Zižkovsky, 1971).
The fact that acetylcholinesterase is the target for organophosphorus poisoning in vertebrates and also in insects raises
the question as to whether the various isozymes differ in their
importance as far as organophosphate poisoning is concerned.
We therefore embarked upon a study of the housefly, *Musca domestica* (L.), in order to find whether these differences were
of importance in the poisoning of the housefly by organophosphates.
    In our early work (Eldefrawi et al., 1970) we analyzed sol-
uble extracts of housefly heads on polyacrylamide gels, then

sliced the gels and assayed AChE by the Ellman method (Ellman et al., 1961). We found four peaks of activity. Evidence that they were all acetylcholinesterases was that butyrylthiocholine was hydrolyzed much more slowly than acetylthiocholine and that they were inhibited by organophosphates and by eserine. Subsequently we developed an improved detection procedure and used only female houseflies, Wilson strain, which were four days old. We shall report information only for the soluble fraction of the acetylcholinesterase, which represents 34% of the whole. Supernatant fractions of the enzyme were separated by polyacrylamide gel electrophoresis and the location of the enzyme bands determined by a modification of the technique of Karnovsky and Roots (1964). The gels were then scanned with an Acta III spectrophotometer.

Figure 1 shows that there are at least seven distinct isozymes, three being found in the thorax and four different ones in the head. We were concerned about the possibility that the different isozymes might each be associated with a subpopulation of the housefly sample rather than representing a group of isozymes in a particular individual. Therefore, in the case of the head isozymes, we performed micro-electrophoresis with individual housefly heads. We were unable to scan these on our spectrophotometer, but it was clear that precisely the same four head isozymes were present in the individual as were present in the whole population which we normally sampled.

The data which we shall present about these seven isozymes derive from the observation of Chiu, Tripathi, and O'Brien (1972) that kinetic studies with substrates and with inhibitors may be performed on the gels directly, without eluting the isozymes. Using this procedure, we showed (Tripathi et al., 1973) that fairly substantial differences existed between the reactivity of the isozymes towards acetylthiocholine and a variety of inhibitors. From now on we shall refer primarily to the isozymes of the head, which is the richest source and the easiest to prepare. Good Lineweaver-Burk plots were obtained for the housefly head, using acetylthiocholine as substrate, and the isozymes varied in their kinetic parameters. In Table I the values of the Michaelis constant and the percent contributions are shown, and it can be seen that the K_m values vary over a 5-fold range. The data also show that all except head isozyme IV make substantial contributions to the total activity. We have noted quite large differences between different housefly strains, as Table 2 shows.

The same technique permits the evaluation of the overall rate constant, k_i, governing the reaction between progressive inhibitors (such as organophosphates) and the individual isozymes on the gel. Table 3 shows the k_i values for three dif-

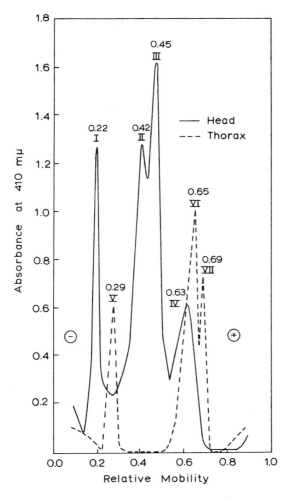

Figure 1. An electropherogram of AChE isozymes of housefly (Wilson strain). The gels were incubated with 1 mM ATCh for 45 min. for head and 90 min. for thorax. The number heading each peak is the R_m value.

ferent inhibitors: the organophosphates malaoxon and Tetram, and the carbamate eserine. Once again, there are easily measurable variations between the isozymes and their sensitivity, but their range is not large, the maximal value being a 2.3-fold range for malaoxon.

It seemed plausible that these modest differences could be coupled with other factors of importance in the intact housefly, particularly the location and availability of the isozyme

TABLE I. MICHAELIS CONSTANTS (K_m) AND MAXIMUM
VELOCITY (V_{max}) OF ATCh (ACETYLTHIOCHOLINE) FOR AChE
ISOZYMES OF HOUSEFLY AT 25°C, pH 6.0

Isozyme	10^4 K_m (M)	Head V_{max} (arbitrary units)	Fractional[a] activity	R_m[b]
I	1.83 (±0.31)	1.89 (±0.07)	0.33	0.22
II	6.87 (±1.42)	1.97 (±0.17)	0.22	0.42
III	4.35 (±0.16)	2.66 (±0.03)	0.38	0.49
IV	7.70 (±0.84)	0.56 (±0.03)	0.07	0.63
		Thorax		
V	1.58 (±0.23)	0.57 (±0.003)	0.25	0.29
VI	6.57 (±0.99)	1.69 (±0.14)	0.40	0.65
VII	3.86 (±1.03)	1.32 (±0.17)	0.35	0.69

Standard errors are shown in parentheses.

[a] Calculated on the basis of relative absorbancy from the stained zymogram with 1×10^{-3} M ATCh.

[b] R_m is expressed as relative mobility of each isozyme, calculated by the ratio of isozyme to the tracking dye migration.

to inhibition. We therefore poisoned houseflies by a variety of organophosphates and traced out the consequences of their inhibition at a variety of times (Tripathi and O'Brien, 1973). Figure 2 shows an example of the results, using the organophosphate paraoxon, which was applied to the tip of the abdomen. As can be seen, the isozymes varied a great deal in their sensitivity to inhibition. One of them (Thoracic VII) was totally inhibited in 80 minutes and did not recover. I should point out that the dosage used was always the LD_{50}, which is to say the dose which caused 50% mortality in 24 hours. Consequently it follows that isozyme VII was fully inhibited in both survivors and victims for the period from 80 minutes to 1280 minutes (21 hours) so that it is apparent that half of the population was able to survive total inhibition of isozyme VII. The implication is that isozyme VII is not very important in poisoning. At the other end of the spectrum, head isozyme III was very little affected in the course of poisoning, so that it is probable that it also was relatively unimportant in the poisoning process. We performed such studies with four different organophosphates, all used at the LD_{50} dose. It seemed

TABLE 2. MICHAELIS CONSTANTS (K_m) AND MAXIMUM VELOCITIES (V_{max}) [*]
OF ATCh FOR AChE ISOZYMES FROM THREE STRAINS
OF HOUSEFLIES, AT 25°C, pH 6.0

Isozyme	KDRO Stock		KDRO-w		CSMA	
	$10^4 \times K_m$	V_{max}	$10^4 \times K_m$	V_{max}	$10^4 \times K_m$	V_{max}
I	1.06 (±0.30)	2.07 (±0.06)	4.07 (±0.42)	1.85 (±0.06)	5.19 (±0.37)	2.09 (±0.06)
II	2.09 (±0.37)	0.78 (±0.03)	3.67 (±0.75)	0.66 (±0.04)	4.96 (±0.41)	0.96 (±0.03)
III	1.93 (±0.29)	1.68 (±0.06)	3.54 (±0.24)	1.13 (±0.02)	3.47 (±0.40)	1.22 (±0.04)
IV	4.13 (±0.97)	0.72 (±0.06)	3.56 (±0.59)	1.53 (±0.08)	3.47 (±0.46)	1.02 (±0.04)

Standard errors are shown in parentheses.

[*] V_{max} in arbitrary units

TABLE 3.   BIMOLECULAR RATE CONSTANTS (k_i) OF AChE ISOZYMES
OF HOUSEFLY HEAD WITH MALAOXON,
TETRAM, AND ESERINE AT 25°C, pH 6.0

Isozyme	Malaoxon	Tetram	Eserine
	$10^{-5} \times k_i$ $(M^{-1} min^{-1})$		
I	7.85 (±0.37)	8.80 (±0.63)	25.46 (±1.22)
II	5.34 (±0.23)	8.50 (±0.17)	22.24 (±1.59)
III	4.86 (±0.22)	7.60 (±0.29)	15.99 (±0.77)
IV	3.42 (±0.45)	5.90 (±0.21)	29.71 (±2.10)

Standard errors are shown in parentheses.

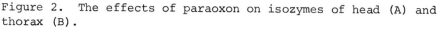

PARAOXON : LD_{50} 4.0 µg/g

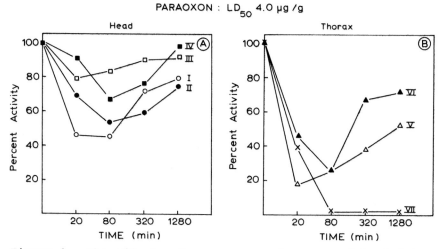

Figure 2.   The effects of paraoxon on isozymes of head (A) and
thorax (B).

plausible that, if one particular isozyme was of great importance, death should be associated with the fact that it is inhibited to some fixed and important extent.   Table 4 shows for
all seven isozymes the extent of inhibitions, expressed as
minimal enzyme activity, for all four compounds.   The column
on the right shows the range of minima, and for most isozymes
the range was rather large; for example for isozyme I the
range was 36 percentage points, which is obtained by subtrac-
ting 18 from the malaoxon column from 54 in the diazinon column.   However, thoracic isozyme V shows minimal variation,
there being only 7 percentage points difference in the extents
of inhibition with these four quite different organophosphates.

TABLE 4. THE MINIMAL PERCENT ACTIVITY OF AChE AFTER POISONING
WITH AN LD_{50} DOSE

Isozyme	Malaoxon	Paraoxon	Diazinon	Dichlorvos	Range[a]
		Head			
I	18	45	54	39	36
II	42	53	28	72	44
III	51	78	32	57	46
IV	28	67	67	70	42
Total[b]	36	61	41	55	25
		Thorax			
V	15	18	22	21	7
VI	5	38	1	16	37
VII	1	1	1	1	0
Total[b]	6	20	7	12	14

[a] Range of minimal activities shown in table.

[b] Equals $\Sigma \underline{fm}$, where $\underline{f}$ is the fractional activity of each isozyme (from Table 1) and $\underline{m}$ is the percentage minimal activity
(from this table).

These limited data therefore suggest that this isozyme is the
most crucial one in organophosphorus poisoning.

At this time we were given a population of houseflies which
exhibited unusual resistance to poisoning by the organophosphate Rabon in the field. Rabon has the structure $(CH_3O)_2P(O)$
$OC(C_6H_4Cl_3)=CHCl$. Groups were selected in the following way.
One group, which we call Cornell susceptible, was bred without contact with Rabon over twenty generations. The other
group, which we call Cornell resistant, was continuously selected at the larval stage over twenty generations with increasing concentrations of Rabon, which was finally at 60 parts
per million in the larval medium. We found that Rabon was
more than 1538 times more toxic to the susceptible than the

resistant group.

Before describing the studies on the sensitivity of the acetylcholinesterase to inhibition, it should be noted that the inhibition process involves two steps as shown in the following scheme.

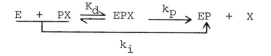

$$E + PX \xrightleftharpoons{K_d} EPX \xrightarrow{k_p} EP + X$$
$$\underbrace{}_{k_i}$$

The first is an affinity step governed by dissociation constant K_d. A very small K_d means that the inhibitor has a high affinity for the enzyme surface. The second is a phosphorylation step, governed by a rate constant, variously called k_p or k_2. A very high k_p means that the ability of bound inhibitor to phosphorylate the enzyme's active site is very high. The measurement of K_d and k_p separately requires rather special techniques (Hart and O'Brien, 1973). For the overall process it is relatively easy to measure the potency by an apparent rate constant, k_i, which in fact can be shown to be the ratio of the values k_p to K_d.

Table 5 shows that the overall potency of five different organophosphates for the isozyme mixture in solution showed the enzyme from resistant flies to be substantially less sensitive than that from susceptible flies and the sensitivity ratio was particularly high for Rabon, for which resistance was maximal. Two lines of evidence show that all the isozymes are converted from a sensitive to a relatively insensitive form. Firstly, if any one isozyme had retained its sensitivity, then the kinetics of inhibition would be extremely nonlinear, because at the high concentrations of inhibitor used the sensitive isozyme would become fully inhibited, and then subsequently the other isozymes would be slowly inhibited. Such nonlinear kinetics were not observed. Secondly, inhibition studies performed directly on the gel showed that every isozyme in the resistant strain had become insensitive to approximately the same degree. Since it is very unlikely that four different genes would all have mutated to produce precisely the same result, we must conclude that the catalytic site of all four isozymes is under the control of a single gene, and that the variety of isozymes is epigenetic rather than genetic.

The isozyme scans of the acetylcholinesterase from the resistant and susceptible strains are shown side-by-side in Figure 3, and although there are some moderate differences in the peak heights, the similarities are far more striking than the differences. A rather more quantitative statement of this

TABLE 5. BIMOLECULAR REACTION CONSTANTS (k_i) FOR
THE INHIBITION OF SUSCEPTIBLE (S) AND
RESISTANT (R) HOUSEFLY BRAIN AChE AT 25°C, pH 7.4

$$k_i \ (M^{-1} \ min^{-1})$$

Compounds	S	R	Ratio	Resistance Factor
Rabon	12.30×10^6	5.96×10^4	206	>1500
Paraoxon	3.19×10^6	0.34×10^5	94	17.0
Dichlorvos	1.16×10^7	0.99×10^5	117	16.0
Diazoxon	0.62×10^8	0.56×10^7	11	9.0
Tetram	2.40×10^7	0.345×10^7	7	-

fact is provided in Table 6, which indicates the Michaelis constant and the maximum velocity for the different isozymes from
the susceptible and resistant strains. The amounts of the

TABLE 6. MICHAELIS CONSTANTS (K_m) AND MAXIMUM VELOCITY (V_{max})
OF ATCh BY SOLUBLE BRAIN AChE FROM SUSCEPTIBLE AND
RESISTANT STRAINS OF HOUSEFLY AT 25°C, pH 7.4

Constant	S	R	Ratio
K_m	0.95×10^{-5} M	3.26×10^{-5} M	3.4
V_{max} (μmoles/hr/mg protein)	18.5	68.3	3.7

enzyme are not drastically changed, nor is the Michaelis constant, and consequently the affinity of the substrate for its
enzyme is probably little changed by the mutation.

Let us now explore in more detail the basis for the difference in sensitivity to inhibition associated with the mutation
which we shall tentatively assume to be the basis of the difference. For the isozyme mixture, we have been able to separate out the effects of k_p, the phosphorylation constant, and
K_d, the affinity constant. Table 7 shows that the effects of
the mutation on the phosphorylation constant are not very
large, being typically a 3-fold increase in sensitivity to the
phosphorylation step which is observed in the resistant enzyme. Tetram behaves in an anomalous way, as we shall discuss

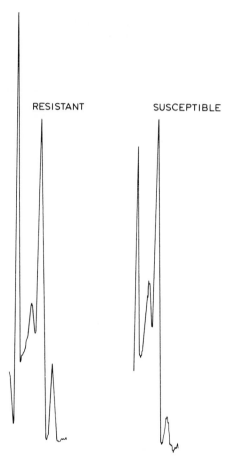

Figure 3. An electropherogram of AChE isozymes from Cornell susceptible and Cornell resistant strains of houseflies.

---

below. When we turn to the changes in dissociation constant (Table 8) it is apparent that this is the sole cause of the enzyme insensitivity which is observed. For the compound Rabon, to which the houseflies of the resistant strain are most resistant, there is an astonishing 573-fold reduction in the affinity of the inhibitor for the enzyme. Similar very large reductions in affinity for paraoxon and dichlorvos are seen and more moderate reductions for diazoxon and Tetram. It is of great interest to note that for the compound Tetram there was substantial affinity change. It should be noted that Tetram is entirely unlike the other organophosphates in that its side group is charged and is quite similar to the natural

TABLE 7. PHOSPHORYLATION CONSTANTS (k_p) OF SUSCEPTIBLE AND RESISTANT HOUSEFLY BRAIN AChE AT 25°C, pH 7.4

Compounds	S	k_p (min-1) R	RATIO
Rabon	0.59	1.64	2.7
Paraoxon	0.87	3.00	3.4
Dichlorvos	0.89	2.94	3.3
Diazoxon	0.83	2.57	3.0
Tetram	0.399	4.84	12.1

substrate acetylcholine. It follows that the mutation which has occurred has had relatively little effect upon the ability to bind acetylcholine (as shown by the previous demonstration of the small effect upon K_m) but quite a large effect upon the binding of Tetram. We may have to discard the earlier simple view that Tetram and acetylcholine bind to similar and presumably anionic sites on the surfaces of the enzyme, whereas the other organophosphates bind to a quite separate area. But it is clear that it is the organophosphate binding site which is primarily affected by the mutation, either by the change in an amino acid residue or by a difference in folding of the enzyme which reduces the effectiveness of that particular binding site.

It is entirely to be expected that results of this kind should be found. It would be remarkable if an important regulatory enzyme like acetylcholinesterase could undergo a mutation which would alter the ability of the enzyme to work on its natural substrate. But numerous inhibitors make obligatory use of binding sites not employed by the natural substrate, and consequently the possibility exists to develop a mutant enzyme which is substantially unaltered as far as the natural substrate is concerned. In view of the fact that the mutation has clearly altered the structure of the enzyme quite profoundly, but has had no substantial effect on the amount of the enzyme as a whole, or of any of the isozymes within it, it is clear that the single gene which mutated is a structural gene and not a regulatory gene.

In summary, acetylcholinesterase isozymes do occur in houseflies, and significant differences do exist among the different isozymes as far as the poisoning process by organophosphates is concerned. A fortunate discovery with respect to resistance has provided evidence that the catalytic sites of all four isozymes, in this instance, involve secondary factors. Possibly the different isozymes represent differing ratios of catalytic subunits with non-catalytic subunits, or

TABLE 8. DISSOCIATION CONSTANTS (K_d) OF SUSCEPTIBLE AND
RESISTANT HOUSEFLY BRAIN AChE AT 25°C, pH 7.4

Compound	Structure	K_d, µM		Ratio
		S	R	
Rabon	$(MeO)_2POC$—⟨ring⟩ Cl, Cl, Cl, CHCl	0.048	27.5	573
Paraoxon	$(EtO)_2PO$ ⟨ring⟩ NO_2	0.27	86.9	322
Dichlorvos	$(MeO)_2POCH=CCl_2$	0.077	29.5	383
Diazoxon	$(EtO)_2PO$—⟨pyrimidine, iPr, CH_3⟩	0.0134	0.463	36
Tetram	$(EtO)_2PSCH_2CH_2NEt_2 \cdot {}^+H$	0.0166	1.404	84

possibly they represent different aggregates of catalytic subunits. This possibility is being explored further.

ACKNOWLEDGEMENTS

We thank Shell Chemical Company for supplying Rabon and
dichlorvos, American Cyanamid for paraoxon and malaoxon, and
CIBA-GEIGY Chemical Corporation for diazinon and diazoxon.
This work was supported by U. S. Public Health Service Grant
ES 00901 and Training Grant ES 98.

REFERENCES

Bajgar, J. and V. Zižkovsky 1971. Partial characterization
of soluble acetylcholinesterase isoenzymes of the rat brain.
J. Neurochem. 18: 1609-1614.

Chiu, Y. C., R. K. Tripathi, and R. D. O'Brien 1972. A gel scanning method for kinetic studies on an acetylcholinesterase isozyme. *Analyt. Biochem.* 45: 480-487.

Eldefrawi, M. E., R. K. Tripathi, and R. D. O'Brien 1970. Acetylcholinesterase isozymes from the housefly brain. *Biochem. Biophys. Acta.* 212: 308-314.

Ellman, G. L., K. D. Courtney, V. Andres Jr., and R. M. Featherstone 1961. A new and rapid colorimetric determination of acetylcholinesterase activity. *Biochem. Pharmacol.* 7: 88-95.

Hall, Z. W. 1973. Multiple forms of acetylcholinesterase and their distribution in endplate and non-endplate regions of rat diaphragm muscle. *J. Neurobiol.* 4: 343-361.

Hart, G. J. and R. D. O'Brien 1973. Recording spectrophotometric method for determination of dissociation and phosphorylation constants for the inhibition of acetylcholinesterase by organophosphates in the presence of substrate. *Biochem.* 12: 2940-2945.

Karnovsky, M. J. and L. Roots 1964. A "direct-coloring" thiocholine method for cholinesterases. *J. Histochem. Cytochem.* 12: 219-221.

Massoulie, J. and F. Rieger 1969. L'acetylcholinesterase des organes électriques de poissons (Torpille et gymnoté); complexes membraniers. *Eur. J. Biochem.* 11: 441-455.

Tripathi, R. K., Y. C. Chiu, and R. D. O'Brien 1973. Reactivity in vitro toward substrate and inhibitors of acetylcholinesterase isozymes from electric eel electroplax and housefly brain. *Pestic. Biochem. Physiol.* 3: 55-60.

Tripathi, R. K. and R. D. O'Brien 1973. Effects of organophosphates in vivo upon acetylcholinesterase isozymes from housefly head and thorax. *Pestic Biochem. Physiol.* 2: 418-424.

Wilson, B. W., J. L. Schenkel, and D. F. Fry 1971. Acetylcholinesterase isozymes and the maturation of normal and dystrophic muscle in *Cholinergic Ligand Interaction* (D. J. Triggle, J. F. Moran, and E. A. Barnard, Eds.) p. 137 Academic Press, New York.

LOCALIZATION OF CREATINE KINASE ISOZYMES IN MUSCLE CELLS: PHYSIOLOGICAL SIGNIFICANCE

HANS M. EPPENBERGER, THEO WALLIMANN,
HANS J. KUHN, and DAVID C. TURNER
Institute for Cell Biology
Swiss Federal Institute of Technology
CH - 8006 Zurich,
SWITZERLAND

ABSTRACT: Further evidence is presented that a small but significant fraction (ca. 7%) of the MM isozyme of creatine kinase (CPK) in chicken skeletal muscle cells is tightly bound to the M-line, a specific region of the myofibrillar contractile apparatus. Unlike chicken skeletal muscle, chicken heart contains almost exclusively the BB-CPK isozyme. About 2% of the BB-CPK in chicken heart binds tightly to the myofibrillar fraction. Indirect fluorescence antibody staining reveals that the BB-CPK bound to heart myofibrils is located at the Z-line regions. Very occasionally, bound CPK can also be detected at the M-lines of heart myofibrils; there are indications that this M-line CPK is the MB isozyme, traces of which have been detected electrophoretically in heart extracts. The possible functions of the different CPK isozymes bound at distinct myofibrillar locations are discussed.

INTRODUCTION

Working muscle cells must continuously regenerate the ATP used in the contraction process. In vertebrates this is accomplished by a phosphotransferase reaction: ADP produced by the action of the myosin ATPase reacts with the phosphagen, phosphocreatine, to form ATP plus creatine. ATP is thus maintained at a high level. The enzyme responsible for catalyzing this important reaction is creatine kinase (CPK).

Most localization studies involving either histochemistry (Sherwin et al., 1969; Khan et al., 1971) or cell fractionation (Kleine, 1965; Ottaway, 1967) have indicated that the bulk of muscle CPK is cytosolic, with lesser amounts associated with mitochondria (particularly in heart) or with the sarcoplasmic reticulum. CPK has thus generally been considered to be a "soluble enzyme". Over the years, however, several observations have suggested that at least some of the CPK in skeletal and heart muscle cells might be bound to elements of the contractile apparatus (Szorenyi and Degtyar 1948; Strohman, 1959; Yagi and Noda, 1960; Yagi and Mase, 1962; Ottaway, 1967;

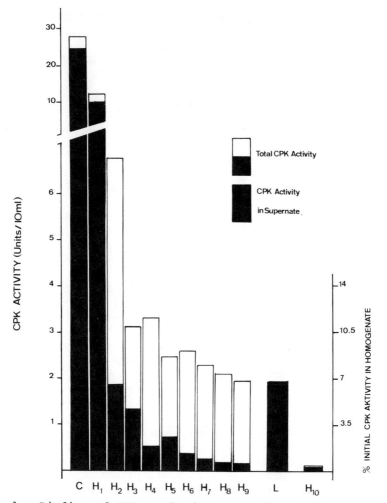

Fig. 1.   Binding of CPK to the low-speed (ca. 1000 x g) pellet
fraction (predominantly myofibrils) from adult chicken skeletal
muscle.   Enzyme activity measured in the total suspension after
each treatment (black plus white) is compared to the activity
subsequently recovered in the supernate (black).   C: crude homogenate (1/100, ^{W}/v) in high ionic strength buffer (0.02 M phosphate buffer containing 0.1 M KCl and 0.001 M dithiothreitol).
H_1-Hg: successive washes in high ionic strength buffer; the
later washes release little CPK into the supernate.   L: extraction (45 min) at low ionic strength (0.005 M Tris-HCl, containing 0.001 M dithiothreitol, pH 7.7); bound CPK is released.
A_{10}:   final high ionic strength wash; little CPK remains, either
bound or free.

Botts and Stone, 1968; Baskin and Deamer, 1970). Intrigued by the possible physiological implications of a close inter-relationship between the ATP-hydrolyzing and ATP-regenerating systems, we decided to reinvestigate the localization of CPK within muscle cells.

We were also aware that a number of myofibrillar proteins of unknown function were being isolated and characterized in several laboratories, and we were "on the lookout" for a myo-fibrillar protein with properties similar to MM-CPK, the only isozyme of CPK found in mammalian and avian skeletal muscle. For further information concerning the different CPK isozymes, see the paper by Turner (1974). MM-CPK had been purified to homogeneity from several mammalian (Kuby et al., 1954; Eppen-berger et al., 1967) and avian (Eppenberger et al., 1967; Roy et al., 1970) sources, and characterized as a dimer having a molecular weight of about 84,000 and possessing two identical subunits. Thus, when Morimoto and Harrington (1972) published their purification and characterization of an 88,000-dalton dimeric protein isolated from the M-line regions of skeletal muscle myofibrils, it was not sheer serendipity that led us to propose that their M-line protein was in fact MM-CPK.

Comparison of other published data for the two proteins (especially amino acid composition and solubility properties) provided strong support for the identity of the two protein preparations (Turner et al., 1973). Furthermore, MM-CPK pur-ified according to our earlier procedure (Dawson and Eppen-berger 1970) was compared directly with M-line protein purified according to the very different procedure of Morimoto and Harrington (1972). The results (summarized in TABLE I) provide conclusive evidence that the M-line protein of Morimoto and Harrington is really MM-CPK (Turner et al., 1973). The pre-sence of the MM-CPK isozyme within the M-line region of each sarcomere was further demonstrated by the indirect immuno-fluorescence method (Turner et al., 1973). Isolated myofibrils reacted with antiserum directed against MM-CPK show a regular pattern of bright fluorescent lines running through the middle of each A-band (Fig. 2a).

Yagi and Mase (1962), Botts and Stone (1968), and Morimoto and Harrington (1972) had shown that MM-CPK (M-line protein) binds to reconstituted myosin filaments. Any protein which binds to myofibrillar myosin only at the M-line regions would be expected to bind to the rod portions of the myosin molecules, since the central regions of the thick filaments (through which the M-lines pass) are devoid of myosin "heads". Recently, a specific interaction of MM-CPK with the rod portion of the myosin molecule has been demonstrated (Houk and Putnam, 1973).

While further studies of the binding of CPK to isolated myofibrillar proteins are obviously of great interest, we

TABLE I

SUMMARY OF EVIDENCE FOR THE IDENTITY
OF M-LINE PROTEIN AND MM-CPK

Similar molecular weight[a][b]
Similar amino acid composition[a][c]
Similar extractability at low ionic strength[a][d][e]
Same subunit composition: two identical subunits[a][b]
Similar specific enzyme activity[e]
Same electrophoretic mobility:
      cellulose acetate[e]
      polyacrylamide[e]
      SDS-polyacrylamide (subunits)[f]
Same pattern of cyanogen bromide fragments[f]
Immunological identity in Ouchterlony test[e]
Same myofibrillar localization[a][e]

---

[a] Morimoto and Harrington (1972)
[b] Dawson et al.,(1967)
[c] Eppenberger et al., (1967)
[d] Ottaway (1967)
[e] Turner et al.,(1973)
[f] Wallimann et al., (1974)

---

turned our attention initially to the specificity of the
interaction of CPK with intact myofibrils. We started with
two questions: 1) Is the observed binding of MM-CPK to myofibrils physiological, or are the M-line regions just "sticky"?
Although our findings (Turner et al., 1973) when taken together with evidence accumulated for M-line protein by others
(Knappeis and Carlsen, 1968; Kundrat and Pepe, 1971; Morimoto
and Harrington, 1972; Eaton and Pepe, 1972) had strongly suggested that MM-CPK is an integral component of the myofibril,
it was still unclear what fraction of the CPK in muscle is
tightly bound to the contractile elements and under what conditions this fraction remained bound. 2) Is the binding of CPK
specific for the MM-CPK isozyme? In electron micrographs of
chicken heart muscle, which contains BB-CPK as the predominant
isozyme at all stages of development, no M-lines are detectable
(Sommer, 1969). But does this mean that no CPK is bound to the
myofibrils in chicken heart muscle? Before attempting to
assess whether myofibrillar CPK might possibly have a function
distinct from that of the bulk muscle CPK, it would obviously
be of interest to know whether myofibrils with no bound CPK

412

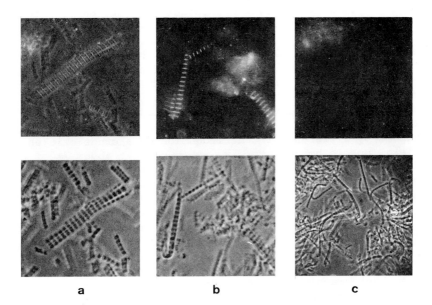

a   b   c

Fig. 2. Detection of MM-CPK in the A-band regions of chicken skeletal muscle myofibrils by the indirect fluorescent anti-body technique (Turner et al., 1973). Fluorescence (top row) and corresponding phase contrast (bottom row) photographs are shown for: a) once-washed, b) 9 times-washed, and c) once-extracted myofibrils. Extensive washing with high ionic strength buffer (0.1 M KCl in 0.02 M phosphate buffer, 0.001 M dithiothreitol, pH 7.0) does not remove the bound CPK (a,b), while brief (45 min) extraction in low ionic strength buffer (0.005 M Tris-HCl, 0.001 M dithiothreitol, pH 7.7) completely removes the bound enzyme (c). Fluorescence and phase contrast exposures have been superimposed in a (top) in order to show that MM-CPK is localized at the middle of each A-band. a,b: x 1700 ; c: x 800.

exist.

METHODS AND RESULTS

FURTHER CHARACTERIZATION OF THE BINDING OF MM-CPK TO SKELETAL MUSCLE MYOFIBRILS

The first steps in the Morimoto-Harrington (1972) procedure for purifying M-line protein from a muscle mince are the

following: 1) exhaustive washing at relatively high ionic
strength (0.1 M KCl in 0.02 M phosphate buffer, pH 7.0) to remove all soluble proteins, hereinafter called "washing"; and
2) extraction of M-line protein at relatively low-ionic
strength (5 mM Tris-CHl buffer, pH 7.7), hereinafter termed
"extraction". Using this procedure we found that approx. 3/4
of the CPK ultimately recoverable from the mince was removed
during the extensive washings, while the remaining 1/4 was extracted in the second step (Turner et al., 1973). When a
similar experiment was performed with the myofibrillar fraction
obtained by differential centrifugation of a crude chicken
skeletal muscle homogenate, it was found that about 7% of the
total CPK activity present in the homogenate remained bound to
myofibrils after extensive washing (Fig. 1). The bound CPK
could be quantitatively released by a single extraction at low
ionic strength (Fig. 1). FIGURES 2a-c show myofibrils stained
for MM-CPK with the indirect fluorescent antibody method after
one (a) and nine (b) washes, as well as after complete extraction by low ionic strength buffer (c). No MM-CPK is detectable
in the M-line region after extraction (Fig. 2c). In Fig. 2a
(upper photograph) phase contrast and fluorescence exposures
have been superimposed, showing that the MM-CPK is bound at
the center of the dark-appearing A-bands, where the M-lines are
known to be located. We have further demonstrated directly,
using the ferritin antibody method, that MM-CPK is found
throughout the disc-shaped M-line visible in the electron microscope (Wallimann et al., 1974). Electron microscope studies
also show that the M-line is no longer visible after extraction
of MM-CPK (Morimoto and Harrington, 1972; Wallimann et al.,
1974). Preincubation of washed myofibrils with anti-MM-CPK
antiserum prevents the extraction of MM-CPK (Wallimann et al.,
1974), confirming a similar finding of Morimoto and Harrington,
(1972) for their M-line protein. Our results thus strongly
indicate, in agreement with the work of others (Knappeis and
Carlsen, 1968; Kundrat and Pepe, 1971; Morimoto and Harrington
1972; Eaton and Pepe, 1972), that a small but significant
fraction of the MM-CPK in skeletal muscle cells regularly contributes to the structure of the myofibrillar M-lines.

The situation in chicken heart muscle. The BB isozyme is,
except for possible traces of MB-CPK (see below), the only CPK
isozyme present in chicken heart. Like MM-CPK, BB-CPK is a
homodimer of 84,000 molecular weight (Dawson et al., 1967).
The two types of subunit, M and B, have very different amino
acid compositions and are clearly the products of different
genes (Eppenberger et al., 1967). Antibodies elicited against
MM-CPK do not cross-react with BB-CPK and vice versa (Eppenberger

et al., 1967), though antisera against either homodimer will
precipitate the heterodimer, MB, small amounts of which have
sometimes been reported to be present in electrophoreses of
chicken heart extracts (Dawson et al., 1965; Eppenberger et
al., 1970). The three CPK isozymes from chicken are readily
separable in a variety of electrophoretic systems (see, e.g.,
Turner, 1974).

When heart myofibrils are washed repeatedly with the same
buffer used for skeletal muscle myofibrils, about 2% of the CPK
activity present in the crude heart homogenate is found to be
tightly bound (Fig. 3). As is the case with skeletal muscle
myofibrils (Fig. 1), this activity can be released by low ionic
strength extraction (Fig. 3). When immunofluorescence with
anti-BB-CPK antiserum was used to localize the BB-CPK bound to
heart myofibrils, a regular pattern of fluorescent cross-
striations was observed - but in the I-bands and not in the
A-bands (Fig. 4a,b). This result was quite unexpected.
Electron micrographs of heart myofibrils reacted in the indirect ferritin antibody procedure for BB-CPK revealed that the
BB-CPK is bound at the Z-line in the center of each I-band.
Extraction (90 min) removes all of the BB-CPK bound at the
Z-lines (Fig. 4c); extraction can be prevented by prior reaction of the bound CPK with anti-BB-CPK antiserum (Wallimann et
al., 1974). Since the Z-line contains no myosin, the binding
of BB-CPK to myofibrils must differ from that of the MM isozyme. Compared to skeletal muscle myofibrils, heart myofibrils
contain only about 1/5 as many active CPK molecules on a weight
basis (Wallimann et al., 1974). Further characterization of
the binding of BB-CPK to heart myofibrils as well as to isolated myofibrillar components will be necessary before an
attempt to explain its physiological significance (if any) can
be made.

Heart myofibrils reacted with anti-BB-CPK antiserum occasionally show fluorescence in the H-bands (where the M-lines
would be if there were any) in addition to the usual Z-line
fluorescence (Fig. 5b, TABLE II). In addition, control incubations with anti-MM-CPK of heart myofibrils infrequently revealed fluorescence in the H-band regions, though never in the
Z-lines (Fig. 5a, TABLE II). The most plausible suggestion
we can offer at this time to explain these exceptional cases
is that MB-CPK, which would react with either antiserum, may
be present in the H-band (M-line) regions of some chicken heart
myofibrils. If this explanation is correct, it remains to be
seen whether this isozyme is specific for a particular heart
region or cell type. It should also be noted that skeletal
muscle myofibrils reacted with anti-BB-CPK antiserum often show
faint fluorescence in the Z-line regions (TABLE II). This may

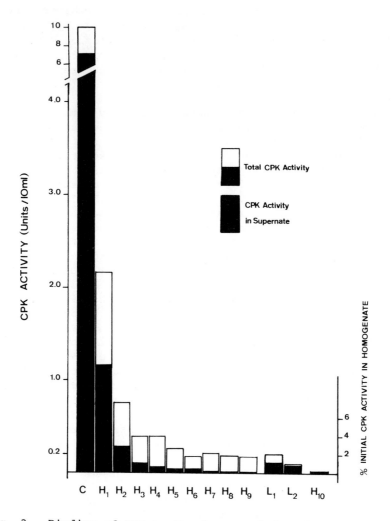

Fig. 3. Binding of CPK to the low-speed (ca. 1000 x g) pellet fraction from adult chicken heart muscle. Measurements and designations as in Fig. 2, except that two low ionic strength extractions of 45 min each (L_1 and L_2) were necessary in order to release the bound CPK.

be due to small amounts of BB-CPK not detected in electrophoreses.

Binding of heterologous isozymes to skeletal and heart muscle myofibrils. Because skeletal and heart muscle in the chicken each contain a single predominant isozyme, our information on

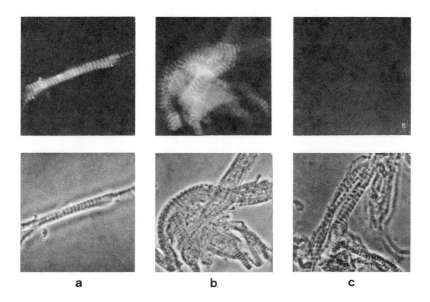

Fig. 4. Detection of BB-CPK bound in a regular pattern to chicken heart myofibrils by the indirect fluorescent antibody technique (Turner et al., 1973). Fluorescence (top row) and corresponding phase contrast (bottom row) photographs are shown for: a) once-washed, b) 9 times-washed, and c) once-extracted myofibrils. Conditions as in Fig. 2. BB-CPK remains bound at the I-band regions even after extensive washing (a,b), but is completely extracted after 90 minutes at low ionic strength (c). x 1700.

the isozyme specificity of binding is still incomplete. Does MM-CPK, when incubated with heart myofibrils, bind at the H-band regions and only there, as we might expect from the experiments already presented? Does BB-CPK bind at the Z-lines of skeletal muscle myofibrils? Before describing the experiments we have carried out, it should be noted that, although MM-CPK was known to bind to isolated skeletal muscle myosin (see above), we knew it to be possible that other proteins (e.g., the second M-line protein described by Eaton and Pepe, 1972) might be required as "linkers" in order to effect binding of either or both isozymes to myofibrils.

Washed, but not extracted, myofibrils from chicken heart or skeletal muscle were incubated with purified BB- or MM-CPK, again washed to remove non-specifically absorbed antigen, and

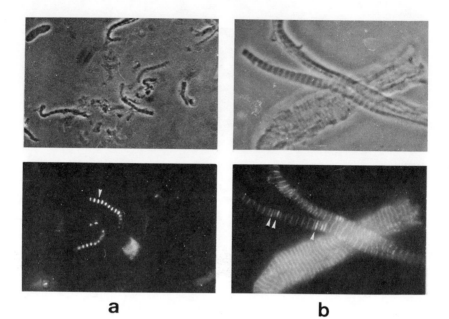

a   b

Fig. 5. Exceptional binding of CPK at the M-line region (center of A-band: arrows) in heart myofibrils, detected by the indirect fluorescent antibody procedure in a small fraction of the isolated washed myofibrils. The M-line CPK reacts with antisera against both MM (a) and BB (b) CPK isozymes, and is therefore suspected to be MB-CPK (see text). a: x 800, b: x 1700.

then reacted with the corresponding antiserum in the indirect fluorescent antibody procedure. The results are summarized in TABLE II. We see that purified MM-CPK does bind to heart myofibrils at the H-band (M-line) region, indicating that heart myofibrils possess binding sites for a CPK isozyme that is never present in heart cells. Electron micrographs of heart myofibrils sectioned and stained after incubation with MM-CPK show electron dense material in the H-bands (Wallimann et al., 1974). In untreated controls no such H-band material could be observed. It is unlikely, however, that these findings reflect formation of a complete M-line structure. Besides the fact that the H-band material is more diffuse than in skeletal muscle M-lines, there is considerable evidence for another M-line protein in addition to MM-CPK (Knappeis and Carlsen, 1968, Eaton and Pepe, 1972). We see further (TABLE II) that BB-CPK binds to skeletal muscle myofibrils at the Z-lines. On the other hand, the results shown in TABLE II also give a first

TABLE II

LOCALIZATION OF CPK-ISOZYMES BY THE INDIRECT IMMUNOFLUORESCENCE METHOD[a] IN ISOLATED WASHED MYOFIBRILS FROM ADULT CHICKEN AND AFTER ADDITIONAL INCUBATION WITH PURE MM- AND BB-CPK

| | Skeletal muscle | | | | Heart muscle | | | |
| | M-line | | Z-line | | M-line region[b] | | Z-line | |
	M-CPK[c]	B-CPK[c]	M-CPK	B-CPK	M-CPK	B-CPK	M-CPK	B-CPK
Native (no incubation)	+++	-	-	+[d]	++[e]	++[e]	-	++
Incubation with purified MM-CPK (1 mg/ml)	+++		+		+++	+	+	
Incubation with purified BB-CPK (1 mg/ml)		+|		++		+		+++

a) Turner et al., (1973).
b) No M-line structure present.
c) M-CPK refers to fluorescence detected after reaction with anti-MM-CPK antiserum; since this antiserum is subunit-specific, the fluorescence could be due to MM- or MB-CPK, or both. Similarly, B-CPK refers to fluorescence detected after reaction with anti-BB-CP antiserum and could reflect the presence of BB- or MB-CPK, or both.
d) Very weak, probably MB-CPK, see fig. 5.
e) Only occasionally; but then strong fluorescence, probably MB-CPK, see fig. 5.

indication that the binding of the MM and BB isozymes is not absolutely specific for the M-line and Z-line regions, respectively. Weak fluorescence is observed both at the Z-lines in heart myofibrils incubated with MM-CPK and at the M-lines in skeletal muscle myofibrils incubated with BB-CPK.

RECONSTITUTION EXPERIMENTS

The foregoing studies were extended to myofibrils of both skeletal and heart muscle from which the bound CPK had been completely extracted (as shown in Figs. 2c and 4c). When such extracted myofibrils were incubated with purified MM- and/or BB-CPK isozymes, a complex pattern of fluorescent cross-striations was found. Though there was preferential binding of MM-CPK to the M-line regions and of BB-CPK to the Z-lines with extracted myofibrils from both heart and skeletal muscle, both isozymes bound at both the M- and Z-lines. Moreover, both glycerinization prior to extraction and the duration of the extraction influenced the subsequent binding of the isozymes. A detailed report of these results will be published elsewhere. (Wallimann et al., 1974).

SUMMARY AND DISCUSSION

Chicken heart myofibrils evidently function without CPK bound to the M-line, just as skeletal muscle myofibrils function without CPK at the Z-line; neither M-line nor Z-line bound CPK, therefore, can be said to be required for muscle function in general. Although the patterns of binding of CPK isozymes to isolated unextracted myofibrils are distinct and highly specific (MM-CPK localized at M-lines in skeletal muscle, BB-localized at Z-lines in heart), the lesser degree of specificity observed in studies of the binding of exogenous isozymes to both washed and fully extracted myofibrils points to the need for detailed examination of interactions of CPK isozymes with myofibrillar components from different sources. A full appreciation of the function of myofibrillar CPK will clearly depend upon better knowledge of such interactions. We must ask ourselves, furthermore, whether all of the binding to myofibrils is physiological. Clearly more experiments are needed. In view of the large "background" of unbound (cytosolic) CPK molecules, however, it will be difficult to demonstrate CPK binding to myofibrils without some preliminary washing procedure. One must try (as we have) to use detection methods that do not introduce "unphysiological" conditions.

WHAT IS THE FUNCTION OF MYOFIBRILLAR CPK?

We have considered the following possibilities: 1) CPK bound at either the M-line or the Z-line allows for more efficient muscle function because of the propinquity of ATP-regenerating and ATP-hydrolyzing sites. 2) MM-CPK at the M-line is important in establishing the proper register of myosin molecules for construction of the thick filament. 3) MM-CPK bound at the M-line is important as a "structural protein" in maintaining the integrity of the already-constructed thick filament. 4) MM-CPK in the M-line, by virtue of its enzymatic activity, participates directly in an ATP-requiring process. 5) MM-CPK bound to myosin has an allosteric effect on myosin ATPase activity.

Several arguments can be advanced in support of 1): (a) A CPK molecule present as a soluble enzyme in the intermyofibrillar space might conceivably be closer to a myosin "head" on the outside of a myofibril than those CPK molecules located at the circumference of the disc-shaped M-line structure would be. But those CPK molecules bound nearer the center of the M-disc would clearly be closer to the myosin "heads" located deep inside the myofibril. An analogous argument would presumably hold for Z-line CPK. (b) Of all the possibilities, only 1) takes into account the fact that CPK can be bound at either the M-line or the Z-line. (c) No additional assumptions need be made about the properties of CPK molecules. CPK would function in its time-honored way—as an enzyme.

The bare central zones of thick filaments formed from myosin *in vitro* are somewhat longer than those of thick filaments *in vivo* (Morimoto and Harrington, 1973), hence the proposal (2) that MM-CPK might be involved in thick filament assembly. In electron micrographs of developing muscle, M-line material is not visible at early stages of myofibril assembly (Etlinger and Fischman, 1972), making this possibility unlikely. The appearance of M-line material later on would, of course, be consonant with a stability role (possibility 3). During each contraction-relaxation cycle, the distance between thick filaments in the hexagonal lattice first increases, then decreases. If MM-CPK acts (perhaps with another M-line protein) as a stabilizing cross-link between the thick filaments, and if the same amount of MM-CPK remains bound throughout the contraction relaxation cycle, then MM-CPK (or the complex of MM-CPK with another M-line protein) must be able to act as a kind of spring or hinge - expanding and contracting or opening and closing with the movements of the thick filaments. However, any such mechanism remains in the realm of speculation as long as it is not known whether a constant amount of MM-CPK is bound throughout the contraction-relaxation cycle. Alternatively, one could postulate that additional CPK molecules are added from the soluble "pool" during contraction and then released during relaxation. Experiments with isolated thick filaments (Morimoto and Harrington, 1973) would appear to constitute a particularly promising approach to the question of a possible structural role for M-line CPK, but at present there is no firm evidence on this point.

There is likewise no evidence for the possibility (4) that processes occurring in the M-line (such as a mediolateral expansion and contraction of M-line material) are energy-requiring, much less that they require direct transfer of ATP from CPK bound at the M-line. The possibility (5) that CPK has an allosteric effect on myosin ATPase activity, originally suggested by the work of Yagi and Mase (1962), should be rein-vestigated, though in view of the binding of MM-CPK to the rod portions of only a few myosin molecules at the center of the thick filament, it seems unlikely that this will turn out to be the function of M-line CPK.

Despite the present dearth of supporting evidence, mech-anisms involving distinct functions for MM- and BB-CPK (as in possibilities 2-5, above) are attractive because, potentially at least, they can account for the existence of CPK isozymes with their stage- and tissue-specific distributions (Turner and Eppenberger, 1974). In common with most workers in the isozyme field, we assume that phenomena such as the CPK devel-opmental transition leading to the presence of high concen-trations of a single isozyme (MM-CPK) in adult skeletal muscle (see Turner, 1974) must be functionally significant. In contrast to many others, however, we tend to think that in this case the adaptive value of the different isozymic forms lies not so much in their catalytic differences (which appear to be minimal) as in their differing abilities to interact with components of the contractile apparatus.

Obviously much more experimentation will be required before we will be able to understand the function of myofibrillar CPK. Still, we felt it worthwhile to discuss a number of possibil-ities, some them highly speculative if for no other reason than to emphasize that CPK isozymes, like the different forms of myosin, may well turn out to be <u>both structural proteins and enzymes</u>.

ACKNOWLEDGEMENTS

We thank Dr. W. F. Harrington for helpful discussions of the possible functions of M-line CPK and Dr. F. Ruch for the use of a fluorescence microscope. Our work has been supported by the Swiss National Science Foundation (grant no. 3.8640.72) and by the Muscular Dystrophy Association of America.

REFERENCES

Baskin, R. J. and D. W. Deamer 1970. A membrane-bound creatine phosphokinase in fragmental sarcoplasmic reticulum. *J. Biol. Chem.* 245: 1345-1347.

Botts, J. and M. Stone 1968. Kinetics of coupled enzymes. Creatine kinase and myosin A. *Biochemistry* 7: 2688-2696.

Dawson, D. M., H. M. Eppenberger, and N. O. Kaplan 1965. Creatine kinase: evidence for a dimeric structure. *Biochem. Biophys. Res. Commun.* 21: 346-353.

Dawson, D. M., H. M. Eppenberger, and N. O. Kaplan 1967. The comparative enzymology of creatine kinases. II. Physical and chemical properties. *J. Biol. Chem.* 242: 210-217.

Dawson, D. M. and H. M. Eppenberger 1970. Creatine kinase. *Methods Enzymol.* 17A; 995-1002.

Eaton, B. L. and F. A. Pepe 1972. M-band protein: two components isolated from chicken breast muscle. *J. Cell. Biol.* 55: 681-695.

Eppenberger, H. M., D. M. Dawson, and N. O. Kaplan 1967. The comparative enzymology of creatine kinases. I. Isolation and characterization from chicken and rabbit tissues. *J. Biol. Chem.* 242: 204-209.

Eppenberger, H. M., M. E. Eppenberger, and A. Scholl 1970. Comparative aspects of creatine kinase isoenzymes. *FEBS Symp.* Vol. 18: 269-279.

Etlinger, G. D., and D. A. Fischman 1972. M and Z band components and the assembly of myofibrils. *Cold Spring Harbor Symp. Quant. Biol.* 37: 511-522.

Houk, T. W., and S. V. Putnam 1973. Location of the creatine phosphokinase binding site of myosin. *Biochem. Biophys. Res. Commun.* 55: 1271-1277.

Khan, M. A., P. G. Holt, G. O. Knight, and B. A. Kakulas 1971. Incubation film technique for the histochemical localization of creatine kinase. *Histochemie* 26: 120-125.

Kleine, T. O. 1965. Localization of creatine kinase in microsomes and mitochondria of human heart and skeletal muscle and cerebral cortex. *Nature* 207: 1393-1394.

Knappeis, G. G. and F. Carlsen 1968. The ultrastructure of the M-line in skeletal muscle. *J. Cell. Biol.* 38: 202-211.

Kuby, S. A., L. Noda, and H. A. Lardy 1954. ATP-creatine transphosphorylase. I. Isolation of the crystalline enzyme from rabbit muscle. *J. Biol. Chem.* 209: 191-201.

Kundrat, E. and F. A. Pepe 1971. The M-band: studies-with fluorescent antibody staining. *J. Cell. Biol.* 48: 340-347.

Morimoto, K. and W. F. Harrington 1972. Isolation and physical properties of an M-line protein from skeletal muscle. *J. Biol. Chem.* 247: 3052-3061.

Morimoto, K. and W. F. Harrington 1973. Isolation and composition of thick filaments from rabbit skeletal muscle. *J. Mol. Biol.* 17: 165-175.

Ottaway, J. H. 1967. Evidence for binding of cytoplasmic creatine kinase to structural elements in heart muscle. *Nature* 215: 521-522.

Roy, B. P., J. F. Laws, and A. R. Thomson 1970. Preparation and properties of creatine kinase from the breast muscle of normal and dystrophic chicken. *Biochem. J.* 120: 177-185.

Sherwin, A. L., G. Karpati, and J. A. Bulcke 1969. Immunohistochemical localization of creatine phosphokinase in skeletal muscle. *Proc. Nat. Acad. Sci.U.S.A.* 64: 171-175.

Sommer, J. R. and A. Johnson 1969. The ultrastructure of frog and chicken cardiac muscle. *Z. Zellforsch.* 98: 437-468.

Strohman, R. C. 1959. Studies on the enzymic interactions of the bound nucleotide of the muscle protein actin. *Biochim. Biophys. Acta* 32: 436-449.

Szorenyi, E. T. and R. G. Degtyar 1948. Phosphocreatineadenosinediphosphateferase of muscle and its correlation with actomyosin. *Ukrain. Biokhim. Zhur.* 20: 234-240.

Turner, D. C., T. Wallimann, and H. M. Eppenberger 1973. A protein that binds specifically to the M-line of skeletal muscle is identified as the muscle form of creatine kinase. *Proc. Nat. Acad. Sci. U. S. A.* 70: 702-705.

Turner, D. C. and H. M. Eppenberger 1974. Developmental changes in creatine kinase and aldolase isoenzymes and their possible function in association with contractile elements. *Enzyme* 15: 224-238.

Turner, D. C. 1974. Isozyme transitions of creatine kinase and aldolase during muscle differentiation *in vitro*. III *Isozymes: Developmental Biology*, C. L. Markert, editor. Academic Press, N. Y.

Wallimann, T., H. J. Kuhn, D. C. Turner, and H. M. Eppenberger 1974. Creatine kinase isoenzymes and myofibrillar structure. (in preparation).

Yagi, K. and L. Noda 1960. Phosphate transfer to myofibrils by ATP-creatine transphosphorylase. *Biochim. Biophys. Acta* 43: 249-259,

Yagi, K. and R. Mase 1962. Coupled reaction of creatine kinase and myosin A-adenosine triphosphatase. *J. Biol. Chem.* 237: 397-403.

REGULATION OF HUMAN GLUTAMINE PHOSPHORIBOSYLPYROPHOSPHATE AMIDOTRANSFERASE BY INTERCONVERSION OF TWO FORMS OF THE ENZYME

EDWARD W. HOLMES, JR.
JAMES B. WYNGAARDEN
WILLIAM N. KELLEY

Departments of Medicine and Biochemistry
Duke University Medical Center
Durham,North Carolina 27710

ABSTRACT.   Human glutamine PP-ribose-P amidotransferase (PP-ribose-P amidotransferase) catalyzes the initial rate-limiting step in purine biosynthesis *de novo*. The catalytic activity of the enzyme is determined by the relative concentrations of purine ribonucleotides (feedback inhibitors) and PP-ribose-P (substrate). Two interconvertible forms of the enzyme with molecular weights of 133,000 and 270,000 have been identified. The small form of the enzyme is converted into the large form in the presence of purine ribonucleotides. The large form of the enzyme is converted into the small form in the presence of PP-ribose-P. Enzyme activity correlates directly with the amount of PP-ribose-P amidotransferase present in the small form of the enzyme. These studies provide a molecular basis for understanding the regulation of purine biosynthesis *de novo* by PP-ribose-P and purine ribonucleotides in man.

Glutamine phosphoribosylpyrophosphate amidotransferase, PP-ribose-P amidotransferase, catalyzes the reaction depicted in Fig. 1.   The substrates of this reaction are glutamine, phosphoribosylpyrophosphate (PP-ribose-P), and water.   The products are glutamate, phosphoribosylamine and pyrophosphate.   Phosphoribosylamine is then converted by a series of reactions to the parent purine ribonucleotide, inosine monophosphate (IMP), and IMP is subsequently converted to guanosine monophosphate (GMP) and adenosine monophosphate (AMP). Data obtained from the study of many species suggest that this reaction catalyzed by PP-ribose-P amidotransferase may be the rate-limiting step in purine biosynthesis *de novo*.

This concept is derived in part from the observation that PP-ribose-P amidotransferase is sensitive to feedback inhibition by purine ribonucleotides (Wyngaarden and Ashton, 1959; Hartman, 1963; Caskey et al., 1964; Nierlich and Magasanik, 1965; Momose, et al, 1965; Rowe and Wyngaarden, 1968; Hill and Bennett, 1969; Rowe et al., 1970; Reem, 1972; Wood et al., 1973; Holmes et al.,

1973a). We have also observed that the kinetics of the reaction
are different when purine ribonucleotides are included in the

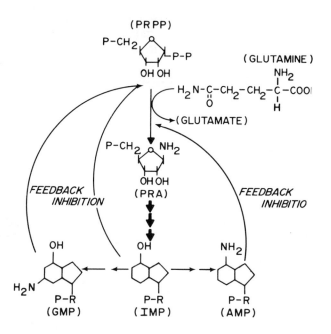

Figure 1. Important steps in purine biosynthesis *de novo*.

assay (Fig. 2). (Holmes et al., 1973a).

The enzyme preparation used for these and all subsequent
studies was purified approximately 20-fold from human placenta.
In the absence of purine ribonucleotides human PP-ribose-P amidotransferase exhibits Michaelis Menten kinetics for the substrate
PP-ribose-P. However, the inclusion of purine ribonucleotides
in the assay system results in a qualitative change in the kinetics from a hyperbolic to a sigmoidal function. The result is
a marked inhibition of amidotransferase by purine ribonucleotides
at physiological concentrations of PP-ribose-P from 10^{-5} to
10^{-6}M (Fox and Kelley, 1971). The inhibition produced by purine ribonucleotides can be reversed by increasing the relative
concentration of PP-ribose-P. These data indicate the importance
of the relative intracellular concentrations of purine ribonucleotides and PP-ribose-P in the control of the activity of human PP-ribose-P amidotransferase.

There are other date available from the study of purine biosynthesis de novo in man which suggest that PP-ribose-P levels
are important in the regulation of this pathway (Fox and Kelley,
1971; Kelley et al., 1970a; Kelley et al., 1970b: Fox et al.,

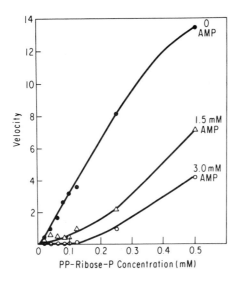

Figure 2. Michaelis-Menten plot with variable PP-ribose-P in the absence and presence of AMP. Assay performed in 50 mM potassium phosphate buffer, pH 7.4, containing 4 mM glutamine, 5 mM MgCl$_2$, and 18 mM β-mercaptoethanol. Equimolar magnesium was added for each concentration of AMP used. No AMP (●—●), 1.5 mM AMP (△—△), and 3.0 mM AMP (o—o). (From Holmes, et al., 1973a).

1970). When PP-ribose-P levels are increased or decreased there is a corresponding change in the rate of purine biosynthesis suggesting that PP-ribose-P amidotransferase is not saturated with this substrate in vivo. The apparent Km for PP-ribose-P of the normal human enzyme is 5 X 10^{-4}M, a value which is 10 to 100 times greater than the intracellular concentration of PP-ribose-P (Fox and Kelley, 1971). Therefore, it is unlikely that PP-ribose-P amidotransferase is saturated with this substrate in vivo.

The molecular basis for the regulation of PP-ribose-P amidotransferase is not readily apparent from the studies cited. The data presented in Fig. 3 suggest a potential mechanism for the regulation of human PP-ribose-P amidotransferase. When a partially purified preparation of the enzyme from human placenta was applied to an 8% agarose gel filtration column the enzyme activity eluted in two distinct peaks. The Stokes radii of the small and large forms were determined to be 54.5 and 65.7 angstroms respectively.

When the same enzyme preparation was applied to an isokinetic sucrose gradient, the major component of enzyme activity was demonstrated to have a sedimentation coefficient of 5.9 (Fig. 4).

In addition there was a faster sedimenting component represented by the shoulder on the major peak. After this same enzyme preparation was passed through a Sephadex G-100 column and
the enzyme activity present in the void volume was discarded,
a single peak of activity was observed on sucrose gradient

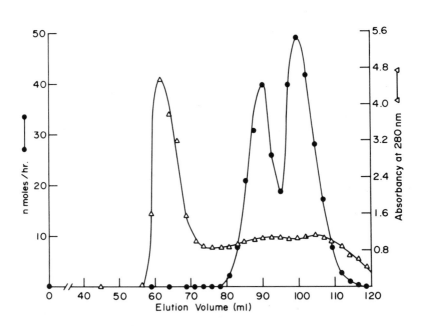

Figure 3. Molecular heterogeneity of human PP-ribose-P amidotransferase demonstrated by gel filtration. Two milliliters
of the enzyme sample were applied to an 8% agarose column equilibrated with 50 mM potassium phosphate buffer, pH 7.4 containing 5 mM $MgCl_2$, 60 mM β-mercaptoethanol, and 0.25 M sucrose.
(From Holmes et al., 1973b).

ultracentrifugation.
   In other experiments to be presented later, it was found
that the faster sedimenting component of PP-ribose-P amidotransferase correspond to the larger form of the enzyme demonstrated
by gel filtration chromatography. The sedimentation coefficient

of this large form of the enzyme was 10.0. A summary of the physical properties of both forms of human PP-ribose-P amidotransferase is presented in Table I.

The following observations suggest a possible relationship between these two forms of PP-ribose-P amidotransferase. When

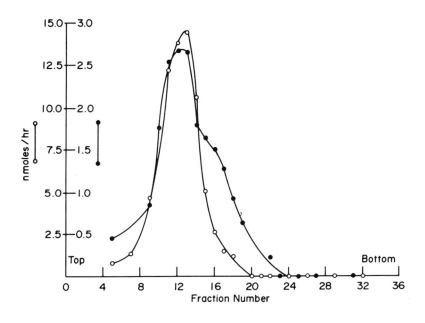

Figure 4. Molecular heterogeneity of PP-ribose-P amidotransferase shown by sucrose gradient ultracentrifugation. The enzyme preparation obtained from the standard purification procedure was applied to one gradient (●——●). This same enzyme preparation was applied to a Sephadex G-100 column and after discarding the enzyme activity in the void volume, the remaining activity was pooled and concentrated by ultracentrifugation. This preparation was applied to the other gradient (o   o) (From Holmes et al., 1973b).

purine ribonucleotides are added to the assay in sufficient concentrations, 100% inhibition of enzyme activity is observed. When PP-ribose-P is added in sufficient concentrations, the inhibition produced by purine ribonucleotides is completely reversed. Since the molecular weight of the large form of PP-ribose-P amidotransferase was found to be twice that of the small form, one explanation for the molecular heterogeneity of PP-ribose-P amidotransferase is an association-dissociation reaction induced by the effector molecules, purine ribonucleotides and PP-ribose-P.

TABLE I

PHYSICAL PROPERTIES OF HUMAN PP-RIBOSE-P AMIDOTRANSFERASE

	Sedimentation Coefficient[a]	Stokes Radius[b]	Molecular Weight
	X 10 ^{13}S	A	
Small Form	5.9 ± 0.4	54.5	133,000
Large Form	10.0 ± 0.4	65.7	270,000

a) The sedimentation coefficient of the small form was determined on 16 occasions and that of the large form on 13 occasions. The value given is the mean ± 1 S.D.

b) The Stokes radius of each form was determined on two occasions and the mean value is reported (From Holmes et al., 1973b).

The studies shown in Fig. 5 were performed with the small form of the enzyme that had been separated from the large form by chromatography on Sephadex G-100. When this enzyme preparation was applied to a sucrose gradient that contained no purine ribonucleotide a single peak of amidotransferase activity was observed as shown in Fig. 4. The sedimentation coefficient of this peak was 5.9, a value which corresponds to a molecular weight of 133,000 (Table I). When this same enzyme preparation was incubated at 37° with purine ribonucleotide (AMP, GMP, or AMP + GMP) and applied to a gradient that also contained purine ribonucleotide, the major peak of enzyme activity had a sedimentation coefficient of 10.0, a value which corresponds to a molecular weight of 270,000 (Table I). Thus the small form of PP-ribose-P amidotransferase is converted to the large form in the presence of purine ribonucleotides.

The large form of the enzyme, that had been separated from the small form by gel filtration on 8% agarose as shown in Fig. 3, was used for the studies depicted in Fig. 6. When this preparation was applied to a sucrose gradient that contained purine ribonucleotide, the sedimentation coefficient of PP-ribose-P amidotransferase was 10.0. If this same preparation was incubated with PP-ribose-P and applied to a gradient that contained PP-ribose-P, the sedimentation coefficient was 5.9. Therefore, the large form of the enzyme is converted to the small form in the presence of PP-ribose-P.

This phenomenon of interconversion between the large and small form of PP-ribose-P amidotransferase can also be demonstrated by repeated passage of the same enzyme preparation

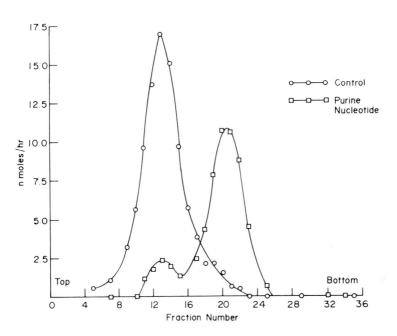

Figure 5. Association of PP-ribose P-amidotransferase induced by purine ribonucleotides. The small form of the enzyme obtained from Sephadex G-100 column chromatography was incubated at 37° for 15 min. with no purine ribonucleotide (o—o) or 5 mM AMP (□——□) and centrifuged into gradients that contained no purine ribonucleotide, or 5 mM AMP, respectively.

through a gel filtration system (Fig. 7). In the top panel of this figure the enzyme preparation obtained from the standard purification procedure was incubated with purine ribonucleotide at 37° and passed through a 10% agarose column that had been equilibrated with purine ribonucleotide. The activity

eluted in one major peak that had a Stokes radius of 65.7 angstroms, a value which corresponds to a molecular weight of 270,0C
(Table I). This activity peak was pooled, concentrated, and
incubated with PP-ribose-P at 37° and applied to the same gel
filtration column that had been re-equilibrated with PP-ribose-
P (bottom panel of Fig. 7) In this instance the activity eluted

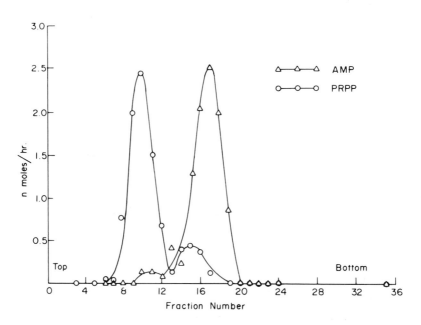

Figure 6. Conversion of the large form of PP-ribose-P amidotransferase to the small form by PP-ribose-P. The large form
of the enzyme was isolated by gel filtration on 8% agarose
(Figure 3). This enzyme preparation, containing 5 mM AMP, was
incubated at 37° for 15 min. in the absence of PP-ribose-P
($\triangle$——$\triangle$) and with 10 mM PP-ribose-P (o——o) and centrifuged
into gradients that contained 5 mM AMP ($\triangle$——$\triangle$) or 1 mM PP-
ribose-P (o——o).

in one major peak with a Stokes radius of 54.5 angstroms, a
value which corresponds to a molecular weight of 133,000 (Table
I). As shown here the large form of the enzyme can be isolated
and subsequently converted to the small form. In other experiments not depicted here the small form was isolated and then
converted to the large form. This would seem to demonstrate
conclusively that the large and small form of human PP-ribose-
P amidotransferase are interconvertible.

In other experiments it was found that L-glutamine alone,
or in combination with PP-ribose-P or with purine ribonucleotide,

had no effect on the molecular size of either the small or large
form of human PP-ribose-P amidotransferase. This is in keeping
with the observation that human PP-ribose-P amidotransferase
exhibited Michaelis-Menten kinetics with glutamine as the variable substrate even in the presence of purine ribonucleotides
if the concentration of PP-ribose-P was saturating (Holmes et
al, 1973a; Holmes et al., 1973b).

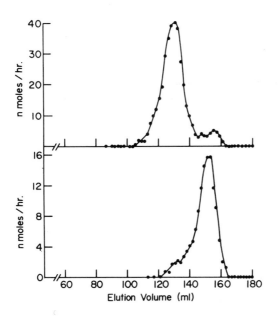

Figure 7. Isolation of the large form of human PP-ribose-P
amidotransferase and subsequent conversion to the small form
by PP-ribose-P. Top, the enzyme preparation from the standard
purification procedure was incubated with 5 mM AMP at 37° for
15 min and eluted from a 10% agarose column also equilibrated
with 5 mM AMP. Bottom, the enzyme activity that was eluted between 122 to 140 ml (top panel) was pooled, concentrated by
ultrafiltration, and then incubated at 37° for 15 min with 10
mM PP-ribose-P. This preparation was applied to the same 10%
agarose column equilibrated with 0.5 mM PP-ribose-P (from Holmes
et al., 1973b).

What is the physiological significance of this interconversion between the small and large form of human PP-ribose-P amidotransferase? The experimental observations described above suggest that the small form of human PP-ribose-P amidotransferase
is the catalytically active species and the large form of the

enzyme is inactive. If this hypothesis is correct, it should
be possible to correlate enzyme activity with the amount of
PP-ribose-P amidotransferase present as the small form of the
enzyme.

The data presented in Fig. 8 is a compilation of two sets of
experiments. Enzyme activity was determined by the routine assay
procedure at a non-saturating concentration of PP-ribose-P. Enzyme activity is expressed as a per cent of the maximal activity

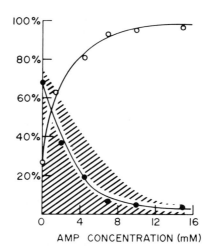

AMP CONCENTRATION (mM)

Figure 8. Correlation between enzyme activity and the small
form of human PP-ribose-P amidotransferase. Catalytic activity was determined at 37° during a 10 min. incubation with 2.5
mM PP-ribose-P, 5 mM MgCl$_2$, 4 mM glutamine, 16 mM β-mercaptoethanol and 0,2,4,7,10 and 15 mM AMP. Activity is expressed in
the cross-hatched area as a per cent of maximal activity observed with 10 mM PP-ribose-P in the absence of AMP. Aliquots
of the same enzyme preparation were incubated at 37° for 10
min. with 2.5 mM PP-ribose-P in the presence of 0,2,4,7,10 and
15 mM AMP. These samples were then applied to sucrose gradients that contained 2.5 mM PP-ribose-P and the corresponding
concentration of AMP. The per cent of amidotransferase present
as the small and large form was determined in each gradient
(small form ●—●; large form o—o). (From Holmes et al, 1973b).

observed with a saturating concentration of PP-ribose-P (cross-
hatched area). This assay was performed in the absence of any
purine ribonucleotide and at several different concentrations
of AMP. As expected, the greater the AMP concentration the lower the enzyme activity. In the other set of experiments performed on the same enzyme sample, the assay conditions just describe

were reproduced as closely as possible and these samples were
then applied to sucrose density gradients in which the assay
conditions had also been reproduced. Following centrifugation
of these samples it was possible to calculate the percentage
of PP-ribose-P amidotransferase present in the small and large
form under each of the experimental conditions used in the
enzyme assays. As shown in Fig. 8 enzyme activity decreased
(cross-hatched area) as the amount of PP-ribose-P amidotransferase present in the small form of the enzyme decreased, while
the amount of PP-ribose-P amidotransferase present in the large
form increased. These observations provide strong support for
an enzyme model in which the small form of human PP-ribose-P
amidotransferase is the catalytically active species and the
large form of the enzyme is catalytically inactive.

Fig. 9 summarizes in schematic fashion the mechanism by which
human PP-ribose-P amidotransferase activity is controlled. PP-
ribose-P causes the large inactive form of human PP-ribose-P

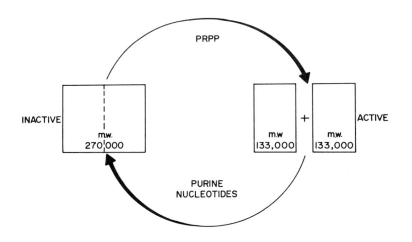

Figure 9. Schematic representation of the mechanism by which
human PP-ribose-P amidotransferase activity is controlled.

amidotransferase to dissociate to the small form of the enzyme
which is the catalytically active species. Purine ribonucleotides cause the small active form of the enzyme to revert to the
large catalytically inactive form of PP-ribose-P amidotransferase. Since the reaction catalyzed by PP-ribose-P amidotransferase is the probable rate-limiting step in the purine biosynthetic
pathway, this model provides an explanation at the molecular
level for the regulation of purine biosynthesis *de novo* by
PP-ribose-P and purine ribonucleotides in man.

REFERENCES

Caskey, C. T., D. M Ashton and J. B. Wyngaarden, 1964. The enzmology of feedback inhibition of glutamine phospho= ribosylpyrophosphate amidotransferase by purine ribonucleotides. *J. Biol. Chem.* 239: 2570-2579.

Fox, I. H., J. B. Wyngaarden and W. N. Kelley, 1970. Depletion of erythrocyte phosphoribosylpyrophosphate in man: A newly observed effect of allopurinol. *New Eng. J. Med.* 283: 1177-1182.

Fox, I. H. and W. N. Kelley, 1971. Phosphoribosylpyrophosphate in man: Biochemical and clinical significance. *Ann. Intern. Med.*, 74: 424-433.

Hartman, S. C., 1963. Phosphoribosylpyrophosphate amidotransferase: Purification and general catalytic properties. *J. Biol. Chem.*, 238: 3024-3035.

Hill, D. L. and L. L. Bennett, Jr., 1969. Purification and properties of 5-phosphoribosylpyrophosphate amidotransferase from adenocarcinoma, 755 cells. *Biochem.*, 8: 122-130.

Holmes, E. W., J. A. McDonald, J. M. McCord, J. B. Wyngaarden and W. N. Kelley, 1973a. Human glutamine phosphoribosylpyrophosphate amidotransferase kinetic and regulatory properties., *J. Biol. Chem.*, 248: 144-150.

Holmes, E. W., J. B. Wyngaarden and W. N. Kelley, 1973b. Human glutamine phosphoribosylpyrophosphate amidotransferase: Two molecular forms inter-convertible by purine ribonucleotides and phosphoribosylpyrophosphate, *J. Biol. Chem.*, 248: 6035-6040.

Kelley, W. N., I. H. Fox and J. B. Wnygaarden, 1970a. Regula- of purine biosynthesis in cultured human cells. I. Effects of orotic acid, *Biochim. Biophys. Acta*, 215: 512-516.

Kelley, W. N., M. L. Greene, I. H. Fox, F. M. Rosenbloom, R. I. Levy and J. E. Seegmiller, 1970b. Effects of orotic acid on purine and lipoprotein metabolism in man. *Metabolism.* 19: 1025-1035.

Momose, H., H. Nishikawa and N. Katsujja, 1965. Genetic and biochemical studies of 5' nucleotide formation. II. Repression of enzyme formation in purine nucleotide biosynthesis in *Bacillus subtilis* and derivation of depressed mutants. *J. Gen. Appl. Microbiol.*, 11: 211-220.

Nierlich, D. P. and B. Magasanik, 1965. Regulation of purine ribonucleotide synthesis by end product inhibition: The effect of adenine and guanine ribonucleotides on the 5'-phosphoribosylpyrophosphate amidotransferase in *Aerobacter aerogenes*. *J. Biol. Chem.*, 240: 358-365.

Reem, G. H., 1972. *De novo* purine biosynthesis by two pathways in Burkitt lymphoma cells and in human spleen. *J. Clin. Invest.*, 51: 1058-1062.

Rowe, P. B. and J. B. Wyngaarden 1968. Glutamine phosphoribosylpyrophosphate amidotransferase. Purification, substructure,
amino acid composition and absorption spectra. *J. Biol.,
Chem.,* 243: 6373-6383.
Rowe, P. B., M. D. Coleman and J. B. Wyngaarden, 1970. Glutamine phosphoribosylpyrophosphate amidotransferase. Catalytic and conformational heterogeneity of the pigeon liver
enzyme. *Biochem.,* 9: 1498-1505.
Wood, A. W. and J. E. Seegmiller 1973. Properties of 5-phospho-
ribosyl-1-pyrophosphate amidotransferase from human lympho-
blasts. *J. Biol. Chem.* 248: 139-143.
Wyngaarden, J. B. and D. M. Ashton, 1959. The regulation of
activity of phosphoribosylpyrophosphate amidotransferase
by purine ribonucleotides: A potential feedback control
of purine biosynthesis. *J. Biol. Chem.,* 234: 1492-1496.

STUDIES ON HUMAN LYSOSOMAL β-D-N-ACETYL
HEXOSAMINIDASE AND ARYLSULFATASE ISOZYMES

MARIO C. RATTAZZI, PATRICK J. CARMODY,
and RONALD G. DAVIDSON
Division of Human Genetics
Department of Pediatrics
Children's Hospital of Buffalo
State University of New York at Buffalo
Buffalo, New York 14222

ABSTRACT. Studies on the structural, functional and genetic relationships among the isozymes of β-hexosaminidase
and arylsulfatase may contribute to the understanding of
the basic biochemical defect in Gm_2 gangliosidosis and metachromatic leukodystrophy.

The hypothesis of a sialyl transferase deficiency as
the cause of the deficiency of β-hexosaminidase A in Tay-
Sachs disease does not seem tenable at the present time,
since the reported effect on β-hexosaminidase A of treatment with crude neuraminidase preparations is most probably
due to merthiolate present in these preparations.

The nature of the basic defect in Gm_2 gangliosidosis
can be investigated in terms of structural relationships
between β-hexosaminidase isozymes by the study of the complementation of β-hexosaminidase A deficiency that takes
place in somatic cell hybrids between fibroblasts from
patients with Sandhoff disease and Tay-Sachs disease.

The differences between plasma and tissue β-hexosaminidase do not appear to be due to interaction with plasma
lipoproteins and the causes of these differences remain
obscure.

A newly developed, sensitive electrophoretic technique
for the study of arylsulfatase isozymes has been applied
to the prenatal diagnosis of MLD and has uncovered a different form of arylsulfatase in amniotic fluid.

INTRODUCTION

In the last few years, a number of human lipid storage
diseases have been characterized as due to, or associated
with, deficient activity of lysosomal enzymes (Bernsohn and
Grossman, 1971; Hers and Van Hoof, 1973). Many of these
lysosomal acid hydrolases have been shown to exist in multiple forms, only one of which may be deficient in a given
disorder. The physiological significance of these multiple
forms of lysosomal enzymes is unknown. The study of a storage

439

disease associated with a lysosomal enzyme deficiency, and of its clinical and biochemical variants, may provide an understanding of the physiological role played by the individual components of the related lysosomal isozyme system in the normal cell. Conversely, the study of the structural and functional relationships among the different isozymes may provide insight into the basic biochemical defect underlying the disease in question. As a result, more accurate diagnostic procedures and therapeutic approaches may become available.

This paper describes studies carried out in our laboratory on two lysosomal enzyme systems, β-D-N-acetyl hexosaminidase [1] and arylsulfatase. Impaired function of these enzymes is associated with two human lipid storage disorders, GM_2 gangliosidosis and metachromatic leukodystrophy, respectively.

METHODS AND MATERIALS

References to established methodologies will be given throughout the text where results of experiments are described. Electrophoretic methods developed in our laboratory for the study of β-hexosaminidase (Rattazzi and Davidson, 1972) and arylsulfatase (Rattazzi, et al., 1973) will be described in some detail in this section.

1. GENERAL PROCEDURE

The support material used for electrophoresis is cellulose acetate gel (Cellogel by Chemetron - Milan, Italy, available through Kalex Scientific Co., Manhasset, N.Y.) in sheets 16 cm X 17 cm, 0.35 mm thick. The gel sheets are thoroughly rinsed with distilled water and equilibrated in the electrophoresis buffer for a few hours. A gel sheet is blotted free of excess buffer and placed, porous face up, in the electrophoretic tank (Gelman, type 51101). The current is applied for 10

Note (1) The following abbreviations are used: β-hexosaminidase, Hex A, Hex B, Hex C: β-D-N-acetyl glucosaminidase A,B,C (EC 3.2.1.30). TSD: Tay-Sachs disease. SJD: Sandhoff-Jatzkewitz disease. ASA, ASB, ASC: Arylsulfatase A,B,C, arylsulfate sulfohydrolase (EC 3.1.6.1). MLD: Metachromatic Leukodystrophy. Gm_2-ganglioside: N-acetylgalactosaminyl-(1→4)-(N-acetylneuraminyl)-(2→3)-galactosyl-(1→4)-glucosyl ceramide. p-NCS: 2-hydroxy-4-nitrophenyl sulfate (*para* nitrocatechol sulfate). BAL: British Anti-Lewisite (2,3 dimercaptopropanol). RDE: Receptor Destroying Enzyme. 4 MU(S):4-methyl umbelliferyl (sulfate).

minutes to stabilize the gel, then the samples are applied as
thin bands (15-20 mm long) with a Lang-Levy micropipette. Five
to 25 µl of cell extract are usually applied and up to 12
samples can be applied to the same gel. Five to 10 minutes
are allowed for absorption and equilibration of the samples.
The current is then applied and the electrophoresis allowed
to proceed for a few hours, usually at room temperature. Constant amperage is always used. At the end of the run, the gel
is marked, cut and dipped 5-6 times in 20 ml of the appropriate
fluorogenic substrate solution on a clean glass plate, blotted
gently with filter paper, and transferred to a glass moist
chamber. After incubation at 37°C for the desired length of
time, the fluorescence of the bands of enzyme activity is
enhanced by placing the gel, porous face downward, for 4 minutes on the surface of the enhancing buffer in a tray. After
very gentle blotting, the bands of enzyme activity are observed by exposing the gel, in the moist chamber, to long wave
(365 nm) ultraviolet light.

2. SPECIFIC PROCEDURES

The following conditions are used for electrophoresis:

a. β-hexosaminidase.
    System A: 0.025M trisodium citrate-citric acid buffer,
pH 5.5, containing 0.01% sodium azide as a preservative.
Electrophoresis for 3 hours at 15 mamp, 100 volts, room temperature. This system is usually employed for quantitation of
the isozymes by excision of the bands and additional incubation in substrate solution.
    System B: 0.036 M sodium barbital - citric acid buffer,
pH 6.0 (0.01% sodium azide). Electrophoresis for 3-4 hours
at 15 mamp, 200 volts, room temperature. This system gives
excellent separation of the isozymes and can be used for
immunoelectrophoresis.
    System C: 0.01 M sodium barbital HCl buffer, pH 8.0
(0.01% sodium azide): Electrophoresis for 2 hours at 7.0
mamp, 400 volts at +4°C. Used mainly for immunoelectrophoresis since immune complexes are readily formed at this pH.
    Substrate solution: 0.1M disodium phosphate - citric
acid buffer pH 4.5, containing 0.1 mg/ml of 4-methyl umbel-
liferyl-N-acetyl β-D-glucosaminide (Pierce Chemical Co.).
Enhancing buffer: 0.25 M sodium carbonate-glycine buffer,
pH 10.
    Cell extracts for electrophoresis are prepared by five
consecutive freezing and thawing cycles (dry ice-acetone mixture) of suspensions of saline-washed cells in 0.05M disodium

phosphate - citric acid buffer, pH 4.5 (1:3, v/v), followed by centrifugation at 40,000 g for 30 minutes at 4°C. Tissue extracts are prepared by ultrasonic disruption at 4°C of minced tissue in the above buffer (1:5, v/v) followed by centrifugation as above.

b. Arylsulfatase.

Electrophoresis buffer: 0.036M sodium barbital - 0.037 M sodium acetate - acetic acid, pH 7.0, containing 1 mM β-merc-aptoethanol and 0.01% sodium azide. Prior to electrophoresis the gels are equilibrated in the same buffer containing 0.05% bovine serum albumin. Electrophoresis is carried out for 3½ hours at 15 mamp, 180 volts. Substrate solution: 0.5 M sodium acetate - acetic acid buffer, pH 7, containing 3 mg/ml of 4 methyl umbelliferyl sulfate, sodium salt (this substrate is prepared by a modification of the method of Rinderknecht et al., 1970, as described by Rattazzi et al., 1973.

Enhancing buffer: same as for β-hexosaminidase.

Cell extracts for electrophoresis are prepared by ultra-sonic disruption at 4°C of suspensions of saline-washed cells in 0.05 M sodium acetate buffer, pH 6.5, containing 0.1% bovine serum albumin and 1 mM,β mercaptoethanol (1:5, v/v), followed by incubation for 30 minutes at 25°C to precipitate ASC and centrifugation at 40,000 g for 30 minutes at 4°C.

Tissue extracts are prepared by ultrasonic disruption at 4°C of minced tissue in distilled water(1:5 v/v) fol-lowed by high speed centrifugation as above. One-tenth volume of 0.5M acetate buffer, pH 5.0, is then added to the high speed supernatant, followed by incubation at 25°C and recentrifugation as above.

3. IMMUNOELECTROPHORESIS

Electrophoresis is carried out as outlined above, in *a,* and *b,* but the samples are applied as dots at 20 mm from one another (center to center). For β-hexosaminidase, Systems B and C are preferred since they afford a better separation of the isozymes. At the end of the run, the antiserum (50-100 µl) is applied to the gel as a thin line equidistant from the sample application points. After absorption of the serum, the gel is incubated for 72 hours in a sealed humid chamber at 27°C. After washing for 8 hours with 0.15 M NaCl solutions containing 0.01% sodium azide, the gels are stained for en-zymatic activity as outlined above. The antigen-antibody complexes retain β-hexosaminidase or arylsulfatase activity when tested with the artificial fluorogenic substrates.

RESULTS AND DISCUSSION

β-HEXOSAMINIDASE AND GM_2 GANGLIOSIDOSIS

1. Characteristics of the isozyme system.

Two major forms of β-hexosaminidase, Hex A and Hex B, are found in human tissues, body fluids, leukocytes and cultured skin fibroblasts. These isozymes can be separated by electrophoresis, ion exchange chromatography and isoelectric focusing, and share a number of biochemical characteristics such as molecular weight, pH optima, Km for several artificial and some natural substrates, and Ki for various inhibitors (Robinson and Stirling, 1968; Okada and O'Brien, 1969; Sandhoff and Wässle, 1971; Srivastava et al., 1974a,b). In addition to the difference in isoelectric point, the two isozymes differ in heat stability, Hex A being more labile at $52^{\circ}C$ (pH 5.5), and in activity towards GM_2 ganglioside, which is apparently degraded *in vitro* only by Hex A (Robinson and Stirling, 1968; Okada and O'Brien, 1969; Sandhoff and Wässle, 1971; Wenger, et al, 1972; Li et al., 1973). The biochemical similarities between the two isozymes suggest that they are structurally related. Further supportive evidence is provided by immunological data, showing that Hex A and Hex B share an antigenic determinant, although Hex A possesses an antigenic determinant not detectable on Hex B (Carroll and Robinson, 1973; Srivastava and Beutler, 1974; Bartholomew and Rattazzi, 1974). In addition, the expression of human Hex A in man-mouse somatic cell hybrids apparently depends on the presence of Hex B (Lalley et al., 1974) suggesting that an interaction of genes or gene products is needed for the expression of the two isozymes.

The two isozymes are believed to be composed of subunits (Srivastava et al., 1974b). It is not known whether the similarities are due to the presence of identical subunits shared by both isozymes, and whether the differences result from different subunits present only in Hex A or from subunits of different structure unique to both Hex A and Hex B (Tateson and Bain, 1971; Robinson and Carroll, 1972; Desnick et al., 1972; Srivastava and Beutler, 1974). Alternatively, since at least β-hexosaminidase purified from human placenta appears to be a glycoprotein (Srivastava et al., 1974 a,b), the differences between Hex A and Hex B may result from a different carbohydrate composition (Robinson and Stirling 1968; Goldstone et al, 1971).

A minor form of β-hexosaminidase, Hex C, has been described (Hoogwinkel et al., 1974; Poenaru and Dreyfus, 1973) but not yet characterized biochemically.

2. GM_2 *gangliosidosis.*

Human GM_2 gangliosidosis, a fatal degenerative disease of the central nervous system, is associated with deficient activity of β-hexosaminidase. In the most common form, TSD, there is deficient activity of Hex A with normal or increased activity of Hex B (Okada and O'Brien, 1969), (Fig. 1). A rare clinical variant of the disease, SJD, has been characterized biochemically as associated with deficient activity of both Hex A and Hex B (Sandhoff et al., 1968; Young et al., 1970; Sandhoff et al., 1971; Okada et al, 1972; Desnick et al., 1972), supporting the hypothesis of a structural relationship between the isozymes. Both diseases are transmitted with an autosomal recessive mode of inheritance.

The inability of extracts of tissues from patients with TSD and SJD to degrade GM_2 *in vitro* (Kolodny, et al., 1969; Sandhoff et al., 1971; Tallman et al., 1972) and the apparent inability of purified Hex B to degrade GM_2 ganglioside *in vitro* (Sandhoff and Wässle, 1971; Sandhoff et al., 1971; Li et al., 1973) support the concept that, *in vivo*, GM_2 **ganglioside** is the natural substrate of Hex A, and that deficient activity of this enzyme, even in the presence of Hex B, results in accumulation of the ganglioside. Several hypotheses have been put forward to explain the deficiency of β-hexosaminidase in TSD and SJD in terms of structural relationships between Hex A and Hex B and these will be discussed in the following sections.

3. *Neuraminidase treatment of* β-*hexosaminidase.*

The observation that treatment of human Hex A with *V. cholerae* neuraminidase preparations results in the conversion of this isozyme to a form electrophoretically similar to Hex B (Robinson and Stirling, 1968) and an analogous observation on rat kidney Hex A (Goldstone et al., 1971) has led these authors to put forward the hypothesis that human Hex A and Hex B are glycoproteins, similar if not identical in polipeptide structure, but differing in the number of sialic acid residues. Deficiency of a sialyl transferase, which normally transfers sialic acid residues to the acceptor Hex B thus converting it to Hex A, has been proposed as the basic biochemical defect in TSD (Goldstone et al., 1971). Both of these related hypotheses have been considered by a number of investigators (Sandhoff et al., 1971; Tateson and Bain, 1971; Paigen, 1971; Snyder et al, 1972; Murphy and Craig, 1972; Carroll and Robinson, 1973; Ikonne and Ellis, 1973; Hayase et al, 1973); however, with one exception (Murphy and Craig, 1973), a number of investigators, including our group, have been unable to reproduce the results of Robinson and Stirling (1968) using purified preparations of *V. cholerae* and *C. perfringens* neuraminidase in a variety of experimental conditions (Carmody and Rattazzi, 1973; Ikonne and Ellis, 1973;

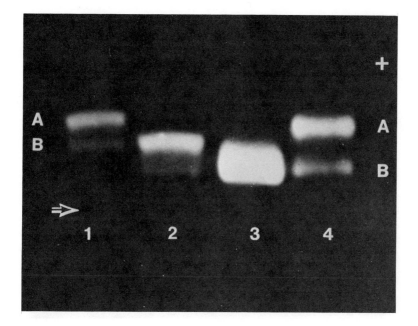

Fig. 1.   Fluorescent bands of β-hexosaminidase activity after electrophoresis at pH 5.5 on cellulose acetate gel and development with fluorogenic substrate. 1: plasma and 4: extract of leukocytes from normal individual. 2: plasma and 3: extract of leukocytes from patient with TSD. A and B: Hex A and Hex B. Arrow: point of application.

Hayase et al., 1973; O'Brien, 1973a; Srivastava et al., 1974b; Swallow et al., 1974; Carmody and Rattazzi, 1974). In contrast,when crude *V. cholerae* culture filtrates are used as a source of neuraminidase human Hex A is converted to a form closely resembling Hex B (Carmody and Rattazzi, 1973; Swallow et al., 1974; Carmody and Rattazzi, 1974). The newly formed Hex B-like enzyme has the same electrophoretic mobility as Hex B in systems A, B, C on cellulose acetate gel (Fig. 2) and on starch gel (Swallow et al., 1974), and it is stable in the conditions (52°C, pH 5.5, 3 hrs.) that result in inactivation of Hex A. By gel filtration on Sephadex G-150 (Rattazzi, 1968) the apparent molecular weight of the conversion product is the same as that of the starting material (Hex A), and the same as that of Hex B (125,000 ± 3,000) (Carmody and Rattazi, 1974).

    Immunological characterization by immunoelectrophoresis

445

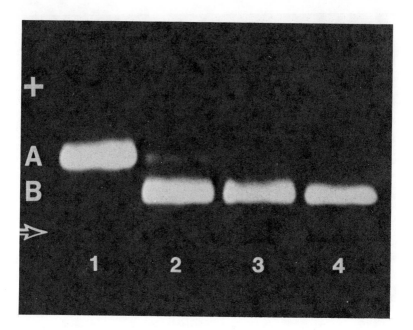

Fig. 2. Effect of *V. cholerae* culture filtrate and merthiolate on Hex A. 1: Hex A incubated for 18 hours at 37°C in
buffer. 2: Hex A incubated for 18 hours at 37°C in the presence of *V. cholerae* filtrate. 3: Hex A incubated for 18 hours
at 37°C in the presence of 1mM merthiolate. 4: Hex B.
Notation as in Fig. 1.

and double diffusion of the conversion product has shown that,
although it retains the antigenic determinant β, common to
Hex A and Hex B, the conversion product no longer exhibits
the specific antigenic determinant α (Bartholomew and Rattazzi,
1974) unique to Hex A. The possibility that the observed
phenomenon may be due to loss of enzymatic activity of the
antigen-antibody complex utilized to detect the immunoprecipitate seems to be ruled out by preliminary inhibition tests.
Characterization of the conversion factor has shown convincing-
ly that it is not neuraminidase, since it is stable after heat
treatment (Fig. 3), not Ca^{++} dependent, dialyzable and resistant to Pronase digestion (Swallow et al., 1974; Carmody
and Rattazzi, 1974) in contrast to the properties of *V. cholerae* neuraminidase (Ada et al., 1961).

The properties of the conversion factor indicated that it

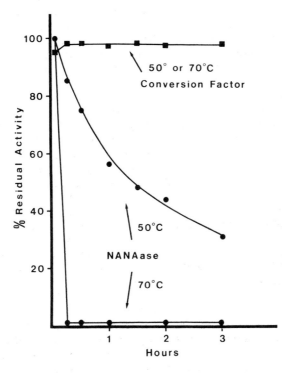

Fig. 3. Effect of heat treatment on the activity of the conversion factor (squares) and of neuraminidase (NANAase,
circles) of *V. cholerae* culture filtrates. Conversion activity was measured by electrophoresis and quantitation of
Hex B-like formed from a known amount of Hex A. Neuraminidase activity was measured in the same samples using sialyl-
lactose as a substrate. Both activities are expressed as a
percent of that present before heat treatment.

was probably a small compound, most probably nonprotein in
nature. Comparison of *V. cholerae* culture filtrates obtained
commercially (Sigma, Burroughs Wellcome) with preparations
obtained from NIH (Lots Wyeth 001, 002, Merck 44936) showed
that only the commercial ones were active with respect to the
conversion of Hex A. After noting that the commercial preparations contained merthiolate, a sulfhydryl reagent used as
a preservative, this compound was tested in the conditions
used to detect *V. cholerae* filtrate conversion activity, and
it was found to convert Hex A to a Hex B-like form (Fig. 2)
indistinguishable from that obtained from *V. cholerae* filtrate treatments. Electrophoretic, heat inactivation and

immunological characteristics of the two conversion products were identical. Furthermore, addition of BAL or cysteine (0.001M) to the incubation mixtures completely inhibited the conversion activity of both merthiolate and *V. cholerae* culture filtrates. These findings strongly support the conclusion that merthiolate and not neuraminidase is the active factor responsible for the conversion of Hex A to the Hex B-like form (Carmody and Rattazzi, 1974).

These results do not exclude the possibility that Hex A and Hex B may be glycoproteins (Srivastava et al., 1974b). However, they indicate that the results obtained by Robinson and Stirling (1968) using Burroughs Wellcome *V. cholerae* culture filtrate (RDE, which contains merthiolate) were probably misinterpreted. Thus, unless direct evidence for the exisence of sialic acid residues on the enzyme molecule is obtained, the existence of a sialyl-transferase deficiency in TSD seems improbable at this point. Whether merthiolate acts on the SH-groups in the conversion of Hex A to Hex B-like, and in the accompanying changes in thermal stability and immunological reactivity, and whether Hex B and Hex B-like are structurally identical is not yet known.

4. *Complementation of Hex A deficiency.*

Assuming that the biochemical defect in TSD and SJD results from two different mutations, Hex A activity might be restored by complementation if the TSD and the SJD genomes are brought together and are active in the same cell. In vivo complementation of biochemical defects has been obtained in different biological systems, including bacteria, fungi, insects (see review by Fincham, 1966). In man, somatic cell hybridization by Sendai virus-mediated fusion of enzyme-deficient fibroblasts, has been used to study complementation of both the intergenic and interallelic types (Siniscalco et al., 1969 ;Silagi et al.,1969; Nadler et al., 1970; Lyons et al., 1973). Ideally, this approach would require that the parental strains be amenable to negative chemical selection, either utilizing some selective media or by making use of variants of the same biochemical defect (Nadler et al., 1970). This allows detection of the enzyme activity restored by complementation in the relatively few heterologous cell hybrids expected from virus-mediated fusion. No such selective system is as yet available in the case of fibroblasts derived from patients with TSD or SJD. However, utilizing the sensitivity and the resolution of cellulose acetate gel electrophoresis, we have been able to demonstrate the complementation of Hex A deficiency in somatic cell hybrids between fibroblasts from individuals with TSD and an individual

448

with SJD (Rattazzi et al., 1974). Two fibroblast lines from unrelated patients with TSD, and one from a patient with SJD were used in three separate experiments. Fusion was obtained by exposure to β-propiolactone-inactivated Sendai virus (1000 HAU units/2.5 x 10^6 cells) of equal proportions of TSD and SJD cells. After incubation at 37°C for 30 minutes in the presence of 30% fetal calf serum, an aliquot of the cell population was frozen, and the remainder was plated in growth medium (Eagle's Basal medium, Grand Island Biological Co. with 10% fetal calf serum and antibiotics). Appropriate controls were provided by separate fusions of homologous cells (TSD x TSD and SJD x SJD) and co-cultivation of heterologous cells not exposed to virus. The cells were harvested at weekly intervals for 5-6 weeks (10-12 cell divisions), and prepared for electrophoresis.

Demonstration of the formation of heterokaryons after virus treatment was obtained by a parallel fusion of SJD cells with TSD cells grown for two passages in medium containing ^3H thymidine. Examination by autoradiography of the virus-treated cell mixture showed that, of the binucleated cells present (11.7% of total), approximately 31.5% (or 3.7% of total) had one labelled and one unlabelled nucleus.

Examination of cell extracts by cellulose acetate gel electrophoresis showed that, in all three fusion experiments, a band of enzyme activity migrating to the position of Hex A from control fibroblasts was visible in TSD-SJD cell-fusion populations harvested after one week of growth (Fig. 4). This band was not present in homogenates of cells harvested immediately after fusion, suggesting that *de novo* synthesis, specified by both genomes, had to occur for the newly formed band to appear. The band was never present in homogenates of TSD or SJD cells separately exposed to Sendai virus, nor in homogenates of cells cultured together in the absence of virus treatment. The newly-formed β-hexosaminidase band became progressively weaker until it was no longer visible in extracts of cells harvested at 4-5 weeks after fusion, presumably due to overgrowth of the TSD-SJD cell hybrids by the parental cells.

The newly formed β-hexosaminidase band was identified as Hex A by several criteria. Its mobility was coincident with that of Hex A after cellulose acetate gel electrophoresis in systems A, B, C (see Methods). After heat treatment of the cell extracts at pH 5.5 for 3 hours at 52°C, the newly formed β- hexosaminidase was inactivated, as was Hex A from normal fibroblast extracts, while Hex B was still fully active in all extracts (Fig. 5.) Immunoelectrophoresis showed that the newly formed enzyme reacted with an antiserum directed against the specific antigenic determinant α of Hex A. Furthermore, fusion

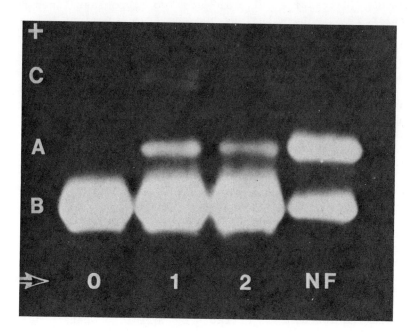

Fig. 4. Hex A activity in cell extracts after exposure to
Sendai virus of mixtures of fibroblasts from a patient with
TSD and a patient with SJD. 0: 30 minutes after virus treatment; 1: one week in culture after virus treatment. 2: two
weeks in culture after virus treatment; NF: extract of fibroblasts from normal individual. Electrophoresis at pH 6.0·
A, B, arrow: same notation as in Fig. 1; C: Hex C, visible
in SJD cells (Roper and Schwantes, 1973).

of the enzymatically active arcs of immunoprecipitation was
observed when partially purified Hex A and extracts of the
hybrid cell population were allowed to react against this
antiserum in double immunodiffusion, suggesting immunological
identity between the two enzymes. These results strongly
support the conclusion that the newly formed β-hexosaminidase
is Hex A, and that complementation has occurred in the TSD-
SJD heterokaryons.

Estimation of the relative activity of Hex A in the hybrid
cell extracts by electrophoresis, elution of the fluorescent
bands and fluorometry (Rattazzi and Davidson, 1972), gave an
average value of 6% Hex A (expressed as percent of Hex B
activity in the same samples) for the three fusion experiments.
This value compares well with the number of labelled TSD-SJD

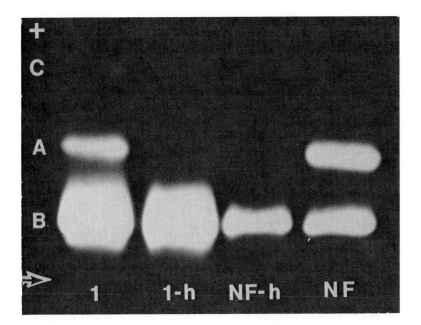

Fig. 5. Heat inactivation of Hex A activity in cell extracts.
1: Sendai virus-treated mixture of TSD and SJD fibroblasts after
one week in culture; 1-h: same extract as 1, heated at 52°C for
3 hours; NF: normal fibroblasts; NF-h: same as NF, heated at
52°C for 3 hours. Electrophoresis and notation as in Fig. 4.
heterokaryons observed by autoradiography in the parallel fusion experiment, which represents 5.8% of all genomes capable
of synthesizing active Hex B (TSD, TSD-TSD homokaryons, and
TSD-SJD heterokaryons. If complementation in these cells restored Hex A activity to a level comparable to that of normal
fibroblasts, in which the ratio of Hex B: Hex A activity is
approximately one, then the Hex A activity of these heterokaryons should indeed amount to about 6% of the total Hex B activity.

A complete biochemical characterization of the newly formed
Hex A, including activity towards GM_2 ganglioside, should establish whether intergenic or interallelic complementation
has taken place, and help in the understanding of the structural relationships between the isozymes and the basic biochemical
defect in the two diseases.

Coincidence of the biochemical characteristics of complemented Hex A with those of control Hex A would be compatible
with intergenic complementation (Fincham, 1966), indicating that
the normal expression of Hex A results from interaction of products coded for by at least two independent genes. General-

451

izing the concept of the sialyl-transferase hypothesis of
Goldstone et al., (1971), this interaction could be of sequential nature. Hex B, coded for by a structural gene (mutated in SJD), would be converted to Hex A by the product of a
modifier gene (mutated in TSD) (Robinson and Carrol, 1972).
Complementation in the TSD-SJD heterokaryons would then result from the normal modifier of the SJD genome acting on the
product of the normal structural gene of the TSD genome. Alternatively, the interaction could be at a more basic structural
level, according to the hypothesis (Desnick et al., 1972; van
Someren and van Henegouwen, 1973; Ropers and Schwantes, 1973;
Srivastava and Beutler, 1974) that Hex A is composed of unlike
subunits $(\alpha\beta)n$, coded for by two genes, one of which, α, is
mutated in TSD, the other, β, in SJD; and that Hex A shares
common β subunits with Hex $B(2\beta)_n$ or $(\beta\gamma)_n$, as inferred from
the deficient activity of both Hex A and Hex B in SJD. Complementation of Hex A activity in TSD-SJD heterokaryons would
then result from assembly of Hex A from normal α subunits
coded for by the SJD genome, with normal β subunits coded for
by the TSD genome.

Differences in kinetic characteristics between complemented
Hex A and control Hex A, on the other hand, would indicate interallelic complementation (Fincham, 1966) and would be ex-
plainable by a hypothesis combining features of the two pre-
ceding ones. Hex B, composed of like subunits, would normally
be converted to Hex A by interaction of a low molecular weight
modifier with a "recognition site" adjacent to the enzyme active site on each subunit. Allelic mutations in TSD and SJD
would affect the subunits of Hex B, altering primarily, but
not exclusively, the recognition site in TSD, thus preventing
conversion to Hex A, and affecting primarily the active site in
SJD, but not necessarily preventing conversion to (inactive)
Hex A in this case. Kinetic abnormalities of Hex B in TSD
(Wenger et al., 1972), absence of enzymatically inactive material
cross-reacting with Hex A in TSD (Bartholomew and Rattazzi,1974:
Carroll and Robinson, 1973; Srivastava and Beutler, 1974), and
enzymatically inactive cross-reacting material with the mobilities of both Hex A and Hex B in one case of SJD (Srivastava and
Beutler, 1974) all support this hypothesis. Complementation of
the interallelic type would then take place in the TSD-SJD heterokaryons according to the "hybrid protein" model of Catcheside
and Overton (1958), with restoration of the recognition site,
and, to a certain degree, of the more complex active site. [2]

---

[2] All three hypotheses (modifier; different subunits; like
subunits and modifier) are complicated by the report, in the
same paper by Srivastava and Beutler (1974), of inactive cross-
reacting material, having the mobility of Hex A, but not of

5. β-Hexosaminidase in plasma.

In human plasma and serum, the electrophoretic mobilities of the major bands of β-hexosaminidase activity are different from those of both Hex A and Hex B from cell and organ extracts (Fig. 6). In TSD and SJD, the activity of these isozymes in plasma and serum is affected in the same way as that of Hex A and Hex B from tissue (O'Brien et al., 1970; Kolodny, 1972). Plasma β-hexosaminidase is thus clearly under the control of the same gene(s) as the tissue isozymes, and one can refer to the plasma isozymes as Hex A and Hex B. A minor β-hexosaminidase isozyme is also present in plasma; it is more evident in TSD (Fig. 1) but also visible in normal plasma after prolonged staining. Since it has the same mobility as Hex B from cell extracts, it may be due to contamination from WBC or platelets.

The difference in electrophoretic mobility between plasma and tissue Hex B had been detected by Stirling (1972) by starch gel electrophoresis. The low sensitivity of the starch gel system had allowed the detection of plasma Hex B only during pregnancy, when the activity of this form increases; therefore, the isozyme had been called "form P" (= pregnancy). Using the more sensitive cellulose acetate gel system, we have consistently observed plasma Hex B in the plasma of individuals of either sex of all age groups, including fetuses as early as 16 weeks of gestation. Hex B of plasma has also been called I (= intermediate mobility) by Price and Dance (1972) from its behaviour on ion-exchange chromatography. In contrast to organ β-hexosaminidase, plasma isozymes are readily affected by short incubation with purified neuraminidase from *V. cholerae* or *Cl. perfringens* (Fig. 7), both Hex A and Hex B moving to a less anodal position after treatment (see also Ikonne and Ellis, 1973; Swallow et al., 1974). Partially purified organ Hex A and Hex B added to plasma, and treated with neuraminidase, do not show any effect even after prolonged treatment, suggesting that sialic acid residues may be present or exposed in the plasma and not in the organ isozymes.

The coincidence of plasma Hex B with the pre-β lipoprotein zone, and of Hex A with the α-lipoprotein zone, by electrophoresis on cellulose acetate gel (system C) the increase of pre-β lipoproteins in pregnancy (Pantelakis et al., 1964) the presence of terminal sialic acid residues in β-lipoproteins

Footnote (2) continued: Hex B, in a second case of SJD. One should postulate in this case a stable Hex A arising from conversion of unstable Hex B by the modifier or, in the unlike-subunits hypothesis, a stable αβ combination and an unstable ββ (or βγ) combination.

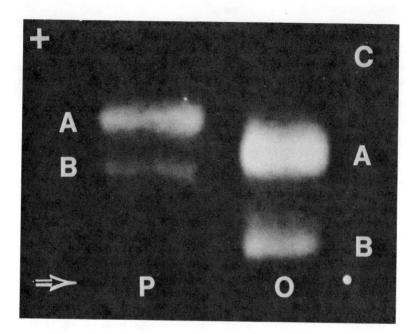

Fig. 6. Different mobility of β-hexosaminidase isozymes from plasma (P) and extract of organ (O). Electrophoresis at pH 6.0, 4 hours. Notation as in Fig. 4. Hex C was faintly visible only in the liver extract.

(Margolis and Langdon, 1966),the report of the association of several enzymes, including at least one lysosomal enzyme, with β-lipoproteins in human plasma (Lawrence and Melnick, 1961; Mermall and McDonald, 1972) prompted us to explore the possibility that the characteristics of plasma β-hexosaminidase could result from the interaction with plasma lipoproteins. However, ultracentrifugal flotation experiments (Havel et al., 1955) followed by assay for β-hexosaminidase (Leaback and Walker, 1961) showed that β-hexosaminidase was sedimenting even after centrifugation for 45 hours at 105,000 g, 16°C at density d=1.226. As expected, all plasma lipoproteins were recovered in the upper layer of the centrifugate. The β-hexosaminidase containing fractions showed the same electrophoretic pattern as untreated plasma, indicating that the observed sedimentation was not due to delipidation during the ultracentrifugal run. Immunoelectrophoresis with anti-α- and anti-β- lipoprotein immune sera also failed to yield substantial evidence that plasma β-hexosaminidase isozymes may

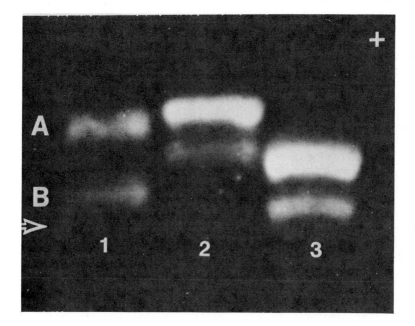

Fig. 7. Effect of treatment with purified *C. perfringens* neuraminidase on plasma β-hexosaminidase isozymes. 1: standard leukocyte extract; 2: plasma incubated at 37°C for 3 hours; 3: plasma incubated at 37°C for 3 hours in the presence of neuraminidase. Electrophoresis at pH 5.5, 4 hours.

be associated with any of the three major classes (High-, Low-, Very Low Density) of plasma lipoproteins. Although the plasma isozymes have a higher apparent molecular weight by gel filtration than the organ isozymes, recent data on molecular weight obtained by ultracentrifugation (Verpoorte, 1974) also seem to exclude association with a large lipoprotein molecule.

An electrophoretic survey of biological fluids other than plasma, including tears (Carmody et al., 1974) cerebrospinal fluid, mother's milk, saliva, urine, bile, and thoracic duct lymph has shown that only in lymph the β-hexosamidase electrophoretic pattern is of the plasma type, all others being of the organ type. Since extracts of intestinal mucosa and of whole small intestine do not show the plasma β-hexosaminidase pattern, it is likely that lymph β-hexosaminidase is derived from plasma, as are most other lymph proteins (Courtice, 1968).

The possibility that the characteristics of plasma β-hexosaminidase may reflect a more "native" configuration of the isozymes, discharged into the circulation by "regurgitation" from secondary lysosomes (Zurier et al., 1973), or an interaction with a plasma constituent after this discharge, is under investigation. Elucidation of the biochemical nature of the differences between plasma and organ β-hexosaminidase may be of importance in the development of replacement therapy in GM_2 gangliosidosis (Desnick et al., 1972; O'Brien, 1973; Hickman et al., 1974).

ARYLSULFATASE AND METACHROMATIC LEUKODYSTROPHY

1. Characteristics of the isozyme system.

Three isozymes of arylsulfatase are present in human tissues: ASA and ASB, which have acidic pH optima and are lysosomal, and ASC, which has a basic pH optimum and is tightly bound to microsomes (Harinath and Robins, 1971a,b; Perumal and Robins, 1973). The structural and functional relationships among these isozymes are not known. The two lysosomal isozymes differ in their isoelectric point, molecular weight, (Harinath and Robins, 1971b), immunological properties,(Neuwelt et al., 1971) activity towards natural substrate (Harzer et al., 1973) and salt sensitivity (Baum et al, 1959). The differential salt sensitivity is exploited for the assay of ASA activity in the presence of ASB, using p-NCS as a substrate (Baum et al., 1959). This differential inhibition assay is not very sensitive, and ASB activity may not be completely inhibited (Porter et al., 1971). Therefore, difficulties may arise when assaying small amounts of ASA activity in the presence of large amounts of ASB activity (Stumpf and Austin, 1971). Increased sensitivity is obtained by the use of the fluorogenic substrate 4 MUS (Rinderknecht et al., 1970) but no differential inhibition assay is available for this substrate. The problem may be overcome by electrophoretic separation of arylsulfatase isozymes on cellulose acetate gel using 4 MUS as substrate, which allows the study of ASA and ASB in small amounts of human tissue, leukocytes, and cultured cells (Rattazzi et al., 1973). In addition to its potential usefulness in the study of structure and function of the arylsulfatase isozymes, the method has immediate applications in clinical medicine.

2. Metachromatic leukodystrophy.

In man, deficient activity of ASA is associated with MLD, a progressive degenerative disorder of the nervous system, transmitted with an autosomal recessive mode of inheritance.

In this disease, accumulation of the sulfatide, cerebroside-3-sulfate, leads to degeneration of myelin and death in the first decade (Austin et al, 1965). Since deficient activity of sulfatide sulfatase coincides with deficient ASA activity in MLD (Jatzkewitz and Mehl, 1969), and since ASB cleaves the sulfatide in vitro at a much slower rate than ASA, (Harzer et al., 1973) it is believed that sulfatide is the natural substrate of ASA and that deficient activity of this enzyme leads to accumulation of sulfatide in MLD. Direct demonstration of ASA deficiency in leukocyte and cultured skin fibroblasts of patients with MLD has been possible using cellulose acetate gel electrophoresis and 4 MUS as a substrate (Fig. 8). The same deficiency of ASA activity can be shown in fibroblasts of patients with the adult form of MLD (Percy and Kaback 1971; Rattazzi et al., 1973), in which the symptoms develop much later in life and the disease has a more protracted course. However, despite the lack of demonstrable ASA activity,fibroblasts from patients with this disease can slowly degrade sulfatide added to the culture medium (Porter et al., 1971). An explanation of this phenomenon could be that ASB is responsible for significant sulfatide degradation *in vivo*. This possibility has been judged doubtful by Harzer et al., (1973) on the grounds that patients with a rare variant of MLD, in which the activity of ASA, ASB, and ASC is deficient (Murphy et al., 1971), live up to their 11th year of age. However, it is possible that allelic mutations may affect the activity of ASB towards sulfatide *in vivo* more severely in infantile MLD, less severely in the "trisulfatase deficiency" and even less severely in the adult-onset form of the disease. In each case, the altered structure of ASB would also result in absence of active ASA, by a mechanism analogous to that proposed for β-hexosaminidase in TSD and SJD (see above). These very cases of "trisulfatase deficiency" apparently due to a single autosomal recessive gene (Murphy et al., 1971), argue for a structural relationship among the three isozymes.

3. *Prenatal diagnosis of MLD.*

By electrophoretic demonstration of deficient ASA activity in leukocytes and cultured skin fibroblasts, we have diagnosed MLD in three patients from two sibships. In one of these families we have recently monitored a pregnancy at risk by examination of cultured amniotic fluid cells obtained by amniocentesis at the 15th week of pregnancy. Electrophoresis of extracts of these cells revealed absence of the ASA band (Fig. 9). Taking advantage of the fact that immune complexes between ASA and antibodies against this enzyme retain enzy-

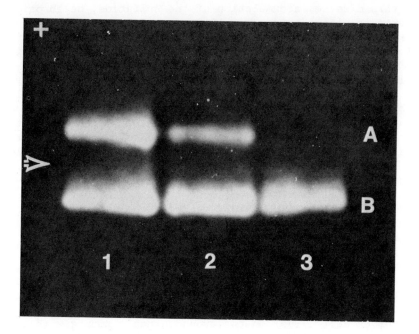

Fig. 8. Fluorescent bands of arylsulfatase activity after electrophoresis of fibroblast extracts on cellulose acetate gel and development with the fluorogenic substrate, 4MUS. 1: normal individual; 2: parent of patient with MLD; 3: patient with MLD. Notice decreased relative activity of band A in 2. A,B:ASA, ASB; arrow: point of application.

matic activity (Neuwelt et al., 1971), we could also confirm absence of ASA activity in these cells by immunoelectrophoresis and immunodiffusion. Although inactive material cross-reacting with ASA has been described in MLD (Neuwelt et al., 1971) the selectivity of immunological methods can still be exploited in this case, since only enzymatically active immune complexes are revealed by incubation with substrate.

Following termination of pregnancy, the diagnosis of MLD in the fetus was confirmed by demonstration of ASA deficiency in extracts of fetal organs examined by electrophoresis, together with suitably matched fetal controls. (Fig. 9). Furthermore, microscopic examination of sections of brain and kidney showed the presence of metachromatic granules of sulfatide already accumulating in the fetal cells.

In an attempt to overcome the problem of the time-lag between amniocentesis and availability of sufficient amounts

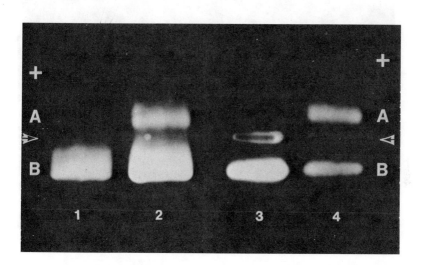

Fig. 9. Arylsulfatase activity in extracts of cultured
amniotic fluid cells and fetal liver. 1: pregnancy at risk
for MLD; 2: normal pregnancy; 3: liver of fetus with MLD
from pregnancy at risk; 4: liver of normal fetus. Activity
at the point of application (arrow) in lane 3 is Hex C, not
completely removed from the preparation of unfrozen fetal
liver. Electrophoresis and notation as in Fig. 8. The
pictures of two electrophoretic runs were combined for this
figure.

of cultured amniotic fluid cells for biochemical analysis, we
also examined the cell-free amniotic fluid from this pregnancy.
Surprisingly, electrophoresis at pH 7.0 of concentrated
amniotic fluid from the fetus with MLD showed an arylsulfatase
pattern indistinguishable from that of other amniotic fluid
samples from normal fetuses: In both normal and MLD amniotic
fluids the ASA band is replaced by a band with lower anodal
mobility. Incubation of gels at alkaline pH showed that this
band is not due to microsomal ASC. The same pattern was also
shown by an amniotic fluid sample from another fetus with MLD,
obtained through the courtesy of Dr. G.H. Thomas (Johns
Hopkins Hospital, Baltimore) (Fig. 10). Most interestingly, only
the slow anodal band of the normal amniotic fluid, and not

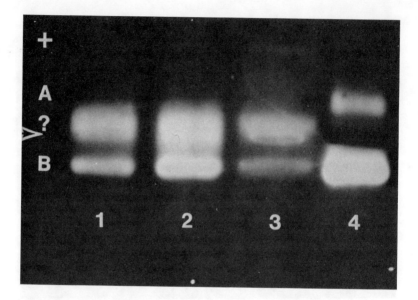

Fig. 10. Arylsulfatase isozymes in amniotic fluid. 1: normal pregnancy; 2: pregnancy in which MLD was diagnosed (same case as 1, 3 in Fig. 9); 3: pregnancy in which MLD was diagnosed by others; 4: standard leukocyte extract. Note band (indicated by ?) just anodal to application point, not coincident with ASA band of leukocyte extract, and present in all amniotic fluid samples. Dark zone in 4 is hemoglobin contaminating the leukocyte preparation.

the corresponding one of the MLD amniotic fluid, gave an enzymatically active immunoprecipitate by immunoelectrophoresis against an anti-human ASA immune serum. Incubation at 37° of cell homogenates prepared in normal amniotic fluid did not lower the mobility of ASA from cells. Conversely, adjusting the pH of the amniotic fluid to that of cell extract by dialysis against the appropriate buffer did not alter the mobility of the slow anodal band of amniotic fluid.

Further biochemical characterization of this arylsulfatase form in amniotic fluid may provide interesting data on the nature of the arylsulfatase isozymes.

CONCLUSIONS

The importance of the concept of isozymes and of the application of the isozyme methodology to clinical medicine is clearly exemplified by β-hexosaminidase (O'Brien, 1973b). Only through the recognition of the existence of multiple forms of this enzyme was it possible to detect the deficiency of Hex A in TSD which was masked in test tube assays by the increase of Hex B activity. Biochemical characterization of the isozymes led to development of reliable diagnostic procedures, especially valuable in prenatal diagnosis, and of methods applicable to screening programs for the detection of individuals heterozygous for the TSD gene. At the same time, biochemical and immunological data on the structural and functional relationships among the isozymes have been gathered, and continue to be obtained, that should lead to the elucidation of the basic biochemical defect in GM_2 gangliosidosis. This in turn should provide a better understanding of several human storage diseases, analogous to GM_2 gangliosidosis, and associated with deficiency of lysosomal acid hydrolases. Most important, knowledge of physiological role of the isozymes could provide data applicable to effective enzyme replacement therapy.

Note added in proof. After this manuscript was submitted for publication, two papers have appeared which are relevant to the data presented above. Tallman et al. (1974) have obtained significant degradation of Gm_2 ganglioside by Hex B in vitro, suggesting that Hex B may be at least in part responsible for the degradation of this glycolipid in vivo. This finding, together with the accumulation of Gm_2 ganglioside in TSD, supports the concept that Hex B may be altered in this disease, as hypothesized in the "like subunits and modifier" model proposed by us. Tallman et al. (1974) have also proposed a "conformational" model of structural relationships between Hex A and Hex B. According to these authors β-hexosaminidase, composed of similar subunits, exists first in a metastable conformation (Hex A), then in a conformation stabilized by disulfide bonds (Hex B). A structural mutation in TSD shifts the equilibrium to the stable Hex B form, with concomitant deficient activity towards Gm_2 ganglioside. A different mutation in SJD affects the activity but not the conformational properties of the enzyme. According to this model, interallelic complementation of Hex A deficiency in TSD-SJD heterokaryons would result from formation of a heteropolymeric β-hexosaminidase molecule, allowing for the existence of an active metastable Hex A.

461

Thomas et al. (1974) have also observed complementation of Hex A deficiency by Sendai virus treatment of TSD and SJD fibroblasts. Although the newly formed β-hexosaminidase activity has not been characterized immunologically, and no direct evidence of formation of heterokaryons has been presented, the results obtained by these authors are in close agreement with ours, and confirm the conclusion that TSD and SJD result from different mutations capable of complementation.

ACKNOWLEDGMENTS

Supported in part by grants from the National Institutes of Health, Public Health Service (GM 15874 and HD 06321) and from the Maternal and Child Health Service, U. S. Department of Health, Education and Welfare (Project 417).

Dr. Rattazzi is a recipient of a Public Health Service Research Career Development Award (K04 GM 70,638) from the Institute of General Medical Sciences.

REFERENCES

Ada, G. L.,E. L. French, and P. E. Lind 1961. Purification and properties of neuraminidase from *Vibrio cholerae*. *J. Gen. Microbiol*. 24: 409-421.

Austin, J., D. Armstrong, and L. Shearer 1965. Metachromatic form of diffuse cerebral sclerosis. *Arch. Neurol*. 13: 593-614.

Bartholomew, W. R., and M. C. Rattazzi 1974. Immunochemical characterization of human β-D-N-Acetyl hexosaminidase from normal individuals and patients with Tay-Sachs disease. I. Antigenic differences between hexosaminidase A and hexosaminidase B. *Int. Arch. Allergy* 46: 512-524.

Baum, H., K. S. Dodgson and B. Spencer 1959. The assay of arylsulphatases A and B in human urine. *Clin. Chim. Acta* 4: 453-455.

Bernsohn, J., and H. Grossman (Editors) 1971. *Lipid storage diseases. Enzymatic defects and clinical implications*. Academic Press, New York.

Carmody, P. J. and M. C. Rattazzi 1973. Is neuraminidase responsible for the *in vitro* conversion of human hexosaminidase A to hexosaminidase B? *Amer. J. Hum. Genet*. 25: 19A.

Carmody, P. J., and M. C. Rattazzi 1974. Conversion of human hexosaminidase A to hexosaminidase "B" by crude *Vibrio cholerae* neuraminidase preparations: merthiolate is the active factor. *Biochim. Biophys. Acta* · (in press).

Carmody, P. J., M. C. Rattazzi, and R. G. Davidson 1973. Tay-Sachs disease - The use of tears for the detection of heterozygotes. *N. Eng. J. Med.* 289: 1072-1074.

Carroll, M., and D. Robinson 1973. Immunological properties of N-acetyl-β-D-glucosaminidase of normal human liver and of Gm_2-gangliosidosis liver. *Biochem. J.* 131: 91-96.

Catcheside, D. G., and A. Overton 1958. Complementation between alleles in heterocaryons. *Cold Spring Harb. Symp. Quant. Biol.* 23: 137-140.

Courtice, F. C. 1968. The origin of lipoproteins in lymph. In: *Lymph and the lymphatic system,* edited by H. S. Mayerson. Charles C. Thomas, Springfield, Ill. p. 89-126.

Desnick, R. J., P. D. Snyder, S. J. Desnick, W. Krivit, and H. L. Sharp 1972. Sandhoff's disease: ultrastructural and biochemical studies. In: *Sphingolipids, sphingolipidoses and allied disorders,* edited by B. W. Volk and S. M. Aronson. Plenum Press, New York, p. 351-371.

Fincham, J. R. S. 1966. *Genetic complementation.* W. A. Benjamin, Inc., New York, p. 62-89.

Goldstone, A., P. Konecny and H. Koenig 1971. Lysosomal hydrolases: conversion of acidic to basic forms by neuraminidase. *FEBS Lett.* 13: 68-72.

Harinath, B. C., and E. Robins 1971,a.Arylsulphatases in human brain: assay, some properties, and distribution. *J. Neurochem.* 18: 237-244.

Harinath, B. C., and E. Robins 1971,b.Arylsulphatases in human brain: separation, purification, and certain properties of the two soluble arylsulphatases. *J. Neurochem.* 18: 245-257.

Harzer, K., K. Stinshoff, W. Mraz and H. Jatzkewitz 1973. The patterns of arylsulphatases A and B in human normal and metachromatic leucodystrophy tissues and their relationship to the cerbroside sulphatase activity. *J. Neurochem.* 20: 279-287.

Havel, R. J., H. A. Eder and J. H. Bragdon 1955. The distribution and chemical composition of ultracentrifugally separated lipoproteins in human serum. *J. Clin. Invest.* 34: 1345-1353.

Hayase, K., S. R. Reisher, and B. F. Miller 1973. Purification and separation of A and B forms of N-acetyl-β-D-hexosaminidase from human aorta. *Prep. Biochem.* 3: 221-241.

Hers, H. G., and F. Van Hoof (Editors) 1973. *Lysosomes and storage diseases.* Academic Press, New York.

Hickman, S., L. J. Shapiro, and E. F. Neufeld 1974. A recognition marker required for uptake of a lysosomal enzyme by cultured fibroblasts. *Biochem. Biophys. Res. Commun.* 57: 55-61.

Hooghwinkel, G. T. M., W. A. Veltkamp, B. Overdijk, and J. J. W. Lisman 1972. Electrophoretic separation of β-N-acetyl hexosaminidase of human and bovine brain and liver and of Tay-Sachs brain tissue. *Hoppe-Seyler's Z. Physiol. Chem.* 353: 839-841.

Ikonne, J. U., and R. B. Ellis 1973. N-Acetyl-β-D-hexosaminidase component A. Different forms in human tissues and fluids. *Biochem. J.* 135: 457-462.

Jatzkewitz, H., and E. Mehl 1969. Cerebroside-sulphatase and arylsulphatase A deficiency in metachromatic leukodystrophy (ML). *J. Neurochem.* 16: 19-28.

Kolodny, E. H. 1972. Sandhoff's disease: studies on the enzyme defect in homozygotes and detection of heterozygotes. In: *Sphyngolipids, sphyngolipidoses and allied disorders*, edited by B. W. Volk and S. M. Aronson. Plenum Press, New York. p. 321-341.

Kolodny, E. H., R. O. Brady, and B. W. Volk 1969. Demonstration of an alteration of ganglioside metabolism in Tay-Sachs disease. *Biochem. Biophys. Res. Commun.* 37: 526-531.

Lalley, P. A., M. C. Rattazzi, and T. B. Shows 1974. Human β-D-N-acetyl hexosaminidases A and B: Expression and linkage relationships in somatic cell hybrids. *Proc. Nat. Acad. Sci.* 71: 1569-1573.

Lawrence, S. H., and P. J. Melnick 1961. Enzymatic activity related to human serum beta-lipoprotein: histochemical, immunoelectrophoretic and quantitative studies. *Proc. Soc. Exp. Biol. Med.* 107: 998-1001.

Leaback, D. H., and P. G. Walker 1961. Studies on glucosaminidase. The fluorometric assay of N-acetyl-β-glucosaminidase. *Biochem. J.* 78: 151-156.

Li, Yu-Teh, M. Y. Mazzotta, C.-C. Wan, R. Orth and Su-Chen Li 1973. Hydrolysis of Tay-Sachs ganglioside by β-hexosaminidase A of human liver and urine. *J. Biol. Chem.* 248: 7512-7515.

Lyons, L.B., R.P. Cox and J. Dancis 1973. Complementation analysis of maple syrup urine disease in the heterokaryons derived from cultured human fibroblasts. *Nature* 243:533-535.

Margolis, S., and R. G. Langdon 1966. Studies on human serum β$_1$-lipoprotein I. Amino acid composition. *J. Biol. Chem.* 241: 469-476.

Mermall, H. L., and H. J. McDonald 1972. Release of enzymic activity from human beta-lipoproteins by sonic radiation. *Proc. Soc. Expt. Biol. and Med.* 141: 735-739.

Murphy, J. V., and L. Craig 1972. Neuraminidase-induced changes in white blood cell hexosaminidase A. *Clin. Chim. Acta* 42: 267-272.

Murphy, J. V., H. J. Wolfe, E. A. Balazs, and H. W. Moser 1971. A patient with deficiency of arylsulfatase A, B, C and steroid sulfatase, associated with storage of sulfatide, cholesterol sulfate, and glycosaminoglycans. In: *Lipid storage diseases. Enzymatic defects and clinical implications.* Edited by J. Bernsohn and H. J. Grossman. Academic Press, New York. p. 67-110.

Nadler, H. L., C. M. Chacko, and M. Rachmeler 1970. Interallelic complementation in hybrid cells derived from human diploid strains deficient in galactose-1-phosphate uridyl transferase activity. *Proc. Nat. Acad. Sci.* 67: 976-982.

Neuwelt, E., D. Stumpf, J. Austin, and P. Kohler 1971. A monospecific antibody to human sulfatase A. Preparation, characterization and significance. *Biochim. Biophys. Acta* 236: 333-346.

O'Brien, J. S. 1973, a. Tay-Sachs' disease and juvenile Gm_2-gangliosidosis. In: *Lysosomes and storage diseases,* edited by H. G. Hers and F. Van Hoff. Academic Press, New York. p. 323-344.

O'Brien, J. S. 1973,b. Tay-Sachs disease: from enzyme to prevention. *Federation Proceedings* 32: 191-199.

O'Brien, J. S., S. Okada, A. Chen, and D. Fillerup 1970. Tay-Sachs disease. Detection of heterozygotes and homozygotes by serum hexosaminidase assay. *New Eng. J. Med.* 283:15-20.

Okada, S., and J. S. O'Brien 1969. Tay-Sachs disease: generalized absence of a Beta-D-N-Acetyl hexosaminidase component. *Science* 165: 698-700.

Okada, S., M. McCrea, and J. S. O'Brien 1972. Sandhoff's Disease (Gm_2 gangliosidosis type 2): clinical, chemical, and enzyme studies in five patients. *Ped. Res.* 6: 606-615.

Paigen, K. 1971. The genetics of enzyme realization. In: *Enzyme synthesis and degradation in mammalian systems,* edited by M. Rechcigl, Jr. University Park Press, Baltimore, p. 1-46.

Pantelakis, S. N., A. S. Fosbrooke, J. K. Lloyd, and O. H. Wolff 1964. The nature and occurrence of pre-beta lipoprotein in diabetic children and pregnant women. *Diabetes.* 13: 153-160.

Percy, A. K., and M. M. Kaback 1971. Infantile and adult-onset metachromatic leukodystrophy. *N. Eng. J. Med.* 285:785-787.

Perumal, A. S., and E. Robins 1973. Arylsulphatases in human brain: purification and characterization of an insoluble arylsulphatase. *J. Neurochem.* 21: 459-471.

Poenaru, L., and J. C. Dreyfus 1973. Electrophoretic studies of hexosaminidases. Hexosaminidase C. *Clin. Chim. Acta* 43: 439-442.

Porter, M. T., A. L. Fluharty, J. Trammell, and H. Kihara 1971. A correlation of intracellular cerebroside sulfatase activity in fibroblasts with latency in metachromatic leukodystrophy. *Biochem. Biophy. Res. Commun.* 44: 660-666.

Price, R. G., and N. Dance 1972. The demonstration of multiple heat stable forms of N-acetyl-β-glucosaminidase in normal human serum. *Biochim. Biophys. Acta* 271: 145-153.

Rattazzi, M. C. 1968. Glucose-6-phosphate dehydrogenase from human erythrocytes: molecular weight determination by gel filtration. *Biochem. Biophys. Res. Commun.* 31: 16-24.

Rattazzi, M. C., and R. G., Davidson 1972. Prenatal detection of Tay-Sachs disease. In: *Antenatal Diagnosis*, edited by A. Dorfman. The University of Chicago Press, Chicago. p. 207-210.

Rattazzi, M. C., J. S. Marks, and R. G., Davidson 1973. Electrophoresis of Arylsulfatase from normal individuals and patients with metachromatic leukodystrophy. *Amer. J. Hum. Genet.* 25: 310-316.

Rattazzi, M. C., J. A. Brown, R. G. Davidson and T. B. Shows 1974. Tay-Sachs and Sandhoff-Jatzkewitz diseases: complementation of β-hexosaminidase A deficiency by somatic cell hybridization. In: *New Haven Conference, Second International Workshop on Human Gene Mapping.Birth Defects:Original Article Series.* The National Foundation, New York (in press).

Rinderknecht, H., M. C. Geokas, C. Carmack, and B. J. Haverback 1970. The determination of arylsulfatases in biological fluids. *Clin. Chim. Acta.* 29: 481-491.

Robinson, D., and M. Carroll 1972. Tay-Sachs disease: interrelation of hexosaminidases A and B. *Lancet.* i: 322-323.

Robinson, D., and J. L. Stirling 1968. N-Acetyl-β-glucosaminidase in human spleen. *Biochem. J.* 107: 321-327.

Ropers, H. H., and V. Schwantes 1973. On the molecular basis of Sandhoff's disease. *Humangenetik* 20: 167-170.

Sandhoff, K., U. Andreae, and H. Jatzkewitz 1968. Deficient hexosaminidase activity in an exceptional case of Tay-Sachs disease with additional storage of kidney globoside in visceral organs. *Life Sci.* 7: 278-285.

Sandhoff, K., K. Harzer, W. Wässle and H. Jatzkewitz 1971. Enzyme alterations and lipid storage in three variants of Tay-Sachs disease. *J. Neurochem.* 18: 2469-2489.

Sandhoff, K., and W. Wässle 1971. Anreicherung und Charakterisierung zweier Formen der menschlichen N-Acetyl-β-D-hexosaminidase. *Hoppe-Seyler's Z. Physiol. Chem.* 352: 1119-1133.

Silagi, S., G. Darlington, and S. A. Bruce 1969. Hybridization of two biochemically marked human cell lines. *Proc. Nat. Acad. Sci.* 62: 1085-1092.

Siniscalco, M., H. P. Klinger, H. Eagle, H. Koprowski, W. Y.
Fujimoto, and J. E. Seegmiller 1969. Evidence for intergenic complementation in hybrid cells derived from two
human diploid strains each carrying an X-linked mutation.
Proc. Nat. Acad. Sci. 62: 793-799.

Snyder, P. D., W. Krivit, and C. C. Sweeley 1972. Generalized
accumulation of neutral glycosphingolipids with Gm_2 ganglioside accumulation in the brain. *J. Lipid Res.*
13: 128-136.

van Someren, H,, and H. B. van Henegouwen 1973. Independent
loss of human hexosaminidases A and B in man-chinese
hamster somatic cell hybrids. *Humangenetik.* 18: 171-174.

Srivastava, S. K., Y. C. Awashti, A. Yoshida, and E. Beutler
1974,a. Studies on human β-D-N-Acetyl hexosaminidases I.
Purification and properties. *J. Biol. Chem.* 249: 2043-
2048.

Srivastava, S. K., A. Yoshida, Y. C. Awashti, and E. Buetler
1974,b. Studies on human β-D-N-Acetyl hexosaminidases II.
Kinetic and structural properties. *J. Biol. Chem.*
249: 2049-2053.

Srivastava, S. K., and E. Beutler 1974. Studies on human β-D-N-
Acetyl hexosaminidases III. Biochemical genetics of Tay-
Sachs and Sandhoff's diseases. *J. Biol. Chem.* 249:
2054-2057.

Stirling, J. L. 1972. Separation and characterization of N-
acetyl-β-glucosaminidases A and B from maternal serum.
Biochim. Biophys. Acta. 271: 154-162.

Stumpf, D., and J. Austin 1971. Metachromatic leukodystrophy
(MLD) IX. Qualitative and quantitative differences in
urinary Arylsulfatase A in different forms of MLD.
Arch. Neurol. 24: 117-124.

Swallow, D. M., D. C. Stokes, G. Corney, and Harry Harris
1974. Differences between the N-acetyl hexosaminidase
isozymes in serum and tissues. *Ann. Hum. Genet., Lond.*
37: 287-302.

Tallman, J. F., W. G. Johnson, and R. O. Brady 1972. The
metabolism of Tay-Sachs ganglioside: catabolic studies
with lysosomal enzymes from normal and Tay-Sachs brain
tissue. *J. Clin. Invest.* 51: 2339-2345.

Tallman, J.F., R.O. Brady, J.M. Quirk, M. Villalba and A.E.
Gal 1974. Isolation and relationship of human hexosaminidases. *J. Biol. Chem.* 249:3489-3499.

Tateson, R., and A. D. Bain 1971. Gm_2 gangliosidoses: consideration of the genetic defects. *Lancet* ii: 612-613.

Thomas, C.H., H.A. Taylor, C.S. Miller, J. Axelman and B.R.
Migeon 1974. Genetic complementation after fusion of
Tay-Sachs and Sandhoff cells. *Nature* 250:580-582.

Verpoorte, J. A. 1974. Isolation and characterization of the major β-N-Acetyl-D-glucosaminidase from human plasma. *Biochemistry*. 13: 793-799.

Wenger, D. A., S. Okada, and J. S. O'Brien 1972. Studies on the substrate specificity of Hexosaminidase A and B from Liver. *Arch. Biochem. Biophys.* 153: 116-129.

Young, E. P., R. B. Ellis, B. D. Lake, and A. D. Patrick 1970. Tay-Sachs disease and related disorders. Fractionation of brain N-acetyl-β-hexosaminidase on DEAE cellulose. *FEBS Lett.* 9: 1-4.

Zurier, R. B., S. Hoffstein and G. Weissman 1973. Mechanism of lysosomal enzyme release from human leukocytes I. Effect of cyclic nucleotides and colchicine. *J. Cell Biol.* 58: 27-41.

DNA-DEPENDENT RNA POLYMERASE SPECIES FROM RAT LIVER TISSUE

B.J. BENECKE, A. FERENCZ AND K.H. SEIFART

Institut für Physiologische Chemie
355 Marburg/L.
Lahnberge
Bundesrepublik Deutschland

ABSTRACT: Multiple forms of DNA dependent RNA polymerase have been isolated from rat liver tissue and HeLa-cells. The subunit structure of the rat liver B enzyme as well as that of the A and B enzymes from HeLa-cells have been investigated. In addition to the enzymes A and B from rat liver, a third enzyme has been described and designated as enzyme C. This enzyme demonstrates an intermediate sensitivity towards the specific inhibitor α-amanitin and is only inhibited by high concentrations of the toxin. Similar characteristics have been described by Roeder for enzyme III from plasmacytoma cells. Studies directed at the biological half-life of the individual polymerase species have demonstrated that these enzymes have extended half-life periods. The conclusion has been drawn that the concentration of RNA polymerase is probably not a regulating factor for the rate of RNA synthesis within the cell and that other factors may be controlling the rate of initiation. The polyanion heparin has been used to study the formation of specific initiation complexes. Enzyme A which is initiated in vivo and isolated as such in association with the rat liver nucleolus is completely resistent toward the polyanion heparin if assayed in vitro. In contrast, purified enzyme A does not form heparin resistent complexes on isolated DNA to a comparable extent and it is concluded that specific initiation complexes may not be formed by the isolated enzyme under these conditions.

INTRODUCTION

Transcription and particularly the understanding of its control has made considerable progress in recent years. The elucidation of the structure of bacterial RNA polymerase (Burgess, Travers, Dunn, and Bautz, 1969), the use of phage genetics and the identification of various positive (Perlman et al., 1970; Zubay et al., 1970) and negative (Gilbert and Müller-Hill, 1967; Zubay et al., 1972; Pirotta et al., 1970) control elements has significantly advanced the understanding of the regulation of RNA synthesis in prokaryotes. However, the mechanism of differential gene-expression in eukaryotic organisms , although

469

pursued with considerable effort, is incompletely understood
at the present time. While structural aspects of the chromosomal template certainly play a role in this connection, the
ubiquitous finding of the occurence of multiple forms of DNA
dependent RNA polymerase in all eukaryotic cells studied thus
far (for a review, see Cold Spring Harbor Symposium, 1970)
implied a possible role for these enzymes in regulating the
transcription of specific genes or groups of genes. This report will deal with the multiplicity of the rat liver RNA
polymerases, their biological half-life, and implications which
this parameter has for the regulation of transcription, and
finally, the question of measuring specific initiation complexes
between RNA polymerase A and DNA.

MULTIPLICITY OF RNA POLYMERASE FROM EUKARYOTIC CELLS

The multiple forms of RNA polymerase from sea urchin and
rat liver have initially been designated by Roeder and Rutter
(1969) as I, II, and III according to their elution-order from
DEAE-Sephadex and by Chambon and his associates (Kedinger et
al. 1970) as A and B for the enzymes from calf thymus according
to their sensitivity toward the mushroom toxin α-amanitin, which
at low concentrations specifically inhibits the B-enzyme (Seifart and Sekeris, 1969; Kedinger et al. 1970). General agreement exists that enzyme A (or I) is structurally associated
with the nucleolus and presumably transcribes the ribosomal
genes, whereas enzyme B (or II) is located in the nucleoplasm.
It has been shown that the B enzyme consists of more than one
sub-class (Weaver et al., 1971; Kedinger et al. 1974) and that
the A-Enzyme may likewise represent a multicomponent group
(Chesterton et al., 1971; Smuckler et al., 1971; Sajdel et al.,
1971) although the functional significance of this finding is
not clear at present.

The subunit structure of a number of eukaryotic RNA polymerase species has been established by different laboratories
and it reveals certain basic similarities to the molecular
architecture of the RNA polymerase from *E. coli* although definite differences have been found for the molecular weight and
stoichiometry of the individual subunits. In agreement with
other laboratories we have previously shown (Seifart et al.
1972) that the B enzyme from rat liver contains subunits of
200, 175, 140, 38, 22, 16 x 10^3 daltons, probably representing
a mixture of B_I and B_{II} (Kedinger et al. 1974). Subsequent
structural studies on HeLa cells have shown (Fig. 1) that the
A enzyme consists of the following subunits: 185, 128, 65, 41,
32 x 10^3 daltons, and that the B enzyme contains high molecular
weight subunits of 210, 165, 140 x 10^3 daltons. The low molecular weight subunits are less clear in this latter case. This is

470

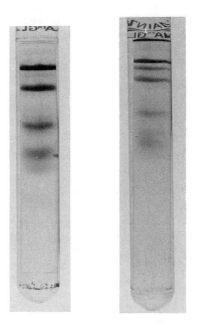

Figure 1. Comparison of the high molecular weight subunits of
RNA Polymerase A (left) and B (right) of Hela Cells on 5% SDS
polyacrylamide gels according to the method of Weber and Osborn
(1969). Electrophoresis was conducted toward the anode for 5
hours at 7 mA per tube, stained for 2 hours in 0.25% Coomasie
brilliant blue, 40% methanol, 7.5% acetic acid and destained
in 20% methanol 7.5% acetic acid.

in fair agreement with values found by other workers (Kedinger
et al., 1974; Sugden and Keller, 1973; Roeder, this volume).
Until the enzymes have been reconstructed it is not completely
clear which of the low molecular weight subunits in fact belong
to the enzymes and which of them may possibly represent low
molecular weight impurities.

Enzyme III initially found by Roeder and Rutter (1969) upon
DEAE Sephadex chromatography and demonstrating resistence toward
α-amanitin up to concentration of 3μg/ml has not been found in
all tissues investigated. It is possible that this may be related to fluctuating quantities of the enzyme in the nuclei of
different organisms, or different states of synthetic activity.
In addition, it appears that this enzyme can be analyzed most
clearly on DEAE Sephadex and may not be resolved upon chromatography on DEAE-cellulose (unpublished results).

It has been observed that the particulate free cytosol of

the liver cell contains an RNA polymerizing activity also reported to occur in thymus and ascites cells (Amalric et al.,
1972) and which has been designated as enzyme C. This enzyme
is a true DNA-dependent RNA polymerase by all catalytic criteria presently available. Its high molecular weight (~ 500,000
daltons) and resistence toward rifampicin render unlikely
that the enzyme is of mitochondrial origin (Reid and Parsons
1971) but do not distinguish it from the nuclear enzymes A
and B. Chromatography on DEAE-cellulose shows, however, that
only a single peak of activity is present in the cytoplasm
(Seifart et al. 1972) eluting at or very close to the elution
point of enzyme A on this exchanger.

Since the initial differentiation of enzymes A and B was
achieved by the inhibitor α-amanitin, titration curves were
conducted with enzymes A, B, and C. The result (Fig. 2) clearly differentiates all three enzymes. Pure preparations of enzyme A (and isolated rat liver nucleoli containing this enzyme)
are completely resistent up to very high concentrations of the
inhibitor, which very effectively inactivates enzyme B at an
enzyme-inhibitor stoichiometry of 1:1 (Seifart et al. 1969,
Chambon et al. 1970). This result renders it very unlikely
that the enzyme in question emanates from the nuclear enzymes
A or B in any variable ratio. This finding is substantiated
by an experiment in which α-amanitin was injected into rats in
vivo for 2 hours at a concentration of 100μg/100g body weight.
It is known that amanitin will not bind to enzyme A, but will
form a very stable complex with enzyme B which is maintained
through the subsequent extraction and purification procedures
(Chambon et al, 1972). Therefore this approach enables one to
freeze the *in situ* situation and allows certain conclusions
concerning the distribution of the enzymes before the preparation disrupts the cellular integrity. If the enzymes are subsequently extracted from nuclei and cytoplasm of such amanitin
treated animals and analyzed on DEAE-cellulose the results in
Fig. 3 Part I, concerning the nuclear enzymes, clearly show
that enzyme B is completely inhibited by amanitin treatment
in vivo, whereas enzyme A is virtually not effected. In contrast enzyme C, (Fig. 3, Part II) although partly inhibited,
is clearly present. This activity cannot be due to enzyme B
since this component was completely eliminated, and it is not
A since that component is not inhibited.

It is therefore fairly certain, that enzyme C represents
an entity distinct from the polymerases A and B. The enzyme
is most readily isolated from the cytoplasm as has been described. It is very likely, however, that this enzyme exerts
its function within the nucleus but very readily equilibrates
with the cytoplasm. The reason for this equilibration and the

472

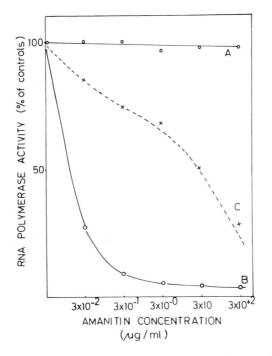

Figure 2.   Effect of increasing concentrations of α-amanitin
on the activity of RNA polymerase species A, B, and C.   Concentrations reflect final concentrations in the assay.   The enzymes
were purified and assayed as described previously (Seifart et
al 1972).

fact that this phenomenon does apparently not apply to the enzymes A and B is not clear at present.   It is possible that the
enzyme is synthesized in great excess over what is required in
the interphase nucleus of the adult rat liver.   In this connection it is interesting to note that 12 hours after partial
hepatectomy, when RNA synthesis is dramatically stimulated but
DNA synthesis has not yet been initiated (Tsukada and Liebermann 1965), the amount of RNA polymerase enzymes A and B extracted from the nuclei and analyzed on DEAE-cellulose chromatography is increased approximately two-fold.   At the same time,
however, the quantity of enzyme C in the cytoplasm is reduced
by 15-20% (unpublished results).   Further experiments must
clarify whether this interesting finding may be related to a
shift of enzyme C into the nucleus under certain physiological
conditions.

   In agreement with our findings (Seifart et al., 1972), Wilhelm at al. (1974) have reported an enzyme from Xenopus oocytes
demonstrating an intermediate sensitivity toward α-amanitin and

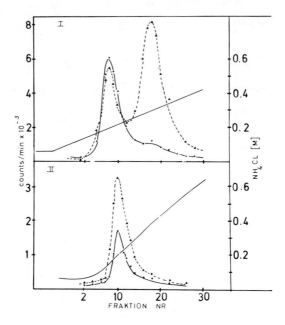

Figure 3. DEAE-cellulose chromatography of RNA polymerase isolated from nuclei (part I) or cytoplasm (part II) from control
(▲– – – –▲) or amanitin treated (●━━━━━●) animals.
Details of the chromatography are as described in the legend
of Figs. 4 and 5.

Roeder (this volume) has made the interesting observation that
polymerase III from mouse plasmacytoma cells demonstrates identical inhibition kinetics toward α-amanitin when compared to
enzyme C from rat liver. Enzyme III from plasmacytoma cells
has been suggested to transcribe the genes for 5 S ribosomal
RNA (Roeder, this volume) although Penman et al. (1974) have
found that enzyme III from HeLa-cells may transcribe t-RNA
genes. This question, as well as possible functional relationships between enzymes III and C, remain to be clarified
by future experiments.

STUDIES CONCERNING THE BIOLOGICAL HALF-LIFE OF RNA POLYMERASES

We attempted to determine the biological half life for the
individual RNA polymerase species since this parameter obviously has very important implications for the regulation of RNA
synthesis. For this purpose protein biosynthesis was blocked
by cycloheximide administered to rats for time periods between
5 min and 24 hr before sacrifice. The antibiotic was admin-

istered at a concentration of 400 µg/100 g body weight, which
resulted in inhibition values of protein biosynthesis of
90% and 75% at 4 and 12 hours respectively. After 24 hours,
protein synthesis had recovered to 66% of control values. We
did not attempt to block peptide formation completely, since
pilot experiments had shown that concentrations of cycloheximide which achieve this purpose will also kill the animals
within 6 hours, thus preventing an assessment of the biological
half-life of slowly turning over proteins.

Nuclei and cytoplasm of control and cycloheximide treated
animals were prepared at the times indicated in Figs. 4 and 5.
Extraction of the enzymes and chromatography on DEAE-cellulose
was conducted as described previously (Benecke et al. 1973).
The results (Fig. 4 and 5) show no significant differences for
the enzymes extractable from control and cycloheximide treated
groups at any of the times investigated, although protein syn-

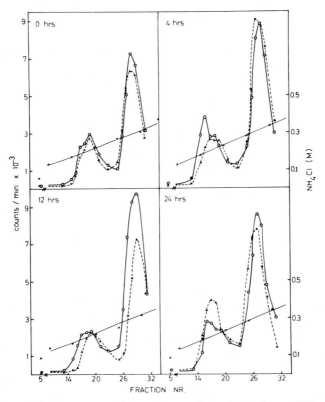

Figure 4. Silmutaneous chromatography of nuclear RNA polymerase activities from control (▲– – – –▲) and cycloheximide

Fig. 4 (con't) treated (O————O) animals on identical 2.5 x 5 cm DEAE-cellulose columns. Elution of approx. 3.5 ml fractions was achieved with a linear gradient from 0.0 to 0.35 NH_4 Cl in buffer (Tris-Cl, 0.05 M, pH 7.9, 20% glycerol, 5 mM mercaptoethanol, 0.25 mM EDTA) pumped from an ultrograd mixer at a flow rate of 40 ml/hr. In order to achieve identical conditions all columns were packed from the same batch of exchanger and were developed in pairs at identical elution rates from one pump (with multiple outlets) from the same elution medium. From Benecke et al. (1973).

---

thesis was dramatically inhibited as described before. Additional experiments (data not shown) with 3.0 mg cycloheximide/ 100 g body weight for 4 hours showed a complete inhibition of peptide synthesis, but enzyme levels in excess of 85% of the control values. It is quite clear that cycloheximide at some point in concentration and time will inhibit the synthesis of RNA polymerase which, as a central biomolecule, must obviously be turning over. The data reported here indicate, however, that this may be a comparatively slow process and that the enzymes studied represent a population of molecules with extended half-life periods. Consequently it is clear, that the enzymes are present in excess under normal conditions and that short-term fluctuations in the rate of RNA synthesis may not be controlled through the factor of enzyme concentration.

It is very clear, that the application of cycloheximide will dramatically inhibit both the accumulation of RNA in vivo (Muramatsu et al. 1970) and RNA synthesis in vitro by nuclei isolated from such animals. Yu and Feigelson (1972) concluded that in vivo administration of cycloheximide resulted in a progressive diminuation in nucleolar polymerase A activity with a $t_{1/2}$ of approximately 1.5 hours. More recent observations (Lampert and Feigelson 1974) seem to show, however, that this measurement rather reflects the amount of polymerase bound within the nucleolus than the total amount of enzyme present. It therefore appears very likely, that the inhibition of RNA synthesis after cycloheximide does not reflect a disappearance of the enzyme but rather demonstrates the dependence on continued protein biosynthesis for the enzyme to express its biological activity in the nucleus or nucleolus. Similar observations have been made in cultures of Chlorella (Wanka and Moors, 1970) where cycloheximide inhibits DNA synthesis, probably without altering the amount of enzyme present. In addition it has been shown by Roeder (1974) that developmental changes of the oocytes of *Xenopus laevis*, which are characterized by dramatic fluctuations in the rate of ribosomal RNA synthesis, are not accompanied by a corresponding fluctuation in the ratio of

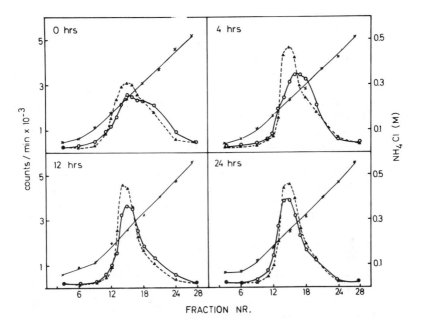

Figure 5. Simultaneous chromatography of the cytosol from control (▲— — — —▲) and cycloheximide treated (O———————O) animals on identical 2.5 x 5 cm DEAE-cellulose columns. Conditions were as described in Fig. 4. In this case the developing gradient was from 0.0 to 0.60 M NH$_4$Cl. From Benecke et al. (1973).

the corresponding enzymes. It is possible that factors that grossly affect the conditions of cellular growth, as is the case in complete tissue replacement after partial hepatectomy as discussed before or during malignant growth (see Roeder, this volume), lead to increased de novo synthesis of enzyme, since preexisting levels may no longer be able to cope with the increased demand. These are long-term effects related to cellular growth, however, and their regulation may be completely different from short-term fluctuations in the rate of transcription.

THE USE OF HEPARIN TO MEASURE SPECIFIC INITIATION COMPLEXES BETWEEN DNA AND RNA POLYMERASE A

The question concerning the amount of RNA polymerase actually engaged in transcription is of extreme importance in connection with the role which the concentration of RNA polymerase

plays in the regulation of transcription. If the enzyme is
present in excess, as has been deduced from the previous section, then clearly the rate of initiation must be a regulating
factor. This is in agreement with conclusions drawn by Cedar
and Felsenfeld (1973) and emphasizes the importance of measuring the initiation reaction. The complicated process of RNA
synthesis can be divided into several steps which have been
well characterized for the bacterial RNA polymerase (eg. Burgess,
1971) and more recently for the enzyme from animal cells (Meilhac and Chambon, 1973). The initial steps of the reaction involve binding of the enzyme to DNA and the formation of an initiation complex. Once bound into such a true initiation complex,
the enzyme from *E. coli* becomes resistent to compounds like
rifampicin(Kerrich-Santo and Hartmann, 1974; Bautz and Bautz
1970), poly I (Bautz et al. 1972) or heparin (Schäfer et al.
1973) which are known to act as inhibitors of free polymerase,
i.e. enzyme which is not bound to DNA. These effects of heparin
have been extended to study the interaction of DNA dependent
RNA polymerase A and DNA under various conditions.

It has been found, that heparin in minute quantities completely inactivates purified RNA polymerase A from rat liver
(Fig. 6). Moreover, the mere binding of the enzyme to DNA does

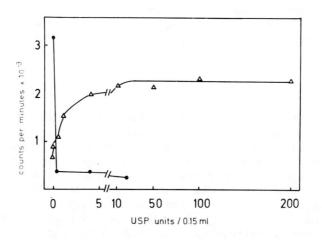

Figure 6. Effect of heparin in vitro on RNA synthesis catalyzed
by enzyme A (●————●) and isolated nucleoli (△————△)
respectively. Heparin was added to the enzyme or nucleoli
before the reaction was started. Nucleoli were isolated from
purified rat liver nuclei by the method of Higashinakagawa et
al. (1972).

not confer resistence toward this inhibitor (Fig. 7.). In
addition it has been found (Ferencz and Seifart 1974) that the
purified A-Enzyme demonstrates only a very limited heparin resistence while transcribing isolated DNA and it is therefore

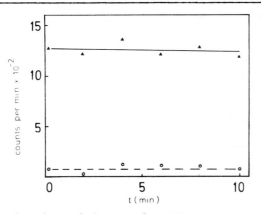

Figure 7. Kinetics of decay of a DNA: enzyme complex in the
presence of heparin (O— — — —O). Purified RNA polymerase A
was incubated with DNA and Mg^{++} for 15 min. at 37° at which
time heparin (0.4 USP units/assay) was added. After the times
indicated, nucleoside triphosphates were added and the incubation was continued for 10 minutes. As a control, RNA polymerase A was preincubated in an identical fashion, but buffer
(▲———————▲) instead of heparin was added, after which the
experimental regime was identical as above.

questionable whether the interaction of the purified enzyme
and isolated DNA represents transcription from true initiation
complexes. In contrast to these results, it was found (Fig.
6) that RNA polymerase A that was initiated in vivo and isolated in association with the nucleolus is resistent to heparin
in vitro. Figure 6 also demonstrates that heparin stimulates
RNA synthesis in isolated nucleoli. This apparently paradoxical effect is due to (i) an inhibition of ribonuclease contained
in the nucleoli (Boctor et al., 1974) and (ii) a dissociation
of chromosomal proteins within the nucleolus leading to a faci-
litation of the elongation reaction of previously bound enzyme.
Both effects collectively cause an increase in the size of the
transcript (Ferencz and Seifart 1974). The significant point
is however, that those polymerase molecules that were initiated
in vivo are not inhibited by heparin in vitro and complete at
least one round of transcription. An interesting aspect concerns the apparent abundance of RNA polymerase A within the
nucleolus. If isolated nucleoli are titrated with increasing

amounts of DNA (Fig. 8) RNA synthesis can clearly be stimulated. This indicates that the nucleolus contains enzyme which is not engaged in transcription and which can be transferred to an exogenously added template. This stimulation of RNA synthesis can clearly be blocked by the prior addition of heparin (Fig. 8). Similar conclusions are reached by time-kinetic analyses which show that nucleolar RNA synthesis plateaus at approximately 20 minutes. From UTP and actinomycin-chase experiments it can be concluded that RNase contributes toward, but is not primarily responsible for this plateau. If exogenous DNA is added after the plateau has been reached, synthesis is resumed (Fig. 9.), possibly because free enzyme in the nucleolus can utilize this template. If nucleoli are preincubated with heparin their initial rate of synthesis and the plateau is higher.

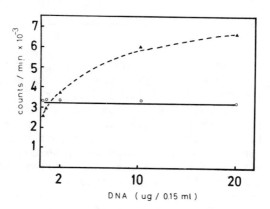

Figure 8. Effect of various DNA concentrations on RNA synthesis of control (▲– – – –▲) and heparin treated (O————O) nucleoli in vitro.

However, addition of DNA to these nucleoli will have absolutely no effect (Fig. 9). This indicates that the free enzyme was complexed by heparin during the preincubation period.

The ability of heparin to quench the DNA-effect depends strongly on the incubation time during which the nucleoli are allowed to react with the exogenous template before heparin is added. To demonstrate this, nucleoli were incubated with either buffer of DNA for various times, after which heparin was added and the resulting synthesis measured for a constant incubation period. The data are summarized in Fig. 10 and represent the difference in RNA synthesis between incubation mixtures containing either nucleoli or nucleoli and DNA as a function of the preincubation time. The results show, that preincubation of nucleoli and DNA at 37° results in a certain amount of

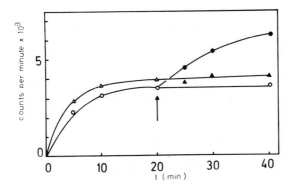

Figure 9. Effect of exogenously added DNA on nucleolar RNA synthesis in the presence (△————△) or absence (O————O) of heparin. Where appropriate, heparin 25 USP units /0.15 ml was added to nucleoli before start of the reaction. In both cases DNA (20 ug/0.15 ml) was added after 20 minutes when synthesis had reached a plateau (arrow).

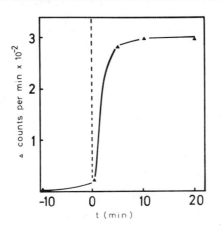

Figure 10. Effect of the preincubation time on the formation of heparin resistent complexes between DNA and nucleolar associated enzyme A. Nucleoli were incubated with either buffer or DNA (20 ug/0.15 ml) for various times at 37°. At these times, indicated in the figure, heparin (25 USP units/0.15 ml) was added to all incubation vials and the subsequent RNA synthesis was measured for a constant time period. The time point: -10 min. indicates, that heparin was added 10 minutes before DNA. The figure represents the difference in RNA synthesis at the appropriate times between nucleoli preincubated with DNA and those preincubated with buffer before heparin was added in either case.

481

heparin-resistent synthesis on DNA which reaches a maximum after a preincubation time of approximately 5 minutes. It therefore seems possible that the surplus enzyme present in nucleoli is still capable of forming such complexes. The exact extent of heparin resistance which is conferred, i.e. what proportion of the free enzyme is actually engaged in such a complex is, however, difficult to assess in a complex system like the intact nucleolus.

Collectively the results of this latter series of experiments allow the following conclusions. The A-enzyme contained in the nucleolus and initiated in vivo is bound into a heparin resistant complex. Such complexes are not formed between isolated nucleolar DNA purified RNA polymerase A, indicating that the enzyme: template interaction occurs in a much less specific fashion under these conditions. The nucleolus apparently contains an excess of enzyme A and preliminary experiments do not rule out that this enzyme is in fact able to form heparin resistant initiation complexes on isolated DNA. It is presently impossible to decide, however, whether heparin resistance is conferred by an integral part of the nucleolar enzyme, which may be lost during extraction and purification, or whether the chromosomal structure, requiring the entire array of chromosomal proteins, is required to direct the polymerase to correct initiation sites.

ACKNOWLEDGMENTS:

We greatfully acknowledge financial support of the Deutsche Forschungsgemeinschaft and the able technical assistence of Doris Schwarz. We thank Prof. Peter Karlson for supplying research facilities and Prof. Th. Wieland (Heidelberg) for a generous gift of α-amanitin. One of us (A.F.) was the recipient of a fellowship by the Alexander von Humboldt Stiftung.

REFERENCES

Amalric, R., M. Nicoloso, and J. P. Zalta 1972. A comparative study of "soluble" RNA polymerase activity of Zajdela Hepatoma Ascites Cells and Calf Thymus. *FEBS Lett.* 22: 67-72.

Bautz, E. K. F. and F. A. Bautz 1970. Initiation of RNA synthesis: the function of σ in the binding of RNA polymerase to promotor sites. *Nature* 226: 1219-1222.

Bautz, E. K. F., F. A. Bautz, and E. Beck 1972. Specificity of sigma-dependent binding of RNA polymerase to DNA. *Mol. Gen. Genet.* 118: 199-207.

Benecke, B. J., A. Ferencz, and K. H. Seifart 1973. Resistance of hepatic RNA polymerases to compounds effecting RNA and protein synthesis in vivo. *FEBS Lett.* 31: 53-58.

Burgess, R. R., A. A. Travers, J. J. Dunn, and E. K. F. Bautz 1969. Factor stimulating transcription by RNA-polymerase. *Nature* 221: 43-46.

Burgess, R. 1971. RNA polymerase. *Ann. Rev. Biochem.* 40: 711-740.

Boctor, A., A. Grossman, and W. Szer 1974. Characterization of an endoribonuclease in rat - liver nucleoli. *Eur. J. Biochem.* 44: 391-400.

Cedar, H. and G. Felsenfeld 1973. Transcription of chromatin in vitro. *J. Mol. Biol.* 77: 237-254.

Chambon, P., F. Gissinger, J. L. Mandel, C. Kedinger, J. Gniazdowski, and M. Meilhac 1970. Purification and properties of Calf Thymus DNA dependent RNA polymerases A and B. *Cold Spring Harbor Symposium*. 35: 693-707.

Chambon, P. F., F. Gissinger, C. Kedinger, J. L. Mandel, M. Meilhac, and P. Nuret 1972. Structural and functional properties of three mamalian nuclear DNA-dependent RNA polymerases. *Acta Endocrinologica (Kbh) Suppl.* 168: 222-246.

Chesterton, J. and P. Butterworth 1971. A new form of mammalian DNA-dependent RNA polymerase and its relationship to the known forms of the enzyme. *FEBS. Lett.* 12: 301-308.

Ferencz, A. and K. H. Seifart 1974. The effect of heparin on RNA synthesis of isolated rat liver nucleoli. *Eur. J. Biochem.* Submitted for publication.

Gilbert, W., and B. Müller-Hill 1972. The lac operator is DNA. *Proc. Natl. Acad. Sci.* U.S. 58: 2415-2421.

Higashinakagawa, T., M. Muramatsu and Sugano, H. 1972. Isolation of nucleoli from rat liver in the presence of magnesium ions. *Exptl. Cell Res.* 71: 65-74.

Kedinger, C., M. Gniazdowski, J. L. Mandel, F. Gissinger, and P. Chambon 1970. α-amanitin: A specific inhibitor of one of two DNA-dependent RNA polymerase activities from Calf Thymus. *Biochem. Biophys. Res. Commun.* 38: 165-171.

Kedinger, C., F. Gissinger and P. Chambon 1974. Animal DNA-dependent RNA polymerases. *Eur. J. Biochem.* 44: 421-436.

Kerrich-Santo, R. and G. Hartmann 1974. Influence of temperature on the action of rifampicin on RNA polymerase in presence of DNA. *Eur. J. Biochem.* 43: 521-532.

Lampert, A. and P. Feigelson 1974. *Fed. Proc.* (Abstract) paper in press.

Meilhac, M. and P. Chambon 1973. Animal DNA-dependent RNA polymerases initiation sites on Calf Thymus DNA. *Eur. J. Biochem.* 35: 454-463.

Muramatsu, M., N. Shimada, and T. Higashinakagawa 1970. Effect of cycloheximide on the nucleolar RNA synthesis in rat liver. *J. Mol. Biol.* 53: 91-106.

Perlman, R., B. Chen, B. de Crombrugghe, M. Emmer, M. Gottes-
man, H. Varmus and J. Pastan 1970. The regulation of lac
operon transcription by cyclic adenosine 3' 5' monophosphate.
Cold Spring Harbor Symp. 35: 419-423.

Penman, S. 1974. personal communication.

Pirotta, V., P. Chadwick, and M. Ptashne 1970. Active form
of two coliphage repressors. *Nature* 227: 41-44.

Roeder, R. and W. Rutter 1969. Multiple forms of DNA-dependent
RNA polymerase in eucaryotic organisms. *Nature* 224: 234-
237.

Roeder, R. 1974. Multiple forms of DNA dependent RNA polymerase
in *Xenopus laevis*. *J. Biol. Chem.* 249: 249-256.

Sajdel, E., and S. Jacob 1971. Mechanism of early effect of
Hydrocortisone on the transcriptional process: Stimulation
of the activities of purified rat liver nucleolar RNA polymerases. *Biochem. Biophys. Res. Commun.* 45: 707-715.

Schäfer, R., W. Zillig and K. Zechel 1973. A model for the
initiation of transcription by DNA-dependent RNA polymerase
from *Escherichia coli*. *Eur. J. Biochem.* 33: 207-214.

Seifart, K. H., and C. E. Sekeris 1969. α-amanitin: A specific inhibitor of transcription by mammalian RNA polymerase.
Z. Naturforsch. 24b: 1538-1544.

Seifart, K. H., B. J. Benecke and P. P. Juhasz 1972. Multiple RNA polymerase species from rat liver tissue. *Arch.
Biochem. Biophys.* 151: 519-532.

Smuckler, E. and J. R. Tata 1971. Changes in hepatic nuclear
DNA-dependent RNA polymerase caused by Growth Hormone and
Triiodothyronine. *Nature* 234: 37-39.

Sugden, B. and W. Keller 1973. Mammalian DNA-dependent RNA
polymerase. *J. Biol. Chem.* 248: 3777-3788.

Tsukuda, K. and I. Liebermann 1965. Liver nuclear ribonucleic
acid polymerase formed after partial hepatectomy. *J. Biol.
Chem.* 240: 1731-1736.

Wanka, F., and J. Moors 1970. Selective inhibition by cycloheximide of nuclear DNA synthesis in synchronous cultures
of chlorella. *Biochem. Biophys. Res. Commun.* 41: 85-90.

Weber, K. and M. Osborn 1969. The reliability of molecular
weight determinations by dodecyl sulfate-polyacrylamide
gel electrophoresis. *J. Biol. Chem.* 244: 4406-4412.

Wilhelm, J., D. Dina, and M. Crippa 1974. A special form of
DNA dependent RNA polymerase from oocytes of *Xenopus laevis*.
Biochem. 13: 1200-1208.

Yu, F. L., and P. Feigelson 1972. The rapid turnover of RNA
polymerase of rat liver nucleolus, and of its messenger
RNA. *Proc. Natl. Acad. Sci.* U.S. 69: 2833-2837.

Zubay, G., D. Schwartz and J. Beckwith 1970. The mechanism
of activation of catabolite sensitive genes. *Cold Spring*

Harbor Symp.   35:   433-435.

Zubay, G., D. E. Morse, W. J. Schrenk and J. H. M. Miller   1972.
   Detection and isolation of the repressor protein for the
   tryptophane operon of E. coli.   *Proc. Natl. Acad. Sci.* U.S.
   69:   1100-1103.

PHYSIOLOGICAL AND PATHOLOGICAL SIGNIFICANCE OF
HUMAN PYRUVATE KINASE ISOZYMES IN NORMAL AND
INHERITED VARIANTS WITH HEMOLYTIC ANEMIA

SHIRO MIWA, KOJI NAKASHIMA, and KENJI SHINOHARA
Third Department of Internal Medicine
Yamaguchi University School of Medicine
1144 Kogushi, Ube-shi, Yamaguchi-ken 755 Japan

ABSTRACT. Eight genetic variants of erythrocyte pyruvate
kinase (PK) and a so-called classical type quantitative
PK deficiency were characterized by electrophoretic,
kinetic, and immunological studies. In classical type PK
deficiency, erythrocyte PK isozymes (PK-R_1 and PK-R_2) were
not detectable but M_2-type PK was present presumably by
compensatory mechanism. The liver lacked L-type PK and
showed only M_2-type PK though no apparent liver dysfunction was demonstrable. Severe hemolytic anemia resulting
from M_2-type PK instability and the loss of erythrocyte
PK production in mature erythrocytes due to the absence
of ribosomes was seen in classical type PK deficiency.
PK "Tokyo I", PK "Nagasaki", and PK "Maebashi" all had
high Km for phosphoenolpyruvate, low Vmax, and urea
instability. Hemolytic anemia was moderate in these cases.
PK "Sapporo" and PK "Tsukiji" showed abnormal nucleotide
specificity, high Km for phosphoenolpyruvate, high Vmax,
and enzyme instability. Hemolytic anemia was moderate to
mild. PK "Tokyo II" showed low Km for phosphoenolpyruvate,
low Vmax, and stable characteristics. Clinically this
case showed mild hemolytic anemia. PK "Ube", detected in
a search for genetic polymorphism in healthy persons, had
abnormal electrophoretic mobility but was functionally
normal.

INTRODUCTION

Pyruvate kinase (PK) (ATP:pyruvate phosphotransferase,
EC 2.7.1.40) is one of the key enzymes in the Embden-
Meyerhof pathway. Tanaka et al (1967) and Imamura and Tanaka
(1972) reported that at least three kinds of electrophoretically distinct PK isozymes existed in mammalian tissues,
namely, L-type PK in the liver, M_1-type PK in the muscle,
and M_2-type PK in the spleen, liver, leukocytes, and other
tissues. Although kinetic, electrophoretic and immunological
characteristics led Bigley et al (1968) to believe erythrocyte PK to be identical to L-type PK, Imamura et al (1973)
and Nakashima et al (1974), using a more advanced electro-

phoretic method, found that erythrocyte PK isozyme showed
different migration from L-type PK. In addition, erythrocyte
PK was later separated into two bands (PK-R$_1$ and PK-R$_2$)
(Nakashima et al, 1974) which changed electrophoretically and
kinetically during erythrocyte aging (Nakashima, 1974).

Since the discovery of erythrocyte PK deficiency as a cause
of hereditary hemolytic anemia by Valentine, Tanaka, and Miwa
(1961), over 135 cases have been documented (Tanaka and Paglia,
1971). It has become more and more apparent that in addition
to the so-called classical type, presumably caused by quantitative PK deficiency, other PK deficiencies may result from
production of genetic PK variants which have abnormal functional characteristics (Tanaka and Paglia, 1971; Brandt and
Hanel, 1971; Staal et al, 1972). 'It is also known that the
clinical manifestations of this disorder are quite heterogeneous, suggesting that the molecular defect(s) is not uniform but rather heterogeneous.

This paper describes the characterization of eight erythrocyte PK variants and discusses the influence of functional
deviation in each PK variant on the clinical severity of
hemolysis. For comparison, clinical features as well as the
characteristics of erythrocyte and liver PKs in two cases with
so-called classical type PK deficiency are presented.

MATERIALS AND METHODS

Details of the patients' clinical histories and laboratory
data have been reported previously (Nakashima et al, 1974;
Miwa et al, 1975).

Tests with venous blood specimens anticoagulated with
heparin were undertaken within 48 hours. Heparinized blood
specimens were passed through a cotton wool column as described
by Busch and Pelz (1966) to remove leukocytes. Filtrated red
cells were washed three times with isotonic saline, centrifuging at 1,200 g for 10 min and removing the buffy coat
after each washing. In certain experiments which required
red cells absolutely free from leukocytes, red cell suspensions
were further passed through a mixed column with sulphoethyl
cellulose and Sephadex G25 (Medium) as described by Nakao
et al (1973). Hemolysate was prepared by saponin hemolysis
and diluted in either water or buffer as required for each
study as described previously (Nakashima et al, 1974).

PK activity was assayed according to Bücher and Pfleiderer
(1955) with slight modification. The standard assay system
employed 1.33 mM phosphoenolpyruvate, 2 mM ADP, 0.15 mM NADH,
8 mM MgSO$_4$, 6 mM EDTA, 75 mM KCl, and 6 U/ml lactate dehydrogenase in 0.05 M triethanolamine-HCl buffer, pH 7.4.

Kinetic studies for phosphoenolpyruvate used 0.2 M Tris-HCl buffer, pH 8.0, instead of 0.05 M triethanolamine-HCl buffer, pH 7.4, to produce clearer sigmoidal curves of allosteric effect (Koster et al, 1971). All assays were performed at 37°C in Gilford 2400-S recording spectrophotometer. An enzyme unit (u) was defined as the number of micromoles of NADH oxidized per min either by red cells containing one g hemoglobin at 37°C (u/g Hb) or by 10^{10} red cells at 37°C (u/10^{10}RBC). The latter expression was used only for PK activity determined by the standard assay system. Tissue extracts from patients were prepared from biopsy specimens at the time of splenectomy and from normal controls at the time of autopsy as described previously (Nakashima et al, 1974).

The thin layer polyacrylamide gel electrophoresis method of Imamura and Tanaka (1972) was employed using the conditions shown in the legends of Fig. 1. Urea stability test was carried out by the modified method of Koster et al (1971) in 0.2 M Tris-HCl buffer, pH 7.4, with 2 M urea at 25°C (Miwa et al, 1975). Normal human erythrocyte PK was purified by the method of Chern et al (1972) to Stage 5. Purified material showed only one protein band as assessed by polyacrylamide disc electrophoresis. Anti-human-erythrocyte-PK serum was obtained by immunization of a rabbit with a mixture of purified PK and Freund's complete adjuvant. Antiserum thus prepared was used for the PK neutralization test. At room temperature, antiserum was serially diluted with 0.5% bovine serum albumin and 5 min after either hemolysate or tissue extract was mixed well with this diluent PK activities were assayed (Miwa et al, 1975). Nucleotide specificity tests were performed by the modified method of Wiesmann and Tönz (1966). Younger and older red cells were prepared by centrifugation of whole blood taking advantage of the higher density of old cells than young cells (Nakashima, 1974).

RESULTS

Pertinent hematological data and physical findings as well as erythrocyte PK activities of 12 cases with PK deficiency are summarized in Table 1. Eight genetic PK variants were differentiated from each other by electrophoresis, kinetic studies for phosphoenolpyruvate, nucleotide specificity test, urea stability test and PK neutralization test by anti-erythro-cyte-PK serum. These PK variants were designated PK "Kiyose", PK "Tokyo I", PK "Nagasaki", PK "Maebashi", PK "Sapporo", PK "Tsukiji", PK "Tokyo II" and PK "Ube". Biochemical characteristics of these variants as well as those of classical type PK deficiency are shown in Table 2. We have defined

TABLE 1

PERTINENT CLINICAL FINDINGS, HEMATOLOGICAL DATA AND
ERYTHROCYTE PYRUVATE KINASE ACTIVITIES ON 12 CASES WITH PYRUVATE KINASE DEFECT

Classification	Age (yr-mo)	Sex	Hemoglobin (g/100 ml)	Reticulocyte (%)	Blood transfusion	Splenomegaly*	Erythrocyte PK activity (u/10^10 RBC)	Consanguinity
Classical type 1	5	F	5.0	13.6	++	3	1.02	–
PK deficiency 2	0-6	M	2.6	18.0	++	4	1.53	–
PK "Kiyose" 3	0-8	M	7.5	11.9	++	3	0.97	–
PK "Tokyo I" 4	14	F	7.3	16.1	+**	9	1.66	+
5	13	F	8.1	18.8	–	5	1.61	+
PK "Nagasaki" 6	5	F	7.7	23.9	–	4	6.80	+
PK "Maebashi" 7	9	M	8.7	9.9	+	3	0.41	+
PK "Sapporo" 8	32	F	8.5	2.7	+	0	5.03	+
PK "Tsukiji" 9	24	F	5.3	5.0	+	5	2.39	+
PK "Tokyo II" 10	33	F	10.4	2.6	–	0	1.61	+
11	18	F	10.4	2.4	–	0	1.25	–
PK "Ube" 12	35	M	15.7	1.6	–	0	5.74	–

Normal range
Mean ± 1 S.D.   4.55 ± 0.69

* cm below the left costal margin

** Blood transfusion only at the time of splenectomy

490

TABLE 2

BIOCHEMICAL CHARACTERISTICS OF EIGHT PYRUVATE KINASE VARIANTS AND A CLASSICAL TYPE PYRUVATE KINASE DEFICIENCY

Classification	Km for PEP	Vmax	Mobility*	Urea stability** 20 min	Urea stability** 60 min	Neutralization by antiserum** x80†	Neutralization by antiserum** x5†	Other remarks
Classical type PK deficiency	1.40	2.12	None (M_2=12)	37	24	103	99	M_2-type in erythrocytes
PK "Kiyose"	3.25	14.53	117	-	-	-	-	double heterozygote
PK "Tokyo I"	4.00	13.02	91	36	13	66	26	increased affinity to GDP, UDP
PK "Nagasaki"	5.25	10.90	90	12	2	-	-	abnormal pH curve, no sigmoidicity in kinetic curve for PEP
PK "Maebashi"	5.85	1.74	100	22	11	71	54	-
PK "Sapporo"	3.25	26.80	110	16	9	63	27	increased affinity to UDP, GDP
PK "Tsukiji"	4.48	32.80	117	48	31	92	82	decreased affinity to ADP
PK "Tokyo II"	1.05	2.92	126	88	84	44	28	2 cases in unrelated family
PK "Ube"	1.73	15.95	110	82	76	22	14	heterozygote
Normal range Mean ± 1 S.D.	1.43-2.23	14.67-18.15	100	80-89	72-84	21-37	12-20	

* Electrophoretic mobility of PK-R_1 (% of normal)

** % residual activity

† Serially diluted antisera were used but only results using x80 and x5 diluents listed in table

In the liver of classical type PK deficiency there is no L-type, in "Kiyose" normal L-type PK is present, and in "Tokyo I" a slow L-type is present.

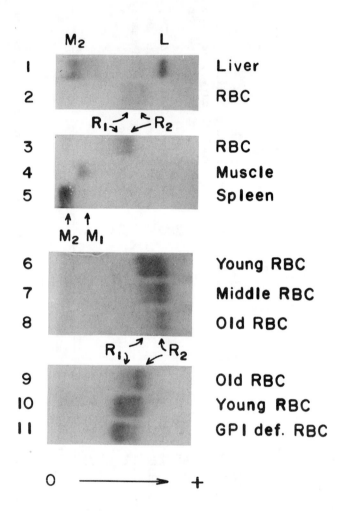

Fig. 1. Photograph of thin layer polyacrylamide gel electrophoresis of liver, muscle, spleen, and erythrocyte pyruvate
kinase. Supporting media: 3.34 % acrylamide gel, 0.1 cm in
thickness. Buffer: 10 mM Tris-HCl, 5 mM $MgSO_4$ 0.5 mM dithiothreitol, 0.5 mM fructose 1,6-diphosphate, pH 8.2. Voltage:
27 V/cm. Time: 4-5 hr. RBC: erythrocytes. GPI def.:
glucosephosphate isomerase deficiency. Liver, spleen, muscle,
and erythrocytes (1-10) except GPI deficiency erythrocytes (11)

(Fig. 1 legend continued) were taken from normal control.
O: origin. M_2: M_2-type PK. L: L-type PK. R_1: PK-R_1.
R_2: PK-R_2. M_1: M_1-type PK.

---

cases which showed no detectable PK-R in the erythrocyte and
no L-type PK in the liver electrophoretically or immunologically as classical type PK deficiency.

Electrophoretic patterns of human PK isozymes are shown in
Fig. 1. The designation of tissue PK isozymes (M_1-, M_2-, and
L-type PK) is that of Imamura and Tanaka (1972) and the designation of the two bands of erythrocyte PK (PK-R_1 and PK-R_2) is
that of Nakashima et al (1974). M_1-type PK had different
migration from M_2-type PK. Electrophoretic patterns changed
in the process of red cell aging. Younger cells had more
PK-R_1 and older cells had less PK-R_1 than normal. This finding is further confirmed by the fact that a case with glucosephosphate isomerase deficiency which had a young mean cell age
revealed more intense PK-R_1 than PK-R_2. In addition to the
electrophoretic change, PK of younger cells had low Michaelis-
Menten constant (Km) for phosphoenolpyruvate and was more
labile while PK of older cells had higher Km for phosphoenolpyruvate and was more stable than normal (Nakashima, 1974).

Results of PK neutralization tests by anti-erythrocyte-PK
serum are shown in Table 2. Normal erythrocyte PK was neutralized markedly while erythrocytes of PK "Tsukiji" and PK
"Maebashi" showed less inhibition. The muscle and the spleen,
which have only M_1-type PK and M_2-type PK respectively, were
not neutralized at all by anti-erythrocyte-PK serum (Miwa et
al, 1975). Since PK of the liver which contains both L-type
and M_2-type PKs showed moderate inhibition, it is reasonable
to assume that L-type PK has similar, if not identical, anti-
genicity to erythrocyte PK.

Typical electrophoretic patterns of some of the abnormal
erythrocyte PK isozymes are shown in Fig. 2. Classical type
PK deficiency had no PK-R in the erythrocytes and no L-type
PK in the liver but instead had M_2-type PK in the erythrocytes
and only M_2-type PK in the liver. Erythrocyte PK of PK
"Kiyose" showed electrophoretically faster moving PK-R_1 and
PK-R_2, while PK "Tokyo I" had slower moving PK-R_1 and PK-R_2
than normal.

Kinetic curves for phosphoenolpyruvate are shown in Fig.
3. Normal erythrocyte PK kinetic curves were sigmoidal but
that of PK "Nagasaki" had no sigmoidisity. PK "Maebashi" and
PK "Tokyo II" had low maximum velocity (Vmax) but PK "Tsukiji"
had markedly high Vmax. It is interesting to note that only
PK "Tokyo II" and PK "Ube" had low Km and normal Km for phosphoenolpyruvate, respectively, while all other PK variants had

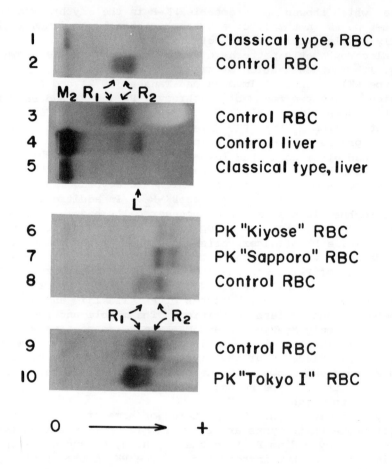

Fig. 2. Photograph of thin layer polyacrylamide gel electrophoresis of erythrocyte and liver pyruvate kinase. Procedure:
same as Fig. 1. RBC: erythrocytes.

---

high Km for phosphoenolpyruvate (Fig. 3; Table 2).

Kinetic curves for ADP determined for PK "Tsukiji" using
both 1 mM and 5 mM phosphoenolpyruvate are shown in Fig. 4.
PK "Tsukiji" was subject to inhibition when ADP concentrations

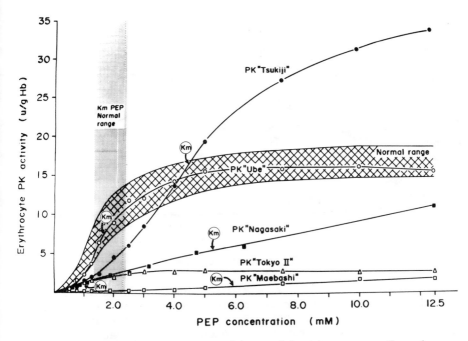

Fig. 3. Erythrocyte pyruvate kinase kinetic curves for phosphoenolpyruvate. Assay system: 2 mM ADP, 0.15 mM NADH, 8 mM
$MgSO_4$, 6 mM EDTA, 75 mM KCl, 6 U/ml lactate dehydrogenase and
varied concentrations of phosphoenolpyruvate (PEP) in 0.2 M
Tris-HCl buffer, pH 8.0 at 37°C. Arrows indicate Michaelis-
Menten constants (Km) of each PK variant.

were increased beyond 0.7 mM in 5 mM phosphoenolpyruvate in
contrast to normal erythrocyte PK which showed no inhibitory
effects under these conditions. PK "Tsukiji" showed very low
Vmax in 1 mM phosphoenolpyruvate.

Case 10 and Case 11 are unrelated but are considered to
have identical PK variants, PK "Tokyo II", as judged by the
several parameters shown in Table 2.

DISCUSSION

Imamura and Tanaka (1972) have presented evidence that in
human tissue, as in rat tissue, at least three forms of PK
(M_1-, M_2-, and L-type) exist and that electrophoretic patterns
of erythrocyte PK are significantly different from those of
L-type PK in the liver. In addition, based on data obtained
in a comparative study of fetal, new-born and adult rat tissue,
they considered M_2-type PK to be the prototype of PK isozymes

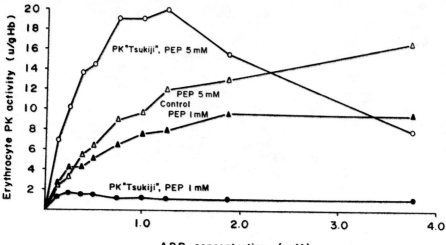

Fig. 4. Erythrocyte pyruvate kinase kinetic curves for ADP from PK "Tsukiji" and normal control. Assay system: 0.15 mM NADH, 8 mM $MgSO_4$, 6 mM EDTA, 75 mM KCl, 6 U/ml lactate dehydrogenase, two concentrations of phosphoenolpyruvate (PEP) (1 mM and 5 mM) and varied concentrations of ADP in 0.2 M Tris-HCl buffer, pH 7.2 at 37°C.

and L- and M_1-types to be differentiated types. Nakashima et al (1974) showed that erythrocyte PK can be separated into two bands (PK-R_1 and PK-R_2) using a slightly modified thin layer polyacrylamide gel electrophoresis technique originally developed by Imamura and Tanaka (1972).

We studied the liver of two cases with classical type PK deficiency, and in both no L-type PK was detected. It is interesting to note that liver dysfunction was not shown by laboratory tests and no morphologically abnormal findings, either microscopically or electronmicroscopically were seen in the liver (Matsumoto et al, 1972). These findings strongly suggest that L-type PK in the liver is a differentiated isozyme which has particular kinetic properties but which can be compensated for by M_2-type PK.

Recently, Imamura et al (1973) suggested from the results of electrophoretic, immunological and kinetic studies, that erythrocyte PK might be a hybrid of L-type PK and M_2-type PK subunits, although definitive proof of this hypothesis is lacking. In contrast to the liver, erythrocyte PK with either quantitative or qualitative defects almost always causes hemolytic anemia. As shown in Fig. 2, classical type PK deficiency had no detectable erythrocyte PK isozyme (PK-R_1 and PK-R_2) but instead had faint M_2-type PK visible in red cells quite free from leukocyte contamination which contained remarkably high reticulocyte counts. In accordance with this electrophoretic finding, erythrocyte PK in classical type PK deficiency was revealed to have typical characteristics of M_2-type PK as judged by low Km for phosphoenolpyruvate, instability to urea, no cross-reaction with anti-erythrocyte-PK serum and marked inhibition by alanine. If the hypothesis which Imamura and Tanaka (1973) have proposed, that M_2-type PK is a prototype PK, is correct, then it is quite possible that even under physiological conditions, red cell precursors produce M_2-type PK and as erythroblasts differentiate, PK-R_1 and PK-R_2 (mainly PK-R_1) are produced by conversion of M_2-type PK. The existence of M_2-type PK in reticulocytes of classical type PK deficiency can be regarded as a compensatory mechanism but further study is necessary in this regard. Urea stability tests revealed that normal M_2-type PK was more labile than normal PK-R_1 and PK-R_2. As mature erythrocytes are no longer able to produce enzyme protein molecules because of the lack of ribosomes, it is quite reasonable to assume that in classical type PK deficiency, mature erythrocytes will die and hemolyze shortly after passing the reticulocyte stage because of increased instability of M_2-type PK in their erythrocytes. Reflecting this, two cases with classical type PK deficiency both had severe hemolytic anemia and required periodic blood transfusions. On the other hand, the liver cells having ribosomes and thus the ability to synthesize M_2-type PK throughout their life cycle, appear to be able to compensate for a lack of L-type PK in classical type PK deficiency.

As reported previously (Nakashima et al, 1974), PK "Kiyose" is a double heterozygote of a maternal classical quantitative deficient PK allele and a paternal qualitative mutant PK allele. This case had had severe hemolytic anemia which necessitated exchange transfusion in the neonatal period and frequent blood transfusions thereafter until a splenectomy was performed. At least in this case, double genetic defects appear to cause synergistic damage to the tetrameric PK molecules, resulting in clinically severe hemolytic anemia.

PK "Tokyo I" (Imamura et al, 1973), PK "Nagasaki" (Nakashima et al, 1974) and PK "Maebashi" (Miwa et al, 1975) all had high Km for phosphoenolpyruvate, low Vmax, and urea instability. Clinically, these cases showed moderate hemolytic anemia. PK "Tokyo I" can be differentiated from PK "Nagasaki" because the former showed abnormal affinity to GDP and UDP, though both had slow moving electrophoretic patterns as compared to normal erythrocyte PK. PK "Maebashi" is clearly different

from PK "Tokyo I" and PK "Nagasaki" because PK "Maebashi"
showed normal electrophoretic mobility.

PK "Sapporo" is an interesting variant because it exhibited
high Vmax, high PK activity, abnormal attitude to nucleotides
and clinically, mild hemolytic anemia. Km for phosphoenolpyruvate was high. In addition, the heterozygotes in the
family had slight anemia. PK "Tsukiji" appears to be similar
to PK "Sapporo" with regard to high Km for phosphoenolpyruvate,
high Vmax, faster migration of PK in electrophoresis, abnormal
nucleotide specificity, enzyme instability as judged by urea
stability tests, and slight hemolytic anemia in the heterozygote. However, while PK "Sapporo" had high normal PK activity by the standard assay method, PK "Tsukiji" had low PK
activity. This fact can be explained by the difference in
affinity to nucleotides as seen in Fig. 4. PK "Tsukiji" was
inhibited by high concentrations of ADP in 5 mM phosphoenolpyruvate and showed extremely low Vmax in 1 mM phosphoenolpyruvate. These findings were not observed in normal erythrocyte PK or PK "Sapporo". The standard assay in our laboratory
uses 1.33 mM phosphoenolpyruvate which is similar to the 1 mM
phosphoenolpyruvate in this experiment. Low phosphoenolpyruvate concentrations reduced activity of PK "Tsukiji" markedly
despite the fact that this variant had high Vmax. Phosphoenolpyruvate concentrations in the erythrocyte are much lower than
1 mM. Hence, this phenomenon seems to reflect the clinical
manifestation of moderate to severe hemolytic anemia, the mechanisms of which remain to be elucidated. It is interesting
to note in this regard that both the mother of PK "Sapporo"
and the parents of PK "Tsukiji" showed mild hemolytic anemia
while heterozygotes for other PK variant genes were asympto-
matic. Homozygotes for PK "Sapporo" and PK "Tsukiji" genes
both had abnormal nucleotide specificity in common. Hence, it
appears reasonable to assume that PK variants which have
abnormal ADP binding sites can cause clinical signs of hemolysis even in heterozygous states. Further study is necessary
to clarify this point.

PK "Tokyo II" had very different characteristics from other
PK variants. This particular variant showed low Km for phosphoenolpyruvate and low Vmax as well as enzyme stability as
judged by the urea stability test. As mature erythrocytes are
unable to produce enzyme moleucles, stable PK enzyme can maintain its total activity in the erythrocytes much longer than
unstable variants such as PK "Tsukiji" and PK "Maebashi" both
of which had moderate to severe anemia. The clinical manifestations of PK "Tokyo II" correlate well with these biochemical findings. Both PK "Tokyo II" patients had mild anemia
with only slight reticulocytosis and had never required blood

transfusions.

An erythrocyte PK electorphoresis screening procedure performed on healthy blood bank donors in a search for genetic polymorphism detected PK "Ube", a healthy and non-anemic PK variant. It is very likely that the propositus is heterozygous for mutant PK gene and normal PK gene. Except for the electrophoretic abnormality, the enzyme was functionally normal and showed no other abnormalities. Heterozygotes of other functionally abnormal mutant PK genes and normal PK gene all showed abnormal kinetic curves. It appears that in PK "Ube", the abnormal molecule has a different electric net charge from the normal, presumably because of a probable single amino acid substitution, but this structural change affects neither the active site nor the antigenic site.

PK of young erythrocytes had more $PK-R_1$ which showed slightly lower Km for phosphoenolpyruvate and was more labile than $PK-R_2$. The latter appeared to be the main component in old erythrocytes. The physiological significance of these findings in the aging process of erythrocytes remains to be further elucidated.

ACKNOWLEDGMENTS

These studies were supported in part by scientific research grants from the Ministry of Education and a cancer research subsidy from the Welfare Ministry, Japan. We are indebted to Miss Hitomi Ogawa for expert technical assistance and to Miss Takako Nishino for assistance in preparing diagrams.

REFERENCES

Bigley, R. H., R. Stenzel, R. T. Jones, J. O. Campos, and R. D. Koler 1968. Tissue distribution of human pyruvate kinase isozymes. *Enzymol. Biol. Clinica* 9: 10-20.

Brandt, N. J., and H. K. Hanel 1971. Atypical pyruvate kinase in a patient with haemolytic anaemia. *Scand. J. Haematol.* 8: 126-133.

Bücher, T., and G. Pfleiderer 1955. Pyruvate kinase from muscle. in *Methods in Enzymology* (Colowick and Kaplan, eds), Vol. I. Academic Press, New York, 435-436.

Busch, D., and K. Pelz 1966. Erythrozytenisolierung aus Blut mit Baumwolle. *Klin. Wsch*. 44: 983-984.

Chern, C. J., M. B. Ritterberg, and J. A. Black 1972. Purification of human erythrocyte pyruvate kinase. *J. Biol. Chem.* 247: 7173-7180.

Imamura, K., and T. Tanaka 1972. Multimolecular forms of pyruvate kinase from rat and other mammalian tissues.

I. Electrophoretic studies. *J. Biochemistry* 71: 1043-1051.

Imamura, K., T. Tanaka, T. Nishina, K. Nakashima, and S. Miwa 1973. Studies on pyruvate kinase (PK) deficiency. II. Electrophoretic, kinetic, and immunological studies on pyruvate kinase of erythrocytes and other tissues. *J. Biochemistry* 74: 1165-1175.

Koster, J. F., G. E. J. Staal, and L. Van Milligen-Boersma 1971. The effect of urea and temperature on red blood cell pyruvate kinase. *Biochim. Biophys. Acta* 236: 362-365.

Matsumoto, N., T. Ishihara, K. Nakashima, S. Miwa, F. Uchino, and M. Kondo 1972. Sequestration and destruction of reticulocyte in the spleen in pyruvate kinase deficiency hereditary nonspherocytic hemolytic anemia. *Acta Haem. Jap.* 35: 525-537.

Miwa, S., K. Nakashima, K. Ariyoshi, K. Shinohara, E. Oda, and T. Tanaka 1975. Four new pyruvate kinase (PK) variants and a classical type PK deficiency. *Brit. J. Haematology* in press.

Nakao, M., T. Nakayama, and T. Kankura 1973. A new method for separation of human blood components. *Nature* (London) 246: 94.

Nakashima, K. 1974. Further evidence of molecular alteration and aberration of pyruvate kinase. *Clin. Chim. Acta* in press.

Nakashima, K., S. Miwa, S. Oda, T. Tanaka, K. Imamura, and T. Nishina 1974. Electrophoretic and kinetic studies of mutant erythrocyte pyruvate kinase. *Blood* 43: 537-548.

Staal, G. E. J., J. F. Koster, and J. G. Nijessen 1972. A new variant of red blood cell pyruvate kinase deficiency. *Biochim. Biophys. Acta* 258: 685-687.

Tanaka, K. R., and D. E. Paglia 1971. Pyruvate kinase deficiency. *Seminars in Hematology* 8: 367-396.

Tanaka, T., Y. Harano, F. Sue, and H. Morimura 1967. Crystallization, characterization, and metabolic regulation of two types of pyruvate kinase isolated from rat tissues. *J. Biochemistry* 62: 71-91.

Valentine, W. N., K. R. Tanaka, and S. Miwa 1961. A specific erythrocyte glycolytic enzyme defect (pyruvate kinase) in three subjects with congenital non-spherocytic hemolytic anemia. *Trans. Ass. Amer. Physicians* 74: 100-110.

Wiesmann, U., and O. Tönz 1966. Investigations of the kinetics of red cell pyruvate kinase in normal individuals and in a patient with pyruvate kinase deficiency. *Nature* (London) 209: 612-613.

ALDEHYDE OXIDIZING ENZYMES IN *LOCUSTA MIGRATORIA*

T. J. HAYDEN and E. J. DUKE
Department of Zoology,
University College,
Belfield,
Stillorgan Road, Dublin 4,
Ireland.

ABSTRACT: Polyacrylamide gel electrophoresis reveals at least seven aldehyde oxidases in *Locusta migratoria L.* All are molybdoproteins and have a molecular weight of about 280,000. Isoelectric focusing confirms that the electrophoretic separation is due to differences in charge density. Attempts to show that the heterogeneity might be due to posttranslational protein modification were unsuccessful. The enzymes appear to fall into three distinct groups on the basis of their ontogenetic patterns and substrate preferences. Phenotypic variations are reported and their possible relationship to subunit structure is discussed. SDS electrophoresis reveals that one band at least is a dimer. The significance of our results for further analysis of the relationships between these forms and the factors which may be involved in the control of their expression are discussed.

INTRODUCTION

The control of enzyme expression in higher organisms may be more complex than in bacteria (Wynngaarden, 1970). Detailed models and reviews of the genetic and epigenetic control of enzymes in higher organisms have been published (Greengard, 1967, Filner et al., 1969, Britton and Davidson 1969, Tomkins et al 1969, O'Malley and Means 1974, Goldberger 1974).

The xanthine : oxygen oxidoreductase (E. C. 1.2.3.2) (XDH, XO) system has been extensively studied in many organisms as a model of the genetic and/or epigenetic regulation of an enzyme (Glassman et al. 1968, Collins et al. 1970, Della Corte and Stirpe 1972, Lee and Fisher 1971). In *Drosophila melanogaster* mutations at several loci have been described, some of which affect both XDH and another molybdo-enzyme, aldehyde: oxygen oxidoreductase (E.C. 1.2.3.1) (aldehyde oxidase, AO) (Glassman et al. 1968). XDH and AO show low substrate specificity with considerable overlap in substrates oxidized (Krenitsky et al. 1972). The structural relationship between XDH, AO and another molybdoenzyme, reduced NADP :

nitrate oxidoreductase (E.C. 1.6.6.3) (nitrate reductase, NR), is further borne out by the assembly of NR activity in a zero mutant of *Neurospora* by *in vitro* complementation with subunits from XO or AO from mammals (Pateman et al. 1964, Ketchum et al. 1970). Other *in vitro* complementation experiments producing active XDH from zero mutants of *Neurospora crassa* and *D. melanogaster* indicates that the molybdoenzymes XDH and AO, particularly, may be structurally conservative to some extent (Duke, unpublished). Furthermore, unpublished observations from this laboratory suggest that, although higher vertebrates may possess only one enzyme functioning as XO, AO, or both, *D. melanogaster* and other invertebrates exhibit XDH and AO which are electrophoretically separable. Multiple forms of both enzymes have been described and have been related to heterozygosity, for XDH (Yen and Glassman 1965, Bianchi and Chessa 1970) and AO (Dickinson 1970) and to proteolytic degradation (Duke et al. 1973). Nutritional factors such as high protein and molybdenum also appear to alter the structure of XDH and AO in *Drosophila* (Glassman et al. 1968, Collins et al. 1970, Duke et al. 1974). These factors also affect the levels of XDH and AO activity in wild-type and mutant strains of *Drosophila*. Wild-type flies cultured on a high protein diet show elevated levels of XDH activity with no apparent increase in enzyme protein, although the enzyme present shows altered electrophoretic mobility (Collins et al. 1970). Low xanthine dehydrogenase (lxd) mutants exhibit low levels of both XDH and AO but can recover to near wild-type levels of activity when fed concentrations of molybdenum which do not affect wild-type flies (Duke et al. 1974). These findings would all appear to indicate that epigenetic mechanisms of enzyme control are operating at the level of enzyme structure whereby changes in activity are achieved.

It was decided, therefore, to investigate the control of these enzymes in another insect, but one such as the locust with a different developmental history. Routine electrophoretic investigations revealed a single band for XDH, but the related AO was seen to be highly heterogeneous making it quite unlike the corresponding enzyme in *Drosophila*. The present paper describes the expression of aldehyde oxidase in *Locusta migratoria*, and preliminary attempts to interpret the system are discussed. This analysis will hopefully provide a basis for future investigations of the nature of enzyme regulatory mechanisms in this organism.

MATERIALS AND METHODS

Animals: Locusts were reared on a diet of fresh grass, bran, and water in glass fronted cages under crowded conditions with

502

constant illumination at 35°C.

Preparation of Extracts: Animals were sacrificed by exposure
to a low temperature of -20°C for 30 minutes. Tissues or
whole animals were homogenized using a Potter-Elvehjem hand
homogenizer in five volumes (w/v) 0.05M TRIS, 10^{-3}M EDTA pH
8.0. The homogenate was centrifuged at 35,600g for 20 minutes
in a Sorvall RC2B centrifuge. The supernatent was fractionated
using ammonium sulphate. The 40-60% precipitate was redissolved in a minimal volume of TRIS-EDTA buffer and was desalted
by dialysis or by elution through Sephadex G-25.

Assay for Aldehyde Oxidase: Assays for aldehyde oxidase were
carried out using a Unicam SP 800 to monitor reduction of
0.005% (w/v) dichlorophenolindophenol in the presence of 1 mM
substrate. In some cases 0.1 mM phenazine methosulphate was
included in the assay.

Polyacrylamide Gel Electrophoresis: Polyacrylamide gel electrophoresis was carried out at various gel concentrations using
the continuous buffer system of Yen and Glassman (1965) in
either the EC-470 slab apparatus of Raymond (1964) or in glass
tubes 80 mm. long and 5 mm. i.d. Two dimensional electrophoresis was carried out according to Raymond (1964). XDH was
detected as previously described (Yen and Glassman 1965). For
aldehyde oxidase hypoxanthine was replaced by 0.1 ml benzal-
delyde and DPN was omitted unless otherwise stated. SDS
electrophoresis was carried out according to Weber et al.
(1972).

Autoradiography: 0.5 - 1.0 m Ci of ^{99}Mo in the form of
ammonium molybdate was fed in 200 ml of a grass homogenate to
locusts for 3 days. Crude extracts were electrophoresed on
a 5% slab gel. The gel was cut into longitudinal strips corresponding to the sample tracks and placed on Perspex support
sheets so that gel and support fit snugly into wetted dialysis
tubing. The gels were dried overnight at 40°C and taped firmly
to Kodak Blue Brand X-Ray film, which was then sandwiched between glass plates. The film was exposed for 14 days. Due
account of the shrinkage which occurred during drying was taken
when the autoradiograph was compared to duplicate strips which
had been stained for enzyme activity.

Gel Filtration Chromatography: Samples were chromatographed
on Sephadex G-200 Columns (2.5 x 50 cm.) and eluted with 0.05M
Tris, 10^{-3}M EDTA, pH 8.0.

Isoelectric Focusing: Isoelectric focusing was carried out in disc gel apparatus containing 4% polyacrylamide gel with 0.054 ml Ampholine, pH3 - 10 (LKB - Produkter AB) per ml gel. The upper cathodal chamber contained 2 ml triethanolamine per 500 ml deionized water adjusted to pH 10.9 by addition of KOH. The lower anodal chamber contained 1 ml conc. H_2SO_4 per 500 ml deionized water. Focusing was carried out at 2 mA/tube until the voltage reached 200 V and then at a constant voltage of 200 for 5 hours. In some cases the amphoteric dyes Fast Green FCF and Congo Red were used as internal markers (Conway - Jacobs and Lewin 1971).

Ion Exchange Chromatography: Samples were first dialyzed against 0.05 M Tris, 10^{-3}M EDTA, pH 8.0 and applied to DE 52 columns (2.5 x 25 cm) which had been equilibrated with the same buffer. Elution was by means of either NaCl (0 - 0.5 M.) or pH (8 - 4) gradients.

Dissociation Studies: Samples of semi-purified band 4 eluted from DE 52 were brought to 10 M with Urea, allowed to stand at $4^{\circ}C$ for 3 hours and then dialyzed against 0.05 M Tris, 10^{-3}M EDTA for 18 hours and then electrophoresed. Other attempts at dissociation were made by bringing aliquots of the same sample to 1 M with NaCl, freezing to $-20^{\circ}C$ and subsequently rethawing. Dialysis for 18 hours was carried out before electrophoresis.

Attempts to Modify Electrophoretic Behaviour of Aldehyde Oxidase by Incubation with Various Reagents: The molybdenum-chelating reagents, sodium dimethyldithiocarbamate, potassium cyanide and disodium 4, 5-dihydroxy - m - benzenedisulphonate, all 10^{-3}M, were added to a crude homogenate and incubated for 60 minutes at $37^{\circ}C$ prior to electrophoresis. In addition 10^{-3}M nicotinamide adenine dinucleotide (NAD), 10^{-3}M mercaptoethanol and 10^{-3} dithiotreitol (Clelands Reagent) were incubated with crude homogenate at $37^{\circ}C$ and $4^{\circ}C$ for 30 minutes and electrophoresed in gels with and without the reagent added before polymerization. Soybean trypsin inhibitor at a range 1 - 10 mg/ml was included in the homogenizing buffer during one extraction sequence in an effort to inhibit possible proteolytic effects.

RESULTS

Electrophoresis. A representative polyacrylamide electrophoretic pattern of the aldehyde oxidizing enzymes

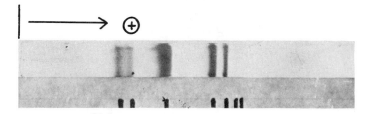

Fig. 1. Aldehyde oxidase electrophoretic pattern from a fifth
instar female locust. A crude whole body extract was electrophoresed at 250 V in a 5% polyacrylamide gel.

of *L. migratoria* is shown in Fig. 1. At least seven aldehyde
oxidizing enzymes are present in whole homogenates. In Fig.
2 XDH can be observed in a more cathodal position to all these
oxidases. These bands have been designated 1 - 8 beginning
with XDH (1) in an anodal direction. In Fig. 2. an autoradiograph of an electrophoretic separation of crude homogenates
from locusts, which had been fed ^{99}Mo, is shown. Radioactivity is detectable in those areas which also contain XDH
and aldehyde oxidase activities (Fig. 2). This would appear
to indicate that all of these enzymes are, in fact, molybdoproteins. Incubation of the homogenate in the presence of a
variety of molybdenum-chelating agents, such as dimethyldithiocarbamate, Tiron, and KCN (all 10^{-3}M) prior to electrophoresis
revealed that only KCN had a significant effect, causing almost total elimination of the activity of all the enzymes as
measured by NBT reduction (Fig. 2). No specific effects on
any of the individual isozymes were observed at the concentrations used.

Molecular Size. All the aldehyde oxidases co-elute with XDH
from G-200 Sephadex indicating that all the electrophoretic
forms are approximately the same size or else they represent
variously dissociated states of an aldehyde oxidase which co-
elutes with XDH and then breaks down on electrophoresis. The
latter possibility is unlikely as all states would have to be
active. Furthermore, two dimensional polyacrylamide gel
electrophoresis (first in 5%, second in 7%) shows that all the
enzymes, including XDH, migrate into positions lying on a
straight line through the origin (Fig. 3). This would seem to
exclude the possibility of dissociation artifacts and confirm
that all forms are of similar molecular size, i.e. 280,000,
as determined by the method of Parish and Marchalonis (1970).

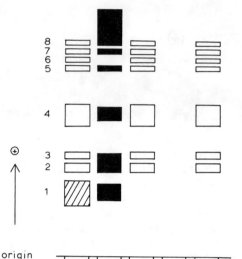

Fig. 2. Diagrammatic representation of polyacrylamide gel electrophoresis of aldehyde oxidases from the locust. The extracts were as follows:- A. Control pattern stained for AO and XDH. B. Autoradiograph of ^{99}Mo-labelled proteins from similar extract as in A. C. Crude extract incubated for 60 minutes at 37°C in the presence of 10^{-3} M sodium dimethyldithiocarbamate before electrophoresis. D. Crude extract incubated for 60 minutes at 37°C with 10^{-3} M potassium cyanide. E. Crude extract incubated for 60 minutes at 37°C with 10^{-3} M 4, 5-dihydroxy-m-benzenedisulphonic acid, disodium salt. Electrophoresis was performed as described in Fig. 1.

In addition, Ferguson plots of mobility against gel concentration yield a set of parallel straight lines which is again in agreement with the above results.

Isoelectric Points. Figure 4a shows the separation obtained when aldehyde oxidases are focused in the first dimension in a 4% gel and then electrophoresed into a 5% gel for a second dimensional separation. The pH profile in Fig. 4b was obtained by measuring the pH after elution of slices of a duplicate first dimension gel with distilled water. The reality of the two dimensional pattern was confirmed by staining the slices of the duplicate gel to relate enzyme location to the pH of the eluate. The pI values obtained for the aldehyde oxidases, in numerical order were 5.55, 5.4, 5.1, 4.8, 4.65, 4.5, 4.4. XDH

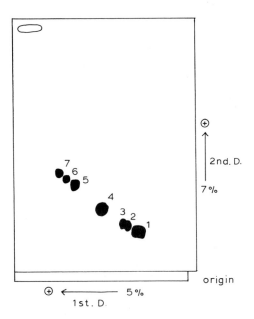

Fig. 3. Two-dimensional electrophoresis of XDH and aldehyde
oxidases in a crude homogenate of *Locusta*. First dimension:
5% gel in the Raymond apparatus. A longitudinal strip corresponding to a sample track was set at right angles in a 7%
gel in the same apparatus for the second-dimensional separation.

focused at pH 5.7. The electrophoretic mobility was found to
be inversely proportional to pI.

Ion Exchange. As expected the aldehyde oxidases elute from
DE 52 anion exchanger in the order expected on the basis of
the pI data (Fig. 5a). Fig 5b shows that the elution pattern
(from DE 52) was different when a sodium chloride gradient was
used as eluant. Isozyme 4 eluted first, followed by 5, 6, and
7 and finally 2 and 3. This indicates that the total charge
on the isozymes does not change in parallel with changes in pH.

Incubations. Incubation with 10^{-3}M NAD, 10^{-3}M mercaptoethanol
or 10^{-3}M dithiothreitol did not alter the electrophoretic pattern. This reduces the possibility that cofactor induced conformational changes or various states of reduction of S-S
groups are responsible for the observed heterogeneity. Inclusion of soybean trypsin inhibitor in the preparation procedure
was also without effect suggesting that the bands are not due

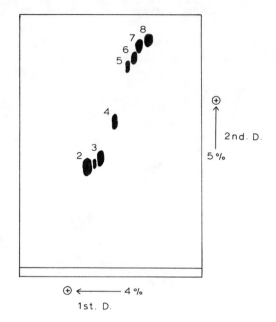

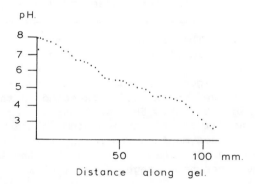

Fig. 4. a. Two dimensional separation of XDH and aldehyde oxidases of crude homogenates of *Locusta* incorporating iso-electric focussing in 4% gel in the first dimension and electrophoresis in the 5% slab gel in the second dimension. b. pH gradient in a duplicate gel to that used in the first dimension of the separation in a. above.

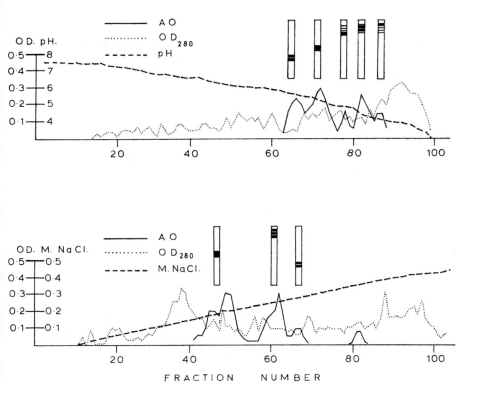

Fig. 5. Elution profile of aldehyde oxidases from a DE 52 anion exchange column with the dimensions of 2.5 x 25 cm. The column was equilibrated with 0.05 M Tris pH 8. a. Elution with pH gradient 8-4. b. Elution with salt gradient 0-0.5 M NaCl. Insets show the electrophoretic pattern associated with each elution peak.

to proteolytic degradation of the native form of the enzyme. Incubation with deoxycholate did not affect the mobility or appearance of bands 2-7. There is a small but consistent change in the mobility of XDH under these conditions. Other data suggest that XDH may be a lipoprotein or have some lipid moiety closely associated with it. Deoxycholate at low concentrations stimulates XDH activity but above 0.5% w/v it acts as an inhibitor. Lipase-treated XDH also shows altered electrophoretic mobility.

Staining of Gels. All the aldehyde oxidizing enzymes displayed oxidase activity when gels were incubated with NBT or

509

INT with benzaldehyde as substrate. Bands 2 and 3 show strong staining under these conditions. Band 4 was fainter and 5, 6, 7, and 8 marginally detectable. Addition of DPN to the medium did not appreciably alter this situation. However, PMS with or without DPN strongly enhanced the staining of all bands.

Substrates. Table 1 shows the relative activities of the three groups of enzymes, i.e. band 2 and 3, band 4 and bands 5, 6, and 7, with a range of aldehyde substrates. In all cases the activity with benzaldehyde was expressed as 100 and other values were expressed relative to this. For each group of enzymes heptaldehyde was the optimum substrate. In general within the range C_3 - C_{12} aldehydes there was an increase in activity up to C_7 and then a decrease. Only the "fast" group showed high levels of DCIP reduction with acetaldehyde as substrate, a fact which correlates with observations on the rates of staining of the bands on gels with these substrates.

Electrophoretic Variants. Fig. 6a shows the apparent polymorphism of bands 5, 6, 7, and 8 as detected on electrophoresis. It has been noted that the intensity of staining often varies between bands and not always binomially as would be expected if random association of subunits, synthesized at the same rate, were involved. The numbers shown refer to frequency of occurrence of the indicated phenotypes without reference to staining intensity. Fig. 6b shows a phenotype detected twice in some 200 individuals examined. All four front bands are displaced anodally when compared to a more normal phenotype. It is of interest to note that the mobility of the other aldehyde oxidases, bands 2, 3, and 4 were normal. It would, therefore, appear unlikely that these latter bands are related to the anodal group and differ only in conformation or arrangement of subunits. Three variants of band 4 have been detected in the following numbers: fast phenotype 13, intermediate 46, and slow 37. The data fit a Hardy-Weinberg equilibrium calculated on the basis of a one locus-two allele hypothesis.

Subunit Structure of Band 4. Fig. 7 shows the results of electrophoresis of band 4 on SDS - polyacrylamide gel (7% acrylamide). The enzyme would appear to be a dimer composed of subunits of 140,000 daltons, the minimal conditions consistent with the electrophoretic variation.

Ontogenesis. Newly hatched first instar nymphs lack band 4 which appears during the first day after hatching. All other stages have the full complement of enzymes except adult males

TABLE 1
RELATIVE ACTIVITIES OF ALDEHYDE OXIDASES FROM
LOCUSTA MIGRATORIA WITH A VARIETY OF SUBSTRATES

Substrate	Isozyme Group		
	Slow (2,3)	Middle (4)	Fast (5,6,7,8)
Acetaldehyde	26	82	230
Propionaldehyde	9	18	33
Butyraldehyde	79	100	66
Valerylaldehyde	92	72	100
Hexaldehyde	180	527	660
Heptaldehyde	215	1090	730
Octaldehyde	157	560	430
Nonaldehyde	157	400	400
Decaldehyde	71	190	200
Dodecyladehyde	20	73	33
Benzaldehyde	100	100	100
Salicylaldehyde	60	41	66
Furfuraldehyde	96	117	100
Pyruvaldehyde	5	18	0
Iso-butyraldehyde	10	18	0

Assay: Reduction of DCIP (Dichloroindophenol). All values
were adjusted so that activity with Benzaldehyde was 100 in
each case.

who lack bands 5, 6, 7, and 8.

Tissue Distribution. The tissue distribution of aldehyde
oxidases reveals that, oesophagus, crop, gizzard, colon,
rectum, fat-body, malpighian tubule, ovary, testis, brain,
central nervous system, and integument in all stages except
adult males possess the full set of isozymes. The activity
of midgut and ileum seems to come mainly from band 4.

DISCUSSION

There is now much evidence to indicate that molybdenum
containing enzymes are subject to diverse genetic and epigenetic controls. The genetic controls of XDH and AO have
been extensively studied in *Drosophila* (Courtright 1967,
Glassman et al. 1968, Dickinson 1970, Chovnick 1974). In addition, nutritional and other factors have been observed to
influence these enzymes in *Drosophila* and other species

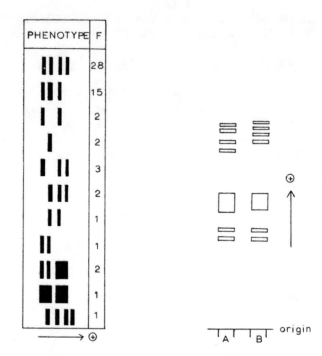

Fig. 6. a. Frequency of occurrence of phenotypic variations of the front set of aldehyde oxidizing enzymes from whole homogenates of *Locusta* as revealed by polyacrylamide gel electrophoresis on 5% gels. F = Frequency. b. Variant showing co-ordinate alteration in electrophoretic mobility of the anodal set of isozymes. A. control. B. Mutant.

(discussed by Duke et al. 1974).

In this paper it has been shown that the expression of aldehyde oxidase in *Locusta migratoria* is quite different from that in *Drosophila* in that the enzyme in the former species is highly polymorphic and seems to consist of three distinct groups of enzymes under different control mechanisms. It is unlikely that this polymorphism is due to the different developmental histories of the two species since some holometabolous insects have multiple electrophoretic forms of AO (unpublished observations). It has been noted in some thirty invertebrate species from the phyla Arthropoda, Annelida, and Mollusca that electrophoretically separable XDH and AO exist, the latter often in more than one form, whereas in those

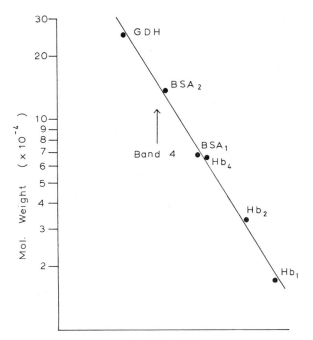

Fig. 7. Determination of the molecular weight of the subunit of isozyme 4 on 7% Polyacrylamide - SDS gel. The marker proteins used were hemoglobin, bovine serum albumin, and glutamic dehydrogenase.

mammals examined a single band stains for both activities (Hayden, unpublished). Attempts to electrophoretically separate AO and XDH/XO in those mammalian tissues, for which both activities have been reported, have so far been unsuccessful. Various authors (Gordon et al. 1940, Felsted et al. 1973) have noted during the purification of AO that XO activity is lost at a certain stage, i.e. after treatment with an organic solvent. It is, however, not clear whether this represents a separation of two proteins or simply an inactivation of the XO component within a single molybdoprotein. What is clear is that with evolution there is a reduction in the molecular polymorphism of molybdoproteins. Little can be said at this stage about the possible physiological basis for this selection process.

The various aldehyde oxidases in *Locusta* would appear to be under the control of different genes. Band 4 is not detectable until about 12 hours after hatching, whereas the other oxidases are present at hatching. Also, the anodal group (5, 6, 7, and 8) is completely absent in all adult males. All the electrophoretic forms have quite clearly the same molecular weight (ca.280,000) and they are fractionated mainly on the basis of their charge densities as seen from the patterns of isoelectric focusing and elution from DE 52 using a pH gradient. The elution pattern with a salt gradient reveals that the titration curves of the isozymes are not parallel because, if this were so, elution would be in the same order as when a pH gradient is used. Some further information on the molecular structures of these oxidases may be obtained by observing electrophoretic variants. Band 4 consists of at least two subunits of equal molecular weight (140,000) when subjected to SDS electrophoresis. This agrees with the interpretation of the 3 observed phenotypes on the basis of a one locus two allele hypothesis.

It is difficult to derive a model to explain the various phenotypic combinations of the anodal group of bands. The co-ordinate anodal displacement of all four isozymes observed in two individuals may indicate the existence of a common subunit or the binding of a common ligand. One possible explanation is that these bands are coded by a system of three loci, the products of two assembling in all possible combinations to form four trimers which then each associate with a single subunit from a third locus. Further experiments are required to test this hypothesis. It is of interest to note that in those two mutants showing co-ordinate displacement of the anodal group of isozymes, bands 2, 3, and 4 remain unchanged. It should also be emphasised that all our efforts to induce epigenetic alterations or conversions of these oxidases have so far failed. In addition it seems unlikely that the heterogeneity is due to proteolytic effects since soybean trypsin inhibitor failed to alter the pattern. In fact, the least heterogeneity is observed in preparations from mid-gut and ileum. Some differences in rates of oxidation of various substrates have been noted although heptaldehyde is the optimum substrate for all three groups of enzymes. Acetaldehyde is oxidized more rapidly by the most anodal group of isozymes both in gels and when assayed spectrophotometrically.

In insects, AO is readily separable from XDH on electrophoresis. Igo and Mackler (1960) state, however, that pig liver AO is a form of XO. This makes the heterogeneity seen in *Locusta* even more interesting. It is intended to further investigate the structural relationships between the various forms. Attempts are being made, also, to analyze what epigenetic factors of physiological significance affect the expression of both the aldehyde oxidases and xanthine dehydrogenase. It is hoped in this way to clarify the relationship between the AO isozymes in *Locusta* and perhaps obtain some insight into the phylogeny of molybdoproteins in general.

REFERENCES

Bianchi, V. and G. Chessa 1970. Allenzimi ad attivita xanthindeidrogenasica in *Anopheles atroparvus*. *Riv. Parasitol*. 31: 299-303.

Britten, R. J. and E. H. Davidson 1969. Gene regulation for higher cells : A theory. *Science*. 165: 349-357.

Chovnick, A., M. McCarron, W. Gelbart and J. Pandey 1974. Electrophoretic variants as a tool in the analysis of gene organization in higher organisms. *IV Isozymes: Genetics and Evolution*, C. L. Markert, Ed. Academic Press. New York.

Collins, J. F., E. J. Duke and E. Glassman 1970. Nutritional control of xanthine dehydrogenase: I The effect in adult *Drosophila melanogaster* of feeding a high protein diet to larvae. *Biochim. Biophys. Acta* 208: 294-303.

Conway-Jacobs, A. and L. M. Lewin 1971. Isoelectric focusing in acrylamide gels: Use of amphoteric dyes as internal markers for determination of isoelectric points. *Anal. Biochem*. 43: 394-400.

Courtright, J. B. 1967. Polygenic control of aldehyde oxidase in *Drosophila*. *Genetics* 57: 25-39.

Della Corte, E. and F. Stirpe 1972. Regulation of rat liver xanthine oxidase: Involvement of thiol groups in the conversion of the enzyme activity from type D into type O and purification of the enzymes. *Biochem. J.* 126: 739-745.

Dickinson, W. J. 1970. The genetics of aldehyde oxidase in *Drosophila melanogaster*. *Genetics* 66: 487-496.

Duke, E. J., P. Joyce and J. P. Ryan 1973. Characterization of alternative molecular forms of xanthine oxidase in the mouse. *Biochem. J.* 131: 187-190.

Duke, E. J., D. R. Rushing and E. Glassman 1974. Nutritional control of xanthine dehydrogenase: II. The effects on xanthine dehydrogenase and aldehyde oxidase of culturing wild-type *Drosophila* and enzyme-deficient mutants on different levels of molybdenum. *Biochem. Genet*. (in press).

Felsted, R. L., A. E. Chu and S. Chakrin 1973. Purification and properties of the aldehyde oxidases from hog and rabbit livers. *J. Biol. Chem.* 248: 2580-2587.

Filner, P., J. L. Wray and J. E. Varner 1969. Enzyme induction in higher plants. *Science* 165: 358-367.

Glassman, E. 1962. Convenient assay of xanthine dehydrogenase in single *Drosophila melanogaster*. *Science* 137: 990-991.

Glassman, E., T. Shinoda, E. J. Duke and J. F. Collins 1968. Multiple molecular forms of xanthine dehydrogenase and related enzymes. *Ann. N. Y. Acad. Sci.* 151: 263-273.

Goldberger, R. F. 1974. Autogenous regulation of gene expression. *Science* 183: 810-816.

Gordon, A. H., D. E. Green and V. Subrahmanyan 1940. Liver aldehyde oxidase. *Biochem. J.* 34: 764-774.

Greengard, O. 1967. The quantitative regulation of specific proteins in animal tissues: Words and facts. *Enzym. Biol. Clin.* 8: 81-96.

Igo, R. P. and B. Mackler 1960. Liver aldehyde oxidase: A form of xanthine oxidase. *Biochim. Biophys. Acta* 44: 310-314.

Ketchum, P. A., H. Y. Cambier, W. A. Frazier, C. H. Madansky, and A. Nason 1970. *In vitro* Assembly of Neurospora assimilatory nitrate reductase from protein subunits of a Neurospora mutant and the xanthine oxidizing or aldehyde oxidase systems of higher animals. *Proc. Natl. Acad. Sci.* (U. S.))66: 1016-23.

Krenitsky, T. A., S. M. Neil, G. B. Elion, G. H. Hitchings 1972. Comparison of the specificities of xanthine oxidase and aldehyde oxidase. *Arch. Biochem. Biophys.* 150: 585-599.

Lee, P. C. and J. R. Fisher 1971. Regulation of xanthine dehydrogenase levels in liver and pancreas of the chick. *Biochim. Biophys. Acta* 237: 14-20.

O'Malley, B. W., and A. R. Means 1974. Female steroid hormones and target cell nuclei. *Science* 183: 610-620.

Parish, C. R. and J. J. Marchalonis 1970. A simple and rapid acrylamide gel method for estimating the molecular weights of proteins and protein subunits. *Anal. Biochem.* 34: 436-450.

Pateman, J. A., D. J. Cove, B. M. Rever and D. B. Roberts 1964. A common cofactor for nitrate reductase and xanthine dehydrogenase which also regulates the synthesis of nitrate reductase. *Nature* 201: 58-60.

Raymond, S. 1964. Acrylamide gel electrophoresis. *Ann. N. Y. Acad. Sci.* 121: 350-365.

Tomkins, G. M., T. D. Gelehrter, D. Granner, D. Martin, Jr. H. H. Samuels and E. B. Thompson 1969. Control of specific gene expression in higher organisms. *Science* 166: 1474-80.

Weber, K., J. R. Pringle and M. Osborn 1972. Measurement of molecular weights by electrophoresis on SDS - acrylamide gel. *Methods in Enzymol.* 26: 3-27.

Wynngaarden, J. B. 1970. Genetic control of enzyme activity in
    higher organisms. *Biochem. Genet.* 4: 105-125.
Yen, T. T. and E. Glassman 1965. Electrophoretic variants of
    xanthine dehydrogenase in *Drosophila melanogaster*.
    Genetics 52: 977-981.

FORMS AND FUNCTIONS OF GLUCONEOGENIC
ENZYMES IN ARTHROPODS

P. W. HOCHACHKA and H. E. GUDERLEY
Department of Zoology
University of British Columbia
Vancouver, B. C., Canada V6T 1W5

ABSTRACT.   Gluconeogenesis can be conveniently discussed
in terms of mechanisms for bypassing key irreversible steps
in glycolysis.   Pyruvate kinase is bypassed by a true enzyme pathway involving pyruvate carboxylase and phosphoenolpyruvate carboxykinase, each catalyzing a reaction involving
high energy phosphate donors.   In gluconeogenic tissues,
such as epidermis, gill, and fat body, pyruvate kinase is
a regulatory enzyme controlled by fructose diphosphate and
other chitin precursors.   Phosphoglycerate kinase is bypassed by a kinetic mechanism involving the production of
a new isozymic form of this enzyme.   The electrophoretic
properties of muscle and epidermal PGK change at molting,
and take on kinetic features that better suit them for
function in the gluconeogenic direction.   Phosphofructokinase is bypassed by fructose diphosphatase (FDPase).   In
crustacean epidermis, this enzyme occurs in two isozymic
forms, termed I and II.   Both forms are strongly inhibited
by AMP, a mechanism that seems both necessary and sufficient
to prevent futile cycling at this point in metabolism.
FDPase II is also regulated by metabolite signals from glycogen and chitin synthesis, as well as by pentose shunt
intermediates.   This makes the FDPase locus responsive to
the demands of the subsequent pathways.   Where these enzymes
occur in muscle, particularly in the highly specialized
insect flight muscle, they may take on unique functions.

In arthropods, gluconeogenesis is a critical process, not
only for the formation of blood sugars and glycogen, but also
for the synthesis of the chitinous exoskeleton.   In order to
grow, arthropods must periodically shed their exoskeleton.
The rapid thickening and hardening of the new cuticle requires
efficient gluconeogenesis.   In insects and crustaceans the
epidermis is most intimately involved with fulfilling this
biosynthetic need, but the events associated with molting
profoundly affect the other tissues.

Growth in insects and crustaceans follows molt cycles, with
those of insects being more complex due to the metamorphoses
they undergo.   The following major changes occur during the
molt cycles.   During intermolt, the animal is actively feeding

and deposits considerable lipid and glycogen. Lipid is stored primarily in the hepatopancreas in crustaceans, and in the fat body in insects. The major glycogen deposits in crustaceans are in muscle, epidermis, blood cells, and the hepatopancreas (Hohnke, 1970; Johnson and Davies, 1972). In insects the major glycogen deposits are in the fat body and muscle (Friedman, 1970). The initiation of the premolt period by hormonal changes brings a separation of the epidermis from the exoskeleton. The new cuticle starts growing under the old, while new gills are forming inside the old and the muscle is decreasing to 60% of its intermolt mass (Skinner, 1966). During late premolt wholesale resorption of chitin from the old cuticle occurs. Insects secrete an enzymatically active molting fluid to facilitate this process. Once the animal has withdrawn from the old cuticle, a rapid thickening and hardening of the new cuticle occurs. This is critical as the soft, immobile animal is highly susceptible to predation. Chitin synthesis is maximal and rapid calcium deposition occurs (in crustaceans) until the cuticle is hardened (Stevenson, 1972; Adelung, 1971). The hexose units for this synthesis come from the resorbed chitin (Speck and Urich, 1971), from gluconeogenesis, as well as from glycogen reserves (Adelung, 1971). Thus, each tissue is faced with changing metabolic demands as a function of the molt cycle.

Gluconeogenesis in arthropods follows basically the same pathway as in mammals, differing in details related to rather unique functional requirements. Its major functions are to supply hexose units for glycogen and chitin synthesis, but in addition it also supplies precursors for blood glucose, trehalose, and mannose. This diversity of end products imposes important modifications on the overall control of the pathway.

The three points of greatest interest in the pathway are those where "downhill" glycolytic reactions require special modifications for their reversal. These are the reactions catalyzed by pyruvate kinase (PK), phosphoglycerate kinase (PGK), phosphofructokinase (PFK)[1], and hexokinase. Hexokinase, phosphofructokinase, and pyruvate kinase have received much attention in mammalian systems as have the mechanisms for their reversal and control. The phosphoglycerate kinase reaction, which proceeds with a large $-\Delta G°$, has been studied mainly in yeast and skeletal muscle (Scopes, 1973; Fritz and White, 1974) where its kinetics and thermodynamics seem to preclude reversal. Possibilities for bypassing this reaction include (1) a true enzymatic bypass (3PGA $\rightarrow$ glycerate $\rightarrow$ glyceraldehyde $\rightarrow$ GAP), (2) a kinetic bypass simply reversing PGK function (Fritz and White, 1974), or (3) a kinetic bypass involving different isozymes of PGK. The latter mechanism was shown to be the case in our investigations of this reaction in various tissues of the crab, *Cancer magister*.

The enzymes unique to gluconeogenesis--phosphoenolpyruvate carboxykinase (PEPCK), pyruvate carboxylase, and fructose 1,6 diphosphatase (FDPase)--often occur as tissue specific isozymes in various arthropod tissues such as gill, epidermis, fat body, flight muscle, and walking leg muscle (Friedman, 1970; Hochachka, 1972; Crabtree et al, 1972). Where these enzymes have been characterized, their regulatory properties fit in well with the requirements of each tissue's metabolic organization, each often playing key, tissue specific functions.

The Pyruvate Kinase Bypass

For gluconeogenesis from alanine, lactate, or pyruvate to occur, the first reaction to be bypassed is the pyruvate kinase reaction. This is usually thought to occur via the pyruvate carboxylase and phosphoenolpyruvate carboxykinase reactions. The two enzymes are present in low levels in crustacean epidermis (Table 1), with PEPCK increasing in specific activity with molting. While enzyme control in the epidermis has not been worked out, the acetyl CoA dependence of the pyruvate carboxylase activity makes an analogous system to that in the liver probable.

Perhaps the most important problem at this branchpoint in the crustacean epidermis is the control of the pyruvate kinase reaction. In both gill and hypodermis, pyruvate kinase levels rise with molting to become 50-100 fold those of PEPCK (Table 1). The molting hypodermis makes both chitin and CO_2 at high rates, having both active gluconeogenesis and glycolysis. The pyruvate kinase reaction thus needs to be under close, but flexible control. Preliminary studies on molting hypodermal PK indicate FDP activation, MgATP inhibition, and FDP reversal of MgATP inhibition. Fig. 1 shows that 1 mM FDP, 5 x 10^{-4}M UDPAG, and 1 mM glucosamine-6-phosphate all decrease the PEP K_m to 5 x 10^{-5}M. MgATP is competitive with PEP, increasing the PEP K_m to 1 mM. The K_i for MgATP is 1.4 mM. Calcium is a potent inhibitor, with 50% inhibition at 5 x 10^{-4}M. This is well within the physiological levels in these tissues (Mackay and Prosser, 1970). Molting hypodermal pyruvate kinase thus resembles the mammalian liver type pyruvate kinase, sensitive to the levels of FDP and other important metabolites in the tissue (Scutton and Utter, 1968).

In insects, the pyruvate kinase reaction has been studied in the flight muscle and fat body of the locust, *Schistocerca gregaria*, by Bailey and Walker (1969). The fat body PK is

TABLE 1

Specific activities in various tissues of *Cancer magister*
expressed in μmoles/min/gm wet weight at 20°C

Enzyme	Intermolt			Freshmolt		
	Hypodermis	Gill	Muscle	Hypodermis	Gill	Muscle
FDPase	0.66	0.44	0.33	1.20	0.559	0.484
PFK	0.182	0.241	5.81	0.375	0.1565	2.122
F6PGAT	0.12	0.12	-	0.15	0.135	0.0025
PGK	3.615	3.278	13.02	5.295	4.34	8.36
PK	6.9	7.7	61.7	19.7	16.2	51.5
PEPCK	0.15	0.08	-	0.228	0.12	-
pyruvate carboxy-lase*	+	+	-	+	+	-

* Demonstrable but not quantifiable due to contaminating enzyme activities.

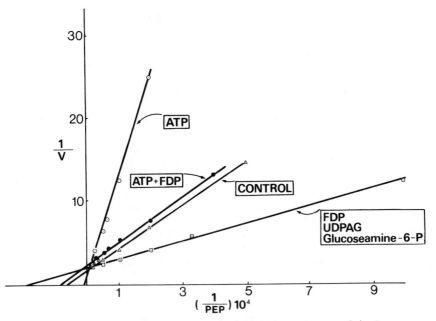

Fig. 1. Lineweaver-Burk plots (1/velocity versus 1/substrate concentration) for molting epidermal pyruvate kinase from *Cancer magister*. Control assay conditions (Δ): 50 mM imidazole-Cl, pH 7.0 with 95 mM K^+, 0.3 mM ADP, 5 mM Mg^{++}, 0.15 mM NADH, PEP to indicated levels, excess LDH. Enzyme in the presence of 0.5 mM UDPAG, 1 mM glucosamine-6-P, and/or 1 mM FDP (□). Enzyme in presence of 6 mM ATP (O). Enzyme in presence of 6 mM ATP and 1 mM FDP (●).

activated by low concentrations of FDP (50% activation at 1 micromolar) and inhibited by ATP. The ATP inhibition is reversed by FDP. FDP also converts the PEP kinetics from sigmoidal to hyperbolic. Flight muscle PK shows no response to FDP at physiological pH, 5 mM ATP causes 50% inhibition, and the PEP kinetics are strictly hyperbolic. The locust tissue specific isozymes show essentially the same regulatory properties as do the pyruvate kinases of mammalian liver and muscle.

The other side of this branchpoint has been utilized for some elegant biochemical specialization in insect flight muscle. Insect flight muscle is one of the most active tissues in nature, being able to sustain a wingbeat frequency of up to 1000/second. In the transition from rest to flight, it undergoes as much as a 100-fold rise in oxygen consumption (Sacktor, 1970). In investigations on the enzymology of this branchpoint in various insect muscles, Crabtree, Higgins, and Newsholme (1972) found little or no PEPCK in any of the flight muscles tested, finding it only in the leg muscles. In contrast, the pyruvate carboxylase activity was found only in the flight muscle, occurring there in levels up to 4 times the level in avian or mammalian liver. A gluconeogenic function seems ruled out by the absence of PEPCK in the same tissue, but a function in providing oxaloacetate is highly reasonable. The flight muscle burns only carbohydrate, and upon initiating flight mobilizes these reserves with a concomitant increase in pyruvate and acetyl CoA levels due to an initial lack of Krebs cycle intermediates. The accumulating acetyl CoA thus serves as a sensitive indicator of the amount of additional oxaloacetate needed for optimal Krebs cycle activity. The flight muscle pyruvate carboxylase shows a pronounced activation by increased acetyl CoA levels. It thus provides oxaloacetate when needed for increased Krebs cycle activity. The high level of this normally gluconeogenic enzyme gives this tissue a particularly sensitive capacity for rapidly stepping up its oxidative metabolism.

Kinetic Mechanisms for PGK Reversal

The reaction catalyzed by PGK proceeds with an overall $\Delta G°$ of about -4.5 kcal/mole and as such would appear to present a fairly large thermodynamic barrier for carbon flow in the gluconeogenic direction. In addition, where the enzyme has been examined, it typically displays a very high affinity for 1,3-DPG and a low affinity for 3-PGA. In mammalian muscle, the K_m for 1,3-DPG is about 2 μM while the K_m for 3 PGA is about 1000 times greater (about 3 mM). Thus, both thermodynamic and kinetic barriers have to be overcome to achieve net gluconeogenic flux through this reaction. Usually, it has been assumed that this barrier is "bypassed" by kinetic mechanisms, but these have never been adequately charted. For this reason, and because there is some confusion in the literature as to the precise pathway of carbon flow in this region of metabolism (Veneziale, Gabrielli, and Lardy, 1970), we examined the problem in some detail in the epidermis and muscle of the crab, *Cancer magister*. Electrophoretic and kinetic analyses reveal distinct muscle and hypodermal isozymes (Fig. 2), both tissues also having intermolt and molting forms of the PGK isozyme. The intermolt muscle PGK moves more slowly than the other forms and displays a higher specific activity than does its molting variant (Table 1). The intermolt epidermal PGK moves ahead of the muscle forms, but slower than its corresponding molting variant. Molting epidermal PGK is present in 1.5-2

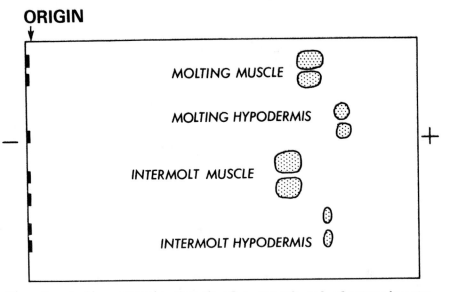

Fig. 2. A diagrammatic sketch of a starch gel electrophoreto-
gram showing zones of PGK activity extracted from crab muscle
and epidermis. Conditions of gel and run: 13% starch gel in
Tris-citrate buffer, pH 7.0, 250 mV, 30 hours. Stain after
Scopes (1968): 4 ml of 50 mM Tris-HCl containing 3-PGA (8 mM),
ATP (8 mM), MgCl$_2$ (5 mM), NADH (3 mM), GAPDH (5 units/ml),
TIM and α-GPDH (2.5 units/ml). After 1 hour of incubation, a
solution of nitroblue tetrazolium (3 mg/ml) and phenazine methosulfate (0.5 mg/ml) was applied to detect unchanged NADH.

fold the intermolt PGK levels. Thus, in both tissues, the
amount and *kind* of isozyme present is varied as a function of
the molt cycle. On examination of the catalytic and regulatory
properties of the PGK isozymes, it becomes clear that the molting hypodermal PGK is better suited for gluconeogenic function
than the intermolt PGK, and that both are in fact better than
the intermolt muscle PGK.

Central to the regulation of PGK for gluconeogenic function
appears to be its affinity for adenylates rather than for
3-PGA, for a comparison of the 4 isozymes reveals a similar
affinity for 3-PGA (Table 2). In contrast, large differences
in enzyme-ATP and enzyme-ADP affinity are evident between the
4 isozymes. In regard to "uphill" function the intermolt muscle PGK I displays the lowest affinity for ATP and the highest
sensitivity to product inhibition by ADP. Indeed this enzyme
is so sensitive to ADP that at all possible physiological
levels of ADP in crustacean muscle (Hochachka et al, 1970) the

TABLE 2
Kinetic parameters of phosphoglycerate kinase
from *Cancer magister*

| | K_m values | |
	ATP	3 PGA
Molting muscle	4.54×10^{-5} M	2.22×10^{-4}
Molting hypodermis	7.6×10^{-5} M	4.84×10^{-4}
Intermolt muscle	1.28×10^{-4} M	2.56×10^{-4}
Intermolt hypodermis	1.21×10^{-4} M	3.24×10^{-4}

| | K_i values for ADP relative to: | |
	ATP	3 PGA
Molting muscle	4.5×10^{-4}	3.46×10^{-4}
Molting hypodermis	2.65×10^{-3}	1.3×10^{-3}
Intermolt muscle	3.5×10^{-5}	6.03×10^{-5}
Intermolt hypodermis	3.16×10^{-4}	3.17×10^{-4}

Assay conditions: 50 mM imidazole-Cl pH 7.0, 2 mM $MgCl_2$,
3 mM 3-PGA, 6×10^{-4} M ATP, 3×10^{-4} M NADH,
excess glyceraldehyde-3-phosphate dehy-
drogenase.
K_i values were determined from Lineweaver-
Burk plots at 3 different ADP levels.

enzyme's uphill function would be kinetically precluded. Although reversible in theory, the enzyme functions in effect
as a unidirectional catalyst (see Barnes et al, 1972, for further discussion of this problem). This high ADP affinity of
course is an eminently sensible design for a PGK that functions
in as highly glycolytic a tissue as crustacean muscle
(Hochachka et al, 1970). However, it is a characteristic that
would be incompatible with gluconeogenic function in the epidermis. Not surprisingly the epidermal isozymes do not show
such an extreme sensitivity to ADP, and of these the lowest
ADP affinity (as judged by K_i values for ADP) is displayed by
the molting hypodermal PGK. The reduced ADP affinity (100-
fold lower than that of the intermolt muscle PGK) coupled with
a five-fold increase in the ATP affinity allows for effective

reversal of the reaction. Thus, in *Cancer magister* the means of reversing the PGK reaction is a kinetic one, expressed by varying the isozyme form present in each tissue in accordance with the demands of the molt cycle.

FDPase - A Mechanism for Bypassing PFK

As far as the available data allow one to assess, the pathway of gluconeogenesis in arthropod epidermis is not at all unusual to the level of the hexosephosphates; at this point, however, the situation becomes rather novel in that a large number of metabolic fates are available for the newly formed hexose phosphates. The first and probably most important branchpoint occurs at the level of F6P, with one metabolic branch leading to chitin biosynthesis and the other branch leading predominantly to glycogen. The first committed metabolite in the pathway to chitin is UDPAG; the first committed metabolite to glycogen is UDPG. In addition, F6P can be used to sustain glucose, mannose, and trehalose reserves. FDPase catalyzes the first and only reaction that (1) is common to this complex end of carbohydrate metabolism, and (2) irreversibly commits carbon flow into this metabolic pool. To our knowledge, the enzyme has been examined in detail in only one arthropod, the arctic king crab, *Paralithodes camtchatica* (Hochachka, 1972).

FDPase in this species occurs in at least 3 isozymic forms. The form predominating in muscle will be discussed further below. The other two forms (Fig. 3), termed FDPase-I (predominant in epidermis) and FDPase-II (predominant in gill) occur in significantly higher activities. When extracted from intermolt crabs, the specific activity (in μM FDP hydrolyzed/ min/gm wet weight at 20°C and optimal assay conditions) is highest (0.6 units) in the epidermis, intermediate (0.3 units) in the gill, and lowest (about 0.1 units) in muscle (Hochachka, unpublished data). These values are similar to those noted for other crustaceans (Table 1; Thabrew et al, 1971). [In an earlier paper, mμM units of FDPase activity were mistakenly reported as μM units (Hochachka, 1972)].

In arthropod epidermis, as in other tissues which contain significant quantities of FDPase and PFK within the same cell compartment, stringent reciprocal control mechanisms are required to prevent significant "futile" cycling at this point in metabolism. In essentially all cases, this is achieved by a single metabolite, AMP, which serves as a potent activator of PFK and a potent inhibitor of FDPase (see Newsholme and Gevers, 1967, for a review). The same control mechanism is operative in the king crab epidermis. Both FDPases present

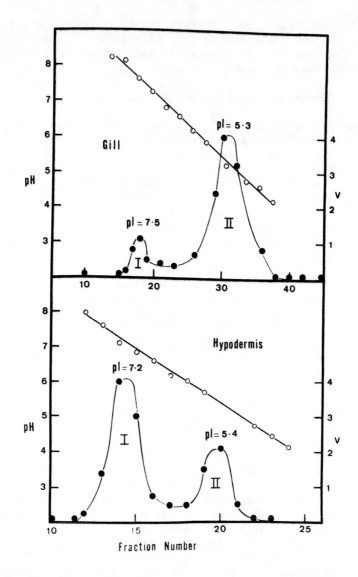

Fig. 3. Isoelectrofocusing of (a) gill, and (b) epidermal FDPase in a pH 3-10 gradient at 5°C. From Hochachka (1972).

in the tissue are sensitive to AMP control, the efficiency of this inhibition being strongly dependent upon K^+, H^+, and Mg^{++} concentrations (Hochachka, 1972). Since all the relevant affinity parameters are within the known in vivo concentration ranges for these compounds, it is evident that interactions

between FDP, AMP, K^+, Mg^{++}, and H^+ on the enzyme could lead
to highly flexible control of catalysis. Yet, given the set
of metabolic pathways operating in this tissue, the above
control mechanisms lack the specificity required for efficient
channelling of carbon through the FDPase "bottleneck" either
towards chitin or towards glycogen (and other storage carbohydrates), but not into both branches simultaneously. What
clearly is required is a set of feedback signals, so that the
activity of FDPase could be regulated relative to the requirements for each of the terminal pathways. At least three such
additional "signals" are now known: UDPAG, the immediate precursor to chitin, UDPG, the immediate precursor to glycogen,
and R5P, a pentose cycle intermediate (Fig. 4 and 5). Inter-

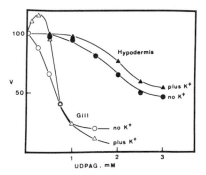

Fig. 4. Effect of UDPAG upon gill (O,Δ) and epidermal (●,▲)
FDPases from the king crab. Assay conditions given Hochachka
(1972). Effect of 100 mM K^+ is shown for gill FDPase in open
triangles, for the epidermal FDPase in closed triangles.

estingly, all three metabolites are inhibitors and all three
have the same "target" isozyme--FDPase II. FDPase I is totally
insensitive to R5P and UDPG, and is only slightly inhibited at
high UDPAG concentrations (Fig. 4 and 5). Thus, these three
metabolites could serve to regulate the supply of a key precursor (F6P) required for their own formation. At the same
time, when their concentrations (in any combination) are so
high as to largely "turn off" FDPase II, the tissue would
still retain the capacity to produce some F6P by FDPase I
catalysis. In fact, the appearance of end product control of
this reaction step can be viewed as setting the evolutionary
stage for an isozyme requirement at this point in metabolism
(Stadtman, 1968), for without more than one isozyme in this
kind of control situation, an over-supply of one end product
(say, UDPAG) could curtail the synthesis of intermediates
required for the alternate branch pathways (Fig. 6).

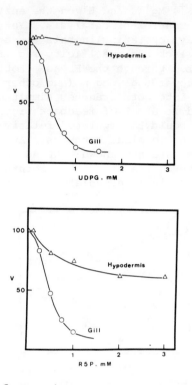

Fig. 5. Effect of UDPG (upper panel) and of R5P (lower panel) on gill (O) and epidermal (Δ) FDPases from the arctic king crab. Assay conditions given by Hochachka (1972).

At the moment we do not know if the relative amounts of FDPase I and II are regulated as a function of the molt cycle, although total FDPase activity clearly is increased during molting (Table 1). We do know, however, that the relative amounts of FDPase I and II differ between tissues such as the gill and epidermis in a manner that fits each tissue's metabolic organization. Thus, in the gill whose prime biosynthetic load is that of chitin synthesis for each filament cover, about 90% of the FDPase activity is accounted for by isozyme II. Almost certainly, UDPAG is a chief regulatory metabolite (Hochachka, 1972); R5P feedback signals from the pentose shunt may also be important since this pathway is highly active in the king crab gill (Hochachka et al, 1970). However, the biosynthetic pathways to glycogen, mannose, and trehalose are essentially inactive in this tissue; hence the need for the "insensitive" FDPase I is lower, and the highly sensitive

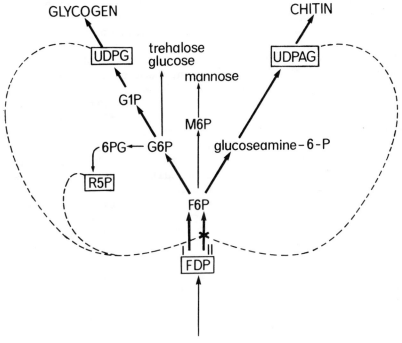

Fig. 6. A summating metabolic map of feedback regulation of FDPase I and II in king crab epidermis, showing metabolite inhibition with a dark cross. AMP control of both FDPases I and II is not shown.

FDPase II satisfactorily fulfills the gill's requirements at this reaction point.

Behrisch (1972), in his work with muscle FDPase from several arctic crabs, has described a form similar to the FDPase II of hypodermis and gill. His muscle FDPase shows a high sensitivity to AMP inhibition, similar to other muscle FDPases, but is activated by PEP. PEP decreases the K_m for FDP and increases the K_i for AMP. The other regulatory properties are similar to the gill form. UDPAG is a potent inhibitor, as are mannose-6-phosphate, and ribulose-5-phosphate. UDPG has no effect on the activity. There is only the one responsive form present in the tissue. Thus any carbon flux through this reaction, whether it be for net gluconeogenesis or for amplification of the PFK regulation, is responsive to signals from chitin synthesis, the pentose phosphate shunt, and mannose metabolism,

as well as to the energy status of the cell. Behrisch was
able to demonstrate pyruvate carboxylase and PEPCK in the muscle of king and tanner crabs, making a gluconeogenic function
for the FDPase more plausible.

All of the above FDPases and in most others which have been
studied, a basic AMP control prevents the simultaneous opera-
tion of PFK and FDPase, preventing the wasteful hydrolysis of
ATP. In bumblebee flight muscle a different situation holds.
Newsholme et al (1972) found that bumblebee flight muscle had
FDPase in levels up to 30 times those in any other muscle.
A gluconeogenic function is ruled out by the absence of PEPCK.
As mentioned before, flight muscle is one of the most efficient
and active in nature. It utilizes carbohydrate as its sole
carbon and energy source. In the bumblebee flight muscle,
PFK and FDPase are in approximately a 1:1 ratio, inversely
correlated with body size. The FDPase differs from any other
previously examined in being refractory to AMP inhibition.
This lack of AMP control allows the futile cycling of carbon,
with concomitant hydrolysis of ATP and release of energy as
heat. This ATP hydrolysis is useful to the bumblebee which
must keep its thorax at over 30°C in order to fly. Bumblebees
are able to function under inclement conditions which keep
honeybees in the hive, thermoregulating on a social scale.
Newsholme et al (1972) postulate that, during flight, the hydrolysis of ATP through muscular contraction and oxidative
phosphorylation of ADP release sufficient heat to maintain the
thoracic temperature. When the bee rests on a flower, gather-
ing nectar, the heat output from contraction diminishes and
the bumblebee, to prevent itself from cooling down, invokes
this cycle which results in significant heat release. Honeybees, which are limited to warm days for foraging, lack these
high FDPase levels. Lardy and his coworkers have measured
the cycling and have established that it occurs experimentally
in response to a drop in the external temperature while the
bumblebee is resting. No cycling occurs during flight, probably due to a potent calcium inhibition of the FDPase (Clark
et al, 1973). This elaboration of a different form of FDPase
is of selective significance, allowing foraging during inclement conditions.

Glucose-6-Phosphatase -
A Mechanism for Bypassing Hexokinase

The glucose-6-phosphatase reaction, the final reaction on
the way to glucose, has not received extensive attention in
arthropods. In crustaceans, it has been measured in the gill
of *Carcinus maenus* where it occurs in high levels and is active

in the release of glucose (Thabrew, Post, and Munday, 1971).
Johnston, Elder, and Davies (1973) confirmed the presence of
glucose-6-phosphatase in blood cells and elucidated the importance of the cells as polysaccharide storage compartments.
In the insect, *Phormia regina*, glucose-6-phosphatase activity
in the fat body was found to be associated with trehalose-6-
phosphatase activity (Friedman, 1970). Increased levels of
trehalose, which inhibit trehalose-6-phosphatase activity,
activate glucose-6-phosphatase activity.

Glucosamine-6-P Synthesis

The formation of glucosamine-6-P is the first step in the
biosynthesis of chitin, and because this is such a key function of early post-molt epidermis, its activity must be carefully integrated with that of gluconeogenesis as a whole. In
arthropods, this key intermediate is usually synthesized by an
amidotransferase reaction. Most amidotransferase reactions
are fully reversible, but in this case, glutamine (rather than
other amino acids or NH_4^+) is the cosubstrate and the amido
nitrogen serves as the donor group, producing a transferase
reaction that in effect is irreversible:

$$F6P + glutamine \longrightarrow glutamate + glucosamine-6-P$$

As a branchpoint enzyme, an irreversible "bottleneck" type
reaction is brought under metabolic control far more readily
than are reversible reactions; hence, nature's choice of
glutamine as the nitrogen donor can be readily appreciated.
Glucosamine-6-P can also be synthesized in a reaction catalyzed
by glucosamine phosphate isomerase:

$$F6P + NH_4^+ \longleftrightarrow glucosamine-6-P + H_2O$$

However, the equilibrium of this reaction, as determined by
in vitro measurements, is in favor of F6P formation.
In crustaceans, both of the above enzymes are present, with
the second probably playing a catabolic role. The F6P-glu-
tamine-amidotransferase increases in specific activity with
molting (Table 1) and the enzyme is present at levels several
fold higher than reported values for mammalian liver
(Winterburn and Phelps, 1971a). Glucosamine-6-phosphate isomerase has been shown to vary with the molt cycle, increasing
in specific activity with molting (Lang, 1971). When both
enzymes are present in the same tissue, it is clear that their
simultaneous function must be prevented in order to avoid glutamine depletion, but at this time nothing is known of mech-

anisms by which this may be achieved. In the mammalian liver
the regulation of the amidotransferase is complex; UDPAG inhibits, with G-6-P potentiating the inhibition. AMP and ATP
further enhance the inhibition, but UTP reverses the inhibition. Substrate inhibition by glutamine which is induced by
UDPAG further controls the enzyme (Winterburn and Phelps,
1971b). Similar controls, if operative, in the crustacean
epidermis are yet to be examined.

In the housefly, F6P-glutamine-amidotransferase does not
occur. Here the isomerase has been modified for synthetic
function. A potent G6P activation of only the aminating
reaction, coupled with N-acetyl-glucosamine-6-phosphate activation, help increase the synthetic rate to 4 times the degradative rate. Furthermore, the presence of the subsequent enzymes on the pathway to chitin synthesis, which would immediately remove the glucosamine-6-P, further favor the biosynthetic reaction (Benson and Friedman, 1970).

CONCLUDING COMMENTS

In this paper we have attempted to give a brief overview
of the functions to which gluconeogenic enzymes are applied
in the arthropods. Although the field is not fully developed,
it is already clear that at many of the "bottlenecks" in
chitin and carbohydrate synthesis, insects and crustaceans
have evolved specific isozymes which integrate carbon flow
through these reactions with the specialized needs of the
tissue. Some of the enzyme variants may occur in response
to changing cyclic requirements, as in the case of the PGK
forms; others, as in the case of the FDPases, may be present
simultaneously giving the tissue a greater metabolic flexibility and responsiveness; still others, as in the case of the
pyruvate kinases, occur in tissue-specific form(s) displaying
functions that are integrated into the overall metabolic organization of the tissue.

Study in this field is far from complete. Does significant
gluconeogenesis occur in muscle or in blood cells at any time
during the molt cycle? What are the control sites in the
branch pathway leading to chitin? In particular, what are the
regulatory properties of the enzymes catalyzing the first and
last steps, two obvious sites of regulation, in the pathway?
How is the reciprocal relationship between chitin and glycogen
synthesis achieved, and what is the role of isozymes in this
control? To date, most such questions remain unanswered. When
new students address themselves to them, they must bear in
mind the cyclic demands of the molt cycle, for this adds a
dimension with both biological implications and experimental
advantages.

534

FOOTNOTE

[1] Abbreviations used:
F6PGAT: fructose-6-phosphate glutamine-amido-transferase
PK: pyruvate kinase
PGK: phosphoglycerate kinase
PFK: phosphofructokinase
PEPCK: phosphoenolpyruvate carboxykinase
FDPase: fructose-1,6-diphosphatase
FDP: fructose-1,6-diphosphate
MgATP: magnesium adenosine triphosphate
UDPAG: uridine diphosphate N-acetyl-glucosamine
PEP: phosphoenolpyruvate
ATP: adenosine triphosphate
ADP: adenosine diphosphate
3-PGA: 3-phosphoglycerate
1,3-DPG: 1,3-diphosphoglycerate
GAP: glyceraldehyde-3-phosphate
Acetyl CoA: acetyl coenzyme A
AMP: adenosine monophosphate
UDPG: uridine diphosphate glucose
R5P: ribose-5-phosphate
F6P: fructose-6-phosphate
G6P: glucose-6-phosphate
UTP: uridine triphosphate
GAPDH: glyceraldehyde-3-phosphate dehydrogenase
TIM: triose phosphate isomerase
αGPDH: α-glycerol phosphate dehydrogenase

REFERENCES

Adelung, D. 1971. Untersuchungen zur Häutungsphysiologie der dekapoden Krebse am Beispiel der Strandkrabbe *Carcinus maenas*. *Helgoländer wiss. Meeres-untersuchungen* 22: 66-119.

Bailey, E. and P. R. Walker 1969. A comparison of the properties of the pyruvate kinases of the fat body and flight muscle of the adult male desert locust. *Biochem. J.* 111: 359-369.

Barnes, Larry O., John J. McGuire, and Daniel E. Atkinson 1972. Yeast diphosphopyridine nucleotide specific isocitrate dehydrogenase. Regulation of activity and unidirectional catalysis. *Biochem.* 11: 4322-4329.

Behrisch, Hans 1972. Regulation by phosphoenolpyruvate of fructose 1,6-diphosphatase in skeletal muscle: evidence for an allosteric activator of the enzyme. *Can. J. Biochem.* 50: 710-713.

Behrisch, Hans, and Craig E. Johnson 1974. Unusual regulatory characteristics of fructose 1,6-diphosphatase from muscle. Studies on the kinetic behaviour of the enzyme from muscle of the arctic tanner crab *Chionocetes bairdi*. *Comp. Biochem. Physiol.* 47B: 427-436.

Benson, Robert, and Stanley Friedman 1970. Allosteric control of glucosamine phosphate isomerase from the adult housefly and its role in the synthesis of glucosamine-6-phosphate. *J. Biol. Chem.* 245: 2210-2228.

Clark, Michael G., David P. Bloxham, P. C. Holland, and Henry Lardy 1973. Estimation of the fructose diphosphatase-phosphofructokinase substrate cycle in the flight muscle of *Bombus affinis*. *Biochem. J.* 134: 589-597.

Crabtree, B., S. J. Higgens, and E. A. Newsholme 1972. The activities of pyruvate carboxylase, phosphoenolpyruvate carboxylase, and fructose diphosphatase in muscles from vertebrates and invertebrates. *Biochem. J.* 130: 391-396.

Friedman, Stanley 1970. Metabolism of Carbohydrates in Insects. in *Chemical Zoology V* (Florkin and Scheer, eds). Academic Press, New York, A: 167-197.

Fritz, Paul J. and E. Lucile White 1974. 3-phosphoglycerate kinase from rat tissues. Further characterization and developmental studies. *Biochem.* 13: 3, 444-449.

Hochachka, P. W. 1972. The functional significance of variants of fructose 1,6-diphosphatase in the gill and hypodermis of a marine crustacean. *Biochem. J.* 127: 781-793.

Hochachka, P. W., G. N. Somero, D. E. Schneider, and J. M. Freed 1970. The organization and control of metabolism in the crustacean gill. *Comp. Biochem. Physiol.* 33: 529-548.

Hochachka, P. W., J. M. Freed, G. N. Somero, and C. L. Prosser 1971. Control sites of glycolysis of crustacean muscle. *Int. J. Biochem.* 2: 125-130.

Hohnke, Lyle and Bradley T. Scheer 1970. Carbohydrate Metabolism in Crustaceans. in *Chemical Zoology V* (Florkin and Scheer, eds). Academic Press, New York, 147-166.

Johnston, Michael, and P. Spencer Davies 1972. Carbohydrates of the hepatopancreas and blood tissues of *Carcinus*. *Comp. Biochem. Physiol.* 41B: 433-443.

Johnston, Michael A., Hugh Y. Elder, and P. Spencer Davies 1973. Cytology of *Carcinus* haemocytes and their function in carbohydrate metabolism. *Comp. Biochem. Physiol.* 46A: 569-581.

Lang, R. 1971. Chitin Synthese bei dem flusskrebs *Orconectes limosus*: Aktivität der Phosphoglucosaminisomerase und einbau von Glucose-U-^{14}C in Chitin. *Z. vergl. Physiologie* 73: 305-316.

Larsson-Raznikiewicz, Mârtha and Lars Arvidsson 1971. Inhibition of phosphoglycerate kinase by products and product homologues. *Eur. J. Biochem.* 22: 506-512.

MacKay, W. C. and C. L. Prosser 1970. Ionic and osmotic regulation in the king crab and two other North Pacific crustaceans. *Comp. Biochem. Physiol.* 34: 273-280.

Newsholme, E. A., B. Crabtree, S. J. Higgins, S. D. Thornton, and Carole Start 1972. The activities of fructose diphosphatase in flight muscles of the bumblebee and the role of this enzyme in heat generation. *Biochem. J.* 128: 89-97.

Passano, L. M. 1960. Molting and Its Control. in *The Physiology of Crustacea, Vol. 1* (Waterman, Talbot, H., eds) Academic Press. p. 473.

Sacktor, Bertram 1970. Regulation of intermediary metabolism with special reference to the control mechanisms in insect flight muscle. *Adv. Insect Physiol.* 7: 268-347.

Scopes, R. K. 1968. Methods for starch gel electrophoresis of sarcoplasmic proteins. *Biochem. J.* 107: 139-150.

Scopes, R. K. 1973. 3-Phosphoglycerate Kinase. in *The Enzymes, Vol. VIII, Part A* (Boyer, Paul., ed) 335-351.

Scrutton, Michael and Merton Utter 1968. The regulation of glycolysis and gluconeogenesis in animal tissues. *Ann. Rev. Biochem.* 37: 249-302.

Skinner, Dorothy M. 1966. Breakdown and reformation of somatic muscle during the molt cycle of the land crab, *Gecarcinus lateralis. J. Exp. Zool.* 163: 115-124.

Speck, Ulrich, Klaus Urich, and Ute Herz-Hübner 1972. Nachweis einer Regulation der Glucosamin Bildung bei dem Flusskrebs *Orconectes limosus* zur zeit der Häutung. *Z. vergl. Physiol.* 76: 341-346.

Speck, U. and K. Urich 1973. Resorption des alten Panzers vor der Häutung bei dem Flusskrebs *Orconectes limosus*: Schicksal des freigesetzten N-Acetyl-glucosamins. *Z vergl. Physiol.* 78: 210-220.

Stevenson, J. Ross 1972. Changing activities of the crustacean epidermis during the molting cycle. *Am. Zool.* 22: 373-380.

Thabrew, Myrtle, P. C. Poat, and Kenneth A. Munday 1971. Carbohydrate metabolism in *Carcinus maenus* gill tissue. *Comp. Biochem. Physiol.* 40B: 531-541.

Veneziale, Carlo, Franco Gabrielli, and Henry A. Lardy 1970. Gluconeogenesis from pyruvate in isolated perfused rat liver. *Biochem.* 9: 3960-3970.

Winterburn, P. J. and C. F. Phelps 1971a. Purification and some kinetic properties of rat liver glucosamine synthetase. *Biochem. J.* 121: 701-709.

Winterburn, P. J. and C. F. Phelps 1971b. Studies on the control of hexosamine biosynthesis by glucosamine synthetase. *Biochem. J.* 121: 711-720.

DIFFERENT MOLECULAR FORMS OF FORMATE DEHYDROGENASE AND NITRATE REDUCTASE SOLUBILIZED FROM THE MEMBRANE OF ESCHERICHIA COLI

J. RÚIZ-HERRERA and E. I. VILLARREAL-MOGUEL
Departamento de Microbiología
Escuela Nacional de Ciencias Biológicas
Instituto Politécnico Nacional
México 17, D. F.
MEXICO

ABSTRACT: Formate dehydrogenase present in the membrane of Escherichia coli is involved in H_2 evolution and nitrate reduction by the bacterium. The kinetic properties, stability, and size of the enzymes involved in the two processes are different. Upon purification, the formate dehydrogenase involved in nitrate reduction dissociated into different molecular aggregates which differed in size and charge. During the initial steps of purification, nitrate reductase was attached to formate dehydrogenase, but they were separated by preparative electrophoresis. Nitrate reductase was solubilized also by SDS and heating. The enzymes solubilized by the three methods differed only in size and charge. The most purified samples of nitrate reductase could be further separated into different molecular forms. Neither formate dehydrogenase nor nitrate reductase required detergents or lipids for activity or stability. It was concluded that formate dehydrogenase and nitrate reductase are present in the form of large aggregates in the membrane. These aggregates were detached by different treatments. Upon purification the aggregates were fragmented.

Escherichia coli metabolizes formate under anaerobic conditions by means of two different multiprotein complexes. One of them, hydrogenlyase, transforms formate into H_2 and CO_2, whereas the other one, nitrate reductase, transfers electrons from formate to nitrate. Nitrate is reduced to nitrite in the process. Both complexes contain formate dehydrogenase. Hydrogenlyase contains also hydrogenase and one or two not yet identified electron carriers (Peck and Gest, 1957) but no cytochromes (Ruíz-Herrera and Alvarez, 1972). The nitrate reductase complex contains cytochrome b_1 components besides nitrate reductase itself (Ruíz-Herrera and De Moss, 1969 and is induced by nitrate (Ruíz-Herrera and De Moss, 1969), whereas hydrogenlyase is induced by formate (Pinsky and Stokes, 1952, Kushner and Quastel, 1953; Ruíz-Herrera et al., 1972).
We measured formate dehydrogenase in cell homogenates

obtained by breaking the cells with lysozyme and freezing and thawing with two different artificial electron acceptors (Peck and Gest, 1957), a one-electron acceptor with low redox potential (benzyl or methyl viologen) and a two-electron acceptor with higher redox potential (phenazine methosulfate or methylene blue). It was found that both activities were membrane-bound and that the level of the one-electron enzyme was increased by growing the cells with formate. The two-electron enzyme increased when the cells were grown in the presence of nitrate (Rúiz-Herrera et al., 1971).

Several mutants unable to reduce NO_3- to NO_2- *in vivo* were found to lack the two-electron formate dehydrogenase only (Ruiz-Herrera and De Moss 1969). Of these mutants, some were unable to evolve H_2. All of these non-gas forming mutants failed to synthesize the one-electron enzyme (Rúiz-Herrera and De Moss, 1969; Rúiz-Herrera and Alvarez, 1972). According to these results it was concluded that there were two formate dehydrogenases or two isozymes of the same enzyme. The pleiotropic effect could be explained by either of two hypotheses: a)- Both enzymes might share the polypeptide responsible for the catalytic action i.e. the one which reacted with formate. This polypeptide may be bound into the membrane to different components, each responsible for the specificity of the electron acceptor. b)- Both enzymes may be genetically unrelated but may require a common protein acceptor to be bound to the membrane. According to this view, both enzymes would be active only if bound to the common protein acceptor. The physiological electron acceptor for the two-electron formate dehydrogenase is cytochrome b_1. However, both phenazine methosulfate and methylene blue can be used as a bypass so that electrons are not required to pass through the cytochromes. This is shown by the fact that mutants lacking cytochrome b_1 can still reduce these two dyes with formate as electron donor. In the case of the one-electron formate dehydrogenase the acceptor is not known, but it might be a non-heme iron containing protein.

We have searched for methods to solubilize formate dehydrogenase and nitrate reductase from the membranes of *E. coli*. We have tested several solvents, chaotropic salts, detergents, and other tensioactive substances at different concentrations and under several conditions. Effective solubilization was achieved only under very few conditions (Table 1). The best method involved the use of the non-ionic detergent BRIJ 36 T, an ether of lauryl alcohol with 10 oxyethylenes. BRIJ 36 T did not inactivate the enzymes, and furthermore the one-electron formate dehydrogenase was stabilized when the membranes were solubilized in BRIJ 36 T. The properties of both formate dehydrogenases were different (Table 2). Kinetic character-

TABLE 1

COMPARISON OF SOLUBILIZATION
METHODS FOR FORMATE DEHYDROGENASE
AND NITRATE REDUCTASE

METHOD*	% PROTEIN SOLUBILIZED	RELATIVE ACTIVITY** IN SUPERNATANT		REMARKS
		TWO-ELECTRON FDHase	NRase	
KSCN 0.15 M	11	17		Not inhibitory
CTAB 0.1mM	25	1	1	Inactivates
TRITON X-100	19	6		Inhibitory
TWEEN- 80 2mg/ml	20	6.2		Inhibitory
SDS 1mM	57	0	63	Selective
DEOXYCHOLATE 1 mg/mg Prot	41	8	4	Inactivates
BRIJ 36 T 5mM	30	100	100	Not inhibitory
HEATING AT ALKALINE	10	0	100	Selective

* Only the best conditions are given
** Taking as 100 the best method.

TABLE 2

COMPARISON OF FORMATE DEHYDROGENASES

PARAMETER	ONE-ELECTRON ENZYME	TWO-ELECTRON ENZYME
CELLULAR LOCATION	CELL MEMBRANE	CELL MEMBRANE
RELATIVE Eo	LOW	HIGH
OPTIMUM pH	6.4	7.3
OPTIMUM T	38°C	30°C
Km FOR ELECTRON DONOR	$2.9 \times 10^{-4}M$	$1.7 \times 10^{-4}M$
Km FOR FORMATE	$1.5 \times 10^{-3}M$	$2.5 \times 10^{-5}M$
TYPE OF KINETICS	NORMAL	BIPHASIC
PHYSIOLOGICAL COMPLEX	HYDROGENLYASE	FORMATE OXIDASE, NITRATE REDUCTASE

istics of the two-electron enzyme were generally maintained
when the enzyme was solubilized and separated in an agarose gel
column.   However, the Km for the electron acceptor increased
when the detergent was removed.   Perhaps this increase in Km
can be attributed to the orientation of the electron acceptor
site of the enzyme, i.e. the one which reacts with formate,

541

towards a hydrophilic environment; the electron donor site is in a hydrophobic environment.

Both formate dehydrogenases were unstable, but the two-electron enzyme was stable if isolated from cells grown in a medium supplemented with ferric ions, molybdate, and selenite, and if dithiothreitol and selenite were added to the buffer employed during the solubilization and purification of the enzyme. In recent experiments using this buffer solution, the enzyme was resolved into several peaks differing in molecular weight (Fig. 1). Similar results were obtained when the detergent was omitted from the elution buffer. Addition of phospholipids did not increase the activity or stability of the enzyme.

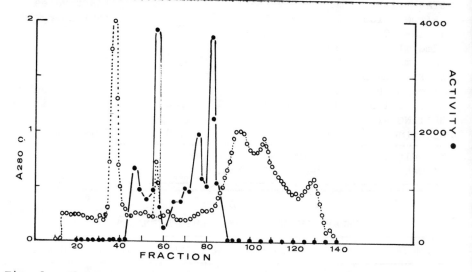

Fig. 1. Chromatography of solubilized formate dehydrogenase in an Agarose 4 B column. Activity is expressed as dpm $^{14}CO_2$ released from radioactive formate per hour of incubation per ml.

The peak having the highest molecular weight was subjected to chromatography in DEAE-cellulose and eluted with a linear gradient of NaCl. Again several peaks of activity were separated. When the largest peak of activity was subjected to preparative electrophoresis in 6% acrylamide gel several peaks were again separated.

Contrary to a previous result (Villarreal-Moguel et al., 1973) it was found that formate dehydrogenase-containing fractions also showed nitrate reductase activity even after purification by agarose gel chromatography or DEAE-cellulose chromatography. These two enzyme activities were separated only after electrophoresis in acrylamide gel. According to this result, both components and perhaps also cytochrome b_1 must form a tight complex which is not separated readily; it should be noted that the active peak contains iron (Villarreal-Moguel et al., 1973). Nitrate reductase was also solubilized by heating at 56° C at pH 8.3 and by treatment with SDS. These treatments did not solubilize the formate dehydrogenase.

The general properties of nitrate reductase solubilized by the three methods were similar except for size and charge (Table 3). Surprisingly, enzyme solubilized with SDS showed the highest molecular weight. The enzymes solubilized by heat and BRIJ 36 T were purified by chromatography in agarose gel, DEAE-cellulose, and Sepadex G-200. In some experiments BRIJ-solubilized enzyme was separated into two peaks during chromatography in Sephadex G-200. Both peaks of activity gave almost the same pattern of protein bands when subjected to electrophoresis in acrylamide gel. Rechromatography of the low molecular weight fraction in Sephadex G-200 at pH 8.3 revealed that the peak of activity was sharp whereas the protein peak was broad. This result suggests the presence of several proteins with a similar but not identical molecular weight. This result may explain the confliciing data in the literature about the molecular weight of nitrate reductase (Taniguchi and Itaoaki. 1960; Showe and De Moss. 1968).

TABLE 3

GENERAL PROPERTIES OF NITRATE REDUCTASE
SOLUBILIZED BY THREE METHODS

| PARAMETER | NITRATE REDUCTASE SOLUBILIZED WITH | | |
	HEAT	SDS	BRIJ 36 T
OPTIMUM TEMPERATURE	45°& 60°C	45°& 60°C	45° & 60°C
OPTIMUM pH	7.3	7.3	7.3
KM METHYL VIOLOGEN	4.3×10^{-7}M	6.2×10^{-7}M	5.5×10^{-7}M
KM FOR FORMATE	8.3×10^{-5}M	12.5×10^{-5}M	8.3×10^{-5}M
VE/VO BIOGEL A-15	2.0	1.87	2.05
VE/VO BIOGEL A-5	1.25	1.03	1.38
RF IN ACRYLAMIDE			
GEL 7%	0.29	0.02	0.03 & 0.16
5%	0.49	0.06	0.29
3.75%	0.83	0.27	0.51

When BRIJ-solubilized enzyme was further purified by preparative acrylamide gel electrophresis, two bands of activity were observed (Fig. 2). Purified nitrate reductase solubilized by both heating and BRIJ gave very similar patterns of proteins at different stages of purification when subjected to electrophoresis in acrylamide gel (Fig. 3). Antibodies prepared

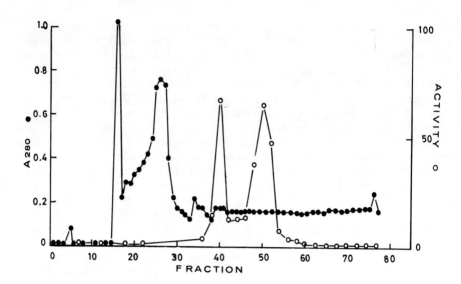

Fig. 2. Preparative electrophoresis in polyacrylamide gel of purified nitrate reductase. Activity expressed as nanomoles nitrate reduced per min per ml.

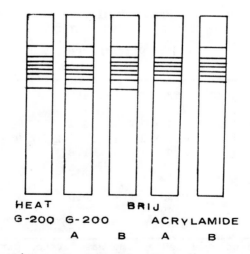

Fig. 3. Schematic representation of electrophoretic patterns in polyacrylamide gel of nitrate reductase at several stages of purification. A and B stand for two different peaks in each case. Peak A from Agarose had the higher molecular weight. Peak A from electrophoresis had the fastest mobility.

against heat-solubilized and purified nitrate reductase gave 3
precipitation bands against this enzyme and one against nitrate
reductase purified beyond the stage of preparative electrophoresis. That antibodies were directed against nitrate reductase
could be demonstrated by use of nitrate reductase-less mutants.
Solubilized membranes of these mutants gave a negative or only
slightly positive precipitation in capillary tubes when mixed
with the antiserum. Several mutants which lacked either or
both formate dehydrogenase and cytochrome b_1 synthesized nitrate
reductase that showed an electrophoretic mobility different from
the wild type (Table 4).

TABLE 4

BEHAVIOUR OF NITRATE REDUCTASE FROM FORMATE DEHYDROGENASE-LESS
MUTANTS IN ACRYLAMIDE GEL ELECTROPHORESIS

STRAIN	BIOCHEMICAL LESION (S)	RF OF NRase IN 7% ACRYLAMIDE GEL
RB-25	fd⁻	0.08
TW-101	fd⁻	0.06
TW-149	fd⁻ cytb⁻	0.13 and 0.30
TW-143	fd⁻ cytb⁻	0.30
WILD Ab-2102		0.03 and 0.12

According to these results we suggest that formate dehydrogenase and nitrate reductase occur as large aggregates in the
membrane of *Escherichia coli*. In these aggregates the active
polypeptide (s) is bound to other polypeptides responsible for
additional properties of the complexes, i.e., recognition of
different electron acceptors or donors, stability, hydrophobic
characteristics, etc. The aggregates can be detached by several
treatments which affect different types of bondings. These
treatments may or may not destroy the integrity and properties
of the normal aggregates but do give rise to different molecular forms of the enzymes. Once in the so called "soluble"
stage, formate dehydrogenase and nitrate reductase and perhaps
other membrane bound enzymes (J. Ruíz-Herrera unpublished
observations) behave as the non-membrane bound enzymes do: that
is, they do not require detergents or lipids for activity or
stabilization.

ACKNOWLEDGEMENTS

This work was carried out with financial support from the
Subvéncion 020 of the CONACYT, México.

REFERENCES

Kushner, D. J. and J. H. Quastel 1953. Factors underlying bacterial enzyme synthesis. *Proc. Soc. Exp. Biol. Med.* 82: 388-392.

Peck, H. D. and H. Gest 1957. Formic dehydrogenase and the hydrogenlyase enzyme complex in coli-aerogenes bacteria *J. Bacteriol.* 73: 706-721.

Pinsky, M. J. and J. L. Stokes 1952. Requirements for formic hydrogenlyase adaptation in nonproliferating suspensions of *Escherichia coli.* *J. Bacteriol.* 64: 151-161.

Ruíz-Herrera, J. and A. Alvarez 1972. A physiological study of formate dehydrogenase, formate oxidase and hydrogenlyase from *Escherichia coli* K-12. Ant. V. Leeuw. *J. Microbiol. Serol.* 38: 479-491.

Ruíz-Herrera, J. and J. A. De Moss 1969. Nitrate reductase complex of *Escherichia coli* K-12: Participation of specific formate dehydrogenase and cytochrome b_1 components in nitrate reduction. *J. Bacteriol.* 99: 720-729.

Ruíz-Herrera, J., A. Alvarez and I. Figueroa 1972. Solubilization and properties of formate dehydrogenases from the membrane of *Escherichia coli.* *Biochim. Biophys. Acta* 289: 254-261.

Showe, M. K. and J. A. De Moss 1968. Localization and regulation of synthesis of nitrate reductase in *Escherichia coli.* *J. Bacteriol.* 95: 1305-1313.

Taniguchi, S. and E. Itagaki 1960. Nitrate reductase of nitrate respiration type from *E. coli.* I. Solubilization and purification from the particulate system with molecular characterization as a metalloprotein. *Biochim. Biophys. Acta* 44: 263-279.

Villarreal-Moguel, E. I., V. Ibarra, J. Ruíz-Herrera and C. Gitler.1973. Resolution of the nitrate reductase complex from the membrane of *Escherichia coli.* *J. Bacteriol.* 113: 1264-1267.

ISOZYMES OF ASPARTATE AMINOTRANSFERASE AND
GLYCOGEN PHOSPHORYLASE IN MUSCLES OF LOWER VERTEBRATES

LYSLOVA, E. M., T. P. SEREBRENIKOVA, N. A. VERZHBINSKAYA
Sechnov Institute of Evolutionary Physiology and Biochemistry
Academy of Sciences of USSR

ABSTRACT. Two isozymes of aspartate aminotransferase and
glycogen phosphorylase b were found in skeletal muscles of
cyclostomes, several species of elasmobranch and teleost
fish, and in amphibians.

The isolated isozymes of both enzymes differ in electrophoretic mobility, affinity to DEAE cellulose, as well as
in apparent Km for substrates. They have different thermostability and different urea denaturation curves. The
isozymes of carp muscle phosphorylase b exhibit a different
Km for the allosteric effector AMP and a different inhibition constant for glucose-6-phosphate. Some species
variation exists in many properties of the isozymes of both
enzymes. One of two isozymes of phosphorylase b, present
in muscles of lower vertebrates (phosphorylase II), was not
found in muscles of higher animals. The other isozyme was
distinctly differentiated in the course of vertebrate
evolution (Km for substrates and for AMP became much lower).

The differences in physico-chemical characteristics of
the AAT isozymes are so pronounced that they probably
originated at an early stage of animal evolution as different proteins, and since then have followed different evolutionary pathways. The mitochondrial AAT isozymes under-
went more profound changes in the course of vertebrate
evolution than did the cytoplasmic isozymes.

INTRODUCTION

The study of isozymes represents a new branch of enzymology and is proving very useful in comparative biochemistry.
New possibilities have arisen for obtaining evidence of molecular evolution of enzymes, and for investigating their origin
and course of differentiation to the level present in higher
animals. This paper presents the results of a comparative
study of two enzymes which play an important role in muscle
metabolism. They are: glycogen phosphorylase (ph-ase, α-
glucanphosphorylase, EC 2.4.1.1.), a key enzyme of glycogenolysis in vertebrate muscles, and aspartate aminotransferase
(AAT, l-aspartate:2-oxoglutarate aminotransferase, EC 2.6.1.
1.), which participates in many processes of muscle metabolism.
We have studied partly purified isozymes of both enzymes

isolated from skeletal muscles of lower vertebrates.

MATERIALS AND METHODS

The species investigated were: Cyclostomes - lamprey (*Lampetra fluviatilis*); Fish - shark (*Squalus acanthias*), skates (*Dasyatis pastinaca, Raja clavata*), marine teleosts - (*Scorpaena porcus, Spicara smaris*), freshwater teleosts - (*Esox lucius, Tinca tinca, Abramis brama, Lucioperca lucioperca, Cyprinus carpio*); Amphibians - frog (*Rana temporaria*).

Lampreys were caught in the autumn during their spawning migration into rivers; during the winter they were kept in fish ponds; marine fish were freshly caught in the Black Sea; freshwater fish were taken from fish ponds.

The isozymes of both enzymes were isolated and partly purified by means of differential sedimentation in ammonium sulfate, differential centrifugation of subcellular fractions, agar gel electrophoresis, and DEAE-cellulose column chromatography (Serebrenikova and Filosofova, 1969; Filosofova and Serebrenikova, 1969; Filosofova, 1970; Filosofova-Lyslova, 1972; Serebrenikova and Khlyustina, 1974).

RESULTS AND DISCUSSION

ISOZYMES OF PHOSPHORYLASE b

Two isozymes of ph-ase were found in muscles of cyclostomes, teleost fish, and amphibians. Two different methods of isozyme separation (agar gel electrophoresis and DEAE column chromatography) yielded equally good results (Fig. 1). In three species - shark, sea bass, and trout (from Sevan lake) we have found only one isozyme of ph-ase, known from the skeletal muscles of higher animals. We found no report of the existence of two isozymes of ph-ase in normal skeletal muscles of higher vertebrates. Only heart muscle, tumors of different origin, and thrombocytes contain two isozymes of ph-ase (Yunis et al., 1962; Davis et al., 1967; Yunis and Arimura, 1966, 1969).

Isozymes of muscle ph-ase differ in electrophoretic mobility - Ph-ase I migrates far toward the anode while ph-ase II remains closer to the origin. They differ in their affinity to DEAE cellulose; ph-ase I was adsorbed on DEAE and could be eluted by 0.04 - 0.2 M KCl; ph-ase II was not adsorbed on DEAE at pH 7.4. There was some species variation in the electrophoretic mobility of both isozymes, and in the affinity of ph-ase I to DEAE.

A more detailed study of ph-ase isozymes was performed using the enzymes of lamprey and carp muscle. The extraction

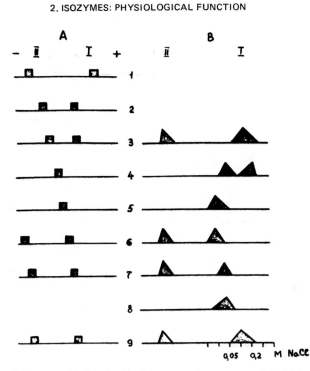

Fig. 1. Patterns of distribution of Ph-ase activities.
Muscle extracts 1:2 w/v , 0.05 M KCl, pH 8.0, 10 h extraction.
A-agar gel electrophoresis, medinal-acetate, pH 8.6; 8 v/cm,
4 h, water-cooling. After elution of protein fractions determination of Ph-ase activity (Cori et al., 1955) and protein
content (Lowry et al., 1951). B-chromatography on DEAE column,
tris-HCl, pH 7.4; elution with 0 - 1 M KCl; zones of Ph-ase
activity shown diagramatically; I and II Ph-ase isozymes. 1-
Frog, 2 - Pike-perch, 3 - Carp, 4 - Scorpionfish, 5 - Sea-bass,
6 - Sting-ray, 7 - Skate (*Raja clavata*), 8 - Dogfish-shark,
9 - Lamprey.

of ph-ase from muscle homogenates and the purification and
separation of the isozymes was performed according to the technique of Metzger et al. (1968). We obtained a 20-fold purification of the enzyme (Serebrenikova and Khlyustina, 1974).

Lamprey and carp muscle contained both isozymes but ph-ase
II was predominant. Without the addition of AMP the purified
enzyme preparation showed no activity. Therefore it may be concluded that both isozymes are molecular varieties of ph-ase b.

We compared the resistance of both isozymes to heat and
urea denaturation. The time/activity curves of heat inactivation of the isozymes were obtained at 50°C, pH 6.3. The
curves differed distinctly for the two isozymes of muscle ph-
ase b. Ph-ase I of lamprey and carp muscles proved to be less

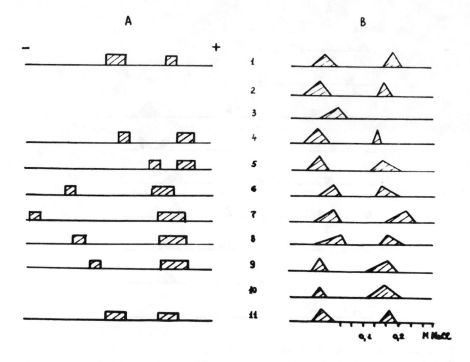

Fig. 2. Patterns of distribution of AAT activities. A - agar
gel electrophoresis, designations as on Fig. 1. B - chromatography on DEAE cellulose, phosphate buffer 0.005 M, pH 7.4,
elution with 0 - 1 M NaCl; zones of AAT activity shown diagramatically (assay by Tonhazy et al., 1950). 1 - Frog,
2 - Golden shiner, 3 - Tench, 4 - Pike-perch, 5 - Carp, 6 -
Sea-bass, 7 - Scorpionfish, 8 - Sting-ray, 9 - Skate (*Raja
clavata*), 10 - Dogfish shark, 11 - Lamprey.

thermostable than ph-ase II (Fig. 4). The difference between
$t_{1/2}$ for the isozymes was statistically significant (p < 0.01).

The course of urea denaturation of isozymes of muscle ph-
ase b differs in some respects from heat denaturation (Fig. 4).
No difference was found between the lamprey ph-ase b isozyme,
but carp ph-ase I and ph-ase II differed markedly. Both isozymes of lamprey muscle ph-ase b were more readily denatured
by urea than were the isozymes from carp muscle.

There is no significant difference in pH optima between the
ph-ase isozymes. The pH optimum of both isozymes in muscle
ph-ase b in lamprey and in carp was 6.3; the same value was
reported for fish muscle ph-ase (Najojama, 1961)--that is,
somewhat lower than the pH-optimum 6.8 known for muscle
ph-ase b of higher vertebrates.

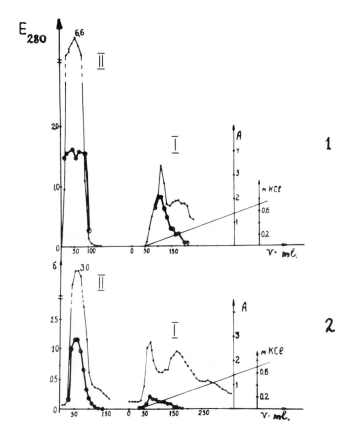

Fig. 3. Chromatography on DEAE cellulose Ph-ase isozymes of lower vertebrates. 1 - Lamprey, 2 - Carp, I and II - Ph-ase isozymes; ordinates: E_{280}; A - micromoles P; M KCl. -.-.-. protein content in samples E_{280} -o-o-o Ph-ase activity, micromoles P/ ml/ min abscissa - volume of eluate in ml. I isozyme - gradient elution with 0 - 0.75 M KCl. II isozyme - elution with β-glycerophosphate buffer 0.002 M. EDTA - 0.001 M, 2-mercaptoethanol - 0.005 M, pH 6.8

For two isozymes of muscle ph-ase b from the lamprey and from the carp, the apparent Michaelis constants K_m were calculated for two substrates--gl-1-P and glycogen; the K_m for the AMP allosteric activator of ph-ase b and the Ki for gl-6-P were also calculated; K_m and Ki were determined graphically following the Lineweaver-Burk method (see Dixon and Webb, 1964).

Isozymes of muscle ph-ase b from the lamprey and from the carp differed significantly in K_m values for both gl-1-P and glycogen (Table I); K_m values for AMP were equal for both

TABLE 1

Km and Ki of muscle Phosphorylase

	Km		Ki		REFERENCES
	Glycogen, mg%	Gl-1-P mM	AMP mM	Gl-6-P mM	
RABBIT	18	6 – 9	0,03		Fisher et al.,1966
COCK	25	6,9	0,04		Jokay et al., 1972
FROG (in spring)	109 ± 13,9(3)	5,9±0,15 (2)	0,142±0,001 (2)		Our own data
CARP ph-I	79 ± 5(5)	18,3±3(4)	0,08±0,002 (3)	0,22±0,04 (2)	Our own data
CARP ph-II	136 ± 3(5)	113±22 (5)	0,15±0,002 (3)	0,15±0,03 (2)	
SHARK	120	24			Cohen et al., 1971
LAMPREY ph-I	1210±80 (3)	47±3 (2)	0,47±0,10 (3)	0,31±0,09 (3)	Our own data
LAMPREY ph-II	310±50 (4)	170±30 (3)	0,47±0,06 (2)	0,36±0,11 (2)	

In parentheses - number of experiments

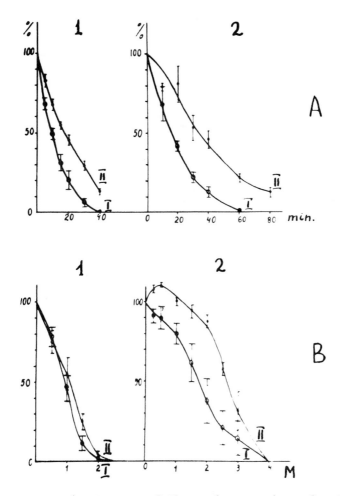

Fig. 4. Denaturation curves of Ph-ase b. A - heat denaturation
B - Urea denaturation. 1 - lamprey, 2 - Carp, -o-o-o Ph-ase
I, -.-.-. Ph-ase II; ordinate - % of initial enzyme activity
($\bar{x} \pm S \bar{x}$) abscissa - incubation time (min) at 50°C [A] urea
concentrations M [B]

isozymes in the lamprey but differed significantly in the
carp. The sensitivity to AMP of carp muscle ph-ase b isozymes
was much higher than for the lamprey. The inhibiting action of
gl-6-P on the AMP activation of ph-ase b was strongly pronounced in both isozymes of muscle ph-ase b in the lamprey and
in the carp. This fact is at variance with the known weak
inhibiting effect of gl-6-P on the AMP activation of muscle

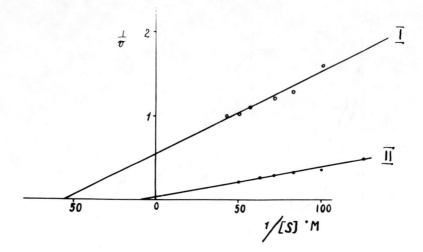

Fig. 5. Relationship of I/V and I/[S] for Ph-ase isozymes
-o-o-o Ph-ase I, -.-.-. Ph-ase II. ordinate - I/V, V - micromoles P/ mg protein/ min abscissa - I/[S], S - gl-I-P,M
the region of direct proportionality 2 - 40 x 10^{-3} M t° - 30°C,
pH 6.3, 1 x 10^{-3}M AMP, 0.5 - I% of glycogen (fin.c.)

ph-ase b from higher vertebrates (Schliselfeld et al., 1970;
Bot et al., 1971; Vereb and Bot, 1972).

Muscle ph-ase b of the frog *R.temporaria* proved to have
interesting characteristics. In our earlier work (Serebrenikova and Filisofova, 1969) we detected two zones of ph-ase
activity on zymograms of frog muscle extracts. Now, having
isolated and purified the muscle ph-ase b from frogs in the
spring (April), we obtained only one isozyme corresponding
to ph-ase I. The K_m of frog muscle ph-ase b showed intermediate characteristics falling between those of fish and
higher vertebrates. Thus K_m for gl-1-P was as low as in
warm-blooded animals, but the K_m for glycogen and for AMP was
very near to values of ph-ase b isozymes from fish. (Table 1).

EVOLUTION OF MUSCLE PHOSPHORYLASE b

Molecular evolution of muscle ph-ase b isozymes has
occurred in vertebrates. Two isozymes of muscle ph-ase b in
the lamprey show high K_m values for both substrates of the
enzyme reaction and also for the allosteric activator, AMP.
That means that they exhibit low affinity for substrates as
well as for the activator. Their further evolution in the
phylum seems to be divergent. Both isozymes still occur in

fish muscles and with their kinetic and molecular characteristics essentially unchanged. On the main road of vertebrate evolution which leads to terrestrial animals, we find only one isozyme of muscle ph-ase b, isozyme - I, which has differentiated. Its affinity for substrates and for the allosteric activator has increased. Amphibians (frogs), occupying an intermediate position between aquatic and terrestrial vertebrates, probably have two isozymes of ph-ase b in their muscle tissue, but the isozyme composition of ph-ase b may undergo seasonal changes. That the seasonal reorganization of energy metabolism occurs in frog muscle is well known (Savina, 1965).

Based on the fact that K_m values of muscle ph-ase b or substrates and for the activator AMP decrease during the course of vertebrate evolution, one may assume that in muscles of warm-blooded animals glycogenolysis proceeds via the AMP activated ph-ase b (Morgan and Parmeggiani, 1969). This is most likely to occur under anaerobic conditions when the AMP concentration in muscle rises, and ATP and gl-6-P decreases.

We do not know the conditions under which ph-ase b functions in the muscles of lower vertebrates; the K_m values for AMP seem too high. But we know that among invertebrates (in molluscs) muscle ph-ase is completely insensitive to AMP (Verzhbinskaya, 1972).

The classification as well as the functional role of the multiple forms of ph-ase is not clear. Are they isozymes in which the subunits are coded by separate loci (Fisher and Krebs, 1966) or do they represent proteins of different degrees of polymerization or, perhaps, different degrees of phosphorylation? These are the questions to be answered. We have designated the two forms of ph-ase b as isozymes because of their different electrophoretic mobilities and different kinetic characteristics.

ISOZYMES OF ASPARTATE AMINOTRANSFERASE

In the muscles of all species investigated we detected two isozymes of AAT--mitochondrial (m-AAT) and cytoplasmic (c-AAT). On zymograms, m-AAT is located in the cathodal zone whereas c-AAT remains near the origin (Fig. 2, 6,). Some species variation in the mobility of both isozyme's may be noted. The position of m-AAT, far from the origin in the cathodal zone of the zymogram, probably shows that in muscles of lower vertebrates m-AAT is positively charged like m-AAT of the heart and skeletal muscles of mammals (Kömendy et al., 1965).

The m-AAT of muscle from lower vertebrates was not adsorbed on DEAE; c-AAT was adsorbed and may be eluted with 0.1 - 0.3 M NaCl. Some species variation was revealed in the affinity

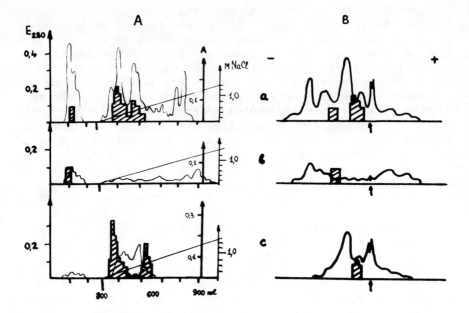

Fig. 6. Fractionation of AAT-isozymes. A - chromatography.
B - electrophoresis, the experimental conditions as in Figs.
1 and 2; A - ordinates: E_{280}; A - micromoles of pyruvate;
M KCl; _____ protein content in samples, shaded areas - AAT
activity in micromoles of pyruvate/ ml/ min. abscissa -
volume of eluate in ml. a - the whole muscle extract (as in
Fig. 1); b - mitochondria, c- cytoplasm (Differential centrifugation of muscle homogenate -1,000 x g, 3,000 x g, and 12,000
x g, 0.44 M mannitol, pH 7.4, mitochondria were disintegrated
with the detergent desoxycholate, then dialyzed against phosphate buffer 0.005 M, pH 7.4 (Filosofova-Lyslova, 1970)

of c-AAT to DEAE cellulose (Filosofova, Serebrenikova, 1969;
Serebrenikova, Filosofova, 1969). On the basis of chromatography data it may be suggested that c-AAT in lower vertebrates contains isozymes (Fig. 6).

Among the poikilothermal vertebrates lampreys showed
extremely high activities of muscle AAT (Fig. 7). Fish
exhibited much lower AAT activity and frog muscle somewhat
larger amounts but never reaching, however, the levels found
in lampreys (Fig. 7). The c-AAT isozyme represents the main
part of muscle AAT; it has a lower specific activity, calculated per mg protein of the fraction, than does m-AAT.

Muscle AAT isozymes of four species belonging to different
classes of vertebrates - cyclostomes, fish, and amphibians-

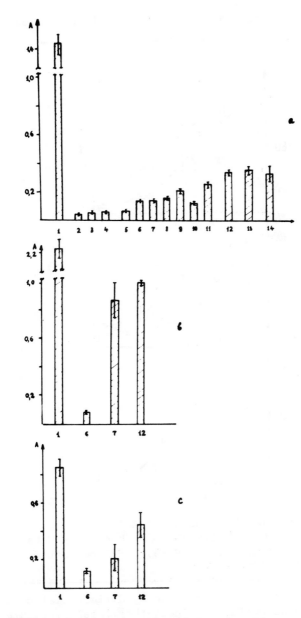

Fig. 7. AAT activities in subcellular fractions of muscle of lower vertebrates. ordinate - AAT activities ($\bar{X} \pm S\,\bar{X}$); (Tonhazy, et al., 1950, and Karmen, 1955 assays) - micromoles of pyruvate, or $NADH_2$/mg protein per min; a, b, c, as on Fig. 6. 1 - Lamprey, 2 - Dogfish-shark, 3 - Skate, 4 - Sting-ray, 5 - Golden shiner, 6 - Pike, 7 - Tench, 8 - Pike-perch, 9 - Carp,

Fig. 7. legend continued. 10 - Scorpionfish, 11 - Sea-bass, 12 - Frog.

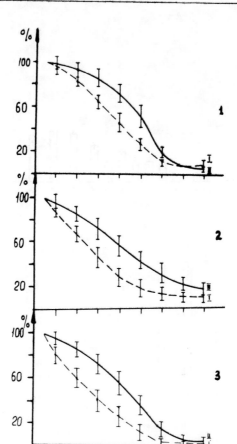

Fig. 8. Urea denaturation curves of AAT isozymes. ordinate - % of initial enzyme activity ($\bar{X} \pm S\bar{X}$ abscissa - urea concentrations, M 1 - Lamprey, 2 - Pike, 3 - Frog, dotted line - c-AAT, continuous line - m-AAT.

were studied after being separated by differential centrifugation of muscle homogenates. Their thermostability, urea denaturation, inhibition by oxaloacetate, pH optima, apparent K_m for aspartate and oxoglutarate were all determined.

The m-AAT of the muscles of lower vertebrates was less stable to heat denaturation than c-AAT (Table 2). The same was reported for heart and skeletal muscle AAT isozymes in

TABLE 2

THERMOSTABILITY OF AAT ISOZYMES

: $t_{1/2}$ °C c-AAT	: $t_{1/2}$ °C m-AAT	: p
LAMPREY	53 $\mp$ 3 (4)	39 $\mp$ 5 (4)	p = 0,05
PIKE	64 $\mp$ 5 (3)	54 $\mp$ 4 (2)	p > 0,05
TENCH	74 $\mp$ 6 (3)	59 $\mp$ 7 (2)	p > 0,05
FROG	65 $\mp$ 4 (4)	47 $\mp$ 3 (4)	p = 0,01

$t_{1/2}$ - temperature at which 50% of initial activity is lost in 15 minutes incubation.

In parentheses - number of experiments.

mammals and birds (Bertland and Kaplan, 1970). Both isozymes of muscle AAT in poikilothermal vertebrates are less thermostable compared to muscle AAT of homeothermal animals.

Urea denaturation curves were obtained (Yamada et al., 1962) at pH 7.4 after a 5 minute treatment with 1 M - 8 M urea. Enzyme inactivation is a function of urea concentration (Fig. 8). The shape of the denaturation curve is different for the two AAT isozymes; c-AAT gives a smoothly sloping inactivation curve, while m-AAT has a distinctly S-shaped curve of inactivation. In our experimental conditions m-AAT of lower vertebrates was more stable to urea denaturation than c-AAT. For heat denaturation the relation was quite the opposite. This result corresponds to that reported by Wada et al. (1968) who found pronounced pH-dependent changes in m-AAT stability to urea treatment. Moreover, at low urea concentrations true denaturation is accompanied by dissociation of dimers into monomers; since monomers are more active (Polianovsky and Ivanov, 1964) they change the slope of the inactivation curve of m-AAT, which dissociates more readily than c-AAT. Both AAT isozymes from muscle of lower vertebrates are very unstable in urea. At pH 7.4 a 5 minute treatment with 8 M urea completely inactivates both isozymes. By contrast, the muscle AAT of mammals retains 85% of the initial activity after 15 minutes of treatment with 8 M urea.

Muscle AAT isozymes of lower vertebrates differ significantly in sensitivity to oxaloacetete. The m-AAT is strongly in-

TABLE 3

INHIBITION OF AAT ISOZYMES BY OXALOACETATE

	c - AAT			m - AAT		
	Spectro-photo-metric assay, Karmen, 1955	Colori-[+/] metric assay Tonhazy et al., 1950	p	Spectro-photo-metric assay Karmen, 1955	Color-[+/] metric assay Tonhazy et al., 1950	p
LAMPREY	0.850 (7)	0.895 (6)	p > 0.05	2.245 (4)	0.598 (2)	p = 0.05
PIKE	0.119 (2)	0.127 (2)	p > 0.05	0.076 (2)	0.042 (2)	p = 0.02
TENCH	0.204 (2)	0.143 (2)	p > 0.05	0.874 (3)	0.256 (3)	p = 0.05
FROG	0.446 (6)	0.348 (10)	p > 0.05	0.996 (7)	0.268 (10)	p = 0.01

+/ Oxaloacetate is accumulated in the course of reaction

in parentheses - number of experiments

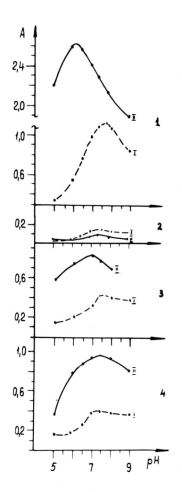

Fig. 9. pH optima curves of AAT isozymes. ordinate - enzyme activities in micromoles $NADH_2$ / mg protein / min abscissa - pH; dotted line - c-AAT; continuous line - m-AAT. 1 - lamprey, 2 - Pike, 3 - Tench, 4 - Frog.

hibited, similarly to m-AAT of heart and liver of mammals (Wada and Morino, 1964), while c-AAT is insensitive to oxalo-acetate (Table 3).

The pH optima are rather alike for both AAT isozymes of lower vertebrates, but still there is a slight difference; the pH optima for m-AAT lies at pH 6.3-7.4, that for c-AAT at pH 7.4-7.8.

Table 4 shows the K_m values for aspartate and oxoglutarate for AAT isozymes from the muscles of lower vertebrates. For comparison, data for heart AAT isozymes of higher animals are

TABLE 4

Km VALUES OF AAT ISOZYMES

| | c - AAT | | m - AAT | |
	Km for l-ASPARTATE	Km for OXOGLUTARATE	Km for L-ASPARTATE	Km for OXOGLUTARATE
		x 10^{-3} M		
PIG heart muscle	2.5	0.3	0.5	1.0
CHICK heart muscle	3.3	0.1	0.6	1.5
FROG skeletal muscles	2.37 ± 0.30 (5)	0.21 ± 0.05 (4)	0.63 ± 0.04 (4)	0.33 ± 0.07 (4)
TENCH skeletal muscles	4.57 ± 2.57 (2)	-	7.18 ± 0.54 (2)	0.66 ± 0.04 (2)
PIKE skeletal muscles	1.78 ± 0.47 (3)	0.17 ± 0.06 (2)	5.79 ± 2.72 (3)	0.13 ± 0.06 (3)
LAMPREY skeletal muscles	5.47 ± 1.57 (4)	0.16 ± 0.03 (4)	0.96 ± 0.06 (3)	0.55 ± 0.02 (4)

TABLE 4 (cont.)

	P of Km s between isozymes		Experimental conditions t°, pH, buffer	REFERENCE
	l-ASPARTATE	OXOGLUTARATE		
PIG heart muscle			t° = 25° Na-phosph. 0.1M pH = 7.5	True Km Wada et al., 1968
CHICK heart muscle			t° = 25° Tris-HCl 0.1M pH = 7.4	Apparent Km Bertland et al., 1970
FROG skeletal muscles	P = 0.01	P > 0.05	t° = 25° Phosphate 0.1M pH=7.6; pH=7.4 c-AAT m-AAT	Apparent Km Our own data
TENCH skeletal muscles	P > 0.05	P > 0.05	t° = 25° Phosphate 0.1M pH=7.5; pH=7.0 c-AAT m-AAT	Apparent Km Our own data
PIKE skeletal muscles	P > 0.05	P > 0.05	t° = 25° Phosphate 0.1M pH=7.6; pH=7.4 c-AAT m-AAT	Apparent Km Our own data
LAMPREY skeletal muscles	P = 0.05	P = 0.01	t° = 25° Phosphate 0.1M pH=7.8; pH=7.0 c-AAT m-AAT	Apparent Km Our own data

included, but we could find no kinetic data for AAT from
skeletal muscles.

In all vertebrates the K_m values for aspartate are high
for the c-AAT isozyme, much higher than for m-AAT. On the
contrary, the K_m values for oxoglutarate are higher for m-AAT
than for c-AAT. This is a common correlation for the AAT isozymes of vertebrates. The c-AAT isozyme shows a similar
correlation between the K_m for two substrates in all vertebrates; K_m for aspartate is nearly 10-fold larger than that
for oxoglutarate (Velick and Vavra, 1962; Wada et al., 1968).

In muscle m-AAT the K_m values for two substrates are distributed in a most complicated manner, the relationship between
two substrates being inverted in the course of vertebrate evolution (Table 4). Muscle AAT isozymes in teleost fish differ
significantly from those other species of lower vertebrates.

In the course of vertebrate evolution the difference
between the AAT isozymes increases. The c-AAT reveals no
significant changes in the phylum of vertebrates; species
variations are small. Beginning at an early stage of vertebrate evolution, the lamprey c-AAT has already differentiated
to resemble the c-AAT isozyme in the muscle tissue of mammals.
More pronounced changes in evolution might be seen in m-AAT
which exhibits markedly different characteristics in different
higher animals.

Profound distinctions in many molecular characteristics of
the AAT isozymes may be detected throughout the course of
vertebrate evolution. These distinctions lead to the conclusion that the AAT isozymes may represent different enzyme
proteins which arose at an early stage of animal evolution and
developed independently in different cell compartments.

REFERENCES

Bertland, L. H., and N. O. Kaplan 1970. Studies on the confor-
mations of the multiple forms of chicken heart aspartate
amino-transferase. *Biochemistry* 9: 2653-2665.

Bot, G., E. Polyk, and G. Vereb 1971. Chemical modification of
various phosphorylases in the presence of allosteric
effectors. *Acta Biochim. Biophys. Acad. Sci. Hung.* 6: 199-
206.

Cohen, Ph., Th. Duewer, and E. H. Fisher 1971. Phosphorylase
from dogfish skeletal muscle. Purification and a comparison of its physical properties to those of rabbit muscle
phosphorylase. *Biochem.* 10: 2683-2694.

Cori, G. T., B. Illingworth, and P. J. Koller, 1955. *Methods
in Enzymology.* 1: 200-205, Pergamon Press, New York,
London.

Davis, C. H., L. H. Schliselfeld, D. P. Wolf, C. A. Leavitt, and E. G. Krebs 1967. Interrelationship among glycogen phosphorylase isozymes. *J. Biol. Chem.* 242: 4824-4833.

Dixon, M., and E. Webb,1964. *Enzymes.* Longmans, Green Co., London, New York, Toronto.

Fisher, E. H., E. G. Krebs 1966. Relationship of structure to function of muscle phosphorylase. *Feder. Proc.* 25: 1511-1520.

Jokay, J., E. Karczag 1972. Purification and properties of cock muscle phosphorylase b. *Proc. Microbiol. Res. Group Hung. Acad. Sci.* 4: 109-123.

Karmen, A. 1955. A note on the spectrophotometric assay of glutamicoxaloacetic transaminase in human blood serum. *J. Clin. Invest.* 34: 131-132.

Lowry, O., N. Rosenbrough, A. Farr, and R. Randell 1951. Protein measurement with the Folin phenol reagents. *J. Biol. Chem.* 193: 265.

Metzger, B., L. Glaser, and E. Helmreich 1968. Purification and properties of frog skeletal muscle phosphorylase. *Biochemistry* 7: 2021.

Morgan, H. E., and A. Parmeggiani 1964. Regulation of glycogenolysis in muscle. III Control muscle glycogen phosphorylase activity. *J. Biol. Chem.* 239: 2440-2445.

Najojama, F. 1961. Enzymatic studies on the glycolysis in fish muscle. *Bull. Japan. Soc. Sci. Fish.* 27: 1022-1025.

Filosofova, E. M., and T. P. Serebrenikova 1969. Combined chromotographic and electrophoretic studies on sarcoplasmic proteins and their enzymic properties in cyclostomes and fishes. In: *Enzymes in Animal Evolution* (in Russian), "Nauka", Leningrad, 59-65.

Filosofova E. M. 1970. Aminotransferases from skeletal muscles of lower vertebrates. *Zh. Evol. Biokhim. Fiziol.* 6: 179-186.

Filosofova-Lyslova, E. M. 1972. Isoenzymes of aspartate and alanine aminotransferases from somatic muscles of some lower vertebrates. *Biokhimiya* 37: 498-506.

Polianovsky, O. L. and V. I. Ivanov 1964. On dissociation of aspartate-glutamate transaminase to subunits. *Biokhimiya* 29: 728-731.

Savina, M. V. 1965. Seasonal changes in respiratory enzymatic activity of mitochondria of somatic frog's muscles. *Citologiya* 7: 247-250.

Schliselfeld, L. H., C. H. Davis, and E. G. Krebs 1970. A comparison of phosphorylase isozymes in rabbit. *Biochemistry* 9: 4959-4965.

Serebrenikova, T. P., and E. M. Filosofova 1969. Fractionation of proteins and an assay of enzymic activities in muscle sarcoplasm of lower vertebrates. In: *Enzymes in Animal Evolution* (in Russian), "Nauka", Leningrad, 50-59.

Serebrenikova, T. P., and T. B. Khlyustina 1974. Isoenzymes of phosphorylase from skeletal muscles of cyclostomes and bony fishes. *Biokhimiya* 39:

Tonhazy, N. E., N. G. White, and W. W. Umbreit 1950. A rapid method of the estimation of the glutamic-aspartic transaminase in tissues and its application to radiation sickness. *Arch. Biochem.* 28: 36-42.

Velick, S. F. and J. Vavra 1962. A kinetic and equilibrium analysis of the glutamic oxaloacetate transaminase mechanism. *J. Biol. Chem.* 237: 2109-2122.

Vereb, G. and G. Bot 1972. The allosteric properties of the isozymes of pig heart phosphorylase. *Acta Biochem. Biophys. Acad. Sci. Hung.* 7: 35-42.

Verzhbinskaya, N. A. 1972. Functional organization of the enzyme system of glycolysis in muscle and nervous tissues of cephalopod molluscs and lower fishes. *Zh. Evol. Biokhim. Fiziol.* 8: 260-268.

Wada, H. and I. Morino 1964. Comparative studies on glutamic-oxalocetic transaminases from the mitochondrial and soluble fractions of mammalian tissues. In: *Vitamins and Hormones,* Academic Press, New York. 22: 411-444.

Wada, H., H. Kayamiyama, and T. Wutanabe 1968. The study of immunochemical properties and structure of aspartate aminotransferase from pig heart. In: *Chemistry and Biology of Pyridoxal Catalysis* (in Russian), "Nauka", Moscow, 75-86.

Yamada, K., S. Sawaki, A. Fukumura, and M. Hayashi 1962. Isozymes of transaminase in rat tissues *J. Vitaminol.* 8: 286-291.

Yunis, A. A., E. H. Fisher, and E. G. Krebs 1962. Purification and properties of rabbit heart phosphorylase. *J. Biol. Chem.* 237: 2809-2815.

Yunis AA., and G. K. Arimura 1966. Heterogenity of glycogen phosphorylase from rat chloroma. *Biochim. Biophys. Acta* 118: 335-343.

Yunis, A. A., and G. K. Arimura 1968. Isoenzymes of glycogen phosphorylase in human leukocytes and platelets: relation to muscle phosphorylase. *Biochim. Biophys. Res. Commun.* 33: 119-128.

SPECTROSCOPIC PROBES OF ENZYME-COENZYME-SUBSTRATE COMPLEXES OF ASPARTATE TRANSAMINASES

M. MARTINEZ-CARRION, S. CHENG,
M. J. STANKEWICZ AND A. RELIMPIO
Department of Chemistry, Biochemistry and Biophysics Program,
University of Notre Dame, Notre Dame, Indiana 46556

ABSTRACT. The relationship of structure and function
among isozymes, in solution, in a given system is of
utmost importance. In glutamate aspartate transaminase the two genetically distinct isozymes, mitochondrial and cytoplasmic, can be probed for variations
in the microenvironment of each protein for substrates,
inhibitor or coenzyme. As natural probes we utilize
the absorbance properties, circular dichroism or the
^{31}p nmr signal of the active center chromophore, pyridoxal phosphate. Covalent enzyme-substrate complexes
with fluorine atoms in specific regions of the substrate can also be prepared. In the latter case the
^{19}F nmr chemical shift sensitivity to its microenvironment acts as the reporter of the properties of
each isozyme in the region of binding this ^{19}F label.
The above techniques have been utilized under equilibria
or dynamic conditions, stopped-flow, to provide evidence
for the subtle differences in what by standard chemical
protein techniques appear to be identical active site
regions. These differences can in certain cases be
used to rationalize some basic functional variances
among the isozymes such as their substrate affinities
and pH optima.

Experimental approaches that provide information on the
correlation of physiological significance to protein structure
are sorely needed for those well documented isozymes of distinct genetic origin. If in principle we accept the pro-
position that catalytic properties and location within the
metabolic structure of the cell are dependent on the physico-
chemical properties of each isozyme, we must agree on the
need for a greater knowledge of the subtleties of the
molecular properties of isozymes.

Thus, it is of utmost importance, if we are to acquire
a clear view of enzyme catalysis in general and the significance of enzymes with different structure but similar
catalytic function, that we understand events resulting from
the interaction in solution of ligands, substrates, inhibitors, or regulatory substances with each isozyme.

Both the cytoplasmic (anionic) and mitochondrial
(cationic) isozymes of glutamate aspartate transaminase,
(GAT) are large proteins of similar molecular weight, each
containing two subunits (Martinez-Carrion and Tiemeier
1966). From a structural point of view they differ in
certain parameters,as discussed below, and in the amino
acid sequence of a peptide from the active site region
(Morino and Watanabe 1969).

The two isozymes catalyze, with similar efficiency,
in vitro, the transfer of an amino group between dicarboxylic
amino and keto acids.

Conventional steady state kinetic approaches can shed
only limited light on overall processes. We have, therefore,
made use of various spectroscopic techniques: a) to study
the interaction of ligands with the active site and b) to
acquire detailed knwledge of the microenvironment provided
by each isozyme to the ligand.

In the specific case of GAT we can take full advantage
of several unique aspects of this system; namely that the
active site chromophore acts as a natural reporter probe of
the effects of ligand interaction, that each isozyme
contains 1 mole of phosphate at the active site region
and that the catalytic mechanism involves a covalent
intermediate.

Active site. The two isozymes contain PLP at the active
site, linked to the ε-amino group of a lysyl residue. Near
the prosthetic group there is also a histidyl residue
(Table I) that is readily destroyed by photoxidation with
methylene blue. The resulting catalytically inactive holo-
isozymes can still bind the substrates. The two other
residues listed in the table are not as critical for
catalytic activity. Thus, modification of the cysteinyl
residues can produce catalytically active S and M isozymes.
The tyrosyl residues of both enzymes can be titrated taking
advantage of the pH dependence of their absorbance at 295
nm. Figure 1 shows the difference in total tyrosine content
between the isozymes and the differences in available
tyrosines between the apo and holoenzyme form of each
enzyme. These data and others from chemical modification
studies, (Riordan and Christen 1970) and circular dichroism
measurements in the ultraviolet region (Martinez-Carrion
et al., 1970) provide the basis for an assignment of
tyrosyl residues to the region of the isozymes responsible
for pyridoxal phosphate (PLP) binding.

TABLE I
Isozymes of GAT-Active Site Properties

Amino acids in active site	Cytoplasmic	Mitochondrial	Function
Primary residues	Lys	Lys	Schiff base formation, probable carboxyl anchor
	His	His	α-proton removal
Secondary residues	Cys	Cys	not essential (steric) hindrance)
	Tyr	?	probable in PLP binding

The prosthetic group. The first impression is that the
basic differences in the overall structural properties
of the isozymes are compensated by an apparent similarity
at the active site. This region can be probed using spectroscopic techniques with varying degrees of success. Absorption spectra, circular dichroism and nuclear magnetic reso -
nance have been used in our laboratory as tools for the
survey of the active site topology, each providing an
increasing degree of resolution.

The binding of the PLP chromophore provides the basis
for both the absorbance and extrinsic dichroicity in the
visible and near ultraviolet regions of the spectrum. Both
isozymes show positive dichroic bands that mimic their
absorption spectra (Martinez-Carrion et al., 1970) and with
similar dissymmetry $\Delta\varepsilon/\varepsilon$, dichroicity/absorbance, value
(Table II). In the presence of substrate the isozymes form
distinctive enzyme-substrate (ES) complexes, due to the
formation of Schiff's base derivatives with the chromophoric coenzyme. The species absorbing at 430 nm yield
optically inactive transitions but those with absorbance at
360 nm, or the ketimine, with absorbance at 335 nm, both
possess dichroicity bands. These bands, which are due to
the ES complexes, vary in $\Delta\varepsilon/\varepsilon$ values for the two isozymes
and this difference in the electronic structure in the

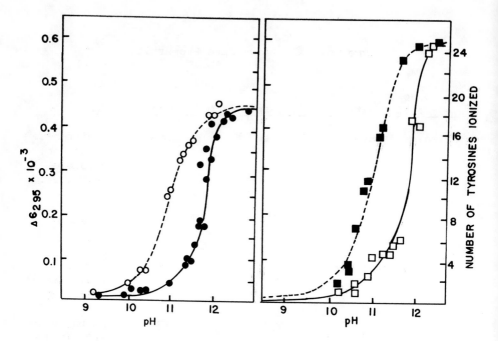

Fig. 1. Spectrophotometric titration of the tyrosyl phenolic hydroxyl groups of the isozymes of GAT. Solvent 0.1 M KCl mitochondrial isozyme(M) KCl. Points indicate equilibrium values. <u>Right</u> M-GAT, <u>left</u> S-GAT, <u>open circles</u>; holoenzyme, <u>full circles</u>; apoenzyme.

electronic structure in the domain of PDP binding in the in the absorbance wave-length for the supernatant (cytoplasmic) isozyme(S) and mitochondrial isozyme(M) isozymes. Thus, they demonstrate minor differences in the molecular envelop in the immediate vicinity of the prosthetic group in each isozyme.

Do these differences translate into variations of the dynamic properties of each isozyme? The answer must be affirmative (Table III). The changes in absorbance of the PLP due to interaction with ligands can be utilized to measure directly the affinity for ligands (Michuda and Martinez-Carrion 1969), the main difference being in their affinity for aspartate and in the range of their pH optima.

Dicarboxylic acids can also form abortive complexes with unique absorbance characteristics. These complexes have a well defined pH dependent absorptivity at 430 nm (Fig.2), which we have used in the past to explain the catalytic activity, pH optima dependence on the degree of

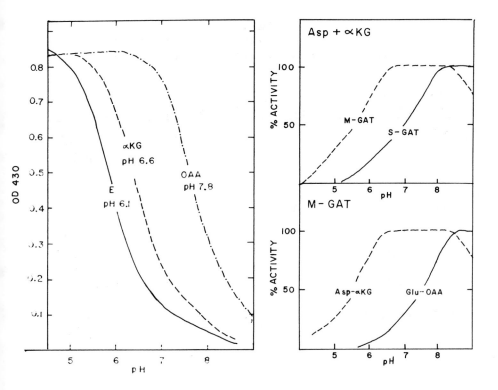

Fig. 2. Effect of dicarboxylic acids on GAT. Left, M-GAT-
pH dependence of the absorbance of the active site prosthetic group in the free enzyme, E, and in the dicarboxylic
acid binary complex with equal amount of α-Ketoglutarate,
αKG, or oxalacetate, OAA. Numbers indicate pH value for the
pKa of the binary abortive complex. Right top-pH-activity
dependence for S and M GAT with aspartate and α-ketoglutarate as substrate pairs, Asp + αKG. Bottom- observed variations in the pH dependence of M-GAT with hight affinity
oxalacetate or low affinity α-ketoglutarate.

---

inhibition by oxalacetate or ketoglutarate in each isozyme
(Michuda and Martinez-Carrion 1969).

Substrate analogues also form complexes with distinct
absorbance and these interactions can be followed under
equilibrium conditions or by stopped-flow kinetics. The
values of the dissociation constants for the amino acid substrates are dependent on the buffer concentration and vary
with the type of buffer (Cheng et al., 1971). Similar

TABLE II

CD Data for Enzyme-Substrate Complex with
Glutamate Aspartate Transaminase Isozymes[a]

Enzyme form or complex	λ_{max} (nm)	$\Delta\varepsilon/\varepsilon \times 10^{-4}$
S-PLP	365	19.2 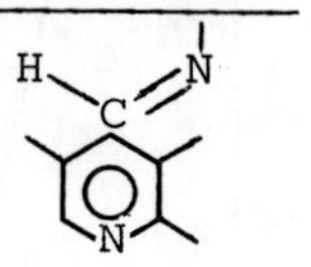
M-PLP	355	17.0
S-PMP	333	12.5 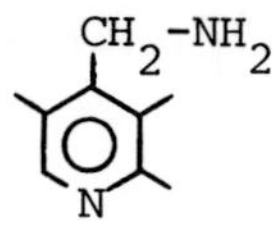
M-PMP	334	10.5
S-Aspartate	430	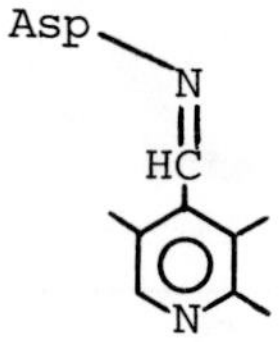 ----
	336	5.87
M-Aspartate	430	----
	335	11.2
S-α-Methylaspartate	430	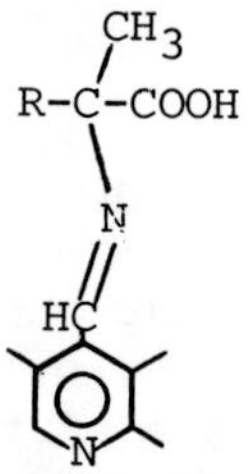 ---
	365	13.9
M-α-Methylaspartate	430	---
	368	8.4
S-Hydroxyaspartate	492	-5.78 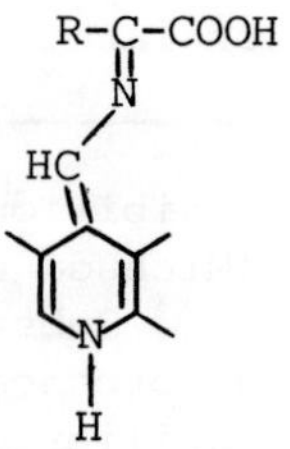
	336	7.15
M-Hydroxyaspartate	494	-4.84
	335	10.5

[a] pH 8.9

studies with α-methyl-aspartate under stopped-flow conditions show (Fig.3) that the rate constant for the formation of the enzyme-amino acid complex can depend on the buffer concentration, expressed as the concentration of tris chloride (Fig.3). The dependence of the dissociation constant on the chloride concentration is indicated in the figure on the right. The stopped-flow experiments also show (Fig. 4) that a pH dependence of amino acid binding is only observed in the presence of ions and that these affect both the velocity and the dissociation constant of interaction of the amino acid with the active center chromophore.

Nuclear magnetic resonance.
   a)   Ligand probes.
   The chromophore is not significantly perturbed by anions but the interaction of anions with the isozymes can be monitored by nuclear magnetic resonance using direct and indirect approaches.
   The direct approach takes advantage in the chemical shift (δ) of the anion trifluoroacetate in the free and enzyme-bound forms (Fig. 5). Variations in the ratio of anion/enzyme can be used to determine the affinity, stoichiometry and pH dependence for this anion. On the basis of these data and results of chemical modification,it appears that one anion binds to the active site of each isozyme (Cheng and Martinez-Carrion 1972). In the cytoplasmic enzyme the histidyl residue is the most likely ligand at the binding site.
   Nmr can also be utilized to study the properties of free and bound inhibitors. Changes in the band-broadening of the nmr signal can be correlated with the affinity of the active site for succinic acid. The obvious dependence of the values of the dissociation constants on ion concentration can also be shown, and the affinity values are found to be pH dependent in both S and M isozymes (Martinez-Carrion et al., 1973). Some of these results are summarized in Table III.
   To illustrate this information a schematic representation of the active site region is shown in Fig. 6. At the top we see the anion bound to the active site histidyl in the holoenzyme; the displacement of the anion and internal covalent complex between the chromophore and substrate is shown in the middle; and the formation of the inhibitory abortive complex is shown in the bottom part of the figure. This diagram is also intended to illustrate the need for chemical studies in solution for evaluation of ligand-binding parameters of proteins. It is quite obvious that a substrate

TABLE III
Isozymes of GAT-Dynamic Properties

Property	Cytoplasmic	Mitochondrial
Substrates (Kd)[a]	mM	mM
Aspartate	4	0.4
Glutamate	13	14
α-Ketoglutarate	0.4	0.7
Oxalacetate	0.02	0.02
pH optimum	8-9 (pH)	6-9 (pH)
Abortive complexes (inhibitory)		
α-Ketoglutarate[a]	50	620
Oxalacetate[a]	100	46
Succinate[b]	3	2
Substrate analogues[b]		
α-Methylaspartate	0.7	0.35
erythro-hydroxy-aspartate	0.02	0.006
Anion Binding[c] (inhibitory)		
Chloride	25 (10)	6 (3)
Trifluoroacetate	29 (15)	∿10
pK for binding of Cl⁻ at the active site[d]	6.2	-----

[a] In 0.1 M Sodium pyrophosphate buffer, pH 8.0.

[b] In the absence of buffer anions, pH 8.2.

[c] Figures are for Kd values with the pyridoxal form of the isozymes, numbers in parentheses refer to Kd values of the pyridoxamine form.

[d] Cheng and Martinez-Carrion (1972).

analogue, succinate, with chemical structure resembling the substrate aspartate can competitively inhibit by binding at the acitve site. However, any inference with respect to enzymatic mechanism to be derived on the basis of structure correlation of this enzyme-inhibitor complex to the mode of binding of the substrate is likely to be misleading.

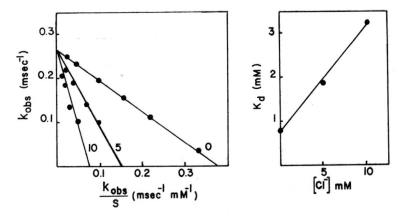

Fig. 3. Stopped-flow measurement of the reaction of α-methylaspartate with M-GAT. <u>Left</u>, plot of the observed rate constant k_{obs} with respect to k_{obs}/S at various Cl^- concentrations, numbers in slopes. <u>Right</u>, secondary plot of the apparent dissociation constant for α-methylaspartate against the Cl^- concentration.

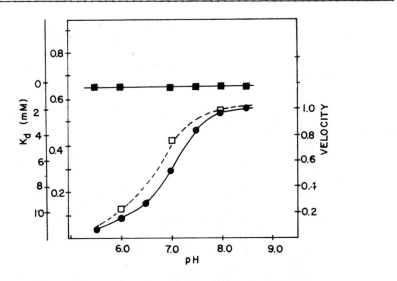

Fig. 4. pH dependence of the dissociation constant of S-GAT-α-methylaspartate complex. 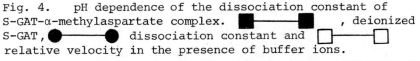 , deionized S-GAT, ●——● dissociation constant and ☐——☐ relative velocity in the presence of buffer ions.

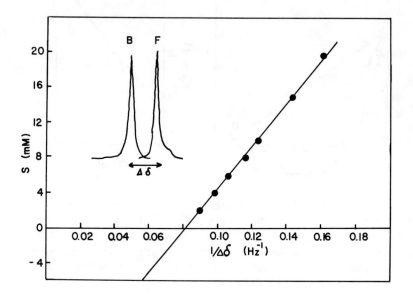

Fig. 5. Chemical shift changes, Δδ, for the trifluoroacetate, S, resonance in association with aspartate transaminase. Insert, B, anion in the presence of GAT; F, free anion.

---

b) Coenzyme probe.

The physical properties of some of the groups involved in the binding regions of the chromophore have also been investigated with the aid of nmr. The ^{31}P (Fig. 7) nmr signal of the phosphate group in PLP or PMP is of great help. The dependence of chemical shift on pH is evident in the free coenzyme, but absent when the PLP or PMP are stoichiometrically bound to the active site. The phosphorus binding site is available in the apoenzyme and can be occupied by inorganic phosphate, which is displaced by the stoichiometric addition of PLP or PMP. Because of this we now express the protein phosphate binding site as a positive pocket where the fully ionized phosphate group fits (Fig. 6).

c) Enzyme-substrate covalent probes.

The covalent nature of the enzyme-substrate complex is currently being exploited in our laboratory to probe the microenvironment provided by each isozyme to the bound substrate. In this case the approach consists of monitoring the spectroscopic properties of specific atoms in the sub-

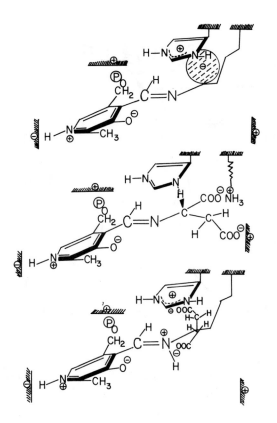

Fig. 6. Schematic representation of the binding of anion, substrate and dicarboxylic acid inhibitor to the active center of GAT.

strate in a 1:1 ES complex. To this end we have prepared a complex of trifluoromethyl menthionine, a substrate for both isozymes, and have covalently attached it to the active site by $NaBH_4$ reduction of the Schiff's base formed between the PLP and the fluorinated amino acid (Fig. 8). Detection of substrate atoms in this complex can be achieved by ^{19}F nmr without interference of other nmr signals arising from the protein, coenzyme or substrate. The appearance of a single signal due to bound amino acid (ES) at a different position from that of an internal control of free substrate (S) is indicative of both the dependence of the chemical

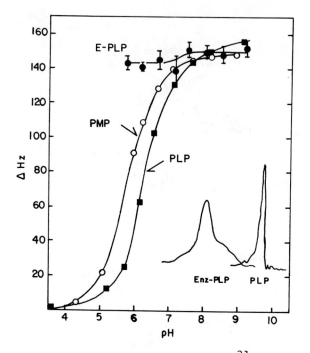

Fig. 7. Chemical shift changes of the ^{31}P nmr signal in free (PLP or PMP) or enzyme bound (E-PLP) phosphate from the coenzyme, O, ■ free coenzyme; ● enzyme bound coenzyme.

---

shift on the tertiary structure microenvironment provided by the protein and of the equivalence of the active site within each dimeric isozyme. Non-equivalence of the sites in the ES complex would lead to a doublet signal for the bound amino acid.

The pH-dependence of the chemical shift quite clearly reflects the sensitivity of the fluorine probe to electric field effects and other changes in the protein environment. The differences in the two isozymes with respect to the environment provided to this part of the substrate is significant. The pK for a group or groups in the protein affecting the distal CF_3 moiety most likely reflect the pKa value of the ε-amino group of the highly reactive lysyl residue. Heats of ionization of over 10,000 calories/mole also agree with this tentative assignment. The mitochondrial isozyme on the other hand shows a marked difference in the

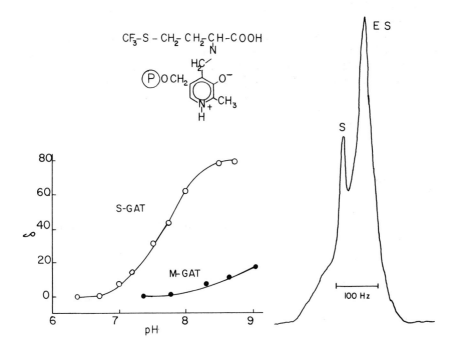

Fig. 8. ^{19}F chemical shift of trifluoromethylpyridoxyl phosphate (I); S, signal due to free compound I when added in stoichiometric excess and used as internal reference in the nmr sample. ES, nmr signal of the stoichiometric addition of compound I to apo GAT. <u>Bottom left</u>, pH dependence of the chemical shift of the ES resonance.

pH dependence of the CF$_3$ group. The titrations with the mitochondrial isozyme (M isozyme) could not be carried to completion due to instability at high pH's, yet the differences in the pKa's are indicative of the difference in the microenvironment provided by each isozyme for the CF$_3$ group of the substrate methionine in an enzyme-substrate complex.

In conclusion, it appears that the S and M isozymes of

GAT are distinct proteins, differing in their gross
structural aspects, yet they are organized so as to present
similar three-dimensional properties around the PLP coenzyme.
This active site region can be examined by a variety of
spectroscopic techniques which take advantage of the properties of the natural chromophore or the topography of
the active center domain can be surveyed by designing
covalently anchored probes. The two isozymes also possess
subtle dissimilarities in the active site regions that impart
the differences in functional behavior, such as differences
in their susceptibility to inhibitors, substrate affinities
and pH optima. In our view, this situation is not unique
to the GAT system and must also occur in other less well
understood isozyme systems.

REFERENCES

Cheng, S., and Martinez-Carrion, M. 1972. The active
    center of aspartate transaminase. A fluorine-19
    nuclear magnetic resonance study of the anion binding
    site. *J. Biol. Chem.* 247: 6597-6602.
Cheng, S., Michuda-Kozak, C., and Martinez-Carrion, M. 1971.
    Effects of anions on the substrate affinities of the
    pyridoxal and pyridoxamine forms of mitochondrial and
    supernatant transaminases. *J. Biol. Chem.* 246: 3623-
    3630.
Martinez-Carrion, M., S. Cheng, and A.M. Relimpio 1973.
    Nuclear magnetic resonance of aspartate transaminase.
    A ^{19}F and ^{1}H investigation of the binding of dicarboxylic
    acids to various forms of each isoenzyme. *J. Biol.
    Chem.* 248: 2153-2160.
Martinez-Carrion, M., and D. Tiemeier 1967. Mitochondrial
    glutamate aspartate transaminase I. Structural com-
    parison with the supernatant isozyme. *Biochemistry,*
    6: 1715-1722.
Martinez-Carrion, M., D.C. Tiemeier, and D.L. Peterson 1970.
    Conformational properties of the isoenzymes of aspartate
    transaminase and enzyme substrate complexes. *Bio-
    chemistry,* 9: 2574-2582.
Michuda, C.M., and M. Martinez-Carrion 1969. Distinctions
    in the equilibrium kinetic constants of the mitochondrial
    and supernatant isozymes of aspartate transaminase.
    J. Biol. Chem. 244: 5920-5927.
Michuda, C.M., and Martinez-Carrion, M. 1970. The isozymes
    of glutamate-aspartate transaminase. Mechanism of
    inhibition by dicarboxylic acids. *J. Biol. Chem.*

245: 262-269.

Morino, Y., and T. Watanabe 1969. Primary structure of
    pyridoxal phosphate binding site in the mitochondrial
    and extramitochondrial aspartate aminotransferase
    from pig heart muscle. *Chymotryptic peptides.*
    Biochemistry, 8: 3412-3417.

Riordan, J.F., and P. Christen 1970. Syncatalytic mod-
    ification of a functional tyrosine residue in asparate
    aminotransferase. *Biochemistry*, 9: 3025-3034.

ISOZYMES OF ADENOSINE DEAMINASE

ROCHELLE HIRSCHHORN
Department of Medicine
New York University School of Medicine
New York, N. Y.

ABSTRACT. Adenosine deaminase (ADA) exists as a polymorphic enzyme in human red blood cells. Additional
forms of this enzyme activity that differ in several
respects from the red blood cell (RBC) form are present
in tissues, including lymphocytes, other than RBC's.
The relationship of RBC ADA to these different tissue
ADA activities has been elucidated. The small molecular
weight, polymorphic RBC isozymes could be converted to
each of four different large molecular weight tissue-
specific isozymes of ADA by incubation with the appropriate tissues. The biochemical and electrophoretic
characteristics of these tissue-specific isozymes, generated by interaction with tissue factors in vitro, were
like those of the various isozymes found naturally in
the different tissues. The catalytic activity of all of
the several forms of the ADA enzyme would thus appear
to reside in a single molecule coded at the same genetic
locus. The various specific tissue isozymes that differ
in electrophoretic mobility and molecular weight are
generated by interaction of this catalytic unit with
factors present in the different tissues.

A proportion of patients with the inherited disorder,
Severe Combined Immunodeficiency of the autosomal recessive type, lack the RBC adenosine deaminase in their
red blood cells. These patients also lack normal tissue
isozymes, although a small amount (< 2% of normal) of
ADA activity could be detected.

Despite the general deficiency of ADA in the tissues
of such patients, we have found residual enzyme activity
(7.84 ± 2.3 vs 22.5 ± 3.9 nmoles inosine/mg protein/min)
in fibroblasts derived from patients with this disorder.
This residual enzyme moves more rapidly toward the anode
in starch gel electrophoresis than the normal tissue
enzyme of fibroblasts but has a molecular weight resembling tissue enzymes (approximately 260,000), considerably larger than that of the red cell enzyme (approximately 35,000). Its K_m is similar to that of the normal
enzyme (5.0 x 10^{-5} M vs 7.1 x 10^{-5} M), but it shows
greater heat stability. It may therefore be regarded
as a mutant of the catalytic unit of the RBC ADA enzyme,

detectable as the tissue isozyme.

Alterations in the proportion of these various isozymes occurs with stimulation of lymphocytes in vitro. Following stimulation with phytohemagglutinin the tissue isozyme is markedly diminished and the RBC isozymes increased. Following stimulation with pokeweed mitogen, a new isozyme of 110,000 molecular weight appears.

INTRODUCTION

Adenosine deaminase is an enzyme of the purine salvage pathway that deminates adenosine to yield inosine. Inosine is then either further degraded to uric acid or salvaged at several points for the biosynthesis of purines (Henderson and Paterson, 1973). This enzyme has been the subject of extensive investigations and several different forms have been described in man (Henderson and Paterson, 1973; Spencer, Hopkins, and Harris, 1968; Edwards, Hopkinson, and Harris, 1971).

In human red blood cells (RBC) (Spencer, Hopkins, and Harris, 1968) adenosine deaminase (ADA) is an enzyme of approximately 35,000 mol. wt. which migrates relatively rapidly toward the anode on starch gel electrophoresis at pH 6.5. This protein is apparently modified after translation to give rise to two secondary enzyme bands, resulting in a typical triple banded pattern. Each of these three bands contains a reactive sulfydryl (SH) group (Hopkinson and Harris, 1969). This protein is also polymorphic in that red blood cells of 90% of individuals demonstrate the phenotype 1-1, while rare individuals demonstrate the phenotype 2-2; heterozygous individuals demonstrate a four banded pattern, presumably representing overlapping of the two-codominant gene products (Spencer, Hopkins, and Harris, 1969; Hopkinson, Cook and Harris, 1969).

Tissues other than red blood cells contain, in addition to the polymorphic RBC isozyme, several other tissue isozymes (3). These have previously been designated a-e according to their electrophoretic mobility. The most common of the tissue isozymes (d) occurs in most tissues, including fibroblasts and in circulating T lymphocytes. This isozyme has a molecular weitht of approximately 260,000. The slower isozyme (e) has been described as typical of kidney and intestine and has a molecular weight of over 480,000. Two minor isozymes (b and c) are variably expressed in most tissues and have been reported to have the same molecular weight as the d isozyme. These "tissue specific" isozymes have been reported not to show the polymorphism observed in the RBC form of

the enzyme in that they do not appear to vary in electrophoretic mobility in tissues of individuals of differing ADA RBC
phenotypes nor do they contain reactive SH groups. Because
of these differences, the several tissue specific isozymes
were considered to be determined at one or more separate
genetic loci which were not allelic with the locus for RBC
ADA.

The absence of adenosine deaminase from red blood cells
has recently been described in a proportion of children suffering from the autosomal recessive form of a disease called
Severe Combined Immunodeficiency (Giblett, Anderson, Cohen,
Pollara, and Meuwissen, 1972; Porter, 1974). This disorder
is a uniformly fatal (if untreated) inherited disease of
infancy which is characterized by a failure of differentiation
of lymphoid cells responsible for the normal development of
both cellular and humoral immunity. The enzyme deficiency
does not appear to be secondary to the disease state in that
it is also transmitted in an autosomal recessive manner with
diminished levels in the obligate heterozygous carrier parents
of the affected children (Scott, Chen, and Giblett, 1974;
Hirschhorn, Rosen, and Parkman, unpublished observations).
We have had the opportunity to examine stored tissues from
several of these patients in order to explore further the
genetic and molecular interrelationships of the various ADA
isozymes.

The normal isozyme pattern of adenosine deaminase in various tissues was examined in homogenates of organs obtained
at autopsy, and stored at -20°C for varying periods of time.
These tissue extracts were electrophoresed in starch gel
(phosphate buffer, pH 6.5, 2.5 V/cm, 18 hrs) and the areas
of adenosine deaminase activity visualized by the coupled
method described by Spencer et al. (1968) in which a blue
formazan dye precipitates at the site of enzyme activity.
The patterns varied from tissue to tissue generally in the
manner described previously by Edwards et al. (1971) and
thus the same nomenclature will be used (Fig. 1). Extracts
of normal liver contained two major isozymes (Fig. 1, channel
1), one with a mobility between that of the first and second
band of the RBC ADA isozymes (isozyme a) and a second band
migrating less rapidly with a mobility similar to that of a
band seen in lymphocytes and fibroblasts (Fig. 1) (isozyme d).
Normal kidney extracts (Fig. 1, channel 3) revealed one
major area of activity with a mobility overlapping and slower than that of isozyme d (isozymes d and e). Normal heart
homogenates (channel 5) contained the slower mobility isozyme
d and the intermediate mobility large molecular weight tissue
isozyme b-c as well as the RBC isozymes. Normal splenic

585

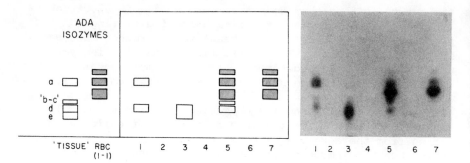

Fig. 1: Tissue homogenates were subjected to starch gel
electrophoresis and stained for adenosine deaminase activity
(Spencer, Hopkins, and Harris, 1968). The various isozymes
found in different tissues are represented diagrammatically
using a previously suggested nomenclature (Henderson and
Paterson, 1973). The isozymes that are specific for RBC's
but present in other tissues are indicated in the diagram as
shaded areas. Channel 1 shows the pattern obtained with
normal liver; channel 2, immunodeficient patient's liver;
channel 3, normal kidney; channel 4, patient's kidney;
channel 5, normal heart; channel 6, patient's heart;
channel 7 contains a normal lymphoid line of RBC ADA phenotype 1-1 for reference.

---

homogenates (Fig. 2, channel 1) showed predominantly an RBC
ADA pattern in this individual (who is a 2-1 phenotype) as
did muscle. Normal intestinal extracts revealed areas of
activity in the a, b, c, d, and e zones. A lymphoid line of
ADA RBC 2-1 phenotype is seen in channel 7 for reference.
Thus, all of the isozymes (a-e and RBC) which have been described previously could be visualized in one or another of
the tissues examined. When homogenates of kidney, liver,
intestines, spleen, heart, and muscle of the affected child
were subjected to electrophoresis in parallel with the homogenates from normal tissues, the red blood cell isozymes, as
seen in Figures 1 and 2 were absent. This confirms the earlier conclusion that the gene controlling transcription of
the protein RBC ADA is expressed in non-hematopoietic tissue
(Edwards, Hopkinson, and Harris, 1971). However, contrary to
prior expectations none of the normal tissue specific isozymes of adenosine deaminase could be detected in any tissue
examined from this patient, although they were easily visualized in normal tissue.
It is unlikely that the absence of these isozymes was
due to complete loss of normal enzyme activity during storage
for four months at -70°C prior to examination since ADA iso-

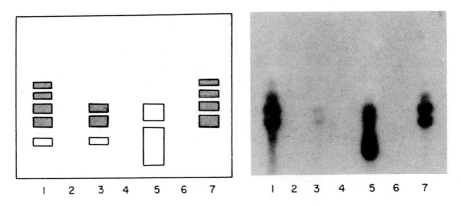

Fig. 2: Tissues were treated in the same manner as those in Fig. 1 but were obtained from an individual of ADA 2-1 phenotype. Channel 1 shows the pattern obtained with normal spleen; channel 2, patient's spleen; channel 3, normal muscle; channel 4, patient's muscle; channel 5, normal intestines; channel 6, patient's intestines; channel 7, normal lymphoid line of RBC ADA 2-1 phenotype.

---

zymes were still detectable in tissues of normal individuals which had been stored either as homogenates or as whole tissues for up to six months at -20°C. Matching samples from the normal and immunodeficient patient contained equal amounts of protein (as determined by the method of Lowry) as well as equivalent amounts of two other enzyme activities, nucleotide phosphorylase and adenylate kinase, as visualized following starch gel electrophoresis. It thus appeared likely that the catalytic unit of the different tissue and RBC ADA enzyme activities were controlled by a single structural genetic locus. Earlier investigations had also suggested a relationship between the red cell and tissue isozymes. Akedo et al. (1972) had demonstrated that normal human tissues contained a heat labile factor which converted a small molecular weight ADA activity to a large molecular weight enzyme activity (Akedo, Nishahara, Shinkai, Komatsu, and Ishikawa, 1972). In order to clarify the genetic and molecular interrelationship of the several ADA isozymes we have examined the interconvertibility of the polymorphic low molecular weight RBC ADA enzyme of differing gentoypes to the various tissue specific high molecular weight isozymes, utilizing both normal tissues and extracts of tissues from children with SCID associated with absent RBC ADA (ADA mutant tissues).

Normal as well as mutant tissues were incubated with RBC ADA for varying periods of time and the results evaluated by starch gel electrophoresis, determination of enzyme activity, and molecular weight determinations.

Extracts of normal kidney, which contained only the very slowly migrating isozymes (d-e) typical of that tissue, were incubated for 5 hrs at 25°C with RBC hemolysates or lymphoid line extracts, which contained only the RBC isozymes of ADA. The mixtures were then subjected to starch gel electrophoresis and the areas of enzyme activity visualized (Fig. 3). Incubation of either the hemolysate or the kidney extract alone did not cause disappearance of activity or alterations in electrophoretic mobility of either the RBC isozymes or the tissue specific isozyme (Fig. 3, channels 1 and 3). Following incubation of the RBC ADA with normal kidney extract, activity with the electrophoretic mobility of RBC ADA was no longer detectable, and only enzyme activity with electrophoretic mobility of the kidney isozyme could be seen (Fig. 3, channel 4). Incubation of RBC ADA with oxidized glutathione, which presumably binds to the SH group of the enzyme protein adding a negative charge and thus increasing the anodal electrophoretic mobility (Hopkinson and Harris, 1969) (Fig. 3, channel 7), did not prevent this apparent conversion of RBC ADA isozyme to kidney isozyme (Fig. 3, channel 6) nor did it alter the mobility of the isozyme generated. Addition of 2-mercaptoethanol also did not prevent this conversion (Fig. 3, channel 5).

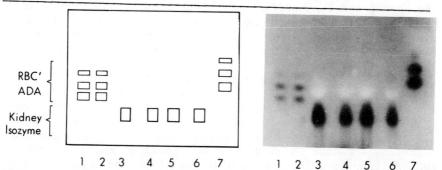

Fig. 3: Conversion of RBC ADA by incubation with normal kidney. Lysates of RBC were incubated alone and with homogenates of normal kidney, the mixture subjected to electrophoresis and the areas of ADA enzyme activity detected as described in the text. Channel 1 contains RBC ADA. Channel 2 contains RBC ADA + 2-mercaptoethanol. Channel 3 contains normal kidney extract. Channel 4 contains RBC ADA + normal kidney extract. Channel 5 contains RBC ADA + 2-mercaptoethanol + normal kidney extract. Channel 6 contains RBC ADA + oxidized glutathione and kidney extract. Channel 7 contains RBC ADA + oxidized glutathione.

These findings suggested that the RBC ADA isozyme had been converted to an isozyme with an electrophoretic mobility indistinguishable from that of tissue isozyme. Clearer evidence for this was obtained when extracts of tissues from children with SCID and lacking RBC ADA, which did not contain any detectable ADA activity on starch gel, were incubated with extracts of normal lymphoid line cells or RBC hemolysates. When extracts of liver from an ADA deficient child, which contained no ADA enzyme activity detectable on gel electrophoresis, were incubated with lymphoid line or RBC extracts (Fig. 4, channels 3 and 7), a portion of the RBC enzyme activity was converted to an enzyme activity with the characteristic electrophoretic mobility of one of the isozymes seen in liver (isozyme d) (Fig. 4, channels 1 and 2). This isozyme has a molecular weight of approximately 260,000 and is also present in T lymphocytes (Hirschhorn and Levytska, 1974) and fibroblasts. Incubation of the same RBC ADA with extracts of the patient's kidney (which by themselves contained no enzyme activity detectable on gel electrophoresis) caused complete conversion of all of the RBC isozyme to an ADA isozyme with the slower mobility characteristic for the normal kidney isozyme (e) (Fig. 4, channels 4, 5, and 6). This isozyme has also been reported to be present in intestine. Conversion by deficient liver and kidney did not require 5 hours incubation but occurred within 5 minutes of mixing. On one occasion it was possible to obtain conversion of RBC isozyme to the labile isozyme a, which is seen in liver and lung, with a mobility intermediate between the 1st and 2nd bands of RBC ADA 1-1 and the 2nd and 3rd bands of heterozygous ADA 2-1 and is of unknown molecular weight. An extract of ADA-deficient liver was incubated with a lymphoid line of 2-1 genotype (Fig. 5, channel 2). Following this incubation no enzyme was seen as the RBC isozyme but instead the major portion of the enzyme activity showed the same electrophoretic mobility as the slower moving d isozyme, the more anodal isozyme a, as well as the isozyme b-c (Fig. 5, channel 1). Similarly, fibroblasts derived from another child with SCID contained a factor which converted the RBC isozyme of normal lymphoid lines to the mobility of a tissue isozyme seen in normal fibroblasts. Thus tissues appear to contain a factor which can convert RBC isozyme to each of the previously described electrophoretically different species of ADA found in the various tissues. The isozyme generated appeared to depend upon the tissue utilized.

There was no loss of enzyme activity during these conversions. When RBC extracts were incubated with liver extracts (Table 1), 10 µl of the RBC hemolysate alone contained 0.575

Conversion of 'RBC' ADA Isozymes to 'Tissue Specific' ADA Isozymes

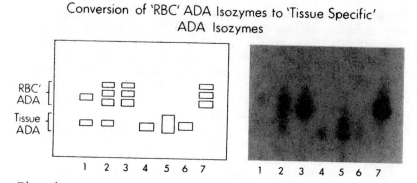

Fig. 4: Conversion of RBC ADA to tissue specific isozymes by incubation with tissues deficient in ADA. Lymphoid line extracts, which contain only RBC ADA, were incubated with extracts of kidney and liver from ADA-deficient patients, which by themselves contained no visible ADA activity. Tissue homogenates and mixtures were subjected to starch gel electrophoresis and stained for ADA activity. Normal liver extract is visualized in channel 1 and normal kidney extract in channels 4 and 6 to serve as markers. RBC ADA is seen in channels 3 and 7. In channel 2 RBC ADA was incubated with liver extract from an ADA-deficient patient, which did not itself contain ADA. ADA activity with the mobility of the slow moving isozyme of normal liver (channel 1) is now visible in addition to the RBC ADA isozymes. In channel 5, RBC ADA was incubated with kidney from an ADA-deficient child, which itself did not contain ADA and virtually all of the enzyme activity is now present as a slow moving kidney-like isozyme.

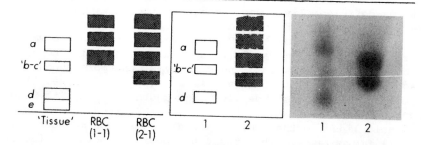

Fig. 5: Conversion of RBC ADA by liver to several different tissue isozymes. RBC ADA was incubated with extracts of liver from an ADA-deficient child, subjected to starch gel electrophoresis and stained for ADA activity. The various isozymes found in different tissues are represented diagrammatically using a previously suggested nomenclature (Edwards, Hopkinson, and Harris, 1971). The isozymes which are specific for RBC's

but present in other tissues are indicated in the diagram
as shaded areas. The RBC isozyme of both a 1-1 and a 2-1
phenotype are diagrammed to aid in identifying the isozymes
generated from the lymphoid line following incubation with
liver extracts. Channel 2 contains an extract of normal lymphocytoid line of a 2-1 ADA genotype. Channel 1 contains an
extract of the same lymphocytoid line that had been incubated
with liver extract devoid of ADA activity. The normal isozyme
a of adenosine deaminase, typical of liver and lung tissue,
was now visible in channel 1 as well as the tissue isozyme
b-c and isozyme d seen in various other tissues.

TABLE I
DETERMINATION OF ADA ENZYME ACTIVITY*

	Enzyme Units (Δ A_{293}/hr)
Red Blood Cells (RBC)	0.575
ADA Deficient Liver	$\underline{-0.068}$
RBC + ADA Deficient Liver (calculated)	0.507
RBC + ADA Deficient Liver (observed)	0.510

* RBC hemolysates and extracts of liver deficient in
ADA were incubated alone and together for 1 hr. and the adenosine deaminase enzyme activity determined. The mixture was
also subjected to gel electrophoresis and staining for enzyme
activity and was found to contain no detectable RBC ADA isozymes but only ADA activity of the mobility of a tissue isozyme.

units of ADA while 10 μl of RBC hemolysate mixed with 10 μl
of liver, which converted all the enzyme to the tissue form,
contained 0.510 units of ADA. The liver alone contained too
little activity to be detected under these conditions of assay. This confirmed that the RBC ADA activity was not destroyed but appeared to be converted without loss of activity.
Although neither iodoacetate nor glutathione interfered
with the apparent conversion of RBC ADA by normal kidney, the
possible role of the free SH groups of the RBC isozyme in the
conversion to tissue isozyme was further investigated using
conditions designed to give a more definitive answer. Liver

tissue from an ADA deficient child was utilized so that the mobility of the isozyme generated could be evaluated.

Iodoacetamide, which binds firmly to SH groups, was utilized rather than glutathione, which binds reversibly. The ability of iodoacetamide to bind the free SH groups of the RBC isozymes was tested by demonstrating that oxidized glutathione could no longer increase the anodal electrophoretic mobility of the RBC's preincubated with iodoacetamide. Conversion of RBC ADA to isozyme d by incubation with ADA-deficient liver proceeded normally under these circumstances and the mobility of the tissue isozyme generated was again not altered by the presence of glutathione.

Tissue isozymes also differ from the RBC isozyme in molecular weight. The isozyme generated in vitro by incubation of RBC's with ADA deficient liver and characterized by electrophoretic mobility was further characterized by estimation of molecular weight. The mixture, containing an excess of RBC enzyme so that not all of the RBC enzyme was converted, as determined by gel electrophoresis, was applied to a calibrated Sephadex G-150 column and the activity of the different fractions determined. The residual RBC activity eluted at the same volume as did RBC ADA alone. In addition, the major portion of the ADA activity was not eluted as a higher molecular weight species with the same elution fraction as the tissue isozyme of normal circulating human lymphocytes (Fig. 6).

In order to compare further the properties of the tissue isozymes generated in vitro as compared to those found naturally, RBC's of individuals who exhibit different polymorphic forms of RBC ADA were used to generate tissue isozymes. Normal tissue isozymes have been described as not expressing the polymorphism detectable in the RBC ADA isozymes. RBC lysates from individuals homozygous for two different genetically determined, polymorphic forms of RBC ADA were obtained (1-1 and 2-2). RBC's demonstrating the more rapid anodal mobility of phenotype 1-1 and those of the electrophoretically slower phenotype 2-2 were each incubated with extracts of liver deficient in ADA. The tissue isozymes generated by such treatment of the electrophoretically different RBC isozymes did not differ from each other in electrophoretic mobility to the extent that the RBC isozymes did. However, the isozyme generated from a 2-2 RBC could reproducibly be shown to migrate slightly less rapidly anodally than that generated from RBC's of a 1-1. Thus when the distance between the slowest bands of RBC ADA 2-2 and RBC ADA 1-1 measured 10 mm, whereas the difference in mobility between the isozymes generated from the RBC ADA 2-2 and RBC ADA 1-1 measured 3.5 mm. This very small difference could easily have been undetected during electro-

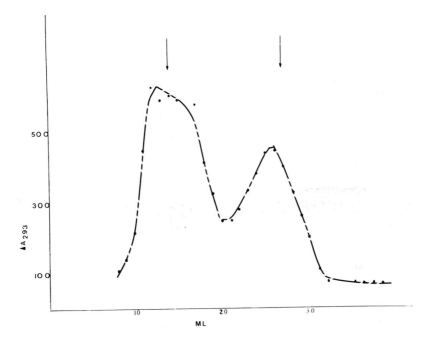

Fig. 6: Conversion of RBC ADA of low molecular weight to a high molecular weight ADA by incubation with ADA-deficient liver. RBC ADA was incubated with liver from an ADA-deficient child, which by itself contained no detectable ADA under these conditions. Gel electrophoresis of an aliquot of the mixture demonstrated partial conversion of a portion of the activity to a tissue isozyme (d). The mixture was applied to a calibrated Sephadex G-150 column (25 x 1.5 cm). The arrow at an elution volume of 28 ml indicated where RBC ADA alone is found. The arrow at elution volume of 14 ml indicates where the tissue isozyme of lymphocytes is found. The conversion mixture contains enzyme activity with the low molecular weight of the original RBC ADA but in addition a high molecular weight activity is now seen.

phoresis of the tissues from rare 2-2 individuals.

These findings suggest that the catalytic unit of all of the different isozymes of ADA is determined by the same genetic locus and that this is the locus affected in SCID with RBC ADA deficiency. In keeping with this hypothesis, we have found evidence that the mutation responsible for absence of the RBC ADA expresses itself in the tissue isozymes. When extracts of fibroblasts from children who do not have the

RBC ADA isozyme in their red cells are concentrated, 2-30% of the ADA activity found in normal fibroblasts can be detected (7.84 ± 2.3 vs 22.5 ± 3.9 nmoles adenosine/mg protein/min). The level of enzyme activity was determined in mutant and normal fibroblasts over a period of six months during which time the fibroblasts were transferred at approximately bi-weekly intervals. No consistent changes in the levels of enzyme activity of the normal or mutant fibroblasts were observed with increasing time in culture.

These normal and mutant fibroblasts were grown in media containing 20% fetal calf serum. Since fetal calf serum exhibits appreciable levels of ADA activity (22 nmoles/ml/min), it was possible that the residual enzyme activity detected in the mutant fibroblasts might represent enzyme endocytosed from the fetal calf serum. Fibroblasts were therefore grown for six days in media containing 20% fetal calf serum and in media containing 20% fetal calf serum which had been previously heated at 56°C for 3½ hours. The heated fetal calf serum contained 4.8 nmoles/ml/min ADA activity as compared with 20.26 nmoles/ml/min in the unheated serum. The level of enzyme activity in both normal and mutant fibroblasts was essentially unchanged in the face of this five-fold difference in the ADA levels in the media. The total ADA activity present in the media in a single tissue culture flask was 9.6 nmole/ml/min, which is less than the enzyme activity of the mutant fibroblasts harvested from that flask (13.5 nmoles/flask/min). Fibroblasts were also grown in media containing 10% of either fetal calf serum or horse serum. The human serum contained 2.77 nmoles/ml/min ADA activity while the activity of the horse serum was less than 1 nmole/ml/min. The levels of endogenous ADA activity of the mutant and normal fibroblasts did not vary in any consistent manner with the levels of exogenous ADA activity in the media. Because of these considerations, it appeared that the residual enzyme activity represented endogenous enzyme synthesized by the mutant fibroblasts.

Lysates of mutant and normal fibroblasts were subjected to electrophoresis on starch gel and the areas of adenosine deaminase activity were visualized. When this activity was visualized after electrophoresis in starch gel, it was found to have a mobility faster than the normal fibroblast isozyme, but different in mobility from the ADA activity of fetal calf serum. The mobility of the mutant ADA activity of the mutant fibroblast was the same whether they had been cultured in human, calf or horse serum, each of which has a different electrophoretic mobility.

Normal fibroblasts, in our hands commonly exhibit two

tissue isozymes, either alone or together in varying proportions. Preliminary data suggest that the slower isozyme appears in very confluent, slow growing cultures. The appearance of this slow tissue isozyme was correlated with the
presence of a shoulder of ADA activity on Sephadex G-200
chromatography of approximately 480,000 molecular weight.
Mutant fibroblasts grown to extreme confluence before harvest-
ing also exhibited two bands of enzyme activity. Each of
these bands was faster in electrophoretic mobility than the
corresponding band of the normal fibroblasts. Most commonly,
mutant fibroblasts demonstrated a single area of activity
with a mobility slightly faster than that of the faster isozyme seen in normal fibroblasts. This mutant activity was
indistinguishable in molecular weight on Sephadex G-200 chromatography (260,000 vs 266,000) or by Km (5.0 x 10^{-5} M) from the
tissue isozyme isolated from normal fibroblasts. However,
this mutant enzyme showed markedly increased heat stability
compared with the normal tissue isozyme. Over 70% of the
activity was present after heating at 56°C for two hours, as
compared with 20% or less for the normal isozyme.

This enzyme activity would appear to be present in tissues
other than fibroblasts since concentrated extracts of liver
and spleen from a SCID patient lacking RBC ADA, analyzed shortly after autopsy, revealed a residual isozyme in liver with
2% of normal activity and mobility similar to the isozyme a
as well as activity in the area of the isozyme d.

These data suggest that a mutation has occurred in the
catalytic unit of the RBC isozyme that results in a protein
with a diminished halflife in the RBC. This catalytic unit
may be stabilized in tissues and fibroblasts after interacting with the conversion factor and appear as a tissue isozyme
with increased anodal mobility.

The catalytic activity of the various isozymes of ADA are
controlled by a single genetic locus, but the distribution of
this catalytic unit between various tissue and RBC isozymes
is under complex control. The relative amounts of the various
tissue RBC isozymes have been observed to change in cultured
fibroblasts (Edwards, Hopkinson, and Harris, 1971).

Thus fibroblasts derived from the same individual harvested at different times may variously express the RBC isozyme
alone, tissue isozymes alone or differing combinations of
each. The factors controlling this variation have not been
elucidated. We have observed a similar phenomenon following
stimulation of lymphocytes. Human peripheral blood lymphocytes can be induced to proliferate in tissue culture by
addition of several different mitogens derived from lectins
including phytohemagglutinin (PHA), Conconavalin A (ConA)

and Pokeweed Mitogen (PWM). Addition of these lectins, which preferentially stimulate different classes of lymphocytes, causes a series of complex metabolic and morphologic altera-tions that culminate in cell division (Hirschhorn and Hirsch-horn, 1974).

When normal resting human peripheral blood lymphocytes are purified in Hypaque/Ficoll gradients and passage through nylon wool to preferentially separate T lymphocytes and extracts subjected to starch gel electrophoresis, two major areas of activity are evident as previously described (Hirschhorn and Levytska, 1974; Wust, 1971). These cell preparations con-tain the polymorphic triple banded RBC isozyme as well as a slower mobility isozyme d. This pattern remains constant over at least five days of tissue culture. When PHA is added to these lymphocyte preparations at the beginning of the culture period, the tissue specific isozyme is markedly diminished after 72 hours. After 112 hours, the tissue iso-zyme is virtually undetectable.

This qualitative shift from a tissue to an RBC form of the enzyme induced by PHA can be quantitated. Extracts of stimulated and non-stimulated lymphocytes were applied to a Sephadex G-150 column and the fractions assayed for ADA activity. It was found that the tissue specific isozyme of lymphocytes was of a large molecular weight represented by the first peak, while the RBC isozyme was retarded with a molecular weight of 35,000 daltons. The distribution of enzyme activity was similar in unstimulated lymphocytes both at 0 time and after 72 hours, whereas following stimulation with PHA for 72 hours, the proportion of isozyme present as the tissue form was markedly diminished. Thirty-five percent of the total activity was present as tissue isozyme at the start of culture and 24% after 72 hours without stimula-tion (Table 2). In contrast, in PHA-stimulated lymphocytes only 8% of the activity was present as the tissue specific isozyme. However, total ADA enzyme activity was unchanged.

We examined the possibility that PHA-stimulated lymphocytes might contain an inhibitor of the tissue isozyme. Fractions containing tissue isozyme isolated by gel chromatography from control lymphocytes, which by themselves had 316 units of ADA activity were mixed with extracts of PHA-stimulated lymphocytes, which by themselves contained 727 units of ADA activity and indeed 983 units of ADA activity were found, demonstrating that extracts of PHA-stimulated lymphocytes do not contain an inhibitor of tissue ADA.

Further alterations in isozymes can be observed with proliferation of another type of cell. When peripheral blood lymphocytes are separated on Hypaque/Ficoll gradients but

TABLE II

ADENOSINE DEAMINASE ACTIVITY OF CULTURED LYMPHOCYTES*

	Total Activity (Δ A$_{293}$/hr/100 X 10^6 cells)	% "Tissue ADA"	% "RBC ADA"
Zero Time	5.548	34.4	65.6
72 hr	5.759	24.3	75.7
72 hr PHA-stimulated	5.364	8.5	91.5

* Human peripheral blood lymphocytes were cultured for 72 hr with and without PHA, and extracts subjected to Sephadex G-150 gel chromatography to separate the "RBC" and "tissue" isozymes of ADA. The enzyme activity present as each of these isozymes was then determined as described in the text.

---

not passed through nylon wool, another type (B) of lymphocyte is obtained in addition to the T lympocyte. This lymphocyte is preferentially stimulated by another mitogen, PWM. The unstimulated lymphocyte population again showed a high molecular weight peak at >260,000 and another peak at ∿35,000. When stimulated with PHA these mixed B and T populations of lymphocytes showed loss of the 260,000 molecular weight peak but there now appeared activity with an estimated molecular weight of 110,000. When PWM, which preferentially stimulates B lymphocytes was used, almost one-third of the total activity was present as the 110,000 isozyme. The remainder of the activity was equally distributed between the 260,000 and 35,000 molecular weight peaks. Thus alteration in the metabolic state of different cell types can give rise to different isozymic forms of the enzyme. In this case T lymphocytes shift their 260,000 molecular weight isozyme to the 35,000 isozyme while B lymphocytes develop an isozyme of 110,000 molecular weight. Differences in the ratio of conversion factor to the catalytic unit arising from changes in rates of synthesis and/or degradation could conceivably control the ratio of one isozyme to another.

In summary, these studies indicate that the catalytic subunit of several of the tissue isozymes of ADA appear to be controlled by the same genetic locus as that determining RBC ADA. The ratio of one of these to another changes in a predictable manner upon stimulation of different classes of lymphocytes. We believe that this arises from the interaction of the catalytic unit with conversion factors which are the product(s) of one or more distinct genetic loci.

ACKNOWLEDGEMENTS

Aided by a grant from the National Institutes of Health (Al-10343) and from the National Foundation.

Rochelle Hirschhorn is a recipient of an N.I.H. Research Career Development Award Al-70254.

REFERENCES

Akedo, H., H. Nishahara, K. Shinkai, K. Komatsu, and S. Ishikawa 1972. Multiple forms of human adenosine deaminase: Purification and characterization of two molecular species. *Biochim. Biophys. Acta.* 276: 257-271.

Edwards, Y. H., D. A. Hopkinson, and H. Harris 1971. Adenosine deaminase isozymes in human tissues. *Ann. Hum. Genet. London.* 35: 207-219.

Giblett, E. R., J. E. Anderson, F. Cohen, B. Pollara, and H. J. Meuwissen 1972. Adenosine deaminase deficiency in two patients with severely impaired cellular immunity. *Lancet.* 11: 1067-1069.

Henderson, J. F. and A. R. P. Paterson 1973. *Nucleotide metabolism.* Academic Press, N. Y. and London.

Hirschhorn, R., F. S. Rosen, and R. Parkman (unpublished observations).

Hirschhorn, K. and R. Hirschhorn 1974. Mechanisms of lymphocyte activation in *Mechanisms of Cell Mediated Immunity* R. J. McCluskey and S. Cohen, eds., John Wiley and Sons, Inc. p. 115.

Hirschhorn, R. and V. Levytska 1974. Alterations in isozymes of adenosine deaminase during stimulation of human peripheral blood lymphocytes. *Cell. Immunol.* 12: 387-395.

Hopkinson, D. A., P. J. L. Cook, and H. Harris 1969. Further data on the adenosine deaminase polymorphism. A report of a new phenotype. *Ann. Hum. Genetics* 32: 361.

Hopkinson, D. A. and H. Harris 1969. The investigation of reactive sulphydryls in enzymes and their variants by starch gel electrophoresis. Studies on red cell adenosine deaminase. *Ann. Hum. Genet. London.* 33: 81.

Porter, Ian, editor, (in press). *Combined Immunological Disease - A Molecular Defect?* Academic Press, New York.

Scott, C. R., S. H. Chen, and E. R. Fiblett 1974. Detection of the carrier state in combined immunodeficiency disease associated with adenosine deaminase deficiency. *J. Clin.*

Invest. 53: 1194-1196.
Spencer, N., D. Hopkins, and H. Harris 1968. Adenosine deaminase polymorphism in man. *Ann. Hum. Genet.* 32: 9-14.
Wust, H. 1971. Adenosine deaminase in lymphocytes and its electrophoretic separation. *Hum. Hered.* 21: 607-713.

ISOZYMES OF SHEEP LIVER THREONINE DEAMINASE

ROBERT S. GREENFIELD and DANIEL WELLNER
Department of Biochemistry
Cornell University Medical College
New York, N. Y. 10021

ABSTRACT. Sheep liver extracts were shown to possess 1, 2, or 3 electrophoretically separable isozymes of threonine deaminase and 5 to 7 isozymes separable by isoelectric focusing. A quantitative study of the enzyme activity in 150 individual livers was carried out. It was found that while the greatest number yielded an extract containing 0.1-0.2 units/ml the activity of some livers was 20 times higher. The distribution was asymmetric and approximated a trimodal curve. Only the livers possessing the highest activity exhibited the three electrophoretic isozymes. Both genetic and physiological factors appear to influence the enzyme activity. Studies of the purified isozymes by gel filtration chromatography suggested that they have a molecular weight of about 53,000 and that they exhibit a tendency to aggregate at high concentrations. Estimates of the subunit molecular weight by gel electrophoresis in the presence of sodium dodecyl sulfate also gave values close to 53,000, indicating that the enzyme is monomeric.

INTRODUCTION

Enzymes that catalyze the non-oxidative deamination of L-threonine (L-threonine hydro-lyases (deaminating), EC 4.2.1. 16, also called threonine dehydrases or threonine dehydratases.) to α-ketobutyrate and ammonia according to equation 1 have been found in microorganisms, plants, and mammalian tissues (Umbarger,

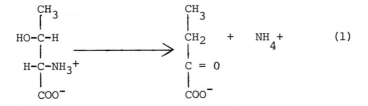

$$
\begin{array}{ccc}
\underset{|}{\overset{CH_3}{|}} & & \underset{|}{\overset{CH_3}{|}} \\
HO-C-H & & CH_2 \quad + \quad NH_4^+ \quad\quad (1) \\
| & \longrightarrow & | \\
H-C-NH_3^+ & & C = O \\
| & & | \\
COO^- & & COO^-
\end{array}
$$

1973). The bacterial enzymes have been divided into two main groups, according to their apparent functions: the "biosynthetic" threonine deaminases, which are believed to be responsible for the formation of α-ketobutyrate used for the synthesis of isoleucine, and the "biodegradative" threonine deaminases, which appear to catalyze the breakdown of threonine to satisfy the energy requirements of the cell. Enzymes belonging to the first group are subject to feedback inhibition by L-isoleucine, whereas the enzymes from the second group require AMP or ADP for maximal activity. Enzymes from both groups require pyridoxal phosphate as coenzyme.

A mammalian enzyme catalyzing reaction 1 has been obtained in crystalline form from rat liver (Nakagawa et al., 1967; Nakagawa and Kimura, 1969) and in partially purified form from sheep liver (Nishimura and Greenberg, 1961). Since the rat liver enzyme is more active toward L-serine than toward L-threonine (Nakagawa et al., 1967), it is more appropriately designated as "serine deaminase" or "serine dehydratase". However, the enzyme from sheep liver, which acts on L-threonine, L-allothreonine, and L-serine, is more active toward L-threonine than L-serine. Furthermore, this enzyme is rapidly inactivated in the presence of L-serine (Nishimura and Greenberg, 1961; Davis and Metzler, 1962; McLemore and Metzler, 1968). On the basis of these properties and in view of the fact that isoleucine is not synthesized in mammalian organisms, it would be reasonable to assume that the physiological function of this enzyme is that of a biodegradative threonine deaminase. It is noteworthy that this enzyme catalyzes the irreversible degradation of a dietary essential amino acid to a product which cannot be utilized for the resynthesis of that amino acid. It would therefore appear that some control of the activity of this enzyme is needed in order not to deplete the threonine reserve needed by the organism for protein synthesis. Mammalian threonine deaminases have not been shown to be allosteric enzymes and have been reported not to be subject to feedback control by nucleotides or isoleucine (Dixon and Davis, 1967; Nakagawa and Kimura, 1969). However, it is possible that the rate of synthesis and degradation of these enzymes are regulated. Thus, a number of studies have shown that the serine deaminase activity of rat liver is strongly influenced by hormonal and dietary influences (see, for example, Jost et al., 1968). Little is known, however, about the factors that regulate the levels of threonine deaminase of sheep liver.

In this paper we report the presence of multiple molecular forms of sheep liver threonine deaminase. It is shown that both the number of isozymes and their amount vary from one animal to another. Estimates of the molecular weight of threonine deaminase were obtained by gel filtration chromatography and polyacrylamide gel electrophoresis in the presence of sodium dodecyl sulfate. The results suggest that the enzyme has a molecular weight of about 53,000 and consists of a single polypeptide chain.

MATERIALS AND METHODS

MATERIALS

Fresh sheep livers were obtained from the abattoir, cooled in ice, and processed within 48 hours. "Ampholine" carrier ampholytes were from LKB Instruments; Sephadex G-150 was from Pharmacia; DE-52 and CM-52 cellulose were from Whatman. Other reagents were of the highest grade available.

DETERMINATION OF THREONINE DEAMINASE ACTIVITY

This was done by the 2, 4-dinitrophenylhydrazine method, essentially as described by Nishimura and Greenberg (1961). The enzyme was incubated for 15 minutes at 37^0 in a final volume of 0.3 ml containing 16.7 mM L-threonine and 0.1 M potassium phosphate buffer, pH 7.4. The reaction was stopped with 0.1 ml of 25% trichloracetic acid and, after centrifugation, 0.1 ml of the supernatant was added to 0.1 ml of 0.1% dinitrophenylhydrazine in 2 N HCl. After 5 minutes, 0.2 ml of ethanol and 0.5 ml of 2.5 N NaOH were added with shaking, and the color was read at 540 nm. One unit of activity is defined as the amount of enzyme required to form 1 μmole of α-ketobutyrate per minute under the above conditions.

POLYACRYLAMIDE GEL ELECTROPHORESIS AND ISOELECTRIC FOCUSING

These were carried out according to the general procedures described by Davis (1964) and Wellner (1971), respectively. Specific conditions are given under each experiment described below. Gel electrophoresis in the presence of sodium dodecyl sulfate was carried out by a modification of the method of Weber and Osborn (1969) described by Cooper and Meister (1972).

ACTIVITY STAIN

Gels were stained for threonine deaminase activity by a modification of the method of Feldberg and Datta (1970). This consisted of incubating the gels at 37^0 in a solution containing L-threonine (20mM), phenazine methosulfate (0.2 mg/ml) and nitroblue tetrazolium (1 mg/ml) in 0.1 M potassium phosphate buffer, pH 8.0. Although threonine deaminase does not catalyze a redox reaction, it appears that one of the reaction intermediates can serve as a reducing agent for the dye. In crude preparations, control experiments in which threonine is left out of the staining solution are needed.

RESULTS

The large variation in the threonine deaminase activity of sheep liver has been reported previously by Nishimura and Greenberg (1961) and by others, but a quantitative study of the distribution of activity has not been published previously. Such a distribution in three batches of about 50 livers each is shown in Fig. 1. It is evident that the distribution curve is not symmetrical. Although approximately 80% of the animals showed an activity of less than 0.5 units/ml in the liver extract, with the greatest number falling between 0.1 and 0.2 units/ml, a few livers yielding 2 units/ml or higher have been observed. Only about 3% of the animals yielded more than 1.5 units/ml of activity. No significant differences were seen between different batches, or between livers obtained in different seasons. Also, there were no differences between different lobes of the same liver. When liver extracts of high activity were mixed with extracts of low activity, an average activity value was obtained. This indicates the absence of threonine deaminase inhibitors in the livers of low activity.

Polyacrylamide gel electrophoresis of crude liver extracts as well as of purified enzyme preparations revealed the existence of a number of isozymes. Fig. 2 shows activity bands obtained with extracts of livers from various regions of the distribution curve. In such preparations, some formazan bands appeared in control experiments in which threonine was left out of the staining solution. In analogy to the "nothing dehydrogenases" reported by others, these bands may be designated as "nothing deaminases". These do not appear in more highly purified preparations. In Fig. 2, only those bands in the portion of the gel marked by arrows represent threonine deaminase. It was found that one or two bands of enzyme appeared in extracts of low activity (< 0.5 units/ml), two bands in extracts of medium activity (0.5-1.5 units/ml), and a third band, with higher mobility, appeared only in extracts of livers with very high activity (> 1.5 units/ml).

The purification procedure we used was a modification of that of Nishimura and Greenberg (1961). Only livers in the medium activity range (0.5-1.5 units/ml and possessing two electrophoretic isozymes) were used in the purification. Except as noted, all steps were carried out at 4^0.

1) Homogenization: Fresh sheep livers were cut into pieces and homogenized at high speed in a Waring blendor for 3 minutes in 2 volumes of 0.1 M potassium phosphate buffer, pH 7.2. The homogenate was centrifuged at 10,000 x g for 15 minutes.

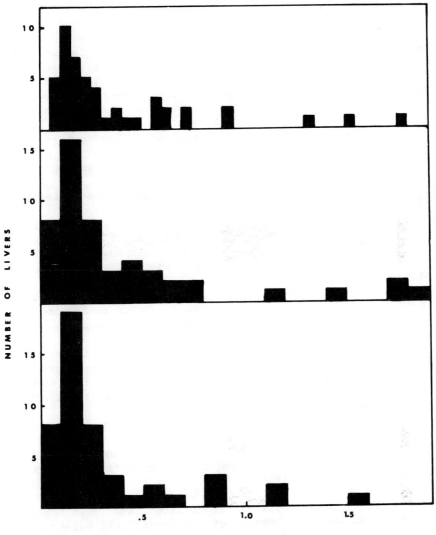

Fig. 1: Distribution of threonine deaminase activity in sheep livers. Three batches of about 50 livers each were analyzed. Activity (units/ml) was measured after the first heat step of the purification procedure. The height of each bar gives the number of livers exhibiting activity in the range represented by its width.

605

2) First heat step: The supernatant from step 1 was heated at 72° for 10 minutes, rapidly cooled, and centrifuged for 15 minutes at 10,000 x g.

3) First ammonium sulfate fractionation: The red supernatant from step 2 was brought to 40% saturation in ammonium sulfate (230 g/l of supernatant). As the salt dissolved, the pH of the solution was maintained at 7.2 by addition of concentrated ammonium hydroxide. After 1 hour, the solution was centrifuged and the precipitate discarded. The supernantant was brought to 60% saturation in ammonium sulfate (by adding 123 g of the salt per liter of supernantant) and centrifuged after 1 hour. The precipitate was redissolved in 0.1 M potassium phosphate buffer, pH 7.2 (1/10 the volume of the homogenate).

4) Second heat step: The above solution was heated to 65° for 10 minutes, cooled, and centrifuged.

5) Second ammonium sulfate fractionation: The supernantant from step 4 was brought to 60% saturation of ammonium sulfate (370 g/l of supernatant) and centrifuged after 1 hour. The precipitate was dissolved in the same buffer as in step 3.

6) Chromatography: The enzyme was chromatographed on DEAE-cellulose as described in Fig. 3. Under these conditions, the activity emerged from the column as a single peak. Fractions 22 to 38 were pooled.

A summary of a typical purification is given in Table 1. Although further purification is achieved by chromatography on carboxymethyl cellulose (see below) a greater than 600 fold increase in specific activity with a 55% yield is achieved up to this point.

Although the existence of two bands of enzyme could be seen in electrophoresis (three bands in the livers of high activity), a greater multiplicity of isozymes could be demonstrated by isoelectric focusing. Thus when the enzyme obtained in step 6 was submitted to isoelectric focusing in polyacrylamide gel, using a pH 3-10 gradient, about 7 bands of activity could be seen (Greenfield and Wellner, unpublished data). In order to obtain an accurate measure of the isoelectric points of these isozymes, an isoelectric focusing experiment was carried out in a sucrose gradient with pH 3-10 carrier ampholytes as shown in Fig. 4. Again, about 7 activity peaks were observed. The isoelectric points, measured at 4°, were 4.50, 5.07, 5.30, 5.65, 6.07, 6.37 and 6.57. In another experiment, using pH 5-8 carrier ampholytes, one of the fractions isolated by carboxymethyl cellulose chromatography (peak I, see below) was resolved by isoelectric focusing into 5 peaks of activity with isoelectric points of 5.26, 5.60, 5.75, 6.03, and 6.6. The

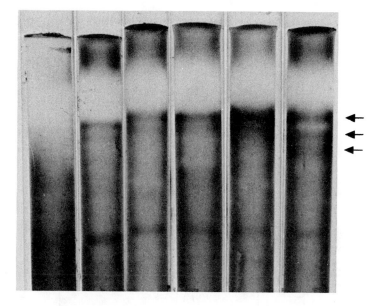

Fig. 2: Polyacrylamide gel electrophoresis of sheep liver
extracts from different regions of the distribution curve
(see Fig. 1). The gels were stained for threonine deaminase
activity. An aliquot of the supernatant from the first heat
step (0.05ml) was applied to each gel. From left to right,
the activities were 0.15, 0.6, 1.1, 1.4, 1.8, and 2.1 units/ml,
respectively. The three activity bands in the last gel are
marked by arrows. Electrophoresis conditions: 200v, 2 h,
5 cm gels, 10% acrylamide, 0.3 M Tris-HCl, pH 8.9.

results of the two experiments are consistent with each other,
and suggest that the shoulder at fraction 58 in Fig. 4 consists
of two unresolved isozymes with isoelectric points of 5.60
and 5.75.

Significant further purification of the enzyme was achieved
by chromatography on a carboxymethyl cellulose column, as
shown in Fig. 5. The elution pattern of enzymatic activity
supports the conclusion that several isozymes are present
in the preparation obtained in step 6. The enzymes in the two
main activity peaks were studied further as described below.
These are designated as peak I (fractions 11-18) and peak II
(fractions 38-50) in Fig. 5.

When the two enzyme fractions were submitted to polyacrylamide gel electrophoresis as shown in Fig. 6, the main protein
band from peak II migrated with a lower mobility than the

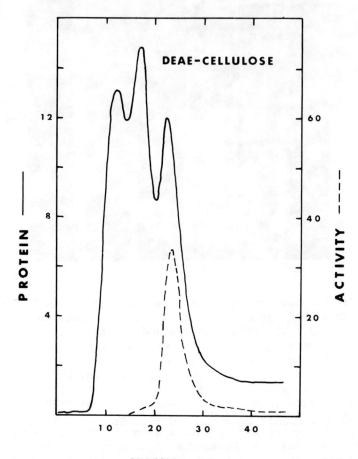

Fig. 3: Chromatography of threonine deaminase on DEAE-cellulose. The enzyme, after the second ammonium sulfate fractionation (step 5), was dialyzed overnight against 2 changes of
0.005 M potassium phosphate buffer, pH 7.2. This solution (70
ml) was applied to a column of Whatman DE-52 (30 x 5 cm) equilibrated with the same buffer, and was washed in with 20 ml of
the buffer. The enzyme was then eluted with a linear gradient
made with 500 ml of 0.05 M and 500 ml of 0.1 M potassium
phosphate buffers, pH 7.2. Fractions of 19 ml were collected
at a flow rate of 1.25 ml/min. Ordinate: protein concentration
(O.D.$_{280}$) and activity (units/ml).

TABLE I

SUMMARY OF PURIFICATION FROM 10 SHEEP LIVERS

FRACTION	VOL. (ml)	ENZ. CONC. (units/ml)	TOTAL ACT. (units)	PROT. CONC. (mg/ml)	TOTAL PROTEIN (g)	SPECIFIC ACTIVITY (units/mg)	YIELD (%)
Homogenate	11,000	.525	5775	150	1650	.0035	100
1st Heat Step 72°C	3,100	1.49	4620	91	282	.0164	80
1st $(NH_4)_2SO_4$ Fractionation 40–60% SAT.	350	11.6	4060	71	24.8	.163	70
2nd Heat Step 65°C	340	11.0	3740	58	20	.189	65
2nd $(NH_4)_2SO_4$ Fractionation 0–60% SAT.	175	19.3	3377	81	14	.238	58
DEAE Cellulose Fractions (22–38)	–	–	3150	–	1.4	2.25	55

609

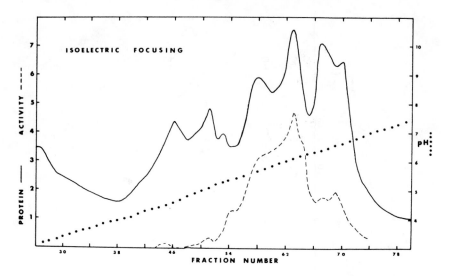

Fig. 4: Isoelectric focusing of threonine deaminase in a sucrose gradient. This was carried out according to Vesterberg and Svensson (1966) in a 440 ml column with pH 3-10 carrier ampholytes. After focusing for 36 hours (final voltage: 800 v), fractions of 3 ml were collected. These were tested for pH, protein (O.D.$_{280}$ x 10) and enzymatic activity (O.D.$_{540}$ x 10, 25 µl aliquots assayed).

one from peak I. The peak I enzyme is therefore the more negatively charged, in agreement with its early elution from carboxymethyl cellulose.

Isoelectric focusing of the same enzyme fractions showed that several isozymes were present in each (see Fig. 7). The material from peak I exhibited two strongly staining bands and a few weaker ones, while that from peak II revealed three strong bands, the third one corresponding to the first strong band of peak I. Some active protein remained at the origin, suggesting aggregation. Gels run in parallel and stained for protein with Coomassie blue revealed a similar pattern.

The two fractions were chromatographed on Sephadex G-150 in order to estimate the molecular weight of the enzyme. As shown in Fig. 8, the peak eluted with the same volume of buffer for both, although the fraction from peak I of the carboxymethyl cellulose column showed more evidence of aggregated material. Comparison of the elution volume of threonine deaminase with that of proteins of known molecular weights (Fig. 9) suggests that the enzyme has a molecular weight of about 53,000. When a crude liver homogenate was chromatographed on the same column, the threonine deaminase

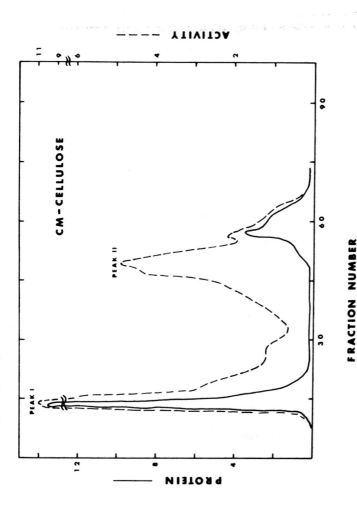
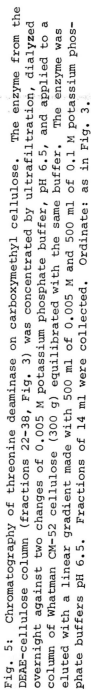

Fig. 5: Chromatography of threonine deaminase on carboxymethyl cellulose. The enzyme from the DEAE-cellulose column (fractions 22-38, Fig. 3) was concentrated by ultrafiltration, dialyzed overnight against two changes of 0.005 M potassium phosphate buffer, pH 6.5, and applied to a column of Whatman CM-52 cellulose (300 g) equilibrated with the same buffer. The enzyme was eluted with a linear gradient made with 500 ml of 0.005 M and 500 ml of 0.1 M potassium phosphate buffers pH 6.5. Fractions of 14 ml were collected. Ordinate: as in Fig. 3.

611

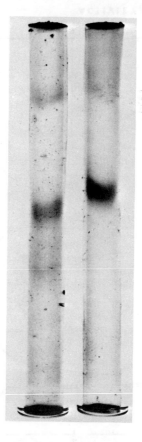

Fig. 6: Polyacrylamide gel electrophoresis of threonine deaminase isozymes. Aliquots from peak I (left) and peak II (right) from the carboxymethyl cellulose column (see Fig. 5) were run under the following conditions: 200 v, 3 hr, 8 cm gels, 6% acrylamide, 0.03 M Tris-HCl, pH 8.9. The gels were stained for protein with amido black in 7% acetic acid. Cathode at top.

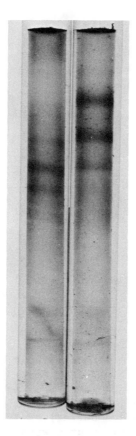

Fig. 7: Isoelectric focusing of threonine deaminase
isozymes in polyacrylamide gel. Aliquots from peak I
(left) and peak II (right) from the carboxymethyl cellulose
column (see Fig. 5) were focused under the following
conditions:16 hr at 100-250 v (max.), 10 cm gels, 7.5%
acrylamide, 1% pH 5-8 "Ampholine" carrier ampholytes.
Cathode solution (top), 0.4% ethanolamine; anode solution
0.2% H_2SO_4. After prefocusing 30 min at 100 v, the enzyme
was applied to the top of the gel in 1% "Ampholine", 5%
sucrose. The gels were stained for threonine deaminase
activity.

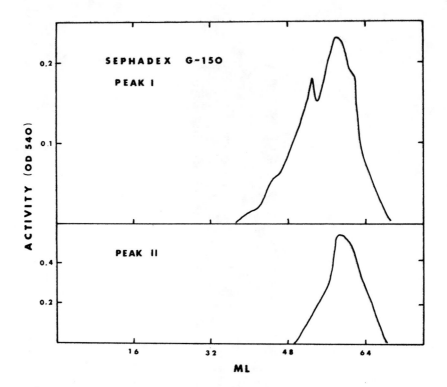

Fig. 8: Gel filtration chromatography of threonine deaminase.
Aliquots of enzyme (0.5 ml) from peaks I and II of the
carboxymethyl cellulose column (see Fig. 5) were applied to
a column (1.5 x 57 cm) of Sephadex G-150 equilibrated with
0.1 M potassium phosphate buffer, pH 7.2, 0.1 M potassium
chloride. The enzyme was eluted with the same buffer. Fractions of 1.6 ml were collected at a rate of 0.1 ml/min.
Activity was measured on 0.1 ml aliquots.

activity eluted in the same position, indicating that no
significant change in molecular weight had taken place during
purification.

When the isozymes isolated by isoelectric focusing were
studied by gel electrophoresis in the presence of sodium
dodecyl sulfate and mercaptoethanol, the main band observed
corresponded in mobility to a polypeptide chain of molecular
weight about 53,000 with a minor component of apparent
molecular weight 160,000 (Fig. 10). Since similar results
were obtained with the isozymes from peak I and peak II of
the carboxymethyl cellulose column, it appears that these

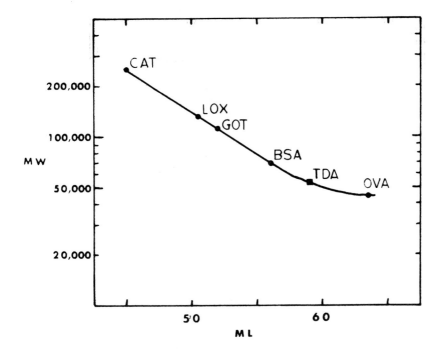

Fig. 9: Molecular weight determination of threonine deaminase by gel filtration chromatography. Standard proteins used were: catalase, L-amino acid oxidase, glutamate-oxaloacetate transaminase, bovine serum albumin, and ovalbumin (circles). The square indicates the position of threonine deaminase. See Fig. 8 for conditions.

isozymes do not differ significantly in molecular weight or in subunit structure. The results of the gel filtration experiments and of gel electrophoresis in the presence of sodium dodecyl sulfate, taken together, suggest that sheep liver threonine deaminase consists of a single polypeptide chain of molecular weight 53,000.

DISCUSSION

The above results show that sheep liver threonine deaminase consists of a number of isozymes, and that both the amount of enzyme and the number of isozymes vary from one animal to another. Although the variation in the amount of enzyme may be the result of either physiological or genetic factors, the finding of an isozyme present only in the livers of highest activity is consistent with a genetic difference.

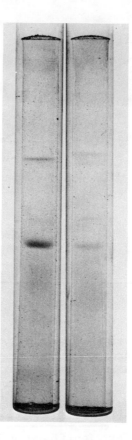

Fig. 10: Polyacrylamide gel electrophoresis of threonine deaminase isozymes in the presence of sodium dodecyl sulfate (SDS).
Peak I (left)and peak II (right). After isoelectric focusing
as described in Fig. 7, the gel positions corresponding to the
principal activity bands (2 for peak I, 3 for peak II) were cut
out, crushed, and extracted with 0.2 ml 0.025 M Tris-acetate
buffer pH 8.4 (3 hr) and 0.3 ml of the same buffer for 16 hours.
The extracts were centrifuged and the supernatants were dialyzed against the same buffer for 16 hours. The samples were
evaporated to dryness, redissolved in 0.1 ml of 1% SDS, 1%
mercaptoethanol, 0.025 M Tris-acetate pH 8.4, and dialyzed
against the same solution for 2 hr Electrophoresis was then
carried out for 3 hr at 100 v, in 8 cm gels, 6% acrylamide, 0.1%
SDS, 0.1% mercaptoethanol, 0.05 M Tris-acetate, pH 8.4. After
fixation in 7% acetic acid, 30% methanol for 16 hours, the gels

Fig. 10 continued: were stained with 0.25% Coomassie blue
in 7% acetic acid, 30% methanol and destained in 7% acetic
acid, 30% methanol. Catalase, bovine serum albumin, and ovalbumin were used as molecular weight standards.

The distribution curve, which approximates a trimodal curve,
is also compatible with a genetic explanation. Thus, a frequent gene (t) coding for an isozyme of low activity and a
rare gene (T) coding for one of high activity may account for
the distribution if it is assumed that the genotype tt results
in low activity, TT in high activity, and tT in intermediate
activity. The experimental curve may be complicated however,
if superimposed on the genetic variation, a variation due to
nutritional or other physiological factors plays an important
part. Thus we have found that the amount of individual isozymes as well as the number of isozymes is greater in livers
of high activity. In rat liver, two isozymes of serine dehydratase have been separated, and these were reported to be
under different types of metabolic control, one being regulated by glucagon and the other by corticosteroids (Inoue
et al., 1971). It is possible that the sheep liver threonine
deaminase levels may also be under hormonal control. Thus,
some of the variation in activity we observed may be due to
differences in conditions under which the animals were kept
before slaughter.

    At this time, nothing is known concerning the structural
differences responsible for the differences in electrophoretic
mobility and isoelectric point of the isozymes. Gel filtration
chromatography indicates a molecular weight of about 53,000.
However, confirmation by other methods is needed for a definitive conclusion in this regard. The enzyme has a tendency
to aggregate as indicated by the Sephadex elution profile and
by the protein remaining at the top of the gel in electrophoresis and isoelectric focusing. Attempts to concentrate the
enzyme by ultrafiltration resulted in the formation of an
insoluble precipitate at concentrations greater than about
1 mg/ml. Preliminary experiments have shown the several
isozymes to be glycoproteins (Greenfield and Wellner, unpublished data). Thus, differences in carbohydrate content may
account for some of the differences in mobility and isoelectric point.

ACKNOWLEDGEMENTS

    This investigation was supported by grant No. CA13259
from the National Cancer Institute, U. S. Public Health
Service.

REFERENCES

Cooper, A. J. L. and A. Meister 1972. Isolation and properties of highly purified glutamine transaminase, *Biochemistry* 11: 661-671.

Davis, B. J. 1964. Disc electrophoresis-II. Method and application to human serum proteins. *Ann. N. Y. Acad. Sci.* 121: 404-427.

Davis, L. and D. E. Metzler 1962. The pH dependence of the kinetic parameters of threonine dehydrase. *J. Biol. Chem.* 237: 1883-1889.

Dixon, E. and L. Davis 1967. Action of sheep liver threonine dehydratase on L-allothreonine and copper salts of L-threonine and L-serine. *Arch. Biochem. Biophys.* 119: 155-158.

Feldberg, R. S. and P. Datta 1970. Threonine deaminase: A novel activity stain on polyacrylamide gels. *Science* 170: 1414-1416.

Inoue, H., C. B. Kasper, and H. C. Pitot 1971. Studies on the induction and repression of enzymes in rat liver. *J. Biol. Chem.* 246: 2626-2632.

Jost, J.-P., E. A. Khairallah, and H. C. Pitot 1968. Studies on the induction and repression of enzymes in rat liver. *J. Biol. Chem.* 243: 3057-3066.

McLemore, W. O.,and D. E. Metzler 1968. The reversible inactivation of L-threonine dehydratase of sheep liver by L-serine. *J. Biol. Chem.* 243: 441-445.

Nakagawa, H., H. Kimura, and S. Miura 1967. Crystallization and characteristics of serine dehydratase from rat liver. *Biochem. Biophys. Res. Commun.* 28: 359-364.

Nakagawa, H. and H. Kimura 1969. The properties of crystalline serine dehydratase of rat liver. *J. Biochem.* 66: 669-683.

Nishimura, J. S. and D. M. Greenberg 1961. Purification and properties of L-threonine dehydrase of sheep liver. *J. Biol. Chem.* 236: 2684-2691.

Umbarger, H. E. 1973. Threonine deaminases. *Adv. Enzymol.* 37: 349-395.

Vesterberg, O. and H. Svensson 1966. Isoelectric fractionation, analysis, and characterization of ampholytes in natural pH gradients. *Acta. Chem. Scand.* 20: 820-834.

Weber, K. and M. Osborn 1969. The reliability of molecular weight determinations by dodecyl sulfate-polyacrylamide gel electrophoresis. *J. Biol. Chem.* 244: 4406-4412.

Wellner, D. 1971. Electrofocusing in gels. *Anal. Chem.* 43: 59A-65A.

PROPERTIES AND PHYSIOLOGICAL SIGNIFICANCE
OF MULTIPLE FORMS
OF MITOCHONDRIAL MONOAMINE OXIDASE (MAO)

MOUSSA B. H. YOUDIM and
GODFREY G. S. COLLINS
MRC Clinical Pharmacology Unit
University Dept. of Clinical Pharmacology
Radcliffe Infirmary
Oxford OX2 6HE, U.K. and
Dept. of Pharmacology
School of Pharmacy
University of London
29/39 Brunswick Square
London WC1N 1AX, U.K.

ABSTRACT. Much indirect evidence has accumulated to
suggest that monoamine oxidase (MAO) may not be a single
enzyme but a number of closely related catalysts with
different substrate and inhibitor specificities. The
enzyme is associated with the outer membrane of mitochondria. Solubilized and purified MAO contains one
mole of covalently bound FAD and eight moles of -SH per
120,000 g enzyme protein. Electrophoresis of the soluble
purified enzyme on polyacrylamide gel exhibits a number
of active cathodic and anodic bands as visualized by
histochemical procedures. The properties of extracted
bands have revealed that they are closely related with
regard to molecular weight, FAD and -SH content, and the
number of subunits and binding sites for the inhibitor
phenelzine. However they exhibit different substrate and
inhibitor specificities, thermal inactivation, and pH
optima. The multiplicity of MAO is not universally
accepted. Because the multiple forms contain varying
amounts of phospholipid material it has been suggested
that they represent artifacts of the purification procedure. However strong evidence is presented that they
are immunologically distinct and associated with different
populations of mitochondria in the liver and the brain.
The physiological role of MAO remains to be elucidated.
Drugs that irreversibly inhibit this enzyme are used in
the chemotherapy of depressive illness. It has therefore been suggested that these drugs owe their beneficial
effect to the inhibition of a particular form of the
enzyme at a specific site in the brain, thereby raising
the level of transmitter amines.

INTRODUCTION

Monoamine oxidase (MAO) (monoamine: O_2 oxidoreductase, deaminating, EC1.4.3.4) is a ubiquitous enzyme occuring in all classes of vertebrates and many invertebrates (Blaschko, 1937). It is distinguished from other amine oxidases by two properties: first, its insensitivity towards the inhibitory activity of unsubstituted acylhydrazines such as semicarbazide, and second, its characteristic substrate specificity pattern. Thus, amines such as the naturally occurring phenylethylamines, dopamine, noradrenaline, β-phenethylamine, and also 5-hydroxytryptamine are readily oxidized whereas diamines are either poorly de-aminated or not metabolized at all (Tipton, 1973; Youdim, 1974a). The reaction catalyzed by MAO may be represented as a two step process

(A) $R-Ch_2-NH_2+O_2 \rightarrow R-CH=NH + H_2O_2$

(B) $R-CH=NH+H_2O \rightarrow RCHO + NH_3$
      the overall reaction being

(C) $R-CH_2-NH_2+O_2+H_2O \rightarrow RCHO + H_2O_2 + NH_3$

HISTORICAL EVIDENCE OF MULTIPLE FORMS OF MAO

It is currently accepted that the microsomal and mito-chondrial fractions account for the total activity of brain and liver homogenates. Although it is now considered that the microsomal MAO is the product of homogenization procedure, can MAO be considered as a single entity? Werle and Roewer (1952) reported more than one type of MAO. These workers were able to separate enzymes from animal sources that were capable of oxidizing only aliphatic monoamines or only aromatic amines. Alles and Heegard (1943) tested a large series of typical substrates with liver extracts from several species and found large species differences as judged by the relative rates of oxidation. They also noted that the enzyme-inhibitor dissociation constant was not independent of the substrate. It was suggested by Satake (1955) that MAO may be a mixture of closely related enzymes which have different substrate specificities and each tissue may contain an enzyme system with a distinct composition. Severina and Gorkin (1963) studying the metabolism of two aralkyl amines, tyramine and benzylamine, were able to inhibit selectively the oxidative deamination of the two amines. The data obtained are ex-plained by assuming the existence either of two different structure-bound amine oxidases in rat liver mitochondria or of two catalytically active sites on the enzyme. In the kinetic studies on the inhibition of the oxidation of sero-tonin and tyramine by rat liver mitochondria (Hardegg and

Heilbron, 1961) it was concluded that the two substrates are
probably oxidized by two different monoamine oxidases. The
activity versus pH curves of MAO in the presence of inhibitors
suggested the presence of two monoamine oxidases not equally
sensitive to tranylcypromine, iproniazid, phenelzine, and
pargyline. Tranylcypromine gave a second maximum at pH
6.5-7.0 (Youdim and Sourkes, 1965). A similar result has
been obtained with the partially heat-inactivated enzyme
(Youdim and Sourkes, 1965) and it was thus concluded that the
two pH maxima could be explained in a number of ways:

(a) Two distinct enzymes with different pH maxima are
present as a mixture.
(b) Active-enzyme-substrate complexes are formed.
(c) An ampholyte inhibitor is present.
(d) A unitary enzyme is present possessing more than one
active site.

The first possibility was favored by the findings with selective enzyme inhibition by drugs (Youdim and Sourkes, 1965).
Horita (1962), studying the influence of pH on serotonin
metabolism in various tissue homogenates, has indicated the
presence of at least two enzyme systems in the heart. Using
serotonin as the substrate it was reported that KCN inhibited
the heart enzyme at pH 9.5 while phenelzine inhibited it at
the physiological pH 7.4. Proflavine acts as a competitive
inhibitor in the case of serotonin oxidation by MAO (Gorkin
et al., 1964) from rat liver. However, a non-competitive
inhibition is observed when tyramine is used as substrate and
the degree of inhibition by proflavine in the presence of
each substrate is significantly different. The main point of
interest in this study is the non-reversibility of inhibition
when tyramine is the substrate as compared to reversibility of
the oxidation of serotonin, supporting the suggestion of
Hardegg and Heilbron (1961) that the enzymes which oxidize
serotonin and tyramine are different.

Gorkin (1963) was able partially to resolve rat liver mitochondrial MAO using a non-ionic detergent OP-10 to solubilize
the mitochondria. After applying the "solubilized" material
to a "Brushite" column, stepwise elution gave two fractions
capable of deaminating either m-nitro-hydroxybenzylamine or
p-nitro-phenthylamine. More recently highly purified "solubilized" enzyme has been subjected to various techniques for
the separation of different forms of MAO. These include the
first polyacrylamide gel electrophoretic separation of liver
(Youdim and Sandler, 1967) and other tissue monoamine oxidases
(see Sandler and Youdim, 1972; Youdim, 1974a).

METHODS AND RESULTS

ELECTROPHORETIC SEPARATION OF THE MULTIPLE FORMS OF MAO

Electrophoresis of the purified enzyme on 5% polyacryl-amide gels using a Tris-HCl, pH 9.1 buffer system, results in the separation of five and four bands of MAO activity from liver and brain, respectively (Youdim et al., 1970). From the liver, three bands (MAO 2, 3,and 4) and from brain, two bands (MAO 2 and 3) migrate from cathode to anode; band one remains at the origin (MAO 1) and a band of activity migrates towards the cathode (MAO R). Enzyme activity in the poly-acrylamide gels can be identified by a histochemical method (Youdim, et al., 1970). It has been shown that after electro-phoresis of such preparations in polyacrylamide gels, sites of enzyme activity detected by tetrazolium histochemical tech-niques do not necessarily coincide in all cases with bands exhibiting MAO activity as measured by more conventional methods such as the radiochemical assays (Lagnado et al., 1971). Furthermore, the intensity of tetrazolium reduction does not necessarily reflect the level of MAO activity found in different forms (Youdim and Lagnado, 1972). These results may be explained in part by the inhibitory property exerted by tetrazolium salts on MAO (Lagnado et al., 1971). Thus, tetrazolium salts used (Shih and Eiduson, 1969; 1971) to identify and to measure the activity of monoamine oxidase as separated electrophoretically on polyacrylamide can result in artifacts of activity depending on the substrate employed (Youdim and Lagnado, 1972; Diaz Borges and D'Iorio, 1973). In our studies, the bands of MAO activity in the gels were separated with a sharp razor, the gels homogenized in 0.05 M phosphate buffer, pH 7.4 to dilute the protein, and the extract centrifuged at 10,000 x g for 10 min. The super-natants which contained almost all the MAO activity were used for further studies with the radiochemical assay of Collins and Southgate (1969).

PROPERTIES OF THE MULTIPLE FORMS OF MAO

Some of the physiochemical properties of the multiple forms of rat liver MAO are shown in Table 1. It is apparent that the various forms of monoamine oxidase are closely re-lated with regard to molecular weight, number of subunits, moles of -SH groups, and FAD content. Differences, however, appear when a comparison is made of the anodic and cathodic forms with regard to pH optima, heat stability, binding of phenelzine, and total phospholipid content (Youdim et al., 1970; Youdim and Collins, 1974). These differences would

622

TABLE I

SOME PHYSIOCHEMICAL PROPERTIES OF THE MULTIPLE FORMS
OF RAT LIVER MONOAMINE OXIDASE

	Anodic Forms			Cathodic Form	
	1	2	3	4	R
Molecular weight [1]	2.95×10^5	3.09×10^5	2.94×10^5	$2.89 \times 10^{(5)}$	–
Moles–SH per $1.5 \times 10^5 g$ [2]	8.0	9.0	9.0	9.0	9.0
Moles FAD per $1.5 \times 10^5 g$ [3]	1	1	1	1	1
pH optima [4]	8.1	8.3	8.1	8.7	7.1
Molecular weight [5] in 8M urea	7.76×10^4	7.73×10^4	8.13×10^4	7.59×10^4	–
Phospholipid phosphorus (μg/mg protein) [6]	0.38	0.20	0.11	0.58	0.015
^{14}C-phenelzine binding (moles per $1.5 \times 10^5 g$) [7]	0.89	0.99	0.90	0.92	0.06

(1)   measured by gel filtration (Youdim and Collins, 1971)
(2)   (Youdim, 1974a)
(3)   (Youdim, 1974a)
(4)   (Youdim et al., 1970)
(5)   measured by gel electrophoresis (Youdim and Collins, 1971)
(6)   (Youdim and Collins, 1974 (in press))
(7)   (Youdim and Collins, 1974 (in press))

suggest that we are dealing with two basically different
monoamine oxidases; for example, the cathodic form is rather
heat stable, has a pH optimum of 7.1, extremely low phospholipid content, and is capable of binding only 0.06 mole of
the irreversible active site-directed inhibitor phenelzine.
In contrast, the anodic forms are heat labile and have much
higher phospholipid content (Youdim and Collins, 1974).   The
binding of [14C]-phenethylhydrazine (phenelzine) to the enzyme was estimated using equilibrium dialysis.   Approximately
one mole of radioactive phenelzine binds to each 150,000 g of
enzyme protein.   However, increasing the concentration of
[14C]-phenelzine substantially increased the amount bound by

form-5, reaching one mole per 150,000 g. We can conclude from the data presented in Table 1 (see also Youdim and Collins, 1971) that each of the multiple forms of rat liver MAO is composed of two subunits of approximately equal molecular weight. However, it was pointed out that as one mole of flavin has been found per mole of intact monoamine oxidase and that only one mole of phenelzine was attached to each mole of native MAO, it would be expected that only one of the subunits possesses a catalytic site (Youdim and Collins, 1971). Recently, Oreland et al., (1973) have confirmed these results and also managed to separate two subunits of equal molecular weight from pig liver MAO by polyacrylamide gel electrophoresis carried out in the absence of a reducing agent.

The molecular basis of the apparent multiple forms of MAO is unclear although polymerization of a basic subunit (Collins and Southgate, 1970; Gomes et al., 1969), conformational differences (Youdim and Collins, 1971), and differences in the amount of bound phospholipid membraneous material (Tipton, 1972; Tipton et al., 1972; Houslay and Tipton, 1973) have all been suggested. Multiplicity of MAO is not universal, however, for in some tissues, including pig brain (Tipton and Spires, 1968), monkey small intestine (Murali and Radhakrishnan, 1970), human platelets (Collins and Sandler, 1971), and tyramine oxidase from *Sarcina lutea* (Yamada et al., 1967) monoamine oxidase appears to be homogeneous. Although indirect biochemical and electrophoretic studies have strengthened the case for the presence of multiple molecular forms of MAO, it has also been argued that they are artifacts of the solubilization and purification procedures (Houslay and Tipton, 1973). In the light of these criticisms, two new approaches have been made to this problem:

   a)   preparation of intact mitochondria, and
   b)   induction of antibody to the purified enzyme.

Both have met with some success. Kroon and Veldstra (1972) separated mitochondria from rat brain on a sucrose density gradient and reported that the various fractions have significantly different MAO activity with four substrates. An objection has been made by Youdim (1973, 1974b) to this procedure, since at high sucrose density, mitochondrial structures are altered. To maintain the mitochondria in their physiological state, Youdim (1973) has separated mitochondria having different monoamine oxidase activities on a gradient consisting of iso-osmotic sucrose with Ficoll added to increase the density and using zonal rotor centrifugation. Two fractions having higher activity for either tyramine or dopamine have been isolated. The induction of antibody to MAO will be discussed in a later section.

CATALYTIC PROPERTIES

The substrate specificity pattern and sensitivity to inhibitors exhibited by the multiple forms of MAO have been of particular interest. Table II shows the Km values and Vmax of the multiple forms of rat liver MAO toward the substrates dopamine, tyramine, and kynuramine. It should be noted that the cathodic form is particularly active against dopamine. The Km values of the anodic forms show wide variations when the different substrates are compared; for example, the Km of form-4 is 210 µM for dopamine but only 10 µM for kynuramine. The Km of the cathodal form for dopamine is particularly low (3.6 µM).

Inhibitors of MAO have been used clinically in the treatment of depressive illness for several years (see Sandler and Youdim, 1972). It has been assumed that the antidepressant properties of these drugs depends on the inhibition of MAO leading to accumulation of amine neurotransmitter substrates at particular sites in the brain (Bevan Jones et al., 1972). Not only do these drugs inhibit the multiple forms of MAO to different extents but there is a correlation between the extent of inhibition achieved in vitro and after administration in vivo (Youdim et al., 1972; Collins et al., 1972). Interest has centered on two new MAO inhibitors, clorgyline (N-methyl-N-propargyl-3 (2,4-dichlorophenoxy) -propylamine hydrochloride, (M and B 9302), and deprenyl (E-250)) which have been reported to be more "specific" against certain of the forms of MAO (see Johnston, 1968; Knoll and Magyar, 1972). Using these agents, two forms of MAO can be distinguished: the "type A" which is inhibited by clorgyline, and "type B" which is more sensitive to deprenyl. In order to attempt to relate enzyme "Types A and B" with the electrophoretically separated forms of MAO we have determined the effect of clorgyline on the multiple forms of MAO using kynuramine and dopamine as substrates (Fig. 1). Forms 1, 2, and 3 are inhibited more readily than forms-4 and -5; this is particularly noticeable when dopamine is the substrate. Whether forms 1, 2, and 3 represent MAO "type A" and form-5 "type B" cannot at present be determined.

INDUCTION OF ANTIBODY TO LIVER MAO

This technique has recently been employed in an attempt to determine whether the MAO present in liver is immunologically identical with that found in brain. Hartman, Yasunobu, and Udenfriend (1971) were the first to induce antibody to whole beef liver MAO and its multiple forms. They obtained a single immunoprecipitant and came to the conclusion that liver enzyme was a single entity; however, using this antibody, not all the

TABLE II
SOME CATALYTIC PROPERTIES OF THE MULTIPLE FORMS
OF RAT LIVER MONOAMINE OXIDASE (MAO)

Substrate	Anodic forms				Cathodic form R
	1	2	3	4	
Dopamine					
Km (μM)	19	100	130	210	3.6
Vmax	0.32(13)	1.00(53)	0.85(108)	0.12(80)	8.97(3139)
Tyramine					
Km (μM)	34	19	22	180	290
Vmax	2.45(100)	1.87(100)	0.79(100)	0.15(100)	0.28(100)
Kynuramine					
Km (μM)	12	19	27	10	46
Vmax	3.11(127)	1.85(99)	0.56(71)	0.28(186)	0.32(114)

Enzyme activity was estimated at 37°C in phosphate buffer (0.05 M, pH 7.4) using either ^{14}C-dopamine and ^{14}C-tyramine (Robinson et al., 1968) or kynuramine as substrate (Kraml, 1965). The Km and Vmax values were estimated from double reciprocal plots and each value is the mean of 3 determinations. The Vmax values are expressed as nmole substrate deaminated/protein/min and have also been expressed (in parenthesis) as percentages of that determined for tyramine (100%).

beef brain monoamine oxidase could be precipitated (Hartman and Udenfriend, 1972) leading to the suggestion that a brain-specific MAO different from that of the liver exists (see also McCawley and Racker, 1973). In Youdim's (1974c) experiments (Youdim and Benda, 1974) antibody was raised to purified rat liver MAO (for methods, see Youdim, 1974c). When soluble MAO from rat liver was tested against antibody induced in rabbits using double diffusion Ouchterlony technique (See Fig. 2), two immunoprecipitin lines of antigenic identity were formed when protein was detected using Amido Black. Identical plates were examined for MAO activity using the histochemical procedure of Glenner et al. (1957). The pattern is identical to that observed for protein staining, showing that each of the immunoprecipitin lines is enzymatically active. Soluble MAO prepared from rat brain when tested against anti-liver MAO also cross-reacted, giving two lines of immunoprecipitate having enzyme activity. The present results indicate that there are no antigenic sites present on liver MAO which are not also present on brain MAO. Furthermore, two immunoprecipitin lines which are enzymatically active are representative of two distinct forms of MAO, supporting the electrophoretic evidence

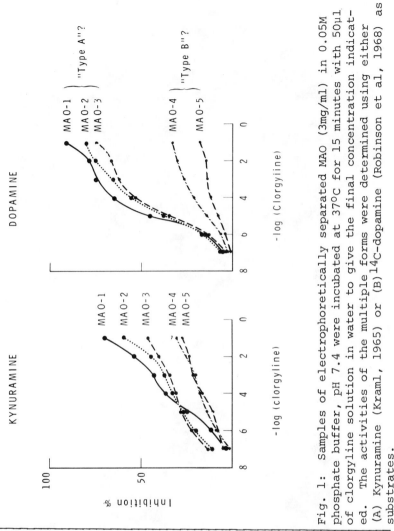

Fig. 1: Samples of electrophoretically separated MAO (3mg/ml) in 0.05M phosphate buffer, pH 7.4 were incubated at 37°C for 15 minutes with 50μl of clorgyline solution in water to give the final concentration indicated. The activities of the multiple forms were determined using either (A) Kynuramine (Kraml, 1965) or (B) ^{14}C-dopamine (Robinson et al, 1968) as substrates.

and the indirect studies with MAO inhibitors. Studies in progress have involved the antigenic properties of electrophoretically separable bands. Our preliminary results would indicate that the two antigens are related to the cathodic and anodic forms of monoamine oxidases as separated electrophoretically. Recently McCauley and Racker (1973) have demonstrated the presence of two antigenically different monoamine oxidases in beef brain having varying substrate and inhibitor specificities. These workers did not rule out the possibility of each form existing in different states.

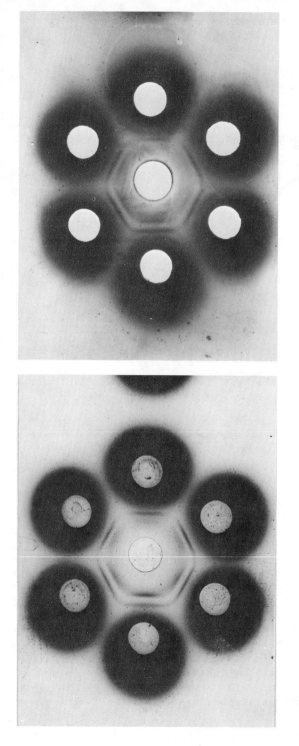

Fig. 2: Immunodiffusion assay of rat liver monoamine oxidase. Ouchterlony double diffusion plates (1.2% Agarose in 0.05M barbitone buffer, pH 8.4) were used. The plates were allowed to diffuse for 48 hours at room temperature. After rinsing in several changes of 0.9% NaCl for 96 hours at 4°C the plates were rinsed with distilled water and stained either (A) with 0.1% Amido Black in 7% acetic acid or (B) for MAO activity with a solution containing 4mg nitroblue tetrazolium, 1mg sodium sulphate and 4mg tryptamine hydrochloride in 5ml of 0.05M phosphate buffer pH 7.4 (Youdim, 1974c).

DISCUSSION

PHYSIOLOGICAL SIGNIFICANCE OF THE MULTIPLE FORMS OF MAO

Evidence for the existence of multiple forms of mitochondrial MAO in brain and liver is becoming overwhelming although absolute proof of their occurrence in vivo will be difficult to achieve. The physiological significance of the different forms of MAO in the regulation of the metabolism of both transmitter and non-transmitter amines can at present only be guessed at. It has been tempting to postulate, on the basis of kinetic data taken under limited and non-physiological conditions, broad theories explaining the importance of multiple forms of monoamine oxidase and their distribution. This point is illustrated in Fig. 3, where marked differences in substrate sensitivity between partially purified rat liver MAO forms-1 to 4 and form-5 at 37°C are shown. On the basis of such striking differences it may be suggested that in vivo substrate inhibition could regulate synthesis of monoamines by feed back end-product inhibition (Hamon et al., 1973). However, levels of biogenic monoamines high enough to inhibit forms 1 to 4 are unlikely to occur in vivo (see Usdin and Snyder, 1974). Furthermore, recent studies in vivo with perfused intact isolated whole liver (Youdim et al., 1974) have shown that substrate inhibition of MAO does not occur even at concentrations as high as 4 mM. Although product inhibition by aldehydes has also been reported in vitro, such phenomena do not occur in vivo (Houslay and Tipton, 1973). Under most conditions, MAO and aldehyde dehydrogenase in the cell are in excess over their substrates. It must be remembered that the MAO assays employ optimal concentrations of reagents very different from concentrations existing in vivo and that rat liver and brain contain many times the concentration of enzyme commonly used in the in vitro assays. The rate-limiting factor in vivo might be the concentration of substrate, either amine or O_2, which under normal circumstances is well below their Km values, thereby rendering the extrapolation from kinetic data derived under optimal conditions in vitro (Houslay and Tipton, 1973b) to conditions in vivo extremely hazardous.

In spite of these difficulties, progress is being made in elucidating the presence and function of multiple forms of MAO in vivo. First, the immunological experiments described in the previous section provide strong evidence that different MAO enzymes exist in vivo. Second, although histochemical procedures for visualizing MAO are imperfect (Youdim, 1974c), Hanker and his workers (1972) have shown heterogeneity of mitochondria with respect to MAO activity when tryptamine and

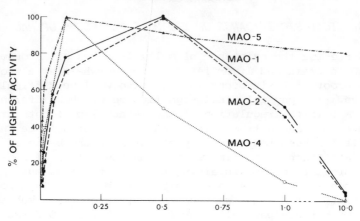

Fig. 3: The effect of various concentrations of kynuramine on the rate of oxidative deamination of kynuramine by the multiple forms of rat liver MAO. Enzyme (3mg/ml) activity was determined at 37^{0}C in 0.05M phosphate buffer, pH 7.4 using the spectrophotofluometric assay described by Kraml (1965). Enzyme Form-3 (MAO_3) which had a very similar pattern of activity as compared to MAO 1 and 2 is not shown for the sake of clarity.

tyramine were used as substrates. More direct studies (Kroon and Veldetra, 1972; Yang and Neff, 1973; Youdim, 1973) involving the subcellular fractionation of liver and brain have confirmed these findings. Third, if enzyme multiplicity were the result of the preparative procedure (Houslay and Tipton, 1973) it might be expected that after administration of MAO inhibitors in vivo all enzyme forms would be inhibited to the same extent; however, this is not the case (Youdim et al., 1972; Collins et al., 1972). In the central nervous system monoamine oxidase is present in the intraneuronal (nerve endings, synaptosomes) as well as extraneuronal (glial cell) mitochondria. A number of recent reports would indicate that the substrate specificities of intra and extra-neuronal monoamine oxidase are different (see Sandler and Youdim, 1972). The reabsorption mechanism (Iversen, 1967) is now considered to be the major non-chemical inactivation of biogenic monoamines following presynaptic release. Assuming that brain MAO exists in multiple forms (Youdim et al., 1969) and that these enzymes differ in their substrate specificity patterns (see Youdim, 1974 a) it seems clear that the role of the enzyme requires re-evaluation. The synaptic MAO constitutes about 10% of the total brain activity (Fahn et al., 1969; Tabakoff et al., 1974). Thus this small amount of enzyme may be responsible for regulating the concentration of free amine at central synapses. Inhibition of intraneuronal MAO rapidly leads to saturation of the storage capacity of the synaptic vesicles and to an increase in the concentration of free amines, which in turn may interfere with uptake of amines released from the nerve endings (Trendelenburg et al., 1972). Evidence has been presented that brain monoamine oxidase activity (Holz-bauer and Youdim, 1973) and its multiple forms may be regulated by endocrine glands (Youdim et al., 1974). The suggestion has been made that the level of amines and amine metabolites may exert an allosteric control of both amine synthesis and metabolism (Collins et al., 1970). On the other hand, the differential intracellular localization of the multiple forms of MAO might also have functional significance (Collins et al., 1970).

In the therapy of depressive disease, drugs are used which inhibit MAO and it is believed that they owe their beneficial effect to this ability and to the resulting accumulation of amines. However, it is found that some compounds are considerably more effective therapeutically than others, and no fully satisfactory explanation for this anomaly has been offered. In 1969 (Youdim et al., 1969) we offered an approach which we thought in better accord with the characteristics of MAO. Suggestions were made that if multiple forms of MAO were also present in human brain, this might explain the varying ability of different MAO inhibitors to combat depression; it should be possible to correlate the therapeutic effect of these compounds with their inhibitory action on a particular form of the enzyme, and the substrate preference of that fraction might even provide a pointer to the biochemical disturbance present in the illness. Human brain MAO, like that of the rat, also exists in multiple forms (Collins et al., 1970) and it seems that the therapeutic usefulness of MAO inhibitors relies on the inhibition of particular forms of the enzyme (Youdim et al., 1972; Bevan Jones et al., 1972). An important finding is that these drugs have varying inhibitory actions when more than one substrate is used to estimate MAO activity. Even though one particular form may be totally inhibited, that of the second may not. Unfortunately many inhibitors of MAO are not as selective in their inhibitory action. The tyramine-hypertensive action that occurs in patients on MAO inhibitors when they eat cheese or tyramine-rich food (Blackwell, 1963) has not been eliminated (Lader et al., 1970). With this in mind it was suggested (Collins et al., 1970) that an approach should be made to synthesize specific selective inhibitors tailored either to an individual form of the enzyme or to the biogenic monoamines, thus eliminating some of the side effects. Recently Yang and Neff (1973; Neff et al., 1974) using the newly synthesized "selective" inhibitors clorgyline (M and B9302) and deprenyl (E-250) have been able to raise selectively the levels of either the neurotransmitter 5-hydroxytryptamine (serotonin) or β-phenyl-ethylamine, respectively, in the rat brain. It is apparent from their studies that the action of these new drugs is the inhibition of the enzyme form active against either 5-hydroxytryptamine or β-phenylethylamine. It is to be hoped that such an approach will in the future lead not only to the more effective treatment of depressive illness but also to an understanding of the underlying chemical lesion.

REFERENCES

Alles, G. A. and E. V. Heegard 1943. Substrate specificity of amine oxidase. *J. Biol. Chem.* 147: 487-494

Bevan Jones, A. B., C. M. B. Pare, W. J. Nicholson, K. Price, and R. S. Stacey 1972. Brain amine concentration after monoamine oxidase inhibitor administration. *Brit. Med. J.* 1: 17-19.

Blackwell, B. 1963. Hypertension crisis due to monoamine oxidase inhibitors. *Lancet* ii: 849-851.

Blaschko, H., D. Richter, and H. Schlossmann 1937. The oxidation of adrenaline and other amines. *Biochem. J.* 31: 2187-2196.

Collins, G. G. S. and J. Southgate 1969. The estimation of monoamine oxidase using ^{14}C-labelled substrates. *Biochem. Pharmacol.* 18: 2285-2287.

Collins, G. G. S. and M. Sandler 1971. Human blood platelet monoamine oxidase. *Biochem. Pharmacol.* 20: 289-296.

Collins, G. G. S., M. Sandler, E. D. Williams, and M. B. H. Youdim 1970. Multiple forms of human brain mitochondrial monoamine oxidase. *Nature* 225: 817-820.

Collins, G. G. S., M. B. H. Youdim, and M. Sandler 1972. Multiple forms of monoamine oxidase: comparison in vitro and in vivo inhibition patterns. *Biochem. Pharmacol.* 21: 1995-1998.

Diaz Borges, J. M. and A. D'Iorio 1973. Polyacrylamide Gel electrophoresis of rat liver mitrochondrial monoamine oxidase. *Can. J. Biochem.* 51: 1089-1095.

Fahn, S., J. S. Rodman, and L. J. Cote 1969. Association of tyrosine hydroxylase with synaptic vesicles in bovine caudate nucleus. *J. Neurochem* 16: 1293-1300.

Glenner, G. C., H. J. Burtner, and G. W. Brown 1957. The histochemical demonstration of monoamine oxidase activity by tetrazolium salts. *J. Histochem. Cytochem.* 5: 591-600.

Gomes, B., I. Igane, H. G. Kloepfer, and K. T. Yasunobu 1969.
Amine Oxidase XIV. Isolation and characterization of the
multiple beef amine oxidase components. *Arch. Biochem.
Biophys.* 132: 16-27.

Gorkin, V. Z. 1963. Partial separation of rat liver mitochondrial amine oxidase. *Nature* 200: 77-78.

Gorkin, V. Z., N. V. Komsarova, M. I. Lerman, and I. V.
Veryovkina 1964. The inhibition of mitochondrial amine
oxidases in vitro by Proflavine. *Biochem. Biophys.* 15:
383-389.

Hamon, M., S. Bourgoin, and J. Glowinski 1973. Feed back
regulation of serotonin synthesis in rat striatal slices.
J. Neurochem. 20: 1727-1745.

Hanker, J. S., C. J. Kusyk, F. E. Bloom, and A. G. E. Pearse
1972. The demonstration of dehydrogenases and monoamine
oxidase by the formation of Osmium Blacks at the sites of
Hatchett's Brown. *Histochemie* 33: 205-230.

Hardegg, W. and E. Heilbronn 1961. Oxydation von serotonin
und tyramin durch rallenlebermitochondrien. *Biochim.
Biophys. Acta* 51: 553-560.

Hartman, B. H. and S. Udenfriend 1972. The use of an immunological technique for the characterization of bovine
monoamine oxidase from liver and brain. *Adv. Biochem.
Psychopharmacol.* 5: 119-128.

Hartman, B. H., K. T. Yasunobu, and S. Udenfriend 1971.
Immunological identity of the multiple forms of beef liver
mitochondrial monoamine oxidase. *Arch. Biochem. Biophys.*
147: 797-804.

Horita, A. 1962. The influence of pH on serotonin metabolism
by rat tissue homogenates. *Biochem. Pharmacol.* II, 147-153.

Houslay, M. D. and K. F. Tipton 1973. The nature of the
electrophoretically separable multiple forms of rat liver
monoamine oxidase. *Biochem. J.* 135: 173-186.

Iversen, L. L. 1967. *Uptake and storage of noradrenaline in
sympathetic nerves.* Cambridge University Press, Cambridge.

Johnston, J. P. 1968. Some observation upon a new inhibitor of
monoamine oxidase in brain tissue. *Biochem. Pharmacol.*
17: 1285-1297.

Knoll, J. and K. Magyar 1972. Some puzzling pharmacological
effects of monoamine oxidase inhibitors. *Adv. Biochem.
Psychopharmacol.* 5: 393-408.

Kraml, M. 1965. A rapid microfluorometric determination of
monoamine oxidase. *Biochem. Pharmacol.* 14: 1684-1686.

Kroon, M. C. and H. Veldstra 1972. Multiple forms of rat
brain mitochondrial monoamine oxidase: subcellular
localization. *FEBS letters* 24: 173-177.

Lader, M. H., G. Sakalis, and M. Tansella 1970. Interaction

between sympathomimetic amines and a new monoamine oxidase inhibitor. *Psychopharmacologia* 18: 118-123.

Lagnado, J. R., M. Okamoto, and M. B. H. Youdim 1971. The effect of tetrazolium salts on monoamine oxidase activity. *FEBS letters* 17: 117-120.

Murali, D. K. and A. N. Radhakrishnan 1970. Preparation and properties of an electrophoretically homogeneous monoamine oxidase from monkey intestine. *Biochim. Biophys. Acta* 206: 61-70.

McCauley, R. and E. Racker 1973. Separation of two monoamine oxidases from bovine brain. *Mol. Cell. Biochem.* 1: 73-81.

Neff, N. H., H.Y. T. Yang, and C. Goridis 1973. Degradation of the transmitter amines by specific types of monoamine oxidases. In *Frontiers in Catecholamine Research*, E. Usdin and S. Snyder, eds., Pergamon Press, Oxford pp. 133-138.

Oreland, L., H. Kinemuchi, and T. Stigbrand 1973. Pig liver monoamine oxidase: studies on the subunit structure. *Arch. Biochem. Biophys*. 159: 854-860.

Robinson, D. S., W. Lovenberg, H. Keiser, and A. Sjoerdsma 1968. Effect of drugs on human platelet and plasma amine oxidase activity in vitro and in vivo. *Biochem. Pharmacol*. 17: 109-119.

Sandler, M. and M. B. H. Youdim 1972. Multiple forms of monoamine oxidase: functional significance. *Pharmacol. Rev*. 24: 331-348.

Satake, K. 1955: *Sakiuno Koso-Kazaku*, 4: 39 cited by H. Okumoro, *J. Bull. Osaka. Med. Sch*. 6: 58 (1960).

Severina, I. S. and V. Z. Gorkin 1963. On the nature of mitochondrial monoamine oxidases. *Biokhimiga* 28: 896-902.

Shih, J. and S. Eiduson 1969. Multiple forms of monoamine oxidase in developing brain. *Nature* 224: 1309-1310.

Shih, J. and S. Eiduson 1971. Multiple forms of monoamine oxidase in the developing brain: tissue and substrate specificities. *J. Neurochem*. 18: 1221-1227.

Tabakoff, B., L. Meyerson, and S. G. A. Alivisatos 1974. Properties of monoamine oxidase in nerve endings from two bovine brain areas. *Brain Res*. 66: 491-508.

Tipton, K. F. 1972. Some properties of monoamine oxidase. *Adv. Biochem. Psychopharmacol*. 5: 11-24.

Tipton, K. F. 1973. Biochemical aspects of monoamine oxidase. *Brit. Med. Bull*. 29: 116-119.

Tipton, K. F. and I. P. C. Spries 1968. The homogeneity of pig brain mitochondrial monoamine oxidase. *Biochem. Pharmacol*. 17: 2137-2141.

Tipton, K. F., M. B. H. Youdim, and I. P. C. Spires 1972. Beef adrenal gland monoamine oxidase. *Biochem. Pharmacol*. 21: 2197-2204.

Trendlelenburg, U., P. R. Draskoczy, and K. H. Graefe 1972. The influence of intraneuronal monoamine oxidase on neuronal net uptake of noradrenaline and on sensitivity to noradrenaline. *Adv. Biochem. Psychopharmacol.* 5: 371-378.

Werle, E. and F. Roewer 1952. Substrate specificity of mono-amine oxidase. *Biochem. J.* 50: 320-326.

Yamada, H., T. Uwajima, H. Kumagai, M. Watanabe, and K. Ogata 1967. Bacterial monoamine oxidase. 1. Purification and crystallization of tyramine oxidase of *Sarcina lutea.* *Agr. Biol. Chem.* 31: 890-896.

Yang, H. Y. T. and N. H. Neff 1973. Monoamine oxidase 1. A natural substrate for type B enzyme. *Fed. Proc.* 31: 797.

Youdim, M. B. H. 1973. Heterogeneity of rat brain and liver mitochondrial monoamine oxidase: subcellular fractiona-tion. *Biochem. Soc. Trans.* 1: 1126-1127.

Youdim, M. B. H. 1974a. Monoamine deaminating system in mammalian tissues. In *MTP International Review of Science,* H. Blaschko, ed., Butterworth Ltd., London (1974) in press.

Youdim, M. B. H. 1974b. Heterogeneity of rat brain mitochon-drial monoamine oxidase. *Adv. Biochem. Psychopharmacol.* *11* (in press).

Youdim, M. B. H. 1974c. The assay and purification of brain monoamine oxidase. In *Methods in Neurochemistry,* N. Marks and R. Rodnight, eds., Plenum Press, New York (in press).

Youdim, M. B. H. and P. Benda 1974. Unpublished results.

Youdim, M. B. H. and J. R. Lagnado 1972. Limitation in the use of tetrazolium salts for the detection of multiple forms of monoamine oxidase. *Adv. Biochem. Psychopharmacol.* 5: 289-292.

Youdim, M. B. H. and G. G. S Collins 1971. The dissociation and reassociation of rat liver mitochondrial monoamine oxidase. *Europ. J. Biochem.* 18: 73-78.

Youdim, M. B. H. and G. G. S. Collins 1974. Some properties of the multiple forms of rat liver mitrochondrial mono-amine oxidase. *Biochem. Pharmacol.* (in press).

Youdim, M. B. H. and M. Sandler 1967. Isoenzyme of soluble monoamine oxidase from human placental and rat liver mitochondria. *Biochem. J.* 105: 43p.

Youdim, M. B. H. and T. L. Sourkes 1965. The effect of heat inhibitors and riboflavin deficiency on monoamine oxidase. *Can. J. Biochem.* 43: 1305-1318.

Youdim, M. B. H., G. G. S. Collins, and M. Sandler 1969. Multiple forms of rat brain monoamine oxidase. *Nature* 223: 626-628.

Youdim, M. B. H., G. G. S. Collins, and M. Sandler 1970. Isoenzyme of a soluble mitochondrial monoamine oxidase In *FEBS Symposium 18, Enzymes and Isoenzymes,* D. Shugar, ed., Academic Press, London. pp. 281-289.

Youdim, M. B. H., M. Holzbauer, and H. F. Woods 1974. In *Neuropsychopharmacology of Monoamines and their Regulatory Enzymes*, E. Usdin, ed., Raven Press, New York, pp. 11-28.

Youdim, M. B. H., G. G. S. Collins, M. Sandler, A. B. Bevan Jones, C. M. B. Pare, and W. J. Nicholson 1972. Human brain monoamine oxidase; multiple forms and selective inhibitors. *Nature* 236: 225-228.

THE USE OF ISOZYMES IN THE
STUDY OF MITOCHONDRIA

DAVID J. FOX

Department of Zoology
The University of Tennessee
Knoxville, Tennessee   37916

ABSTRACT. The use of marker enzymes has greatly facilitated the study of both structural and functional aspects of
various subcellular organelles. This investigation shows
that the usefulness of marker enzymes may be further extended by taking into account the subcellular distribution
of the different isozymic forms of many enzymes. In particular, malate dehydrogenase, NAD- and NADP-dependent isocitrate dehydrogenases, and glutamate-oxaloacetate transaminase have been used to investigate the phenomenon of mitochondrial heterogeneity. Crude mitochondrial preparations
from rat cerebrum are heavily contaminated with synaptosomes
(vesicles derived from nerve endings during homogenization)
which contain quantities of entrapped cytoplasm causing an
overestimate of the heterogeneity for some enzymes. The
removal of most of the synaptosomes has been accomplished
and isozymes have been used to aid in the interpretation
of enzyme profiles when mitochondria are centrifuged to
isopycnicity in sucrose density gradients.

INTRODUCTION

Mitochondria isolated from different tissues or organs are
known to differ both in morphology and in their enzymic composition. The activities of NADP-dependent isocitrate dehydrogenase (IDH-NADP)(Fox & Abacherli, 1971), Glycerophosphate
oxidase (GP-OX), Glutamate dehydrogenase (GDH) (Pette et al.,
1962) and Malic enzyme (ME) (Brdiczka and Pette, 1971), for
example, vary over a wide range depending upon the tissue of
mitochondrial origin. Other enzymes [Malate dehydrogenase
(MDH), Succinate dehydrogenase (SDH), Glutamate oxaloacetate
transaminase (GOT), Cytochrome-c (Cyt-c) (Pette et al., 1962)
and to some extent NAD-dependent isocitrate dehydrogenase (IDH-
NAD) (Goebell and Klingerberg, 1963)], however, exist in constant or near-constant proportion groups regardless of the
tissue of mitochondrial origin. Even within single tissues
the mitochondrial populations have been reported to be heterogeneous: Swick et al. (1967) have shown by low speed sucrose
density gradient centrifugation that rat liver Ornithine-retoacid
aminotransferase (OKAAT) activity is associated with mitochondria

of lower mean diameter than are Aspartate aminotransferase (AAT), GDH, and IDH-NADP; the specific activities of SDH, NADH-dependent cytochrome-c reductase (NADH-cyt-c-red), and Beta-hydroxybutyrate dehydrogenase (β-HBDH) have been shown by Katyare et al. (1970) to be different for light, fluffy, and heavy rat liver mitochondria; mitochondria isolated from intestinal crypt and villar cells are different in their abilities to oxidize succinate and glycerol-l-phosphate (Hülsmann et al., 1970); Bullock et al. (1971) found great diversity in the respiratory control ratios for rat muscle mitochondria isolated according to size on a discontinuous sucrose gradient; many differences are found in the enzymic complement of mitochondria isolated from rat brain nerve endings (synaptosomes) and glia + neurons (Salganicoff & Koeppe, 1968); in a study involving the centrifugation of crude rat brain homogenates through a sucrose density gradient in a zonal rotor, Blokhuis and Veldstra (1970) concluded that there exist "many populations of mitochondria, each with its own enzyme complement...".

In response to the use of crude rat brain homogenates by Blokhuis and Veldstra (1970), Fox (1973) examined crude and partially purified mitochondria from rat cerebral hemispheres for heterogeneity when centrifuged to isopycnicity in sucrose density gradients. He concluded that there is sufficient contamination of washed primary rat brain mitochondria with cytoplasmic enzymes entrapped in synaptosomes to cause possible misinterpretation of the extent of heterogeneity after isopycnic sucrose density gradient centrifugation, when based solely on enzyme activity measurements. Since many enzyme activities are found in both the cytoplasm and mitochondria, care must be taken not to attribute to mitochondria, activity which may be of extramitochondrial origin. Primary brain mitochondrial preparations being heavily contaminated with synaptosomes (vesicles derived from nerve endings containing large quantities of entrapped cytoplasm and even organelles such as mitochondria) make this problem particularly troublesome. The present investigation confirms and extends the conclusions of Fox (1973) and uses isozymes to aid in the interpretation of enzyme profiles when mitochondria are centrifuged to isopycnicity in sucrose density gradients.

MATERIALS AND METHODS

Preparation of Mitochondria

All mitochondrial isolations were made from 10% homogenates of finely minced rat cerebral hemispheres, heart, liver or kidneys or mouse heart, liver or kidneys in med A [20mM Triethanolamine, 0.25 M sucrose, 1 mM Ethylenediamine tetraacetic acid (EDTA), pH 7.2 (Kadenbach, 1966)] by successive centrifugations at 1,000g (10 min) and 12,000g (15 min). In most cases the 12,000g pellet was resuspended in med B [3% Ficoll, 0.12 M mannitol, 0.03 M sucrose, pH 7.4 (Clark & Nicklas, 1970)] and further centrifuged at 1,000g and 12,000g. This second 12,000g pellet was resuspended (with or without removal of the fluffy white layer as indicated in the Text and figure legends) in 1-2 ml med B, layered over 9 ml med C (2X med B) and centrifuged at 12,000g for 30 min. The pellet of purified mitochondria obtained was resuspended in either CM cellulose elution medium (see below) or in 1-2 ml of med C.

Isopycnic Sucrose Density Gradient Centrifugation

Gradients of 0.8-1.7 M sucrose (30 ml) were generated in 1" x 3" nitrocellulose centrifuge tubes (Beckman) using an Isco model 570 gradient former. Two ml of purified rat brain mitochondria resuspended in med C were carefully layered over the sucrose gradients and in turn were overlayered with 2 ml of med B. The gradients were centrifuged for 18 hr at 25,000 rpm in the SW 25.1 rotor of a Beckman L3-50 centrifuge, following which 30, 20 drop fractions were collected using an Isco model 183 density gradient fractionator. Each fraction was brought to 0.5% Triton X-100, incubated for at least 30 min and assayed spectrophotometrically for protein (O.D. 280); MDH (at 340 nm in 0.1M Tris HCl, pH 8.0, 10 mM NAD, 20 mM L-malate); IDH-NAD (Goebell and Klingenberg, 1964); Lactate dehydrogenase (LDH) (Clark and Nicklas, 1970); IDH-NADP (at 340 nm in 0.1M Tris-HCl, pH 8.25, 0.5 mM NADP, 1 mM Mg SO_4 and 2 mM D-L isocitrate); GOT (modified from DeLorenzo and Ruddle, 1970).

Electrophoresis

Electrophoresis was performed in 13% starch gels (Connaught) using a Buchler Vertical gel Electrophoresis apparatus at 300 v and 20 mA for 16 hr using 40 mM Tris-citrate as gel buffer and 0.1x CBT (citrate-borate-Tris, Fox, 1971) in the electrode reservoirs. Alternatively electrophoresis was performed for 1 hr at 300 v and 1mA/strip on cellulose acetate using 40 mM Tris-citrate pH 8.0 containing 5 mM β-mercaptoethanol. Enzyme activity (MDH, LDH, IDH-NADP) was located within the sliced gel or strip by incubating it in the appropriate spectrophotometric assay mixtures containing 2 ml of Nitroblue tetrazolium (Sigma, 6 mg/ml) and 2 ml of phenazine methosulfate (Sigma, 0.25 mg/ml) per 20 ml of reaction mixture. IDH-NAD is not stable to electrophoresis and GOT was localized by the method of DeLorenzo and Ruddle (1970).

Mitochondrial content of cytoplasmic IDH-NADP

The extent of cytoplasmic IDH-NADP which is found associated with heart, liver, and kidney mitochondria from rats and mice was determined after separation of the cytoplasmic and mitochondrial isozymes by C M cellulose (Whatman) column chromatography. Centrifuged (30 min at 30,000 g) sonicates (3 x 15 sec at a setting of 40 on a Bronwill Biosonik made by Blackstone Ultrasonics) of mitochondria prepared through media B and C as described above were applied to columns of C M cellulose equilibrated in 18 mM Tris titrated to pH 6.0 with solid citric acid, 1 mM EDTA, 5 mM β-mercaptoethanol, 10% glycerol, 1 mM $MgSO_4$, and 1 mM α-ketoglutarate (to stabilize the enzyme), eluted with 5 ml of the same buffer to bring off the cytoplasmic isozyme and further eluted with 10 ml of 0.25 M NaCl in the same buffer to bring off the mitochondrial isozyme (see Fig. 4 for details of the apparatus used).

RESULTS AND DISCUSSION

When crude primary rat cerebral hemisphere mitochondria (2nd 12,000g pellet, i.e. centrifuged once through med B) are subjected to isopycinc density gradient centrifugation two main peaks of protein are seen (Fox, 1973), one at the interface of the gradient and the loading zone (consisting mostly of soluble proteins and myelin) and another somewhat asymetrical peak deep into the gradient. By further purifying the mitochondria through med C, Fox (1973) was able to resolve the asymetrical peak into two distinct, yet partially overlapping, peaks and to greatly reduce the amount of protein at the loading interface. The results of a similar experiment are seen in Fig. 1B. In order to gain some insight into the origin of the two peaks, the crude mitochondrial (second 12,000 g) pellet was separated into its two components, by gently washing away the white fluffy layer with 2 ml of med C, leaving the brown firmly packed pellet. The two components were layered in med C (but not having been centrifuged through med C) onto separate sucrose gradients; their protein profiles may be seen in Fig. 1A. In Fig. 1C is seen the results of fraction by fraction addition of these two peaks, which closely approximates the profile reported for crude mitochondria by Fox (1973). It therefore appears that the less dense (peak at tube 20) of the two peaks is contributed primarily by the fluffy layer and the peak of greater density (peak at tube 23) by the firmly packed portion of the crude pellet. By comparing Figs. 1, A, B, and C it can be seen that the result of centrifugation through med C is to largely remove the protein content of the upper half of the gradient and to reduce somewhat the less dense of the two main peaks.

The profiles of the mitochondrial marker enzyme IDH-NAD

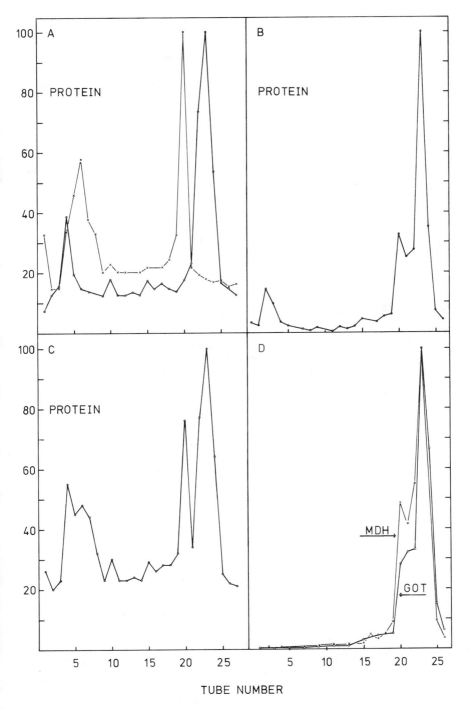

Figure 1

Figure 1 (legend). A. Composite of separate centrifugations of the fluffy (X-X) and the packed (0-0) portions of the second 12,000g pellet of rat cerebral hemisphere mitochondria assayed for protein at 280 nm. Peaks are at tube 20 (fluffy) and at tube 23 (packed). Centrifugation was as described in Fig. 1B. B. Distribution of protein (O.D. 280) in a 0.8-1.7 M sucrose gradient after 18 hr of centrifugation at 25,000 rpm in a Beckman 25.1 rotor of rat cerebral hemisphere mitochondria purified through med C. Peaks are at tubes 2, 20 and 23. C. Fraction by fraction summation of the two protein distributions seen in Fig. 1A to simulate the pattern obtained by centrifugation of the fluffy and packed portions of the second 12,000g pellet together in the same gradient. D. Composite of activity distributions of GOT and MDH determined from the same sucrose density gradient (0.8-1.7 M) of med C-purified rat cerebral hemisphere mitochondria. Same gradient and conditions as Fig. 1B and Fig. 2A and B.

(Pycock and Nahorski, 1971) may be seen in Fig. 2 A and C. The shoulder to the left of the main peak in med C-treated mitochondria (Fig. 2A) is obviously contributed by mitochondria from the fluffy layer (Fig. 2C, peak at tube 20).

The situation becomes more complex, however, when one considers the profiles (Figs. 2 B and D) for the cytoplasmic marker enzyme LDH. As for the protein profile, treatment of the crude mitochondria with med C results in the elimination of most of the LDH (Fig. 2B) in the upper half of the gradient (only a small peak at tube 16 remains). Removal of the fluffy layer from the second 12,000g pellet resulted in a shift of one tube to the lighter side (peak now at tube 22) of the denser of the two LDH peaks (Fig. 2D). The IDH-NAD and LDH peaks from the fluffy layer are superimposable at tube 20 (compare Fig. 2C and D).

Interestingly, the well-packed brown pellet from the second 12,000g centrifugation is not as homogeneous as one might suppose, as can be seen by the quantities of LDH associating with bodies of lower density filling the upper half of the gradient (Fig. 2D). This material is less dense than the main peak (at tube 22), but doubtless is of much larger size else it would not have been found in the packed portion of the second 12,000g pellet.

If we may use LDH as indicative of synaptosomal and IDH-NAD of mitochondrial localization within the gradient, several interesting observations follow. There is a main peak of synaptosomes (1 EU/ml of LDH) at tube 20 (Fig. 2D) which is contributed by the fluffy layer and is greatly reduced (tube 20, Fig. 2B) by centrifugation through med C. There is a denser

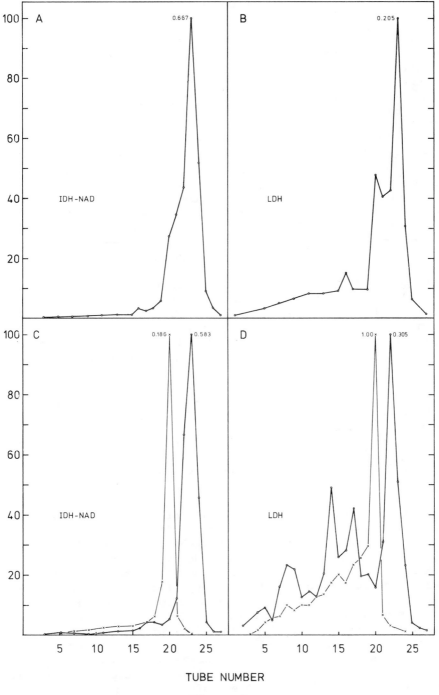

TUBE NUMBER

Figure 2

Figure 2 (Legend). A. Distribution of the mitochondrial marker enzyme IDH-NAD in a 0.8-1.7 M sucrose gradient after 18 hr of centrifugation at 25,000 rpm in a Beckman 25.1 rotor of rat cerebral hemisphere mitochondria purified through med C. Percent maximal activity (0.667 EU/ml, tube 23) vs. tube number (less to more dense, left to right). B. Same as A except that LDH was the enzyme being assayed. Peaks are at tubes 16, 20, and 23. C. Composite of separate centrifugation of the fluffy (X-X) and the packed (0-0) portions of the second 12,000g pellet of rat cerebral hemisphere mitochondria assayed for the mitochondrial marker enzyme IDH-NAD. Peak activities are at tube 20 (fluffy) and tube 23 (packed). Centrifugation was as in A. D. Same as C except that LDH was the enzyme being assayed. Peaks are at tubes 20 (fluffy) and 22 (packed).

---

peak (but of less magnitude, 0.305 EU/ml of LDH) of synaptosomes (tube 22, Fig. 2D) which is almost coincident with the mitochondrial peak at tube 23 (Fig. 2C) and which is also at least partially removed by med C. as judged by the coincidence of the LDH and IDH-NAD peaks at tube 23 (Fig. 2A and 2B). Electron microscopy has shown that many synaptosomes actually contain entrapped mitochondria (Clark and Nicklas, 1970; Salganicoff and Koeppe, 1968). It is not surprising, therefore, to find coincidence of the LDH (Fig. 2D) and IDH-NAD (Fig. 2C) peaks at tube 20. This may in fact represent a homogeneous population of synaptosomes, all of which contain entrapped mitochondria. The less dense synaptosomal peaks (Fig. 2D) found in the upper half of the gradient obviously contain no mitochondria (as judged by their lack of IDH-NAD). It is not possible to say whether the dense peaks for LDH (tube 22) and IDH-NAD (tube 23) represent partially overlapping populations of pure synaptosomes and pure mitochondria or whether some of the synaptosomes actually contain some of the mitochondria.

In interpreting the heterogeneity of brain mitochondria several facts must be noted. Salganicoff and Koeppe (1968) have reported that synaptosomal mitochondria derive from nerve endings and that free mitochondria derive from neuronal cell bodies and glial cells. When assessed separately for their enzyme contents Salganicoff and Koeppe found that seven of eleven enzymes examined were disproportionately distributed between these two kinds of mitochondria. It is no wonder, then, that Blokhuis and Veldstra (1970) find massive heterogeneity when using not just crude mitochondria, but whole brain homogenates.

Separation of the two populations of mitochondria, perhaps through several washings of the fluffy (nerve ending mitochondria) and packed (glial + neuronal mitochondria) portions of the second 12,000g pellet, would be expected to reduce the apparent extensive heterogeneity to levels more in keeping with those of liver, (Swick et al., 1967; Katyare et al., 1970) for example.

The problem of cytoplasmic contamination of the two mito-chondrial peaks remains. Using the cytoplasmic and mitochon-drial isozymes of MDH, Fox (1973) has shown that the proportion of cytoplasmic to mitochondrial forms, as judged by cellulose acetate electrophoresis, increases as samples are taken from dense to progressively less dense portions of the MDH activity peak, even when using med C-purified mitochondria. In this investigation that result is confirmed not only for MDH, but for GOT as well (Fig. 3). The activity profiles for both MDH

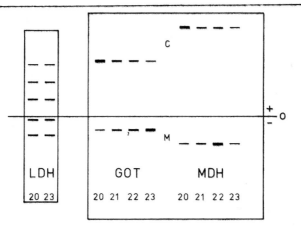

Figure 3. Interpretive composite schematic representation of cellulose acetate (left) zymogram in which samples from tubes 20 and 23 of figure 2B were electrophoresed for 1 hr at 1mA at 300v and stained for LDH activity. Interpretive composite schematic representation of starch gel zymogram (right) in which samples from tubes 20, 21, 22, and 23 were electrophoresed at 300v for 16 hr at 20mA and stained for GOT and MDH. Note that the synaptosomal (20) and mitochondrial (23) peaks yield the same LDH pattern. For both GOT and MDH the cytoplasmic isozyme stained heavier for the synaptosomal (20) peak while the mitochondrial isozyme stained heavier for the mitochondrial (23) peak.

and GOT may be seen in Fig. 1D. Not only may LDH be seen in both peaks (Fig. 2B) on the basis of activity, but all five isozymes with equal distribution of staining intensities are found in both peaks (Fig. 2B tubes 20 and 23) (see zymogram in Fig. 3). It was judged safe to use the two isozymes of MDH and GOT since rat liver mitochondria contain none of the cytoplasmic isozyme.

This is not true, however, for IDH-NADP. Table 1 shows that both rat and mouse mitochondria from heart, liver, and kidney are associated, at least, with some of the cytoplasmic form. By separating the IDH-NADP cytoplasmic and mitochondrial isozymes (Fig. 4) from centrifuged sonicates of mitochondria it was possible to estimate the percent contribution of the cytoplasmic isozyme to the total IDH-NADP activity. This enzyme

TABLE 1

Percent of total IDH-NADP activity

Mouse	Cyt.	Mito.
Heart	2.7%	97.3%
Liver	34.8%	65.2%
Kidney	15.8%	84.2%
Rat	Cyt.	Mito.
Heart	0.6%	99.4%
Liver	16.9%	83.1%
Kidney	1.5%	98.5%

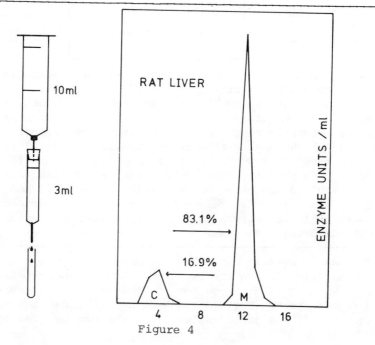

Figure 4

Figure 4 (Legend). A. Schematic representation of the CM cellulose columns used for the separation of the cytoplasmic and
mitochondrial forms of IDH-NADP. One ml of centrifuged mitochondrial sonicate was applied to the 3 ml column of CM cellulose. The 10 ml reservoir was then attached and filled first
with 5 ml of elution buffer (see M & M) followed by 10 ml of
elution buffer containing 0.25 M NaCl. B. Typical elution profile of the two IDH-NADP isozymes from CM cellulose. The one
shown is of rat liver.

---

would therefore be of little value in serving as a marker for
cytoplasmic and mitochondrial compartments. Whether the cytoplasmic isozyme is merely adherant to the outer mitochondrial
membrane or whether it exists in the intermembrane space or
even in the matrix is not known. However, as can be seen in
Fig. 5 the cytoplasmic isozyme is more easily released from
the mitochondria than is the mitochondrial form, suggesting
that it may be loosely adherant to the outer membrane or may
be located in the intermembrane space.

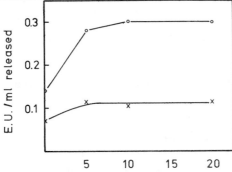

SECONDS of SONICATION

Figure 5. Med C-treated mouse liver mitochondria were resuspended in the hypotonic medium used for CM column elution and
divided into five 1 ml aliquots which received either 0, 5,
10, or 20 seconds of sonication. The membranes were removed
by centrifugation for 30 min at 30,000g and the supernatant
was applied to the CM column. Fifteen drop fractions were
collected, each was assayed for IDH-NADP activity and the relative amounts of c IDH-NADP (X-X) and total IDH-NADP (O-O)
were calculated. The results indicate that suspension of mitochondria in hypotonic medium alone is sufficient to release
most of the c IDH, whereas sonication is necessary to release
the bulk of the m IDH.

REFERENCES

Blokhuis, G. G. D., H. Veldstra 1970. Heterogeneity of mitochondria in rat brain. *FEBS-Letters* 11: 197-199.

Bridiczka, D., D. Pette 1971. Intra- and extramitochondrial isozymes of (NADP) malate dehydrogenase. *Europ. J. Biochem.* 19: 546-551.

Bullock, Gillian, E. E. Carter, and A. M. White 1971. The heterogeneity of muscle mitochondria demonstrated by discontinuous density-gradient centrifugation. *Biochem. J.* 125(4): 107.

Clark, J. B., W. J. Nicklas 1970. The metabolism of rat brain mitochondria. Preparation and characterization. *J. Biol. Chem.* 245: 4724-4731.

DeLorenzo, R. J. and F. H. Ruddle 1970. Glutamate oxalate transaminase (GOT) genetics in *Mus musculus:* Linkage, polymorphism and phenotypes of the GOT-2 and GOT-1 loci. *Biochem. Genet.* 4: 259-273.

Fox, David J. 1971. The soluble citric acid cycle enzymes of *Drosophila melanogaster.* I. genetics and ontogeny of NADP-linked isocitrate dehydrogenase. *Biochem. Genet.* 5: 69-80.

Fox, David J. 1973. Contribution of cytoplasmic enzymes to apparent heterogeneity of rat brain mitochondria. *Wilhelm Roux' Archiv* 172: 75-79.

Fox, David J., E. Abacherli 1971. The use of enzymes in the study of mitochondrial heterogeneity. *Amer. Zool.* 11: 690.

Goebell, H., M. Klingenberg 1963. DPN linked isocitrate dehydrogenase in near constant proportion to the respiratory chain. *Biochem. Biophys. Res. Commun.* 13: 213-216.

Goebell, H., M. Klingenberg 1964. DPN-spezifische Isocitrat-Dehydrogenase der Mitochondrien. *Biochem. Z.* 340: 441-446.

Hülsmann, W. C., W. G. J. Iemhoff, J. W. O. van den Berg, A. M. De Pijper 1970. Unequal rates of development of mitochondrial enzymes in rat small intestinal epithelium. *Biochim. Biophys. Acta (Amst.)* 215: 553-555.

Kadenbach, B. 1966. Synthesis of mitochondrial proteins: Demonstration of a transfer of proteins from microsomes into mitochondria. *Biochim. Biophys. Acta (Amst.)* 134: 430-442.

Katyare, S. S., P. Fatterpaker, A. Sreenivasan 1970. Heterogeneity of rat liver mitochondrial fractions and the effect of tri-iodothyronine on their protein turnover. *Biochem. J.* 118: 111-121.

Pette, D., M. Klingenberg, Th. Bücher 1962. Comparable and specific proportions in the mitochondrial enzyme activity pattern. *Biochem. Biophys. Res. Commun.* 7: 425-429.

Pycock, C. J., S. R. Nahorski 1971. The validity of mitochondrial marker enzymes in rat heart. *J. Molec. Cell. Cardiol.* 3: 229-241.

Salganicoff, L., R. E. Koeppe 1968. Subcellular distribution of pyruvate carboxylase, diphosphopyridine nucleotide and triphosphopyridine nucleotide isocitrate dehydrogenases, and malic enzyme in rat brain. *J. Biol. Chem.* 243: 3416-3420.

Swick, R. W., J. L. Strange, S. L. Nance, J. F. Thomson 1967. The heterogeneous distribution of mitochondrial enzymes in normal rat liver. *Biochem.* 6: 737-744.

ISOZYMES OF PHOSPHOENOLPYRUVATE METABOLISM

CHRIS I. POGSON*, IAN D. LONGSHAW, and DAVID M. CRISP
Department of Biochemistry
University of Bristol
The Medical School
University Walk, Bristol, U.K.

ABSTRACT. Fructose diphosphate-sensitive pyruvate kinases are associated with gluconeogenic tissues, fructose diphosphate-insensitive activities with tissues with high glycolytic flux rates. Thus, in liver, PK-L is associated exclusively with the parenchymal cell fraction, whilst PK-M activity is non-parenchymal. In adipose tissue, glycerophosphate synthesis through oxalacetate may play an important role in vivo. The properties of the interconvertible forms of adipose tissue pyruvate kinase are correlated with the presence of potentially substantial glycolytic and glyceroneogenic capacities within the same cell. The use of the relatively specific transaminase inhibitor, aminooxyacetate, as a tool for the investigation of the relative roles of phosphoenolpyruvate carboxykinase isozymes is discussed. Evidence is presented to show that a substantial part of the total gluconeogenic flux in guinea pig kidney involves the cytosolic isozyme. The possibility is discussed that the increase in gluconeogenesis in the acidotic rat kidney cortex may be associated with the presence of a higher molecular weight species of phosphoenolpyruvate carboxykinase.

INTRODUCTION

A large number of papers have been devoted to consideration of the regulation of gluconeogenesis in mammals and other organisms. It is therefore initially somewhat surprising that we have no real understanding of the molecular mechanisms involved in this process. It is known that certain reactions are candidates as the foci of control phenomena, and also that the enzymes catalyzing these reactions viz., glucose-6-phosphatase, fructose 1,6-diphosphatase, phosphoenolpyruvate carboxykinase, and pyruvate carboxylase, are, in several instances, subject to long-term regulation through changes in the rates of their synthesis and degradation. One outstanding and re-

---

* present address: Biological Laboratory
University of Kent
Canterbury, CT2 7NJ, U.K.

markable fact however is that we know little about the short-term controls on any of these systems; indeed, one is tempted to think that a single co-ordinated mechanism may be involved in the regulation at all of these loci.

Since it is probably true that tissues with gluconeogenic capacity also always contain the enzymes of glycolysis, one important characteristic of such tissues is the presence of "energetically-wasteful" metabolic cycles. Although there is now evidence that a degree of "cycling" may occur physiologically (Clark et al, 1974; Friedman et al, 1971), and may be advantageous from a control standpoint (Newsholme and Start, 1973), it is clear that the extent of this process must be subject to strict regulation.

We have been concerned over the past few years with the mechanisms involved in one of these cycles, namely, that connected with the synthesis and metabolic fate of phosphoenol-pyruvate in various mammalian tissues. It is clear from these studies that isozymic patterns may play a significant role in the eventual understanding of the underlying control mechanisms.

The Regulation of Pyruvate Kinase

a) Liver and kidney

Liver is composed of a number of cell types. The "typical" liver cell, the parenchymal cell, constitutes approximately 60% of the total cell number, and, by virtue of its large size, 90-95% of the total cell mass (disregarding the appreciable proportion of extracellular material). The residue of cellular material is made up of Kupffer and other structural cells (bile duct, anteriolar walls, etc.).

It was reported in 1967 that liver PK can be resolved electrophoretically into at least two distinct fractions (Tanaka et al, 1967). One of these, PK-M, exhibits properties apparently similar in many respects to those of the skeletal muscle enzyme; investigations of the other PK-L, revealed a sigmoid rate dependency with increasing PEP concentrations, and a striking sensitivity to activation by FDP (Carbonell et al, 1973), reminiscent of that noted earlier for the yeast enzyme (Hess et al, 1967).

Using a modification of the techniques introduced by Berry and Friend (1969), we have prepared "viable" parenchymal cells in good yield from mouse liver (Crisp and Pogson, 1972). Examination of these cells reveals that they contain exclusively PK-L. Conversely, the non-parenchymal cell fractions contain only PK-M.

Parenchymal cells also contain all liver glucose-6-phos-
phatase and glucokinase activity; very little (Werner et al,
1972), if any, hexokinase is present (Crisp and Pogson, 1972).
This is consistent with the postulate that gluconeogenesis is
characteristic of only one hepatic cell type. These findings
also shed some light on the regulation of carbohydrate utilization; one may now speculate that the specific enzymes glucokinase, fructokinase, and galactokinase are parenchymal, the
less-specific hexokinase being predominantly non-parenchymal.

In kidney the situation is more confused. Although gluconeogenesis is entirely a cortical function, there is again a
problem regarding cell heterogeneity. Thus both PEPCK (Guder,
in press) and the acidosis-responsive phosphate-dependent glutaminase (Curthoys and Lowry, 1973) are associated only with
the proximal tubular fractions. This may in part explain the
confusing number of PK activities which appear on cellulose
acetate electrophoresis of tissue extracts (Fig. 1); some at
least of these activities are FDP-sensitive (Jimenez de Asua
et al, 1971; Pogson, unpublished results).

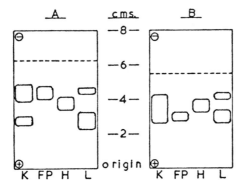

Fig. 1. *Cellulose acetate electrophoresis of mammalian PK
activities* (from Pogson, 1968a).
Activities were separated in two buffered (A - 15 mM imidazole,
0.5 mM FDP, pH 7.4; B - 15 mM imidazole, 5 mM EDTA, 10 mM
2-mercaptoethanol, pH 6.8), and were visualized by fluorescence quenching under short wave UV light following spraying
with medium containing excess lactate dehydrogenase, NADH and
appropriate substrates. K, kidney; FP, fat pad; H, heart;
L, liver.

b) Epididymal adipose tissue

Adipose tissue is unable to synthesize glucose from C_3 or
C_4 precursors. It does, however, possess a modified pathway

in that, while both glucose-6-phosphatase and fructose 1,6-diphosphatase are undetectable, both pyruvate carboxylase and PEPCK are present in appreciable amounts (Ballard et al, 1967; Wise and Ball, 1964). The function of this modified pathway is "glyceroneogenic" (Ballard et al, 1967; Reshef and Shapiro, 1970), that is, the production of L-glycerol-3-phosphate under conditions where glucose conversion to this metabolite may be limiting. Under "normal" conditions, when glucose is readily available, adipose tissue is "glycolytic", and glucose may provide the carbon skeletons of both the glycerol and fatty acid moieties of newly-synthesized triglyceride (Margolis and Vaughan, 1962). When pyruvate is the sole carbon source in vitro, glyceroneogenesis ensures that sufficient L-glycerol-3-phosphate is present to esterify at least a proportion of the fatty acid available (Reshef et al, 1970).

In comparison with liver, therefore, adipose tissue might be regarded as being more glycolytic than gluconeogenic, while still possessing a capacity for substantial synthetic flux. The properties of adipose tissue pyruvate kinase in turn differ from those of the liver enzyme (Pogson, 1968a, b; Carbonell et al, 1973) and may be a reflection of the different metabolic balance in the two tissues.

Initial investigations into the properties of rat adipose tissue PK revealed that the observed kinetic behavior towards PEP is substantially affected by the composition of the extracting medium (Table 1). When imidazole alone is used, the extracted enzyme exhibits characteristics similar to those described for muscle PK including a hyperbolic rate response to increasing PEP concentrations. When, however, a magnesium chelator (e.g. citrate, EDTA or ATP) is present, the kinetic properties resemble more closely those of the liver enzyme, e.g., a sigmoid response towards increasing PEP with a Hill coefficient of 1.75 (Pogson, 1968a, b). The two forms have been preliminarily designated PK-A (sigmoid with PEP) and PK-B (hyperbolic with PEP). FDP affects adipose tissue PK in two ways. In the first place, PK-A resembles liver PK-L in that it is activated by FDP, although both the apparent affinity and the degree of activation are lower than in the liver. Secondly, incubation of PK-A with FDP at low concentrations (5×10^{-5} M) leads to conversion to PK-B (Table 2). This is opposed by EDTA and increased by Mg^{2+}, although this ion alone has no effect on the interconversion. This is consistent with a mechanism involving the binding of FDP to the enzyme either directly as its magnesium chelate or as part of a ternary system with Mg^{2+} as the third partner.

Both PK-A and PK-B are inhibited by ATP and activated by K^+ and NH_4^+. The kinetic constants for these ligands appear to be similar for both forms of the enzyme (Table 3).

The two forms are clearly differentiable by their electrophoretic behavior (Pogson, 1968a). PK-B activity shows a mobility similar to that of the muscle enzyme, while PK-A

TABLE 1

EFFECT OF EXTRACTION MEDIUM ON ADIPOSE
TISSUE PYRUVATE KINASE ACTIVITY

The basal extraction medium contained 20 mM-imidazole, pH
6.8, with additions as indicated. Activities were measured
with 2.5×10^{-4} M and either 3×10^{-5} M (*low*) or 10^{-3} M (*high*)
PEP.

Additions to medium	Ratio of activities (high/low)
None	3.1
10 mM 2-mercaptoethanol	3.4
5 mM EDTA	52
5 mM EDTA, 10 mM 2-mercaptoethanol	52
5 mM EDTA, 20 mM $MgCl_2$	6.4
5 mM EGTA	3.6
10 mM Citrate	45

once more resembles liver PK-L (Fig. 1). Using the ultracentrifugal technique devised by Yphantis (Yphantis and Waugh,
1956), we were able to show that the two forms also differ
with respect to their apparent sedimentation coefficients,
PK-A sedimenting at 5.3 - 5.6S and PK-B at 7.2 - 7.3S (Pogson,
1968a). These compare with the values of 7.2S for the possibly
related red cell enzyme (Koler et al, 1964).

In contrast to liver PK-L, adipose tissue total PK activity
is not affected by either acute alloxan-diabetes or 3-day
starvation. The activity is increased, however, from 0.45 ±
0.03 to 0.66 ± 0.06 units/mg protein in rats fed a high carbohydrate diet (Pogson and Denton, 1967).

The presence in vitro of two interconvertible forms of adipose tissue PK suggests that such interconversion may be of
physiological import. A muscle-type PK activity is clearly
"advantageous". Where glucose is in short supply, however,
and the dicarboxylate shuttle becomes active, then it is important that flux through PK be severely curtailed to avoid
"cycling". This could be achieved by the presence of a high
Km form of the enzyme, such as PK-A. The relative physiological concentrations of PEP and FDP (Halperin and Denton,
1969) and the Ka for the latter are consistent with such a
role.

At present FDP and ATP are the only metabolites known to
affect the interconversion of the two forms at physiological
concentrations. Although these are possible antagonists, it

TABLE 2

EFFECT OF INCUBATION MEDIUM COMPOSITION ON THE ACTIVITY OF ADIPOSE TISSUE PYRUVATE KINASE

The percentage of PK-A was calculated from the activities measured at two PEP concentrations as described by Pogson (1968b).

Extraction Medium	Additions	Time at 34° (min)	% PK-A
20 mM - imidazole pH 6.8 } 5 mM - EDTA } 10 mM - 2-mercaptoethanol }	-	0	99
"	-	10	91
"	16.7 mM EDTA	10	99
"	0.5 mM FDP	10	1
"	0.05 mM FDP	10	33
"	0.05 mM FDP, 33.3 mM EDTA	10	80
20 mM - imidazole pH 6.8 (preincubated for 10 min)	-	0	64
"	-	10	76
"	0.05 mM EDTA	10	88
"	16.7 mM EDTA	10	98
"	0.05 mM FDP	10	28
"	0.05 mM FDP, 16.7 mM EDTA	10	58

TABLE 3

KINETIC CONSTANTS OF PK ACTIVITIES IN MAMMALIAN TISSUES

Data from Pogson (1968b)

	Muscle	Liver	Adipose Tissue	
	PK-M	PK-L	PK-A	PK-B
Km (mM)				
PEP	0.075	0.84	0.6	0.067
ADP	0.27	0.1	0.33	0.67
Ka (mM)				
Mg^{2+}	0.45		2.3 - 4.3	1.4 - 2.7
K^+	11		20-50	10-25
NH_4^+	11		9-34	4-10
FDP	-	0.001	0.03	-
Ki (mM)				
ATP	0.15		1.5	1.5

is unlikely that ATP levels vary significantly in this tissue (Halperin and Denton, 1969). There is evidence, however, that the interconversion may be an in vivo phenomenon. Thus, although there is no overall change in enzyme activity during starvation, the extracted enzyme (in imidazole alone) contains only 10% PK-B in contrast to that from control animals where the PK-B content, although variable, is consistently higher (20-70%).

PEPCK Isozymes

The livers of very many of a growing number of species studied contain two PEPCK activities. One of these is mitochondrial, the other cytoplasmic. The properties of the two enzymes differ markedly, however, from one species to another (Hanson and Garber, 1973; Table 4), and also vary within one species according to the developmental and dietary status of the animal. This latter arises from the relative "inducibility" of cytoplasmic PEPCKs when compared with the mitochondrial enzymes. In species where the PEPCK activity in both cellular compartments is appreciable, the following questions arise: (1) Do both enzymes contribute simultaneously to the total gluconeogenic flux? (2) Do both enzymes contribute to gluconeogenic flux, but not act simultaneously i.e. is one pre-

TABLE 4

INTRACELLULAR DISTRIBUTION OF *PEPCK* IN VARIOUS SPECIES

Species	% of total	
	Mitochondria	Cytosol
rat } mouse } hamster }	5	95
guinea pig	50	50
man	60	40
cow	50	50
sheep	35	65
rabbit } pigeon }	95	5

dominant under certain dietary or hormonal states, and the other under different conditions?   (3) Does only one of the enzymes function as a component of the gluconeogenic sequence? If so, what is the role of the other, and can any explanation be readily adduced to account for the inter-specific variations in enzyme distribution?

While it is at present difficult to give categorical answers to all of these points, it is now possible at least to clarify the situation by ruling out certain alternatives.

We have investigated the effects of the inhibitor amino-oxyacetate (AOA) on glucose synthesis in kidney cortex slices (Longshaw et al, 1972). AOA, a derivative of hydroxylamine, is a relatively specific inhibitor of pyridoxal phosphate-dependent enzymes (Roberts et al, 1964; Kun et al, 1964). It is to be expected therefore that gluconeogenic sequences involving transamination should be inhibited by AOA, while those independent of transamination should be unaffected. In the rat where PEPCK activity is essentially all cytosolic, gluconeogenesis from pyruvate, citric acid cycle and equivalent substrates involves transport of reducing equivalents from the mitochondrian to the cytosol via the malate porter system. With lactate, cytosolic NADH is generated directly and oxalacetate transport involves the transaminase-dependent porter (Fig. 2). Thus AOA inhibition should be demonstrable

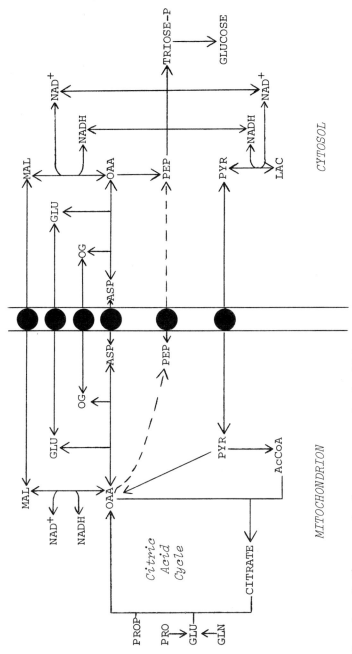

Fig. 2. *Pathways of gluconeogenesis in kidney cortex.*
The dotted line indicates carbon flow where PEPCK activity is mitochondrial. Details of mitochondrial porter systems have been omitted for the sake of clarity. AcCoA, acetyl coenzyme A; ASP, aspartate; GLN, glutamine; GLU, glutamate; LAC, lactate; MAL, malate; OAA, oxalacetate; OG, oxoglutarate; PRO, proline; PROP, propionate; PYR, pyruvate.

when lactate is provided as a gluconeogenic carbon source; this has been shown to be true (Rognstad and Katz, 1970).

In the contrasting case, where PEPCK is exclusively mito-chondrial, PEP may be transported from mitochondrial to cyto-sol in exchange for citrate or malate (Robinson, 1971a, b); lactate gluconeogenesis should thus be insensitive to AOA. Where the substrate is relatively oxidized (i.e. pyruvate or alanine), or is metabolized directly in the mitochondrion (e.g. citric acid cycle intermediates, glutamate), the trans-port of reducing equivalents to the cytosol via the malate transporter results in a superfluity of oxalacetate in the cytosol; this can return to the mitochondrion, and complete the cycle via the aspartate system, involving an AOA-sensitive step. It is clear therefore that the pattern of AOA inhibition of glucose synthesis from a range of substrates may give an indication as to which of the isozymes of PEPCK is involved. Using this technique we have confirmed earlier reports with rat kidney cortex slices, and have shown that slightly lower but comparable inhibition by AOA of lactate gluconeogenesis exists in guinea pig kidney cortex. We could find no evi-dence of inhibition when other substrates were added (Table 5). These findings indicate that, under the prevailing experimental conditions, cytosolic PEPCK plays a dominant role in guinea pig kidney gluconeogenesis. Although it is possible that the mitochondrial enzyme plays some part in the overall process, it appears that the level of its activity cannot be greatly increased to compensate for the inhibition of the "cytoplasmic PEPCK" pathway.

It has recently been elegantly demonstrated that the effec-tiveness of AOA as an inhibitor of lactate gluconeogenesis in guinea pig liver is dependent on the prevailing intra-mito-chondrial redox state (Arinze et al, 1973). These experiments are consistent with the thesis that both PEPCK isozymes play a role in gluconeogenesis, the proportional contribution of each being controlled by dietary and hormonal considerations. In preliminary experiments with guinea pig kidney cortex slices, we find that addition of acetoacetate and β-hydroxy-butyrate in varying proportions does not affect the response towards AOA (N. Stamper and C. I. Pogson, unpublished results). These results do not, however, necessarily conflict with the liver data, but may indicate that different control mechanisms are effective in the two tissues.

Gluconeogenesis in rat kidney cortex is increased during metabolic acidosis (Goodman et al, 1966). Where acidosis is prolonged (i.e. for 6 hr or more), this increase may be cor-related with an elevation of total assayable PEPCK activity (Alleyne and Scullard, 1969). Immunotitration data have con-

TABLE 5

EFFECT OF *AOA* ON GLUCONEOGENESIS IN
RAT AND GUINEA PIG KIDNEY CORTEX

Rats were starved overnight; guinea pigs were allowed to feed *ad libitum*. All substrates were 10 mM; AOA was 0.1 mM. Rates of gluconeogenesis are expressed as nmoles/hr/mg dry wt with the number of observations in parentheses.

Substrate	AOA	Glucose production	
		Rat	Guinea pig
L-Lactate	−	65.8 ± 4.5(5)	23.0 ± 3.9(5)
	+	9.2 ± 1.8(5)*	6.4 ± 0.7(5)*
Pyruvate	−	169.3 ± 9.8(6)	30.8 ± 3.3(5)
	+	167.8 ± 5.5(6)	31.1 ± 4.5(5)
L-Glutamate	−	60.7 ± 5.5(6)	
	+	64.0 ± 4.9(6)	
L-Malate	−		53.6 ± 7.5(5)
	+		53.6 ± 6.0(5)
Succinate	−		74.5 ± 4.1(5)
	+		82.0 ± 10.3(5)

* P < 0.01. All other differences insignificant.

firmed that this rise is due to increased PEPCK protein rather than activation of existing enzyme (Longshaw and Pogson, 1972). It is clear, however, that glucose synthesis begins to increase at a time before that at which PEPCK "induction" is detectable (Alleyne, 1970). Metabolite measurements indicate that PEPCK may be the rate-limiting enzyme of gluconeogenesis under these conditions (Alleyne, 1968; Hems and Brosnan, 1971), implying that the observed increases in flux through this step are due to some form of activation of existing enzyme.

To date, no simple physiological effectors of PEPCK have been discovered. We have therefore investigated the possibility of different molecular forms of PEPCK. The molecular weight of the kidney enzyme in supernatant fractions is approximately 60,000 as determined from gel filtration data; sucrose density gradient fractionation gives a value of 67,000 for "normal" kidney PEPCK. These values are close to those reported earlier for the rat liver enzyme (Ballard and Hanson, 1969; Chang and Lane, 1966). Ultracentrifugal data with undiluted cytosol (Amberson et al, 1964), again employing the moving partition cell (Yphantis and Waugh, 1956), and measurement of appropriate marker enzymes reveals that the molecular weights of "control" and "acidotic" PEPCKs are 83,000 and 128,000 respectively (Fig. 3, Longshaw and Pogson, 1972). These data are consistent with the presence of two forms of the enzyme which may differ in catalytic activity. Similar measurements with diluted extracts agree with the lower molecular weights determined by other techniques. The mechanisms regulating the possible interconversion of PEPCK activities are currently under study.

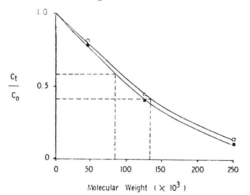

Fig. 3. Typical partition cell distribution pattern obtained with undiluted extracts of acidotic (open symbols) and normal (closed symbols) rat kidney cortices. The molecular weights of the markers were as follows: catalase 247,000; lactate dehydrogenase 126,000; and triosephosphate isomerase 50,000. Dotted lines indicate C_t/C_o ratios (ratio of enzyme activity in top compartment to that in the original extract) obtained for PEPCK activities in the two cases (from Longshaw and Pogson, 1972).

ABBREVIATIONS

AOA, amino-oxyacetate; FDP, fructose 1,6-diphosphate; PEP, phosphoenolpyruvate; PEPCK, phosphoenolpyruvate carboxykinase (EC. 4.1.1.32); PK, pyruvate kinase (EC. 2.7.1.40).

ACKNOWLEDGMENTS

This work was supported by grants from the British Diabetic Association and the Medical Research Council to Professor P.J. Randle, whom we also thank for his interest and encouragement throughout.

REFERENCES

Alleyne, G. A. O. 1968. Concentrations of metabolic intermediates in kidneys of rats with metabolic acidosis. *Nature* 217: 847-848.

Alleyne, G. A. O. 1970. Renal metabolic response to acid-base changes. II. The early effects of metabolic acidosis on renal metabolism in the rat. *J. Clin. Invest.* 49: 943-951.

Alleyne, G. A. O. and G. H. Scullard 1969. Renal metabolic response to acid-base changes. I. Enzymatic control of ammoniagenesis in the rat. *J. Clin. Invest.* 48: 364-370.

Amberson, W. R., A. C. Bauer, D. E. Philpott, and F. Roisen 1964. Proteins and enzyme activities of press juices obtained by ultracentrifugation of white, red and heart muscles of the rabbit. *J. Cell. Comp. Physiol.* 63: 7-21.

Arinze, I. J., A. J. Garber and R. W. Hanson 1973. The regulation of gluconeogenesis in mammalian liver: the role of mitochondrial phosphoenolpyruvate carboxykinase. *J. Biol. Chem.* 248: 2266-2274.

Ballard, F. J. and R. W. Hanson 1969. Purification of phosphoenolpyruvate carboxykinase from the cytosol fraction of rat liver, and the immunochemical demonstration of differences between this enzyme and the mitochondrial phosphoenolpyruvate carboxykinase. *J. Biol. Chem.* 244: 5625-5630.

Ballard, F. J., R. W. Hanson, and G. A. Leveille 1967. Phosphoenolpyruvate carboxykinase and the synthesis of glyceride-glycerol from pyruvate in adipose tissue. *J. Biol. Chem.* 242: 2746-2750.

Berry, M. N. and D. S. Friend 1969. High yield preparation of isolated rat liver parenchymal cells: A biochemical and fine structural study. *J. Cell. Biol.* 43: 506-520.

Carbonell, J., J. E. Feliu, R. Marco, and A. Sols 1973. Pyruvate kinase. Classes of regulatory isoenzymes in mammalian tissues. *Eur. J. Biochem.* 37: 148-156.

Chang, H. C. and M. D. Lane 1966. The enzymatic carboxylation of phosphoenolpyruvate. II. Purification and properties of liver mitochondrial phosphoenolpyruvate carboxykinase. *J. Biol. Chem.* 241: 2413-2420.

Clark, M. G., D. P. Bloxham, P. C. Holland, and H. A. Lardy 1974. Estimation of the fructose 1,6-diphosphatase-phosphofructokinase substrate cycle and its relationship to gluconeogenesis in rat liver in vivo. *J. Biol. Chem.* 249: 279-290.

Crisp, D. M. and C. I. Pogson 1972. Glycolytic and gluconeogenic enzyme activities in parenchymal and non-parenchymal cells from mouse liver. *Biochem. J.* 126: 1009-1023.

Curthoys, N. P. and O. H. Lowry 1973. The distribution of glutaminase isoenzymes in the various structures of the nephren in normal, acidotic, and alkalotic rat kidney. *J. Biol. Chem.* 248: 162-168.

Friedman, B., E. H. Goodman, Jr., H. L. Saunders, V. Kostos, and S. Weinhouse 1971. Estimation of pyruvate recycling during gluconeogenesis in perfused rat liver. *Metabolism* 20: 2-12.

Goodman, A. D., R. E. Fuisz, and G. F. Cahill, Jr. 1966. Renal gluconeogenesis in acidosis, alkalosis and potassium deficiency; its possible role in regulation of renal ammonia production. *J. Clin. Invest.* 45: 612-619.

Halperin, M. L. and R. M. Denton 1969. Regulation of glycolysis and L-glycerol 3-phosphate concentration in rat epididymal adipose tissue in vitro; role of phosphofructokinase. *Biochem. J.* 113: 207-214.

Hanson, R. W. and A. J. Garber 1973. Phosphoenolpyruvate carboxykinase. I. Its role in gluconeogenesis. *Amer. J. Clin. Nutr.* 25: 1010-1021.

Hems, D. A. and J. T. Brosnan 1971. Effects of metabolic acidosis and starvation on the content of intermediary metabolites in rat kidney. *Biochem. J.* 123: 391-397.

Hess, B., R. Haeckel, and K. Brand 1967. FDP activation of yeast pyruvate kinase. *Biochem. Biophys. Res. Commun.* 24: 824-831.

Jimenez de Asua, L., E. Rozengurt, and H. Carmmiath 1971. Two different forms of pyruvate kinase in rat kidney cortex. *FEBS Lett.* 14: 22-24.

Koler, R. D., R. H. Bigley, R. T. Jones, D. A. Rigas, P. van Bellinghen, and P. Thompson 1964. Pyruvate kinase: Molecular differences between human red cell and leucocyte enzyme. *Cold Spring Harbor Symp. Quant. Biol.* 29: 213-221.

Kun, E., J. E. Ayling, and B. G. Baltimore 1964. Studies on specific enzyme inhibitors. VIlI. Enzyme-regulatory mechanisms of the entry of glutamic acid into metabolic pathways in kidney tissue. *J. Biol. Chem.* 239: 2896-2904.

Longshaw, I. D. and C. I. Pogson, 1972. The effects of steroids and ammonium chloride acidosis on phosphoenolpyruvate carboxykinase in rat kidney cortex. I. Differentiation of the inductive process and characterization of enzyme activities. *J. Clin. Invest.* 51: 2277-2283.

Longshaw, I. D., N. L. Bowen, and C. I. Pogson 1972. The pathway of gluconeogenesis in the cortex of guinea pig kidney: use of amino-oxyacetate as a transaminase inhibitor. *Eur. J. Biochem.* 25: 366-371.

Margolis, S. and M. Vaughan 1962. α-Glycerophosphate synthesis and breakdown in homogenates of adipose tissue. *J. Biol. Chem.* 237: 44-48.

Newsholme, E. A. and C. Start 1973. *Regulation of Metabolism.* John Wiley & Sons, London and New York.

Pogson, C. I. 1968a. Two interconvertible forms of pyruvate kinase in adipose tissue. *Biochem. Biophys. Res. Commun.* 30: 297-302.

Pogson, C. I. 1968b. Adipose tissue pyruvate kinase. Properties and interconversion of two active forms. *Biochem. J.* 110: 67-77.

Pogson, C. I. and R. M. Denton 1967. Effect of alloxan diabetes, starvation and refeeding on glycolytic kinase activities in rat epididymal adipose tissue. *Nature* 216: 156-157.

Reshef, L., R. W. Hanson, and F. J. Ballard, 1970. Possible physiological role for glyceroneogenesis in rat adipose tissue. *J. Biol. Chem.* 245: 5979-5984.

Reshef, L. and B. Shapiro 1970. The physiological function and regulation of glyceroneogenesis in adipose tissue. in *Adipose Tissue: Regulation and Metabolic Functions* (Jeanrenaud and Hepp, eds). Academic Press, New York.

Roberts, E., J. Wein, and D. G. Simonsen 1964. γ-Aminobutyric acid (γABA), vitamin B_6 and neuronal function - A speculative synthesis. *Vitams. Hormones.* 22: 503-559.

Robinson, B. H. 1971a. Transport of phosphoenolpyruvate by the tricarboxylate-transporting system in mammalian mitochondria. *FEBS Lett.* 14: 309-312.

Robinson, B. H. 1971b. Role of the tricarboxylate-transporting system in the production of phosphoenolpyruvate by ox liver mitochondria. *FEBS Lett.* 16: 267-271.

Rognstad, R. and J. Katz 1970. Gluconeogenesis in the kidney cortex: Effects of D-malate and amino-oxyacetate. *Biochem. J.* 116: 483-491.

Tanaka, T., Y. Harano, F. Sue, and H. Morimura 1967. Crystallization, characterization and metabolic regulation of two types of pyruvate kinase isolated from rat tissue. *J. Biochem.* 62: 71-91.

Werner, H. V., J. C. Bartley, and M. N. Berry 1972. Glucose-adenosine 5'-triphosphate 6-phosphotransferases of isolated rat liver parenchymal cells. *Biochem. J.* 130: 1153-1155.

Wise, E. M. and E. G. Ball 1964. Malic enzyme and lipogenesis. *Proc. Natl. Acad. Sci.* 52: 1255-1262.

Yphantis, D. A. and D. F. Waugh 1956. Ultracentrifugal characterization by direct measurement of activity. II. Experimental. *J. Phys. Chem.* 60: 630-636.

POSSIBLE ROLE FOR SUPEROXIDE AND SUPEROXIDE DISMUTASE IN G6PD TYPE HEMOLYTIC ANEMIA

ROSANNE M. LEIPZIG
GEORGE J. BREWER
FRED J. OELSHLEGEL JR.
Department of Human Genetics
University of Michigan Medical School
1137 E. Catherine
Ann Arbor, Michigan 48104

ABSTRACT. An enzyme which has been studied extensively by electrophoresis under the various names of achromatic regions, indophenol oxidase, and tetrazolium oxidase, is now known to be superoxide dismutase (SOD). The SOD enzyme catalyzes the conversion of the superoxide radical (O_2^-) to hydrogen peroxide (H_2O_2). In this paper we have investigated the possibility that O_2^- and the H_2O_2 produced from it by SOD may be involved in the mechanism of glucose-6-phosphate dehydrogenase (G6PD) deficiency type hemolysis.

Deficiency of G6PD results in inadequate reduction of NADP to NADPH, which in turn leads to inadequate maintenance of reduced glutathione (GSH). The levels of GSH are maintained through glutathione reductase, with NADPH as a co-factor. The inability to reduce NADP and maintain adequate levels of GSH make the red cell susceptible to oxidant injury.

Our studies of the possible involvement of O_2^- in the mechanism of G6PD type hemolysis include the following: 1) O_2^- is capable of directly oxidizing NADPH and NADH in a buffer system, and it is more reactive than peroxide in effecting this oxidation. 2). O_2^- generated outside intact red cells results in decreased GSH levels in G6PD deficient but not in normal red cells. 3). Three drugs, menadione, APH, and hydroquinone, hemolytic to G6PD deficient cells, all produce O_2^- in vitro.

From this work we hypothesize that drugs and other factors such as infection and acidosis which lead to hemolysis in G6PD deficiency may do this by producing O_2^-. The O_2^- produced may damage the G6PD deficient red cell by virtue of direct oxidation of NADPH and NADH, or by dismutation through SOD to H_2O_2 which can then adversely affect GSH through glutathione peroxidase.

BACKGROUND

ACHROMATIC REGIONS, SUPEROXIDE, AND SUPEROXIDE DISMUTASE (SOD)

Our laboratory has been very involved for years in starch gel electrophoresis, and has studied thousands of human hemolysates in which a number of enzymes have been stained with tetrazolium-linked systems, Along with the bluish bands indicating the presence of the enzymes being studied, light bands, or achromatic regions, have constantly been observed. These bands are seen not only with human hemolysates but are also seen with extracts from other human tissues and tissues from other species (Brewer 1967).

The banding pattern for these achromatic regions were always constant in the numerous human hemolysates tested, until in 1967 we discovered a variation in the pattern in the hemolysate from one patient. We then began to study the protein responsible for these bands. It was soon seen that neither substrates nor co-enzymes were needed during staining to induce the appearance of these bands. A system consisting of phenazine methosulfate (PMS), Nitroblue tetrazolium (NBT), and exposure of the gel to light caused the gel to develop a bluish background containing the light bands.

Upon the death of the patient mentioned above with the variant achromatic pattern, tissues were obtained from the liver, heart, kidneys, skeletal muscle, skin, and intestines. These tissues, along with control tissues for each organ obtained during autopsy of an accident victim, were subjected to electrophoresis. In these non-erythrocytic tissues two banding regions were observed, one migrating anodally and identical to that seen in the hemolysates (designated region A) and one nearer the origin (designated region B). The pattern exhibited in region B was similar for both the patient and the control. Region A, however, showed the same variant pattern in all the patient's tissues as was present in his red cells, whereas region A of the control tissues showed the pattern seen in normal hemolysates. It therefore appears that non-erythrocytic tissues have an additional protein(s), not present in hemolysates, capable of generating achromatic regions. These two regions, A and B, appear to be controlled by different genes, since genetically determined variation in region A did not produce any variation in the bands of region B. It can be concluded that the subunit producing the variation in region A is not shared by the proteins of region B. Through family studies of our patient we have determined that the subunit producing the variants in region A is determined by an autosomal gene (Brewer, 1967).

In our early work we speculated that the bluish background seen in the system described (PMS, NBT, and light) was due to the reduction of the NBT catalyzed by light. The achromatic regions were thought to indicate the presence of a protein capable of either protecting the NBT from reduction in the presence of light and PMS or a protein capable of reoxidizing the reduced NBT (Brewer 1967). The first of these turned out to be the correct explanation. The protein was classified as an indophenol oxidase and has been called by others tetrazolium oxidase.

Lippit and Fridovich (1973) subsequently theorized that photo-reduction of PMS causes, upon reoxidation, the formation of superoxide radicals (O_2^-). These radicals cause the reduction of tetrazolium to blue formazan. This reaction can be prevented by the action of superoxide dismutase (SOD), which scavenges O_2^- and produces hydrogen peroxide (H_2O_2) according to the reaction:

$$2H + O_2^- + O_2^- \xrightarrow{\quad SOD \quad} H_2O_2 + O_2$$

Lippit and Fridovich have shown that the achromatic region enzyme described by us is identical in electrophoretic mobility and staining properties to SOD. We have confirmed this identity (Brewer 1974). Lippit and Fridovich also found that this enzyme does not produce achromatic regions in the presence of reduced NBT, and therefore must act to protect the tetrazolium from reduction by scavenging the $O_2^·$ produced during reoxidation of photo-reduced PMS.

Superoxide, as a free radical of molecular oxygen, is capable of acting as either a reducing agent or as an oxidizing agent. Superoxide reduces NBT, tetranitromethane, and cytochrome c, while oxidizing epinephrine (McCord and Fridovich 1969A; Beauchamp and Fridovich 1971). The addition of SOD to any reaction containing one of these compounds and O_2^- inhibits that portion of the redox reaction due to O_2^-. It is therefore possible to use any of these compounds in a quantitative assay for O_2^- by comparing the amount of reduction (oxidation) in the absence of SOD to the amount of reduction (oxidation) in the presence of SOD.

Several systems have been shown to generate superoxide in vitro. Xanthine oxidase (XO), a flavin-containing enzyme, causes a univalent reduction of oxygen to O_2^- (McCord and Fridovich 1969A). The auto-oxidation of sulfite to sulfate (McCord and Fridovich 1969B), and the exposure of riboflavin to light in the presence of oxygen (Massey 1969), have also been shown to generate O_2^-. Superoxide is also available commercially in the form of a potassium salt.

The sites of in vivo O_2^- production are not yet well defined. The production of O_2^- has been demonstrated in phagocytizing and resting leukocytes (Babior 1973). It is also formed during the auto-oxidation of hemoglobin to methemoglobin

(Misra 1972; Wever 1973). It is most likely that O_2^- is also generated in vivo during the normal activity of enzymes such as XO, NADPH Oxidase (Curnette 1974), and cytochrome P450 (Strobel 1971).

GLUCOSE-6-PHOSPHATE DEHYDROGENASE (G6PD) DEFICIENCY

A hemolytic anemia due to G6PD deficiency is characterized by the sensitivity of G6PD deficient erythrocytes to a variety of diverse agents (Carson 1956). The classical precipitating agents are drugs, including the 8-amino-quinolines, the nitro-furans, menadione, hydroquinone, acetylphenylhydrazine (APH) and many others, but fever, infection (Mengel 1967), and acidosis (Gant 1961) have also been implicated in causing hemolytic reactions in G6PD deficient individuals. The G6PD deficient red cells exposed to a hemolytic agent undergo large decreases in their reduced glutathione (GSH) levels, and Heinz bodies are formed within the cells. The activity of G6PD is greater in young cells than in old, and a greater tendency toward hemolysis is seen among older cell populations (Piomelli 1968). The gene determining G6PD is located on the X chromosome.
The mechanism by which G6PD deficiency leads to hemolytic anemia in the presence of these various precipitating agents is uncertain. Cells which are G6PD deficient have a decreased ability to regenerate NADPH, which is important in the conversion of oxidized glutathione (GSSG) to GSH. The G6PD deficient red cells have a lower than normal content of GSH (Beutler 1955), which decreases further upon exposure to various hemolytic agents (Flanagan 1958). Reduced glutathione constitutes the major source of free sulfhydryl groups in red cells and is possibly important in protecting sulfydryl dependent enzymes, hemoglobin, and other cellular constituents against oxidative damage. It has been suggested that a diminution in the content of GSH in G6PD deficient red cells during administration of a hemolytic agent may result in inactivation of one or more vital sulfhydryl-dependent glycolytic enzymes, leading to a decreased production of energy and subsequent lysis of the cell (Mohler 1961). It has also been suggested that membrane oxidation might result due to the lack of NADPH and GSH. Cohen and Hochstein (1964) have shown that several of the hemolytic drugs produce H_2O_2. Hydrogen peroxide, spontaneously and in a reaction catalyzed by glutathione peroxidase, oxidases GSH, and therefore might be an agent responsible for hemolysis. Although several possibilities such as the above have been suggested, the actual mechanism of hemolysis remains unclear.
We have hypothesized that O_2^- might be involved in the

mechanism of G6PD deficient type hemolysis (Leipzig 1974). In the remainder of this paper we present data bearing on this hypothesis. First, we have looked at the production of O_2^- by certain known hemolytic drugs. Second, we have looked at the oxidation of NADPH and NADH by O_2^-. Third, we have looked at the effect of O_2^- on GSH levels of intact normal and G6PD deficient red cells.

MATERIALS AND METHODS

All chemicals except SOD were obtained from Sigma Chemical Co., St. Louis, Mo.

All optical densities were recorded on a Gilford Model 2400-S Recording Spectrophotometer.

Superoxide dismutase was obtained through the generosity of Dr. Donald Hultquist (Reed et al., 1970).

ESTIMATES OF SUPEROXIDE PRODUCTION

Superoxide production in whole blood was estimated by comparing NBT reduction in the absence and presence of SOD. Reduction of NBT in the absence of SOD was measured in a 4.6 ml solution containing in final concentration, 2.7×10^{-4}M NBT, 3.0 mls whole blood, and 0.6 ml of 0.5 M Tris-0.145M NaCl buffer. Reduction of NBT in the presence of SOD was measured using the same solution and replacing the 0.6 ml buffer with 0.1 ml SOD and 0.5 ml buffer. After incubation for 30 minutes at 37°C each tube was spun in a Sorvall Refrigerated Centrifuge RCB-2 for 6 minutes at 12,000 rpm, and the supernatant diluted 1:5 in the 0.5 M Tris-NaCl buffer. Reduced NBT was determined by absorbance at 595 nm.

Whole blood O_2^- production in the presence of menadione and hydroquinone was assayed using the same system as above. Reduction of NBT by whole blood, whole blood plus drug, and whole blood, drug, and SOD were compared. The 0.5 ml of buffer was replaced with the drug to yield a final concentration in the cuvette of 1×10^{-3} M.

Thus far we have been unable to measure the effect of APH on producing O_2^- in whole blood because of technical difficulties. However, we have utilized a buffer system to demonstrate O_2^- production by APH. In this system, reduction of NBT by APH was measured by the addition of 5 mg/ml APH to 3.0 mls of 0.05 M Tris buffer containing a final concentration of 2.7×10^{-4}M NBT. This was incubated for 3 hrs. at 25°C and the absorbance determined at 515 nm. Reduction of NBT in the presence of SOD was measured using the same system and replacing 0.1 ml buffer with 0.1 ml SOD.

OXIDATION OF PYRIDINE NUCLEOTIDES

To study the effect of varying concentrations of H_2O_2 on NADPH, solutions containing final concentrations of 0.74 mM NADPH and 1.4 M H_2O_2, 0.14 M H_2O_2, 0.014 M H_2O_2, or 0.14 mM H_2O_2 were made up in 0.5 M Tris, pH 7.2. These solutions were incubated for 30 min. at 25°C.

To study the effect of O_2^- on NADPH or NADH we generated O_2^- using 27.5 mM xanthine ($\bar{X}$), .0162 units XO, and .121 mM NADPH (or .121 mM NADH) in 3.1 mls. final volume 0.1 M phosphate buffer, pH 7.4. The change in absorbance at 340 nm was recorded over 35 min. at 30°C. To measure the amount of inhibition by SOD of NADPH oxidation by O_2^-, 0.1 ml SOD was substituted for 0.1 ml buffer.

To examine the effect of O_2^- produced outside whole cells on GSH levels within red cells, O_2^- was generated in a 5.13 ml solution containing 4 ml whole blood, 0.584 M X, and 0.0216 units XO. Incubation was carried out at 37°C for 2 hours. To test if this could be inhibited by SOD and/or catalase, 100 λ SOD and/or 320 units of catalase were added. Glutathione was assayed according to the method of Prins and Loos (1969).

RESULTS

HEMOLYTIC DRUGS AND SUPEROXIDE PRODUCTION.

Menadione and hydroquinone were added to whole blood and O_2^- production measured as a function of the inhibition of NBT reduction in the presence of SOD (Figure 1A). Whole blood reduction of NBT was inhibited 55% in the presence of SOD (left hand set of bars in Figure 1A). Menadione added to whole blood increased NBT reduction by 20% and 58.6% of the total NBT reduction was inhibited by SOD.

Acetylphenylhydrazine in a buffer system (Figure 1B) caused a twenty fold increase in NBT reduction over the buffer control, and 79% of this reduction was inhibitable by SOD.

OXIDATION OF PYRIDINE NUCLEOTIDES

We have found that O_2^- generated in a buffer system is capable of directly oxidizing both NADH and NADPH (Figure 2). This system has a calculated potential for O_2^- generation of 0.567 μ moles based upon the activity of XO. Under these conditions .0063 μ moles of NADPH were oxidized. This

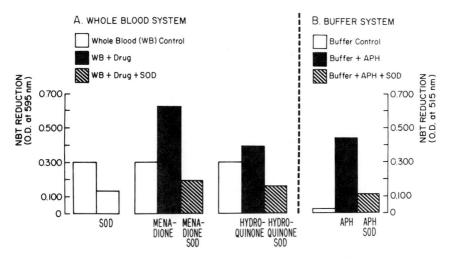

Fig. 1. This figure shows the production of O_2^- by whole blood and by G6PD deficiency type hemolytic drugs. On the far left, the comparison of NBT reduction in whole blood in the presence and absence of SOD is presented. APH was studied in a buffer system rather than whole blood because of technical difficulties. All blood samples were from non-G6PD deficient individuals.

oxidation was 92% inhibitable in the presence of SOD. The same amount of O_2^- caused the oxidation of .004 µ moles of NADH and this reaction was totally inhibited in the presence of SOD (Figure 2).

The direct oxidation of NADPH by H_2O_2 over a comparable time period was also studied. Table 1 demonstrates that H_2O_2 is capable of oxidizing NADPH; however, amounts of H_2O_2 comparable to those of O_2^- generated previously (Figure 2), are unable to oxidize NADPH.

ACTION OF SUPEROXIDE ON NORMAL AND G6PD DEFICIENT RED CELLS

Superoxide generated outside whole cells caused a decrease in the GSH levels of G6PD deficient red cells of 31-39%, whereas normal cells were unaffected (Figure 3). These results were not affected by the addition of SOD or catalase.

DISCUSSION

Glucose-6-phosphate dehydrogenase is responsible for reducing NADP to NADPH in the red cell. The NADPH produced maintains GSH in the reduced form in a reaction catalyzed by

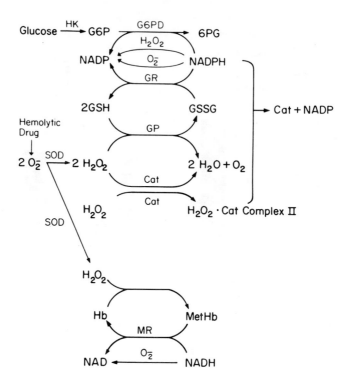

Fig. 2. This figure shows the effect of O_2^- on NADPH and NADH. On the right, the number of μ moles oxidized in 35 minutes. Details of the system are given in the text.

glutathione reductase. Glucose-6-phosphate dehydrogenase deficiency, by limiting the amount of NADPH and GSH produced, makes these cells more susceptible to oxidant injury. In this paper we have examined the possibility that O_2^- may be produced by drugs hemolytic to G6PD deficient cells and the O_2^- may interact unfavorably with NADPH, NADH, and GSH.

We have found that O_2^- is capable of directly oxidizing NADPH and NADH. We have also found the H_2O_2 is capable of directly oxidizing NADPH. It might be argued that the O_2^- effect on NADPH and NADH is due to H_2O_2 formed from O_2^-, either by SOD or spontaneously. However, two of our observations suggest that this is not the case. 1) 0.0063 μ moles of NADPH were oxidized by the estimated 0.567 μ moles of O_2^- produced in the system summarized in Figure 2. Comparable amounts of H_2O_2 caused no apparent oxidation (Table 1). 2) SOD inhibits NADPH oxidation by O_2^-. Since H_2O_2 is generated by SOD, one might expect that addition of SOD would increase or at least maintain the amount of NADPH oxidized. These two observations

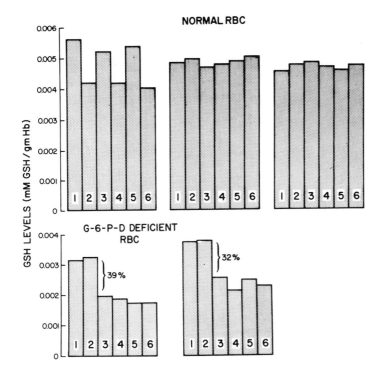

Fig. 3. This figure shows the effect of O_2^- generated outside red cells on GSH levels in normal and G6PD deficient cells. Key: 1 is an incubated control with no additions; 2 is an additional incubated control containing X; 3 contains the O_2^- generating system (X,XO); 4 contains X, XO, and SOD; 5 contains X, XO, and catalase; and 6 contains X, XO, SOD, and catalase.

TABLE 1

OXIDATION OF NADPH BY VARIOUS CONCENTRATIONS OF H_2O_2

μ MOLES H_2O_2 USED	μ MOLES NADPH OXIDIZED
4,890	0.0197
489	0.0092
48.9	0.0023
0.489	0.0

lead us to conclude that O_2^- is more reactive than H_2O_2 in oxidizing NADPH.

The effects of O_2^- described on NADPH and NADH are in buffer systems. Our results in intact G6PD deficient red cells suggest biological relevance of these studies. The generation

675

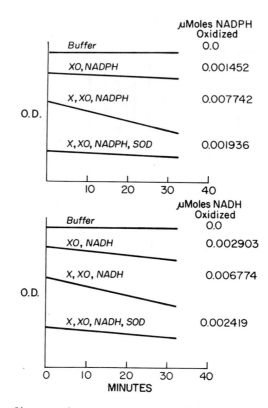

Fig. 4. This figure shows a summary of possible pathways by which O_2^- can interfere with normal red cell metabolism. Key: GR is glutathione reductase; GP is glutathione peroxidase; cat is catalase; MR is methemoglobin reductase; and HK is hexokinase. All other abbreviations are explained in the text.

Since NADPH is required to maintain red cell GSH levels, any agent which oxidized NADPH may jeopardize the ability of the red cell to remain intact. The O_2^- produced by a hemolytic agent can directly oxidize NADPH, thereby diminishing red cell GSH levels. Superoxide is also spontaneously and enzymatically converted to H_2O_2, which can also oxidize NADPH directly, although probably at a slower rate than O_2^-. H_2O_2 is capable of oxidizing hemoglobin to methemoglobin, a process by which O_2^- is generated. Hydrogen peroxide also combines with catalase to form an inactive H_2O_2 catalase complex II which requires the oxidation of NADPH for activation and thereby further depletes the supply of NADPH.

---

of O_2^- in whole blood caused a decrease in the GSH level of deficient but not of normal cells. It thus appears that O_2^-

is capable of entering a red cell and may thereby cause hemolysis in G6PD deficient individuals.

The hypothesis that O_2^- is involved in G6PD deficient type hemolysis is strengthened by our findings that three hemolytic drugs, menadione, hydroquinone, and APH are capable of producing O_2^- in vitro. Cohen and Hochstein (1964) found that several hemolytic drugs, including menadione and hydroquinone, produce H_2O_2 in vitro. We speculate that at least some of the H_2O_2 production seen by Cohen and Hochstein may not have arisen directly from the drugs but rather from O_2^- produced by the drugs. We further speculated that O_2^- production might be a common feature of the various G6PD type hemolytic agents. Consideration of another hemolytic agent, infection (Mengel, 1967), can be used in support of this latter speculation. Babior, et al., (1973) have recently reported that phagocytizing polymorphonuclear leukocytes produce O_2^-. This may be the source of the H_2O_2 these cells are also known to produce. Thus G6PD deficient red cells during infection may be affected by O_2^- released from phagocytizing leukocytes.

In Figure 4 we present a summary of the possible relationships between O_2^- and G6PD deficiency. The production of O_2^- by a hemolytic agent can cause loss of GSH and NADPH in a G6PD deficient cell. This loss could occur through the direct action of O_2^- and/or the conversion of O_2^- to H_2O_2 by SOD. We therefore suggest that O_2^- production may be a common pathway by which various hemolytic agents produce a G6PD type hemolytic anemia.

REFERENCES

Babior, B. M., R. S. Kipnes, J. T. Curnette 1973. Biological defense mechanisms: The production by leukocytes of superoxide, a potential bactericidal agent. *J. Clin. Invest.* 52: 741-744.

Beauchamp, C. O., and I. Fridovich 1971. Superoxide Dismutase: Improved assays and an assay applicable to acrylamide gels. *Anal. Biochem.* 44: 276-287.

Beutler, E., R..J. Dern, C. L. Flanagan, A. S. Alving 1955. The hemolytic effect of primaquine. VII. Biochemical studies of drug-sensitive erythrocytes. *J. Lab. Clin. Med.* 45: 286-295.

Brewer, G. J., 1967. Achromatic regions of tetrazolium stained starch gels: Inherited electrophoretic variation. *Am. J. Hum. Genet.* 19: 674-680.

Brewer, G. J., 1974. *The Red Blood Cell* Vol. I. 2nd Edition (D. MacN. Surgenor, ed.), Academic Press, N. Y. Chapter 9-General Red Cell metabolism. pp. 387-426.

Carson, P. E., C. L. Flanagan, C. E. Ickes, and A. S. Alving 1956. Enzymatic deficiency in primaquine-sensitive erythrocytes. *Science* 124: 484-485.

Cohen, G., and P. Hochstein 1964. Generation of hydrogen peroxide in erythrocytes by hemolytic agents. *Biochemistry* 3: 895-900.

Curnette, J. T., D. M. Whitten, and B. M. Babior 1974. Defective superoxide production by granulocytes from patients with chronic granulomatous disease. *N. Engl. J. Med.* 290: 593-597.

Flanagan, C. L., S. L. Schrier, P. E. Carson, and A. S. Alving 1958. The hemolytic effect of primaquine. VIII. The effect of drug administration on parameters of primaquine sensitivity. *J. Lab. Clin. Med.* 51: 600-608.

Gant, F. L., and G. F. Winhster 1961. (abst) Primaquine sensitive hemolytic anemia complicating diabetic acidosis. *Clinical Research* 9: 27.

Leipzig, R. M., G. J. Brewer, and F. J. Oelshlegel, Jr. 1974. (abst) Superoxide and superoxide dismutase in G6PD type hemolysis. *Clinical Research* 22: 397A.

Lippit, B., and I. Fridovich 1973. Tetrazolium oxidase and superoxide-dismutase: Evidence for identity. *Arch. Biochem. Biophys.* 159: 738-741.

McCord, J. M., and I. Fridovich 1969A. Superoxide dismutase: An enzymatic function for erythrocuprein (hemocuprein). *J. Biol. Chem.* 244: 6049-6055.

McCord, J. M., and I. Fridovich 1969B. The utility of superoxide dismutase in studying free radical reactions. *J. Biol. Chem.* 244: 6056-6063.

Massey, V., S. Strickland, S. G. Mayhew, L. G. Howell, P. C. Engel, R. G. Matthews, M. Schuman, and P. A. Sullivan 1969. The production of superoxide anion radicals in the reaction of reduced flavins and flavoproteins with molecular oxygen. *Biochem. Biophys. Res. Commun.* 36: 891-897.

Mengel, C. E., E. Metz, and W. S. Yancy 1967. Anemia during acute infections: Role of glucose-6-phosphate dehydrogenase deficiency in Negroes. *Arch. Intern. Med.* 119: 287-290.

Misra, H. P., and I. Fridovich 1972. The generation of superoxide radicals during auto-oxidation of hemoglobin. *J. Biol. Chem.* 247: 6960-6962.

Mohler, D. N., and W. J. Williams 1961. The effect of phenylhydrazine on the adenosine triphosphate content of normal and glucose-6-phosphate dehydrogenase-deficient human blood. *J. Clin. Invest.* 40: 1735-1742.

Piomelli, S., L. M. Corash, D. D. Davenport, J. Miraglia, and E. L. Amorosi 1968. In vivo ability of glucose-6-phosphate

dehydrogenase in GD^{A-} and $GD^{Mediterranean}$ deficiency.
J. Clin. Invest. 47: 940-948.

Prins, H. K., and J. A. Loos, 1969. In: *Biochemical Methods in Red Cell Genetics* (J. J. Yunis, ed) Academic Press,N.Y., p. 127.

Reed, D. W., P. G. Passon, and D. E. Hultquist 1970. Purification and properties of a pink copper protein from human erythrocytes. *J. Biol. Chem.* 245: 2954-2961.

Strobel, H. W., and M. J. Coon 1971. Effect of superoxide generation and dismutation on hydroxylation reactions catalyzed by liver microsomal cytochrome P-450. *J. Biol. Chem.* 246: 7826-7829.

Wever, R., B. Oudega, and B. F. VanGelder 1973. Generation of superoxide radicals during the auto-oxidation of mammalian oxyhemoglobin. *Biochem. Biophys. Acta.* 302: 475-478.

FUNCTIONAL ASPECTS OF "NOTHING DEHYDROGENASE" POLYMORPHISM
IN THE PIKE-PERCH (*LUCIOPERCA LUCIOPERCA* L.)

WILHELMINA DE LIGNY, H. HOGENDOORN, B. L. VERBOOM,
and J. WILLEMSEN
Netherlands Institute for Fishery Investigations
Ymuiden, The Netherlands

ABSTRACT. Starch and polyacrylamide gel electrophoresis
of sera and tissue homogenates from pike-perch (*Lucioperca
lucioperca* L.) revealed variable zones with tetrazolium
reducing activity in the absence of added substrate and
coenzyme. This activity could not be ascribed to the action
of dehydrogenases such as ADH and LDH. The frequency distribution of the electrophoretic patterns suggested that
the polymorphism is under the control of three major alleles. Analysis of four year-classes of pike-perch from the
Yssel Lake showed that the distribution of the phenotypes
may differ significantly from the Hardy-Weinberg distribution in some year-classes, but not in others. In aquarium experiments in which the fish were exposed to a rapid
rise in temperature from 21°C upwards, it was found that
phenotypes containing the fast moving electrophoretic zone
survived longer. In vitro experiments investigating the
effect of temperature on the reducing activity of sera
showed that the phenotypes differ in thermostability, and
that the fast moving zone is the more thermostable. In
the temperature range from 7° to 32° C phenotypes containing
the fast moving zone however have less reducing activity.
It is suggested that the polymorphism may provide the pike-
perch with a balanced mechanism for survival in the fluctuating climatic conditions of shallow freshwater lakes.
Attempts to find what the physiological significance of the
system is have so far provided no answers.

INTRODUCTION

"Nothing dehydrogenase" (NDH) polymorphism in pike-perch
was encountered during a general search for enzyme polymorphism
in this species, which is both a popular sports fish and a
commercial catch in inland Dutch waters. Catches are subject
to large yearly fluctuations, dependent on the recruitment of
strong or weaker year-classes. Environmental influences, in
particular temperature, are thought to be of importance in
determining the recruitment (Deelder and Willemsen, 1964),
but no unequivocal relationship with climatic conditions has
been established so far.

681

The study of enzyme polymorphisms was undertaken with the object of investigating the possibility of differential mortality under the influence of environmental variablity and its eventual bearing on the fluctuations in year-class strength.

MATERIAL AND METHODS

Pike-perch, 1 to 2 years of age, were obtained from the Yssel Lake and the Westeinder Lake in the west of the Netherlands. Tissues and blood-clots, taken from fresh-frozen fish, were homogenized in equal volumes of 0.1 M Tris-HCl, pH 7.5, containing 1 % Triton X-100, using a motor-driven glass rod and purified sand. Erythrocytes and plasma were obtained by centrifugation from heparinized blood taken from live fish by cardiac puncture. The erythrocytes were washed twice with isotonic saline prior to homogenization. Leucocytes were separated from whole blood by density gradient centrifugation (Bøyum, 1968).

Tissue and blood-cell homogenates were centrifuged at 3000 rpm, and the supernatant was used for electrophoresis. The plasma was electrophoresed or stored frozen at - 38°C for spectrophotometric activity measurements.

Electrophoresis was carried out horizontally in starch gel or polyacrylamide gel slabs (Bijlsma and Van Delden, 1974); a temperature of 4°C was maintained by using cooling plates. The starch gels were prepared using 11 % Connaught starch and the following buffers:

a. Tris-EDTA-borate, pH 8.6 (Nance et al., 1963);
b. Citric acid - Tris, pH 7.5 (Clayton et al., 1972);
c. Phosphate, pH 6.8 or adjusted to pH 7.4 (after Bengtsson and Sandberg, 1973).

Samples were inserted on Whatman 3 MM filter paper. A voltage gradient of 10 to 20 V/cm was applied for 4 to 4½ hr. Acrylamide gels were prepared using 6% Cyanogum. The buffers employed were Tris-citric acid, pH 6.8 (Bijlsma and Van Delden, 1974) or phosphate pH 6.8 (Bengtsson and Sandberg, 1973). Following a 2 hr prerun to remove the ammonium persulphate used for polymerization of the gels the samples were inserted in premade slots and electrophoresed at 20 V/cm for 2-2½ hr.

NDH zones, unless otherwise stated, were developed in a staining solution containing 15 mg MTT (Serva) and 10 mg PMS (BDH) per 100 ml 0.1 M phosphate, pH 8.0. Alcohol dehydrogenase (ADH), glutamate dehydrogenase (GDH), lactate dehydrogenase (LDH), and malate dehydrogenase (MDH) zones were developed using ethanol (or propanol), glutamate, lactate, and malate as substrates and NAD as coenzyme at pH 8.5, 8.0, 8.0,

and 8.6 for the respective staining solutions. For the development of isocitrate dehydrogenase (IDH) zones, isocitrate
and NADP were used at pH 8.5. For the detection of monoamine
oxidase and aldehyde oxidase activity, tryptamine and benzaldehyde were used as substrate in a pH 8.0 staining buffer.

Spectrophotometric measurements of NDH activity were made
using a reaction mixture consisting of 0.1 M phosphate pH 7.4
or 8.0, containing 1mM Na_4-EDTA, 1 mg bovine serum albumin/ml,
0.1 to 0.25 mM PMS, and 0.08 mM DCIP (2,6-dichlorophenolindo-
phenol). DCIP reduction was measured at 600 nm (King, 1963).
Substrates were added to the reaction mixture at a concentration of 0.5 - 2 mM.

Oxidative activity of NDH was also screened by eye in tubes
using a reaction mixture (Robinson and Lee, 1967) consisting
of 0.7 ml 0.1 M phosphate, pH 7.4 or 8.0, 0.1 ml 3-amino-9-
ethyl-carbazole (AEC) (40 mg/10 ml dimethyl formamide), 0.1
ml peroxidase (Boehringer Grade II, 10 mg/10 ml 0.1 M phosphate pH 7.3), 0.1 ml of 10 to 40 mM substrate solution, and
0.1 ml of the sample to be tested. H_2O_2 liberated by oxidative
activity in the presence of peroxidase forms a colored product
by oxidation of AEC.

RESULTS AND DISCUSSION

ELECTROPHORETIC PATTERNS

Figure 1 shows the NDH and LDH patterns from sera of pike-
perch and other freshwater fishes electrophoresed in starch
gel. In the absence of added substrate (upper part) NDH bands
with tetrazolium reductase activity are revealed in the sera
of perch and pike-perch but not pike and Cyprinids. In the
pike-perch sera the bands are variable. The NDH bands are
also seen in addition to the more cathodal LDH bands, in the
lower gel stained for LDH.

In decreasing order with respect to staining activity, homogenates of kidney, heart, stomach, intestine, liver, and
spleen of pike-perch revealed the characteristic NDH patterns
shown in the sera. Homogenates of the lens of the eye showed
very weakly staining NDH bands, whereas homogenates of erythrocytes and leucocytes exhibited no NDH bands.

Detection of other dehydrogenases in gels of the tissue
homogenates revealed intensly stained MDH isozymes in most
organs. In heart and liver an intensly stained IDH band was
found along with specific ADH and GDH bands. In all staining
solutions the NDH bands were also revealed. They were least
pronounced in the MDH staining solution at pH 8.6. They were
separated equally well in the three buffer systems used in
the starch gels and always migrated anodally ahead of the ADH,

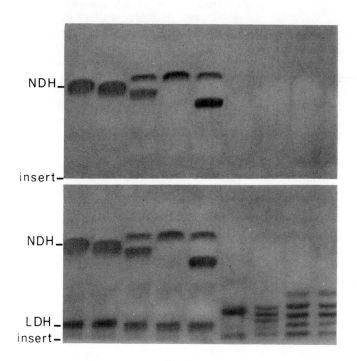

perch   pike-perch   pike Cyprinids

Fig. 1. "Nothing dehydrogenase" and lactate dehydrogenase
patterns of serum from perch, pike-perch, pike, and cyprinids
(from left to right: roach, silver bream, ide) in starch gel
in phosphate pH 6.8. Upper part: gel stained for "nothing
dehydrogenase"; lower part: gel stained for lactate dehydrogenase.

GDH, and MDH bands. In acrylamide gels in Tris-citric acid,
pH 6.8, the IDH band appeared between the variable NDH bands.
    Zones with tetrazolium oxidase activity were revealed in
all gel systems by use of the dehydrogenase staining mixtures
and were separated distinctly from the NDH bands. Incubation
of starch gel slices, in which serial dilutions of pike-perch
serum had been run, in staining mixtures for monoamine oxidase and aldehyde oxidase resulted in a decrease of the activity of the NDH bands, compared to control slices of the
same gel. Overnight dialysis of pike-perch serum against 0.1
M phosphate buffer to remove any metabolite that could serve

as substrate for the NDH isozymes did not affect their staining activity.

In screening serum samples and tissue homogenates from representatives of 16 additional species (both freshwater and marine fish) including seven cyprinids, pike (*Esox lucius*), eel (*Anguilla anguilla*), herring (*Clupea harengus*), mackerel (*Scomber scombrus*), cod (*Gadus morhua*), haddock (*Melanogrammus aeglefinus*), whiting (*Merlangius merlangus*), plaice (*Pleuronectes platessa*, and sole (*Solea solea*), NDH zones so far have been limited to perch (*Perca fluviatilis*) and pike-perch. In a total of 15 perch there was no indication of NDH polymorphism.

NDH PHENOTYPES IN PIKE-PERCH OF THE YSSEL LAKE

A total of 709 pike-perch, belonging to four successive year-classes, were examined for NDH patterns. Electrophoresis of frozen sera or whole blood obtained from frozen fish was performed routinely in acrylamide gels using the Tris-citric acid buffer system, pH 6.8. All except one of the 709 fish were classified into one or another of six phenotypes consisting either of one (phenotypes FF, MM, and SS) or two bands (phenotypes FM, FS, and MS), where F, M, and S stand for the F(ast), M(edium) and S(low) anodal mobility of the individual bands. The exceptional fish possessed a very slow band, designated V, together with the S band. The six regularly occurring phenotypes are shown in Figure 2. There were no obvious differences between males and females with respect to the six phenotypes and their distribution.

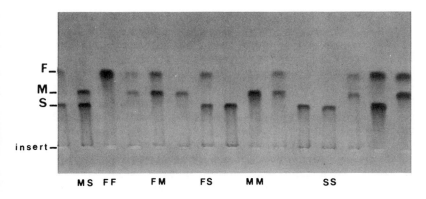

Fig. 2. "Nothing dehydrogenase" phenotypes of pike-perch sera in polyacrylamide gel in Tris-citric acid pH 6.8.

The six regularly occurring phenotypes can be accounted for on the basis of three codominant, autosomal alleles, F, M, and S. Obviously, a fourth allele, V, would be necessary in conjunction with allele S, to account for phenotype SV in the one fish.

The distribution of the six regularly occurring phenotypes is shown by year-class in Table 1. Also shown are the expected Hardy-Weinberg distributions based on the gene frequencies estimated from each year-class.

With the exception of year-class 1972, there were significant deviations (Table 1) of observed from expected numbers. The deviations were not consistent from year to year with respect to the phenotypes, and there was no overall excess of homozygous types as has been frequently observed with respect to other enzyme polymorphisms in fish (Utter et al., 1970; Jamieson et al., 1971; Koehn et al., 1971).

Comparison with data on the strength of the year-classes, derived from assessments of larvae and young fish, indicates that deviations from genetic equilibrium were not characteristic for weak year-classes. Among the year-classes sampled the 1970 year-class was considered a very good year-class and the 1972 year-class very weak.

Differences in allele frequencies were not significant for the 1970-1971-1972 year-classes (Chi-square: 1.43, $P_{df\ 4} >$ 0.80), but the 1973 year-class deviates significantly from the total of the samples (Chi-square: 9.01, $P_{df\ 2} < 0.025$) and appears characterized by the higher frequency of the S allele. This year-class can be considered as the direct offspring of the 1970 and 1971 year-classes.

Altogether, these data suggest differential viability of pike-perch connected with their NDH phenotypes.

AQUARIUM EXPERIMENTS; INITIAL OBSERVATIONS

During the course of investigating the upper lethal temperature of 1-year old hatchery reared pike-perch under aquarium conditions (Willemsen, unpublished) the following observations were made. When the temperature was raised gradually at a rate of approximately 0.5° to 1°C per day from an initial temperature of about 15°C no mortality occurred up to 32°C. A rapid increase in temperature from 22°C to 31°C in 4 hr resulted in heavy mortality, as did an increase of 3°C in 4 hr for fish that had been acclimatized gradually to a temperature of 32°C. In a final experiment with 20 survivors of the 60 with which the experiments were started and that gradually had been brought up to 33°C, further gradual increase of the temperature to 35.6°C killed the remaining fish gradually

TABLE I

OBSERVED AND EXPECTED NDH PHENOTYPE DISTRIBUTION IN FOUR YEAR-CLASSES OF PIKE-PERCH FROM THE ŸSSEL LAKE

Year-class	Age		Phenotypes							N	Chi-square	Gene-frequencies		
			FF	FM	FS	MM	MS	SS	SV		df = 2	F	M	S
1970	2	O[x]	28	40	47	30	41	8	1	195	6.45 P < 0.05	0.367	0.362	0.269
		E[x]	26	52	39	26	38	14						
1971	2	O	10	35	18	5	15	9		92	9.69 P < 0.01	0.397	0.326	0.277
		E	14	24	20	10	17	7						
1972	1	O	32	70	47	36	43	19		247	1.20 P > 0.50	0.366	0.375	0.259
		E	33	68	47	35	48	16						
1973	1	O	14	57	59	19	39	24		212	9.71 P < 0.01	0.340	0.316	0.344
		E	24	46	50	21	46	25						

x$_O$: observed numbers; E: expected numbers.

TABLE 2
"NOTHING DEHYDROGENASE" PHENOTYPES IN PIKE-PERCH FROM
FINAL TEMPERATURE ACCLIMATIZATION EXPERIMENT

	Phenotypes						Total
	FF	FM	FS	MM	MS	SS	
Fish dying first				10			10
Fish dying last	2	3		5			10

over a period of 4 days. NDH phenotypes were determined in these fish and grouped according to the order in which they had died. The distribution of the phenotypes is given in Table 2.

In spite of the small number of fish involved in this experiment longer survival of the FF and FM types compared to the MM types was observed. Also noticed was the absence of fish possessing the S band in this group surviving the total series of acclimatization experiments.

HEATING EXPERIMENTS

A total of 212 one-year old pike-perch was collected from the Yssel Lake and divided among tanks with running tapwater at about 12° C. During a 2-week adaptation period the temperature increased to 14°C. Feeding was initiated and the NDH phenotypes were determined using a minute piece of the ventral fin. No mortality occurred during this period.

The temperature then was raised at a rate of 1° to 1.5°C per day to 21°and 25°C in Experiment 1 and Experiments 2 and 3 respectively. During this period a total of 35 fish died.

The remaining fish then were exposed to a further more rapid increase in temperature in the three separate experiments. The initial and final temperatures and the rate of increase are summarized in Table 3. In Experiments 1 and 2 fish started dying after 32 hr and all fish had died after 99 and 72 hr respectively. The temperatures at this time were 31.4°C and 32.4°C in Experiments 1 and 2 respectively. In Experiment 3 mortality started after 5 hr and by 33.6°C all fish had died.

Fish were collected in the order in which they had died, and frozen rapidly for analysis of the NDH phenotypes. The results are presented in Table 3.

In all three experiments, fish with the phenotypes FF, FM, and FS were more frequent in the group that survived longest, while the phenotypes MM, MS, and SS had a higher frequency in the group that died first. The difference between the distribution of the phenotypes containing the F band compared

TABLE 3

MORTALITY OF "NOTHING DEHYDROGENASE" PHENOTYPES OF PIKE-PERCH DURING WARMING-UP EXPERIMENTS

| | Phenotypes | | | | | | Total | Phenotypes with | | Chi-square |
	FF	FM	FS	MM	MS	SS		F	non-F	F vs non-F
Initial sample	14	57	59	19	39	24	212	130	82	0.33
Mortality during	4	10	9	6	5	1	35	23	12	
slow warming-up	2.3	9.4	9.7	3.1	6.4	4.0		21.5	13.5ˣ	
Experiment 1*										
fish dying first		7	7	3	5	4	26	14	12	
fish dying last	2	6	13	1	3	1	26	21	5	4.28
Experiment 2*										
fish dying first	1	4	2	3	4	5	19	7	12	
fish dying last	2	7	4		3	2	18	13	5	4.74
Experiment 3*										
fish dying first		3	6	3	7	3	22	9	13	
fish dying last	1	6	10	1	3	1	22	17	5	6.02
Total experiments 1, 2, and 3										
fish dying first	1	14	15	9	16	12	67	30	37	
fish dying last	5	19	27	2	9	4	66	51	15	14.66

ˣDistribution expected according to the distribution in the initial sample

*Experiment 1: Warmed up from 21° to 31.4°C at a rate of 2.3°C/24 hr ;
Experiment 2: Warmed up from 25° to 32.4°C at a rate of 3.1°C/24 hr ;
Experiment 3: Warmed up from 25° to 33.6°C at a rate of 8.6°C/8 hr.

to those not containing the F band for the total of the three experiments was highly significant (P <0.001). The differences in the separate experiments were significant at the 5% and 1% level, and all results were similar in showing a higher frequency of the phenotypes containing the F band in the group of fish surviving longest. In contrast the distribution of F and non-F phenotypes among the fish that died during the preliminary slow warming-up period was not significantly different from that in the initial sample (P >0.50).

In general the distribution of the phenotypes at the end of the experiment corresponded well with the distribution that was determined prior to the experiment. Ten exceptions were ascribed to misclassification during the preliminary testing, because of the minimal amount of finray tissue employed.

Throughout the warm-up experiments the oxygen content and nitrite content of the water were regularly determined. Oxygen values varied mostly between 70 and 90% of the saturation value at a given temperature. The nitrite content exceeded a value of 1 mg/liter twice in two of the tanks, during a period in which no mortality occurred and during the further experiment remained well below this value.

OXYGEN DEPLETION EXPERIMENT

After the initial period of slow warming-up to 25°C of one of the tanks the temperature was kept constant while the oxygen supply was diminished. When the oxygen content had reached a value of 2.1 mg/liter, or 25% of the saturation value, it was entirely cut off. Over a period of 2 hr all fish died while the oxygen content decreased to 1.8 mg/liter. The fish were collected in the order in which they died and the NDH phenotypes were analyzed. The results, as presented in Table 4, show that there was no difference between the distribution of the phenotypes in the groups that died first or

TABLE 4

MORTALITY OF "NOTHING DEHYDROGENASE" PHENOTYPES OF PIKE-PERCH DURING OXYGEN DEPLETION EXPERIMENT

	Phenotypes						
	FF	FM	FS	MM	MS	SS	Total
Fish dying first	2	8	2	1	4	5	22
Fish dying last	1	6	6	1	5	3	22

later (Chi-square F vs non-F bands: 0.09; P >0.80).

It is evident from these experiments that NDH phenotypes have a different survival value when the temperature is increased rather rapidly from temperatures of 21°C upwards. The phenotypes containing the F band survived better. On the other hand presence or absence of the F band did not affect mortality due to oxygen depletion. This indicated that the differential mortality observed during the heating experiments was not due to differences between the phenotypes in their efficiency to use oxygen, or in their need for oxygen, although at higher temperatures this may become limiting relative to the increased demand.

SPECTROPHOTOMETRIC MEASUREMENTS

Spectrophotometric measurements of NDH activity were carried out using 2,6-dichlorophenolindophenol (DCIP) instead of a tetrazolium dye, of which the reaction product is insoluble. The reduction of DCIP is accompanied by decoloration of the blue dye and measured by the decrease of extinction at 600 nm. The use of this system for the assay of succinate oxidase is described by King (1963).

The measurements were carried out using serum or eluates of NDH bands from starch gels. The sera were thawed immediately before use, but no appreciable decrease of activity was observed during the course of a day. Serum samples were diluted with isotonic saline varying from 1 in 2 to 1 in 7 depending on their activity to obtain complete reduction of 0.08 mM DCIP in approximately 10 to 20 minutes in the presence of 0.25 mM PMS in 0.1 M phosphate, pH 7.4, at a temperature of 20°C. This standardized procedure was used in the following measurements unless otherwise stated.

The activity of sera with different NDH phenotypes was compared in 13 samples. The NDH activity (measured in 0.1 M phosphate, pH 8.0, and 0.1 mM PMS) decreased in the following order: MS, FS, FM, FF. The activities of the individual MS sera were 97.7, 96.0, and 89.8% of the activity of the most active MS serum. The activities of the other sera relative to this MS serum were 87.5, 85.8, and 85.8% for the FS sera, 88.6, 72.7, and 60.2% for the FM sera and 52.8, 43.7, and 33.5% for the FF sera.

The influence of the reaction temperature was studied over a range from 7°C to 32°C. Sera of the phenotypes FF, FS, MM, and MS were used in a dilution, ranging from 1 in 2 to 1 in 7, in which they had approximately equal activities when measured at 20°C. All phenotypes showed a similar increase in activity over the range from 7°C to 32°C (Figure 3a).

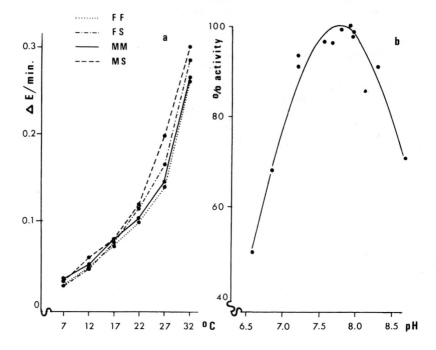

Fig. 3a: Activity of sera of the phenotypes FF, FS, MM, and MS as a function of the reaction temperature. Sera were tested in a dilution of 1 in 2, 1 in 4, 1 in 4, and 1 in 7 respectively.

Fig. 3b. Activity of serum of the phenotype MS as a function of pH of 0.1 M phosphate buffer, expressed in percent of the maximum activity.

The effect of pH on NDH activity was studied for serum of the phenotype MS over a range from pH 6.6 to 8.7. The results indicated a pH optimum at about pH 7.7 to 7.9 (Figure 3b).

The effect of prolonged in vitro exposure to higher temperatures on the NDH activity of sera of the phenotypes FF, FM, FS, MM, and MS, was investigated. The activity of the sera was measured after incubation for 30 and 60 minutes in a water bath at 35°C, 45°C, and 55°C and compared to the activity of the same samples kept on ice. The results are illustrated in Figure 4.

Incubation for 30 minutes at 35°C stimulated the activity of the FS phenotype and to a lesser extent the FF phenotype, while the activity of the FM and MM phenotypes was impaired. After incubation for 30 or 60 minutes at 45°C, FF and FM

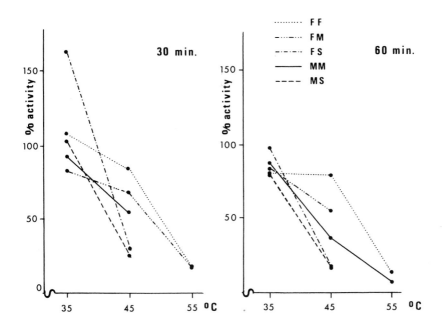

Fig. 4. Activity of sera of the phenotypes FF, FM, FS, MM, and MS after incubation at 35°, 45°, and 55°C for 30 and 60 minutes, expressed in percent of the activity of sera kept on ice.

phenotypes had considerable residual activity. These sera also retained some activity after incubation at 55°C. In contrast the FS and MS phenotypes showed a sharp decrease in activity after 30 or 60 minutes incubation at 45°C. The MM phenotype appeared to occupy an intermediate position.

After the incubation experiment part of the serum was electrophoresed in starch gel in citric acid-Tris, pH 7.5, and the gel was stained for NDH activity. Figure 5 shows that the reduction of the activity of the phenotype FS after incubation at 45°C was predominantly due to reduction of the activity of the S band. After incubation at 55°C the F band still retained some activity. In the phenotype MS incubation at 45°C affected the M band less than the S band but relatively more than the F band of the FS phenotype.

These experiments show that the NDH isozymes differ in thermostability, F being the most thermostable and S being the least thermostable. This observation explains the greater thermostability of the FF and FM phenotypes over the MM and

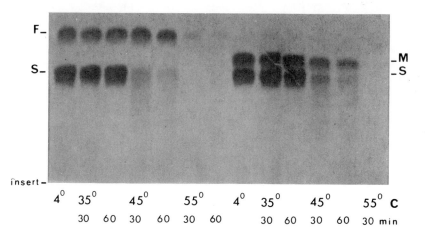

Fig. 5. "Nothing dehydrogenase" patterns in starch gel in citric acid-Tris pH 7.5 of sera of the phenotypes FS and MS after incubation at 35°, 45° and 55°C for 30 and 60 minutes compared to sera kept on ice.

MS phenotypes.

The behavior of the NDH phenotypes on incubation in vitro parallels the results of the aquarium experiments with regard to the superiority of the F variant phenotypes at higher temperatures. The difference in the activity of the phenotypes at 20°C that appears to be maintained at temperatures from 7°C to 32°C is in the other direction with the phenotype MS having a higher activity than the phenotypes containing the F band, and the FF homozygote in particular.

It is tempting to speculate that the "nothing dehydrogenases" are involved in a physiological process providing pike-perch with a balanced mechanism essential for their maintenance in the variable climate of a shallow freshwater lake. This will remain speculative until the nature and function of the NDH isozymes are understood. Attempts to solve this question so far have provided predominantly negative answers, indicating what they are not.

ATTEMPTS AT IDENTIFICATION OF PIKE-PERCH NDH

Pearse (1972), in summarizing the histochemical literature, described the activity of "nothing dehydrogenase" as leading to the production of formazan by the reduction of tetrazolium

salts "in the absence of any substrate" but "in systems containing NAD (or NADP)". The activity increases progressively
from pH 7.0 to pH 9.0. It is inhibited by sulfhydryl inhibitors making it likely that protein-bound SH is responsible for
this reaction but "not excluding the participation of smaller
molecules like glutathione".

NDH of pike-perch differs from the "nothing dehydrogenase"
described by Pearse (1972) by having a pH optimum in the
range of pH 7.7 - 7.9. Furthermore, addition of the sulfhydryl
inhibitors p-chloromercuribenzoic acid (PCMB) and iodoacetate
to sera prior to electrophoresis or to the reaction mixture
in the spectrophotometric measurements (at final concentrations
of 5mM in the sera and 1mM in the reaction mixture) did not
affect the appearance of the NDH bands or the reducing activity. Participation of non-protein molecules, like glutathione,
as suggested by Pearse (1972), appears unlikely in view of the
fact that the activity of the serum was not affected by prolonged dialysis.

A second source of "nothing dehydrogenase" activity, particularly in gels, has been brought to light by the work of
Shaw and Koen (1965), who showed that alcohol dehydrogenase
may be responsible for apparent "nothing dehydrogenase" zones
in starch gels. Reduction of tetrazolium salts by lactate
dehydrogenase acting on non-specific substrates has been reported by Falkenberg et al. (1969). This and other possibilities for "nothing dehydrogenase" activity, like endogenous or
bound substrate, have been discussed by Ferguson (1971).

Unlike ADH and LDH, pike-perch NDH does not require the
participation of NAD (or NADP). Participation of endogenous
substrate is only feasible if it were bound to the enzyme or
some other non-dialyzable component and remained bound during
dialysis, or if it were a non-dialyzable component. The
possibility that a bound metabolite may act as substrate for
pike-perch NDH was investigated by measuring the effect of
substrates for known oxidases on the PMS-DCIP reduction by a
serum of the phenotype MM. The following substances in a
final concentration of 0.5 to 2mM did not influence the reducing activity: glucose, lactate, succinate, urate, hypoxanthine, and L-amino acids. Addition of amines including
aliphatic, aliphatic heterocyclic, and aromatic mono- and
diamines, however, impaired the reducing activity to a varying extent, depending on the complexity of the carbon skeleton
but not the basic properties of the amines. It was thought
that the inhibitory effect of the amines may involve binding
to NDH, analogous or similar to substrate binding, but no reaction was measurable under the conditions of the experiment.
Therefore, sera were incubated with the various amines in a

reaction mixture containing 3-amino-9-ethylcarbazole (AEC) and peroxidase. No reaction, indicative of liberation of H_2O_2 by oxidative deamination of the amines, was obtained.

Participation of a non-dialyzable component of relatively small molecular weight was suggested by an experiment in which serum of the phenotype FS was concentrated using an ultrafiltration-membrane of 15,000 MW retentivity. 60% of the reducing activity of the serum was lost in the process. The possibility that cytochrome c may be involved was examined by addition of potassium cyanide in a final concentration of 1mM to the PMS-DCIP reaction mixture. This did not affect the reducing activity of sera of the phenotypes FM and FS. The NDH reaction therefore does not involve electron transport via cytochrome c and cytochrome c oxidase.

Thus, our attempts at identification of pike-perch NDH have provided no evidence that it is involved in a catalytic reaction. In further investigations the possibility will be considered that the NDH in the serum may be part of an enzyme or enzyme system present in the tissues and may not exhibit the catalytic properties of the complete system.

ACKNOWLEDGEMENTS

The stimulating interest of Dr. J. Visser, Agricultural University, Wageningen, and the help received from Dr. C. Stormont, University of California, Davis, in preparing the manuscript are gratefully acknowledged.

Participation of the first author in the Third International Isozyme Conference was made possible by a grant from the Netherlands Organization for the Advancement of Pure Research (Z.W.O.).

REFERENCES

Bengtsson, S. and K. Sandberg 1973. A method for simultaneous electrophoresis of four horse red cell enzyme systems. *Anim. Blood Grps. Biochem. Genet.* 4: 83-87.

Bijlsma, K. and W. van Delden 1974. Polymorphism at the G6PD and 6PGD locus in *Drosphila melanogaster*. *Drosoph. Inf. Serv.* 50 (in press).

Bøyum, A. 1968. Separation of leucocytes from blood and bone marrow. *Scand. J.Clin. Lab. Invest.* 21, Suppl. 97: 109 pp.

Clayton, J. W. and D. N. Tretiak 1972. Amine-citrate buffers for pH control in starch gel electrophoresis. *J. Fish.Res. Bd. Can.* 29: 1169-1172.

Deelder, C. L. and J. Willemsen 1964. Synopsis of biological data on pike-perch *Lucioperca lucioperca (Linnaeus)* 1758.

F. A. O. Fish. Synopsis 28: 52 pp.

Falkenberg, F., F. G. Lehman, and G. Pfleiderer 1969. Die LDH-Isoenzyme als Ursache für unspezifische Tetrazoliumsalz-Anfärbungen in Gelzymogrammen ("nothing dehydrogenase"). *Clinica Chim. Acta* 23: 265-278.

Ferguson, E. E. Jr. 1971. Tetrazolium staining of lactate dehydrogenase isoenzymes: reduction of tetrazolium salts in the absence of added 1-lactic acid. *Enzymologia* 40: 81-98.

Hitzeroth, H., J. Klose, S. Ohno, and U. Wolf 1968. Asynchronous activation of parental alleles at the tissue-specific gene loci observed on hybrid trout during early development. *Biochem. Genet.* 1: 287-300.

Jamieson, A., W. de Ligny, and G. Naevdal 1971. Serum esterases in mackerel, *Scomber scombrus* L. *Rapp. P. -v. Réun. Cons. Perm. Int. Explor. Mer.* 161: 109-117.

King, T. E. 1963. Reconstitution of respiratory chain enzyme systems. XI. Use of artificial electron acceptors in the assay of succinate-dehydrogenating enzymes. *J. Biol. Chem.* 238: 4032-4036.

Koehn, R. K., J. E. Perez, and R. B. Merritt 1971. Esterase enzyme function and genetical structure of populations of the freshwater fish, *Notropis stramineus*. *Am. Nat.* 105: 51-69.

Nance, W. E., A. Claflin, and O. Smithies 1963. Lactic dehydrogenase: genetic control in man. *Science* 142: 1075-1076.

Pearse, A. G. E. 1972. *"Histochemistry"*, 3rd. ed., Churchill Livingston, Edinburgh, 1518 pp.

Robinson, J. C. and G. Lee 1967. Preparation of starch gel zymograms: peroxide-producing enzymes and ceruloplasmin. *Arch. Biochem. Biophys.* 120: 428-433.

Shaw, C. R. and A. L. Koen 1965. On the identity of "nothing dehydrogenase". *J. Histochem. Cytochem.* 13: 431-433.

Utter, F. M., C. J. Stormont, and H. O. Hodgins 1970. Esterase polymorphism in vitreous fluid of Pacific hake, *Merluccius productus*. *Anim. Blood Grps. Biochem. Genet.* 1: 69-82.

SUBCELLULAR LOCALIZATION OF RAT KIDNEY GLUTAMINASE
ISOZYMES AND THEIR ROLE IN RENAL AMMONIAGENESIS

NORMAN P. CURTHOYS
Department of Biochemistry
University of Pittsburgh
School of Medicine
Pittsburgh, Pennsylvania 15261

ABSTRACT. Rat kidney contains two distinct glutaminase isozymes; one of which is phosphate dependent, the other is
unaffected by phosphate but is strongly activated by
maleate. The phosphate dependent glutaminase is contained
in mitochondria. Its activity cofractionates with mitochondrial marker activities during differential centrifugation and on sucrose gradients. Submitochondrial fractionation by digitonin-Lubrol treatment in the presence of
borate indicated that this glutaminase isozyme is contained
in the inner mitochondrial membrane. This conclusion was
confirmed by a swell-shrink sonication procedure carried
out in the absence of borate. The contribution of this
glutaminase to renal ammoniagenesis could effectively be
regulated by modulation of mitochondrial transport of glutamine in response to metabolic acidosis. In contrast, the
bulk of the phosphate independent glutaminase was recovered
in the microsomal fraction and correlated with luminal membrane markers during isopycnic centrifugation. This glutaminase activity is released selectively from intact kidney cells by treatment with papain, suggesting that it is
located on the external cell surface. The phosphate independent glutaminase and Y-glutamyl transpeptidase reactions appear to be catalyzed by the same enzyme. Both
activities exhibit a close correlation in the various zones
of the kidney and during purification. Maleate appears to
increase glutaminase activity by converting the transpeptidase into a specific hydrolase. The physiological equivalent to maleate could function during acidosis to regulate
the contribution of this enzyme to renal ammoniagenesis.

INTRODUCTION

In response to metabolic acidosis, the mammalian kidney
exhibits increased synthesis of ammonia (Balagura-Baruch,
1971). A large proportion of this ammonia is excreted in the
urine and thereby increases the kidney's capacity for excreting
anions and titratable acid without depleting sodium and potassium reserves. The primary source of urinary ammonia is the

amide and amine nitrogens of glutamine which the kidney extracts from the plasma (Van Slyke et al, 1943). Glutaminase
isozymes catalyze the first reaction in what is thought to be
the primary pathway of renal ammoniagenesis (Goldstein, 1967).
In addition to the classical phosphate dependent glutaminase,
Katunuma et al (1967) have reported the occurrence and the
separation of a second glutaminase which is not affected by
phosphate but is activated strongly by maleate. The two isozymes have different pH optima, K_m for glutamine, and heat
stabilities. They also have a very complementary distribution
in the various tubular structures of the kidney (Curthoys and
Lowry, 1973). Both isozymes are particulate and, therefore,
a characterization of their subcellular localization was undertaken in order to obtain information basic to understanding
their role in renal ammoniagenesis.

METHODS

Kidneys were homogenized in a buffer containing 0.33 M
sucrose, 20 mM N-2-hydroxyethylpiperazine-N'-2-ethanesulfonic
acid and 0.2 mM EDTA, pH 7.5 and the crude homogenate was fractionated by differential centrifugation (Hogeboom et al, 1948).
Isopycnic centrifugation was performed by layering 0.5 ml of
either mitochondrial fraction or heavy microsomal fraction
(resuspended pellet obtained by centrifuging the post-mito-
chondrial supernatant at 40,000 x g for 30 min) on a 28 ml
linear sucrose gradient prepared in the homogenate buffer and
centrifuging in an SW 25.1 rotor at 25,000 rpm for 5 hours.
Submitochondrial localization of phosphate dependent glutaminase was achieved by the digitonin-Lubrol procedure of
Schnaitman and Greenawalt (1968) and Chan, Greenawalt and
Pederson (1970) except that the inner membrane-matrix particle
was resuspended in 10 mM sodium borate, 100 mM potassium phosphate, 100 mM potassium pyrophosphate buffer, pH 8.9 and incubated at room temperature for 10 min before addition of Lubrol-
WX. The swell-shrink sonication procedure of Sottocassa et al
(1967) for mitochondrial fractionation was also used, but the
resulting inner membrane was separated by isopycnic centrifugation. A preparation of intact kidney cells was obtained by
a modification (Curthoys, unpublished results) of the procedure
of Nagata and Rasmussen (1970). Regional localization of
enzyme activities within the kidney was determined by the procedure of Waldman and Burch (1963).

Glutaminase isozymes were assayed by the procedure of
Curthoys and Lowry (1973). Cytochrome oxidase, malate dehy-
drogenase, and adenylate kinase were assayed as described by
Schnaitman and Greenawalt (1968). The procedures of Sottocassa

(1967) were used to assay for TPNH cytochrome c reductase.
Acid phosphatase (Schachter et al, 1970) and alkaline phosphatase (Ray, 1970) were assayed using p-nitrophenyl phosphate
as substrate. γ-glutamyl transpeptidase was assayed by following the disappearance of glutathione using alanine as an
acceptor amino acid (Palekar et al, 1974) or by following the
appearance of p-nitroanaline from a solution containing 5 mM
L- γ-glutamyl-p-nitroanalide, 0.2 mM EDTA and 50 mM imidiazole,
pH 7.2.

RESULTS AND DISCUSSION

The two glutaminase activities fractionated differently
during differential centrifugation (Table 1). Three-fourths
of the phosphate dependent glutaminase activity is recovered
in the mitochondrial fraction. The percent recovery of the
phosphate dependent glutaminase activity in each of the fractions corresponds well with the recovery of cytochrome oxidase
activity. But, in contrast, only one-third of the phosphate
independent glutaminase was recovered in the mitochondrial
fraction and the remaining two-thirds was recovered in the
microsomal fraction.

TABLE 1
SUBCELLULAR FRACTIONATION OF GLUTAMINASE ISOZYMES

Fraction	PDG[b]		PIG	
	units	%	units	%
Nuclear[a]	4.6	22	1.2	8
Mitochondria	16.0	76	4.4	28
Microsome	0.5	2	9.7	63
Cytosol	0.0	-	0.2	1

[a] The crude homogenate was centrifuged at 600 x g for 10 min
to remove nuclear material. Mitochondria were prepared by
subsequent centrifugation at 8000 x g for 10 min. The
resulting supernatant was then centrifuged at 100,000 x g
for 1 hour to separate microsomes (pellet) and cytosol
(supernatant).

[b] Abbreviations are: PDG, phosphate dependent glutaminase;
PIG, phosphate independent glutaminase.

Localization of the Phosphate Dependent Glutaminase

Following isopycnic centrifugation of the mitochondrial fraction, the phosphate dependent glutaminase activity banded in two distinct regions with mean densities of 1.185 and 1.166 (Fig. 1). These densities correspond well with those previously reported for two types of rat kidney mitochondria

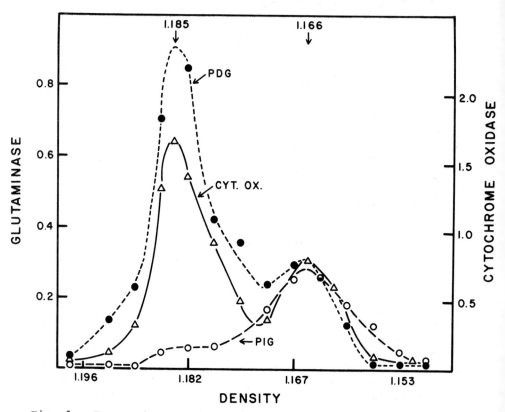

Fig. 1. Isopycnic centrifugation of mitochondrial fraction obtained by differential centrifugation. Enzyme activities are expressed as μmoles ml^{-1} min^{-1}. Abbreviations are as in Table 2.

(Ch'ih and Devlin, 1971). The phosphate dependent glutaminase activity correlates well with cytochrome oxidase and rotenone insensitive DPNH cytochrome c reductase activities (markers for inner and outer mitochondrial membranes, respectively). This correlation confirms that the phosphate dependent glutaminase is located within the mitochondria. In contrast, the

phosphate independent glutaminase bands only in the region of
the light mitochondria. But, sedimentation velocity experiments indicate that this glutaminase sediments much slower
than either form of mitochondria and therefore, none of the
phosphate independent glutaminase is contained in mitochondria,
but instead is contained in some organelle or membrane fragment which has a density similar to light mitochondria but
is much smaller in size.

The submitochondrial localization of the phosphate dependent glutaminase was accomplished by sequential treatment with
digitonin and Lubrol-WX (Table 2). Following treatment of
mitochondrial fraction with digitonin to selectively strip off
the outer membrane and to release proteins from the intra-
cristate space, the bulk of the phosphate dependent glutaminase was recovered in the pellet obtained by low speed centrifugation. This pellet corresponds to the inner membrane-matrix
particle. Subsequent addition of Lubrol resulted in greater
than 70% loss of phosphate dependent glutaminase activity.
Klingman and Handler (1958) have reported that addition of
borate stabilized glutaminase activity towards extraction with
organic solvents. Addition of borate to the inner membrane-
matrix preparation resulted in a 50-100% increase in phosphate
dependent glutaminase activity. This treatment also resulted
in increased cytochrome oxidase and malate dehydrogenase
activities. These increased activities may indicate that the
phsophate dependent glutaminase is latent. Following treatment with borate, subsequent fractionation with Lubrol could
be accomplished with excellent recovery of glutaminase activity. The validity of this fractionation is attested to by the
fact that 50-80% of each of the marker activities are recovered
in their appropriate fraction. The fact that 75% of the phosphate dependent glutaminase activity is recovered in the inner
mitochondrial membrane fraction and the close correlation
between percent recoveries for glutaminase and cytochrome oxidase activities in the various fractions indicates that the
phosphate dependent glutaminase is also contained in the inner
mitochondrial membrane.

This result was confirmed by the procedure of Sottocassa
et al (1967). The mitochondrial fraction was subjected to
swelling in hypotonic Tris-phosphate, followed by shrinking
of the inner membrane by addition of sucrose and MgATP and
short bursts of sonication. The mitochondria were then subjected to isopycnic centrifugation in a sucrose gradient which
was designed so that both the solubilized proteins (malate
dehydrogenase) and the less dense outer membrane (rotenone
insensitive DPNH cytochrome c reductase) did not enter the
gradient (Fig. 2). But, the inner membrane (cytochrome oxi-

TABLE 2

DIGITONIN-LUBROL SUBFRACTIONATION OF MITOCHONDRIA

Fraction	Total Enzyme Activity (μmoles min^{-1})					
	PDG[a]	RIDCR	Aden K	Cyt Ox	Mal D	
Mitochondria	15.3	19.0	39.4	87.2	79.2	
Outer Membrane	3.4 (12)	8.1 (50)	1.2 (18)	25.3 (19)	10.8 (4)	
Intracristate	0.3 (1)	4.0 (24)	3.6 (54)	0.3 (1)	88.4 (29)	
Inner Membrane	21.8 (75)	2.8 (17)	0.7 (10)	105.0 (79)	6.5 (2)	
Matrix	3.5 (12)	1.4 (9)	1.2 (18)	1.7 (1)	198 (65)	
Recovery	29.0 (100)	16.3 (100)	6.7 (100)	132 (100)	303 (100)	

Numbers in parentheses are percent of total activity recovered.

[a] Abbreviations are: PDG, phosphate dependent glutaminase; RIDCR, rotenone-insensitive DPNH cytochrome c reductase; Aden K, adenylate kinase; Cyt Ox, cytochrome oxidase; Mal D, malate dehydrogenase.

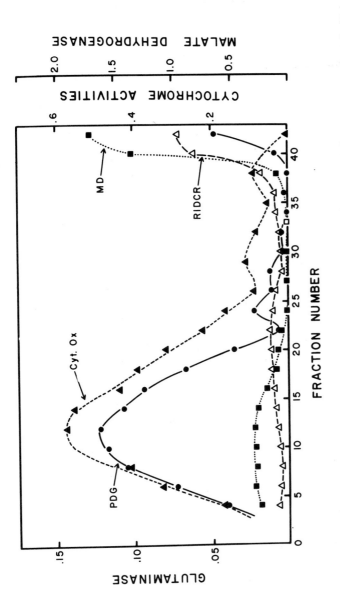

Fig. 2. Isopycnic centrifugation of swell-shrink sonicated mitochondria. Enzyme activities are expressed as μmoles ml^{-1} min^{-1}. Abbreviations are: MD, malate dehydrogenase; the others are as given in Table 2.

dase) banded at a density greater than intact mitochondria.
The close correlation between phosphate dependent glutaminase
and cytochrome oxidase confirms that this glutaminase is contained in the inner membrane and indicates that the results of
the previous experiment was not an artifact of treatment with
borate.

If the phosphate dependent glutaminase is latent (contained
on the inner surface of the inner membrane) then its access to
substrate is limited by a specific permeability barrier.
Alteration in mitochondrial transport of glutamine in response
to metabolic acidosis could very effectively regulate the contribution of this glutaminase activity to renal ammoniagenesis.
Alternatively, because of its submitochondrial localization,
the phosphate dependent glutaminase could be a component of the
mitochondrial transport system for glutamine. This is consistent with the experiments of Kovacevic et al (1970) which have
shown that rat kidney mitochondria rapidly take up glutamine,
but that during glutamine uptake only glutamate accumulated
within mitochondria.

Localization of Phosphate Independent Glutaminase

Following isopycnic centrifugation of the heavy microsomal
fraction, phosphate independent glutaminase bands in a sharp
symmetrical peak with a mean density (1.163) similar to that
observed for light mitochondria (Fig. 3). This glutaminase
activity does not correlate with markers for lysosomes (acid
phosphatase) or for microsomes (TPNH cytochrome c reductase),
but it does show an excellent correlation with alkaline phosphatase and γ-glutamyl transpeptidase activities. Both of
these latter activities have been shown by histological staining (Jacobsen et al, 1967; Glenner et al, 1962) to be localized in the luminal membrane (brush border) of rat kidney proximal tubules.

A preparation of intact kidney cells results in release of
20% of the phosphate independent glutaminase activity but less
than 10% of the lactate dehydrogenase and cytochrome oxidase
activities (Fig. 4). Subsequent incubation of the cell preparation with papain resulted in continued release of phosphate independent glutaminase activity from the cells. After
4 hours greater than 60 % of this glutaminase activity was
released whereas less than 20% of the lactate dehydrogenase
and 10% of the cytochrome oxidase activities were released.
The results of these experiments suggest that the phosphate
independent glutaminase is contained primarily on the external
surface of the luminal membrane of the proximal tubule cells
of the kidney.

706

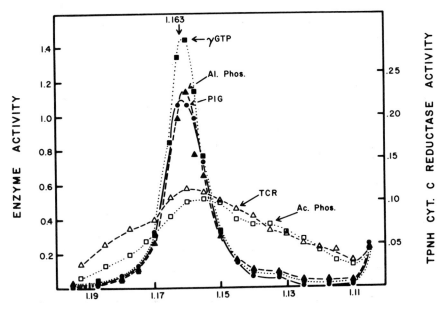

Fig. 3. Isopycnic centrifugation of heavy microsomal fraction obtained by differential centrifugation. Enzyme activities are expressed as μmoles ml^{-1} min^{-1}. Abbreviations are : TCR, TPNH cytochrome c reductase; Ac Phos, acid phosphatase; the others are as given in Fig. 5.

Characterization of the Phosphate Independent Glutaminase

The phosphate independent glutaminase has a broad substrate specificity. It can hydrolyze a number of glutamine analogs including γ-glutamylhydroxamate. Because of this broad specificity and its close correlation with γ-glutamyl trans-peptidase, the possibility that both reactions were catalyzed by the same enzyme was investigated. The only inconsistency in the literature with this possibility was the report of Glenner et al (1962) that histological staining indicates that γ-glutamyl transpeptidase is localized in the luminal membrane of both proximal convoluted and proximal straight tubule cells of the rat kidney, whereas quantitative micro analysis (Curthoys and Lowry, 1973) has shown that the phosphate inde-pendent glutaminase is localized primarily in the proximal straight tubule cells. To resolve this, both activities were assayed in the various regions of kidney tissue (Fig. 5). A close correlation between phosphate independent glutaminase

707

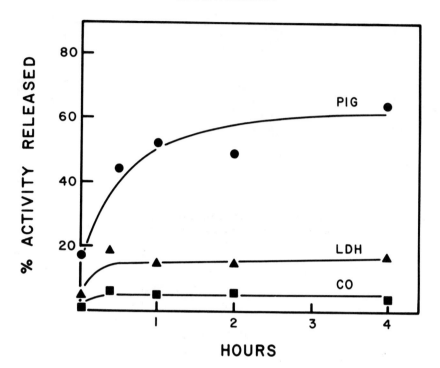

Fig. 4. Release of phosphate independent glutaminase from intact rat kidney cells. Both kidneys were perfused in situ with a collagenase, hyaluronidase solution containing 0.2 mg/ml papain. The kidneys were then excised, minced finely and incubated in the same enzyme solution for 30 min at 37°. The cell preparation was obtained by centrifugation at 150 x g for 1 min and washed twice. At zero time, this preparation was incubated at 37° in a solution containing 1.25 mg/ml papain. At various times aliquots were taken and centrifuged to remove cells. Both the cell preparation and the resultant supernatant were assayed for phosphate independent glutaminase (PIG), lactate dehydrogenase (LDH) and cytochrome oxidase (CO) activities.

and γ-glutamyl transpeptidase activities was observed. Both activities were maximal in outer stripe region of medulla, consistent with the idea that both enzyme activities were localized primarily in the proximal straight tubule cells. In contrast, the alkaline phosphatase activity was found to be maximal in both cortex and outer stripe regions consistent with its histological localization in the luminal membranes of both the proximal convoluted and proximal straight tubule

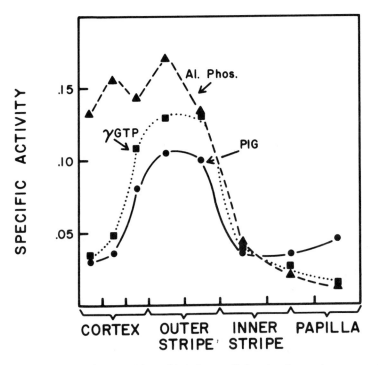

Fig. 5. Regional distribution of luminal membrane activities in rat kidney. A cone of kidney tissue was cut in such a way that its base consisted solely of cortical tissue and as one progressed towards the apex of the cone, one progressed through outer stripe and inner stripe regions of the medulla and finally the apex consisted solely of papillary region. Consecutive slices were then cut from the base of the cone, homogenized and assayed for phosphate independent glutaminase (PIG), γ-glutamyl transpeptidase (γ GTP) and alkaline phosphatase (Al Phos) activities. Specific activity is expressed as μmoles min^{-1} mg^{-1}.

cells (Jacobsen et al, 1967).

A comparison of phosphate independent glutaminase and γ-glutamyl transpeptidase in various preparations is shown in Table 3. In crude homogenates the γ-glutamyl transpeptidase activity when assayed by following the disappearance of glutathione, is 20 times greater than phosphate independent glutaminase. The microsomal preparation was obtained by differential centrifugation and was heated to 50° for 10 min. These conditions selectively destroy the phosphated dependent glutaminase activity (Katunuma et al, 1967). The purified preparation was obtained by solubilization of enzyme by treatment

TABLE 3

COMPARISON OF PHOSPHATE INDEPENDENT GLUTAMINASE
AND γ-GLUTAMYLTRANSPEPTIDASE ACTIVITIES

| | Specific Activity (μmoles min^{-1} mg^{-1}) | | |
	PIG[a]	γGLUpNA	GLUTH
Crude Homogenate	0.063	0.084 (1.3)	1.3 (20)
Heated Microsomes	0.32	0.41 (1.3)	6.0 (19)
Purified Preparation	10.4	13.4 (1.3)	167 (16)

[a] Abbreviations are: PIG, phosphate independent glutaminase;
γGLUpNA and GLUTH, γ-glutamyl transpeptidase assayed with
γ-glutamyl-p-nitroanalide and glutathione, respectively.

Numbers in parentheses are the ratio of γ-glutamyl transpeptidase to glutaminase activity.

of microsomes with papain and subsequent gel filtration and
chromatography on QAE-Sephadex and hydroxylapatite (Curthoys,
unpublished results). This preparation produces one major
band on polyacrylamide gels which is coincident with both
activities and which stains extensively for carbohydrate with
periodic acid Schiff's reagent. The ratio of γ-glutamyl
transpeptidase and glutaminase activities remains constant in
all of these preparations.

The phosphate independent glutaminase is activated by
maleate (Katunuma et al, 1967). To a lesser extent, maleate
also activates the γ-glutamyl transpeptidase activity. But,
the half maximal activation of both reactions occurs at the
same maleate concentration. With glutamine as a substrate,
maleate has a greater activating effect on the formation of
glutamate than it does on the formation of ammonia. This is
because in the absence of maleate, the phosphate independent
glutaminase produces a greater amount of ammonia than glutamate.
The addition of maleate stimulates both activities but now
stoichiometric amounts of ammonia and glutamate are formed.
It appears that the presence of maleate makes this enzyme a
more specific glutaminase. The chromatogram shown in Fig. 6
indicates that maleate has a similar effect on the product
specificity with γ-glutamyl-p-nitroanalide as a substrate.
In the absence of maleate, the major product migrates more
slowly than γ-glutamyl-p-nitroanalide and is probably γ-glu-
tamyl- γ-glutamyl-p-nitroanalide. But, in the presence of
maleate the major product is glutamate. Therefore, maleate

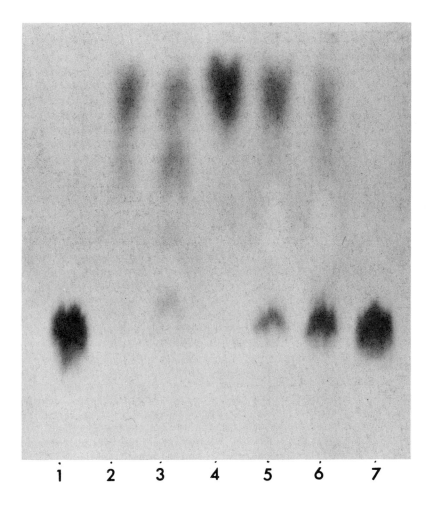

Fig. 6. Effect of maleate on products formed by phosphate independent glutaminase- γ-glutamyl transpeptidase enzyme. Approximately, 0.03 units of highly purified enzyme were added to 125 μl of the γ-glutamyl-p-nitroanalide assay mix which contained either no maleate (2 and 3) or 60 mM maleate (5 and 6). The reactions were spotted by addition of acid at 3 min (2 and 5) and 10 min (3 and 6). Then 10 μl were stopped on Whatman #2 paper. Glutamate was spotted in positions 1 and 7 and in position 4 unreacted γ-glutamyl-p-nitroanalide was spotted. The chromatogram was run in a phenol:H_2O (75:25) solvent adjusted to pH 7.2 with concentrated NH_4OH and developed with ninhydrin.

binds either to a third site or to the receptor site of the
γ-glutamyl transpeptidase and blocks transfer, making this
enzyme a specific glutaminase.

These experiments suggest that the γ-glutamyl transpeptidase and phosphate independent glutaminase are reactions
catalyzed by the same enzyme which is localized on the external
surface of the luminal membrane of the proximal straight tubule
cells. The extracellular localization is consistent with the
proposed role for γ-glutamyl transpeptidase in amino acid
transport (Orlowski and Meister, 1970). But, physiological
experiments on amino acid readsorption suggests that the bulk
of amino acid transport occurs in the initial portion of the
proximal tubule. Our finding that this activity is localized
in the most distal or straight portion of the proximal tubule
is therefore inconsistent with the proposal that it functions
in readsorbing the bulk of the filtered amino acids. The γ-
glutamyl cycle may function as a reserve mechanism to recover
amino acids not transported by other processes. A single turn
of the γ-glutamyl cycle requires hydrolysis of 3 ATPs. This
would provide sufficient energy to concentrate amino acids
against a very large concentration gradient and could account
for the observation that under normal conditions no amino
acids can be detected in the urine, even thought the urine
represents a 100 fold concentration of the initial glomerular
filtrate.

It is unlikely that maleate is a physiological activator of
phosphate independent glutaminase activity. But the occurrence of this effect suggests the possibility of a physiological counterpart to maleate. Katunuma et al (1968) have reported the extraction of a kidney factor which stimulates phosphate
independent glutaminase activity and Alleyne and Roobol (1974)
have demonstrated the presence of a factor in serum of acutely
acidotic rats which stimulates ammonia synthesis in kidney
slices. The appearance of such a factor in the tubular lumen
in response to acidosis could convert the γ-glutamyl transpeptidase into a specific glutaminase. Also, the glutaminase
activity has a lower pH optimum than the γ-glutamyl transpeptidase activity. And therefore, the acidification of the
urine which occurs during acidosis could also promote glutaminase activity over transpeptidase. Therefore, the possibility
that the phosphate independent glutaminase- γ-glutamyl transpeptidase enzyme may contribute to renal ammoniagenesis in
response to acidosis warrants further consideration.

ACKNOWLEDGMENTS

This investigation was supported in part by research grants AM 16651 from National Institute of Arthritis, Metabolism and Digestive Diseases and P-26 from Health Research and Services Foundation, Pittsburgh, Pa.

Many of the experiments in this paper were carried out with the excellent technical assistance of Robert F. Weiss and Theresa Kuhlenschmidt.

REFERENCES

Alleyne, G. A. O. and A. Roobol 1974. Regulation of renal ammoniagenesis. I. Stimulation of renal cortex ammonia-genesis in vitro by plasma isolated from acutely acidotic rats. *J. Clin. Invest.* 53: 117-121.

Balagura-Baruch, S. 1971. Renal metabolism and transfer of ammonia. in *The Kidney, vol. 3* (Rouiller and Muller, eds). Academic Press, New York, 253-327.

Chan, T. L., J. W. Greenawalt, and P. L. Pederson 1970. Biochemical and ultrastructural properties of a mitochondrial inner membrane fraction deficient in outer membrane and matrix activities. *J. Cell Biol.* 45: 291-305.

Ch'ih, J. J. and T. M. Devlin 1971. Protein synthesis in two mitochondrial populations isolated by isopycnic centrifugation from normal and renoprival kidney. *Biochem. Biophys. Res. Commun.* 43: 962-967.

Curthoys, N. P. and O. H. Lowry 1973. The distribution of glutaminase isoenzymes in the various structures of the nephron in normal, acidotic and alkalotic rat kidney. *J. Biol. Chem.* 248: 162-168.

Glenner, G. G., J. E. Folk, and P. J. McMillan 1962. Histochemical demonstration of a gamma-glutamyl transpeptidase like activity. *J. Histochem. Cytochem.* 10: 481-489.

Goldstein, L. 1967. Pathways of glutamine deamination and their control in the rat kidney. *Amer. J. Physiol.* 213: 983-989.

Hogeboom, G. H., W. C. Schneider, and G. E. Pallade. 1948. Cytochemical studies of mammalian tissues. I. Isolation of intact mitochondria from rat liver; some submicroscopic particulate material. *J. Biol. Chem.* 172: 619-636.

Jacobsen, N. O., F. Jorgensen, and A. C. Thomsen 1967. On the localization of some phosphatases in three different segments of the proximal tubules in the rat kidney. *J. Histochem. Cytochem.* 15: 456-469.

Katunuma, N., A. Huzino, and I. Tomino 1967. Organ specific control of glutamine metabolism. *Advan. Enz. Reg.* 5: 55-69.

Katunuma, N., T. Katsunuma, I. Tomino, and Y. Matsuda 1968. Regulation of glutamine activity and differentiation of the isoenzymes during development. *Advan. Enz. Reg.* 6: 227-242.

Klingman, J. D. and P. Handler 1958. Partial purification and properties of renal glutaminase. *J. Biol. Chem.* 232: 369-380.

Kovacevic, Z., J. D. McGiven, and J. B. Chappell 1970. Conditions for activity of glutaminase in kidney mitochondria. *Biochem. J.* 118: 265-274.

Nagata, N. and H. Rasmussen 1970. Renal gluconeogenesis: effects of Ca^{+2} and H^{+}. *Biochim. Biophys. Acta* 215: 1-16.

Orlowski, M. and A. Meister 1970. The γ-glutamyl cycle: a possible transport system for amino acids. *Proc. Natl. Acad. Sci.* 67: 1248-1255.

Palekar, A., S. S. Tate, and A. Meister 1974. Formation of 5-oxoproline from glutathione in erythrocytes by the γ-glutamyltranspeptidase-cyclotransferase pathway. *Proc. Natl. Acad. Sci.* 71: 293-297.

Ray, T. K. 1970. A modified method for the isolation of the plasma membrane from rat liver. *Biochim. Biophys. Acta* 196:1-9.

Schachter, H., I. Jabbal, R. L. Hudgin, L. Pinteric, E. J. McGuire, and S. Roseman 1970. Intracellular localization of liver sugar nucleotide glycoprotein glycosyltransferase in a Golgi-rich fraction. *J. Biol. Chem.* 245: 1090-1100.

Schnaitman, C. and J. W. Greenawalt 1968. Enzymatic properties of the inner and outer membranes of rat liver mitochondria. *J. Cell Biol.* 38: 158-175.

Sottocassa, G. L., B. Kuylenstierna, L. Ernster, and A. Bernstrand 1967. An electron-transport system associated with the outer membrane of liver mitochondria. *J. Cell Biol.* 32: 415-438.

Waldman, R. H. and H. B. Burch 1963. Rapid method for study of enzyme distribution in rat kidney. *Amer. J. Physiol.* 204: 749-752.

Van Slyke, D. D., R. A. Phillips, P. B. Hamilton, R. M. Archibold, P. H. Futcher, and A. Hiller 1943. Glutamine as source of urinary ammonia. *J. Biol. Chem.* 150: 481-482.

ROLE OF ADENYLYLATED GLUTAMINE SYNTHETASE ENZYMES AND URIDYLYLATED REGULATORY PROTEIN ENZYMES IN THE REGULATION OF GLUTAMINE SYNTHETASE ACTIVITY IN *ESCHERICHIA COLI*

E.R. STADTMAN, J.E. CIARDI, P.Z. SMYRNIOTIS
A. SEGAL, A. GINSBURG AND S.P. ADLER
Laboratory of Biochemistry, National Heart and Lung
Institute, National Institutes of Health,
Bethesda, Maryland   20014

ABSTRACT. Regulation of glutamine synthetase in *Escherichia coli* is facilitated by an elaborate cascade system involving two kinds of nucleotidylylation reactions which are modulated by the ratios of various metabolites. The activity of glutamine synthetase is controlled by the covalent attachment and detachment of an adenylyl group to one or more of the enzyme's 12 subunits. The adenylylation and deadenylylation reactions are both catalyzed by a single adenylyltransferase (ATase), whose activity is specified by a regulatory protein. The capacity of the regulatory protein to stimulate adenylylation on the one hand or deadenylylation on the other is specified by the covalent attachment of a uridylyl group to one or more of its four identical subunits. Because glutamine synthetase and the regulatory protein are multimeric proteins composed of several identical subunits, the epigenetic modifications by nucleotidylylation lead to multimolecular enzyme forms. Hybrid molecules of glutamine synthetase (i.e., enzymes composed of both adenylylated and unadenylylated subunits) prepared in vitro by dissociation and reassociation of mixtures of fully adenylylated and fully unadenylylated forms, are indistinguishable from enzymes produced in vivo containing the same average number of adenylylated subunits. Whereas these enzymes cannot be physically separated from one another by chromatography or by electrophoresis, they can be detected on the basis of differences in their catalytic potential and in their susceptibility to inactivation by 4 M urea. Heterologous subunit interactions in hybrid molecules affect the capacities of adenylylated and unadenylylated subunits to catalyze the biosynthesis of glutamine in the presence of Mn^{2+} and Mg^{2+}, respectively, as well as their apparent affinities for the substrate glutamate. From the non-linear relationship between Co^{2+}-supported biosynthetic activity and the average state of adenylylation, it is suggested that adenylylation of a subunit causes inactivation of that subunit as well as those unadenylylated subunits that are in direct contact with it. Hybrids produced by reversible dissociation of a mixture containing equal amounts of fully unadenylylated and succinylated fully adenylylated glutamine synthetase can be separated from either parental type by electrophoresis; these hybrids are mixtures of partially adenylylated-succinylated molecules with 4 to 9 adenylylated subunits per dodecameric aggregate. In contrast to glutamine synthetase enzymes, the uridylylated and unmodified forms of the regulatory protein can be separated by disc gel electrophoresis.

Activity of glutamine synthetase in *Escherichia coli* is regulated by several different mechanisms including: a) repression and derepression of its synthesis in response to fluctuations in the concentrations of various nutritional factors (Woolfolk, et al., 1966; P. Senior, unpublished data); b) the interconversion of inactive and active configurations of the enzyme in response to variations in the concentrations of divalent cations (Kingdon, et al., 1968; Shapiro and Ginsburg, 1968); c) cumulative feedback inhibition by a number of different end products of glutamine metabolism (Woolfolk and Stadtman, 1967); d) a highly sophisticated cascade system involving the adenylylation of glutamine synthetase (Mecke, et al., 1966; Shapiro, et al., 1967; Kingdon, et al., 1967; Shapiro, 1969; Anderson and Stadtman, 1970); and the uridylylation of a small regulatory protein (Brown, et al., 1971; Mangum, et al., 1973; Adler, et al., 1974).

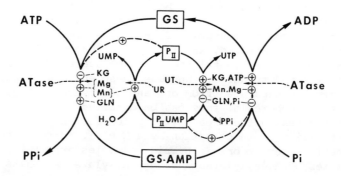

Fig. 1. Interrelationship between the uridylylation of the P_{II} regulatory protein and the adenylylation of glutamine synthetase, and the control of these modifications by metabolites. Abbreviations: UR=uridylyl removing enzyme, UT=uridylyltransferase; ATase=adenylyltransferase; KG=α-ketoglutarate; GLN=glutamine.

Figure 1 illsutrates the interrelationship between adenylylation and deadenylylation of glutamine synthetase and the uridylylation and deuridylylation of the regulatory protein, P_{II}. The outer circle in Fig. 1 shows that the reaction of glutamine synthetase (GS) with ATP leads to the formation of adenylylated enzyme (AMP-GS) and PPi; whereas deadenylylation of the enzyme involves a phosphorolytic cleavage of the AMP-GS to form ADP and GS. In the adenylylation reaction, the adenylyl group of ATP becomes attached in phosphodiester linkage to the phenolic hydroxyl group of a unique tyrosyl residue in each subunit of the enzyme.

Since the adenylylation and deadenylylation reactions are catalyzed by the same adenylyltransferase (ATase), the activity of this enzyme must be rigorously controlled to avoid futile coupling that would result only in senseless phosphorolysis of ATP and PPi (Anderson, et al., 1970). Such senseless coupling is avoided by the action of the regulatory protein, P_{II}, which specifies whether the ATase catalyzes, adenylylation or deadenylylation of glutamine synthetase (Brown, et al., 1971). As is shown in Fig. 1 the unmodified form of P_{II} specifically stimulates the capacity of ATase to catalyze the adenylylation of GS; this capacity is dependent on the presence of a divalent cation and is stimulated by glutamine, and is inhibited by α-ketoglutarate and orthophosphate. In contrast, the uridylylated form of P_{II} (i.e., UMP-P_{II}) can only stimulate the capacity of ATase to catalyze deadenylylation of AMP-GS; moreover, this deadenylylation reaction is inhibited by glutamine and Pi and is dependent on the presence of ATP, α-ketoglutarate, and a divalent cation. The inner circle of Fig. 1 shows that the interconversion of P_{II} between its uridylylated and unmodified states is catalyzed by the joint action of a uridylyltransferase (UT) and a uridylyl removing enzyme (UR). The uridylyltransferase catalyzes attachment of a uridylyl group from UTP to one of two tyrosyl residues on each subunit of P_{II} (Adler, Purich and Stadtman, 1974); this requires also the presence of ATP, α-ketoglutarate and a divalent cation, and is inhibited by glutamine and Pi. Figure 1 shows also that Mn^{2+} alone may activate the deuridylylation of UMP-P_{II}, but under certain conditions the UR-catalyzed deuridylylation reaction is stimulated by either Mn^{2+} or Mg^{2+} provided α-ketoglutarate and ATP are also present (Adler, et al., 1974). It is evident therefore that in *Escherichia coli*, regulation of glutamine synthetase activity is achieved by an elaborate cascade system involving two kinds of nucleotidylylation reactions which are modulated by the ratios of various metabolites. These nucleotidylylation reactions are pertinent to the present symposium discussion because they involve epigenetic modifications of protein structure which lead to multiple molecular forms of enzymes.

Since glutamine synthetase is composed of 12 identical subunits (Woolfolk, et al., 1966; Shapiro and Ginsburg, 1968) which are arranged in two superimposed hexagonal rings (Valentine, et al., 1968), and since all 12 of the subunits can be adenylylated (Kingdon, et al., 1967), it is obvious that the enzyme can exist in molecular forms that differ from each other with respect to the number (0-12) and orientation of adenylylated subunits within single molecules. M.S. Raff and W.C. Blackwelder (personal communication) have calculated that 382 molecular forms of glutamine synthetase are possible. However, unlike other isozymes which have generally been detected as separable electrophoretic or chromatographic bands, the forms of glutamine synthetase cannot be resolved by any of the conventional chromatographic or electrophoretic techniques (J. Ciardi, et al., 1974). Among other possibilities, adenylylation of glutamine synthetase could be associated with concomitant protonation of the enzyme to maintain a similar net charge. Nevertheless, it has been possible to show that hybrid enzyme moelcules (i.e., molecules containing both adenylylated and unadenylylated subunits) do exist and that heterologous interactions between adenylylated and unadenylylated subunits affect the catalytic parameters and stability of the enzyme.

PREPARATION OF PARTIALLY ADENYLYLATED ENZYME

For comparative studies, enzyme preparations containing an average of 0-12 adenylyl groups per mole have been prepared by the following procedures.

1. *Direct isolation*. So called <u>natural</u> enzyme preparations containing different amounts of covalently bound adenylyl groups per mole were isolated from different batches of cells grown under different nutritional conditions. Unadenylylated enzyme $(GS_{\bar{0}})$[1] is readily obtained from bacteria harvested in the stationary phase following growth on a glucose or glycerol medium with a limiting concentration of NH_4^+ as the sole source of nitrogen (Kingdon and Stadtman, 1967; Shapiro and Stadtman, 1970), or from cells grown in the presence of 100 µM Mn^{2+} under <u>conditions of Mg^{2+}</u> limitation (P. Senior, unpublished data).

[1]The average number of adenylylated subunits per mole of enzyme, hereafter referred to as the "state of adenylylation", is designated by the subscript $\bar{n}$ following the standard abbreviation for glutamine synthetase, GS. Since GS contains 12 subunits the value of $\bar{n}$ can vary from $\bar{0}$-$\overline{12}$; thus, unadenylylated enzyme is $GS_{\bar{0}}$, fully adenylylated enzyme is $GS_{\overline{12}}$, and an enzyme preparation containing an average of 7.5 equivalents of adenylyl groups per mole is $GS_{\overline{7.5}}$.

718

Adenylylated enzyme forms $GS_{\overline{11}}$ to $GS_{\overline{12}}$) are obtained from cells harvested in the stationary phase following growth on glucose or glycerol containing an excess of glutamate as the sole nitrogen source (Shapiro and Stadtman, 1970). Enzyme preparations at intermediate states of adenylylation ($GS_{\overline{1.0}}$ to $GS_{\overline{11}}$) are obtained from cells harvested at various times during the transition from log phase to stationary phase of growth.

2. *In vitro adenylylation or deadenylylation.* Enzyme at any average state of adenylylation can be obtained either by incubating $GS_{\overline{0}}$ for various periods of time with ATP, glutamine, Mg^{2+} and the adenylyltransferase (Ginsburg et al., 1970) or by controlled hydrolysis of $GS_{\overline{12}}$ with snake venom phosphodiesterase (Stadtman, et al., 1970; Stadtman, 1969).

3. *Subunit dissociation and reassociation of mixtures of $GS_{\overline{0}}$ and $GS_{\overline{12}}$.*
As is illustrated in Fig. 2, when mixtures of unadenylylated and fully adenylylated enzyme are treated with 7M urea at 0°, both forms are completely dissociated to form a heterogeneous pool of adenylylated (closed circles) and unadenylylated subunits (open circles).

Fig. 2. Preparation of reconstituted hybrid forms of glutamine synthetase by dissociation and reassociation of subunits in mixtures of adenylylated and unadenylylated enzyme. The open circles and filled circles refer to unadenylylated and adenylylated subunits. For simplicity, only one hexagonal ring of glutamine synthetase is illustrated. See Ciardi, et al.(1973) for details.

Ten-fold dilution of the dissociated subunit mixture with Tris-HCl buffer containing KCl, Mg^{2+}, Mn^{2+} and 2-mercaptoethanol (pH 7.5) results in reassociation of the subunits to produce a 55-65% yield of <u>reconstituted</u> <u>enzyme</u> consisting of catalytically active dodecameric aggregates with an average state of adenylylation approximately equal to that calculated from the ratio of $GS_{\overline{0}}$ and $GS_{\overline{12}}$ in the original mixture (Ciardi, et al., 1974). These <u>reconstituted</u> enzyme preparations are indistinguishable from <u>natural</u> enzyme preparations at the same state of adenylylation with respect to catalytic potential and physical characteristics.

4. *"Mixed" enzyme preparations.* For comparative purposes, $GS_{\overline{0.8}}$ and $GS_{\overline{12}}$ were mixed in various proportions to yield enzyme preparations of the desired average state of adenylylation. For example, a mixture containing 56% $GS_{\overline{0.8}}$ and 44% $GS_{\overline{12}}$ would

have an average of 5.7 adenylylated subunits per mole of enzyme and is referred to as "GS$_{\overline{5.7}}$ mix". Unlike <u>natural</u> or <u>reconstit-</u> <u>uted</u> enzyme preparations of intermediate states of adenylyla-tion, the <u>mixed</u> enzyme preparations do not contain hybrid forms (i.e., enzyme molecules composed of both adenylylated and un-adenylylated subunits (Stadtman, et al., 1970; Ginsburg, et al., 1970; Ciardi, 1970).

EVIDENCE FOR THE EXISTENCE OF HYBRID FORMS OF GLUTAMINE SYNTHETASE

The observations that adenylylation of glutamine synthetase is accompanied by marked changes in divalent cation specificity (Kingdon, et al., 1967; Ginsburg, et al., 1970) and in sus-ceptibility to inactivation by urea (Stadtman, et al., 1970; Ciardi, et al., 1973) are the basis of kinetic studies which disclose the existence of hybrid forms of glutamine synthetase.
1. *Relationships between state of adenylylation and the apparent affinity for glutamate.*
The biosynthetic activity of GS$_{\overline{12}}$ has an absolute requirement for Mn^{2+} whereas GS$_{\overline{0}}$ can be activated by either Mg^{2+} or Co^{2+} but <u>not</u> by Mn^{2+}. Data in Table I show that in the presence of Mn^{2+}, when all activity is presumably due to adenylylated sub-units, the concentration of glutamate required for half-maximal activity the $S_{\overline{0.5}}$ value) of <u>natural</u> enzyme preparations (see above) increased from 1.3 to 7.6 as the state of adenylylation varied from GS$_{\overline{2.3}}$ to GS$_{\overline{11.8}}$. In contrast the $S_{\overline{0.5}}$ value mea-sured in the presence of Mg^{2+} (unadenylylated subunit activity) decreased from 5.0 to 2.0 as the state of adenylylation was increased from GS$_{\overline{2.3}}$ to GS$_{\overline{9.0}}$ (Denton and Ginsburg, 1970). If the <u>natural</u> enzyme preparations used in these experiments were simply mixtures of GS$_{\overline{0}}$ and GS$_{\overline{12}}$, then the $S_{\overline{0.5}}$ values for glu-tamate should be independent of the average state of adenylyl-ation. It is evident, therefore, that the <u>natural</u> enzymes con-tain hybrid enzyme species and that heterologous interaction between adenylylated and unadenylylated subunits within these hybrid molecules affect either the affinity of the subunits for glutamate or lead to altered divalent cation specificity.
2. *Relationship between average state of adenylylation and the catalytic potential.*
The data in Fig. 3 show that in the presence of Mg^{2+} the spe-cific activity of enzyme preparations obtained by mixing GS$_{\overline{12}}$ and GS$_{\overline{0}}$ in various proportions in a linear inverse function of the average state of adenylylation. This is as expected from the fact that GS$_{\overline{12}}$ is totally inactive in the presence of Mg^{2+}. In contrast, the specific activities of partially adenylylated enzyme preparations obtained by adenylylation of E$_{\overline{0}}$ in vitro are complex sigmoidal functions of the average state

TABLE I

THE EFFECT OF ADENYLYLATION STATE ON THE CONCENTRATION
OF L-GLUTAMATE REQUIRED FOR HALF-MAXIMAL BIOSYNTHETIC
ACTIVITY IN THE PRESENCE OF Mg^{2+} OR Mn^{2+} [a]

Enzyme	$S_{0.5}$ (mM)	
	Mg^{2+}	Mn^{2+}
$GS_{\overline{2.3}}$	5.0	1.3
$GS_{\overline{9.0}}$	2.0	5.0
$GS_{\overline{11.8}}$		7.6

[a] The $GS_{\overline{2.3}}$ and $GS_{\overline{9.0}}$ preparations were isolated directly from cells and the $GS_{\overline{11.8}}$ preparation was derived from $GS_{\overline{2.3}}$ by in vitro adenylylation. $S_{\overline{0.5}}$ refers to the concentration of L-glutamate required for half-maximal activity (pH 7.1, 25°) in the presence of Mg^{2+} or Mn^{2+} as indicated. The data are from Denton and Ginsburg (1970).

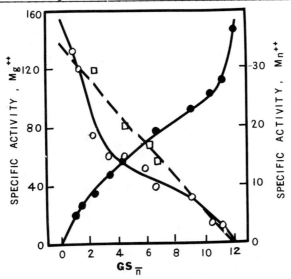

Fig. 3. Relationship between the state of adenylylation and the biosynthetic activity of glutamine synthetase preparations. Solid lines refer to the activity of enzyme preparations obtained either by direct isolation from various batches of cells or by in vitro adenylylation of $GS_{\overline{2.3}}$. Open circles refer to

721

(Fig. 3 legend continued)
the Mg^{2+} -dependent activity; closed circles refer to the Mn^{2+} -dependent activity. The dotted line refers to the Mg^{2+} - dependent activity of preparations obtained by mixing $GS_{\bar{0}}$ with $GS_{\overline{12}}$ in proportions to yield average states of adenylylation as indicated by the square symbols. The figure is from Gins- burg, et al. (1970) where the experiment is described in detail. of adenylylation, whether assayed in the presence of Mg^{2+} or Mn^{2+} (Ginsburg, et al., 1970). Similar results were obtained (data not shown) with natural enzyme preparations of intermed- iate states of adenylylation. Therefore, enzymatically der- ived partially adenylylated enzyme preparations produced either in vitro or in vivo are not simply mixtures of $GS_{\bar{0}}$ or $GS_{\overline{12}}$, but contain hybrid enzyme forms; moreover, heterologous inter- actions between adenylylated and unadenylylated subunits in the hybrid molecules affect their catalytic potential.

A somewhat similar, though less complex, relationship ex- ists between the average state of adenylylation and the specific activity of glutamine synthetase measured in the presence of Co^{2+}. As is illustrated in Fig. 4, the specific activities of mixed enzyme preparations are an inverse linear function of the average state of adenylylation, reflecting the fraction of $GS_{\bar{0}}$ present; $GS_{\overline{12}}$ is completely inactive in the presence of Co^{2+}. In contrast, the specific activity of partially adenyl- ylated natural enzyme preparations and of enzyme adenylylated in vitro are curvilinear functions of the average state of adenylylation. These results offer further evidence for the existence of hybrid enzyme forms. Of several different pos- sibilities considered, it is found that the specific activity of these hybrids correlates well with the average number of subunits in a given hexagonal ring that are joined on both sides by unadenylylated subunits (Fig. 5). In arriving at this relationship, the number of unique, nonsuperimposable configurations of glutamine synthetase for each state of aden- ylylation was determined, assuming that adenylylation occurs by a random mechanism. For each configuration, the number of unadenylylated subunits not in contact with adenylylated sub- units within the same hexagonal ring was determined. The pro- cedure used in this analysis is illustrated in Fig. 6 which shows the distribution of adenylylated subunits that are pos- sible when a molecule contains only 2 adenylylated subunits . The good correlation between specific activity and the number of unadenylylated subunits not in contact with an adenylylated subunit (shown in Fig. 5) suggests that adenylylation of a sub- unit leads not only to its loss of ability to utilize Co^{2+} for activity but also to a loss in the ability of neighboring un- adenylylated subunits to be activated by Co^{2+}.

722

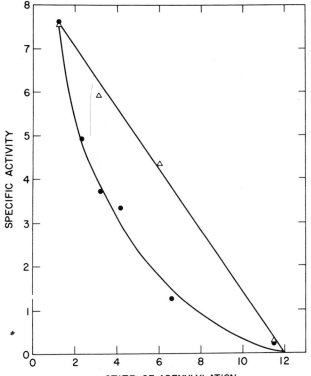

Fig. 4. Effects of the state of adenylylation on the Co-dep-
endent activity of glutamine synthetase. Native enzyme preparations (0) $E_{(2.3)}$ $E_{(6.6)}$, $E_{(11.8)}$ were prepared by the
method of Woolfolk, et al. (1966); $E_{(3.2)}$, $E_{(4.2)}$ were prepared by the method of Ginsburg, et al. (1970). Mixtures of
$E_{(1.2)}$ and $E_{(11.8)}$ (Δ) had average states of adenylylation of
3.2 and 5.8. Specific activity was determined under the following conditions: L-glutamate, 50 mM; NH_4Cl, 50 mM; $CoCl_2$,
7.5 mM; ATP, 7.5 mM. The pH was 7.0. To obtain significant
data, 83 µg of $E_{(11.8)}$ were used and 5.0 µg or less of all
other enzymes were used.

3. *Relationship between state of adenylylation and susceptibility of glutamine synthetase to inactivation by urea.*
Data in Fig. 7 show that $GS_{\bar{0}}$ is perfectly stable when incubated in a buffer containing 4M urea and 0.5 mM ADP, whereas the
activity of $GS_{\overline{12}}$ is completely lost after only a few minutes.
When a mixture of $GS_{\bar{0}}$ and $GS_{\overline{12}}$ having an average state of adenylylation of $GS_{\overline{7.5}}$ was incubated under these conditions, there
was a rapid loss of about 60% of the initial activity, attributable to the fraction of $GS_{\overline{12}}$ present, and then no further change.

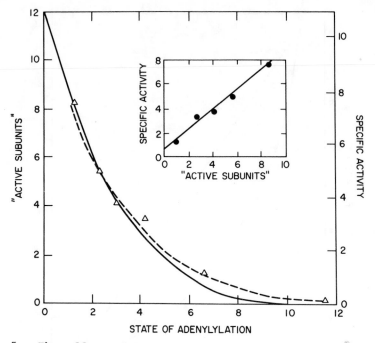

Fig. 5. The effect of heterologous subunit interactions in
hybrid glutamine synthetase molecules on the Co^{2+}-supported
biosynthetic activity. The average number of unadenylylated
subunits adjacent only to other unadenylylated subunits within
each ring of the double hexagon structure was calculated
("active subunits") for various average states of adenylylation
as described in Fig. 6 and plotted against the average state of
adenylylation (solid curve). The open triangles (dashed line)
are measurements of Co^{2+}-supported biosynthetic activity per
milligram of enzyme for different enzyme preparations. The
insert indicates the relationship between the Co^{2+}-supported
biosynthetic activity of the enzyme and the calculated number
of unadenylylated subunits that are adjacent only to other
unadenylylated subunits. The figure is from Seqal and Stadtman (1972), where the experimental details are described.
State of adenylylation assays (Stadtman, et al., 1970) showed
that the stable activity remaining after 30 minutes was all
due to $E_{\bar{0}}$. In contrast, when a natural enzyme preparation
of approximately the same state of adenylylation ($GS_{\overline{7.2}}$) was
incubated under these conditions, the rate of inactivation
was only about one-half as rapid, and the stable activity
that remained after 30 minutes was about $GS_{\overline{5.0}}$; an almost
identical result was obtained when a reconstituted enzyme
preparation (prepared by subunit dissociation and reassociation

724

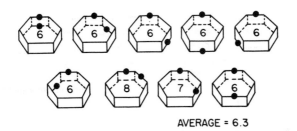

AVERAGE = 6.3

Fig. 6. Determination of the average number of unadenylylated subunits adjacent only to unadenylylated subunits. The black dots represent positions of 5'-AMP moieties on the adenylylated subunits. The unique nonsuperimposable configurations GS_2 are displayed. The number in the center of each molecule is the number of unadenylylated subunits adjacent only to other unadenylylated subunits. To determine the average number of unadenylylated subunits not adjacent to an adenylylated subunit, the number of such subunits for each of the nine configurations was summed and divided by the number of configurations to give 6.3.

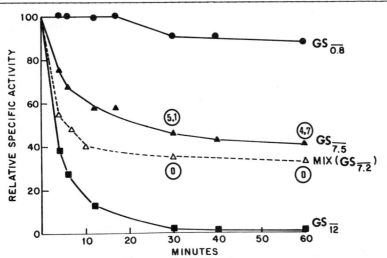

Fig. 7. Detection of glutamine synthetase hybrid forms by the kinetics of inactivation in 4 M urea. Enzyme preparations (0.24 mg/ml) were incubated at 37 , pH 7.0, in a reaction mixture containing 0.187 M mixed imidazole buffer (composed of equal concentrations of imidazole, 2,4-dimethylimidazole, and 2-methylimidazole), 0.5 mM ADP, 25 mM L-glutamine, 25 mM potassium arsenate, and 4 M urea. At times indicated, aliquots were removed and assayed for transferase activity and for the average state of adenylylation. The numbers in the circles indicate the average states of adenylylation of the $GS_{\overline{7.5}}$

(Fig. 7 legend continued)
(large circles) and mix $GS_{\overline{7.2}}$ (small circles) after 30 and 60 min. exposure to 4 M urea, as indicated. The $GS_{\overline{0.8}}$ and $GS_{\overline{7.5}}$ preparations were isolated directly from different batches of cells. The $GS_{\overline{12}}$ prepared by ATase-catalyzed adenylylation of the $GS_{\overline{7.5}}$ preparation. The "Mix $GS_{\overline{7.2}}$" was prepared by mixing $GS_{\overline{0.8}}$ and $GS_{\overline{12}}$ in proper proportions.

of a mixture of $GS_{\overline{0.8}}$ and $GS_{\overline{12}}$ was incubated under these inactivation conditions. These results show that natural and reconstituted enzyme preparations of intermediate states of adenylylation contain hybrid molecules and that heterologous interaction between adenylylated subunits in the hybrids stabilize adenylylated subunit activity.

ELECTROPHORETIC SEPARATION OF RECONSTITUTED HYBRIDS CONTAINING SUCCINYLATED-ADENYLYLATED SUBUNITS

It has not been possible to separate adenylylated from unadenylylated enzyme by conventional electrophoretic or chromatographic techniques.

As noted above, reconstituted enzyme preparations obtained by subunit dissociation and reassociation of mixtures of $GS_{\overline{0}}$ and $GS_{\overline{12}}$ as described in Fig. 2 apparently contain hybrid molecular forms, since they exhibit inactivation kinetics in 4 M urea that are similar to those of natural enzyme preparations of comparable states of adenylylation (Fig. 7). Unfortunately the production of hybrids could not be verified by direct analysis because until now it has not been possible to detect marked differences in the physical-chemical characteristics of adenylylated and unadenylylated enzymes. Unlike other isozymes these forms are not separable by conventional electrophoresis techniques or by chromatography (Ciardi, et al., 1973). To prepare chemically modified variants of adenylylated enzyme that could be separated from $GS_{\overline{0}}$, a sample of $GS_{\overline{12}}$ was succinylated by reaction with succinic anhydride by the procedure of Meighen and Schachman, (1970). As illustrated in Fig. 8, three different reconstituted enzyme preparations were prepared by dissociation and reassociation of $GS_{\overline{0.8}}$, succinylated $GS_{\overline{12}}$, and a mixture of $GS_{\overline{0.8}}$ and succinylated $GS_{\overline{12}}$. Figure 9 shows that all three reconstituted preparations are widely separated from one another by electrophoresis on cellulose acetate, with the reconstituted enzyme derived from a mixture of $GS_{\overline{0.8}}$ and succinylated $GS_{\overline{12}}$ migrating as a rather diffuse band of intermediate mobility. These results show that succinylation of $GS_{\overline{12}}$ produces a chemically modified variant that is easily separated from unmodified $GS_{\overline{0.8}}$ and offers direct proof that reconstituted enzyme preparations obtained by the dissociation

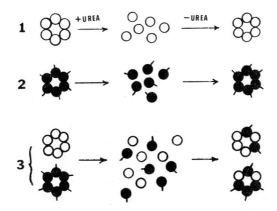

Fig. 8. Preparation of <u>reconstituted</u> enzymes from (1) unadenylylated enzyme, (2) succinylated-adenylylated enzyme and
(3) a mixture of unadenylylated and succinylated-adenylylated
enzyme. For symbols see legend to Fig. 2. Details are described by Ciardi, et al., 1973.

and reassociation of mixtures of $GS_{\overline{0.8}}$ and $GS_{\overline{12}}$ are composed
of hybrid species of enzyme. The diffuseness of the electro-
phoretic band obtained with the <u>reconstituted</u> enzyme prepared
from the mixture of $GS_{\overline{0.8}}$ and $GS_{\overline{12}}$ suggests that this preparation consists of a mixture of hybrid forms at different states
of adenylylation. This is confirmed by the data in Fig. 10
showing that a diffuse band obtained by disc gel electrophoresis
of the <u>reconstituted</u> enzyme mixture is heterogeneous. From the
distribution of succinylylated-adenylylated subunits and unadenylylated subunits in this band it could be calculated
that the average state of adenylylation varied from $GS_{\overline{9.0}}$ on
the side closest to the anode to $GS_{\overline{4.0}}$ on the side closest to
the cathode.

P_{II}-REGULATORY PROTEIN

As shown in Fig. 1 modulation of adenylylation and deadenylylation of glutamine synthetase is facilitated by the
uridylylation and deuridylylation of the P_{II} regulatory protein. the P_{II} protein is a small protein of about 44,000 M.W.
and is composed of 4 identical subunits of 11,000 M.W. each
(Adler, Purich and Stadtman, 1974). Since uridylylation
involves attachment of a UMP group in phosphodiester linkage
to a tyrosyl residue on each subunit, at least 5 different
species of the regulatory protein are possible. Preliminary
studies show that fully uridylylated P_{II} (i.e., P_{II} $(UMP)_4$)
is readily separated from unmodified P_{II} by disc gel electro-

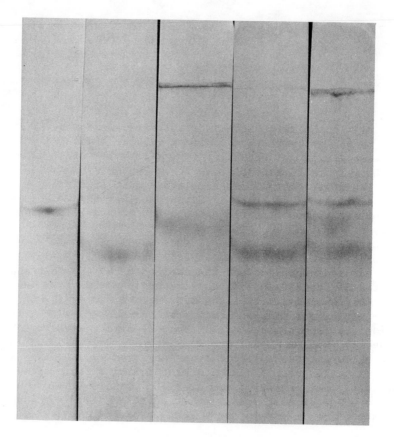

Fig. 9. Separation of reconstituted hybrid glutamine synthetases from the reconstituted succinylated-adenylylated ($E_{\overline{12}}$)
and unadenylylated $E_{\overline{0.8}}$ enzymes from which they were prepared.
Succinylated $E_{\overline{12}}$, unmodified $E_{\overline{0.8}}$, and a mixture containing
equal amounts of both were each subjected to dissociation and
reassociation conditions (Fig. 8). Two to six micrograms of
reconstituted enzyme protein were applied to cellulose acetate
strips and electrophoresis carried out. Protein was stained
with Ponceau S and Nigrosin. The bands of the strips, from left
to right, are: (1) $E_{\overline{0.8}}$, (2) succinylated $E_{\overline{12}}$, (3) partially
succinylated hybrids, (4) $E_{\overline{0.8}}$ and succinylated $E_{\overline{12}}$, and (5)
all three glutamine synthetase preparations. This **Fig.** is from

(Fig. 9 legend continued)
Ciardi, et al., 1973.

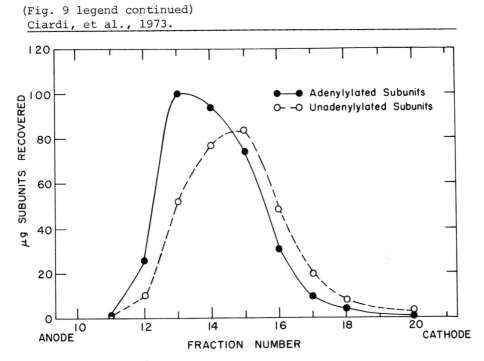

Fig. 10. Distribution of succinylated-adenylylated and unadenylylated subunits in reconstituted hybrid enzymes following reversible dissociation. Succinylated [^{14}C] adenylyl-E_{12}
(with 25% residual transferase activity) was mixed with an
equal amount of $E_{\overline{0.8}}$. The mixed enzymes were dissociated into
subunits and the subunits subsequently reassociated. The
resultant reconstituted enzyme mixture was subjected to electrophoresis on an acrylamide disc gel at pH 7.2. The gel was then
sliced into 1 mm sections and the enzyme eluted with 10 mM
imidazole-HCl-1 mM $MnCl_2$, pH 7.0. Aliquots of each fraction
were assayed for Mg^{2+}-dependent γ-glutamylhydroxamate activity
as a measure of unadenylylated subunits and radioactivity was
measured by liquid scintillation to estimate adenylylated
subunits. The amount of each type of subunit was calculated
for each fraction. This Fig. is from Ciardi, et al., 1973.
phoresis. It remains to be determined if hybrid forms containing 1-3 uridylylated subunits can be separated from one another
by electrophoresis, and whether or not heterologous interaction
between uridylylated and nonuridylylated subunits in hybrid
molecules affect either the capacity of the P_{II} protein to
direct the activity of the adenylyltransferase or its susceptibility to control by various metabolites.

REFERENCES

Adler, Stuart P., J.H. Mangum, G. Magni and E.R. Stadtman 1974. Uridylylation of the P_{II} regulatory protein in cascade control of *Escherichia coli* glutamine synthetase. *Third International Symposium on Metabolic Interconversion of Enzymes*, Springer Verlag, New York, pp. 221-233.

Adler, Stuart P., D. Purich and E.R. Stadtman 1974. Cascade control of *E. coli* glutamine synthetase: Properties of the P_{II} regulatory protein and the uridylyltransferase--uridylyl removing enzyme. *FASEB* 33:1427.

Anderson, W.B., S.B. Hennig, A. Ginsburg and E.R. Stadtman 1970. Association of ATP: Glutamine synthetase adenylyl-transferase activity with the P_I component of the glutamine synthetase deadenylylation system. *Proc. Natl. Acad. Sci. U.S.A.* 67:1417-1424.

Anderson, Wayne B. and E.R. Stadtman 1970. Glutamine synthetase deadenylylation: A phosphorolytic reaction yielding ADP as nucleotide product. *Biochem. Biophys. Res. Communs.* 41:704-709.

Brown, M.S., A. Segal and E.R. Stadtman 1971. Modulation of glutamine synthetase adenylylation and deadenylylation is mediated by metabolic transformation of the P_{II}-regulatory protein. *Proc. Natl. Acad. Sci. U.S.A.* 68:2949-2953.

Ciardi, J.E., F. Cimino and E.R. Stadtman 1973. Multiple forms of glutamine synthetase. Hybrid formation by association of adenylylated and unadenylylated subunits. *Biochemistry* 12:4321-4330.

Denton, M.D. and A. Ginsburg 1970. Some characteristics of the binding of substrates to glutamine synthetase from *Escherichia coli*. *Biochemistry* 9:617-632.

Ginsburg, A., J. Yeh, S.B. Hennig and M.D. Denton 1970. Some effects of adenylylation on the biosynthetic properties of the glutamine synthetase from *Escherichia coli*. *Biochemistry* 9:633-649.

Kingdon, H.S., J.S. Hubbard and E.R. Stadtman 1968. Regulation of glutamine synthetase, XI. The nature and implications of a lag phase in the *Escherichia coli* glutamine synthetase reaction. *Biochemistry* 7:2136-2142.

Kingdon, H.S., B.M. Shapiro and E.R. Stadtman 1967. Regulation of glutamine synthetase, VIII. ATP: glutamine synthetase adenylyltransferase, an enzyme that catalyzes alterations in the regulatory properties of glutamine synthetase. *Proc. Natl. Acad. Sci. U.S.A.* 58:1703-1710.

Kingdon, H.S. and E.R. Stadtman 1967. Regulation of glutamine synthetase, X. Effect of growth conditions on the susceptibility of *Escherichia coli* glutamine synthetase to feedback inhibition. *J. Bacteriol.* 94:949-957.

Mangum, John H., G. Magni and E.R. Stadtman 1973. Regulation of glutamine synthetase adenylylation and deadenylylation by enzymatic uridylylation and deuridylylation of the P_{II} regulatory protein. *Arch. Biochem. Biophys.* 158:514-525.

Mecke, D., K. Wulff, K. Liess and H. Holzer 1966. Characterization of a glutamine synthetase inactivating enzyme from *Escherichia coli*. *Biochem. Biophys. Res. Communs.* 24:542-558.

Meighen, E.A. and H.K. Schachman 1970. Hybridization of native and chemically modified enzymes. I. Development of a general method and its application to the study of the subunit structure of aldolase. *Biochemistry* 9:1163-1176.

Segal, A. and E.R. Stadtman 1972. Effects of cobaltous ion on various catalytic parameters and on heterologous subunit interactions of *Escherichia coli* glutamine synthetase. *Arch. Biochem. Biophys.* 152:356-366.

Shapiro, B.M. 1969. The glutamine synthetase deadenylylating enzyme system from *Escherichia coli*. Resolution into two components, specific nucleotide stimulation and cofactor requirements. *Biochemistry* 8:659-670.

Shapiro, B.M. and A. Ginsburg 1968. Effects of specific divalent cations on some physical and chemical properties of glutamine synthetase from *Escherichia coli*. Taut and relaxed enzyme forms. *Biochemistry* 7:2153-2167.

Shapiro, B.M., H.S. Kingdon, and E.R. Stadtman 1967. Regulation of glutamine synthetase VII. Adenylylglutamine synthetase: a new form of the enzyme with altered regulatory and kinetic properties. *Proc. Natl. Acad. Sci. U.S.A.* 58:648-649.

Shapiro, B.M. and E.R. Stadtman 1970. Glutamine synthetase from *Escherichia coli*. *Methods in Enzymology*, H. Tabor and C.W. Tabor, eds., Academic Press, New York. pp. 910-922.

Stadtman, E.R. 1969. On the structure and allosteric regulation of glutamine synthetase from *Escherichia coli*. In: *The Role of Nucleotides for the Function of Conformation of Enzymes*. H.M. Kalckan, H. Klenow, M. Ottensen, A. Munch-Petersen and J.H. Thaysen, eds., Munksgaard, Copenhagen. pp. 111-137.

Stadtman, E.R., A. Ginsburg, J.E. Ciardi, J. Yeh, S.B. Hennig, and B.M. Shapiro 1970. Multiple molecular forms of glutamine synthetase produced by enzyme catalyzed adenylylation and deadenylylation reactions. *Advan. Enzyme Reg.* 8:99-118.

Valentine, R.C., B.M. Shapiro and E.R. Stadtman 1968. Regulation of glutamine synthetase XII. Electron microscopy of the enzyme from *Escherichia coli*. *Biochemistry* 7:2143-2152.

Woolfolk, C.A., B.M. Shapiro and E.R. Stadtman 1966. Regulation of glutamine synthetase, I. Purification and properties of glutamine synthetase from *Escherichia coli*. *Arch. Biochem. Biophys.* 116:177-192.

SELECTION MECHANISMS MAINTAINING ALCOHOL DEHYDROGENASE POLYMORPHISMS IN *Drosophila melanogaster*

RICHARD AINSLEY and G. BARRIE KITTO
Department of Zoology and
Clayton Foundation Biochemical Institute
Department of Chemistry, University of Texas
Austin, Texas 78712

ABSTRACT. Alcohol dehydrogenase (ADH) is an important detoxification mechanism in fruit flies which live as larvae in fermenting fruits, thus in high alcohol concentration. Many habitats are probably lethal for fruit flies lacking an active ADH detoxification system--which accounts for nearly complete absence of null-activity ADH phenotypes in wild populations. Selection maintaining the two common polymorphic forms seems to be based on subtle differences in the affinities of ADH's for ethanol, propanols and butanols. All ADH's have an active site apparently with greater affinity for propanols and butanols than for ethanol (which is probably the most important toxin)--and by such a large factor that non-toxic levels of propanol and butanol act as competitive inhibitors of ethanol which is at lethal concentration. Interestingly, one form of ADH most efficiently oxidizes ethanol in the absence of other alcohols but is not the form most efficient in the presence of other alcohols. Since ethanol is a toxin that causes delayed development or death, this subtle difference in kinetic properties of the ADH's produced by different phenotypes may result in considerable selection for efficient ethanol oxidizers. However, since the most efficient ethanol-oxidizing phenotype will vary from one habitat to another and also vary in time, ADH polymorphism is itself an important adaptation of *Drosophila melanogaster*.

Recently, several studies have been carried out to understand mechanisms which maintain genetic polymorphisms at various loci which code for proteins. Alcohol dehydrogenase (*Adh*) of *Drosophila melanogaster* is one of the most studied enzyme systems. The reason for this emphasis on ADH is that the function of ADH is less obscure than that of most other isozymes. This is so because *D. melanogaster* larvae live inside fermenting fruits which naturally contain a rather high concentration of alcohols. Thus, most workers have assumed ADH to have a role in an alcohol detoxification system, i.e., a role in removing toxic alcohols at least from body tissues. Proceeding from this assumption, they found associations be-

733

tween gene frequency changes and temperature or alcohol
stresses (Gibson, '70; Bijlsma-Meeles and van Delden, '74).
Others reported results of wild population surveys in which
ADH gene frequencies show clines correlating with latitude
(Pipkin, et al., '74; Johnson and Schaffer, '73; Vigua and
Johnson, '73). In the case of surveys, a selection mechanism
was assumed to be responsible for the observed gene frequency
cline. The mechanism was linked to differences in kinetic
properties of the different ADH morphs, under the assumption
that increased ethanol oxidation efficiency is a property
that is maximized by natural selection. It is desirable,
however, to make a direct test of the assumption that ADH
plays a role in detoxification of alcohols. A direct approach to this problem is feasible since *D. melanogaster*
stocks are available which lack ADH activity (Sofer and Hatkoff, '72), and since we have developed a culture technique
which allows careful regulation and monitoring of alcohol
concentrations in media. The experiments here attempt first
to define the physiological role of ADH, and then propose
selection mechanimsms for maintaining *Adh* polymorphisms which
are directly related to the physiological role of ADH.

MATERIALS AND METHODS

Drosophila melanogaster Stocks

Two mutant lines of *D. melanogaster* kindly made available
by Dr. W. H. Sofer have both high fecundity and high viability
on potato-Kalmus media. One line is homozygous for the white
eyes (w; 1-1.5), and for Adh^n_1 mutant (2-50.1); the other
homozygous for black body (b; 2-48.5), purple eyes (pr;
2-54.5), and cinnabar eyes (cn; 2-57.5), as well as Sofer's
Adh^n_2.

A mutant stock homozygous for Adh^{fast} (referred to as
Adh-F in this report and Adh_4 by Johnson and Schaffer, '74
Adh^{fast} by Grell et al., '65 and Adh-I by Pipkin et al., '74)
and black body was constructed by crossing a multiply-marked
second chromosome line with a wild type line homozygous for
Adh-F, and selecting the appropriate recombinants in the F_3
generation. Wild type lines homozygous for Adh-F and Adh-S
(Adh-S in this report corresponds to Adh^{slow} of Grell et al.,
'65, Adh-II of Pipkin et al., '74, and Adh^6 by Johnson and
Schaffer, '73) were obtained by Dr. Y. Tobari repeated
pair mating of wild type flies which were originally collected from a Texas population.

TABLE I

POTATO-KALMUS MEDIA FORMULATION

Standard liquid part

$(NH_4)_2SO_4$ -------------------------------------	2.0 g
$MgSO_4 \cdot 7H_2O$ -------------------------------	0.5 g
KH_2PO_4 --	1.5 g
Tartaric acid -------------------------------	6.0 g
Sodium propionate -------------------------	6.0 g
Sterile distilled water --------------------	1 liter

Standard solid part

Killed Brewers yeast -----------------------	15 g
Ground instant mashed potatoes -------------	85 g

Modifications of Liquid Part for Alcohol Treatments

Alcohol Treatment	Amount of treatment alcohols added to standard liquid (final volume is 100 ml)
Ethanol	3 ml
Ethanol 2-butanol	3 ml .25 ml
1-propanol	2 ml
2-propanol	2 ml
1-propanol 2-propanol	2 ml 2 ml

Final Mixture - Mix 3 ml of liquid part with 0.5 g solid part. Mix well before mixture hardens.

Culture Media

A modified Kalmus-yeast media (Kalmus, '43) was used for all cultures in this experiment (see Table I for formulation). Instant mashed potatoes were substituted for sucrose which also eliminated the need for agar. Sodium proprionate was added to the media after tartaric acid was found to be unable to control frequent mold infections.

Alcohol Stress Techniques

A flowchart of the techniques is given in Fig. 1. The first step is to start a larval culture from a cohort of approximately 300 eggs which are collected in the shortest possible time period (3 to 12 hours). When the culture is two days old, second instar larvae are separated from the culture media by first diluting the media with cold water and trapping the larvae in a strainer. The larvae are then transferred either en masse by a spatula or one at a time with forceps to the appropriate test media in a one-inch dia-

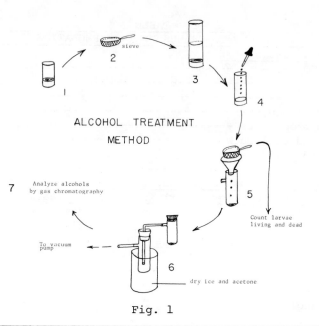

Fig. 1

meter shell vial. After the larvae are transferred, the test
chamber is sealed with cellophane tape to another shell vial
making an air tight seal. More recently the shell vial with
the test media has been plugged with sterile cotton and
placed in a half-liter plastic box containing not more than
one other test culture. The sealed cultures are opened 12 to
36 hours later and the test media dissolved by adding 5 ml of
cold distilled water after which the larvae are quickly separated from the media by a strainer. Larvae are counted and
scored as living if they move either their mouth parts or
posterior spiricles in response to being touched by forceps.
The dissolved media was placed in a test tube with a side
arm. This test tube was then conncected to a cold trap and
the alcohols and water were separated by vacuum sublimination
as shown in step six of Fig. 1. Lastly, the alcohol solution
was analyzed on a varion prepmatic gas chromatograph using a
column 3 feet long and 3/8 inches diameter composed of
chromosorb-p solid support and carbowax 20M liquid phase with
a He flow rate of 100 cc/min. The concentration of alcohols
was quantitatively determined by numerically integrating the
curve heights to approximate the area under the curves which
is proportional to concentration. In order to minimize error, at least three separate curves were obtained from the
same culture extraction and the average of the three areas
recorded. Control vials were carried through all steps

except that no larvae were transferred into them.

Since *D. melanogaster* is collected from banana rots and cactus rots in the wild, samples of these rots were analyzed for alcohol content. One banana ferment was initiated with an inoculation of Fleischman's yeast and another allowed to ferment "naturally." A cactus (*Opuntia* sp.) rot, collected from Travis County, Texas, was also analyzed. The alcohol content of the rots was determined by extracting water and alcohols using the vacuum sublimination technique described above. The alcohols were identified by comparison of elution times on the gas chromatograph described above and chemically identified as alcohols by the specific color change of dichromate ions in nitric acid when the effluent of the gas chromatograph was bubbled through the solution (Walsh and Merritt, '60).

Measurement of Larval Viability

Second instar larvae of each ADH stock were placed in test media consisting of Potato-Kalmus-Yeast media supplemented with 3% ethanol. The larvae were removed after 28 hours and scored as living or dead. In another larger experiment second instar larvae from all the laboratory stocks except Adh^n_2 were used. For each stock fifty larvae were trans-

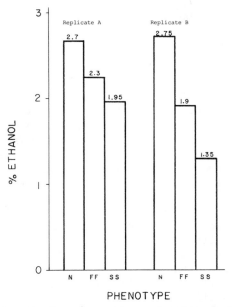

Fig. 2. *In vivo* rates of ethanol oxidation.

737

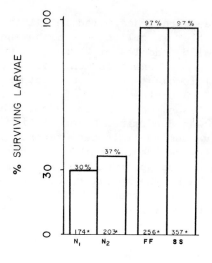

PHENOTYPE

Fig. 3. Larval viability in 3% ethanol treatments.

ferred into each of eight replicate vials containing 3% ethanol supplemented test media and each of eight control vials
containing standard Potato-Kalmus-Yeast media. The vial
were stored in air tight containers until the adults emerged.
Percent viability was determined after counting all the
emerged adults.

RESULTS

Analysis of fermenting bananas determined that only the
alcohol ethanol was present in high concentrations. The concentration varied between 2 and 4 percent by volume of the
liquid volatiles in the fermenting bananas. A cactus rot,
which is a larval habitat for *D. melanogaster* in Texas, was
also analyzed. Unfortunately, it was not possible to obtain
an active rot; however, the dormant rot analyzed had 1 percent ethanol by volume in the liquid volatiles of the cactus
(*Opuntia* sp.) rot.

Alcohol oxidation rate experiments are summarized in Figs.
2, 4, and 5. In each case a large qualitative difference in
the efficiency of flies homozygous for different naturally
occurring *Adh* alleles was detected.

Larval viability in the presence of 3% ethanol by volume
of the liquid volatiles in Kalmus-Potato-Yeast was measured
and found to be as illustrated in Fig. 3 after 28 hours.
Subsequent studies indicate that 3% ethanol is lethal for

both negative ADH stocks while the viability of all active
ADH phenotypes is between 85% and 90% (N=400 for each stock
used).

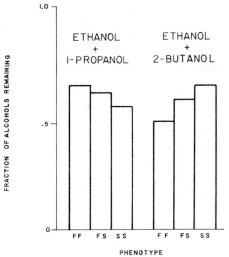

Fig. 4. *In vivo* rates of ethanol oxidation when other
alcohols are present.

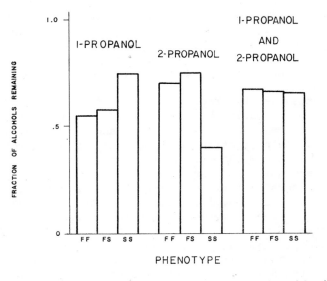

Fig. 5. *In vivo* rates of propanol oxidation.

DISCUSSION

Before selelction mechanisms can be proposed for maintaining genetic polymorphisms, it is necessary to understand the
adaptive value of all the phenotypes involved. Therefore,
first consider the comparison of wild type stocks which have
normal ADH activity and Adh negative stocks which lack ADH
activity. If the physiological role of ADH is to oxidize
alcohols as a step in the detoxification of alcohols, it
should be possible to select against Adh^n alleles with concentrations of ethanol in the order of those found in fermenting bananas and cacti, i.e., 2-3 percent by volume ethanol. This experiment was done using 3 percent ethanol and
the results are summarized in Fig. 2. Note that nearly two-
thirds of the larvae homozygous for Adh^n alleles died under
the same conditions that only three percent of both $Adh-FF$
and $Adh-S$ larvae died. Further, if there was no strong membrane barrier to keep alcohols from diffusing into the bodies
of the larvae, it should be possible to measure alcohol oxidation indirectly by measuring the amount of alcohol remaining after larvae have been placed in a sealed culture containing a fixed amount of alcohol. The results of this experiment using 3 percent ethanol are illustrated in Fig. 3.
Note that both wild type stocks are efficient ethanol oxidizers while the mutant negative Adh stocks are inefficient.
One problem remains with this experiment, namely that ethanol
is recovered in lower yields than expected from homozygous
Adh^n cultures. This probably results from dilution of the
media by metabolic water since freshly prepared homogenates
of homozygous Adh^n larvae lack detectable ADH activity and
since the larvae are able to noticeably digest potato starch.
At any rate, ADH seems to be clearly involved in the detoxification of alcohols.

Next, we attempted to predict in $vivo$ efficiency of alcohol oxidation by comparing kinetics obtained from partially
purified ADH enzymes from $Adh-FF$, $Adh-SS$ and $ADH-FS$ stocks.
Unfortunately, the predictions were not very strong. This
was true because the Michaelis constants for the different
enzyme forms are quite similar although specific activity of
the enzymes differs in purified preparations as well as in
crude homogenates (Gibson, '70; Day and Clarke, '74, and
Rasmuson et al., '66). However, our work indicates that the
different enzyme forms do not purify equally well and that
the efficiency problems arise very early in the purification
procedures. Therefore, it is not reasonable to compare efficiencies of the different ADH enzymes using kinetic data
from partially purified enzymes--or even completely purified

enzymes since the amount of enzyme per individual cannot be determined. However, using the alcohol treatment technique described above, it has been possible to detect qualitative differences in alcohol oxidation efficiencies of the three common active ADH phenotypes even in conditions of constant temperature. Note in Fig. 4 that Adh-SS is the most efficient ethanol oxidizer when only ethanol is present, but that it is the least efficient when sub-lethal doses of 2-butanol are added. Adh-FF is most efficient when 2-propanol is the substrate but Adh-SS is most efficient when 1-propanol is the substrate (see Fig. 5). All ADH morphs are equally efficient on the average when equal amounts of 1-propanol and 2-propanol are added to the test cultures.

Thus it appears that concentrations of alcohols which can be expected to be realized in the wild can select for one or the other ADH morph. Since *Drosophila melanogaster* live in small isolated pockets as larvae, it is reasonable to assume that these isolated pockets are coarse grained habitats for larvae and are different in their alcohol contents probably as functions both of time and space. Thus, ADH polymorphism may be the result of disruptive selection taking place both in time and space. We are currently following up this possibility with laboratory selection experiments.

ACKNOWLEDGEMENTS

This work was supported in part by the Clayton Foundation Biochemical Institute and by PHS Grant No. 5T01 GM 0037-14 from the National Institute of General Medical Sciences. This investigation is partial fulfillment of the Ph.D. requirements of R. Ainsley at The University of Texas at Austin.

REFERENCES

Bijlsma-Meeles, E. and W. van Delden 1974. Intra- and inter-population selection concerning the alcohol dehydrogenase locus of *Drosophila melanogaster*. *Nature* 247: 369-371.

Day, T. and B. Clarke 1974. The relative quantities and catalytic activities of enzymes produced by alleles at the alcohol dehydrogenase locus in *D. melanogaster*. *Biochem. Genet.* 11: 155-166.

Gibson, J. 1970. Enzyme flexibility in *Drosophila melanogaster*. *Nature* 227: 959-960.

Grell, E. H., K. B. Jacobson, and J. B. Murphy 1965. Alcohol dehydrogenase in *Drosophila melanogaster* and genetic variants. *Science* 149: 80-83.

Johnson, F. M. and H. E. Schaffer 1973. Isozyme variability of the genus *Drosophila*. VII. Genotype - Environment relationships in populations of *D. melanogaster* from the eastern United States. *Biochem. Genet.* 10: 149-163.

Kalmus, H. 1943. A factorial experiment on the mineral requirements of a *Drosophila* culture. *Am. Naturalist* 77: 376-380.

Pipkin, S. B., C. Rhodes, and N. Williams 1974. Adh^{II} frequency in relation to temperature in natural populations of *D. melanogaster*. *Isozyme Bull.* 7: 53.

Rasmuson, B., L. R. Nilson, M. Rasmuson, and E. Zeppezaur 1966. Effects of heterozygosity on alcohol dehydrogenase (ADH) activity in *Drosophila melanogaster*. *Hereditas* 56: 313-316.

Sofer, W. H. and M. A. Hatkoff 1972. Chemical selection of alcohol dehydrogenase negative mutants in *Drosophila*. *Genetics* 72: 545-549.

Vigue, C. L. and F. M. Johnson 1973. Isozyme variability in the genus *Drosophila*. VI. Frequency-property-environment relationships of allelic alcohol dehydrogenases in *D. melanogaster*. *Biochem. Genet.* 9: 213-227.

Walsh, J. T. and C. Merritt 1960. Qualitative functional group analysis of gas chromatography effluents. *Analytical Chem.* 32- 1378-1381.

γ-GLUTAMYL TRANSPEPTIDASE: PROPERTIES IN RELATION
TO ITS PROPOSED PHYSIOLOGICAL ROLE

SURESH S. TATE
Department of Biochemistry, Cornell University
Medical College, New York, N.Y. 10021

ABSTRACT. Two rat kidney γ-glutamyl transpeptidase isozymes
have been purified. Both contain about 30% carbohydrate;
although the amino acid composition is similar, the isozymes
show different mobilities on polyacrylamide gels. The isozymes have been examined with respect to their γ-glutamyl
donor and acceptor specificities. Both exhibit similar
specificities. Several glutathione analogs and derivatives
and many γ-glutamyl amino acids were active as γ-glutamyl
donors. Glutathione disulfide was about 5% as active as
glutathione. Of the amino acids tested, L-glutamine and
L-methionine were the best acceptors of the γ-glutamyl group.
A number of dipeptides were active as acceptors and several
of these were more active than L-glutamine. Aminoacylglycine
derivatives were, in the examples studied, more active than
the corresponding free NH_2-terminal amino acids, probably
reflecting their affinity for the "cysteinylglycine" site
of the enzyme. These and other properties of the transpeptidase isozymes are discussed in relation to the proposed function of the γ-glutamyl cycle in amino acid and
peptide transport (Orlowski and Meister, 1970; Meister, 1973).
It has also been shown that maleate inhibits transfer of
the γ-glutamyl moiety to an acceptor and markedly increases
hydrolysis of the γ-glutamyl donor. Thus, glutamine, a
poor substrate for transpeptidase as compared to glutathione,
is rapidly hydrolysed by transpeptidase in the presence of
maleate. Evidence has been obtained suggesting that the
previously described "maleate-stimulated phosphate-independent glutaminase" is a catalytic function of transpeptidase.

Glutathione (L-γ-glutamyl-L-cysteinyl-glycine) occurs in
virtually all living cells. The major pathway of its breakdown involves a reaction, catalyzed by γ-glutamyl transpeptidase, in which the γ-glutamyl moiety is transferred to an
amino acid acceptor:

Glutathione + L-amino acid $\rightleftharpoons$ L-γ-glutamyl-L-amino acid
                                     + L-cysteinyl-glycine

This is followed by conversion of γ-glutamyl amino acid to 5-
oxoproline and free amino acid, a reaction catalyzed by

743

γ-glutamyl cyclotransferase. Cysteinyl-glycine is hydrolyzed to its component amino acids, and 5-oxoproline is cleaved to glutamate in an ATP-dependent reaction. Glutamate, cysteine, and glycine are then utilized for the two step synthesis of glutathione. The reactions which degrade glutathione together with those which lead to its synthesis consititute a series of enzyme-catalyzed reactions which has been termed the γ-glutamyl cycle (Orlowski and Meister, 1970: Van Der Werf, et al., 1971; Meister, 1973).

γ-Glutamyl transpeptidase, a membrane-bound enzyme, is abundant in the kidney where it has been found to be localized in the brush border of the proximal convoluted tubules (Albert et al., 1961; Glenner and Folk, 1961; Glenner et al., 1962; Glossman and Neville, 1972; George and Kenny, 1973). Other locations of high transpeptidase activity include epithelial cells intimately involved in transport processes and the se - cretion of specialized body fluids, such as the epithelia of the jejunal villi, choroid plexus, salivary glands, bile duct, seminal vesicles, epididymis, and ciliary body (see Meister, 1973; and Tate and Meister, 1974a, for a comprehensive litera- ture survey.) It has been suggested that the transpeptidase and other enzymes of the cycle are involved in the transport of amino acids or peptides across such epithelial surfaces (Orlowski and Meister, 1970, Meister, 1973, 1974). The trans- peptidase, according to this postulate, functions in the trans- location of amino acids across cell membranes in a process in which the γ-glutamyl moiety of glutathione functions as a carrier. The studies reported here describe the properties of rat tissue transpeptidases, with particular emphasis on the kidney enzymes, and discuss these in relation to their proposed role in amino acid and peptide transport as well as other physiological phenomena.

PURIFICATION OF RAT KIDNEY γ-GLUTAMYL TRANSPEPTIDASES

A number of investigations have been carried out on the properties of crude and parially purified γ-glutamyl transpep- tidase preparations from kidney (Hanes et al., 1952; Fodor et al., 1953; Hird and Springell, 1954; Revel and Ball, 1959; Goldbarg et al., 1960; Binkley, 1961; Szewczuk and Baranowski, 1963; Orlowski and Meister, 1965; and Richter, 1969); highly purified preparations have been obtained from the kidneys of beef (Szewczuk and Baranowski, 1963), hog (Orlowski and Meis- ter, 1965), and human (Richter, 1969). In the present work, the rat kidney enzyme was purified by a modifiaction of the procedure used for the isolation of the bovine kidney enzyme by Szewczuk and Baranowski (1963). Fresh kidneys (from

Sprague-Dawley rats, 300-350 g) were homogenized in a Waring
Blendor with four volumes of 0.9% sodium chloride at 4°(step
1). The homogenate was treated with 0.5 volume of acetone
(precooled to -15°), and the mixture was centrifuged at 16,000
x g for 30 min. The pellet was homogenized with 0.1 M Tris-
HCl buffer (pH 8) containing 1% sodium deoxycholate and the
suspension was stirred at 4° for 4 hours and then centrifuged
at 16,000 x g for 60 min (step 2). Solid ammonium sulfate
was added to the clear supernatant solution to achieve 70% of
ammonium sulfate saturation and the solution was allowed to
stand at 4° for 18 hours. The precipitate obtained after centrifugation was dissolved in 0.01 M Tris-HCl buffer (pH 8) and
the solution was dialyzed against 60 volumes of the same buffer
(step 3).

The dialyzed solution was treated at 25° with 0.4 volume
of n-butanol added dropwise with constant stirring over a period
of 15 min. The mixture was incubated at 37° for 15 min and
then centrifuged. The aqueous layer was collected and treated
with 1.5 volume of acetone. The precipitate was collected by
centrifugation and dissolved in 0.01 M Tris-citrate buffer
(pH 8.9) and dialyzed against 100 volumes of this buffer (step
4).

The dialyzed solution was applied to a DEAE-cellulose
column (Whatman DE-52, microgranular, previously swollen)
equilibrated with the Tris-citrate buffer. The column was
treated with the buffer until the eluate was free of protein;
about 50% of the enzyme applied emerged in the buffer and the
most active fractions were pooled (fraction A). The column
was then eluted with 0.1 M sodium acetate buffer (pH 5.5); a
second peak of enzyme activity emerged (fraction B), which
accounted for about 15% of the applied enzyme activity (step
5). Both fractions A and B were separately precipitated by
addition of solid ammonium sulfate to 70% of ammonium sulfate
saturation; the pellets were dissolved in a minimal volume of
0.01 M Tris-HCl buffer (pH 8) containing 0.02 M NaCl, and then
chromatographed separately on Sepharose 4B-200 columns (Sigma;
1.5 x 100 cm. Step 6). Table 1 summarizes the results of a
typical purification procedure from 140 g of kidney (from Tate
and Meister, 1974a).

DETERMINATION OF γ-GLUTAMYL TRANSPEPTIDASE ACTIVITY

γ-Glutamyl transpeptidase catalyzes the transfer of the
γ-glutamyl moiety of a γ-glutamyl donor to a variety of acceptors, which may be amino acids or peptides. A number of enzymatic procedures have been used in studies reported here.
The methods are described in detail elsewhere (Tate and Meister,
1974 a). In assays with glutathione, the procedures used

TABLE I

PURIFICATION OF γ-GLUTAMYL TRANSPEPTIDASE FROM RAT KIDNEY

Step	Volume	Protein		Activity[1]		
		Concen-tration	Total	Total Units	Specific	Yield
	ml	mg/ml	mg	units/mg		%
1. Homogenate	650	39.0	25,400	34,200	1.3	100
2. Deoxycholate extract	300	19.8	5,930	29,100	4.9	85
3. Ammonium sulfate precipitation	230	17.3	3,970	28,700	7.2	84
4. Butanol treatment and acetone precipitation	82	12.8	1,050	14,900	14.2	44
5. Chromatography on DEAE-cellulose:						
Fraction A	58	3.0	174	6,850	39.4	20
Fraction B	18	1.9	34	1,980	58.2	6
6. Chromatography on Sepharose 4B-200:						
Fraction A	10.4	4.0	41.6	4,650	112.0	14
Fraction B	6.6	1.4	9.2	1,100	120.0	3

[1]Activity was determined in the presence of γ-glutamyl-p-nitro-anilide (2.5 mM) and glycyl-glycine (20 mM) at pH 8.0 (Tris-HCl buffer) and 37°. Specific activity is defined in terms of μmoles of p-nitroaniline released/min/mg of protein (Tate and Meister, 1974a).

---

involved determination of the disappearance of glutathione (Ball, 1966), or the formation of γ-glutamyl product in systems containing ^{14}C amino acids. In addition to glutathione, a number of other γ-glutamyl derivatives can function as γ-glutamyl donors. A convenient method involves the use of L-γ-glutamyl-p-nitroanilide (Orlowski and Meister, 1963); the enzyme catalyzes the following reaction in the presence of added acceptors, γ-Glutamyl-p-nitroanilide + acceptor $\longrightarrow$

γ-glutamyl-acceptor + p-nitroaniline

The formation of p-nitroaniline is readily determined from the increase in absorbance at 410 nm.

New procedures for the assay of transpeptidase have also been developed in which S-substituted glutathiones are employed; the products of the enzymatic reactions include the corresponding S-substituted derivatives of cysteinyl-glycine which exhibit characteristic spectral properties. These procedures are based on the finding by Avi-Dor (1960) that incubation of S-pyruvoyl-glytathione with a kidney particulate preparation led to a marked increase in absorbance at 300 nm which was associated with the formation of the S-pyruvoyl derivative of cysteinyl-glycine. It was suggested that this reaction is catalyzed by γ-glutamyl transpeptidase. In the present work, large changes in absorbance were observed when the purified transpeptidase preparation was incubated with the S-pyruvoyl, S-acetophenone, and S-propanone derivatives of glutathione (Figure 1). Particularly useful are the S-pyruvoyl and S-acetophenone derivatives of glutathione; the

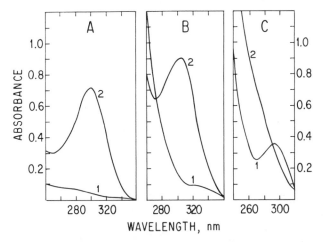

Fig. 1. Spectral changes associated with the action of γ-glutamyl transpeptidase on S-substituted derivatives of glutathione. The reaction mixtures (final volume, 1 ml) contained 0.1 M triethanolamine-HCl buffer (pH 8), 20 mM gly-gly, the glutathione derivative and 1.8 µg enzyme A. Spectra were recorded before (curve 1) and 15 min after addition of the enzyme (curve 2). A. S-pyruvoyl-glutathione (0.125 mM) B. S-acetophenone derivative of glutathione (0.2 mM) C. S-propanone derivative of glutathione (2 mM). (From Tate and Meister, 1974a).

reaction pathways and the structures tentatively assigned to the S-substituted cysteinyl-glycine products, which exhibit high absorbance in the ultraviolet region, are shown in Figure 2 (Tate and Meister, 1974a). The spectral changes observed upon addition of enzyme can be readily monitored spectrophotometrically thus providing convenient assay techniques.

SOME CHARACTERISTICS OF RAT KIDNEY γ-GLUTAMYL TRANSPEPTIDASES A AND B

The two forms of the enzyme, A and B (Table 1) exhibit similar specific activities towards γ-glutamyl-p-nitroanilide, glutathione, and its derivatives. Both forms contain about 30% carbohydrate of which 7-10% is sialic acid. The amino acid compositions of the isozymes are similar (Table II). The notable features are the relatively low content of half-cystine,

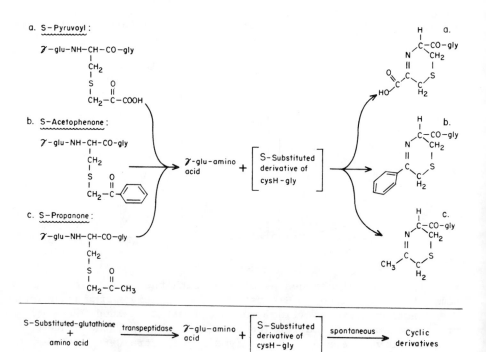

Fig. 2. Action of γ-glutamyl transpeptidase on S-substituted derivatives of glutathione (From Tate and Meister, 1974a).

TABLE II

AMINO ACID COMPOSITION OF RAT KIDNEY γ-GLUTAMYL
TRANSPEPTIDASE ISOZYMES

Amino Acid	Residues/1000 Total Residues	
	Enzyme A	Enzyme B
Asp	92	101
Thr	59	69
Ser	82	85
Glu	104	104
Pro	52	58
Gly	98	99
Ala	84	83
1/2 Cys	2-4	2-4
Val	73	71
Met	14	11
Ileu	49	50
Leu	82	77
Tyr	35	37
Phe	47	45
His	20	22
Lys	53	49
Arg	26	24

and high content of the basic amino acid residues. The two
forms, however, exhibit different mobilities on polyacrylamide
gels, form B migrating (toward the anode) more rapidly at pH
8.5 than form A (Figure 3). It should be noted that these enzymes are very poorly stained by either Coomassie Blue or
Buffalo Black.

The relatively high carbohydrate content of the isozymes
presents considerable difficulties in the determination of the
molecular weights of either the native enzyme or the dissociated monomers. Studies are now in progress on the subunit
structure of the two forms. Like other glycoproteins (Bret-
sher, 1971; Segrest et al., 1971; Segrest and Jackson, 1972),
the transpeptidase isozymes behave anomalously during sodium
dodecyl sulfate (SDS) polyacrylamide gel electrophoresis.
This and the fact that the monomers are very poorly stained
with Coomassie Blue has prevented reasonable estimates of the
subunit molecular weights. Both isozymes, however, yield
similar monomer bands on SDS gels.

The transpeptidase isozymes are extremely stable in so-

A B

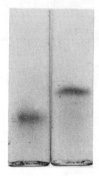

Fig. 3. Polyacrylamide gel electrophoresis of rat kidney γ-glutamyl transpeptidase isozymes A and B. Electrophoresis was run in 4% gels at pH 8.5 (0.05 M Tris-acetate buffer) at 25°.

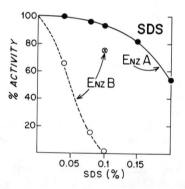

Fig. 4. Effect of sodium dodecyl sulfate (SDS) on the activities of γ-glutamyl transpeptidase isozymes. The isozymes (7 μg of each) were preincubated at 25° for 20 min. in solutions (final volumes, 50 μl) containing 50 mM Tris-HCl buffer (pH 8.0) and SDS as shown. Aliquots (2 μl) were withdrawn and assayed for residual transpeptidase activity at 37° in solutions (final volume, 1 ml) containing 2.5 mM γ-glutamyl-p-nitroanilide, 50 mM glycyl-glycine, and 50 mM Tris-HCl buffer (pH 8.0). ⊗, Solution also contained 10 mM γ-glutamyl-α-aminobutyrate.

TABLE III

ACTIVITY OF RAT KIDNEY γ-GLUTAMYL TRANSPEPTIDASE ISOZYMES
TOWARD VARIOUS AMINO ACID ACCEPTORS

Acceptor	Activity with glutathione	
	Enzyme A	Enzyme B
None (Control)	6.3	6.2
L-Glutamine	20.4	19.7
L-Methionine	18.3	19.3
L-Asparagine	10.9	10.5
L-α-Aminobutyrate	11.2	11.4
L-Alanine	12.0	11.8
Glycine	9.3	8.8
L-Phenylalanine	9.3	9.0
L-Threonine	5.2	5.9
L-Leucine	10.1	10.9
L-Isoleucine	6.9	7.2
L-Glutamate	10.9	12.0
L-Aspartate	6.3	6.2
L-Arginine	10.9	11.3
L-Histidine	5.3	6.0

Taken in part from Tate and Meister (1974a). Values given are
specific activities (μmoles of glutathione utilized per min
per mg enzyme) with glutathione (5 mM) as the γ-glutamyl donor;
the acceptor concentration was 20 mM.

lution both as purified preparations and in partially purified
fractions; indeed, the entire purifiaction procedure described
in Table 1 can be carried out at 25° with no noticeable decrease in yields. The two isozymes, however, exhibit different susceptibility towards SDS. Thus, the isozyme A is considerably more resistant to the denaturing action of SDS than
isozyme B (Figure 4). Considerable protection was afforded
when a γ-glutamyl amino acid, a substrate of the enzyme, was
present in incubation mixtures containing isozyme B and SDS.
Further studies are required to explain the difference in the
effect of SDS on the two isozymes.

*ACTIVITY OF RAT KIDNEY γ-GLUTAMYL TRANSPEPTIDASE ISOZYMES
TOWARDS VARIOUS AMINO ACIDS AND PEPTIDES AS ACCEPTORS*

Table III and IV summarize in part the extensive studies
(Tate and Meister, 1974a) carried out on the ability of vari-

TABLE IV

ACTIVITY OF RAT KIDNEY γ-GLUTAMYL TRANSPEPTIDASE ISOZYMES
TOWARD VARIOUS PEPTIDE ACCEPTORS

Acceptor	Activity with glutathione	
	Enzyme A	Enzyme B
Glycine	9.3	8.8
Gly-gly	50.5	46.2
Gly-L-ala	39.0	35.4
Gly-D-ala	8.0	8.0
Gly-L-asn	16.8	16.4
Gly-L-val	7.1	7.5
Gly-L-pro	15.6	14.2
Gly-D-phe	5.9	5.5
Gly-gly-gly	10.7	10.1
L-Alanine	12.0	11.8
L-Ala-gly	29.5	29.2
L-Ala-L-ala	7.9	7.1
L-Methionine	18.3	19.3
L-Met-gly	26.9	27.0
L-Met-L-ala	11.3	11.1
L-Met-L-met	26.2	23.4
L-Met-L-pro	11.0	11.1
L-Met-L-ala-L-ser	10.0	-
L-Glutamate	10.9	12.0
L-Glu-gly	22.0	20.1
L-Glu-L-ala	8.0	8.0
L-Glu-L-α-amino-butyrate	31.8	-

In part from Tate and Meister (1974a). Values given are
specific activities with glutathione as the γ-glutamyl donor.

ous amino acids and peptides to serve as acceptors of the γ-
glutamyl moiety. The data shown are for activities with glutathione. Similar relative activities were observed for the
two isozymes with either γ-glutamyl-p-nitroanilide or the S-
derivatives of glutathione as the γ-glutamyl donors. Both
forms of the enzyme exhibit, in general, similar amino acid
and peptide acceptor, as well as γ-glutamyl donor, specificities. Although most amino acids serve as γ-glutamyl acceptors,
by far the best substrates (as noted also in previous studies

(Fodor et al., 1953; Hird and Springell, 1954; Revel and Ball, 1959; Goldbarg et al., 1960; Binkley, 1961)) were neutral amino acids such as L-glutamine, L-methionine, L-asparagine, and L-alanine. Branched chain amino acids, L-threonine, L-aspartate, and L-histidine were relatively poor acceptors, whereas L-proline, D-amino acids and α-methyl amino acids did not exhibit detectable activity. There appears to be no absolute requirement for a free αcarboxylate group in the amino acid acceptor, since L-isoglutamine, glycinamide, and L-methionine methyl ester are relatively good substrates.

In agreement with previous findings (Fodor et al., 1953; Revel and Ball, 1959; Goldbarg et al., 1960; Binkley, 1961; Orlowski and Meister, 1965) glycyl-glycine was the best acceptor; high activity was also observed with a number of other peptides (Table IV). Several dipeptides showed significantly more activity than the free amino acid corresponding to the amino-terminal residue, especially when glutathione was the donor. Thus, the activities with glutathione and glycyl-glycine, glycyl-L-alanine, and glycyl-L-asparagine were 5.4, 4.2, and 1.8 times greater than that found with glycine. Similarly, certain methionyl peptides were much more active than methionine, but others were less active. L-Alanyl-glycine was more active than L-alanine, while L-alanyl-L-alanine was less active. While the nature of the amino acid side chain clearly affects acceptor activity (Table III), such effects are significantly modified when the carboxyl group is bound by peptide linkage to another amino acid residue. Attachment of a C-terminal glycine residue to an amino acid always increased acceptor activity. Peptides containing D-amino acid residues (glycyl-D-phenylalanine and glycyl-D-alanine) were inactive.

The high activities found with the amino acyl-glycines may reflect interaction of these peptides with the same enzyme site that binds cysteinyl-glycine moiety of glutathione. This site probably binds other dipeptides as well as amino acid acceptors. Thus, it appears that the acceptor site is relatively unspecific. It is notable, for instance, that glycyl-L-proline and L-methionyl-L-proline are active while L-proline itself is not.

It should be emphasized that the substrate specificities of the transpeptidase A and B are virtually identical. Most studies, however, were carried out with isozyme A. Several of the intermediate fractions (as well as the homogenate) obtained during the purification procedure (Table 1) were also studied with respect to γ-glutamyl donor and amino acid and peptide acceptor specificities; they were found to exhibit specificity similar to that shown by isozymes A and B.

ACTIVITY OF RAT KIDNEY γ-GLUTAMYL TRANSPEPTIDASE
TOWARDS VARIOUS γ-GLUTAMYL DONORS.

Data on transpeptidation between several glutathione analogs and derivatives and methionine are summarized in Table V.

TABLE V

ACTIVITY OF γ-GLUTAMYL TRANSPEPTIDASE TOWARD
ANALOGS AND DERIVATIVES OF GLUTATHIONE

γ-Glutamyl Donor	Specific Activity	Relative Activity [to glutathione]
Glutathione	16.0	[100]
Glutathione disulfide	0.8	5
L-γ-glutamyl-L-α-amino-butyryl-glycine (ophthalmic acid)	13.0	81
γ-(α-methyl-DL-glutamyl)-L-cysteinyl-glycine	0.6	4
S-Methyl-glutathione	9.4	59
S-Propanone-glutathione	24.8	155
S-Acetamido-glutathione	27.5	172
S-Acetophenone-glutathione	29.6	185
S-Pyruvoyl-glutathione	9.9	62

From Tate and Meister (1974a). Transpeptidase activity was
determined with 5 mM γ-glutamyl donor and 20 mM L-[14C] methionine. Specific activity is expressed as the μmoles of γ-glutamyl-methionine formed per min per mg of enzyme A. Essentially
similar results were obtained with enzyme B.

Glutathione disulfide was relatively inactive, whereas several
other analogs and derivatives were effective as donors, in
several instances more active than glutathione. A number of
γ-glutamyl amino acids can also serve as effective γ-glutamyl
donors; the most active of those tested are the γ-glutamyl
derivatives of glutamine, methionine, cysteine, α-aminobutyrate,
and serine (Tate and Meister, 1974a). Thus, it appears that
effective interaction with the enzyme can occur with compounds
in which the moiety attached to the γ-glutamyl group varies
considerably in structure.

THE PHYSIOLOGICAL ROLE OF γ-GLUTAMYL TRANSPEPTIDASES

The membraneous localization of transpeptidase in epithelial cells intimately involved in transport processes and in
the secretion of specialized body fluids is consistent with

the concept that the enzyme plays a role in one or more aspects of these processes. Although the question of its physiological role has not been settled, it seems probable that the enzyme functions in transport phenomena involving amino acids, peptides, or both. That it might function in amino acid transport was suggested many years ago by Hird (1950), Binkley (1951, 1954), Springell (1953), and Ball et al. (1953); the idea has often been mentioned again in the literature (Knox, 1960; Orlowski and Szewczuk, 1961, Orlowski, 1963; Kokat et al. 1965; Meister, 1965; Greenberg et al., 1967; Orlowski et al., 1969). The idea gained strength from the findings by Meister and coworkers which led to the formulation of the γ-glutamyl cycle (Orlowski and Meister, 1970; Van Der Werf et al., 1971; Meister, 1973, 1974). According to this thesis, the membrane-bound γ-glutamyl transpeptidase functions in the translocation of the amino acid by interacting with extracellular amino acid and intracellular glutathione yielding a γ-glutamyl amino acid. The amino acid is released within the cell from its γ-glutamyl carrier by the action of cyclotransferase. The ATP-dependent decyclization of 5-oxoproline and the synthesis of glutathione catalyzed by the two synthetases are viewed as the energy-requiring recovery steps needed for regeneration of the carrier precursor.

If transpeptidase does indeed mediate the transport of amino acids, the broad specificity of the enzyme suggests that it may translocate whole groups of amino acids performing transport functions of the "high capacity-low specificity" type (Young and Freedman, 1971). The possibility that there are transpeptidase isozymes of different specificities (Orlowski and Meister, 1970) which account for group-specific transport systems known to function in the proximal renal tubules (Young and Freedman, 1971; Segal and Thier, 1973) now seems unlikely. Thus several enzyme fractions during purification of rat kidney transpeptidase as well as the purified isozymes A and B exhibit similar amino acid acceptor specificity. Kinetic studies show that the two isozymes also exhibit similar affinity towards various amino acids (Tate and Meister, 1974a). Transport specificity might be imposed by the existence of specific binding proteins (recognition sites) on the membrane surface. It is also possible that the transpeptidase may perform a special function in the transepithelial transport of methionine and glutamine, which are by far the best substrates. In this regard, it is of interest that γ-glutamyl-methionine and γ-glutamyl-glutamine are excellent substrates of γ-glutamyl cyclotransferase (Orlowski et al., 1969; Orlowski and Meister, 1973).

These considerations regarding the possible role of kidney

transpeptidase in transport processes also seem to apply to the enzyme in other tissues. Thus, partially purified γ-glutamyl transpeptidases of rat jejunal epithelial cells, lung, and liver were examined with respect to amino acid and peptide acceptor specificities (Table VI). Although certain differences are observed, especially in the case of activity of the lung and liver preparations towards arginine and methionyl-glycine, in general the relative activities towards various dipeptides and amino acids seem to be similar to those exhibited by kidney transpeptidases. Similar specificities have also been noted for enzyme preparations from rat brain, seminal vesicle, and epididymis (Linda W. DeLap, unpublished data). These similarities point to a common function for the transpeptidases at these various cell surfaces. Further work is needed to identify the isozymes, if any, present in tissues other than the kidney.

TABLE VI

γ-GLUTAMYL TRANSPEPTIDASES OF SOME RAT TISSUES: SUBSTRATE SPECIFICITY

Acceptor	Enzyme Source			
	Kidney Enzyme A	Jejunal Epithelial Cells	Lung	Liver
Gly-gly	100	100	100	100
Gly-L-ala	77	50	80	81
L-Ala-gly	58	63	40	37
L-Met-gly	53	63	20	26
L-Methionine	34	50	21	30
L-α-Aminobutyrate	22	20	20	15
L-Alanine	24	30	14	23
Glycine	18	21	10	13
L-Phenylalanine	19	14	8	13
L-Arginine	22	20	34	30

Values given are the relative activities (relative to gly-gly) obtained with the acceptors shown and γ-glutamyl-p-nitroanilide. The data are for purified kidney isozyme A, and partially purified enzyme preparations from other tissues (unpublished data).

It is notable that rabbit erythrocyte γ-glutamyl transpeptidase is inactive towards γ-glutamyl-p-nitroanilide and its acceptor specificity with glutathione as the donor is quite distinct from that shown by either the rabbit kidney or rat kidney enzyme (Palekar et al., 1974). These findings raise the possibility that glutathione-specific transpeptidases are present in other tissues also; their detection might be facilitated by the use of the S-derivatives of glutathione described here.

The possibility that γ-glutamyl transpeptidases of the kidney and other rat tissues function in dipeptide transport must also be considered. This idea is especially attractive in view of the high activity of the transpeptidases toward several dipeptides. Very active peptide transport systems exist in the small intestine and in the kidney (Milne, 1971; Smyth, 1972; Matthews, 1972; Ugolev, 1972) and it has been suggested that dipeptidases are involved in the handling of peptides at these sites. Such peptidases seem to be primarily cytosolic enzymes (Peters, 1972; Das and Radhakrishnan, 1973). It is possible that the transpeptidase performs a translocation function for peptides analogous to that considered above for amino acids. Release of the dipeptide from the γ-glutamyl carrier would require additional transpeptidation since γ-glutamyl cyclotransferase does not act on γ-glutamyl dipeptide derivatives (Orlowski et al., 1969). Further information is required regarding the orientation of the enzyme molecules within the cell membranes in order to better define the mechanism whereby the transpeptidase might mediate transport. The possible role of the transpeptidase in other physiological phenomena also requires consideration. Thus the high activity of the kidney enzyme (and possibly of enzymes from other tissues) toward S-substituted glutathiones would be consistent with a function in mercapturic acid formation (Boyland and Chasseaud, 1969).

Finally, a possible role of γ-glutamyl transpeptidase in renal ammoniagenesis should also be considered. Curthoys, at this Conference, showed that the rat kidney phosphate-independent glutaminase preparation also catalyzed release of p-nitroaniline from γ-glutamyl-p-nitroanilide and the disappearance of glutathione (Curthoys, 1974). With γ-glutamyl-p-nitroanilide, what appeared to be $(\gamma$-glutamyl$)_2$-p-nitroanilide was formed; in the presence of maleate, however, the major product was glutamate. Maleate has been shown to be a strong activator of phosphate-independent glutaminase (Katunuma et al., 1966; 1967; 1973). On the basis of these and other properties such as insolubility and stability, Curthoys suggested that the phosphate-independent glutaminase and γ-glutamyl transpeptidase activities were catalyzed by the same enzyme.

These findings have been confirmed and extended in our laboratory (Tate and Meister, 1974b). Tables VII and VIII

TABLE VII

EFFECT OF MALEATE ON THE INTERACTION OF γ-GLUTAMYL TRANSPEPTIDASE WITH S-METHYL-GLUTATHIONE AND METHIONINE

Substrates	Maleate present + absent 0	Products, nmoles formed			
		Glutamate	γ-glu-S-methyl-glutathione	S-methyl-cySH-gly	γ-glu met
S-Methyl-glutathione	0	84	100	183	
S-Methyl-glutathione	+	655	<20	638	
S-Methyl-glutathione + Methionine	0	(40)	10	372	335
S-Methyl-glutathione + Methionine	+	625	0	735	161

Taken from Tate and Meister (1974b). The reaction mixtures (final volume, 0.1 ml) contained Tris-HCl buffer (50 mM; pH 8.0), isozyme A (2 μg), S-methyl-glutathione (10 mM), and, as indicated, L-methionine (10 mM), and sodium maleate (50 mM). Incubations were for 20 min. at 37°.

TABLE VIII

EFFECT OF MALEATE ON THE INTERACTION OF γ-GLUTAMYL TRANSPEPTIDASE
WITH GLUTAMINE AND METHIONINE

Substrates	Maleate present: + absent: 0	Products formed, nmoles		
		γ-glu-gln	γ-glu-met	glutamate
Glutamine	0	21	-	33
Glutamine	+	79	-	407
Glutamine + methionine	0	16	25	26
Glutamine + methionine	+	65	97	322

Taken from Tate and Meister (1974b). The experimental conditions were similar to those described in Table VII; L-glutamine (10 mM) and L-methionine (10 mM). Incubations were for 60 min. at 37°.

759

show the effect of maleate on the interaction of transpeptidase isozyme A with S-methyl-glutathione and glutamine, respectively. Similar results were obtained with isozyme B. Maleate greatly stimulates the utilization of both compounds with S-methyl-glutathione; either in absence or presence of the acceptor L-methionine, maleate strongly inhibits the formation of trans-peptidation products. On the other hand, hydrolysis to gluta-mate is greatly stimulated. Similar results were obtained with γ-glutamyl-α-amino-butyrate either in absence or presence of methionine (Tate and Meister, 1974b). Glutamine is a poor sutstrate of transpeptidase acting both as a donor and acceptor of γ-glutamyl moiety. Its utilization is stimulated several fold by the presence of maleate (Table VIII), the principal product being glutamate, hydrolysis again being preferred to transpeptidation.

Furthermore, transpeptidase catalyzes the formation of γ-glutamyl hydroxamate from a number of γ-glutamyl compounds (including glutamine) and hydroxylamine at relative rates which closely approximate those found for transpeptidation reactions with amino acids (Tate and Meister, 1974b). This γ-glutamyl transfer reaction was stimulated 4- to 5-fold in the presence of maleate (about 14-fold stimulation was observed with glutamine). Both the maleate-stimulated hydrolysis of glutamine and the formation of γ-glutamyl hydroxamate are strongly inhibited by borate and L-serine, a reagent combination that has long been known to inhibit γ-glutamyl transpeptidase (Revel and Ball, 1959). More recently, the rat kidney phos-phate-independent glutaminase has been purified according to the procedure of Katunuma et al.(1968) and has been shown to possess catalytic properties identical to those of the γ-glutamyl transpeptidase preparations used here (Tate and Meister, manuscript in preparation).

Figure 5 summarizes the reactions catalyzed by γ-glutamyl transpeptidase and the effect of maleate on these. Previous kinetic studies (Tate and Meister, 1974a) and the fact that the enzyme catalyzes γ-glutamyl hydroxamate formation from γ-glutamyl compounds (Tate and Meister, 1974b) are consistent with the view that the transpeptidation reaction involves intermediate formation of a γ-glutamyl enzyme. Interaction of maleate with the enzyme favors hydrolysis of the γ-glutamyl donor and decreases transpeptidation. Such an effect, if operative in vivo would be expected to interfere with the proposed function of the γ-glutamyl cycle in amino acid trans-port. It is thus of interest that administration of maleate to animals lead to extensive aminoaciduria (Harrison and Harrison, 1954; Rosenberg and Segal, 1964; Bergeron and Vade-boncoeur, 1971). These findings, although consistent with
f

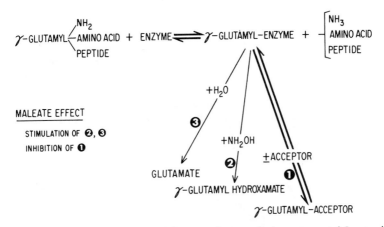

Fig. 5. Reactions catalyzed by γ-glutamyl transpeptidase (from Tate and Meister, 1974b).

function of the transpeptidase and of the γ-glutamyl cycle in amino acid transport, must, however, be viewed with caution since administration of maleate also leads to increased excretion of phosphate and glucose (Harrison and Harrison, 1954; Berliner et al., 1950). Also, there is evidence that maleate may interfere with other enzymes (Webb, 1966) and that this reagent can react with glutathione as well as other sulfhydryl compounds (Morgan and Friedmann, 1938; Webb, 1966).

Maleate, which converts transpeptidase to a glutaminase or a glutathionase, is unlikely to be a physiological modulater of this enzyme. It would be of great interest from the point of view of its possible role in renal ammoniagenesis to investigate whether a counterpart of maleate exists in vivo.

ACKNOWLEDGEMENTS

I am indebted to Dr. Alton Meister for the valuable advice and support given to all phases of this work. This research was supported in part by grants from the American Cancer Society (No. BC-85E) and the National Science Foundation (GB 27454).

REFERENCES

Albert, Z., M. Orlowski, and A. Szewczuk 1961. Histochemical demonstration of γ-glutamyl transpeptidase. *Nature* 191: 767-768.
Avi-Dor, Y. 1960. A study of the effect of particle-bound γ-glutamyltranspeptidase on the product of interaction of Fluoropyruvate with glutathione. *Biochem. J.* 75: 370-374.

Ball, E. G., O. Cooper, and E. C. Clarke 1953. On the hydrolysis and transpeptidation of glutathione in marine forms.
Biol. Bull. 105: 369-370.

Ball, C. R. 1966. Estimation and identification of thiols in
rat spleen after cysteine or glutathione treatment:
Relevance to protection against nitrogen mustards. *Biochem.
Pharm.* 15: 809-816.

Bergeron, M., and M. Vadeboncoeur 1971. Microinjections of L-
leucine into tubular and peritubular capillaries of the
rat. *Nephron* 8: 367-374.

Berliner, R. W., T. J. Kennedy, and J. G. Hilton 1950. Effect
of maleic acid on renal function. *Proc. Soc. Exp. Biol.
Med.* 75: 791-794.

Binkley, F. 1951. Metabolism of glutathione. *Nature* 167: 888-
889.

Binkely, F. 1954. In: *Symposium on Glutathione* (eds., Colowick
et al.). Academic Press, Inc., New York, N. Y. p. 160.

Binkley, F. 1961. Purification and properties of renal glutathionase. *J. Biol. Chem.* 236: 1075-1082.

Boyland, E., and L. F. Chasseaud 1969. The role of glutathione
and glutathione S-transferases in mercapturic acid biosynthesis. *Advan. Enzymol.* 32: 173-219.

Bretscher, M. S. 1971. Major human erythrocyte glycoprotein
spans the cell membrane. *Nature, New Biol.* 231: 229.

Curthoys, N. P. 1975. Subcellular localization of rat kidney
glutaminase isozymes and their role in renal ammoniagenesis.
II. Isozymes: Physiology and Function. C. L. Markert,
editor. Academic Press. N. Y.

Das, M., and A. N. Radhakrishnan 1973. Glycyl-L-leucine hydrolase, A versatile 'master' dipeptidase from monkey small
intestine. *Biochem. J.* 135: 609-615.

Fodor, P. J., A. Miller and H. Waelsch 1953. Quantitative
aspects of enzymatic cleavage of glutathione. *J. Biol.
Chem.* 202: 551-565.

George, S. G., and A. J. Kenny 1973. Studies on the enzymology
of purified preparations of brush border from rabbit
kidney. *Biochem. J.* 134: 43-57.

Glenner, G. G., and J. E. Folk 1961. Histochemical demonstration of a γ-glutamyl transpeptidase-like activity. *Nature*
192: 338-339.

Glenner, G. G., J. E. Folk, and P. J. McMillan 1962. Histochemical demonstration of a γ-glutamyl transpeptidase-like
activity. *J. Histochem. Cytochem.* 10: 481-489.

Glossman, H. and D. M. Neville, Jr. 1972. γ-glutamyl transferase in kidney brush border membranes. *FEBS Letters*
19: 340-344.

Goldbarg, J. A., O. M. Freidman, E. P. Peneda, E. E. Smith,

R. Chatterji, E. H. Stein, and A. M. Rutenburg 1960. The colorimetric determination of γ-glutamyl transpeptidase with a synthetic substrate. *Arch. Biochem. Biophys.* 91: 61-70.

Greenberg, E., E. E. Wollaeger, G. A. Fleisher, and G. W. Engstrom 1967. Demonstration of γ-glutamyl transpeptidase activity in human jejunal mucosa. *Clin. Chim. Acta* 16: 79-89.

Hanes, C. S., F. J. R. Hird, and F. A. Isherwood 1952. Enzymic transpeptidation reactions involving γ-glutamyl peptides and α-amino-acyl peptides. *Biochem. J.* 51: 25-35.

Harrison, H. E., and H. C. Harrison 1954. Experimental production of renal glycosuria, phosphaturia, and aminoaciduria by injection of maleic acid. *Science* 120: 606-608.

Hird, F. J. R. 1950. The γ-glutamyl transpeptidation reaction. Doctoral dissertation, Cambridge University, England.

Hird, F.J.R. and P.H. Springell 1954. The enzymic reaction of amino acids with glutathione. *Biochem. J.* 56: 417-425.

Hird, F.J.R. and P.H. Springell 1954. The enzymic hydrolysis of the γ-glutamyl bond in glutathione. *Biochim. Biophys. Acta* 15: 31-37.

Katunuma, N., I. Tomino, and H. Nishino 1966. Glutaminase isozymes in rat kidney. *Biochem. Biophys. Res. Commun.* 22: 321-328.

Katunuma, N., A. Huzino, and I. Tomino 1967. Organ specific control of glutamine metabolism. In: *Advances in Enzyme Regulation* (ed., Weber, G.). Pergamon Press, N. Y., Vol. 5, pp. 55-69.

Katunuma, N., T. Katsunuma, I. Tomino, and Y. Matsuda 1968. Regulation of glutaminase activity and differentiation of the isozyme during development. In: *Advances in Enzyme Regulation* (ed., Weber, G.). Pergamon Press, N. Y., Vol. 6, pp. 227-242.

Katunuma, N., T. Katsunuma, T. Towatari, and I. Tomino 1973. Regulatory mechanisms of glutamine catabolism. In: *The Enzymes of Glutamine Metabolism*, (eds., Prusiner, S., and Stadtman, E. R.). Academic Press, N. Y., pp. 227-258.

Knox, W. E. 1960. In: *The Enzymes* (2nd Edition), 2: 253-294. Academic Press, Inc., New York, N. Y.

Kokot, F., J. Kuska, and H. Grzybek 1965. γ-Glutamyl transpeptidase in the urine and intestinal contents. *Arch. Immunol. Therap. Exp.* 13: 549-556.

Matthews, D. M. 1972. Rates of peptide uptake by small intestine. In: *Peptide Transport in Bacteria and Mammalian Gut*. (CIBA Foundation Symposium), 71-92. Elsevier.

Meister, A. 1965. *Biochemistry of the Amino Acids* (2nd Edition), Vol. 1, pp. 475-482. Academic Press, Inc., N. Y.

763

Meister, A. 1973. On the enzymology of amino acid transport. *Science* 180: 33-39.

Meister, A. 1974. Glutathione: Metabolism and function via the γ-glutamyl cycle. *Life Sciences* 15: 177-190.

Milne, M. D. 1971. Transport of amino acids and peptides in the gut and the kidney. *The Scientific Basis of Medicine Annual Reviews* 161-177.

Morgan, E. J., and E. Friedmann 1938. Interaction of maleic acid with thiol compounds. *Biochem. J.* 32: 733-742.

Orlowski, M. 1963. The role of γ-glutamyltranspeptidase in the internal diseases. *Clinic. Arch. Immunol. Ther. Exp.* 11: 1-61.

Orlowski, M., and A. Meister 1963. γ-glutamyl-p-nitroanilide: A new convenient substrate for determination and study of L- and D-γ-glutamyltranspeptidase activities. *Biochim. Biophys. Acta* 73: 679-681.

Orlowski, M., and A. Meister 1965. Isolation of γ-glutamyl transpeptidase from hog kidney. *J. Biol. Chem.* 240: 338-347.

Orlowski, M., and A. Meister 1970. The γ-glutamyl cycle: A possible transport system for amino acids. *Proc. Natl. Acad. Sci. U. S.* 67: 1248-1255.

Orlowski, M., and A. Meister 1973. γ-glutamyl cyclotransferase: Distribution, isozymic forms, and specificity. *J. Biol. Chem.* 248: 2836-2844.

Orlowski, M., P. G. Richman, and A. Meister 1969. Isolation and properties of γ-L-glutamylcyclotransferase from human brain. *Biochemistry* 8: 1048-1055.

Orlowski, M., and A. Szewczuk 1961. Colorimetric determination of γ-glutamyl transpeptidase activity in human serum and tissues with synthetic substrates. *Acta. Biochimica Polonica* 8: 189-200.

Palekar, A. G., S. S. Tate, and A. Meister 1974. Formation of 5-oxoproline from glutathione in erythrocytes by the γ-glutamyltranspeptidase-cyclotransferase pathway. *Proc. Natl. Natl. Acad. Sci. U. S.* 71: 293-297.

Peters, T. J. 1972. Subcellular fractionation of the enterocyte with special reference to peptide hydrolases. In: *Peptide Transport in Bacteria and Mammalian Gut.* (CIBA Foundation Symposium), 107-122. Elsevier.

Revel, J.P. and E.G. Ball 1959. The reaction of glutathione with amino acids and related compounds by γ-glutamyl transpeptidase. *J. Biol. Chem.* 234: 577-582.

Richter, R. 1969. Some properties of γ-glutamyl transpeptidase from human kidney. *Arch. Immun. Therap. Exp.* 17: 476-495.

Rosenberg, L. E., and S. Segal 1964. Maleic acid-induced inhibition of amino acid transport in rat kidney. *Biochem. J.* 92: 345-352.

Segal, S., and S. O. Thier 1973. Renal handling of amino acids. In: *Handbook of Physiology;* Section 8, Chapter 15 (eds., Geiger, S. R., Orloff, J., and Berliner, R. W.). pp. 653-676.

Segrest, J. P., R. L. Jackson, E. P. Andrews, and V. T. Marchesi 1971. Human erythrocyte membrane glycoprotein: A re-evaluation of the molecular weight as determined by SDS polyacrylamide gel electrophoresis. *Biochem. Biophys. Res. Commun.* 44: 390-395.

Segrest, J. P., and R. L. Jackson 1972. Molecular weight determination of glycoproteins by polyacrylamide gel electrophoresis in sodium dodecyl sulfate. *Methods in Enzymol.,* 28B: 54-63.

Smyth, D. H. 1972. Peptide transport by mammalian gut. In: *Peptide Transport in Bacteria and Mammalian Gut.* (CIBA Foundation Symposium), 59-70, Elsevier.

Springell, P. H. 1953. Amino acid metabolism with special reference to peptide bond transfer. Doctoral Thesis, Melbourne University, Australia.

Szewczuk, A., and J. Baranowski 1963. Purification and properties of γ-glutamyl transpeptidase from beef kidney. *Biochem. Zeit.* 338: 317-329.

Tate, S. S., and A. Meister 1974a. Interaction of γ-glutamyl transpeptidase with amino acids, dipeptides, and derivatives and analogs of glutathione. *J. Biol. Chem.* 249: (in press).

Tate, S. S. and A. Meister 1974b. Stimulation of the hydrolytic activity and decrease of the transpeptidase activity of γ-glutamyl transpeptidase by maleate; Identity of a rat kidney maleate-stimulated glutaminase and γ-glutamyl transpeptidase. *Proc. Natl. Acad. Sci.* U. S. 71: 3329-3333.

Ugolev, A. M. 1972. Membrane digestion and peptide transport. In: *Peptide Transport in Bacteria and Mammalian Gut.* (CIBA Foundation Symposium), 123-143, Elsevier.

Van Der Werf, P., M. Orlowski, and A. Meister 1971. Enzymatic conversion of 5-Oxo-L-proline to L-glutamate coupled with cleavage of ATP to ADP, a reaction in the γ-glutamyl cycle. *Proc. Natl. Acad. Sci.* U. S. 68: 2982-2985.

Webb, J. L. 1966. In: *Enzyme and Metabolic Inhibitors* (Academic Press, New York), Vol. 3, pp. 285-335.

Young, J. A., and B. S. Freedman 1971. Renal tubular transport of amino acids. *Clin. Chem.* 17: 245-266.

COMPARATIVE PHYSICAL, CHEMICAL, AND ENZYMIC PROPERTIES
OF THE ISOZYMIC FORMS OF AVIAN ALDOLASES
AND FRUCTOSE DIPHOSPHATASES

RONALD R. MARQUARDT

Department of Animal Science
University of Manitoba
Winnipeg, Manitoba R3T 2N2, Canada

ABSTRACT. The distribution of chicken *(Gallus domesticus)*
tissue (muscle, oviduct, heart, brain, kidney, and liver)
aldolase (EC 4.1.2.13) isozymes was established electrophoretically. Each tissue, being composed of varying proportions of six isozymes, has a distinct isozyme pattern.
Brain, oviduct, and heart tissue possess five isozymes of
aldolase: 1,2,3,4, and 5. Leg and breast muscle of adults
possess aldolase 5. Kidney and liver predominantly possess
aldolase 6 with traces of isozymes 1-5. Aldolase 1 corresponds to aldolase C of mammals, 5 to A and 6 to B.
     Antibodies to 5 (A) react with 2,3, and 4, but not with
1 (C) and 6 (B). Aldolase 5 (C) hybridizes in vitro with
1 (A) to form 2,3, and 4. Aldolase 6 also hybridizes in
vitro with 1 to form three intermediate hybrids, but these
hybrids have not been detected in tissues. Pure preparations of aldolases 1 (C), 5 (A), and 6 (B) were found to
have similar molecular properties but to have different immunochemical properties, amino acid compositions, electrophoretic mobilities, and enzymatic properties.
     Two distinct isozymic forms of fructose diphosphatases
(EC 3.1.3.11) have been shown to occur in avian tissue, one
present in liver and kidney and another in skeletal muscle.
Comparison of the purified forms of the liver and muscle
enzymes demonstrate that they are similar with regard to
certain molecular properties but have different electrophoretic mobilities, isoelectric points, immunochemical
reactivities, thermal stabilities, and enzymatic properties.
In general, the muscle enzyme appears to be considerably
more sensitive to cofactor requirements and inhibitors than
the liver enzyme.

INTRODUCTION

The objective of this paper is to summarize the present
status of knowledge concerning the comparative distribution
and properties of chicken *(Gallus domesticus)* aldolases (EC
4.1.2.13) and fructose diphosphatases (EC 3.1.3.11). Three
distinct aldolases, A, B, and C, from chicken muscle, liver,

and brain (Marquardt, 1969a, b, and c; Marquardt, 1970; and
Marquardt, 1971a and b) and two distinct fructose diphosphatases from chicken muscle and liver (Olson and Marquardt, 1972;
Marquardt and Olson, 1974a; and Marquardt and Olson, 1974b)
have been isolated in pure form. These studies complement
those of Rutter and co-workers (Blostein et al., 1963; Rutter
et al., 1968; Penhoet et al., 1966; Penhoet, et al., 1969a; and
Penhoet et al., 1969b) with mammalian aldolases and those of
Horecker and co-workers (Black et al., 1972; Enser, et al.,
1969; Horecker, 1974; Pentremoli et al., 1970; Sia et al., 1969;
Traniello et al., 1971; and Van Tol et al., 1972) with mammalian
fructose diphosphatases.

METHODS

The methods used in these experiments, including the purification to homogeneity of the predominant forms of avian muscle (aldolase 5 or A), brain (aldolase 1 or C), and liver (aldolase 6 or B) aldolases, and muscle and liver fructose diphosphatases, were as described previously (Marquardt 1969a, b, and
c; Marquardt 1970; Marquardt 1971a and b; Marquardt and Olson,
1974a; Marquardt and Olson, 1974b; and Olson and Marquardt,
1972). Hybridization of native and chemically modified fructose diphosphatase was based on the general method of Meighen
and Schachman 1970.

RESULTS

COMPARATIVE PROPERTIES OF AVIAN ALDOLASES

Tissue Isozyme Patterns: Typical aldolase isozyme patterns
for the 14-day-old chick and the mature hen (Fig. 1) reveal
the existence of multiple molecular forms of aldolase in various tissues. Of the tissues tested, only breast muscle aldolase (aldolase 5 or A) migrated as a single band. Hen leg muscle is similar to breast muscle, being composed predominantly
of aldolase 5. In chick leg muscle, in addition to isozyme 5,
considerable amounts of isozyme 4 and traces of 1,2, and 3 are
also evident. The intensity of isozymes 1-4 in leg muscle is
related to the relative proportions of red muscle tissue included in the homogenate. Heart, brain, and oviduct, in contrast
to muscle tissues, contain five readily discernible isozymes.
The proportions of isozymes, as indicated by intensity of bands,
are quite different. Heart predominantely contains aldolase
2 and 3, whereas in brain tissue aldolases 1 (C) and 2 predominate with lesser amount of the other three isozymes. Oviduct,
in contrast, has about equal activities of all five isozymes.
Liver and kidney each contain six isozymes. Isozymes 1-5 are
the same as those found in other tissues. Isozyme 6 (aldolase
B), which is predominant in these tissues, is found exclusively

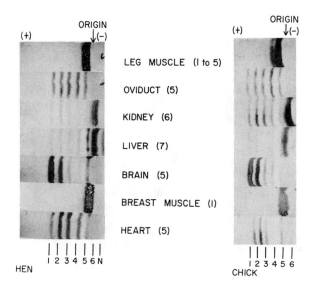

Figure 1. Cellulose-polyacetate electrophoresis of selected chicken tissue homogenates to reveal multiple forms of aldolase. Typical patterns for a mature laying hen (left) and 14-day-old chick (right). The numbers in brackets refer to number of isozymes associated with each tissue. Tissue preparations and electrophoresis was carried out as described earlier (Marquardt, 1969c). The strips were stained for aldolase activity by the method of Penholt et al. (1966).

in kidney and liver. The band N in liver, which developed in the absence of fructose diphosphate, was a nonspecific dehydrogenase. Studies by Herskovitz et al (1967) and Lebhertz and Rutter (1969) are consistent with the above observations.

Immunological and Hybridization Studies: The immunological properties of aldolases from various tissues and species were examined by double diffusion and precipitin analysis. The results presented in Fig. 2 show that there is a strong reaction between anti-liver aldolase and pure liver aldolase and the enzyme present in liver and kidney extracts. No cross-reaction was observed with the adolases from either brain or muscle. Other studies with anti-chicken breast muscle aldolase (anti-A) reveal a strong cross-reaction between anti-A and pure aldolase A but little or no reaction between anti-A and aldolase B or C. Anti-A in addition to precipitating adolase A also is able to

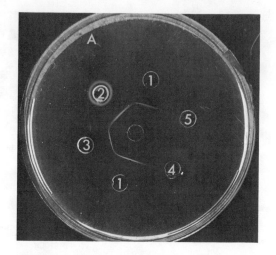

Figure 2.   Two-dimensional immunodiffusion of various aldolases.
The center well contained chicken liver aldolase rabbit antiserum.   Side wells contained: (1) pure chicken liver aldolase,
(2) chicken liver extract, (3) chicken kidney extract, (4)
chicken muscle extract, and (5) chicken brain extract.   See
Marquardt (1971b) for further detail.

---

precipitate other forms of the enzyme including isozymic bands
2, 3, and 4 (Fig. 3).   In this experiment aldolase 1 (C) or
aldolase 6 (B) were not precipitated by anti-A.   These latter
observations are consistent with the above observations.

The extent of cross-reaction between anti-liver aldolase
and aldolase from liver extracts of several vertebrate species
(Fig. 4) indicate that chicken and turkey aldolases are similar
in that both are completely precipitated.   In contrast, rabbit
and rat liver aldolases are not precipitated with the antiserum.

On the basis of these studies it may be concluded that in
chickens, as in mammals, there are three parental forms of aldolase with each tissue having a different distribution of
isozymic forms.   The predominant forms which are immunologically
distinct are aldolase A or 5 from muscle, aldolase B or 6 from
liver or kidney and aldolase C or 1 from brain.   Other forms
of aldolase found in various tissues represent hybrid forms

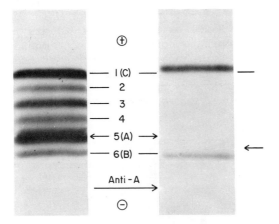

Figure 3. Effect of anti-aldolase A on certain aldolase isozymes. The zymogram on the left is that obtained from a mixture of isozymes isolated from brain, muscle, and liver, and
that to the right, after anti-aldolase A was added. Preparation of isozymes, cellulose-polyacetate electrophoresis and
enzyme staining were as described by Marquardt (1969c).

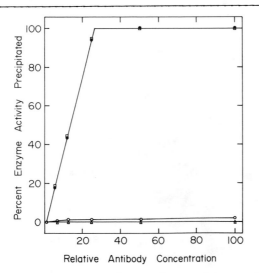

Figure 4. Precipitation of vertebrate liver aldolases with
chicken liver antiserum. (●) Chicken liver extract, (□)
turkey liver extract, (O) rat liver extract, and (△) rabbit
liver extract. See Marquardt (1971b) for further detail.

of the parental aldolases. The ability of parental aldolases
to form hybrid sets of isozymes is depicted in Fig. 5. When
aldolase C (1) and A (5) or C (1) and B (6) are mixed and incubated under appropriate conditions five membered isozyme
sets form. The electrophoretic mobilities of the A–C (5-1)
hybrids, produced in vitro, correspond to those detected in

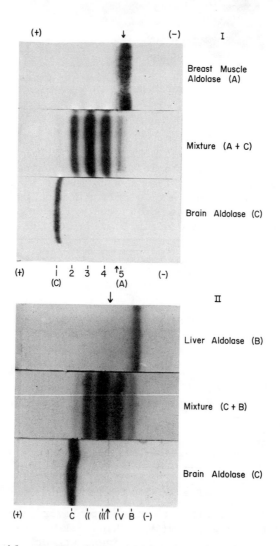

Figure 5. Acid-catalyzed hybridization of avian aldolases.
Upper frame (I), chicken brain aldolase C (1) and chicken

Fig. 5 (con't) breast muscle aldolase A (5); lower frame (II), chicken brain aldolase C (1) and chicken liver aldolase B (6). Hybridization procedures and the detection of the hybridized products by cellulose-polyacetate electrophoresis have been described by Marquardt (1969c). The middle zymogram in each frame represents the hybridized proteins while the exterior strips are the parental isozymes.

---

tissue extracts. The formation of five-membered hybrid sets from pairs of parental aldolases is most readily explained in the basis that aldolases are composed of tetrameric subunits. The naturally occurring 1-5 hybrids observed in tissue extracts probably result from the synthesis of two different kinds of aldolases in the same cell with the consequent formation of multiple combinants. Hybrids of aldolase B (6) and aldolase C (1) were not detected suggesting that aldolase B (6) may not be accompanied by other aldolases in the same cells.

Comparative Properties: In order to obtain an objective comparison of the amino acid composition of avian and rabbit aldolase isozymes a deviation function was calculated (Marquardt, 1971b). The results summarized in Table I suggest there is a difference in amino acid composition between homologous aldolases from different species and between heterologous aldolases within the same species, the divergence being larger in the latter comparisons. The more distantly related yeast aldolase has a sufficiently low divergence value to suggest an interrelationship among the various aldolases. All of the aldolase values are much lower than those for the two unrelated proteins.

A summary of the properties of avian aldolases A, B, and C is presented in Table 2. The proteins have similar molecular properties but differ with regard to amino acid composition, electrophorectic mobilities, immunochemical reactivities and catalytic properties. Aldolase A has a higher rate of fructose diphosphate cleavage, a lower fructose 1-phosphate cleavage rate, and a higher Km for both substrates than aldolase B. Aldolase A and C has similar Km values but different V_{max} values.

COMPARATIVE PROPERTIES OF AVIAN FRUCTOSE DIPHOSPHATASES

Electrophoretic and Molecular Properties: An Electrophoretic comparison of pure forms of the liver and muscle enzymes show that the two proteins have a different migration rate in pH 8.8 buffer (Fig. 6). Sedimentation velocity analysis of the two enzymes reveal a single symmetrical peak with an $S°_{20W}$ value of 7.0 S for the muscle enzyme and 6.8 S for the liver enzyme. Meniscus depletion sedimentation equilibrium analysis also indicated homogeneous species with molecular weights of

TABLE I

Divergence of amino acid composition of rabbit and
chicken aldolases[a]

Description	Value
Comparison of the same enzymes	
Rabbit liver x rabbit liver	1.0
Rabbit muscle x rabbit muscle	<u>1.3</u>
Average	1.2
Comparison of isozymes from the same species	
Chicken liver x chicken muscle	3.3
Chicken liver x chicken brain	3.8
Rabbit liver x rabbit muscle	3.1
Rabbit liver x rabbit brain	<u>4.2</u>
Average	3.6
Comparison of homologous isozymes from different species	
Chicken liver x rabbit liver	3.4
Chicken brain x rabbit brain	3.0
Chicken muscle x rabbit muscle	<u>2.2</u>
Average	2.9
Comparison of rabbit liver aldolase with yeast aldolase and unrelated proteins	
Rabbit liver x yeast aldolase	4.9
Rabbit liver x chymotrypsin	12.0
Rabbit liver x human cytochrome c	13.9

[a] Divergence = D x 100. See Marquardt (1971b) for further details.

136,000 for liver and 146,000 for muscle fructose diphosphatase.
The subunit structure of fructose diphosphatase was investigated using a hybridization procedure. Liver fructose diphosphatase was first succinylated so as to increase its mobility and was then hybridized with the normal liver enzyme. Hybridization results in the formation of five activity bands having different mobilities. Comparison of these results with the controls leads to the conclusion that three intermediate bands

TABLE II

Properties of aldolase variants from chicken

	Muscle (A,5)	Liver (B,6)	Brain (C,1)
Amino acid composition	Different		
Electrophoretic mobilities	Different		
Percent aldolase precipitated with anti-A	100	0	0
Percent aldolase precipitated with anti-B	0	100	0
Number of subunits	4	4	4
s°_{20W}	7.9	8.0	8.0
Stokes radius (mu)	4.7	4.6	4.6
Molecular weight	158,000	154,000	155,000
FDP/FMP Activity Ratio	30	2.2	30
Vmax (FDP cleavage)[a]	24	1.1	8
Km (M), fructose diphosphate	4×10^{-5}	1×10^{-6}	7×10^{-5}
Km (M), fructose diphosphate	2×10^{-2}	9×10^{-4}	3×10^{-2}

[a] umoles substrate cleaved per min per mg protein at 30° C.

arose from mixing of subunits from succinyl-fructose diphosphatase and fructose diphosphatase. This type of pattern which
develops for both the liver and muscle enzymes would be expected
for the hybridization of two tetrameric molecules each composed
of four similar chains (Marquardt and Olson 1974).

Immunochemical Analyses: Immunodiffusion patterns of antiserum
to chicken liver fructose diphosphatase results in a strong reaction between the antiserum and the corresponding enzyme but
only a weak reaction between the antiserum and the muscle enzyme (Fig. 7). Quantitative precipitin patterns (Fig. 8.)
demonstrate that pure liver fructose diphosphatase and the enzymes from liver and kidney extracts are readily precipitated
with the antiserum. The antiserum, however, precipitates only
a small amount of the enzyme from muscle, particularly at low
antiserum concentrations.

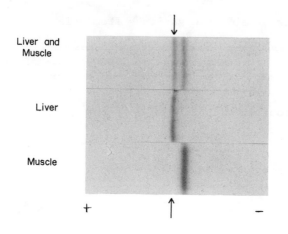

Figure 6. Cellulose-polyacetate electrophoresis of avian liver and muscle fructose diphosphatases. Electrophoresis was carried out at pH 8.8 and the strips were stained for protein with Ponceau S. See Olson and Marquardt (1972) for further detail.

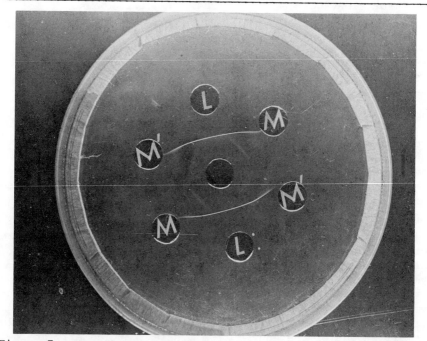

Figure 7. Two dimensional immunodiffusion of avian fructose diphosphatases. L, liver fructose diphosphatase and M, muscle

Fig. 7 (con't) fructose diphosphatase. The center well contained chicken liver fructose diphosphatase antiserum. See
Olson and Marquardt (1971) for further detail.

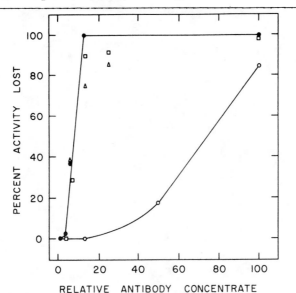

Figure 8. Precipitation of cystalline chicken liver fructose
diphosphatase (●) and the fructose diphosphatases from extracts
of chicken liver (□), kidney (Δ) and breast muscle (O) with
antiserum to chicken liver fructose diphosphatase. See Olson
and Marquardt (1971) for further detail.

A comparison of the immunodiffusion pattern of liver fructose diphosphatases from several species (Fig. 9) indicates
that there is a good cross-reaction between the antiserum and
crystalline chicken liver fructose diphosphatase and the liver
fructose diphosphatases of budgerigar and turkey. The contin-
uity between the precipitin bands of chicken liver fructose
diphosphatase and turkey liver extract fructose diphosphatases
suggest a fairly high degree of similarity between the two enzymes. Spur formation between the preciptin bands of budgerigar
liver extract fructose diphosphatase and chicken and turkey
liver fructose diphosphatases suggest that budgerigar liver
enzyme is immunologically different from the enzyme present
in the liver of chicken and turkey. There is, however, little
or no cross-reaction with liver fructose diphosphatases of rat
and rabbit. Quantitative precipitation results are consistent
with the above results.

Thermal stabilities: Both liver and muscle fructose diphospha-

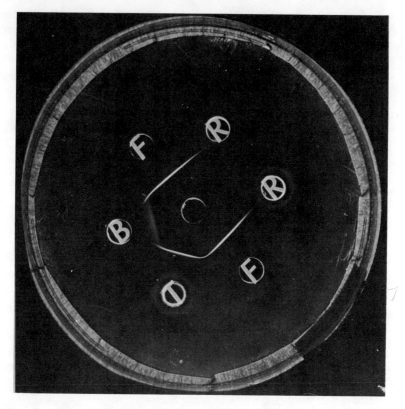

Figure 9. Two dimensional immunodiffusion of various fructose diphosphatases. The center well contained chicken liver fructose diphosphatase rabbit antiserum. The outer wells contained: F, crystalline chicken liver fructose diphosphatase; R, rabbit liver extract; R', rat liver extract; B, budgerigar liver extract; and T, turkey liver extract. See Olson and Marquardt (1971) for further detail.

---

tases are more stable in Tris buffer when stored at 20°C than at 2°C (Marquardt and Olson, 1974). In both cases the muscle enzyme decays at a faster rate than the liver enzyme. The approximate half life at pH 7.5 of the liver and muscle enzymes are 2.0 and 1.5 days at 2°C and longer than 6 and 3 days at 20°C, respectively. In the presence of added $MgCl_2$ and fructose diphosphate both enzymes are stable when stored at either temperature. Even after a 10-day storage period no appreciable loss in enzyme activity occurs. In the absence of added cofactors maximum stability of the two enzymes is obtained at approximately pH 6.0. Both enzymes are considerably less stable at pH 8.6 as compared to pH 6.0.

The results depicted in Fig. 10 demonstrate that both muscle and liver fructose diphosphatases are markedly stabilized by certain compounds. Individually fructose diphosphate (FDP) provided the greatest degree of protection for the muscle enzyme while $MgCl_2$ is most effective with the liver enzyme. Combinations of $MgCl_2$ and FDP or $MgCl_2$, FDP and AMP in the case of muscle enzyme, markedly increased enzyme stability. In the

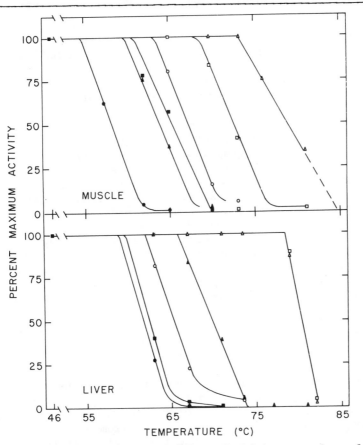

Figure 10. Thermal inactivation of chicken muscle and liver fructose diphosphatases as affected by certain compounds. Stock muscle and liver enzymes were diluted to 3 and 2 ug/ml respectively, with 0.1 M Tris, 2mM EDTA, 10 mM 2-mercaptoethanol and 1 mg/ml bovine serum albumin, pH 7.5 (●) containing 0.5 mM AMP (■); 2 mM fructose diphosphate (FDP, O); 10 mM $MgCl_2$ (▲); 10 mM $MgCl_2$ and 2 mM FDP, or 10 mM $MgCl_2$, 2 mM FDP, and 0.1 mM AMP (△). The enzyme solutions were incubated at the temperature indicated for 15 min, immediately placed in

Fig. 10 (con't) an ice bath for 3 min., centrifuged at 50,000 g for 10 min. and assayed.

---

presence of the three compounds, initial inactivation of the muscle and liver enzyme does not occur until temperatures of 74 and 78°C, respectively, are reached. This compares with initial inactivation temperatures in the absence of added cofactors of 55°C for muscle enzyme and 59°C for the liver enzyme.

Kinetic Properties: Avian liver and muscle fructose diphosphatases have an absolute requirement for Mg^{++} or Mn^{++} and the pH profiles of both enzymes are dependent upon concentration of Mg^{++}. At high pH levels increasing concentrations of Mg^{++} inhibits both enzymes whereas at lower pH levels the opposite pattern occurs. The apparent Km for $MgCl_2$ at pH 7.5 is 0.5 mM for the muscle enzyme and 5 mM for the liver enzyme. (Marquardt and Olson, 1974). Fructose diphosphate inhibits both enzymes. At pH 7.5 the inhibitor constants (Ki) are 0.18 and 1.3 mM for muscle and liver fructose diphosphatases, respectively. The muscle enzyme is considerably more sensitive to AMP inhibition than the liver enzyme. At pH 7.5 and in the presence of 1 mM $MgCl_2$; 50% inhibition of muscle and liver fructose diphosphatases occur at AMP concentrations of 3.0×10^{-8} and 2.3×10^{-6} M, respectively. EDTA activates both enzymes. The degree of activation is time and concentration dependent. The degree of EDTA activation decreases with increasing $MgCl_2$ concentration. Ca^{++} is a very potent inhibitor of both liver (Ki 1×10^{-4} M) and muscle (Ki 1×10^{-5} M) fructose diphosphatases. Ca^{++} appears to be a competetive inhibitor with regard to Mg^{++}. There is, however, a positive homeotropic interaction among Mg^{++} sites of both enzymes in the presence of Ca^{++} (Marquardt and Olson, 1974).

An overall summary of some of the comparative properties of the two enzymes (Table III) demonstrate that muscle and liver fructose diphosphatases differ with regard to certain properties. In general, the muscle enzyme appears to be considerably more sensitive to cofactor requirement and inhibition than the liver enzyme. These results suggest that there are at least two isozymic forms of avian fructose diphosphatases.

TABLE III

Comparative properties of liver and muscle
fructose diphosphatases

	Liver	Muscle
Electrophoretic mobilities	Different	
pI	8.1	8.6
S°_{20W}	6.9	7.0
Molecular weight	136,000	146,000
Number of subunits	4	4
Immunological reaction with anti-liver FDPase	+	−
Thermal stabilities	Different	
Km, fructose diphosphate	Different	
Km (mM) Mg^{++}	5	0.5
Ki (mM) Ca^{++}	0.1	0.01
Ki (mM) fructose diphosphate	1.3	0.18
Ki (M) AMP	2×10^{-6}	3×10^{-8}

REFERENCES

Black, W. J., A. Van Tol, J. Fernando, and B. L. Horecker
1972. Isolation of a highly active fructose diphosphatase
from rabbit muscle: Its subunit structure and activation
by monovalent cations. *Arch. Biochem. Biophys.* 151: 576-
590.

Blostein, R., and W. J. Rutter 1963. Comparative studies of
liver and muscle aldolase II. Immunochemical and chromatographic differentiation. *J. Biol. Chem.* 238: 3280-3285.

Enser, M., S. Shapiro, and B. L. Horecker 1969. Immunological
studies of liver, kidney, and muscle fructose 1, 6-diphos-
phatases. *Arch. Biochem. Biophys.* 129: 377-383.

Herskovits, J. J., C. J. Masters, P. M. Wassarman, and N. O.
Kaplan 1967. On the tissue specificity and biological
significance of aldolase C in the chicken. *Biochem. Biophys. Res. Comm.* 26: 24-29.

Horecker, B. L. 1974. Biochemistry of Isozymes. *I. Isozymes:
Molecular Structure* C. L. Markert, editor, Academic Press,
New York.

Lebhertz, H. G., and W. J. Rutter 1969. Distribution of fructose diphosphate aldolase variants in biological systems. *Biochem.* 8: 109-121.

Marquardt, R. R. 1969a. Multiple molecular forms of avian aldolases. I. Crystallization and physical properties of chicken *(Gallus domesticus)* breast muscle aldolase. *Can. J. Biochem.* 47: 517-526.

Marquardt, R. R. 1969b. Multiple molecular forms of avian aldolases. II. Enzymic properties and amino acid composition of chicken *(Gallus domesticus)* breast muscle aldolase. *Can. J. Biochem.* 47: 527-534.

Marquardt, R. R. 1969c. Multiple molecular forms of avian aldolases. III. Tissue distribution and properties of the predominant isozymes from chicken *(Gallus domesticus)*. *Can. J. Biochem.* 47: 1187-1194.

Marquardt, R. R. 1970. Multiple molecular forms of avian aldolases IV. Purification and properties of chicken *(Gallus domesticus)* brain aldolase. *Can. J. Biochem.* 48: 322-333.

Marquardt, R. R. 1971a. Multiple molecular forms of avian aldolases. V. Purification and molecular properties of chicken *(Gallus domesticus)* liver aldolase. *Can. J. Biochem.* 49: 647-657.

Marquardt, R. R. 1971b. Multiple molecular forms of avian aldolases. VI. Enzymic properties and amino acid composition of chicken liver aldolase and comperative immunochemical properties. *Can. J. Biochem.* 49: 658-665.

Marquardt, R. R., and J. P. Olson 1974a. Thermal inactivation of avian liver and muscle fructose 1,6-diphosphatase in the presence of fructose 1,6-diphosphate, AMP and divalent cations. *Can. J. Biochem.* 52: in press.

Marquardt, R. R., and J. P. Olson 1974b. Comparative enzymatic properties of avian liver and skeletal muscle fructose 1,6-diphosphatases. Submitted for publication.

Meighen E. A., and H. K. Schachman 1970. Hybridization of native and chemically modified enzymes. I. Development of a general method and its application to the study of the subunit structure of aldolase. *Biochem.* 9: 1163-1176.

Olson, J. P., and R. R. Marquardt 1972. Avian fructose-1,6-diphosphatases I. Purification and comparison of physical and immunological properties of the liver and breast muscle enzymes from chicken *(Gallus domesticus)*. *Biochim. Biophys. Acta.* 268: 453-467.

Penhoet, E. E., T. Rajkumar, and W. J. Rutter 1966. Multiple forms of fructose diphosphate aldolase in mammalian tissues. *Proc. Nat. Acad. Sci.* 56: 1275-1282.

Penhoet, E. E., M. Kochman, and W. J. Rutter 1969a. Isolation of fructose diphosphate aldolases A, B, and C. *Biochem.* 8: 4391-4395.

Penhoet, E. E., M. Kochman, and W. J. Rutter 1969b. Molecular and catalytic properties of aldolase C. *Biochem.* 8: 4396-4402.

Pontremoli, S. and B. L. Horecker 1970. Fructose 1,6-diphosphatase from rabbit liver. *In Current Topics in cellular Regulation* (Horecker, B. L. and E. R. Stadtman, eds). Vol. 2: pp 173-199, Academic Press, New York.

Rutter, W. J., T. Rajkumar, E. Penhoet, M. Kochman, and R. Valentine 1968. Aldolase variants: Structure and physiological significance. *Ann. N. Y. Acad. Sci.* 151: 102-117.

Sia, C. L., S. Tranillo, S. Pontremoli, and B. L. Horecker 1969. Studies on the subunit structure of rabbit liver fructose diphosphatase. *Arch. Biochem. Biophys.* 132: 325-330.

Traniello, S., S. Pontremoli, Y. Tashima, and B. L. Horecker 1971. Fructose 1,6-diphosphatase from liver: Isolation of the native form with optimal activity at neutral pH. *Arch. Biochem. Biophys.* 146: 161-166.

Van Tol, A., W. J. Black, and B. L. Horecker 1972. Activation of rabbit muscle fructose diphosphatase by EDTA and the effect of divalent cations. *Arch. Biochem. Biophys.* 151: 591-596.

REGULATION OF MULTIPLE FORMS OF AMINO ACID CATABOLIZING ENZYMES IN RAT LIVER

HENRY C. PITOT, ROBERT LYONS, and JESUS RODRIGUEZ
McArdle Laboratory for Cancer Research
The Medical School, University of Wisconsin
Madison, Wisconsin 53706

ABSTRACT. That multiple forms of enzymes in rat liver are under regulatory control has been known for some time. Patterns of enzyme synthesis and regulation are complex. Single genes control multiple forms of the same enzyme, and multiple alleles for the same enzyme exist in different species. The synthesis of multiple forms of serine dehydratase and of tyrosine aminotransferase is regulated by one or more different hormones. Serine dehydratase occurs in two forms, one regulated by glucagon and the other by cortisone. In the case of tyrosine aminotransferase, one form appears to be comparable to that reported by others and different from the inducible tyrosine aminotransferase. The other three forms exhibit several hormonal controls, the synthesis of one form clearly being regulated by insulin. Ornithine aminotransferase occurs in soluble and intramitochondrial forms as distinguished by rate of labeling which indicates a precursor-product relationship between the soluble and the intramitochondrial form.

The possibility that the multiple forms of serine dehydratase and tyrosine aminotransferase may be the result of different states of oxidation of sulfhydryl and disulfide groups within the molecule is discussed.

Multiple molecular forms of enzymes occurring within the same organism and having the same catalytic activities have been known for some years. The chairman of this conference, Dr. Clement L. Markert, with his colleagues were among the first to recognize the significance and initiate the description of isozymes (Markert and Møller, 1959). In the 15 years since that report numerous investigations, many of which are presented or referred to in these volumes, have shown the importance in biology of multiple molecular forms of enzymes, termed isozymes.

The original findings utilizing model isozymic systems such as lactate dehydrogenase (Wieland and Pfleiderer, 1961, Cahn et al., 1962) have now been extended to a variety of other enzymes exhibiting isozymic forms. That different proteins possessing the same catalytic activity and subject to different regulatory mechanisms, both at the level of the enzyme protein

itself and in the synthesis of these various isozymes has now
been well documented (Stifel and Herman, 1972). In the case of
of the model system, lactate dehydrogenase, the levels of the
various isozymic forms of this enzyme in different organs
have been reported to be the result of differential rates of
degradation of each of the two separable subunits (Fritz et
al., 1969).

As knowledge of multiple forms of enzymes increases, the
operational term "isozyme" can be made more specific and ex-
planatory. For example, the presence of protein modifications
occurring post-transcriptionally have now been reported (Holzer
and Duntze, 1971; Soffer, 1971; Segal, 1973). The studies reported in this paper are directed towards investigations of
closely related multiple forms of three amino acid catabolizing
enzymes of rat liver and hepatoma--serine dehydratase, tyrosine
aminotransferase, and ornithine aminotransferase. While the
experimental evidence is as not yet conclusive, the working
hypothesis to explain the existence of these multiple forms is
that they are post-transcriptional products resulting from
hormonal mechanisms in the case of the first two, and, the
multiple forms of ornithine aminotransferase reflect the requirement for the transfer of finished protein from one cell
compartment to another.

REGULATION OF MULTIPLE FORMS OF SERINE DEHYDRATASE

Several studies from this laboratory (Pitot and Peraino,
1964; Jost et al., 1968) have demonstrated that the rat liver
enzyme, serine dehydratase, can be induced quite readily in
rat liver by feeding dietary protein or by the administration
of amino acids, glucagon, and/or cortisone. More recently
Inoue et al. (1971) demonstrated that serine dehydratase of
rat liver and several hepatocellular carcinomas could be
separated into two forms by gel electrophoresis or DEAE
chromatography.

In Figure 1 is seen the separation of the two forms of
serine dehydratase from crude extracts of rat liver utilizing
DEAE-cellulose column chromatography. The two peaks of enzyme
activity on the left were obtained from rats that had been
given tryptophan five hours prior to sacrifice. The enzyme
pattern on the right was from an animal receiving glucagon
five hours prior to sacrifice. Both animals had previously
been maintained on a laboratory chow diet. The two forms of
the enzyme found in the animal administered tryptophan have
been purified and crystallized. They exhibit no differences
in molecular weight, immunochemistry, phosphate content, or
spectral and kinetic characteristics. While minor questionable

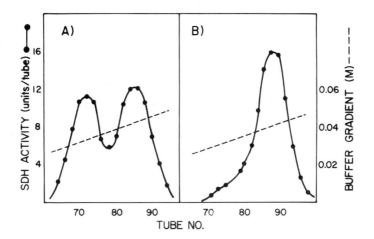

Fig. 1. DEAE-cellulose column chromatography of serine dehydratase in crude extracts from rat liver. A, extract obtained from liver of rat which had been administered tryptophan (100 mg) 5 hours prior to sacrifice. B, enzyme pattern from liver extract of rat receiving glucagon (0.5 mg) 5 hours prior to sacrifice. See Inoue, et al.,(1971) for details.

differences were found in the amino acid composition of the two forms, peptide analyses did show one or two differences in the peptide "maps" of the two forms (Pitot et al., 1972). The most striking characteristic of the two forms were the differences in their occurrence which was noted after the administration of glucagon. Under these circumstances only form I was evident. In Table 1 is seen the effect of various dietary and hormonal circumstances on the occurrence of the two forms of serine dehydratase. From these data, diabetic, newborn, and animals fed a chow diet or given glucagon exhibit almost all of their serine dehydratase in form II. Animals fed a high protein diet or given tryptophan show roughly equal contents of the two forms. On the other hand adrenalectomized animals exhibit form II only, unless cortisone is administered to them. These studies suggest that the synthesis of form I is regulated by cortisone while that of form II is regulated by glucagon.

In attempting to determine both metabolic and chemical differences between the two forms of this enzyme, Inoue et al. (1971) were able to demonstrate that neither form was the direct precursor of the other. On the other hand, these studies did not rule out the possibility that both forms could originate from a common precursor. These and other possibilities will be discussed later in this paper.

TABLE 1

The Levels of the Two Isozymes of Serine Dehydratase in Liver and Hepatoma Under Various Dietary and Hormonal Conditions

Treatment	Type of SDH %		Activity of SDH
	I	II	(units/g liver)
Intact rats			
Chow	0	100	11.5
90% protein diet (1 week)	55	45	105.0
Diabetes (1 week)	2	98	78.4
Newborn (1 week)	0	100	7.8
0% Protein diet (5 days)	--	---	0.9
0% Protein diet + glucagon (10 hr)[*]	3	97	8.2
0% Protein diet + tryptophan (9 hr)[*]	56	44	4.0
Adrenalectomized rats			
12% Protein diet (3 days)	0	100	8.4
12% Protein diet + tryptophan (9 hr)[+]	0	100	9.2
12% Protein diet + cortisone (6 hr)[+]	31	69	19.1
Morris hepatoma 7800			
Chow	80	20	35.0
Adrenalectomy (1 week)	57	43	10.5
90% Protein diet (1 week)	79	21	52.0

[*] Rats were fed on protein-free diet for 5 days, and then starved for 10 hr. Glucagon (0.5 mg/rat, intraperitoneally) and tryptophan (100 mg/rat, intubation) were given at 0 and 6 hr.

[+] Rats were fed on 12% protein diet for 3 days and then starved for 10 hr. Tryptophan (40 mg/rat) and cortisone (10 mg/rat) were only given at 0 hr.

Rats were killed at the times shown in the Table.

REGULATION OF MULTIPLE FORMS OF TYROSINE AMINOTRANSFERASE

In 1962 Kenney reported that during the purification of tyrosine aminofransferase from rat liver, it was possible to separate the enzyme into four distinct fractions by chromatography on hydroxylapatite. More recent studies by Holt and Oliver (1969) demonstrated the separation of at least three forms of tyrosine aminotransferase using gel electrophoresis. These studies suggested that the amount of each form varied with certain hormonal conditions in the host especially in relation to cortisone, insulin, or cyclic nucleotides.

In this laboratory Iwasaki et al. (1973) repeated the earlier work of Kenney (1962) separating and purifying four forms of tyrosine aminotransferase from rat liver. In Table 2 is seen data concerning the regulation of the four forms of tyrosine aminotransferase in heart, kidney, and liver of adult male adrenalectomized rats. From these data it can be noted that form I is ubiquitous in all organs studied and does not change in liver as a result of the dietary or hormonal conditions under which the animal is placed. On the other hand, the administration of cortisone causes a significant rise in forms II and III while glucagon appears to favor form III and epinephrine, form II. In the intact animal cortisone administration does cause a significant increase in all three forms.

In the adrenalectomized, diabetic animal, form IV virtually disappeared. Upon administration of insulin to this animal, a low but significant increase in the total level of the enzyme occurs which is due almost entirely to an increase in form IV.

As with the multiple forms of serine dehydratase no immunochemical, molecular weight, heat stability, or pH dependency differences could be noted in forms II, III, and IV. Form I appears to be a different enzyme in that it is quite heat labile, exhibits a significantly greater K_m for tyrosine, and acts on other aromatic amino acids as well as pyruvate and oxaloacetate (Iwasake et al., 1973). Recently, Johnson et al. (1973) have also described three forms of tyrosine aminotransferase separable on CM-Sephadex. The level of these forms also changed upon the administration of cortisone. However, these authors were able to interconvert the several forms isolated on CM-Sephadex by incubating liver supernatant at 25° or treating the enzyme with potassium cyanate. These authors suggested that the multiple forms may represent products of degradation of the enzyme. Further studies on the forms separable by CM-Cephadex in our laboratory have confirmed that changes occur in their levels on incubation in vitro.

TABLE 2

REGULATION OF 4 FORMS OF TYROSINE AMINOTRANSFERASE (TAT)
IN HEART, KIDNEY, AND LIVER OF ADULT MALE ADRENALECTOMIZED RATS

Tissue and Treatment	Number of Animals	Form of TAT in %				Total TAT[+] Activity (μmole/g/hr)
		I	II	III	IV	
Heart;						
60% protein diet. Killed at 1400 hours	(5)	100	0	0	0	13.1 ± 4.0
Kidney;						
60% protein diet. Killed at 1400 hours	(4)	100	0	0	0	14.7 ± 3.5
Liver;						
12.5% protein diet. Killed at 1400 hours	(5)	21	20	38	21	32.0 ± 3.0
12.5% protein diet. Killed at 2400 hours	(5)	12	20	40	28	78.5 ± 11.0
60% protein diet. Killed at 1400 hours	(5)	10	24	42	23	135 ± 11.0
60% protein diet. Killed at 2400 hours	(5)	3	25	40	31	618 ± 44.0
12.5% protein diet + cortisone (10 mg/100 BW). Killed 6 hrs later	(4)	2	40	43	15	891 ± 61.2

TABLE 2 (continued)

12.5% protein diet + glucagon (0.15 mg/100 BW). Killed 3 hrs later	(5)	5	35	44	16	287 ± 24.3
12.5% protein diet + cAMP (5 mg/100 BW). Killed 90 min later	(4)	12	52	30	5	122 ± 10.0
12.5% protein diet + epinephrine (250 r/100 g BW) 4 hrs and 2 hrs before sacrifice	(3)	8	47	35	10	114 ± 11.0
12.5% protein diet + 4 mg of alloxan i.v. sacrificed 6 days later	(3)	13	55	27	3	96 ± 7.5
12.5% protein diet alloxan-diabetic 6 days + 1 unit of insulin per 100 g BW. Killed 3 hrs later	(3)	12	24	25	38	121 ± 9.6

+ Values represent mean ± standard error. The total TAT activity is that determined on the total liver extract. The % of each form was determined after chromatography (Iwasaki et al., 1973).

Iwasaki et al. (1973) were unable to demonstrate any differences in the rate of decay of forms II, III, and IV after induction by hydrocortisone and labelling in vivo with subsequent determination of loss of label from the enzyme by immunochemical techniques. In preliminary studies in our laboratory it appears that there are significant differences between the forms separable on CM-Sephadex and those separated on hydroxylapatite.

In addition, Spencer and Gelehrter (1974) demonstrated in cultured hepatoma cells the presence of a minor form of tyrosine aminotransferase having different electrophoretic properties from that inducible by corticosteroids. A similar electrophoretic difference in tyrosine aminotransferase isozymes was also reported by Mertvetsov et al. (1973). While the relationship between this minor, more cathodal form and form I as described by Iwasaki et al., is not yet clear, there are certain similarities.

Unfortunately, the studies thus far do not tell us what the differences between multiple forms of serine dehydratase and tyrosine aminotransferase are. Recent investigations in this laboratory have demonstrated that form I of serine dehydratase can be converted to form II by incubation for four hours in 0.1 M mercaptoethanol (unpublished observations). In addition, incubation of form III of tyrosine aminotransferase in the presence of glutathione-insulin transhydrogenase suggests the conversion of form III to form IV. If these experiments are valid one might suggest that the differences between the several forms of these enzymes reside in their sulfhydryl-disulphide interrelationships. On the other hand, the differences may be more subtle such as those described for rabbit muscle phosphoglucose isomerase by Noltmann and his associates (Blackburn et al., 1972). While the structure of serine dehydratase and tyrosine aminotransferase are as yet unknown, it is possible that these two forms do contain some disulfide bridges. Serine dehydratase contains six 1/2-cystines of which at least two represent sulfhydryl groups. If such is the case then extensive reduction and denaturation in guanidinium hydrochloride should eliminate the differences between the multiple forms of both enzymes.

COMPARTMENTALIZED FORMS OF ORNITHINE AMINOTRANSFERASE

Along with the early studies on serine (threonine) dehydratase (Pitot and Peraino, 1964) dietary induced increases were demonstrated in the mitochondrial enzyme, ornithine aminotransferase. At that time attempts to demonstrate multiple forms of this enzyme had not met with reproducible

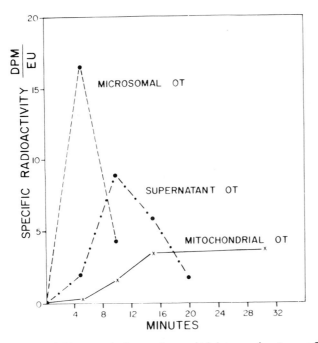

Fig. 2. Specific activity of ornithine aminotransferase (OT)
isolated from microsomal, supernatant and mitochondrial
fractions of liver from rats given 25 microcuries of (U)
^{14}C-L-leucine at 0 time.

success. Yip and Collins (1971) have shown that the kidney
and liver ornithine aminotransferase appear to be identical.
    In recent preliminary experiments in our laboratory we
have been able to demonstrate the presence of small amounts of
ornithine aminotransferase in the soluble fraction of the
hepatic cell. In crude preparations the heat stability of this
soluble form of the enzyme is considerably greater than that
of the mitochondrial form. In order to study further the metabolic difference between these two forms, pulse-labelling with
^{14}C-valine was carried out in vivo. The results of this
experiment are seen in Figure II. It may be noted that the
specific activity of ornithine aminotransferase isolated
immunochemically from the microsomal fraction was extremely
high five minutes after the administration of the label. The
specific activity of this supernatant enzyme peaked five minutes later while that of the mitochondrial enzyme did not reach
a steady state of labeling until 15 minutes after administration
of the radioactive amino acid. These data suggest that ornithine aminotransferase is synthesized on the endoplasmic

reticulum of the liver cell and then transferred from the supernatant to the mitochondria. On the basis of earlier experiments by Whiting and Elliott (1972) with mitochondrial δ-aminolevulinic acid synthetase, one would expect the extra- and intramitochondrial forms to be somewhat different. These workers showed that the molecular weight of the enzyme in the two compartments appear to be different, probably due to aggregation.

Thus, our studies although preliminary in nature do present evidence for at least two forms of hepatic ornithine aminotransferase, the differences between the two forms being related to their compartmentalization within the cell.

CONCLUSIONS

On the basis of the studies reported in this paper, multiple forms of enzymes involved in amino acid catabolism may readily be demonstrated. Studies by other workers on arginase (Gasiorowska et al., 1970) and alanine-glyoxylate aminotransferase (Snell and Walker, 1971) indicate further examples of both multiple forms of an enzyme in the cytosol and two distinct forms related to compartmentalization of an enzyme normally found in the mitochondria. While the evidence is by no means overwhelming, our working hypothesis at the present time suggests that these multiple forms are the result of post-translational modification of these enzymes with the potential for the disulphide interchange enzyme being involved in the multiple forms of serine dehydratase and tyrosine amino transferase. If this latter conjecture is born out, then the effects of steroid hormones on the disulphide-interchange enzyme that had previously been reported by Rabin and his associates (Sunshine et al., 1971) must be extended to include corticosteroids and even peptide hormones within mammalian liver.

REFERENCES

Blackburn, M. N., J. M. Chirgwin, G. T. James, T. D. Kempe, T. F. Parsons, A. M. Ragister, K. D. Schnaekerz, and E. A. Noltmann 1972. Pseudoisoenzymes of rabbit muscle phosphoglucose isomerase. *J. Biol. Chem.* 247: 1170-1179.

Cahn, R. D., N. O. Kaplan, L. Levine, and E. Zwilling 1962. Nature and development of lactic dehydrogenase. *Science* 136: 962-969.

Fritz, P. J., E. S. Vesell, E. L. White, and K. M. Pruitt 1969. The roles of synthesis and degradation in determining tissue concentrations of lactate dehydrogenase-5. *Proc. Natl. Acad. Sci., U. S. A.* 62: 558-565.

Gasiorowska, I., Z. Porembska, J. Jachimowicz, and I. Mochnacka 1970. Isoenzymes of arginase in rat tissues. *Acta Biochem. Polon* 17: 19-30.

Holt, P. G., and I. T. Oliver 1969. Multiple forms of tyrosine aminotransferase in rat liver and their hormonal induction in the neonate. *FEBS Letters* 5: 89-91.

Holzer, H., and W. Duntze 1971. Metabolic regulation by chemical modification of enzymes. *Ann. Rev. Biochem.* 40: 345-374.

Inoue, H., C. B. Kasper, and H. C. Pitot 1971. Studies on the induction and repression of enzymes in rat liver. VI. Some properties and the metabolic regulation of two isozymic forms of serine dehydratase. *J. Biol. Chem.* 246: 2626-2632.

Iwasaki, Y., C. Lamar, K. Danenberg, and H. C. Pitot 1973. Studies on the induction and repression of enzymes in rat liver. Characterization and metabolic regulation of multiple forms of tyrosine aminotransferase. *Europ. J. Biochem.* 34: 347-357.

Johnson, R. W., L. E. Roberson, and F. T. Kenney 1973. Regulation of tyrosine aminotransferase in rat liver. X. Characterization and interconversion of the multiple enzyme forms. *J. Biochem.* 248: 4521-4527.

Jost, J-P., E. A. Khairallah, and H. C. Pitot 1968. Studies on the induction and repression of enzymes in rat liver. V. Regulation of the rate of synthesis and degradation of serine dehydratase by dietary amino acids and glucose. *J. Biol. Chem.* 243: 3057-3066.

Kenney, F. T. 1962. Induction of tyrosine α-Ketoglutarate transaminase in rat liver II. Enzyme purification and preparation of antitransaminase. *J. Biol. Chem.* 237: 1605-1612.

Markert, C. L., and F. Møller 1959. Multiple forms of enzymes: Tissue, ontogenetic, and species specific patterns. *Proc. Natl. Acad. Sci., U. S.* 45: 753-763.

Mertvetsov, N. P., V. N. Chesnokov and R. I. Salganik 1973. Specific changes in the activity of tyrosine aminotransferase isoenzymes in rat liver after cortisol treatment and partial hepatectomy. *Biochim. Biophys. Acta* 315: 61-65.

Pitot, H. C., Y. Iwasaki, H. Inoue, C. B. Kasper, and H. Mohrenweiser 1972. Regulation of the levels of multiple forms of serine dehydratase and tyrosine aminotransferase in rat tissues. *Gann Monograph* 13: 191-204.

Pitot, H. C., and C. Peraino 1964. Studies on the induction and repression of enzymes in rat liver. I. Induction of threonine dehydrase and ornithine- -transaminase by oral intubation of casein hydrolysate. *J. Biol. Chem.* 239: 1783-1788.

Segal, H. L. 1973. Enzymatic interconversion of active and inactive forms of enzymes. *Science* 180: 25-32.

Snell, K., and D. G. Walker 1971. The regulation and development of the isoenzymes of rat liver L-alanine-glyoxylate aminotransferase. *Biochem. J.* 125: 68p.

Soffer, R. L. 1971. Enzymatic modification of proteins. IV. Arginylation of bovine thyroglobulin. *J. Biol. Chem.* 246: 1481-1484.

Spencer, C. J. and T. D. Gelehrter 1974. Pseudoisozymes of hepatic tyrosine aminotransferase. *J. Biochem.* 249: 577-583.

Stifel, F. D., and R. H. Herman 1972. Role of isozymes in metabolic control. *Am. J. Clin. Nutr.* 25: 606-611.

Sunshine, G. H., D. J. Williams, and B. R. Rabin 1971. Role for steroid hormones in the interaction of ribosomes with the endoplasmic membranes of rat liver. *Nature New Biol.* 230: 133-136.

Whiting, M. J., and W. H. Elliott 1972. Purification and properties of solubilized mitochondrial amino levulinic acid synthetase and comparison with the cytosol enzyme. *J. Biochem.* 247: 6818-6826.

Wieland, T., and G. Pfleiderer 1961. Chemical differences between multiple forms of lactic acid dehydrogenases. *Ann. N. Y. Acad. Sci.* 94: 691-700.

Yip, M. C. M., and R. K. Collins 1971. Purification and properties of rat kidney and liver ornithine aminotransferase. *Enzyme* 12: 187-200.

THE HEXOKINASE ISOZYMES: SULFHYDRYL CONSIDERATIONS IN THE REGULATION OF THE PARTICLE-BOUND AND SOLUBLE STATES

HOWARD M. KATZEN and DENIS D. SODERMAN
Merck Institute for Therapeutic Research
Rahway, New Jersey 07065

ABSTRACT. The presence in types I and II hexokinases of sulfhydryls required for catalysis was demonstrated by the inactivating effects of N-ethylmaleimide (N.E.M.) and p-chloromercuribenzene-sulfonate (PCMBS) on the soluble isozymes. Interaction of soluble type I with oxidized glutathione and incubation of intact tissues with N.E.M. produced catalytically-active electrophoretic variants (I_S) of type I, indicating the presence in type I of sulfhydryl(s) distinct from those supporting catalysis.

Incubation of intact diaphragm, skeletal and heart muscles in vitro with cell-permeable thiol inhibitors (N.E.M. or iodoacetate) paradoxically *increased* the yields (total and specific activities) of type II in resultant soluble extracts by greater than 10 fold, while type I (except for the formation of I_S) was unaffected, and G-6-PD and the LDH isozymes were inactivated. The effect on type II was found to be attributable to a translocation of the isozyme from the particulate-bound to the soluble state.

The evidence suggests that this translocation reflects an intracellular redistribution of type II, possibly brought about by an inhibition by N.E.M. of rebinding of type II during a natural equilibrium between the bound and soluble states of the isozyme. Since extracts of "latent-bound" type I were not inactivated by the inhibitors, it is suggested that resistance of type I to inactivation *in situ* reflects its existence in this latent state in the intact cell. It is also suggested that the sulfhydryl(s) not implicated in the catalytic activities of type II may be involved in binding to membrane components of the cell.

INTRODUCTION

The mammalian hexokinase isozymes (E.C. 2.7.1.1), first discovered in human cells in culture and rat liver by kinetic analyses and gel electrophoresis (Katzen et al, 1965) and in rat liver by DEAE-chromatography (Gonzalez et al, 1964) and shortly afterwards found in other tissues of the rat (Katzen and Schimke, 1965), have since been shown in all reported tissues (Katzen, 1967) and species (Grossbard et al, 1966) to exist at least as two, and in most cases as three, distinct isozymes. These have been designated types I, II, and III according to their sequence of migration on gel electrophoresis and elution from DEAE-cellulose chromatography. In addition, liver contains type IV, the fastest electrophoretically migrating form previously called "glucokinase". Because the level of type II has been found to be readily subject to hormonal and nutritional influences, it may be designated as the regulatory isozyme of the three "low K_m" hexokinases (types I to III). In this regard it appears similar to type IV, the "high K_m" liver isozyme. Type I has appeared insensitive to such influences while there is evidence that the level of type III is significantly elevated in several "insulin-sensitive" tissues from diabetic rats. These and other previous findings on the properties, regulation, and role of the isozymes have already been reviewed elsewhere (Katzen, 1967) and discussed in other extensive studies by Katzen and Schimke (1965), Grossbard and Schimke (1966), Gonzalez et al (1967), Katzen et al (1970), and Bessman and Gots (1974).

Since the latter part of this period, attention has been directed to studies of the subcellular distribution or "compartmentalization" of the isozymes in various tissues. Of particular interest has been the significance attached to glucose-6-PO_4 and other factors potentially capable of regulating the reversible release and binding of hexokinase. These aspects have been discussed in some detail in reports by Rose and Warms (1967), Katzen et al (1970), and Gumaa and McLean (1971) and recently reviewed by Anderson et al (1971) and Purich et al (1973). From these and other studies of the regulation of the isozymes (e.g., see Kosow et al, 1973), the individuality of each isozyme has been established. This was previously exemplified by the finding of Katzen et al (1970) that the "latent" state of hexokinase found associated with mitochondrial and microsomal fractions of various tissues (Teichgraber and Beisold, 1968; Wilson, 1968; Katzen et al, 1970) is attributable only to type I.

Despite the great attention paid to the binding and release (solubilization) of hexokinase, little is known of the binding sites on either the intracellular (or cellular) membranes or the isozymes. Rose and Warms (1967) previously demonstrated with liver extracts that interaction of hexokinase with mitochondria depends upon a component of the enzyme structure that is distinct from the elements that support catalysis, as shown by effects of chymotrypsin and liver enzymes in causing complete loss of binding capacity without affecting the catalytic properties. They point out that this suggests a specific role for this part of the hexokinase molecule (isozyme not specified).

Although nothing has thus far been reported on any sulfhydryl involvement in the interaction between subcellular particulates and hexokinase (Katzen, Soderman, and Wiley, in
preparation), a few brief reports have suggested that sulfhydryl groups in at least types I and II may be required for
their catalytic activities, maintainence of structures and
hormonal regulation. Murakami et al (1973) have most recently
observed the complete inactivation, by the thiol inhibitor
p-chloromercuribenzene sulfonate (PCMBS), of soluble types I
and II from several tissues. Redkar and Kenkare (1972) demonstrated the presence of numerous sulfhydryls in bovine brain
hexokinase. We previously reported (Katzen, 1966; Katzen et
al, 1970) that the deficiency of type II hexokinase activity
that we observed in insulin-sensitive tissues from diabetic
rats appeared less pronounced when mercaptoethanol or reduced
glutathione were included in the extraction and assay media.
In addition, when thiols were not added, type II frequently
appeared as two electrophoretically separable bands, designated
types II_a and II_b. These studies suggested to us that the
interconversion of these type II-related forms may have been
the result of a thiol-catalyzed disulfide-interchange reaction.
Interestingly, it appeared that diabetes resulted in an apparently preferential loss of II_a that preceded the subsequent
preferential loss of II_b in the more chronic diabetic rat.
However, the significance of these observations remains
unclear.

The purpose of the present study was to determine what role,
if any, the isozyme--and other intracellular--thiols may play
in the intracellular distribution and regulation of the hexokinases. The present findings suggest that, in addition to
the sulfhydryls required for catalytic activity, there exists
in some of the isozymes separate and distinct sulfhydryl residues not required for catalytic activities, but necessary for,
or involved in, the binding to membrane components of the cell.
In addition, we wish to stress the potential importance of
sulfhydryl factors in studies of the regulation of the hexokinases and in interpretations of the significance of the isozyme pattern observed in tissue extracts.

RESULTS AND DISCUSSION

Evidence for Presence in Hexokinase Isozymes of Sulfhydryl Residues Distinct from Those That Support Catalysis

During the course of our starch gel electrophoretic studies,
it became apparent that at least type I hexokinase contained
one or more sulfhydryl groups in its molecule. Initially we

found that incubation of soluble tissue extracts with oxidized glutathione (GSSG) produced a slow moving variant of type I in addition to the usual band of type I at concentrations of GSSG (about 1 mM) that had little, if any, apparent effect on the activity of any of the isozymes. At higher concentrations of GSSG, the activity of type I was completely abolished. An analogous phenomenon could be observed when N-ethylmaleimide (N.E.M.) was used as the thiol inhibitor. In Fig. 1 it can be seen that treatment of intact diaphragm tissue in vitro with N.E.M. consistently resulted in the appearance in the soluble extracts of an active, but relatively slow-migrating form of type I, designated I_S. The great increase seen here in type II due to the N.E.M.-treatment will be discussed later in this article.

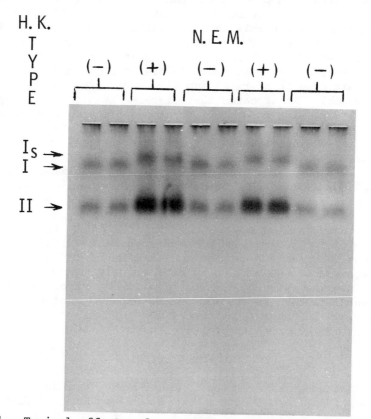

Fig. 1. Typical effects of treatment of intact rat diaphragm with N-ethylmaleimide (N.E.M.) on hexokinase isozymes according to starch gel electrophoresis of soluble extracts. Tissues were incubated in Krebs Ringer-PO_4 buffer, pH 8.0, containing

(Fig. 1 legend continued) EDTA (5 mM) and glucose (0.1M) with (+) or without (-) N.E.M. (10 mM) for 7 min at 34°. Tissues were then washed well with buffer containing 2-mercaptoethanol (2-M.E.) (5 mM) but excluding N.E.M., and then minced and homogenized (at 4°) in equal volumes of 0.1M Tris buffer, pH 7.4, containing EDTA (5 mM) and 2-M.E. (5 mM). 105,000 x g supernatant (soluble) extracts, electrophoresis, hexokinase activity staining and other details are as previously described (Katzen et al, 1970).

The variant form of type I resulting from the GSSG-treatment, referred to above, was presumably due to a "mixed disulfide" between the isozyme and GSSG. Similarly, the I_S seen in Fig. 1 likely represents type I alkylated with N.E.M. via the same sulfhydryl group(s).

The retention of activity associated with these variants indicates the presence in type I of a cysteine residue(s) not necessary for catalytic activity. In addition, the extensive loss of activity of both types I and II after treatment of these *soluble* isozymes with higher concentrations of GSSG (not shown), and with PCMBS and N.E.M. (Fig. 2), indicates the blocking of one or more sulfhydryls implicated in the catalytic activities of both isozymes. Gel electrophoretic assays confirmed the abilities of GSSG, PCMBS or N.E.M. to readily inactivate both type I and II hexokinases (not shown). In view of the retention of activities in the variants of type I, the "non-catalytic" sulfhydryl sites in this isozyme would appear to be distinct from the sites required to support catalysis.

Previous evidence (Katzen, 1966; Katzen et al, 1970) suggests that type II hexokinase may also contain "non-catalytic" cysteine(s) in addition to those shown here to be required for catalysis. This could help explain the existence of the thiol-related catalytically-active types II_a and II_b referred to in the Introduction. Thus, II_b, II_a or both, could represent active disulfide oligomers produced from non-catalytic sulfhydryls in these isozymes. Accordingly, exogenous (Katzen, 1966) or endogenous thiol factors in the cell could regulate these interconvertible forms of type II.

Paradoxical and Specific Increase of Type II Hexokinase from Intact Tissues Exposed to Thiol Inhibitors

Because of the possible significance attached to the relationship between particle-bound and free (soluble) activities (reviewed by Anderson et al, 1971 and Purich et al, 1973), it was determined to examine further the localization of the iso-

801

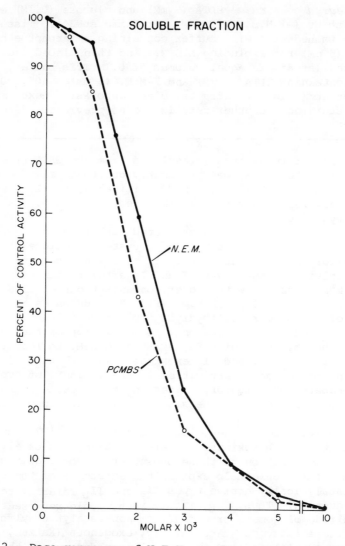

Fig. 2. Dose-response of N.E.M.- and p-chloromercuribenzene-sulfonate (PCMBS)-treatments of soluble extract of diaphragm on net hexokinase activity. Aliquots of the 105,000 x g supernatant fraction (Katzen et al, 1970) were each incubated separately in the designated final concentrations of inhibitors in 0.1 M Tris buffer, pH 8.0, containing 0.07 M glucose and 5 mM EDTA, for 7 min at 25°, and resultant solutions immediately assayed for hexokinase by the spectrophotometric method as in Table 1.

zymes *in situ*. Inasmuch as it has been suggested that the particulate-bound "latent" state of hexokinase (type I) may be
due to the isozyme buried within, or otherwise masked by, subcellular membranes and therefore inaccessible to, or otherwise
inactive to, exogenous substrates (Katzen et al, 1970), we considered it possible that this isozyme and possibly the particulate-bound overt forms (I and II), may also be resistant in
the intact cell to exogenous thiol inhibitors. Thus, the susceptibility of the isozymes in intact tissues to "cell-permeable" thiol inhibitors, as compared to their susceptibilities
in solubilized and particle-bound extracts, may reflect the
native states of these isozymes.

Jacob and Jandl (1962) had earlier demonstrated that the
intact erythrocyte cell membrane is readily permeable to the
sulfhydryl alkylating agent N.E.M., but not to PCMBS, another
thiol inhibitor. In the present experiments, conditions for
exposure of intact tissues to the inhibitors (e.g., Fig. 1)
were selected so as to be compatible with those that led to
rapid and complete inactivation of both types I and II isozymes
in their soluble states (see Fig. 2). Additionally, after
incubation with the inhibitors, the tissues were extensively
washed so as to minimize contamination, by the inhibitors, of
the media used for extraction and assay purposes. Isolated
intact diaphragm was initially selected as the target tissue
because of its easily observable content of both types I and
II isozymes and the sensitivity of its type II to hormonal
regulation (Katzen et al, 1970).

Contrary to our expectation that hexokinase in the native
state (i.e., in intact tissue) would be either insensitive or
inhibited by a cell-permeable thiol inhibitor, incubation of
diaphragm with high concentrations of N.E.M. consistently resulted in considerable *increases* in the type II isozyme present
in the solubilized extracts (Fig. 1), even after 90 min of
incubation (not shown). While spectrophotometric assays of
the net activity of all of the hexokinase isozymes in the
soluble extract indicated an enhancement of over 4-fold due to
N.E.M. (Table 1), staining intensities of the isozymes separated on starch gel electrophoresis revealed this net increase
to be attributable solely to type II (Fig. 1). Since the activity of type I consistently appeared virtually unaffected by
this N.E.M.-treatment, the selective increase of type II was
significantly greater than 4-fold. As seen in Table 1, the
specific increase was about 10 times control values.

Although the catalytic activity of type I appeared unaffected by the N.E.M.-treatment, the consistent appearance of the
electrophoretic variant (I_s) (Fig. 1) indicates that this isozyme nevertheless was exposed to some degree to the N.E.M. in

TABLE 1

EFFECT OF N.E.M.-TREATMENT OF INTACT RAT DIAPHRAGM
ON YIELD OF SOLUBILIZED HEXOKINASES

"Pre-incub." refers to designated supplements added to Krebs-
Ringer - PO_4 buffer, pH 8.0 (containing 0.05 M glucose and
5 mM EDTA) incubated for 15 min at 34° before addition of
tissue, after which tissue was added and incubation continued
for 7 min as in Fig. 1. The tissue was then removed and
washed 3 times with buffer containing supplements designated
under "post-incub.", and incubated in the buffer with these
supplements for 15 min at 34°. The tissue was then washed
with buffer, homogenized in 0.1 M Tris buffer, pH 7.5, containing 5 mM 2-M.E. and 0.1 M glucose, and the 105,000 x g
supernatant assayed for units (u) (also per mg. protein) of
net hexokinase activity by the G-6-PD-coupled spectrophotometric assay previously described (Katzen et al, 1970). Intensities of staining of electrophoretic isozymic bands were
quantitatively estimated relative to the lightest band assigned
an intensity of "1". N.E.M. was at 3 mM and reduced glutathione (GSH) at 30 mM. Asterisks refer to variant type I (I_s).
Other details are in Fig. 1.

Supplements to		Net Hexokinase		Electrophoretic	
Pre-Incub.	Post-Incub.	u x 10^3/ml	u x 10^3/mg	type 1	type II
None	None	181	9.2	1	4
N.E.M.	None	707	30.3	1*	40
N.E.M.+GSH	None	179	8.3	1	4
GSH	None	225	11.8	1	4
N.E.M.	GSH	725	35.8	1*	45
None	GSH	247	12.2	1	6

the intact tissue. Further evidence for the selectivity of
N.E.M. in enhancing total solubilized type II is demonstrated
by the comparable N.E.M.-induced increase in the *specific*
activity (units per mg of total solubilized protein) of net
solubilized hexokinase (Table 1).

This paradoxical and striking effect of N.E.M. is further
illustrated by comparing the dose-response "activating" effect
of N.E.M. on net hexokinase activity in intact diaphragm (Fig.
3) with the dose-response inhibitory effect when the isozymes
in the soluble extract were treated with the thiol inhibitor
(Fig. 2).

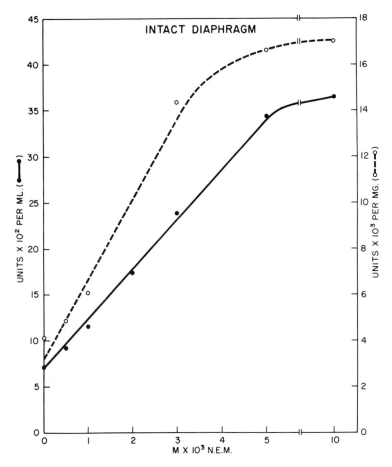

Fig. 3. Dose-response of N.E.M.-treatment of intact diaphragm on yields of net soluble hexokinase activity. Details of tissue incubation and preparation of soluble extracts are given in Fig. 1; hexokinase assay is as in Table 1.

That N.E.M. does, in fact, permeate the diaphragm muscle cell membrane is demonstrated by the ability of N.E.M. to completely inactivate intracellular glucose-6-PO_4-dehydrogenase and significantly inactivate at least 2 of the 5 isozymes of LDH (Table 2) under conditions identical to those that elevated type II. This is consistent with the known sensitivity of these cytoplasmic enzymes to thiol inhibitors (Pfleiderer et al, 1959; Jacob and Jandl, 1962), and further demonstrates the selectivity toward type II of the intracellular effect of N.E.M.

805

TABLE 2

EFFECTS OF N.E.M.-TREATMENT OF DIAPHRAGM
ON YIELDS OF SOLUBILIZED GLUCOSE-6-PO$_4$ DEHYDROGENASE
AND LACTIC DEHYDROGENASES

Tissue was incubated with or without N.E.M. as in Fig. 1, and
total soluble activities extracted as in Table 1. Intensities
of bands were compared with type I hexokinase under comparable
conditions as in Table 1. G6PD was assayed as in the hexokinase procedure in Table 1 except using G6P as substrate and
excluding added G-6-PD and ATP; LDH was assayed according to
Markert (1963). Other details are as in Fig. 1.

± N.E.M.	Net Enzyme Activity		Electrophoretic Bands				
	u/ml	u/mg	5	4	3	2	1
	G6PD						
(−)	0.79	0.038	5	−	−	−	−
(+)	0.08	0.006	0	−	−	−	−
	LDH						
(−)	280	17.3	12	10	5	10	15
(+)	246	14.8	12	10	5	5	8

In a preliminary study of the tissue specificity of the action of N.E.M., we found that, according to gel electrophoretic
assays, yields of type II in soluble extracts of heart and
gastrocnemius muscle were also significantly elevated after
N.E.M.-treatment of the intact tissue in vitro. In contrast,
the activities of type I, II, and III hexokinases in intact
liver appeared unaffected, while the isozymes seen in epididymal adipose tissue (types I and II) were completely inactivated.

In order to demonstrate further that the effects of N.E.M.
on intact tissues were actually due to intracellular rather
than extracellular actions, the effects of PCMBS, the pre-
viously-noted cell-impermeable thiol inhibitor, were tested.
While PCMBS-treatment of soluble types I and II hexokinase was
shown to lead to their complete inactivation, treatment of
intact diaphragm with this inhibitor had no significant effect
on either the yields or electrophoretic mobilities of types I
or II hexokinase in resultant soluble extracts (Table 3). In
addition, dithiobisnitrobenzoic acid (DTNB), another sulfhydryl blocking agent to which the cell membrane is poorly
permeable (Hellman et al, 1973), also had no effect on either

isozyme, while iodoacetate, a cell membrane-permeable thiol
inhibitor, mimicked N.E.M.

TABLE 3

TREATMENT OF INTACT DIAPHRAGM WITH OTHER THIOL INHIBITORS

Tissues were treated with 5,5'-dithiobisnitrobenzoate (DTNB),
iodoacetate (IA) and p-chloromercuribenzenesulfonate (PCMBS)
at final conc. of 10 mM and soluble extracts prepared and
assayed for hexokinases as in Fig. 1. Other details are as
in Table 1.

| Treatment | Net Hexokinase | | Electrophoretic | |
	$u \times 10^3/ml$	$u \times 10^3/mg$	type I	type II
None	145	10.1	1	3
DTNB	151	10.8	1	3
IA	428	33.6	1	12
PCMBS	139	9.8	1	3

Further evidence that the N.E.M. action described above was
intracellular, as well as irreversible, but can be blocked by
a sulfhydryl-agent is also shown in Table 1. Reduced glutathione (GSH) effectively neutralized this action of N.E.M. However, GSH could not reverse the effect of N.E.M., an alkylating
agent known to form stable addition products with sulfhydryl
compounds (Smyth et al, 1960). Inasmuch as incubation of
diaphragm with excess GSH added *after* the N.E.M. exerted its
action did not reverse the effect of N.E.M. ("Post-incubation"
in Table 1), this "activating" action of N.E.M. must have
transpired before, rather than after homogenization of the
tissue. It is interesting to note that GSH itself consistently
led to slight but significant increases in yields of solubilized hexokinase activity (Table 1). This also hints at a
role for sulfhydryl factors in the regulation of hexokinase.

Effect on Type II Hexokinase of N.E.M.
Added During Homogenization

Since treatment of intact muscle tissues with thiol inhibitors led to elevated type II activity in the soluble extracts,
while similar treatment of the soluble fraction had inactivated
this isozyme, we examined the degree of intactness of the cell
required to observe the transition from one effect to the
apparently opposite one. As seen in Table 4, addition of

TABLE 4

EFFECT OF STEP OF N.E.M. ADDITION DURING EXTRACTION OF
SOLUBILIZED DIAPHRAGM HEXOKINASE

N.E.M. was at a final conc. of 10 mM in incubation of intact
tissue, and at 1 mM for homogenates (higher concentrations led
to less activity). Tissue was minced before homogenization;
"semi-homogenate" refers to condition of mince after a single
stroke of glass pestle through glass tissue grinder tube,
after which homogenization (Katzen, 1967) was completed.
When N.E.M. was added to completed "homogenate", additional
strokes were required to homogeneously disperse the aliquot
of N.E.M.

Step of N.E.M. Addition	Net Hexokinase $u \times 10^3$/ml	$u \times 10^3$/mg	Electrophoretic type I	type II
None	119	7.1	1	5
Intact tissue	479	28.7	1	40
"Semi-homogenate"	222	15.6	1	15
Homogenate	154	10.7	1	8

N.E.M. at progressively advanced stages of the disintegration
of diaphragm resulted in a concomitantly decreased expression
of the "activating" effect. The degree of contribution of
unbroken cells surviving the early stages of homogenization
and thereby accounting for the enhancing effect of N.E.M. cannot be ascertained. However, it is important to note that the
cell-impermeable PCMBS had only an inhibitory effect at all
concentrations tested. Nevertheless it is likely that during
the homogenization the observed activities reflected the net
effect of enhancement by N.E.M. of yields of solubilized type
II opposing the effect of inactivation of the solubilized
isozymes. That opposing actions of N.E.M. occurred is evidenced by the response to increasing concentrations of N.E.M.
accompanying the homogenization of the tissue (Table 5).

A study of the tissue specificity of the effect of N.E.M.
added during the homogenization of the tissue is shown in
Table 6. The specificity exhibited here imitated that seen
using intact tissues, as previously noted. The effect of
N.E.M. in elevating yields of type II in the soluble fraction
appears limited to, or most effective in, muscle tissues (i.e.,
diaphragm, gastrocnemius and heart).

TABLE 5

DOSE-RESPONSE EFFECT OF N.E.M. ADDED DURING
HOMOGENIZATION OF DIAPHRAGM

N.E.M. was added to "semi-homogenate" described in Table 4.

N.E.M.	Net Hexokinase	
	u x 10^3/ml	u x 10^3/mg
0	238	8.5
10^{-5}M	285	10.7
10^{-4}M	379	13.6
10^{-3}M	382	14.1
10^{-2}M	295	11.4

Lack of Effect of Thiol Inhibitors
on Type I Hexokinase Activity in Intact Tissue

Because particulate-bound latent hexokinase (type I) had
previously only been demonstrated in fractions of tissue
extracts (Teichgraber and Biesold, 1968; Wilson, 1968; Katzen
et al, 1970), the present experiments were designed to shed
some light on the nature of this isozyme in the intact cell.
Thus, the susceptibility of type I in the intact tissue to
thiol poisons was compared to the susceptibility of latent
hexokinase in the particulate extract.

Accordingly, we found that the particulate-bound latent
activity in diaphragm extracts [in which, according to gel
electrophoretic assays, we confirmed our previous finding
(Katzen et al, 1970) that this activity is attributable only
to type I], appeared completely resistant to either N.E.M.,
iodoacetate or PCMBS. After unmasking this activity as particulate-bound "overt" activity, or solubilizing it with detergents and salts (Katzen et al, 1970), the previously-latent
type I became readily susceptible to inactivation by the thiol
inhibitors. It is important to note that despite the apparent
predominance of type II in this tissue (Katzen, 1967), none
of it could be found to exist in the latent state.

On the basis of the finding, previously mentioned, that the
catalytic activity of type I appears unaffected in intact
muscle and liver treated with cell-permeable thiol inhibitors,
it is suggested that the sulfhydryl group(s) required for the
catalytic activity of this isozyme (and possibly the others)
exists in the native state protected from, or inaccessible to,

TABLE 6

TISSUE SPECIFICITY OF EFFECT OF N.E.M. ACCOMPANYING HOMOGENIZATION

Conditions are as in Table 5. "Gastroc" (gastrocnemius muscle) and other rat tissues refer to "semi-homogenates". (*) refers to a trace of type III and (**) refers to the sum of types IIa and IIb isozymes that appeared.

Tissue	N.E.M.	Net Hexokinase		Electrophoretic Types			
		u x 10^3/ml	u x 10^3/mg	I	II	III	IV
Gastroc.	(−)	128	9.2	1	5	0*	0
	(+)	216	13.1	1	15	0*	0
Heart	(−)	199	9.5	5	2	0*	0
	(+)	387	16.3	5	6	0*	0
Liver	(−)	910	23.6	6	2	3	3
	(+)	996	23.4	6	2	3	4
Adipose	(−)	185	23.2	6	10**	0	0
	(+)	22	3.1	1	0	0	0

those inhibitors. The possible existence of type I in the latent state *in situ* could explain these findings. That the thiol inhibitors did enter the cell in sufficient concentrations to render them capable of otherwise interacting with type I without loss of catalytic activity was shown by the appearance of the I_S variant with N.E.M. This would also argue against the presence of intracellular factors capable of neutralizing the action of the thiol inhibitor.

Evidence for Thiol Inhibitor-Induced Redistribution of Type II Isozyme between Particulate-Bound and Soluble States

In order to explain the apparently paradoxical thiol inhibitor-induced increase in yields of type II in the soluble extracts, several possibilities were considered. Inasmuch as type II is the most unstable of the predominant two isozymes from muscle tissues (Katzen and Schimke, 1965; Grossbard and Schimke, 1966), it was considered possible that select concentrations of the thiol-blocking agents might act by preferentially inhibiting intracellular sulfhydryl factors capable of degrading or inactivating this isozyme. However, at no concentration of N.E.M. (see Fig. 3) were we able to observe any effect on the soluble isozyme other than an inhibitory one. Moreover, N.E.M. was found to be incapable of protecting type II from heat, proteolytic or other inactivation in soluble extracts. Finally, according to assays of the yields of solubilized isozyme, we found the slow rate of loss of type II with time and heat in isolated intact diaphragm to be unaffected by N.E.M.-treatment of the tissue.

However, when we examined the effect of N.E.M.-treatment of diaphragm on the ratio of soluble to particle-bound hexokinase in resultant homogenates, it became apparent that N.E.M. caused a distinct *decrease* in particle-bound type II concomitant to the observed increase of this isozyme in the soluble (supernatant) fraction (Table 7). A significant decrease in type II associated with the particulate fraction of the homogenate can be seen here to have resulted from N.E.M.-treatment of the intact tissue whether the insoluble fraction was assayed as a suspension with type II bound to the particulates (absence of solubilizing agent) or with isozyme solubilized by the detergent-salt treatment. Consistent with this reciprocal relationship was the lack of observable effect on the total activity in the homogenate, according to assays conducted either before or after solubilization of the homogenate by the detergent-KCl treatment. As expected, the detergent-salt treatment exposed latent (type I) activity as reflected by the solubilizing agent-induced increase in total (homogenate) activity.

In confirmatory experiments, wherein the tissue was directly homogenized in the presence of the solubilizing agents,
total (particulate plus soluble) activity appeared unaffected
by prior N.E.M. treatment of the tissue. Moreover, the supernatant fraction derived from this homogenate also appeared
unaffected by this N.E.M. treatment, since the solubilizing
agent solubilized virtually all of the activity from the particulates. Clearly then, the N.E.M.-induced elevation in
yields of soluble type II hexokinase was due to a shift in the
distribution of this isozyme from the particulate to the
soluble fraction.

Implications of Sulfhydryls in Type I and II Isozymes and of Effect of Thiol Inhibitors on Particulate-Soluble Distribution of Type II

The present findings clearly demonstrate the presence in
both types I and II hexokinase of sulfhydryl group(s) required
for catalytic activities. All of the sulfhydryl blocking
agents tested had in common the ability to inactivate completely both of these isozymes in their soluble states. The
finding that interaction of type I with N.E.M. or oxidized
glutathione in the intact tissue led to the formation of
electrophoretic variants of this isozyme with retention of
catalytic activity also indicates the presence of an additional
thiol group(s) in this isozyme that is distinct from the
thiol(s) necessary for catalysis.

It is interesting to note that type I in the native state
in most of the tissues examined appeared protected from the
inhibitory effects of the thiol inhibitors. This would suggest
that *in situ*, the sulfhydryl(s) required for catalysis (in
contradistinction to those not required, and coupled to N.E.M.
and GSSG) are not free as they are in soluble extracts. It
was previously demonstrated (Katzen et al, 1970), and confirmed
here, that only type I represents the latent-bound state of
hexokinase earlier found in particulate fractions. The resistance, or protection, of type I from inactivation by thiol
inhibitors likely reflects the native state of this isozyme.
Latent-bound hexokinase in particulate extracts also was
demonstrated here to be resistant to inactivation by thiol
inhibitors, analogous to its insusceptibility to exogenous
substrate (Wilson, 1968; Craven et al, 1969). It is therefore
suggested that the intracellular resistance of type I to inactivation by thiol poisons indicates that this isozyme also
exists in the native state in the latent-bound condition.
However, although it was demonstrated here that this resist-

TABLE 7

EFFECT OF N.E.M.-TREATMENT OF INTACT DIAPHRAGM ON HEXOKINASE
YIELDS IN SOLUBILIZED AND PARTICLE-BOUND EXTRACTS AND HOMOGENATES

Pooled tissues were treated, as in Fig. 1, with and without N.E.M. as noted under "Tissue" column, and equal aliquots of the extracts from this pool were treated with or without the solubilizing agent (0.5% Triton X-100 in 0.9 M KCl) (Katzen et al, 1970) as noted under "Extract" column. Assays of "particulate" and "homogenate" extracts were made on their suspensions (Katzen et al, 1970). Other details are as in Fig. 1 and Table 1.

Treatment of		Net Hexokinase		Electrophoretic	
Tissue	Extract	u x 10^3/ml	u x 10^3/mg	type I	type II
		Supernatant			
Buffer	Buffer	138	8.8	3	1
N.E.M.	Buffer	410	29.6	3	15
		Particulate			
Buffer	Buffer	469	7.5	–	–
N.E.M.	Buffer	142	–	–	–
Buffer	Triton-KCl	550	8.1	5	13
N.E.M.	Triton-KCl	176	3.0	4	2
		Homogenate			
Buffer	Buffer	334	7.7	–	–
N.E.M.	Buffer	347	6.2	–	–
Buffer	Triton-KCl	459	8.6	5	5
N.E.M.	Triton-KCl	418	7.2	4	6

813

ance could not be accounted for by intracellular neutralization of N.E.M., the possibility that the native conformation of the intracellular type I molecule, or the presence of intracellular stabilizing factors, may protect the "catalytic-related" sulfhydryl in type I, cannot be entirely ruled out.

In addition to type I, previous evidence suggested that type II hexokinase may also contain cysteine(s) discrete from those that support catalysis. The thiol-required interconversion of the two catalytically active forms of type II (Katzen, 1967; Katzen et al, 1970) suggests the involvement of a disulfide-interchange interaction to explain these forms, and the presence of such discrete sulfhydryl components.

In view of the present demonstration that treatment of intact muscle tissues with sulfhydryl blocking agents resulted in a shift in the distribution of type II from the particulate to the soluble fractions, it would be tempting to speculate that the putative sulfhydryl(s) site distinct from the catalytic site in this isozyme may be involved in its binding to intracellular membranes.

Several possibilities, implicating sulfhydryls in a membrane binding site, may be involved to explain this redistribution. Since treatment of the tissue with thiol inhibitors increased solubilized type II parallel to a decrease in particulate activity, it is possible that interaction of the inhibitors with intracellular membrane or other sulfhydryls initiated the *release* of particle-bound isozyme into the soluble phase before or during the extraction steps. However, we were unable to observe any release of bound hexokinase by treatment of isolated particulate fractions with N.E.M. Moreover, as seen in Table 4, addition of N.E.M. to homogenized extracts had a much lesser effect than addition of thiol inhibitor to the intact tissue.

The most likely possibility is that the thiol inhibitor acts by blocking a sulfhydryl-required *rebinding* of type II to intracellular membranes in the intact tissue. The site(s) of this block could be a sulfhydryl on the isozyme, the particle, or both. This possibility stipulates that the inhibitor blocks the rebinding step during a reversible release of type II thereby altering a natural equilibrium between the bound and free isozyme in the intact cell. Less likely is the alternative possibility that the inhibitors preferentially block the binding of the isozyme during the extraction procedure without significantly inhibiting the isozyme itself. This implies that homogenization is required for extensive sulfhydryl-required redistribution of the natural unbound (soluble) state of the isozyme, as has been reported for cytochrome c (Beinert, 1951).

The present evidence that a sulfhydryl site(s) on type II hexokinase distinct from the catalytic site may be responsible for the capacity of the isozyme to bind cellular components is consistent with the observations of Rose and Warms (1967). They reported the destruction (by liver enzymes or chymotrypsin) of the capacity of tumor hexokinase to bind liver mitochondria without significant loss of the hexokinase activity. They suggested that a specific role likely exists for a part of the hexokinase molecule that is involved in the interaction of the enzyme with mitochondria but is discrete from the elements that support catalysis.

The present evidence also points out the danger in using hexokinase activities found in soluble and particulate extracts as a reflection of the natural intracellular distribution of the hexokinases. In addition to previously reported factors capable of regulating the bound and free isozymes, the present data indicate that sulfhydryl factors should also be considered.

REFERENCES

Anderson, J. W., R. H. Herman, J. Tyrrell, and R. Cohn 1971. Hexokinase: a compartmented enzyme. *Amer. J. Clin. Nutr.* 24: 642-650.

Beinert, H. 1951. The extent of artificial redistribution of cytochrome C in rat liver homogenates. *J. Biol. Chem.* 190: 287-292.

Bessman, S. P. and R. E. Gots 1974. The mechanism of insulin action: Mitochondrial acceptor theory. *Intra-Science Chem. Rept.* 8: 73-84.

Craven, P. A., P. Goldblatt, and R. E. Basford 1969. Brain hexokinase. Preparation of inner and outer mitochondrial membranes. *Biochem.* 8: 3525-3532.

Gonzalez, C., T. Ureta, J. Balbul, E. Rabajille, and H. Niemeyer 1967. Characterization of isoenzymes of adenosine triphosphate: D-hexose 6-phosphotransferase from rat liver. *Biochem.* 6: 460-468.

Grossbard, L. and R. Schimke 1966. Multiple hexokinases of rat tissues. Purification and comparison of soluble forms. *J. Biol. Chem.* 241: 3546-3560.

Grossbard, L., M. Weksler, and R. Schimke 1966. Electrophoretic properties and tissue distribution of hexokinases in various mammalian species. *Biochem. Biophys. Res. Commun.* 24: 32-38.

Gumaa, K. A. and P. McLean 1969. Sequential control of hexokinase in ascites cells. *Biochem. Biophys. Res. Commun.* 35: 824-831.

Hellman, B., L. Idahl, A. Lernmark, J. Sehlin, and I. Taljedal 1973. Role of thiol groups in insulin release: studies with poorly permeating disulfides. *Mol. Pharmacol.* 9: 792-801.

Jacob, H. S. and J. H. Jandl 1962. Effects of sulfhydryl inhibition on red blood cells. I. Mechanism of hemolysis. *J. Clin. Invest.* 41: 779-792.

Katzen, H. 1966. The effect of diabetes and insulin in vivo and in vitro on a low K_m form of hexokinase from various rat tissues. *Biochem. Biophys. Res. Commun.* 24: 531-536.

Katzen, H. 1967. The multiple forms of hexokinase and their significance to the action of insulin. *Advan. Enz. Reg.* 5: 335-356.

Katzen, H. and R. Schimke 1965. Multiple forms of hexokinase in the rat: Tissue distribution, age dependency and properties. *Proc. Natl. Acad. Sci.* 54: 1218-1225.

Katzen, H. M., D. D. Soderman and H. Nitowsky 1965. Kinetic and electrophoretic evidence for multiple forms of glucose-ATP-phosphotransferase activity in human cell cultures and rat liver. *Biochem. Biophys. Res. Commun.* 19: 377-382.

Katzen, H., D. D. Soderman, and C. Wiley 1970. Multiple forms of hexokinase. Activities associated with subcellular particulate and soluble fractions of normal and streptozotocin diabetic rat tissues. *J. Biol. Chem.* 245: 4081-4096.

Kosow, D., F. Oski, J. Warms, and I. A. Rose 1973. Regulation of mammalian hexokinase: Regulatory differences between isoenzyme I and II. *Arch. Biochem. Biophys.* 157: 114-124.

Markert, C. L. 1962. Lactate dehydrogenase isozymes: dissociation and recombination of subunits. *Science* 140: 1329-1330.

Murakami, K., Y. Imamura, and S. Ishibashi 1973. Difference between hexokinase iso-enzymes in sensitivity to sulfhydryl inhibitor. *J. Biochem.* 74: 175-177.

Pfleiderer, G., D. Yeckel, and T. Wieland 1959. Factors affecting the activity of lactate dehydrogenase. *Arch. Biochem. Biophys.* 83: 275-281.

Purich, D. L., H. J. Fromm, and F. B. Rudolph 1973. The hexokinases: Kinetic, physical, and regulatory properties. *Advan. Enzymol.* 39: 249-326.

Redkar, V. and U. Kenkare 1972. Bovine brain mitochondrial hexokinase. *J. Biol. Chem.* 247: 7576-7584.

Rose, I. A. and J. Warms 1967. Mitochondrial hexokinase. Release, rebinding and location. *J. Biol. Chem.* 242: 1635-1645.

Smyth, D. G., A. Nagamatsu, and J. S. Fruton 1960. Some reactions of N-ethylmaleimide. *J. Amer. Chem. Soc.* 82: 4600-4608.

Teichgraber, P. and D. Biesold 1968. Properties of membrane-bound hexokinase in rat brain. *J. Neurochem.* 15: 979-989.

Wilson, J. E. 1968. Brain hexokinase: A proposed relation between soluble-particulate distribution and activity in vivo. *J. Biol. Chem.* 243: 3640-3647.

PHYSIOLOGICAL SIGNIFICANCE OF PHOSPHOFRUCTOKINASE ISOZYMES

MICHAEL Y. TSAI, FRESIA GONZALEZ, and ROBERT G. KEMP
Department of Biochemistry
Medical College of Wisconsin
Milwaukee, Wisconsin 53233

ABSTRACT. Considerable metabolic diversity in terms of regulation is observed in carbohydrate metabolism among various tissues of a given animal. Previous work has pinpointed the enzyme phosphofructokinase as the principal regulatory site in glycolysis. This enzyme is subjected to regulatory control by a great variety of metabolites and the regulatory diversity among various tissues can be achieved by differences in effector concentrations in the tissues or by the presence of tissue specific isozymes with differing sensitivities to the metabolic effectors. It is proposed that both mechanisms are operating.

Three different types of phosphofructokinase subunits are found in rabbit tissues; isozyme A of heart and skeletal muscle, isozyme B of liver and erythrocytes, and isozyme C present as hybrids with isozyme A in the brain. The regulatory properties of purified rabbit phosphofructokinase from muscle, liver, and brain have been compared. These data suggest that creatine phosphate, citrate, and inorganic phosphate are the most important effectors of phosphofructokinase in muscle and that 2,3-diphosphoglycerate is an important regulator of erythrocyte phosphofructokinase. Phosphofructokinase of liver and brain is apparently chiefly controlled by the relative amounts of adenine nucleotides and inorganic phosphate.

In general, it is observed that specific isozymes are present in the various tissues that are particularly sensitive to those metabolites that reflect the energy state of the cell.

The metabolism of glucose or its storage form, glycogen, is the exclusive source of energy for red blood cells, for the brain under most conditions, and for muscle during exercise. Furthermore, the enzymes of the glycolytic sequence are present, without exception, in all mammalian tissues. Despite the common occurrence of the enzymes to catalyze all of these reactions in all tissues, there does not appear to be a single mode of regulation of the pathway in all tissues. Compare, for example, brain, skeletal muscle, and liver. The brain has a fairly constant demand for glucose that does not vary greatly at any level of mental activity. Glycolysis in skeletal muscle, on the other hand, varies greatly. In resting muscle, glycolysis is almost completely shut down, but a tremendous surge in glycolytic flux occurs during and immediately following exercise. The liver runs a very modest amount of glucose through this pathway and, in fact, it usually runs the pathway in reverse to synthesize glucose.

How is such diversity in regulation achieved? We must look to the individual enzymes of the glycolytic pathway. Most work in recent years has pinpointed the enzyme phosphofructokinase as the principal regulatory site in glycolysis (see Bloxham and Lardy, 1973, or Mansour 1972, for summaries of the evidence). Phosphofructokinase shows complicated regulatory properties as evidenced by the long list of metabolites that affect its activity (Table 1). Such a list provides us with the first clue to how diversity of regulation among tissues can be achieved. The table shows the known activators and inhibitors of phosphofructokinases that have been purified and studied from different mammalian tissues. The inhibitors include ATP, which is one of the two substrates of this enzyme, a whole series of phosphorylated compounds, and citrate. The inhibition of PFK by phosphorylated compounds and citrate is synergistic with the action of ATP. Inhibition can be reversed by substrate activators such as fructose-6-P and fructose 1-6-di-P, adenosine nucleotides such as ADP, AMP, and cAMP, and by inorganic phosphate. What provides for regulatory diversity is that these different metabolites are known to be present at very different concentrations within the cells of different organs (Lowry et al., 1964; Minakami and Yoshikawa, 1965; Imai et al., 1964; Williamson et al., 1969; Ruderman et al., 1971). The level of ATP in muscle is about 5 to 7 mM as compared to about 2-3 mM in erythrocytes, liver, and brain. P-creatine is present in skeletal muscle at a concentration of about 20 mM whereas the level of this metabolite in brain is about 2 mM and none is present in liver and erythrocytes. In contrast, 2,3-di-P-glycerate exists in appreciable amounts only in the erythrocytes. Obviously the differences in the levels of metabolites among tissues can provide much of the diversity in regulation that we observe, but yet another mechanism to account for this diversity would be the existence of tissue specific isozymes of phosphofructokinase with differing sensitivities to the various metabolic effectors. It is this mechanism that is the subject of this communication.

This work will be divided into two parts: first, evidence will be presented that there are indeed isozymes of phosphofructokinase, and second, that the isozymes have differing

TABLE I

METABOLIC EFFECTORS OF MAMMALIAN PHOSPHOFRUCTOKINASE[a]

Activators	Inhibitors
Fructose-6-P	ATP
Fructose-1,6-diP	Citrate
Glucose-1,6-diP	P-enolpyruvate
AMP	Creatine-P
ADP	3-P-glycerate
Cyclic AMP	2-P-glycerate
K^+	2,3-diP-glycerate
NH_4^+	
$PO_4^=$	

[a] Compiled from the reviews of Bloxham and Lardy (1973) and of Mansour (1972).

sensitivities to metabolic effectors in a manner consistent with the regulatory mode of tissues containing those isozymes.

To provide evidence for the existence of phosphofructo-kinase isozymes, several approaches have been employed by our laboratory and by other investigators (Layzer and Conway, 1970; Tanaka et al., 1971; Kurata et al., 1972) employing cellulose acetate electrophoresis, direct purification of the isozymes, and the use of antisera prepared against purified isozymes. In rabbit tissues, heart PFK migrates upon electro-phoresis as a single band similar to the enzyme found in skeletal muscle. Erythrocyte and liver PFK move most rapidly as a single band towards the anode. In extracts of other rabbit tissues such as adipose, lung, and stomach, there appear often three equally spaced bands of activity in addi-tion to the bands that correspond to the liver types and muscle type PFK, although in some cases, one of the three bands is missing in the extracts of adipose tissue (Fig. 1). In no case was there a band of activity that moved faster than the liver type PFK or slower than the muscle type PFK. Knowing that under the conditions we used for electrophoresis on cellulose acetate strips rabbit muscle PFK exists as a tetramer composed of four identical subunits with molecular weights of 80,000-90,000, we suspected that the electrophor-etic species with mobilities intermediate to the muscle type PFK and liver type PFK were hybrid isozymes. To test this, hybridization of the purified rabbit liver and muscle PFK was carried out (Tsai and Kemp, 1972). Dissociation was carried out by lowering the pH of each enzyme solution to 5.5, mixing an equal portion of each dissociated enzyme solution,

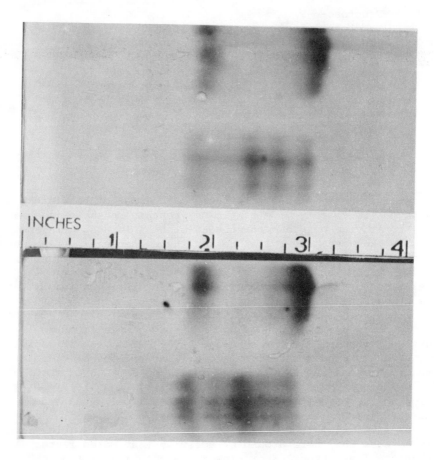

Fig. 1: Electrophoresis of phosphofructokinase in extracts of adipose and stomach tissues. Composite of two separate electrophoreses. From left to right: a mixture of purified muscle and liver PFK, adipose extract, a mixture of purified muscle and liver PFK, lung extract. Scale indicates inches from origin. Electrophoresis was carried out for 3 hours at 340 volts. Anode at the top.

and raising the pH of the mixture to above 7.0 in the presence of fructose 1,6-di-P and dithiothreitol to achieve reassociation. This treatment results in recovery of 90% of the total activity and upon electrophoresis on cellulose acetate strips, three equally spaced intermediate species, in addition to the muscle PFK and liver PFK, were found. These results are shown in Fig. 2. In contrast, the control sample, which was prepared by mixing muscle PFK and liver PFK solutions after the pH's of the individual enzyme solutions have been returned to 7.0, showed only two bands. We have designated muscle PFK as type A PFK and liver PFK as type B PFK (Tsai and Kemp, 1973). A complete set of AB hybrids is found in rabbit lung, adipose, and stomach tissues. In all the other species we have studied (rabbit, rat, mouse, and guinea pig) the phosphofructokinase from skeletal muscle migrates as a single band on cellulose acetate electrophoresis. This most slowly migrating band is in contrast to the fastest moving band which is always found in liver and in erythrocytes (Gonzalez et al., 1974). Incomplete sets of hybrid isozymes are found in several mouse and guinea pig tissues.

To further characterize the rabbit isozyme patterns, immunochemical techniques were employed (Tsai and Kemp, 1973). Antisera to either purified skeletal muscle PFK or to purified liver PFK was prepared by repeated subcutaneous injections into guinea pigs of the respective enzymes mixed with Freund's complete adjuvant. The sera were collected and a fixed amount of diluted purified enzyme or tissue extracts were incubated with varying amounts of antisera and then centrifuged to remove the antigen-antibody complex. The enzyme activity was measured in the supernatant solution. Table II summarizes the end points of the titration curves of phosphofructokinase activities with either of the two antisera. Anti-liver PFK serum, which does not crossreact with PFK-A, precipitated about 95% of the activities of purified liver PFK solution as well as PFK in crude extracts of liver and erythrocytes, thus confirming the electrophoresis patterns that erythrocyte PFK and liver PFK are both type B enzyme. Anti-muscle PFK serum precipitated more than 90% of the skeletal muscle and heart activities, again confirming the fact that heart and skeletal muscle both have type A enzyme. Both anti-B serum and anti-A serum precipitate more enzyme activity from a mixture of AB hybrids produced from equal amounts of type A and type B enzyme than from an unhybridized mixture of the two enzymes. The precipitation data on extracts of lung, stomach, and adipose tissue also showed patterns similar to the in vitro generated hybrid mixture. In some tissues, such as the brain and thymocytes, even a combination of excess anti-A type PFK

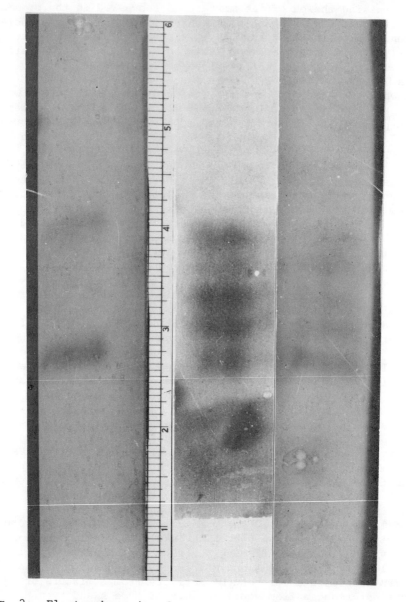

Fig. 2: Electrophoresis of a mixture of hybridized liver and muscle types of PFK. From left to right: a mixture of liver and muscle PFK after each enzyme has been dissociated and re-associated separately, and samples from two different attempts at hybridization.

TABLE II

PRECIPITATION OF PHOSPHOFRUCTOKINASE (PFK) BY ANTISERA

The data given are the results obtained by the addition
of 25 µl of antiserum to liver phosphofructokinase and
50 µl of muscle phosphofructokinase antiserum. Extracts
prepared in 30 mM KF, 4 mM EDTA, 1 mM dithiothreitol
(pH 7.5). Data taken from Tsai and Kemp (1973).

| | % Activity removed by | |
Enzyme Sources	Liver PFK antiserum	Muscle PFK antiserum
Purified liver PFK	96	32
Purified muscle PFK	3	98
Purified muscle PFK and liver PFK[a]	46	69
Purified muscle and liver PFK hybridized mixture[b]	68	82
Heart extract	2	92
Skeletal muscle extract	4	92
Liver extract	93	25
Erythrocyte hemolysate	96	32
Lung extract	58	78
Stomach extract	58	67
Adipose extract	64	58
Cerebrum extract	11	50
Thymocyte sonicate	27	21

[a]Mixture of an equal number of enzyme units of each
enzyme.

[b]Mixture of equal number of units of each enzyme was
subjected to hybridization conditions as described
in text.

and anti-B type PFK sera does not precipitate all the enzyme
activity. Because of the potential significance of brain PFK,
this tissue was examined more extensively. Fig. 3 shows a
cellulose acetate electrophoresis of PFK from a brain extract
and the activity that remained after it was precipitated with
a combination of anti-type A PFK sera and anti-type B PFK sera.
Two observations can be made from this zymogram. First, the
three to four bands that are present in the brain extract are
more tightly spaced than those found in the hybridized mixture
of type A and type B enzymes; second, a single band of activity
remained after the brain extract was precipitated with both
anti-type A sera and anti-type B sera. This remaining band

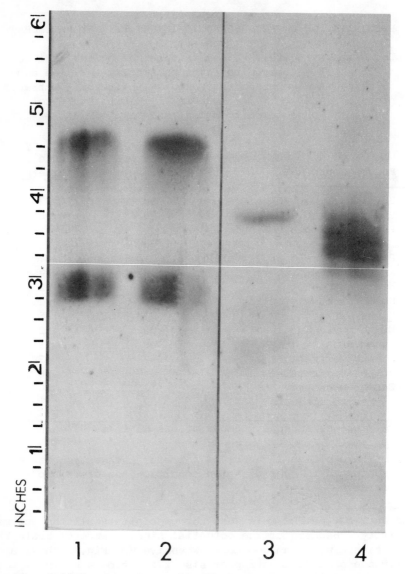

Fig. 3: Cellulose acetate electrophoresis of brain extract and antisera treated brain extract. For details see legend of Fig. 2. 1 and 2 are mixtures of muscle and liver PFK. 4 is brain extract and 3 is brain extract after treatment of 0.1 ml extract with 0.05 ml of anti-muscle PFK serum and 0.025 ml of anti-liver PFK serum followed by centrifugation at 18,000 xg for 20 min. Taken from Tsai and Kemp (1973).

was designated type C phosphofructokinase. When PFK was purified from brain and dissociated and reassociated by pH changes, five bands were seen with the slowest moving band corresponding to that of PFK A, and the fastest moving band corresponding to that of PFK C.

Thus there appears to be three types of phosphofructokinase monomer, with some tissues containing hybrids of the three types. The next question to ask is whether the isozymes have different kinetic properties and whether these properties explain previously mentioned regulatory diversity in carbohydrate metabolism. PFK A, the sole isozyme of skeletal muscle and heart, is readily prepared in homogeneous form (Kemp and Forest, 1968). PFK B was purified from rabbit liver to a high degree of purity in our laboratory (Kemp, 1971) and from rabbit erythrocytes by Tarui et al. (1972). We undertook the purification of brain phosphofructokinase, but in this case we were not purifying a single electrophoretic species but the entire set of isozymes present in the brain. We were successful in obtaining a highly purified preparation of brain PFK that contained all of the component enzymes that are observed in the crude extract of brain. No differences between the purified rabbit brain enzyme and activity in crude extract of brain were found by means of cellulose acetate electrophoresis and by the relative amount of activity precipitated by anti-type A PFK and anti-type B PFK sera. As suggested for a number of other isozymic systems, the various molecular forms of phosphofructokinase may be the products of different genes. A comparison of the regulatory properties of the brain phosphofructokinase with those of isozyme A, the only form of phosphofructokinase found in rabbit skeletal and cardiac muscle, and those of isozyme B, the predominant isozyme species in liver and erythrocytes, may yield information that will indicate the genetic origin of the isozymes that are present in each cell of these different tissues.

ATP, one of the two substrates of the enzyme phosphofructokinase, acts as an inhibitor of the enzyme beyond certain optimum concentrations. The optimum concentrations are different for enzymes from different tissues, presumably reflecting in part the varying levels of ATP in different tissues, and in part the different glycolytic fluxes in tissues. This is shown in Fig. 4. Fig. 4A shows that at pH 7.1 with a relatively low level of fructose-6-P, the liver enzyme is completely inhibited at 0.1 mM ATP, whereas 50% inhibition of the muscle and brain enzymes was achieved at 0.4 mM and 0.7 mM ATP respectively. Raising the level of fructose-6-P to 4mM decreases the inhibition by ATP (Fig. 4B). The K_i for ATP for the enzymes from brain, muscle, and liver are 1.2 mM, 1.0 mM,

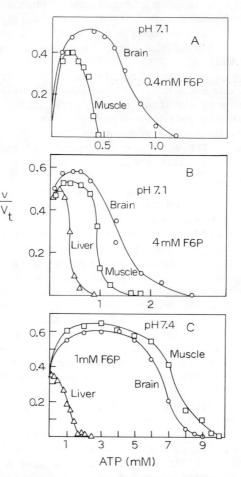

Fig. 4: ATP inhibition of phosphofructokinases. Vt is defined
as the velocity observed at pH 8.2 in the glycylglycine-glycero-
phosphate buffer in an assay system as described elsewhere.
The data in this and all the following figures **were** obtained
with a Gilford Model 2000 spectrophotometer at 26° and in 3 ml
of a medium containing 50 mM Tes titrated with KOH to the desired pH, and a final concentration of 150 mM potassium ions
by the addition of KCl. Other ingredients of the assay included 1 mM EDTA, 0.2 mM DPNH, 1 mM dithiothreitol, 0.6 units of
aldolase, 0.3 units of glycero-P dehydrogenase, and ATP and
fructose-6-P at the indicated concentrations. $MgCl_2$ was always
present 5 mM in excess of the concentration of ATP. All enzymes were dialyzed extensively before use to remove ammonium
sulfate. Rates were determined 4-6 minutes after starting the

(Fig. 4 continued)   reaction by the addition of an amount of
fructose-6-P as indicated in the figures.

and 0.4 mM, respectively. When the pH of the medium is 7.4
(Fig. 4C) inhibition of PFK by ATP occurs at considerably
higher concentrations. The Ki's are 2.8 mM, 3.1 mM, and 0.7
mM for the enzymes from brain, muscle, and liver, respectively. Significantly, it is seen that the enzyme from brain is
inhibited at lower levels of ATP than the muscle PFK, the
reverse of what was found at pH 7.1. This implies a more
dramatic response of muscle PFK to changes in pH in vivo.
Indeed increases in intracellular muscle pH has been suggested as a consequence of creatine phosphate hydrolysis, while
lactate accumulation would lower the pH. The levels of ATP
in muscle, brain, and liver were reported around 6 mM, 3.0 mM,
and 2.7 mM, respectively. The amounts of ATP required to
inhibit these enzymes in vitro are therefore within the range
of physiological concentrations. AMP, cyclic AMP, ADP, and
Pi can reverse the inhibition of PFK by ATP. Fig. 5 shows
the activation of PFK by AMP. The assays are done under conditions where all three enzymes are inhibited to 95% and
activations were measured by the activities shown after the
addition of a known amount of activator. Under these conditions, the Ka for all these enzymes are very similar to
each other (0.2 mM) and to the physiological range of concentrations of AMP in these tissues.

     The turning on and off of glycolysis in skeletal muscle,
however, cannot be accounted for entirely by the changing
levels of adenine nucleotides. It was shown by Helmreich and
Cori (1965) that despite the large increase in energy expen-
diture during muscle contraction, the level of ATP was decreased by only a fraction of the total amount, and even the
change in the ratio of ATP/AMP in muscle could not account
for the profound increase in glycolytic flux during contraction. The ratio of ATP/AMP is presumably maintained fairly
constant during contraction at the expense of phosphocreatine
through the action of creatine phosphokinase. It is most
likely that the signal to turn on glycolysis during contraction
is the falling level of P-creatine, an inhibitor of PFK, and
the concomitant rise in the level of Pi, an activator of the
enzyme. Fig. 6 shows the effect of P-creatine on the activities of PFK from all three tissue sources. Here the specificity is quite striking, for only the activity of skeletal
muscle PFK was affected by this metabolite. The Ki's of the
muscle PFK for creatine phosphate are 1.9 and 8.4 mM at pH
7.1 and pH 7.4, respectively. The Ki is of course profoundly
influenced by the concentration of ATP. In contrast, neither
the brain nor the liver enzymes are affected even under con-

829

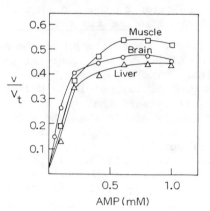

Fig.5: Activation of phosphofructokinases by adenosine
monophosphate at pH 7.4. Data taken from Tsai and Kemp (1974).
Vt and details of the assay are described in Fig. 4. 1 mM
fructose-6-P and sufficient ATP to decrease the activity of
each enzyme to about 5% of the maximum activity at pH 7.4
were present. This amount is

      ○ brain PFK, 8.5 mM ATP
      □ muscle PFK, 9.5 mM ATP
      △ liver PFK, 2.0 mM ATP

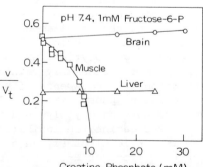

Fig. 6: Inhibition of phosphofructokinases by creatine
phosphate. Assay conditions were described in legend to
Fig. 4. Final concentrations of ATP were 2.0 mM for brain
and muscle PFK and 0.5 mM for liver PFK. Data from Tsai
and Kemp (1974).

ditions where the ATP levels are inhibitory. The concentration of P-creatine is about 20 mM in skeletal muscle and is
much lower in brain. The specificity of this metabolite for
the A type PFK indicates the important role P-creatine plays
during muscular contraction.

On a molecular basis, it is also interesting to note that
whereas the brain PFK consists mainly of type C PFK and hybrids containing both C type and A type subunits, a potent
inhibitor of the A type isozyme does not seem to affect these
hybrid isozymes. The A monomers apparently cannot operate
independently. As noted previously, muscle contraction not
only lowers the content of creatine phosphate, which is an
inhibitor for PFK, but also increases the content of Pi, a
potent activator of PFK. Fig. 7 shows the effect of Pi as an
activator of phosphofructokinase from muscle, liver, and brain.
This figure shows the increasing activity of PFK with increasing concentrations of Pi, tested under conditions where an
amount of ATP is present such that each enzyme is about 95%
inhibited. The muscle enzyme is the least sensitive of the
three preparations to the activating effect of inorganic
phosphate, but these activating levels of Pi are in the physiological range.

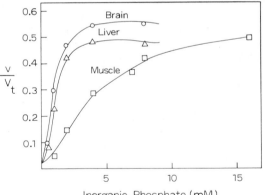

Fig. 7: Activation of phosphofructokinase by inorganic
phosphate at pH 7.4. Experimental conditions were identical
to those described in the legend of Fig. 5. Data from Tsai
and Kemp (1974).

Citrate represents still another important effector of
phosphofructokinase. It is known that cardiac tissue will
preferentially utilize fatty acids of ketone bodies and that
the mechanism for the inhibition of glucose metabolism is
thought to result from increased levels of citrate produced

by this mechanism. The effect of citrate as an inhibitor is most pronounced with the type A or muscle PFK as shown in Fig. 8. Here the muscle and brain preparations are compared and it can be seen that the A isozyme of skeletal and cardiac muscle is more sensitive to citrate inhibition than is brain PFK. A previous comparison of muscle and liver phosphofructo-kinase has shown that the liver enzyme is also less sensitive to the inhibitory action of citrate (Kemp, 1971).

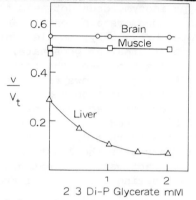

Fig. 8: Inhibition of PFK by 2,3-di-P-glycerate at pH 7.4. Vt is defined in Fig. 4. Fructose-6-P was present at 1 mM and ATP at 0.4 mM. Data from Tsai and Kemp (1974).

Another important regulatory metabolite of PFK is 2,3-diphosphoglycerate which is known to be present in high concentrations in mammalian erythrocytes. This metabolite can be considered an energy storage compound as suggested by the fact that it is the first metabolite to be depleted during blood preservation (Bartlett, 1973). Fig. 9 shows that 2,3-diphosphoglycerate is a specific inhibitor of the type B PFK which is present not only in the liver but also in erythrocytes. The specific inhibitory effect of this effector on the B type PFK and its high concentration in erythrocytes suggests a critical regulatory role for this metabolite in erythrocytes.

One can generalize the foregoing by simply stating that the activity of phosphofructokinase is inversely related to the energy state of the cell. The energy charge or state of all cells is not determined simply by the level of ATP or by the ratio of ATP/ADP and AMP. Creatine phosphate is an energy reserve in skeletal and cardiac muscle, and 2,3-diphospho-glycerate can play a similar role in erythrocytes. We have seen that specific isozymes are present in the various tissues that are particularly sensitive to those metabolites that reflect the energy state of the cell. The type A isozyme of

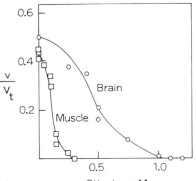

Fig. 9: Inhibition of PFK by citrate at pH 7.1. For details, see legend to Fig. 4. Fructose-6-P was present at 1 mM and ATP at 0.4 mM. Data from Tsai and Kemp (1974).

muscle and heart is very strongly inhibited by creatine phosphate whereas the other isozymes are insensitive to this effector. The type B isozyme found in erythrocytes is particularly responsive to 2,3-diphosphoglycerate concentrations. Of course, all of the enzymes are sensitive to the inhibitory action of ATP, although to varying extents. In liver and brain where the rates of glycolysis change less than in the skeletal muscle, glycolytic fluxes are probably regulated mainly by the changing ratio of ATP to AMP. Diversity of the regulation of glycolysis in different tissues can thus be accounted for both by the varying concentrations of the different effectors of PFK in different tissues and by the presence of tissue-specific isozymes that have differing sensitivities toward the various effectors, particularly those that reflect the energy state of the cell.

ACKNOWLEDGMENT

The authors are indebted to Miss Alice Hsu for excellent technical assistance. This work was supported by a grant from the U. S. Public Health Service (AM 11410).

REFERENCES

Bartlett, G. R. 1973. Changes of red cell phosphate compounds during storage of human blood in ACD, in *Erythrocytes. Thrombocytes, leukocytes,* E. Gerlach, Georg Thieme, eds. Stuttgart, pp. 139-143.

Bloxham, D. P. and H. A. Lardy 1973. Phosphofructokinase, in *The Enzymes*, Vol. IX, Third Edition, P. D. Boyer, ed., Academic Press, New York, pp. 239-278.

Gonzalez, F., M. Y. Tsai, and R. G. Kemp 1974. Distribution of phosphofructokinase isozymes in several species. Submitted for publication.

Helmreich, E. and C. F. Cori 1965. Regulation of glycolysis in muscle. *Advan. Enzyme. Regul.* 3: 91-107.

Imai, S., A. L. Riley, and R. M. Berne 1964. Effect of Ischemia on adenine nucleotides in cardiac and skeletal muscle. *Circ. Res.* 15: 443-450.

Kemp, R. G. and P. B. Forest 1968. Reactivity of the sulfhydryl groups of muscle phosphofructokinase. *Biochemistry* 7: 2596-2603.

Kemp, R. G. 1971. Rabbit liver phosphofructokinase: Comparison of some properties with those of muscle phosphofructokinase. *J. Biol. Chem.* 246: 245-252.

Kurata, N., T. Matsushima, and T. Sugimura 1972. Multiple forms of phosphofructokinase in rat tissues and rat tumors. *Biochem. Biophys. Res. Commun.* 48: 473-479.

Layzer, R. B. and M. M. Conway 1970. Multiple isozymes of human phosphofructokinase. *Biochim. Biophys. Res. Commun.* 40: 1259-1265.

Lowry, O. H., J. V. Passonneau, F. X. Hasselberger, and D. W. Schulz 1964. Effect of ischemia on known substrates and cofactors of glycolytic pathway in brain. *J. Biol. Chem.* 239: 18-30.

Mansour, T. E. 1972. Phosphofructokinase. *Current Top. Cell. Reg.* 5: 1-46.

Minakami, S. and H. Yoshikawa 1965. Thermodynamic considerations on erythrocyte glycolysis. *Biochem. Biophys. Res. Commun.* 18: 345-349.

Ruderman, N. B., C. R. S. Houghton, and R. Hems 1971. Evaluation of the isolated perfused rat hindquarter for the study of muscle metabolism. *Biochem. J.* 124: 639-651.

Tanaka, T., T. An, and Y. Sakaue 1971. Studies on multimolecular forms of phosphofructokinase in rat tissues. *J. Biochem.* 69: 609-612.

Tarui, S., N. Kono, and K. Uyeda 1972. Purification and properties of rabbit erythrocyte phosphofructokinase. *J. Biol. Chem.* 247: 1138-1145.

Tsai, M. Y. and R. G. Kemp 1972. Hydridization of rabbit muscle and liver phosphofructokinases. *Arch. Biochem. Biophys.* 150: 407-411.

Tsai, M. Y. and R. G. Kemp 1973. Isozymes of rabbit phosphofructokinase: Electrophoretic and immunochemical studies. *J. Biol. Chem.* 248: 785-792.

Tsai, M. Y. and R. G. Kemp 1974. Rabbit brain phosphofructo-
    kinase: Comparison of regulatory properties with those
    of other phosphofructokinase isozymes. Submitted for
    publication.
Williamson, J. R., R. Scholz, and E. T. Browning 1969.
    Interactions between fatty acid oxidation and the citric
    acid cycle in perfused rat liver. *J. Biol. Chem.*
    244: 4617-4627.

SUBSTRATE SPECIFICITIES OF PLANT PEROXIDASE ISOZYMES

EDWIN H. LIU
Department of Biology
University of South Carolina
Columbia, S. C. 29208

ABSTRACT. Peroxidase (E. C. 1.11.1.7) utilized hydrogen peroxide to oxidize a wide range of hydrogen donors. Because this enzyme system consists typically of a large number of isozymes in plants, the possibility exists that the peroxidase system consists of a family of homologous enzymes which have different substrate specificities and perform different physiological functions in the cell.

The biochemical properties of two horseradish peroxidase isozymes were compared. One isozyme (No. 5) which migrates cathodally has an isoelectric point of 8.5 and a buoyant density of 1.326 was compared to the most anodally migrating peroxidase (No. 1) which has an isoelectric point of 3.0, and a larger proportion of carbohydrate, as evidenced by its buoyant density of 1.379. When the substrate homovanillic acid is used as the hydrogen donor for peroxidase, the anodic isozyme has a specific activity 37.6 times greater than the cathodic isozyme. Ammonia has been shown to quantitatively enhance the activity of peroxidase by binding reversibly to a ligand on the enzyme which is not the hydrogen peroxide binding site. While ammonia acts to stimulate quantitatively the cathodic isozyme, no ammonia enhancement of activity is observed with the anodic peroxidase isozyme. These results suggest differences in the reactivity of these isozymes which may be based on differences in the active site.

Two zymogram stains for peroxidase have been developed utilizing substrates of possible physiological significance. These are eugenol, a lignin precursor, and tyrosine. These stains do not recognize the same peroxidase isozymes as the conventional benzidine stain, and in fact isozymes are visualized which are not reactive with conventional peroxidase substrates.

INTRODUCTION AND DISCUSSION

The enzyme peroxidase (donor: H_2O_2 oxidoreductase; E. C. 1.11.1.7) is capable of utilizing hydrogen peroxide to oxidize a wide variety of hydrogen donors, such as phenolic substances, aromatic, primary, secondary and teriary amines, leuco-dyes, certain heterocyclic compounds such as ascorbic acid and indole, and certain inorganic ions, particularly the iodine ion (Saunders, 1964.

Peroxidases are heme-containing proteins. While they as a class are not particularly specific as to hydrogen donor, they are very specific in their requirement for hydrogen peroxide. Peroxidases are thought to be monomers with a molecular weight of about 40,000. Peroxidases consist of an apoenzyme which contains both carbohydrate and protein, combined with an iron porphyrin which has been identified as protohemin IX (Keilen, 1951).

The first observation that peroxidase activity consisted of more than one electrophoretically distinct species was made by Theorell in 1942. By electrophoresis of a pure horseradish peroxidase at pH 7.5 he could resolve two components; the anodally migrating he called "true peroxidase" and the cathodally migrating he called "para-peroxidase". He considered the para-peroxidase to be a derivative of true peroxidase. Keilen and Hartree (1951) observed that when horseradish peroxidase was stored at 0°C, a new component appeared.

Recognition that peroxidase activity actually exists in multiple forms came from Jermyn (1954), who observed by electrophoresis on filter paper that there were five components of peroxidase isozymes.

In all higher plants investigated, there exists a multitude of peroxidase isozymes. The number and relative concentration of these peroxidases varies between different tissues and with the developmental stage of an organ (Scandalios, 1964; 1969). Because there is so much peroxidase activity in a cell and changes in its activity are inversely proportional to growth in plants, this enzyme system has been used as a model to study hormonal control of growth processes in plants (Galston, 1969).

It is characteristic of peroxidase that its isozyme pattern is complicated and difficult to interpret. Even in the most thoroughly studied peroxidase system, that isolated from horseradish roots, there is no clear agreement as to the actual number of peroxidase isozymes present. Shannon, et al., (1966) reported seven horseradish peroxidase isozymes which could be recognized and resolved by ion exchange chromatography. Starch gel electrophoresis of commercially prepared horseradish peroxidase contains eleven isozymes (Klapper and Hackett, 1965). Thin layer isoelectric focusing is reported to resolve twenty isozymes of horseradish peroxidase (Delincee, 1970). This complexity of isozyme expression in peroxidase must be recognized and dealt with in any serious study of the biochemistry and physiology of this enzyme.

The electrophorectic mobility of peroxidase isozymes can be altered considerably with no loss of activity merely by storage at cold room temperatures, at pH's of 7.0 or higher (Liu and Lamport, 1973).

One of the most important reasons for ambiguity in the interpretation of peroxidase isozymes is that individual peroxidase isozymes have different reactivities with the various substrates used to measure their activity. The resolved isozymes of horseradish peroxidase exhibit specific activities which differ ten fold when O-dianisidine was used to assay peroxidase activity. When these same isozymes were assayed for the oxidation of oxaloacetate, the differences in specific activity of individual isozymes did not fit the pattern established with O-dianisidine (Kay, et al, 1967). In the dwarf tomato plant, an anodally migrating peroxidase isozyme stains very well when benzidine is the hydrogen donor, but reacts very little with guaiacol (Evans, 1968). This same relationship is found for one of the isozymic components of horseradish peroxidase (Jermyn and Thomas, 1954). Differences in reactivity of horseradish peroxide isozymes with other substrates such as ascorbic acid, phloro-glucinol and mesidine have also been reported (Chmielmicks, et al., 1973) Peroxidase isozymes in tobacco tissue cultures react differently with phenolic compounds such as ferulic acid (Pickering, et al., 1973).

The differences in reactivity with substrates may be an indication that the peroxidase isozymes must have subtle differences around the enzyme active site which direct the specificity of the enzyme for particular hydrogen donors.

This variability in the substrate specification of individual peroxidase isozymes underscores one of the basic problems concerning peroxidase research: no clear cut physiological role can yet be assigned for this enzyme in higher plants.

In those plant systems where genetic analysis of peroxidase isozymes has been attempted using conventional redox stains, e.g., in maize (Brewbaker, 1974) tomato (Rick, et al., 1974), and Datura (Smith, 1974), it is apparent that many genetic loci are responsible for the expression of peroxidase isozymes.

In the horseradish peroxidase system, one isozyme, the most anodally migrating, is distinct because its buoyant density of 1.379 suggests that it contains a greater proportion of carbohydrate than the other isozymes. The zymogram technique provides a second dimension to cesium chloride density gradient centrifugation fractionations, and allows the determination of densities of individual isozymes even though these isozymes have overlapping distributions. (Quail and Varner, 1970). When the fractions from a cesium chloride gradient are subjected to electrophoresis and stained for peroxidase activity, the individual isozymes are resolved in Gaussian distributions (Figure 1). With a densitometer trace, one can locate the peak of each isozyme and determine its density. The more anodally migrating isozyme (No. 1) has the greater density, which is 1.379. This is contrasted with the density of the major cathodic isozyme

(No. 5) which is calculated to be 1.326.

The biochemical properties of horseradish peroxidase isozyme (No. 1), and peroxidase (No. 5) were compared (Table I).
The homovanillic peroxidase assay (Guilbault, 1968) was used.

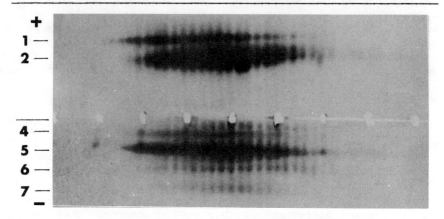

Fig. 1. Zymogram showing the distribution of individual peroxidase isozymes in the cesium chloride density gradient fractionation of horseradish root extract. 5 μl samples of the odd
numbered fractions are sequentially applied to paper wicks,
then subjected to starch gel electrophoresis. The zymogram
is then stained for peroxidase activity with benzidine. The
gaussian distribution of each isozyme in the gradient can be
seen. Isozymes with greater bouyant densities, e.g. No. 1,
have distributions which are shifted to the left.

This assay differs from the spectrophotometric measurement of
the oxidation of redox dyes such as O-dianisidine because it
is fluorescent and involves a one-electron transfer to form
a free radical. In the case of isozyme-5, its specific activity was 6.75 FU/min/OD^{275}. Isozyme-1, however, has a specific
activity of 254 FU/min/OD^{275}. The dense, anodic peroxidase
isozyme thus has a specific activity 37.6 times greater than
the less dense cathodally migrating peroxidase.

The pH optimum of the O-dianisidine reaction is 5.6 and the
optimum for the homovanillic acid assay is pH 8.5; the reactions
were thus carried out at these pH's. It is interesting that
at the optimum of pH 5.6 for the O-dianisidine reaction the
apparent K_m for H_2O_2 is 4×10^{-3}M for the anodic isozyme, but
for the same binding of H_2O_2 to the heme moiety, the apparent
K_m is two orders of magnitude lower, 1.5×10^{-5}M at the pH 8.5
optimum of the homovanillic acid assay. This two orders of
magnitude difference in the binding of H_2O_2 to the enzyme may
be due to the pH difference in optima for these substrates.

At high pH, and using the spectrophotometric O-dianisidine assay,

TABLE I. BIOCHEMICAL PROPERTIES OF HORSERADISH PEROXIDASE ISOZYMES

Sample	pI	Homovanillic Sp. Act. Fu/min/OD$_{275}$ protein	Rel. Sp. Act.	K_{app} (H_2O_2)
Anodic Peroxidase (1)	3.0	254	37.6	1.5×10^{-5}M.
Cathodic Peroxidase (5)	8.0	6.75	$\equiv 1$	---
		o-Dianisidine Sp. Act.* U/min mg. protein		K_{app} (H_2O_2)*
Anodic Peroxidase (A-3)		0.9		4.0×10^{-3}M
Cathodic Peroxidase (C)		0.1		

*From Kay, Shannon, and Lew (1967)

ammonia has been shown to quantitatively enhance the peroxidase reaction (Fridovich, 1963). When the anodic and cathodic per- oxidase isozymes are tested for this ammonia enhancement at pH 9.3 (Table II), the cathodic form showed a quantitative increase of activity with added ammonia. The anodic band however, had no response to added ammonia in the incubation medium. Ammonia did not enhance the homovanillic acid assay for either isozyme. Kinetic evidence has shown that the ammonia enhancement effect is due to the reversible binding of ammonia to a ligand on peroxidase which is not the hydrogen peroxidase binding group (Fridovich, 1963). Since the anodic isozyme has a greater density and presumably a larger carbohydrate moiety, it is con- ceivable that the carbohydrate group could sterically inter- fere with the approach of the ammonium ion to the reactive ligand in peroxidase. Another explanation could be that the reactive ligand is missing in the anodic peroxidase.

These differences in reactivity with homovanillic acid and response to ammonia suggest that these isozymes may have differ- ent active site architecture, and different reactivities with substrates.

If peroxidase isozymes have different reactivities with sub- strates, then the possibility exists that in the investigation of this enzyme in plant systems, the use of any one substrate for peroxidase will result in a failure to recognize and identi- fy isozymes which are not reactive with that hydrogen donor.

One way to determine whether peroxidase isozymes have differ- ent substrate specificities would be to develop peroxide zymo- gram stains which are based on substrates with possible physio- logical significance, rather than the standard redox dyes such as benzidine or guaiacol. Crude extracts could then be simul- taneously stained for peroxidase activity with several substrate. Differences in the isozyme banding pattern of the same extract when stained with different substrates would indicate the exist- ence of isozymes which were preferentially reactive with a par- ticular substrate, and could thus possibly perform different physiological functions.

In an investigation of the possible substrate specificities of peroxidase isozymes, it would be most meaningful to assay those activities which are definitely known to occur in vivo. There are at least three classes of chemical compounds in organ- isms whose formation is linked to the enzymic activity of per- oxidase. The first are halogenated compounds such as iodotyro- sine and diiodotyrosine which are formed after a peroxidase cata- lyzed oxidation of iodide. However, plant peroxidases, either as a purified protein or as an extract of the mascerated tissue, do not catalyze this halogenation reaction (Ljunggren, 1966). Peroxidase is reactive with tyrosine, and can form dityrosine, which is the condensation product of tyrosine free radicals.

TABLE II. THE EFFECT OF AMMONIA ON THE ACTIVITY
OF ISOLATED HORSERADISH PEROXIDASE ISOZYMES

Isozymes	NH4OH conc	pH	Peroxidase Activity* ΔOD^{460}/min
No. 5	-	9.3	.148
No. 1	-	9.3	.107
No. 5	.75mM	9.3	.269
No. 1	.75mM	9.3	.110
No. 5	1.50mM	9.3	.435
No. 1	1.50mM	9.3	.092
No. 5	3.00mM	9.3	.628
No. 1	3.00mM	9.3	.105
No. 5	6.00mM	9.3	1.05
No. 1	6.00mM	9.3	.110

*Peroxidase is assayed spectrophotometrically with 0-dianisi-
dine.

Peroxidase can act on both peptide-bound tyrosine, and the
free acid, since dityrosine is found as a protein crosslink in
resilin (Anderson, 1964), elastin (LaBella, et al, 1967), and
in the gel form of collagen (LaBella, et al., 1968). It is
not known whether dityrosine crosslinks are found in plant proteins. The third class of compounds whose formation is catalyzed by peroxidase is lignin, which is found in all higher
plants. Lignin is formed by the condensation of phenyl alcohol
free radicals.

Because of their possible physiological significance as substrates of peroxidase, eugenol (2-methoxy-4-allyl phenol) which
is a lignin precursor (Siegel, 1956) and tyrosine were used to
develop new zymogram stains which allow the visualization of
peroxidase isozymes after starch gel electrophoresis.

METHODS AND RESULTS
EUGENOL ZYMOGRAM STAIN FOR PEROXIDASE

Peroxidase in the presence of hydrogen peroxide catalyzes
the formation of free radicals of eugenol, and these radicals
condense and form a white precipitate. The precipitate is not
a lignin compound as such, but when the reaction is performed
in the presence of a preformed cellulose matrix such as cellulose powder or paper, the resultant product tests positively
for lignin (Siegel, 1956).

Since the eugenol reaction results in a precipitate, and a
precipitation reaction is a requisite for good zymogram stains,
it appeared that eugenol might be a good choice as a substrate
to visualize peroxidase isozymes. Furthermore, since the peroxidase catalyzed eugenol product is essentially a biphenyl

compound, it is probable that it would be fluorescent.

The following peroxidase zymogram stain was developed:

75 mg eugenol (2-methoxy-4-allyl phenol) ($\sim$ 60 µl)
100 ml 0.05 M sodium phosphate buffer, pH 6.0
10 µl 30% H_2O_2-- final concentration is 9 x 10^{-4}M
Stir for 30 minutes, room temperature.

Peroxidase isozymes are observed by flooding the cut surface of a starch gel with the reaction mixture, then viewing under short-wave ultraviolet light. Blue bands of fluorescence appear within one minute.

Eugenol can also be used as a substrate in a quantitative peroxidase assay. This is accomplished by measuring the increase in fluorescence 430 nm with excitation at 323 nm (Table III).

TABLE III. CONTINOUS SPECTROPHOTOMETRIC ASSAY OF PEROXIDASE ACTIVITY USING TYROSINE AS A SUBSTRATE. ALL REACTIONS WERE RUN AT ROOM TEMPERATURE, IN .05M PHOSPHATE BUFFER, pH 6.0.

Enzyme	H_2O_2	ΔOD^{260}/min	OD^{260}/min per mg. prot.
100 µg	9 x 10^{-4}M	0.349	3.49
25 µg	9 x 10^{-4}M	0.073	2.92
100 µ	none	0.000	--
none	9 x 10^{-4}M	0.000	--
100 mg*	9 x 10^{-4}M	0.018	0.18

*Incubated at 90°C for 10 minutes

TYROSINE ZYMOGRAM STAIN FOR PEROXIDASE ACTIVITY

The peroxidase catalyzed formation of fluorescent dityrosine does not occur below pH 7.0 and optimum rate of formation is not reached until pH 8.5. However, peroxidase is reactive at pH's below 7.0.

A peroxidatic reaction with tyrosine as a substrate has been observed at pH 6.0 (Table IV). The reaction is monitored by measuring the increases in absorbance at 260 nm with time when peroxidase is added to a reaction mixture containing 0.5 mg/ml tyrosine and H_2O_2 at a concentration of 9 x 10^{-4} M. The nature of the product is unknown except that it is nonfluorescent, and absorbs ultraviolet light. It is possible that this compound is a dityrosine which is not anion stablized, and thus not fluorescent.

The following peroxidase zymogram stain utilizing tyrosine as a substrate was developed:

    1 mg tyrosine

    100 ml 0.5 M sodium phosphate buffer, pH 6.0

    10 μl 30% H_2O_2 -- final concentration is 9 x 10^{-4}M

    Stir for 30 minutes, room temperature.

Peroxidase isozymes are observed by following procedures as for the eugenol assay. Dark fluorescent absorbing bands appear within one minute.

TABLE IV. CONTINUOUS FLUORESCENT ASSAY OF PEROXIDASE ACTIVITY USING EUGENOL AS A SUBSTRATE. ALL REACTIONS WERE RUN AT ROOM TEMPERATURE, IN .05M Tris-HCl, pH 8.0.

Enzyme	H_2O_2	ΔF/min.	ΔF/min./mg.protein
10 μg	9 x 10^{-4} M	.320	32.0
25 μg	9 x 10^{-4}M	.820	32.8
0 μg	9 x 10^{-4}M	0	0
25 μg	0	0	0

When tyrosine is used as a peroxidase zymogram substrate at pH 8.5, dityrosine is readily formed on the gel surface. However, blue fluorescence is observed over the whole gel, and no bands can be distinguished, since this is apparently not a precipitating reaction.

The tyrosine and eugenol peroxidase stains were judged to be specific for peroxidase for the following reasons: (1) bands were not recognized on the cut surface of the gel before the application of the tyrosine or eugenol reaction mixture; (2) when H_2O_2 is omitted from the reaction mixture, no bands develop; (3) no bands develop when heat inactivated peroxidase is used as a source of enzyme.

Horseradish petiole extracts were subjected to starch gel electrophoresis, sliced transversely into three identical pieces; each slice was stained for peroxidase activity with either benzidine, eugenol, or tyrosine (Figure 2). Differences could be seen in the isozymes recognized by each stain, their relative intensities, and their migration. These stains also indicate that peroxidase isozymes exist in the cell which are undetectable by conventional redox dye staining procedures.

The eugenol and benzidine substrates were used to study the developmental expression of peroxidase isozymes in the garden pea *(Pisum sativum)*. This was done to determine whether certain eugenol specific isozymes might be expressed independently

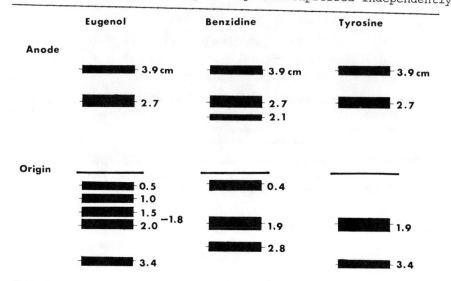

Fig. 2. Schematic of peroxidase isozymes visualized from a single zymogram of horseradish petiole extract when different substrates are used. Samples were subjected to starch gel electrophoresis. The developed gel was sliced transversely into three 2 mm slices, and each slice was stained for peroxidase activity with either eugenol-H_2O_2, benzidine-H_2O_2, or tyrosine-H_2O_2. Figures to the right of each isozyme are the migration distances (cm.) from the origin.

from other benzidine specific peroxidases.

The expression of pea peroxidase isozymes was studied by taking Alaska peas (Carolina Biological Supply Co.) in light grown conditions, and homogenizing roots, shoots and leaves. Extracts were subjected to starch gel electrophoresis at pH 8.3. Transverse sections of the same gel were stained for peroxidase activity using either the benzidine or eugenol staining method.

In the comparison of peroxidase isozymes in developing pea tissues using the benzidine and eugenol substrates (Figure 3), it is apparent that while several isozymes are reactive with both substrates, other peroxidases are specific for one substrate, and that these isozymes are often expressed at different stages of development.

These data, from horseradish and peas, indicate that peroxidase isozymes can have large differences in their reactivity

with substrates, and that peroxidase isozymes specific for one
substrate can go undetected if another substrate is used in
assay procedures. In the case of peas, the developmental expression of certain eugenol specific isozymes appear to be

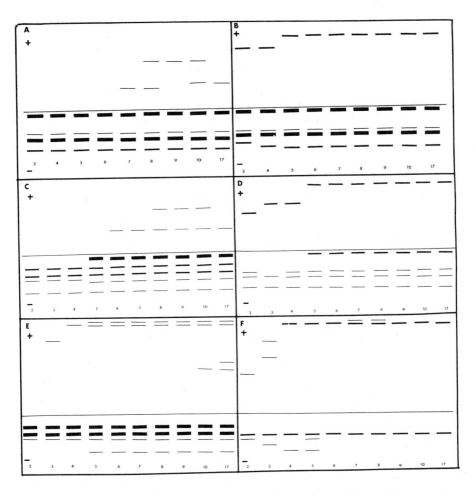

Figure 3. Schematic of peroxidase isozymes of pea tissues during early development. Numerals refer to the age of the plant
in days. Tissues were homogenized, and extracts were subjected
to starch gel electrophoresis. The gel was then sliced transversely into two slices. One slice was stained with benzidine-
H2O2, and the other with eugenol-H2O2. Isozymes are drawn to
scale, and are thus directly comparable. A. Leaf extracts
stained with benzidine; B. Leaf extracts stained with eugenol:
C. Root extracts stained with benzidine; D. Root extracts

Figure 3. Cont'd. stained with eugenol; E. Shoot extracts stained with benzidine; F. Shoot extracts stained with eugenol.

---

under distinct temporal control as compared to other benzidine specific isozymes.

The peroxidase system may thus represent a family of enzymes whose reactivities with different substrates are distinct, and which are expressed independently at different times of development.

ACKNOWLEDGEMENTS

A portion of this research was conducted at the AEC Plant Research Laboratory, Michigan State University as part of a Ph.D. dissertation under contract AT(11-1-1338. The author gratefully acknowledges the assistance of Ms. Donna M. Gibson who performed the developmental study on peas.

REFERENCES

Anderson, S. O., 1964. The cross-links in reslin identified as dityrosine and trityrosine. *Biochim. Biophys. Acta.* 93: 213-215.

Brewbaker, J. L. and Y. Hasegawa, 1974. Polymorphisms of the 12 major peroxidases of maize. *III. Isozymes: Developmental Biology.* C. L. Markert, editor, Academic Press, New York.

Chmielnicka, J., P. Ohlssen, K. Paul and T. Stijbrad 1971. Substrate specificity of plant peroxide. *FEBS Letters* 17: 181-184.

Delincee, H. and B. J. Radola 1970. Thin layer isoelectric focusing on Sephadex layers of horseradish peroxidase. *Biochim. Biophys. Acta.* 200: 404-407.

Evans, J. L. 1968. Peroxidase from the extreme dwarf tomato plant: Identification, isolation and partial purification. *Plant Physiol.* 43: 499-503

Fridovich, I. 1963. The stimulation of horseradish peroxidase by nitrogenous ligands, *J. Biol. Chem.* 238: 3921.

Galston, A.W. and P. J. Davies 1969. Hormonal regulation in plants. *Science* 163: 1288-1297.

Guilbault, G. C., P. Brignac, and M. Zimmer 1968. Homovanillic acid as a fluorimetric substrate for oxidative enzymes. *Anal. Chem.* 40: 190-196

Jermyn, M. A. and R. Thomas 1954. Multiple components in horseradish peroxidase. *Biochem. J.* 56: 631-639.

Junggren, J. 1966. Catalytic effect of peroxidase on the iodination of tyrosine in the presence of H_2O_2. *Biochim. Biophys. Acta.* 113: 71-78.

Kay, E., L. M. Shannon, and J. Y. Lew 1967. Peroxidase isozymes from horseradish roots. II. Catalytic properties. *J. Biol. Chem.* 242: 2470-2473.

Keilin, D. and E. F. Hartree 1951. Purification of horseradish peroxidase and comparison of its properties with those of catalase and methaemoglobin. *Biochem. J.* 49: 88-104.

Klapper, M. H. and D. P. Hackett 1965. Investigation on the multiple components of commercial horseradish peroxidase. *Biochim. Biophys. Acta.* 96: 272-282.

LaBella, F., F. Keeley, S. Vivian and D. Thornhill 1967. Evidence for dityrosine in elastin. *Biochem. Biophys. Res. Commun.* 26: 748-53.

LaBella, F., et al. 1968. Formation of insoluble gels and dityrosine by the action of peroxidase on soluble collagens. *Biochem, Biophys. Res. Commun.* 30: 333-8

Liu, E. H., and S. T. A. Lamport 1973. The pH induced modification of the electrophoretic mobilities of horseradish peroxidase. *Arch. Biochem. Biophys.* 158: 822-826.

Pickering, J. W., B. L. Powell, S. H. Wender and E. C. Smith 1973. Ferulic Acid: A substrate for two isoperoxidase Nicotiana tobacum tissue cultures. *Phytochemistry* 12: 2639-2643.

Quail, P. H. and J. E. Varner 1971. Combined gradient - gel electrophoresis procedures for determining buoyant densities or sedimentation coefficients of all multiple forms of an enzyme simultaneously. *Anal. Biochem.* 39: 344-355.

Rick, C. M., R. W. Zobel and J. F. Fobes 1974. Four peroxidase loci in red fruited tomato species: Genetic and geographical distribution. *PNAS* 71: 835-839.

Saunders, B. C., A. G. Holmes-Siedle and B. P. Stark 1964. Peroxidase -- *The Properties and Uses of a Versatile Enzyme and Some Related Catalysts.* Butterworth, London.

Scandalios, J. G. 1964. Tissue specific isozyme variation in maize. *J. Heredity.* 55: 281-285.

Scandalios, J. G. 1969. Genetic control of multiple moleular forms enzymes in plants. *Biochem. Genet.* 3: 37-79.

Shannon, L. M., E. Kay and J. Lew 1966. Peroxidase isozymes from horseradish roots. I. Isolation and physical properties. *J. Biol. Chem.* 241: 2166-2172.

Siegel, S. M. 1956. The biosynthesis of lignin: Evidence for celluloses as sites for oxidative polymerization of eugenol. *J. Am. Chem. Soc.* 78: 1753-1755.

Smith, H. H. and M. E. Conklin 1974. Effects of gene dosage on peroxidase isozymes in *Datura stramonium* trisomics. *III. Isozymes: Development Biology,* C. L. Markert, editor, Academic Press, New York.

THE ISOPEROXIDASES OF *CUCURBITA PEPO* L.[*]

D. W. DENNA and MERRY B. ALEXANDER
Department of Horticulture, Colorado State University
Fort Collins, Colo. 80521

ABSTRACT. Using a technique for the simultaneous separation of anodic and cathodic isozymes in polyacrylamide gel slabs, 351 accessions of *Cucurbita pepo* squash were surveyed for peroxidase differences. At pH 9 four different anodic and three different cathodic banding patterns were observed involving a total of 14 different isoperoxidase bands. Three of the anodic patterns were found to be due to combinations of two codominant alleles designated Px_1^1 and Px_1^2. The Px_1^2 allele produced a single peroxidase band while Px_1^1 produced a cluster of four. None of the isoperoxidase differences were associated with detectable morphological differences.

The isoperoxidases varied as to when they first became active in germinating seeds, their relative activities in various vegetative organs, and the extent to which they were bound to cell fragments. Two of the isoperoxidases were found to be strongly bound by noncovalent bonds to cell fragments, and an additional relatively small amount of peroxidase activity was released from cell fragments by an enzyme other than cellulase present in crude extracts from *Trichoderma viridae*.

At least two isoperoxidases not bound to cell fragments were found to be capable of acting in vitro as a lignin polymerase utilizing eugenol and hydrogen peroxide.

INTRODUCTION

Plant **peroxidases** are particularly useful for isozyme studies because of their probable presence in all plant groups, enzymatic stability, and ease of detection (Saunders et al., 1964). Peroxidases can be separated by gel electrophoresis into bands referred to as isoperoxidases or peroxidase isozymes (Siegel and Galston 1967). Isoperoxidase banding patterns have been shown to vary with plant genotype, developmental stage, organ, tissue, and location in cells (Pierce and Brewbaker 1973; Ridge and Osborne 1972).

The purpose of this study was primarily to characterize genetic isoperoxidase differences in *Cucurbita pepo* L. and secondarily to investigate possible developmental and

---

[*]Paper No. 1889 of the Colorado Experiment Station.

physiological effects upon them.

MATERIALS AND METHODS

PLANT MATERIALS AND GROWING CONDITIONS

Forty named cultivars (Table 1) were obtained from various seed companies in the United States, Canada and England. The remaining 311 accessions were provided by the U. S. D. A. Plant Introduction Stations at Geneva, New York and Ames, Iowa.

TABLE 1

The Classification of Some Named Cultivars of *Cucurbita pepo* On the Basis of Their Isoperoxidase Patterns

Class I	Class II	Class III
Black Green	Benning's Green Tint	Seneca Prolific Hybrid
Blackini	Black Beauty	
Burpee's Bush	California Field	
Table Queen	Early Sugar	
Butterbar	Little Pumpkin	
Caserta	New Spaghetti	
Chefini	Sugar	
Cheyenne Bush		
Cocozella		
Connecticut		
Field		
Corn Field		**Class IV**
Cozini		Cornell Bush Table Queen
Early Prolific		Delicata
Straightneck		Early White Bush Scallop
Gold Bar		Ebony
Greyzini		Ranger
Marrow Bush		Summer Crookneck
Marrow Trailing		Table Queen
Small Sugar		
St. Pat's		
Scallop		
Sweetnut		
Tricky Jack		
Tuckerneck		
Young's Beauty		
Zucchini		
Zucchini Hybrid		
7 X 7		

The seeds were planted in wooden flats in a mixture of top soil, sand, and peat (2:1:1) and kept on a greenhouse mist bench for seven days to maximize germination and developmental uniformity. All plant materials were selected for uniformity and modality except in the genetic studies where it was necessary to use all the plants. In the 14 day developmental study a new flat of seeds was started every 24 hours for two weeks. The experiments were carried out during the summer to assure normal growth and development. The greenhouse (evapo-cooled) temperatures averaged (mean) 25°C during the day and 21°C at night. In the survey involving the original 351 accessions, because of the large number of plants and our primary interest in determining the maximum number of isoperoxidase bands in *C. pepo*, up to six individual plants representing each variety or plant introduction were harvested in bulk. The field grown plants were begun in peat pots in the greenhouse May 14th and subsequently transplanted to the field.

ENZYME EXTRACTION

Preliminary experiments demonstrated no advantage in extracting peroxidases at or near 0°C rather than at room temperature (22°C); thus, extractions were carried out at room temperature. Equal fresh weight quantities of plant tissues were washed, ground in a mortar, squeezed through two layers of cheese cloth, and the juice centrifuged at 1,800 x g for five min. The resulting "raw juice" supernatant was used directly for all of the isoperoxidase analyses except those involving bound peroxidases. In the binding studies a given filtrate pellet was repeatedly resuspended in distilled water and recentrifuged at 14,000 x g for 10 min until no visible peroxidase activity remained in the supernatant. Peroxidases bound by secondary bonds to cell fragments were removed by treating pellets with 1M NaCl, 8M urea, or 3% sodium cholate, respectively. For the covalently bound peroxidases the pellets were repeatedly resuspended in 1M NaCl and recentrifuged at 14,000 x g for 10 min until no visible peroxidase activity remained in the supernatant upon reacting an aliquot with a solution of 10^{-3}M tolidine and 10^{-2}M H_2O_2 (10:1 v/v). Five gram quantities of pellet material were incubated with various "cellulase" preparations in 5 ml of 0.01M, pH 5 citrate buffer for 24 and 48 hrs respectively. The cellulase treatments were 20 u of cellulase II (*Trichoderma viridae*) (Worthington Biochem., Freehold, N. J.), 15,000 μ of crystalline cellulase (*Trichoderma viridae*) from Seikagaku Kogyo Co., Tokyo, Japan, and a crude cellulase preparation obtained from *Myrothecium verrucaria* according to the method of Whitaker and Thomas 1962 and provided by Dr. P. Hanchey of the Department of Botany and Plant Pathology. The cellulase activities of all three preparations were confirmed by the cellulose-azure assay (Calbiochem, San Diego, Calif.).

GEL PREPARATION AND ELECTROPHORESIS

All of the equipment used for casting the gels and their electrophoresis was purchased from Ortec Inc., Oak Ridge, Tenn. The chemicals for the gels were obtained from Polysciences, Inc., Warrington, Pa. Acrylamide source is critical to high resolution. The equipment and gel casting techniques are described in the Ortec Operating and Service Manual for Models 4010/4011 Electrophoresis System. The formulation of the gels and their electrophoresis has been previously described by Denna and Alexander 1973.

Starting at the bottom of the cell in the casting stand successive layers of 8% (38 mm), 6% (10 mm), and 4-1/2% (4 mm) acrylamide in pH 9.0, 0.375 M tris-sulfate buffer were cast. The 12 sample wells (10 mm deep) were formed in an 8% (10 mm) acrylamide, pH 9.0, 0.075 M tris-sulfate gel. One ml of a given peroxidase containing raw juice was mixed with one ml of 80% sucrose in pH 9.0, 0.018 M tris-sulfate buffer, and a 20 μl sample of the mixture pipetted into a well. The wells were capped with an 8% acrylamide, pH 9.0, 0.075 M tris-sulfate gel, followed by a final 8% (20 mm) acrylamide, pH 9.0, 0.375 M tris-sulfate gel layer. This last layer allowed for cathodic protein migration.

The gel slab was developed in a tris-borate pH 9.0, 0.065 M buffer system with an Ortec 4100 pulsed power supply at a constant 325 v. The power settings were 5 min at 75 pulses per second (pps) (45 mA), 5 min at 140 pps (87 mA), 5 min at 225 pps (105 mA), and 40 min at 300 pps (115 mA). At the completion of the electrophoretic run the gel was removed from the cell and incubated with a solution of 10^{-3} M tolidine and 10^{-2} M H_2O_2 (10:1 v/v). The isoperoxidases appeared as brown bands in the transparent gel. The presence of lignin was detected by the phloroglucinol test as described by Jensen 1962.

RESULTS

A total of 351 cultivars and plant introduction accessions were examined for possible isoperoxidase differences. Four anodic and three cathodic banding pattern differences were detected (Fig. 1). A total of 14 different isoperoxidase bands were observed.

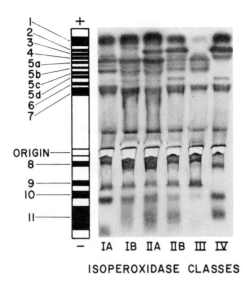

Fig. 1. Isoperoxidase banding pattern classes and band enumeration. Diagrammatic zymogram depicts modal band widths and locations within the gels. A and B within a class illustrate the frequently observed nongenetic band variation.

INDIVIDUAL PEROXIDASE BANDS

Some of the minor isoperoxidase bands showed visible qualitative and quantitative variability between individual plants within the same cultivar. Typical variability is seen in the A and B representatives of pattern classes I and II in Fig. 1. All of the cultivars examined, with the exception of a deviant of Seneca Prolific Hybrid (Fig. 1, class III), possessed a strong isoperoxidase band 1. Band 2 was usually not detected. Isoperoxidase 3 is a major band that is present in classes II and IV and absent from I and III. The **isoperoxidase** 5 cluster (5a, 5b, 5c, and 5d) appeared to be the products of a **single** gene; evidence for this will be presented in the discussion. Band 6 exhibited great variability in intensity but at least traces of it were observed in all plants. Isoperoxidase band 7 was present in all cultivars.

The 4 cathodic isoperoxidase bands frequently lacked definition making classification difficult. Bands 8 and 9 were individually present in some cultivars and accessions, and absent from others. For all samples having relatively well defined cathodic banding, isoperoxidases 10 and 11 were always found to be present.

ISOPEROXIDASE PATTERN CLASSES

Among 351 cultivars examined, taking into account experimental variability, four different anodic and two different cathodic pattern classes were observed (Fig. 1). The primary difference between the class I and II anodic patterns is the absence of band 3 from the former and its presence in the latter. The anodic class III pattern was obtained from a single seed lot of the cultivar Seneca Prolific Hybrid; two other seedlots of the cultivar produced only the class I pattern. The only essential difference between the class III and the class I pattern appears to be the greatly reduced intensity of band 1 in the former. Class IV at first glance appears to be quite different anodically from class I, but when the isoperoxidase 5 complex is considered as a unit, the differences become simpler. The class I individuals have the isoperoxidase 5 cluster in place of band 3, while the reverse is true for the class IV individuals. This point will be further clarified when considering the data on the inheritance of pattern differences.

Three different cathodic patterns were observed. Either band 8 or 9 was absent or both were present together; no samples were observed lacking both bands 8 and 9.

INHERITANCE STUDY

Preliminary studies indicated that the class I and IV anodic isoperoxidase patterns were the products of isoperoxidase homozygotes. The F_1's resulting from crosses between plants representative of the two classes gave a class II pattern; mixtures of equal quantities of extracts from the two classes of homozygotes also gave a class II pattern. The F_2 segregation patterns fit the hypothetical ratio of 1 class I:2 class II:1 class IV (Table 2 and Fig. 2) as expected for two codominant alleles; the backcross data further supported the hypothesis.

DISTRIBUTION OF ISOPEROXIDASE CLASSES

A majority (62.5%) of 40 named cultivars (Table 1) possessed a class I anodic pattern. The class IV anodic pattern was of relatively low (17.5%) frequency, as was the class II (17.5%). As previously mentioned, only a single cultivar (2.5%) had a class III anodic pattern. The cathodic patterns were not always sufficiently clear to classify accurately.

The 311 plant introduction accessions were originally collected from a wide range of geographical areas, most of

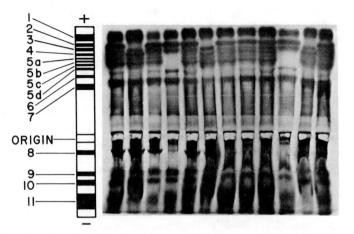

Fig. 2. Typical class I x class IV F_2 samples showing segregation for the alleles PX_1^1 and PX_1^2 their associated isoperoxidases.

TABLE 2

ISOPEROXIDASE SEGREGATION PATTERNS FOR CROSSES
INVOLVING INDIVIDUALS HOMOZYGOUS FOR CLASS I
AND CLASS IV PATTERNS RESPECTIVELY

Cross	Phenotypes Observed			Expected Ratio	X^2	P
	I	II	IV			
Class IV x Class I, F_2						
C.B. Table Queen [1] x Cheyenne Bush	27	56	25	1:2:1	.14	.90
E.W.B. Scallop[2] x Zucchini	31	55	36	1:2:1	.66	.85
Class I x Class IV, F_2						
Cheyenne Bush x C. B. Table Queen	25	46	23	1:2:1	.15	.90
Zucchini x E.W. Bush Scallop	28	53	21	1:2:1	1.12	.75
(Class IV x Class I) x Class IV						
(E.W.B. Scallop x Zucchini) x E.W.B. Scallop	--	52	49	1:1	.10	.75
(Class I x Class IV) x Class I						
(Zucchini x E.W.B. Scallop) x E.W.B. Scallop	43	47	--	1:1	.19	.70

[1]Cornell Bush Table Queen
[2]Early White Bush Scallop

them in the northern hemisphere. The percentage distribution according to class was: 87.1% class I, 10.6% class II, 2.3% class IV, and none in class III. No noteworthy deviations from these averages were noted for any geographical locales represented.

DEVELOPMENTAL STUDY

A preliminary study established that, under the conditions of enzyme extraction and electrophoresis, the dry seed failed to yield any detectable soluble peroxidase, and that the isoperoxidase banding patterns for two, three, and four week old seedlings were indistinguishable. In order to pinpoint the developmental pattern dynamics more precisely extracts of class IV germinating seeds and seedlings were examined every 24 hrs for the first 14 days beginning with imbibition (Fig. 3). Extracts of seed that had imbibed water for one or two days, like the dry seed, failed to show any visible peroxidase activity in the gel. On the third day isoperoxidase 4 (trace), 6 and 7 became visible followed by band 3 on the fourth day. Isoperoxidase 1 appeared for the first time on the fifth day followed by the remaining isoperoxidase bands 2 and 8 on the sixth day. A diminuation of band 7 was observed from the fifth through the seventh days followed by a resurgence in activity from the eight through eleventh days; no similar quantitative pattern change in this isoperoxidase was observed in four replicates. Seeds and seedlings of class I plants showed the same developmental changes except for the substitution of cluster 5 for isoperoxidase band 3.

PLANT PART DIFFERENCES

A study of class IV seedling plant part isoperoxidase differences involving 21 and 120 day old plants was carried out to determine if there were any organ specific differences. No clear-cut qualitative banding differences were apparent although differences in band intensities (quantitative) were observed for different plant parts and tissues of different ages. Again, class I plants gave the same results as class IV except for the substitution of cluster 5 for isoperoxidase 3.

BOUND PEROXIDASES

One M NaCl, 8 M urea, 3% sodium cholate, and 8 M urea plus 3% sodium cholate all resulted in the release of mainly isoperoxidase bands 7 and 10 (Fig. 4). Preliminary experiments

A

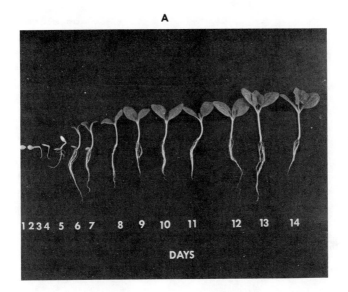

B

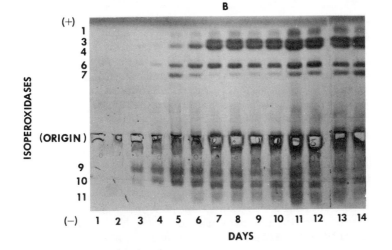

Fig. 3. Developmental study of the isoperoxidases in class IV seeds and seedlings of *Cucurbita pepo*. A. Seedling growth stages. B. Isoperoxidase patterns obtained from the seeds and seedlings on successive days.

demonstrated that the three reagents did not interfere with the normal raw juice gel patterns. The three reagents in the concentrations used were roughly equivalent in their ability

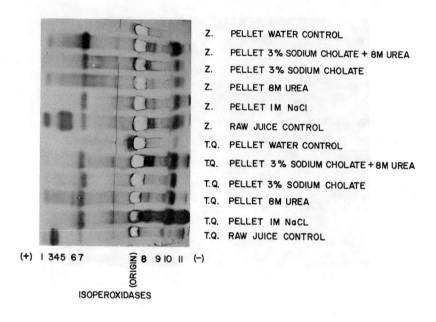

Z.   PELLET WATER CONTROL

Z.   PELLET 3% SODIUM CHOLATE + 8M UREA

Z.   PELLET 3% SODIUM CHOLATE

Z.   PELLET 8M UREA

Z.   PELLET IM NaCl

Z.   RAW JUICE CONTROL

T.Q. PELLET WATER CONTROL

T.Q. PELLET 3% SODIUM CHOLATE + 8M UREA

T.Q. PELLET 3% SODIUM CHOLATE

T.Q. PELLET 8M UREA

T.Q. PELLET IM NaCL

T.Q. RAW JUICE CONTROL

(+) I 345 67   8 9 10 II (−)

(ORIGIN)

ISOPEROXIDASES

Fig. 4. The release of soluble peroxidases from class IV
(Table Queen - T.Q.) and class I (Zucchini - Z) cultivars of
Cucurbita pepo cell fragments (pellets) treated with several
secondary bond breaking reagents.

to release noncovalently bound peroxidase from the pellets.

Three cellulase preparations were tested for their ability
to release covalently bound peroxidase from cell fragments.
The cellulase activity of each was confirmed using the cellulose azure assay, and each was shown not to change typical
raw juice banding patterns over incubation periods of 48 hrs.
The crystalline cellulase from *Trichoderma* (Seikagaku) was
completely ineffective in releasing detectable amounts of
peroxidase from the cell fragments (pellets). However, the
crude *Trichoderma* cellulase preparation (Worthington) and the
Myrothecium cellulase released a relatively small amount of
peroxidase activity, which produced mainly diffuse zones of
activity in the gels near the anodic and cathodic fronts. It
was not possible to say whether any of the bands corresponded
to those described for the raw juices. Cellulase incubation
periods of 24 and 48 hrs gave the same results, and no
differences were observed for cell fragments from plants with
a type I or IV raw juice patterns.

LIGNIN POLYMERASE ACTIVITY

The various isoperoxidases were tested for possible lignin polymerase activity using eugenol and hydrogen peroxide as substrates and phloroglucinol for the detection of lignin-like products. Most of the lignin polymerase activity was found in a band corresponding to isoperoxidase 3. Smaller amounts of activity were associated with the gel areas occupied by bands 5a, 7 and 11. No phloroglucinol positive products were visible in the gels containing the previously covalently bound peroxidases. However, eugenol plus hydrogen peroxide did produce white or pale yellow banding patterns that appeared the same as those produced by tolidine.

DISCUSSION AND CONCLUSIONS

The small amount of isoperoxidase variability among the accessions of *Cucurbita pepo* is in marked contrast to the high degree of variability found in *Zea mays* (Hamill and Brewbaker 1969). The lack of any obvious phenotypic differences between plants differing in their isozyme patterns corroborates the work of Loy 1972.

It is suggested that the genetic locus submitted to genetic analysis be designated Px_1 and the respective codominant alleles Px_1^1 and Px_1^2. Px_1^1 refers to the allele controlling the cluster of four bands (5a, b, c, and d). Px_1^2 designates the allele responsible for isoperoxidase band 3. The clustering of isoperoxidase bands in sets of four are seen in other species of *Cucurbita* (Alexander 1972) and peas (Siegel 1966). In *Cucurbita* species the cluster is frequently shifted up or down relative to *C. pepo* indicating that the individual isoperoxidases in the clusters are probably initially products of the same gene. This interpretation is further supported by the observation that the cluster is developmentally always present as a unit.

The experiment concerned with the developmental appearance · of isoperoxidases demonstrated that high levels of peroxidase activity are not necessary for early seedling development. The results also suggested that different peroxidase genes were being turned on at different developmental times. The possibility remains, however, that some of the developmentally related changes could be epigenetic, such as the attachment of polysaccharide. For genetic studies the results indicate that genetically identical plants between one and 12 weeks old can be expected to yield the same isoperoxidase patterns

The plant organ survey, which demonstrated that all organs contain the same isoperoxidases, albeit in different activity ratios, indicates that peroxidase function(s) is probably not causally related to differential organogenesis but relates to some process(es) essential to the development of at least all vegetative organs. For genetic sampling purposes, where it is important to retain the reproductive capability of the plant in question, the utilization of a combination of leaves and petioles can be expected to yield an isoperoxidase pattern representative of the whole plant.

The studies concerned with the cellular binding of peroxidases showed that mainly two out of potentially 14 isoperoxidases were strongly bound by secondary bonds to cell fragments. The fact that reagents known to break electrostatic, hydrogen, and hydrophobic bonds, respectively, could each by themselves bring about the release of the two isoperoxidases indicated that the peroxidases were bound by a combination of the three kinds of bonds. Covalently bound peroxidases were released by two out of three cellulase preparations. Significantly the highly purified, crystalline cellulase from *Trichoderma viridae* failed to release any peroxidase from the cell fragments (pellets), while the crude preparation from the same organism did. These results indicate that the covalently bound peroxidase is not attached directly to the cellulose but to some other cell wall or membrane component susceptible to degradation by an as yet unidentified enzyme in the crude "cellulase" preparations. The diffuse peroxidase bands that result following incubation with the crude cellulase preparations probably represent peroxidases with varying lengths of fragments from the unknown cell wall or membrane attachment component. The fact that the 48 hr incubation period did not appreciably alter the resulting electrophoretic isoperoxidase pattern as compared to the 24 hr treatment suggests that the unknown component is not being randomly degraded. One would expect random degradation in time to result in a narrowing of at least sections of the original diffuse bands. The possibility that proteolytic or polysaccharide degrading enzymes were working directly on the various peroxidases was largely discounted by the observation that the crude cellulase preparations had no detectable effect upon the isoperoxidase patterns of raw juice extracts.

Over the years peroxidases have been implicated as lignin polymerases on the basis of various correlative (Siegel 1962) and biochemical (Harkin and Obst 1973) studies. The hypothesis tends to be supported in a general way in our studies in that high peroxidase activity does not appear to be essential in very young seedlings and is present in all vegetative organs. One intriguing aspect of lignification is that it is carried on outside the cell, and therefore the lignin polymerase must be transported through the plasma membrane and subsequently stabilized at the site where lignification is to occur. The fact that at least some peroxidases are glycoproteins (Shannon et al., 1966), and that the polysaccharide moiety could be associated with an exocytotic transport mechanism mediated by the Golgi apparatus also tends to support the idea that at least one of the peroxidases could act in vivo as a lignin polymerase.

In our studies it appears that the polyacrylamide matrix of the gels was able to substitute for the cellulose matrix requirement for in vitro lignification reported by Siegel (1962). Lignin polymerase activity was found primarily in an area of the gel corresponding to isoperoxidase band 3, and to a lesser degree in an area corresponding to the isoperoxidase cluster 5. It is to be recalled that these bands are determined by different codominant alleles. These results, however, must be considered as of a preliminary nature because only one hydrogen donor, eugenol, was tested under one set of conditions. As noted earlier there are no obvious phenotypic differences between plants homozygous for either of the alleles. It is a discomfiture that the covalently bound peroxidases failed to produce any visible phloroglucinol positive product after incubation with the substrates for over 24 hrs, while the very soluble isoperoxidases 3 and 5 did. It would seem likely in light of the matrix requirement that isoperoxidases 3 and 5 are at least weakly bound in vitro, and possibly also in vivo. Clearly additional experimentation concerning in vitro lignification mediated by isoperoxidases is needed.

REFERENCES

Alexander, Merry B. 1972. Peroxidases and development in *Cucurbita pepo* L. Master's Thesis, Colorado State University, Fort Collins, Colo.

Denna, D. W. and Merry B. Alexander 1973. Simultaneous electrophoretic separation of anionic and cationic peroxidase isozymes in polyacrylamide slabs. *Techniques for High Resolution Electrophoresis*. ORTEC Incorp., Oak Ridge, Tenn.

Hamill, D. E. and J. L. Brewbaker 1969. Isoenzyme polymorphism in flowering plants. IV. The peroxidase isoenzymes of maize (*Zea mays*). *Physiol. Plant.* 22: 945-958.

Harkin, J. M. and J. R. Obst 1973. Lignification in trees: Indication of exclusive peroxidase participation. *Science* 180: 296-298.

Jensen, W. A. 1962. *Botanical Histochemistry*. W. A. Freeman and Co., San Francisco.

Loy, J. B. 1972. A comparison of stem peroxidases in bush and vine forms of squash (*Cucurbita maxima* Duch. and *C. pepo*), *J. Exp. Bot.* 23: 450-457.

Peirce, L. C. and J. L. Brewbaker 1973. Application of isozyme analysis in horticultural science. *Hort. Sci.* 8: 17-23.

Ridge, Irene and Daphne Osborne 1972. Role of peroxidase when hydroxyproline-rich protein in plant cell walls is increased by ethylene. *Nature* 229: 205-209.

Saunders, B. C., A. G. Holmes-Siedle, and B. P. Stark 1964. *Peroxidase*. Butterworth, Washington.

Shannon, L. M., E. Kay, and J. Y. Lew 1966. Peroxidase isozymes from horseradish roots. I. Isolation and physical properties. *J. Biol. Chem.* 241: 2166-2172.

Siegel, Barbara Z. 1966. The molecular heterogeneity of plant peroxidases. Doctoral Thesis, Yale University, New Haven, Conn.

Siegel, Barbara Z. and A. W. Galston 1967. Indoleacetic acid oxidase activity of apoperoxidase. *Science* 157: 1557-1559.

Siegel, S. M. 1962. *The Plant Cell Wall*. Macmillan Co., New York.

Whitaker, D. R. and R. Thomas 1962. Improved procedures for preparation and characterization of *Myrothecium* cellulase. *Canad. J. Biochem. and Phy.* 41: 667-696.

TWO PAIRS OF ISOZYMES IN THE AROMATIC BIOSYNTHETIC AND CATABOLIC PATHWAYS IN *NEUROSPORA CRASSA*

NORMAN H. GILES and MARY E. CASE
Department of Zoology
University of Georgia
Athens, Georgia 30602

ABSTRACT. In *Neurospora crassa* two early reactions in the constitutive common aromatic (polyaromatic) synthetic pathway exhibit interesting interrelationships with comparable early reactions in the inducible quinate-shikimate (hydroaromatic) catabolic pathway. Genetical and biochemical evidence indicates that this organism possesses two pairs of isozymes in the two pathways--two dehydroquinases and two shikimate oxidoreductases. Normally, one member of each pair functions in the biosynthetic pathway and the other in the catabolic pathway. However, under certain conditions, the two DHQases can substitute for one another in vivo, while the catabolic (NAD) oxidoreductase can substitute for the biosynthetic (NADP) oxidoreductase. Even mutants lacking both biosynthetic enzymes can (under certain conditions) utilize the two catabolic isozymes to grow without a polyaromatic supplement. These results indicate interesting functional similarities between the two pairs of enzymes and suggest possible evolutionary relatedness. However, at this stage in the evolution of the *N. crassa* genome, the two pairs of activities are encoded in quite separate genetic material--two unlinked gene clusters (the *arom* and *qa* clusters) and the two biosynthetic activities are part of a multienzyme complex (the *arom* complex). At present no evidence exists concerning possible amino acid homologies between the two pairs of isozymes.

INTRODUCTION

The common aromatic (polyaromatic) biosynthetic pathway in *Neurospora crassa* involves seven reactions between the two precursors, phosphoenolpyruvic acid and erythrose-4-phosphate, and the branch point compound, chorismic acid. This pathway is constitutive in *Neurospora crassa* and reactions two through six are catalyzed by five enzymes which are encoded in five structural genes in the *arom* gene cluster and comprise the *arom* multienzyme complex (Giles et al, 1967a). The related hydroaromatic pathway involves the catabolism of quinate and shikimate. The initial reactions in this pathway are catalyzed by three inducible enzymes which are encoded in three structural genes in the qa gene cluster. These three enzymes are under the positive control of a regulatory protein encoded in the $qa-1^+$ gene which is also part of the qa gene cluster. $qa-1$ mutants are non-inducible for all three qa enzymes. (Giles, Case, and Jacobson, 1973).

In this paper we discuss two pairs of isozymes in these two pathways. Under certain conditions (typically, the presence of a mutant gene resulting in the formation of an inactive enzyme) one member of an enzyme pair can substitute for the other member. The pertinent interrelationships of the two pathways are shown in Fig. 1.

THE TWO DEHYDROQUINASE ISOZYMES

The first pair of isozymes catalyzes the conversion of dehydroquinic acid (DHQ) to dehydroshikimic acid (DHS). This reaction occurs in both pathways and involves two distinct dehydroquinases: a biosynthetic, constitutive activity (5-dehydroquinate dehydratase, EC 4.4.1.10) (B-DHQase) and a catabolic inducible dehydroquinase (C-DHQase).

The first indication that $Neurospora$ might contain two dehydroquinases having similar specificities stemmed from our initial failure to recover mutants lacking only biosynthetic dehydroquinase activity among many hundreds of polyaromatic auxotrophs (Giles et al, 1967a). The recovery of other mutants lacking all five activities present in the $arom$ multienzyme complex (aggregate), including biosynthetic dehydroquinase, indicated that the $arom$ gene cluster should contain a gene encoding that activity. Biochemical studies soon indicated the presence in $N.$ $crassa$ of two clearly distinguishable dehydroquinases (Giles et al, 1967b). Additional studies also provided evidence for a previously undetected hydroaromatic pathway involving the catabolism of quinate and shikimate which has one step in common with the biosynthetic pathway, i.e., the conversion of DHQ to DHS (Fig. 1).

The hypothesis has been proposed (Giles et al, 1967a) that a primary function of the $arom$ multienzyme complex is to channel intermediates common to the two pathways, e.g., DHQ and DHS, which would otherwise act as inducers of the potentially competing catabolic pathway. The three quinate-shikimate catabolic activities, which are encoded in a second gene cluster--the qa cluster--are present in wild type (WT) at only very low levels in the absence of an added external inducer such as quinic acid (QA). However, induction does occur if an internal inducer accumulates as a result of a mutational block in the biosynthetic pathway, e.g., in $arom-1$ mutants which lack DHS reductase activity and accumulate DHS and DHQ.

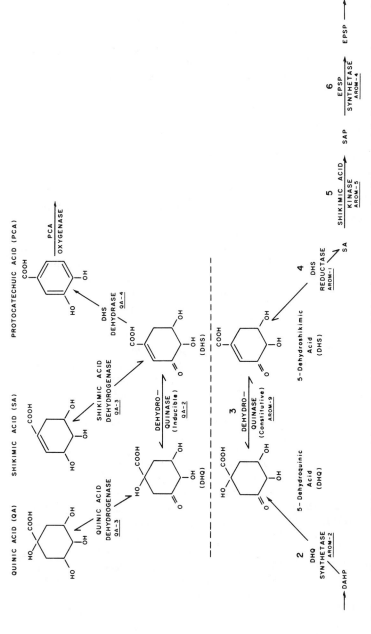

Fig. 1. Diagram of the reactions in the constitutive common (polyaromatic) synthetic pathway (below) and in the inducible quinate-shikimate catabolic pathway (above), indicating the metabolic interrelationships in *Neurospora crassa*.

By contrast, *arom-2* mutants, which lack DHQ synthetase, do not accumulate DHQ and have only wild type levels of C-DHQase. These observations indicated an interrelationship between the two pathways. Further evidence for an interrelationship came from the demonstration that the two isozymes, which have been shown to carry out the same catalytic reaction in vitro (i.e., the conversion of DHQ to DHS) can, under certain conditions, substitute for one another in vivo, despite their normal effective compartmentalization in wild type.

As mentioned previously, initial efforts to obtain mutants in the *arom* cluster lacking only B-DHQase were unsuccessful. The discovery of an inducible C-DHQase appeared to explain this observation since the assumption could be made that mutants lacking only B-DHQase would accumulate DHQ, induce the C-DHQase which could substitute for B-DHQase in vivo, and thus not be recovered. Proof of this assumption required the development of a method for recovering *arom* mutants lacking only B-DHQase. The fortunate detection of a mutant non-inducible for C-DHQase (as well as for the other two *qa* enzymic activities) did, in fact, make possible the recovery of such mutants (Rines, Case, and Giles, 1969). These *arom-9* mutants lack B-DHQase, have high levels of C-DHQase and can grow well on minimal medium (with no polyaromatic supplement). This result suggested that C-DHQase can substitute for B-DHQase, thus permitting the biosynthesis of the end products of the common aromatic pathway. Final proof for this conclusion came with the recovery of mutants specifically non-inducible for C-DHQase activity. Utilizing *arom-9* mutants, it was possible to select by filtration enrichment on sucrose minimal medium for mutants in the quinate pathway lacking C-DHQase activity (Rines, 1968a). Two types of mutants were recovered: (1) *qa-1* mutants--pleiotropic types non-inducible for C-DHQase and for the other two enzymes in the early part of quinate-shikimate pathway; (2) *qa-2* mutants--non-inducible for C-DHQase only. Additional experiments employing filtration enrichment on QA as a sole carbon source resulted in the recovery of additional *qa-1* and *qa-2* mutants, and of two additional types: (3) *qa-3* mutants--non-inducible for both QDHase and SDHase, and (4) *qa-4* mutants--non-inducible for dehydroshikimate dehydrase (DHS-Dase) (Chaleff, 1974) (Fig. 1).

The double mutants *arom-9 qa-2* (and also *arom-9 qa-1*) lack in vitro activity for both dehydroquinase isozymes and fail to grow on sucrose minimal medium (and also on QA as a sole carbon source). These results supply the final proof that C-DHQase can substitute in vivo for B-DHQase in the common aromatic pathway.

The reciprocal relationship also holds, since *qa-2* mutants lacking C-DHQase can grow on QA as a sole carbon source if B-DHQase activity is present ($arom-9^+$). On the basis of comparative growth data, C-DHQase is more efficient in substituting for B-DHQase than the reverse (Table 1), since the three-day dry weight of $arom-9$ mutants on minimal medium is essentially equivalent to that of wild type, while $qa-2$ mutants exhibit only ca 15% of the three-day dry weight of wild type when grown on QA as a sole carbon source.

TABLE 1

Comparative growth (dryweight) as per cent of wild type of an $arom-9$ mutant (lacking B-DHQase), a $qa-2$ mutant (lacking C-DHQase), and a $qa-1$ mutant (lacking all three qa enzymes).*

| Strain | Enzymic Activity | | Growth Medium | |
	B-DHQase	C-DHQase	Sucrose Minimal	Minimal Quinic Acid
WT	+	+	100	100
$arom-9$ $qa-2^+$	0	+	92	110
$arom-9^+$ $qa-2$	+	0	84	15
$arom-9$ $qa-2$	0	0	0	0
$arom-9^+$ $qa-1$	+	0	116	0

* Strains were grown in duplicate in 20 ml of Fries minimal medium in 125 ml Erlenmeyer flasks at 25° for 3 days; sucrose = 1.5%; Q.A. = 0.3% as sole carbon source.

Despite the fact that these two dehydroquinase isozymes can catalyze the same reaction in vitro and substitute for one another in vivo, they can be clearly distinguished on the basis of several different criteria: physical, biochemical, immunological, and genetical. They are clearly very different proteins which separate completely upon purification. The biosynthetic activity is one of the five enzymic components of the $arom$ biosynthetic multienzyme complex which has a molecular weight of ca 230,000 (Burgoyne, Case, and Giles, 1969; Jacobson et al, 1972). The catabolic activity, which is not physically associated (at least in vitro) with the two enzymes catalyzing adjacent reactions in quinate-shikimate catabolism, has an apparent molecular weight of ca 160,000. On the basis of present evidence (Jacobson, Case, and Giles, 1973), the catabolic enzyme is a multimer consisting of a single type of subunit having a remarkably low molecular weight. The C-DHQase is very heat stable, which serves as a useful method of distinguishing it from B-DHQase, and has greatly facilitated its purification. The Michaelis constants for the two isozymes have been determined under standard assay conditions from Linewever-Burke plots by Rines (1968b). The apparent Km value of the catabolic enzyme (0.36 mM) was found to be almost twenty times higher than that of the biosynthetic enzyme (0.02M). Antibodies formed to purified preparations of the catabolic enzyme do not cross react with purified *arom* aggregate protein possessing high B-DHQase activity (Hautala, Jacobson, and Giles, unpublished).

The two isozymes are encoded in quite unrelated structural genes: the *arom-9* gene within the *arom* gene cluster in linkage group II encodes the biosynthetic activity, whereas the *qa-2* gene within the *qa* gene cluster in linkage group VII encodes the catabolic activity. The two isozymes are also under very different genetic regulatory control. The biosynthetic activity is constitutive and to date there is no clear evidence as to the genetic mechanisms involved in regulating either the synthesis or repression of this enzyme. By contrast, the catabolic activity is highly inducible and its synthesis is under the positive control of a regulatory protein encoded in the $qa-1^+$ gene, which is also within (at one end of) the *qa* cluster (Giles, Case, and Jacobson, 1973).

THE TWO SHIKIMATE OXIDOREDUCTASE ISOZYMES

The second pair of isozymes is involved in the metabolism of shikimic acid (SA). In the biosynthetic pathway the enzyme DHS reductase (DHS-Rase) (shikimate:NADP oxidoreductase, EC 1.1.1.25) catalyzes the conversion of DHS to SA, whereas in the catabolic pathway the enzyme shikimate dehydrogenase (SDHase) (quinate [shikimate]: NAD oxidoreductase EC 1.1.1.24), which functions as both a shikimate and a quinate dehydrogenase, catalyzes the conversion of SA to DHS. For both activities, assays are conducted in the direction of shikimate dehydrogenation, and the enzymes can also be referred to as catabolic shikimate dehydrogenase and anabolic (or biosynthetic) dehydroshikimate reductase (measured as a shikimate dehydrogenase).

Evidence that the catabolic activity (SDHase) can substitute in vivo for the biosynthetic activity (DHS-reductase) was an unexpected result of studies of spontaneous revertants occurring with a high frequency on sucrose minimal medium without a complete polyaromatic supplement (lacking tryptophan) in *arom-1* mutants (which lack DHS reductase). Earlier studies of

Gross and Fein (1960) had indicated that many *arom-1* revertants
were not back mutants but resulted from mutations at an unlinked locus, thus representing some type of "suppressor"
mutant. Combined genetical and biochemical studies (Case,
Giles, and Doy, 1972) have shown that these *arom-1* "revertants"
are actually mutants within the *qa-4*+ gene (which encodes DHS-
Dase) and are thus double mutants (*arom-1 qa-4*) lacking both
DHS reductase and DHS-Dase activities (cf Fig. 1). Despite
the absence of both these activities, the double mutants are
prototrophs and can grow on sucrose minimal medium without a
polyaromatic supplement. This situation is true for double
mutants selected as revertants as well as for double mutants
obtained as recombinants from crosses of the two single
mutants. The data indicate that, although various *arom-1 qa-4*
double mutants selected as "revertants" may differ markedly in
their growth rates on sucrose minimal medium compared to wild
type, all doubles tested to date (including ones obtained as
recombinants) grow more slowly than does wild type (Table 2).

TABLE 2

Comparative growth (dry weight) on sucrose minimal medium
as per cent of wild type (WT) of single *arom-1* mutants and
qa-4 mutants (lacking DHS-dehydrase: DHS-Dase) and of various *arom-1 qa-4* double mutants.*

Strains	Enzymic Activity		Growth as % of WT	
	DHS-Rase	DHS-Dase	2 Days	3 Days
WT	+	+	100	100
arom-1+ *qa-4* (M18)	+	0	108	96
arom-1 qa-4+	0	+	0	0
arom-1 qa-4	0	0		
1135-R3†			2	14
1183-R2†			2	38
7655-R6†			44	90
1135-M18-7.7‡			16	78

* Strains were grown in duplicate on 20 ml of Fries minimal
  medium (1.5% sucrose) in 125 ml Erlenmeyer flasks at 25°C.

† Double mutants selected as *arom-1* prototrophic revertants.

‡ Double mutant obtained from an *arom-1* x *qa-4* (M18) cross.

The results just described suggest that the double mutants, although lacking DHS-reductase activity, can convert DHS to SA in the biosynthetic pathway since they require no tryptophan supplement. If this suggestion is correct, how does the *loss* of an enzymic activity in the shikimate catabolic pathway permit mutants lacking an enzyme in the common aromatic pathway to grow without a complete polyaromatic supplement? All available evidence indicates the SDHase is the enzymic activity which is replacing DHS reductase in these double mutants. The major initial evidence for this conclusion came from reversion studies with various *arom-1 qa* double mutants. As shown in Table 3, an active *qa-3*[+] gene (which encodes SDH-QDHase) must be present in such double mutants in order for *arom-1* reversions to be recovered. The presence of either a *qa-3* mutant (which lacks SDHase activity) or of a *qa-1* regulatory mutant (in which all three *qa* catabolic enzymes are non-inducible) prevents the recovery of *arom-1* revertants. By contrast, an *arom-1 qa-2* double mutant reverts with a high frequency and combined genetical and biochemical evidence indicates that these "revertants" are, in fact, triple mutants of the genotype *arom-1 qa-2 qa-4*.

The ability of SDHase to substitute for DHS reductase in double mutants of the genotypes *arom-1 qa-4* can be interpreted on the basis of the intermediates expected to accumulate in such double mutants. Single *arom-1* mutants are known to accumulate DHS which induces and is metabolized by DHS-Dase to protocatechuic acid (PCA) (Gross and Fein, 1960). *Arom-1 qa-4* double mutants should result in a still greater accumulation of DHS since little or no conversion to PCA can occur (depending on the leakiness of the mutants). This greater accumulation of DHS causes the induction of a high level of SDHase activity and by mass action results in a net conversion of DHS to SA by SDHase, which thus substitutes for DHS reductase and supplies the aromatic requirement of the *arom-1* mutants.

Especially strong support for the interpretation just presented comes from reversion studies with other mutants in the *arom* gene cluster. In addition to *arom-1*, the only other auxotrophic *arom* mutants which yield prototrophic revertants of the "suppressor" type (*qa-4* doubles) are the pleiotropic A-C mutants which lack both DHS-reductase and B-DHQase activities. In *arom A-C qa-4* double mutants, the *qa-4* mutation results in protrophy since the excess accumulation of DHQ (and/or DHS) leads to the induction of both C-DHQase--which substitutes for B-DHQase, as discussed in the first part of the paper--and SDHase--which substitutes for DHS reductase. In this remarkable strain both biosynthetic isozymes are substituted for by their respective catabolic enzymes although growth is consider-

TABLE 3

Recovery of spontaneous prototrophic revertants on sucrose minimal medium without a complete polyaromatic supplement in *arom-1* and various *arom-1 qa* double mutants

Strain	Growth on minimal	Recovery of reversions
arom-1 qa-3[+]		
1135	0	++
1183	0	+
7655	0	++
arom-1 qa-3 (M16)		
1135	0	0
1183	0	0
arom-1 qa-1 (*qa-3*[+] inactive)		
1135	0	0
1183	0	0
arom-1 qa-3[+] *qa-2*		
1135	0	++
1183	0	+
arom-1 qa-4	+	--

ably less than in comparable *arom-9* and *arom-1 qa-4* strains. All other auxotrophic *arom* mutants are blocked at one or more other steps in the common aromatic pathway indicating that these reactions cannot be substituted for by "suppressors" of the *arom-1* type and that an alternative aromatic biosynthetic pathway is not present in *Neurospora*.

Contrary to the situation with the dehydroquinase isozymes, present evidence indicates that a reciprocal relationship does not exist with the shikimate oxidoreductase isozymes, i.e., that the normal biosynthetic enzyme (DHS reductase) cannot substitute in vivo for the normal catabolic enzyme (SDHase). This conclusion is based on data indicating that *qa-3 arom-1*[+] mutants do not exhibit significantly greater growth on shikimate as a sole carbon source than do *qa-3 arom-1* double mutants, which are only slightly leaky at most.

The evidence discussed previously indicates that the two shikimate oxidoreductase isozymes are functionally related in

873

that the C-DHQase can substitute for the B-DHQase in vivo, at least under certain special conditions. Despite this fact, it is clear that, as in the case of the dehydroquinase isozymes, the two enzymes are clearly distinguishable on the basis of several criteria. The two proteins separate completely upon purification. DHS reductase is a component of the *arom* multi-enzyme complex and is encoded in the *arom-1* gene within the *arom* gene cluster. SDHase, which is encoded in the *qa-3*[+] gene, has not been substantially purified but appears not to be associated physically, at least in vitro, with the other two enzymes encoded in the *qa* gene cluster. It has a relatively low molecular weight (ca 30,000) and may occur as a multimer, although the biochemical evidence and that from allelic complementation tests with the available *qa-3* mutants have not yet provided definitive evidence on this point. Both biochemical (partial purification) and genetical (mutational) studies indicate that SDHase and QDHase activities are present on the same protein. To date, there is no evidence to indicate that more than one active site is present. Normally, the catabolic enzyme (SDHase) used NAD as a co-factor and the bio-synthetic enzyme (DHS-reductase) used NADP. Thus an interesting biochemical feature of these results is that in the *arom-1* *qa-4* prototrophic double mutant, an NAD-specific enzyme (SDH-Dase) is being used for reductive biosynthesis. Since NADPH is normally used for reductive anabolic reactions, including the normal aromatic biosynthetic pathway in *N. crassa*, it is noteworthy that this organism can utilize an NADH enzyme instead. The reverse possibility appears not to exist, since, on present evidence, the NADP-specific enzyme appears incapable of substituting for the NAD enzyme in the oxidative catabolism of SA.

In conclusion, these studies provide evidence for interesting interrelationships between the constitutive common aromatic synthetic and the inducible quinate-shikimate catabolic pathways in *Neurospora crassa*. This organism possesses two pairs of isozymes in the two pathways--two dehydroquinases and two shikimate oxidoreductases. Normally, one member of each pair functions in the biosynthetic pathway and the other in the catabolic pathway. However, under certain conditions, the two DHQases can substitute for one another in vivo, while the catabolic (NAD) oxidoreductase can substitute for the biosynthetic (NADP) oxidoreductase. These results indicate interesting functional similarities between the two pairs of enzymes and suggest possible evolutionary relatedness. However, at this stage in the evolution of the *N. crassa* genome, the two pairs of activities are encoded in quite separate genetic material--two unlinked gene clusters (the *arom* and *qa* clusters) and the two biosynthetic activities are part of a multienzyme complex (the *arom* complex). Obviously, much additional evidence--some of which will, no doubt, be difficult to obtain--will be required to determine whether significant amino acid sequence homologies exist between the two pairs of enzymes.

ACKNOWLEDGMENTS

This research was supported by Atomic Energy Commission Contract AT(38-1)-735. The authors wish to express their appreciation to Dr. Robert Metzenberg for a critical reading of the manuscript.

LITERATURE CITED

Burgoyne, L., M. E. Case, and N. H. Giles 1969. Purification and properties of the aromatic (*arom*) synthetic enzyme aggregate of *Neurospora crassa. Biochim. Biophys. Acta* 191: 452-462.

Case, M., N. H. Giles, and C. H. Doy 1972. Genetical and biochemical evidence for further interrelationships between the polyaromatic synthetic and the quinate-shikimate catabolic pathways in *Neurospora crassa. Genet.* 71: 337-348.

Chaleff, R. S. 1974. The inducible quinate-shikimate catabolic pathway in *Neurospora crassa.* I. Genetic organization. *J. Gen. Microbiol.* 81: 337-355.

Giles, N. H., M. E. Case, C. W. H. Partridge, and S. I. Ahmed 1967a. A gene cluster in *Neurospora crassa* coding for an aggregate of five aromatic synthetic enzymes. *Proc. Natl. Acad. Sci.* 58: 1453-1460.

Giles, N. H., C. W. H. Partridge, S. I. Ahmed, and M. E. Case 1967b. The occurrence of two dehydroquinases in *Neurospora crassa*, one constitutive and one inducible. *Proc. Natl. Acad. Sci.* 58: 1930-1937.

Giles, N. H., M. E. Case, and J. W. Jacobson 1973. Genetic Regulation of quinate-shikimate catabolism in *Neurospora crassa. Molecular Cytogenetics* (Hamkalo and Papaconstantinou, eds), Plenum Press, 309-314.

Gross, S. R. and A. Fein 1960. Linkage and function in *Neurospora. Genet.* 45: 885-904.

Jacobson, J. W., B. A. Hart, C. H. Doy, and N. H. Giles 1972. Purification and stability of the multienzyme complex encoded in the *arom* gene cluster of *Neurospora crassa. Biochim. Biophys. Acta* 289: 1-12.

Jacobson, J. W., M. E. Case, and N. H. Giles 1973. Purification and properties of catabolic dehydroquinase produced by wild type and various mutants in the *qa* gene cluster of *Neurospora crassa*. *Genet*. 74: S125 Abs.

Rines, H. W. 1968a. The recovery of mutants in the inducible quinic acid catabolic pathway in *Neurospora crassa*. *Genet*. 60: 215 Abs.

Rines, H. W. 1968b. Genetical and biochemical studies on the inducible quinic acid catabolic pathway in *Neurospora crassa*. Ph.D. thesis (unpublished).

Rines, H. W., M. E. Case, and N. H. Giles 1969. Mutants in the *arom* gene cluster of *Neurospora crassa* specific for biosynthetic dehydroquinase. *Genet*. 61: 789–800.

SEPARATION OF ATYPICAL AND USUAL FORMS OF HUMAN SERUM CHOLINESTERASE BY AFFINITY CHROMATOGRAPHY

B.N. LA DU AND Y.S. CHOI
Guttman Laboratory for Human Pharmacology and Pharmacogenetics
Department of Pharmacology, New York University
New York, New York   10016

ABSTRACT. Complete separation of a mixture of usual and atypical human cholinesterases has been achieved using an affinity chromatographic column containing a diamino-caproyl-phenyltrimethyl ammonium ligand. Resolution of the two cholinesterases was due to qualitative differences in the active sites of the usual and the variant esterases for the column ligand. Serum cholinesterase from an individual heterozygous for the atypical esterase gene separated quite differently than it did from a mixture of atypical and usual sera. Rather than eluting as two distinct peaks of esterase activity, the natural heterozygous sample separated as a series of esterases with progressively higher dibucaine numbers. From the pattern and properties of the component esterases, it is concluded that the major molecular form of serum cholinesterase in serum, C_4, is a tetramer, and the serum of heterozygotes contains five tetrameric molecular species, produced by random combination of usual and atypical esterase subunits. Evidence for the identity of such tetramers of mixed subunit composition was obtained by rechromatography of the pooled fractions believed to contain mainly A_2U_2. This component eluted as a single species at the expected position, and had a dibucaine number of 60, as predicted.

Affinity chromatography may be useful in the resolution of normal and variant enzymes from heterozygous individuals for further kinetic investigations, particularly in instances where a mixture of usual and variant enzymes cannot be separated by electrophoresis, or by Sephadex chromatography.

INTRODUCTION

Atypical serum cholinesterase, a variant form of the enzyme, is associated with an inherited sensitivity to the muscle relaxant drug, succinylcholine. Approximately 1:2,500 individuals is homozygous for the atypical cholinesterase gene and would develop paralysis of the respiratory muscles and require artificial respiration for several hours after an operation if given the usual dose of succinylcholine (Kalow, 1962).

The atypical esterase differs from the usual serum cholinesterase in several respects: it has a lower apparent affinity for choline ester substrates; it also has a lower affinity for a number of competitive inhibitors, including dibucaine; and it has a lower turnover number with benzoylcholine as the substrate. These differences suggest that the variant enzyme differs in structure at, or near the active center, perhaps by only a single amino acid substitution. The atypical esterase exists in the same pattern of multiple molecular forms as the usual enzyme, (one major component and three minor bands) and these can be separated by either starch gel or acrylamide electrophoresis. However, the zymogram pattern of atypical-usual heterozygous serum shows only 4 bands of esterase activity, rather than 8, as would be expected if the molecular species of usual and atypical esterases differed appreciably in their electrophoretic migration under standard conditions. A mixture of usual and atypical esterases cannot be separated by the usual column chromatography techniques. It was once reported (Liddell et al., 1962) that at very high pH, separation of usual and atypical cholinesterases was achieved by paper chromatography and by chromatography on DEAE-cellulose; but other laboratories, including our own, have not been able to confirm these results. It is possible that slight differences in net charge between usual and atypical esterases would permit some resolution of these enzymes if experimental conditions could be selected to take advantage of these charge differences.

MATERIALS AND METHODS

Sepharose (Pharmacia) 4B was activated with cyanogen bromide and coupled with the ligand, N(6-aminocaproyl-6'-aminocaproyl)-p-aminophenyl -trimethylammonium bromide hydrobromide as described by Rosenberry et al., 1972. For the experiments described here, Sepharose-ligand with a ratio of 1.0 μmole of ligand per ml of gel was used. We gratefully acknowledge samples of ligand-Sepharose supplied by Dr. Rosenberry, and of ligand generously provided by Dr. John J. Burns of Roche Research Laboratories, Hoffmann-La Roche Inc., Nutley, New Jersey.

Serum samples (4-6 ml) were dialyzed for 48 hours against 0.02 M phosphate buffer, pH 6.9, before application to a 1 x 10 cm column containing approximately 10 ml of the ligand-Sepharose gel in a 3-5° cold room. After the dialyzed serum

was applied to the affinity column, the system was allowed
to equilibrate overnight before addition of elution buffer.
Elution of the proteins by 0.02 M potassium phosphate buffer,
pH 6.9, was followed by the stepwise addition of the same
buffer containing added NaCl to increase the ionic strength.
Flow rate was approximately 4 ml per 30 min, fractions of
approximately 1.5 ml were collected, and protein elution was
monitered by following the 280 nm absorbance of each fraction.
Cholinesterase activity was assayed by the spectrophotometric
method of Kalow and Lindsay (1955) and the dibucaine numbers
were determined by a microadaptation of the method of Kalow
and Genest (1957).

RESULTS

SEPARATION OF USUAL AND ATYPICAL CHOLINESTERASES

Dialyzed sera from usual and atypical homozygous individuals (2.5 ml of each) were mixed and immediately applied to
a ligand-Sepharose column, equilibrated overnight, and elution
aliquots of approximately 1.0 ml were collected the next day
(Fig. 1). Coincident with the elution of the major protein
peak was a large peak of cholinesterase activity with atypical
esterase characteristics (dibucaine number, 18). After the
protein concentration in the eluates had decreased nearly to
baseline values, the elution buffer was changed to one containing 0.7 M NaCl. A second peak of cholinesterase activity
followed, and this enzyme had the characteristics of usual
cholinesterase (dibucaine number, 78). Over 85% of the total
esterase activity applied to the column was recovered, and
the kinetic properties of the fractions identified as atypical
and usual esterase peaks agreed very closely with those of
the corresponding original sera (Table 1).

The observation that atypical cholinesterase eluted
before the usual cholinesterase was not unexpected in view of
the lower apparent affinity of atypical esterase for choline
ester substrates and for most inhibitors (Harris, 1970).

Although in the above experiment the separately dialyzed
sera were mixed just before application to the column, we
obtained the same resolution when the mixture was stored for
5 days at 4^0 before chromatography. More vigorous treatments
to try to produce molecular species of mixed composition,
however, were not attempted.

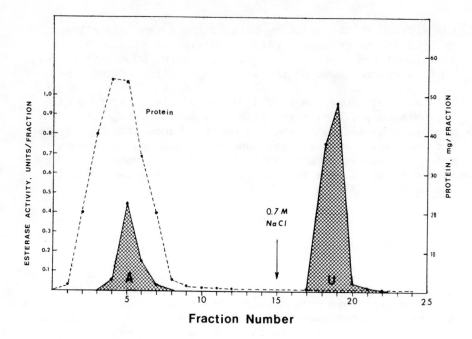

Fraction Number

Fig. 1. Separation of usual (U) and atypical (A) cholinesterase activity from a mixture of dialyzed sera (2.5 ml of each)
from homozygous U and homozygous A individuals. The dibucaine
number of the mixture was 65; for the A serum, 17; and for the
U serum, 78. The dibucaine numbers of the elution fractions
were as follows: fraction 4=20, 5=18, 6=20; and fractions
18=79, 19=78, 20=73 and 21=75.

---

*ELUTION PATTERN OF CHOLINESTERASE ISOZYMES FROM
AU-HETEROZYGOTE SERUM*

Dialyzed serum (4.0 ml) from an individual known to be
heterozygous for the atypical cholinesterase gene by pedigree
and by dibucaine and fluoride inhibition characteristics
(dibucaine number, 60; fluoride number, 52), was applied to
a ligand-Sepharose column, equilibrated overnight, and eluted
as described under Methods except that 0.2 M NaCl was used
in the eluant buffer to displace the more tightly bound esterase during the later phase of the chromatographic separation
(Fig. 2).

It was apparent that serum from a natural AU heterozygote
gave a distinctly different elution pattern than an artificial

TABLE I

RECOVERY AND KINETIC PROPERTIES OF USUAL AND ATYPICAL
CHOLINESTERASES SEPARATED BY AFFINITY CHROMATOGRAPHY

Before mixing sera	Activity Units	K_m for benzoylcholine mM
2.5 ml usual serum	2.35	0.0036
2.5 ml atypical serum	0.725	0.0238

Recovery from column			
Tubes	Activity Units	%	K_m for benzoylcholine mM
Usual activity (18-21)	1.94	82.6	0.0036 (tube 19)
Atypical activity (4-7)	0.79	109.3	0.0261 (tube 5)

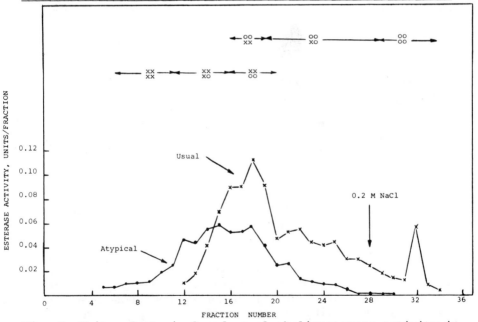

Fig. 2. Units of atypical and usual cholinesterase activity in
each fraction eluted from the column after applying 4.0 ml of
dialyzed serum from an AU-heterozygous individual with a dibucaine number of 60. Calculated proportions of A and U activity were based upon the dibucaine number of the fraction and
the assumption that dibucaine numbers of 20 and 78 represent
100% A and U activity, respectively. The proposed zones for

Legend for Fig. 2, continued.
different tetrameric species (O=U subunit, X=A subunit) shown
above were calculated as explained in the text.

---

mixture of atypical and usual sera. The dibucaine numbers of
subsequent eluates increased gradually from about 15 to 78,
and most of the esterase activity was eluted before the NaCl-
buffer addition. The natural heterozygote elution profile
could be explained by the existence of tetrametric species of
the enzyme composed of usual and atypical subunits in various
proportions. Calculations of the relative percentage of
atypical and usual esterase activity present in each eluant
fraction was made from the observed dibucaine number of each
tube, based upon dibucaine numbers of 20 and 78 representing
100% A and U activity, respectively.

The units of A and U activity in each fraction, determined from the dibucaine number and total activity per fraction,
are shown in Fig. 2. If it were assumed that the five possible tetrameric species (A_4, A_3U, A_2U_2, AU_3, U_4) were formed
by random combination of A and U subunits and the latter were
about equal in number, A activity should be distributed within
four of the tetrameric species (A_4, A_3U, A_2U_2, AU_3) in a ratio
of 1:3:3:1. A similar distribution of U activity would be
expected among the tetramers U_4, U_3A, U_2A_2, and UA_3. Division
of the total recovered A and U activity in accord with the
above distribution pattern (1:3:3:1) gave the same transition
zones between tetrameric species (Fig. 2). Furthermore, a
dibucaine number could be calculated for each possible tetramer,
and these could be inserted between each transition zone to
illustrate a model in which each tetrameric form was completely separated from the others, if they were sequentially eluted
from the column. This theoretical model is given with the
experimental data (dibucaine number for each tube) in Fig. 3.
Except for some obvious deviation in A_3U, the experimental
values are in reasonable accord with a proposed tetrameric
structure for serum cholinesterase. Other possible isozymic
patterns, such as dimers, or tetramers with only two active
sites, have been considered. While such other possibilities
cannot be completely excluded, the experimental data are in
closest agreement with the tetrameric distribution. The
suggestion that human serum cholinesterase is a tetramer was
proposed earlier (La Du and Dewald, 1971) and evidence for
the tetrameric structure was obtained by Scott and Powers
(1972).

In accord with the expectation that discrete tetramers,
for example, A_2U_2, are present in the fractions, we would

presume that fractions 17 through 21 (Fig. 3) contain primarily this particular tetramer. Those fractions, 17-21,

THEORETICAL AND EXPERIMENTAL DIBUCAINE NUMBERS

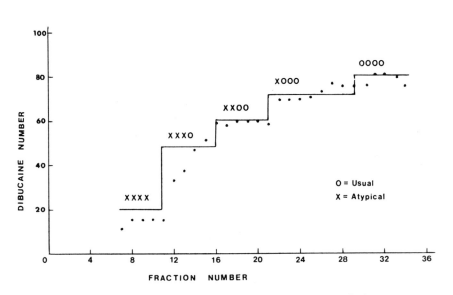

Fig. 3. Theoretical dibucaine numbers and transition zones
for the 5 possible tetramers (O = U subunit, X = A subunit)
from Figure 2, compared with the observed dibucaine number
for each fraction.

were pooled and rechromatographed to see whether the dibucaine
number would be relatively constant or vary over a wide range,
like the heterozygous AU serum sample. As can be observed
in Fig. 4, over 90% of the eluted cholinesterase activity
distributed over 29 fractions had a dibucaine number of approximately 60, the theoretical dibucaine number for A_2U_2
tetramer.

In order to demonstrate the sequential elution of different tetramers more clearly, a larger dialyzed serum sample
(10 ml) from the same AU heterozygous individual was chromatographed on a longer column (1 X 30 cm) containing the same
ratio of ligand per ml of gel as used before. The proportion
of the esterase activity in each fraction assigned to each
of the respective tetramers was calculated from the observed
dibucaine number and the theoretical dibucaine numbers for

883

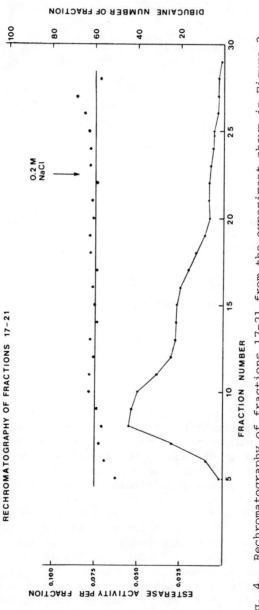

Fig. 4. Rechromatography of fractions 17-21 from the experiment shown in Figure 2. The expected dibucaine number of 60 for the A₂U₂ tetramer is indicated by the horizontal line. Actual dibucaine number and cholinesterase activity are plotted for each fraction.

each tetrameric species, based upon the values for the terminal homologous tetramers, U_4 and A_4, found to be 78 and 15, respectively (Fig. 5).

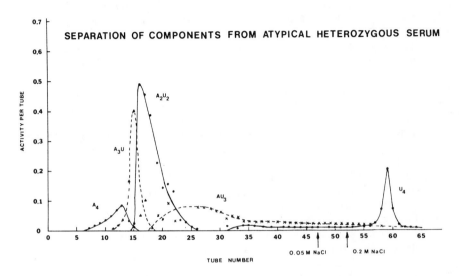

Fig. 5. Pattern of tetramer elution from 10 ml of the same AU-heterozygous serum as used in Figure 2 on a longer affinity chromatography column (1 X 30 cm). Assignment of the tetrameric composition of each tube was based upon the dibucaine number and total activity of each fraction and the calculated dibucaine numbers for each possible tetramer. The homologous tetramers at the beginning and end, A_4 and U_4, had dibucaine numbers of 15 and 78, respectively.

Although the tetrameric patterns are illustrated more distinctly by this experiment, it is apparent that better resolution can be achieved by minor modifications of the affinity chromatographic technique. By rechromatography of selected fractions it should now be possible to isolate and characterize the specific isozymic components of serum cholinesterase.

ACKNOWLEDGMENTS

This project was supported by PHS Grant GM-17184.

REFERENCES

Harris, H. 1970. *The Principles of Human Biochemical Genetics*. North Holland Publishing Co., London. pp. 109-120.

Kalow, W. 1962. *Pharmacogenetics. Heredity and the Response to Drugs*. W.B. Saunders Co., Philadelphia. pp. 69-93.

Kalow W. and K. Genest 1957. A method for the detection of atypical forms of human cholinesterases. Determination of dibucaine numbers. *Canad. J. Biochem. Physiol.* 35: 339-346.

Kalow, W. and H.A. Lindsay 1955. A comparison of optical and manometric methods for assay of human serum cholinesterase. *Canad. J. Biochem.* 33:568-574.

La Du, B.N. and B. Dewald 1971. Genetic regulation of plasma cholinesterase in man. *Adv. Enzyme Regulation.* 9:317-332.

Liddell, J., H. Lehmann, D. Davies, and A. Sharih 1962. Physical separation of pseudocholinesterase variants in human serum. *Lancet.* 1:463-464.

Rosenberry, T.L., H.W. Cahng, and Y.T. Chen 1972. Purification of acetylcholinesterase by affinity chromatography and determination of active site stoichiometry. *J. Biol. Chem.* 247:1555-1565.

Scott, E.M. and R.F. Powers 1972. Human serum cholinesterase, a tetramer. *Nature, New Biology.* 236:83-84.

Index

887

A 5
B 6
C 7
D 8
E 9
F 0
G 1
H 2
I 3
J